"十三五"国家重点图书
2013~2025年国家辞书编纂出版规划

英汉染整词汇
（增补版）

English-Chinese Dictionary of Dyeing & Finishing
（Revised and Enlarged Edition）

岑乐衍　主编

中国纺织出版社

图书在版编目(CIP)数据

英汉染整词汇:增补版/岑乐衍主编.—北京:中国纺织出版社,2017.1

"十三五"国家重点图书 2013~2025年国家辞书编纂出版规划

ISBN 978-7-5180-1653-2

Ⅰ.①英… Ⅱ.①岑… Ⅲ.①染整-词汇-英、汉 Ⅳ.①TS19-61

中国版本图书馆 CIP 数据核字(2015)第 107327 号

策划编辑:秦丹红　　责任编辑:秦丹红　朱利锋
责任校对:寇晨晨　　责任设计:何　建　　责任印制:何　建

中国纺织出版社出版发行
地址:北京市朝阳区百子湾东里A407号楼　邮政编码:100124
销售电话:010—67004422　传真:010—87155801
http://www.c-textilep.com
E-mail:faxing@c-textilep.com
中国纺织出版社天猫旗舰店
官方微博 http://weibo.com/2119887771
北京新华印刷有限公司印刷　各地新华书店经销
2017年1月第1版第1次印刷
开本:850×1168　1/32　印张:26.25
字数:1064千字　定价:168.00元

凡购本书,如有缺页、倒页、脱页,由本社图书营销中心调换

编委会

主　编　岑乐衍
编　委　庄淑娟　黄平章　金远同　汪　澜
　　　　钱　蕾　岑安庆　纪淑仪　李爱民

增补版前言

原《英汉染整词汇》是2003年出版的，距今已有十余年，是我国染整行业唯一正式出版的染整英汉专业词汇，它为我国染整行业的发展做出了贡献。但随着世界科技及印染行业的快速发展，该工具书有必要进行增补修订。

作者今年90多岁，已步入耄耋之年。自大学毕业，就先后到印染厂、纺织科研院所、纺织工业局等从事各项工作，其中，从事染整事业已历经50年之久。在工作之余也曾编写过不少印染著作并撰写论文，退休后还常年阅览各种染整图书和杂志，而且也经常翻译国外染整技术资料和手册等，留心记录专业新词汇。到目前为止，已积累新词汇近万条（原书词汇量约为45000条）。增添的内容涉及环保、节能、生态、激光、纳米、等离子体、数码、生化等新技术，以及新纤维、染整新技术、新设备和计算机技术。并且还增加了染整新技术缩略语以及国外纺织、化学有关团体名称的缩略语。这些内容在一般缩略语词典中并无涉及。在专业词汇方面，一些新名词还加注说明，以便读者查考，如新型纤维、新型染料等的化学结构及优良性能等。词目中，译意力求准确、简明、实用。

修订部分主要是将已不存在的国外染料化工厂商名删去，防止误导；将上一版中某些有错误的词条予以改正，并增加或备注了一些缩略语，使其应用范围更为广泛。

另外，考虑到我国近年与世界各国、各地区的交流更为广泛，故专门搜集到各个国家和地区的英汉名词对照作为附录，使本书实用性更强，希望为读者所欢迎。

岑乐衍谨识
2016.5.1

第1版前言

在步入21世纪的新经济时代中,纺织染整工业的现代化建设,更需要用高新的科学技术来武装,为此我们编写了这本词汇,其目的是力求从国际高新技术、工艺、设备以及染化料水平等方面来表达其技术含量和价值,以便能在努力赶超国际先进水平的长河中,做出应有的贡献。

本词汇包括复合词、缩写词在内,共收词目约45000条,内容以漂、染、印(雕刻)、整工艺;设备;染化料;测配色;环保等为主,并突出收集近年发展的生物酶、生态工程、绿色工程,功能性处理,数码微机控制,纳米,等离子体,微胶囊等高新技术词目。染料、助剂品种,除保留部分普遍和常用的外,绝大部分为国外化工厂商近年开发的新品种,且属无害化或易生物降解的化学品。词目中译意力求正确、简明、实用;部分词目除译名外,还做了必要的解释。

本词汇供纺织染整工业广大科技人员和高等院校师生以及化工、染料、助剂厂商的技术、营销人员查阅,对其他领域的科技人员也有参考价值。

本词汇的编写工作,从1997年开始酝酿,组成编写班子,历时五年。编写组同志在收集新词目方面做了不少努力,参考了很多有价值的文献资料,付出了辛勤的劳动。但由于水平有限,缺点、错误和疏漏之处在所难免,恳切欢迎广大读者批评指正。

岑乐衍谨识
2002年6月

编写说明

1. 本词汇不论单词、缩略语，均按英文字母顺序排序；拼法相同但有大小写差别的，小写在前，大写在后。

2. 以希腊字母 α、β 等，斜体英文字母 N、o、m、p、d、l 等，阿拉伯数字 1、2、3 等开头的词条，以其后主词条的英文字母顺序排序。

3. 方括号"[ ]"、圆括号"( )"、尖括号"〈 〉"及其中内容，以及特殊符号"&"、"""、"/"、连字符"-"、逗号","、点"."等均不参加排序。

4. 英文拼写方式以英式为主。为了便于读者查阅，本词汇中将常见词的美式拼写方式也列入书中，如"color【美】＝colour"。

5. 同一词目有几个译名时，意义相同或相近的译名用逗号","分开，意义不同的译名用分号"；"分开。

6. 括在方括号"[ ]"内的外文或中文，在使用时可省略；括在圆括号"( )"内的中文是注解或说明语。

7. 缩略语词目圆括号"( )"内的英文表示全称；当同一缩略语代表不同的外文全称时，本词汇分别给予解释。括在尖括号"〈 〉"内的外文表示该词目的缩略语。本词汇中，缩略语与英文全称可互查，便于读者使用；不是英文语种的缩略语，本词汇中未列出全称。

8. 中文解释前括在"【 】"内的中文表示该词目的语种。

9. 括在圆括号"( )"内等于号"＝"后面的外文，表示与该词目可通用。

10. 注解中的"商名"表示词目是商标名称或商业名称,有的词目在"商名"后还标出国名与公司名,便于读者查找、应用。

11. 专业略语仅在必要时加注,含义比较明显的专业共用词,一般不注专业略语。专业略语用"《　》"表示。本词汇所用的专业略语有:《化纤》、《毛》、《丝》、《麻》、《针》(针织)、《漂》、《染》、《印》、《整》、《雕刻》、《测色》(测色配色)、《试》(试验测试)、《电》(电气和电子)、《环保》(环境保护和三废治理)、《机》(机械)、《自动》(自动控制)、《计算机》、《裁》(裁剪)、《数》(数学)、《化》(化学、化工)、《物》(物理)、《生化》。

12. 词汇中收集了一批国外著名染料、助剂品种的商名、制造厂及其技术内容,这对染整企业在使用国外染料、助剂时,将有所帮助。

13. 书后附有各个国家与地区名单,便于从事国际贸易的读者查寻和使用。

目　录

词汇正文 …………………………………………（1～817）
主要参考书目 ……………………………………（818）
附录　国家与地区名单（包括参加奥运会及
　　　邮票发行）……………………………（819～827）

A

AA（acrylic acid）丙烯酸
AAC（automatic addition control）自动加入控制
AAC metering system 自动加入控制计量系统（用于活性染料吸尽染色，德国制）
AAFA（American Apparel & Footwear Association）美国服饰与鞋类协会
AAid（acrylamide）丙烯酰胺
aal 阿尔（红色的植物染料）
AAMA（American Apparel Manufacturers Association）美国服装制造商协会
AAS（American Academy of Sciences）美国科学院
AAS（atomic absorption spectrometry，atomic absorption spectroscopy））原子吸收光谱测定法
AATCC（American Association of Textile Chemists and Colourists）美国纺织化学家和染色家协会
AATCC Fading Unit〈AFU〉AATCC［制定的］光照褪色单位
AATCC Fastness Tests 美国纺织化学家和染色家协会［制定的］纺织品牢度测试方法
AATCC Fibre Method 美国纺织化学家和染色家协会［制定的］纤维测试方法
AATT（American Association for Textile Technology）美国纺织工艺协会
abacterial 无细菌的
abating 缩合软化

Abbe refractometer 阿贝折射计
Abbot-Cox process 艾博特—考克斯染色法（还原染料悬浮体染色法）
Abbot dyeing process 艾博特染色法（多孔轴芯染色法）
abbreviated analysis 简略分析
abbreviated drawing 简图
abbreviated formula 简式
abbreviated system 短流程工艺
ABCM（Association of British Chemical Manufacturers）英国化学制造商协会
ABC process（accelerated biological chemical process）加速生物化学法
Aberdeen finish 阿伯丁整理
aberrant 变种；异常的
ABFA（azobisformamide）偶氮二甲酰胺（发泡剂）
abietic acid 松香酸
abietyl alcohol 松香醇
ability 能力；性能
abiotic 无生命的；合成的
abirritation 减缓，抑制
abluent 洗涤的；洗涤剂
ablution 清洗，净洗；吹除
abnormal brilliance 异常光泽
abnormal fibres 异型纤维
abnormality 反常，不规则
ABNT 巴西技术标准协会
aborted 异常中止的
abortive design 不完整的花纹
above floor level 地平面上（机器安装位

置）
abradability 磨蚀性,磨损性
abradant 磨蚀剂,磨料
abrade 磨蚀,磨损;磨光
abraded wool 磨损羊毛
abrader（＝abraser，abrasimeter） 耐磨试验仪
abrasion 磨蚀,磨损,擦伤
abrasion coefficient 磨损系数
abrasion ghosting 磨损虚印;转移印花虚印(转印纸打卷时造成的搭色虚印)
abrasion index 磨蚀指数
abrasion machine 耐磨试验仪
abrasion mark 擦伤痕
abrasion proof 耐磨
abrasion resistance 抗磨损性
abrasion resistance finish 抗磨损整理
abrasion test 耐磨试验
abrasion tester 耐磨试验仪
abrasion wear 磨耗
abrasion wear test machine 耐磨试验仪
abrasion weight loss 磨损质量损失
abrasive fabric 砂布
abrasive paper 砂纸
abrasive resistance 耐磨[牢]度
abrasive roller 磨毛辊(磨毛机的主要部件,辊筒表面包绕砂布带)
abridged spectrophotometer 简装光谱光度计,滤色光度计
abruption 拉断,破坏,断裂;断路
abs.（absorption） 吸收[作用]
ABS（acrylonitrile-butadiene-styrene） 丙烯腈—丁二烯—苯乙烯共聚物
ABS（advanced biosystems） 先进生化系统
ABS（aerodynamic brake spinning） 气动控制纺纱

ABS（alkyl benzene sulphonate） 烷基苯磺酸盐
ABS copolymer 丙烯腈—丁二烯—苯乙烯共聚物
absinthe green 苦艾绿[色](暗黄绿色)
absinthe yellow 苦艾黄[色](暗绿光黄色)
absolute adsorption 绝对吸着
absolute alcohol 无水酒精
absolute dry condition 绝对干燥状态
absolute dry weight 绝对干重
absolute humidity 绝对湿度
absolute intensity 绝对强度
absolutely lubrication-free chain 绝对免润滑传动链(用于拉幅烘干机)
absolute pressure 绝对压力
absolute sensitivity 绝对灵敏度
absolute strength 绝对强力
absolute temperature 绝对温度
absolute value 绝对值
absolute viscosity 绝对黏度
absorbability 吸附能力;被吸附
absorbable organic halogen〈AOX〉 可吸附的有机卤素《环保》
absorbance 吸光度
absorbed water 吸附水
absorbency 吸附性,吸收能力;吸水性,吸液性
absorbency index 吸收指数
absorbent 吸收剂;吸声材料
absorbent compounds 吸湿性化合物
absorbent cotton 脱脂棉
absorbent fabric 吸湿性织物
absorbent filter 吸收性过滤器
absorbent filtering medium 吸收过滤介质
absorbent finishing 吸湿整理(提高吸湿性能)

absorbent gauze 绷带纱布,药用纱布
absorber 吸收器;减震器;吸收剂
absorbility 吸收性,吸收能力
absorbing agent 吸收剂
absorbing column 吸收柱
absorbing duct 吸声管
absorbtion value 吸收值
absorptance 吸收比,吸收系数;吸收性,吸收能力
absorptiometer 吸光测定计;吸收比色计;[液体]溶气计
absorption〈abs.〉吸收[作用]
absorption band 吸收光带,吸收光谱带
absorption base 吸收基
absorption cell 吸收池,比色皿
absorption coefficient 吸收系数
absorption coefficient/scattering coefficient〈K/S〉吸收系数/散射系数(用以表示颜色深度的值)
absorption colour 吸收色(光谱补色)
absorption curve 吸收曲线
absorption equilibrium 吸收平衡
absorption fabric 吸湿性织物(经化学处理的平纹粗棉布或粗麻布,可吸收气体或蒸汽)
absorption factor 吸收因子
absorption gauge 织物吸湿率测试仪
absorption isotherm 吸附等温线
absorption maximum 最大吸收
absorption of moisture 吸湿
absorption of perspiration 吸汗性,耐汗性
absorption paper 吸收纸
absorption photometry 吸光光度学;吸收测光法
absorption promoting agent 吸收促进剂
absorption rate 吸收速度

absorption spectro-photometer 吸收分光光度计
absorption spectro-photometry 吸收分光光度测定法
absorption spectroscopy 吸收光谱法;吸收光谱学
absorption spectrum 吸收光谱
absorption test 吸收试验;吸水试验
absorptive capacity 吸收能力
absorptivity 吸附性;吸收率;吸光系数
ABS resin 丙烯腈—丁二烯—苯乙烯树脂,ABS 树脂
abstergent 洗涤剂,净洗剂
abstract 提取物;抽象的
abstract design 抽象图案;局部花纹图案
abstract pattern 抽象图案
abutment screw 止动螺钉
ABWAG 【德】德国废水处理法规
A. C.(alternative current) 交流[电]
ACAE(amino carboxylic acid ester) 氨基羧酸酯
A. C. arc weld 交流电弧焊
acari[d] 螨
acaricide 杀螨剂
acaristatic action 抗螨虫性
acaroid gum 禾木胶,草树胶
ACC(automatic combustion control) 自动燃烧控制,燃烧自控系统
accelerant 促染剂;加速剂,催化剂,促进剂
accelerant incubation 快速保温试验
accelerated ageing 加速老化,加速陈化
accelerated ageing test 加速老化试验
accelerated biological chemical process〈ABC process〉加速生物化学法
accelerated characterization 催化特性,加

速特性
accelerated deep layer aeration process 深层加速曝气法《环保》
accelerated experiment 快速试验法
accelerated exposure and weather machine 加速暴晒气候牢度测试机
accelerated process 加速流程
accelerated soiling method ［地毯］快速沾污试验法
accelerated storage 加速存储老化
accelerated surface aeration process 表面加速曝气法《环保》
accelerated test 快速试验
accelerated weathering test 加速风蚀试验，加速老化试验
accelerating agent 促染剂；催化剂，促进剂
accelerating potential 加速电压
acceleration 加速度；催速作用
accelerator 加速剂，催化剂，促进剂；加速器
Accelerotor 埃克西来罗试验仪（测织物耐磨和起球用，商名）
accelo-filter 加速过滤器
accent colour 强调色
acceptability 合格度，可接受程度
acceptable defect level〈ADL〉允许疵点标准
acceptable limit 允许限度
acceptable quality level〈AQL〉合格质量标准，验收合格标准
acceptable tolerance range 合格容差幅度
acceptance 验收；认可，肯定
acceptance check 合格检验，验收
acceptance test 验收试验
acceptance tolerance 验收公差
accepted product 合格产品

access 存取《计算机》
access door 移门（安装在烘房等设备上，供操作人员或维修人员进出处理故障用）
access hole 人孔，入口
accessibility 可及度，可及性，可接近性（指机件在装配、拆卸、检查、调整时是否便于伸手操作）
accessories 附件；附属设备；服饰品（包括纽扣、戒指、耳环、眼镜、手表、提包、帽子、手套、鞋等）
accessories for baby clothing 婴儿衣物的服饰品
accessories for clothing 衣物的服饰品
accessory ingredient 助剂
accessory pigment 辅［助］色素
accident 故障，事故
accidental colour 偶生色（凝视有色物体后将物体移开时产生的一种色觉）
accidental error 偶然误差，随机误差
accidental shutdown 事故停车
accident error 偶然误差，随机误差
acclimate 使服水土，使适应环境
acclimation（＝acclimatization）气候适应［作用］，［环境］适应性；驯化作用《环保》
acclimatize 使服水土，使适应环境
accommodation area 占地面积
accommodation cost 占地面积成本
accommodator 调节器
A. C. commutator motor 交流整流子电动机（一种调速电动机）
accommodation point 调节点（常指热塑性纤维熔点）
accompanying drawing 附图
accompanying dyestuff 伴染着色染料（指

防染、拔染着色染料）
accompanying fabric ［测试沾色用的］贴衬织物
accordion fabric 单面提花针织物
accordion pipe 波纹管
accordion pleated skirt 百褶裙
accumulation 聚积，积累，堆积
accumulation of cold 蓄冷
accumulation of heat 蓄热
accumulation ratio 堆积率，浓缩率《环保》
accumulative effect 累积效应
accumulator 存布器；储蓄器；蓄电池
accumulator scray 储布装置
accuracy 准确性，精确度，精度
acenaphthene 苊，萘嵌戊烷
ACESA (Australian Commonwealth Engineering Standards Association) 澳大利亚联邦工程标准协会
acetal 乙缩醛，乙醛缩二乙醇；缩醛，醛缩醇
acetalating agent 缩醛剂
acetaldehyde 乙醛
acetaldehyde sulfoxylate 乙醛合次硫酸盐
acetal fibre 缩醛纤维
acetalization 缩醛化作用
acetalizing degree 缩醛化度
acetal linkage 缩醛键
acetal reactants 乙醛缩二乙二醇反应物
acetal resin 缩醛树脂
acetamide 乙酰胺
acetamido(＝acetamino) 乙酰氨基
acetanilide 乙酰替苯胺
acetate 乙酸盐，醋酸盐；乙酸酯，醋酸酯；醋酯纤维
acetate cellulose fibre 醋酯纤维
acetate crepe 醋酯绉绸

acetate dyes 醋酯纤维染料，醋纤染料
acetate fibre 醋酯纤维，醋酸纤维
acetate fibre solvent 醋酸纤维溶媒
acetate filament[yarn] 醋酯长丝
acetate rayon 醋酯人造丝
acetate rayon staple 醋酯短纤维
acetate resist 醋酯纤维防染
acetate silk 醋酯人造丝，醋酯长丝
acetate/silk blends 醋纤与丝的交织物
acetate staple fibre 醋酯短纤维
acetic acid 乙酸，醋酸
acetic acid amide 乙酰胺
acetic aldehyde 乙醛
acetic anhydride 无水醋酸，醋酸酐
acetic ether 乙酸乙酯
acetin 甘油醋酸酯，醋精
acetoacetanilide 乙酰乙酰苯胺
acetoacetic ester 乙酰乙酸乙酯
acetone 丙酮
acetone collodion 丙酮火棉胶
acetone formaldehyde resin finishing agent 丙酮甲醛树脂整理剂
acetone solubility test 丙酮溶解度测试（检验醋酯等纤维用）
acetyl 乙酰基
acetylacetone 乙酰丙酮
acetylacetone colorimetric test 乙酰丙酮比色法（树脂整理游离甲醛测定法）
acetylacetone method 乙酰丙酮法，又名戊二酮法（测定免烫整理织物游离甲醛含量）
acetylated cotton 醋酯棉
acetylating agent 乙酰［化］剂
acetylation 乙酰化作用
acetylene 乙炔，电石气
acetylene diurea 乙炔化二脲

acetylene generator 乙炔发生器
acetylene tetrachloride 四氯乙烷
acetylene welding 乙炔焊接
acetyl ethyl acetate 乙酰乙酸乙酯
acetyl guanidine 乙酰胍
acetylides 乙炔化合物
acetylization 乙酰化作用
acetyl number 乙酰[基]值
acetyl value 乙酰[基]值
achiote 胭脂树红
achromatic 消色差的;无色的;非彩色的
achromatic colour 无彩色(指灰、黑、白),非彩色;消色差色
achromatic objectives 消色差物镜
achromatic pigment 非彩色颜料
achromatic point 非彩色点(白点,理想白)
achromatism 消色差性;非彩色
achromatopsy 色盲,全色盲
achromic 消色的;无色的
achromic effect 消色效应
acid accelerator 酸性促进剂
acid acceptor 吸酸剂;受酸体
acid ageing 酸老化;酸陈化
acid ager 酸陈化器
acid alizarine dyes 酸性媒染染料
acid anhydrides 酸酐
acid azo dyes 酸性偶氮染料
acid azo pigment 酸性偶氮颜料
acid-base catalysis 酸碱催化
acid-base indicator 酸碱指示剂
acid-base titration 中和滴定,酸碱滴定
acid-base titration detector 酸碱滴定检测器
acid bath 酸浴
acid bleaching 酸性漂白

acid brightening 酸性增光整理《丝》
acid buffer 酸缓冲
acid catalysis 酸催化
acid catalyst 酸催化剂,酸性催化剂;酸催化器
acid catalyzed 酸催化的
acid cellulase unit〈ACU〉酸性纤维素酶活力单位
acid chlorination 酸性氯化(羊毛防缩整理)
acid chrome dyes 酸性铬媒染料
acid colloid method 酸性胶体整理法(用于毛织物防缩整理,或用于棉织物防腐整理)
acid colour dip dyeing 酸性染料浸染
acid colour discharge 酸性染料拔染
acid colour printing 酸性染料印花
acid colour resist dyeing 酸性染料防染
acid converted starch 酸转化淀粉
acid cracking 酸裂解法(用于羊毛脂回收)
acid cure 酸固化
acid damage 酸破坏
acid degradation 酸降解
acid degraded finish 酸降解整理
acid degumming 酸性脱胶
acid desizing 酸退浆
acid developing 酸性显色
acid deweighting finishing 酸减量整理
acid discharge 酸拔染《印》
acid donor 析酸剂,供酸体
acid-dyeable fibres 酸性染料可染型纤维
acid dye for Nylon 锦纶用酸性染料
acid dyeing metal complex dyes 酸性金属络合染料
acid dye lake 酸性染料色淀

acid dyes 酸性染料
acid end group 末端酸基
acid equivalent 酸当量
acid fading 酸气褪色
acid-fast bacteria 耐酸细菌
acid-fastness 耐酸牢度；耐酸性
acid felting 酸毡缩，酸缩呢，酸缩绒
acid finish 浓硫酸整理法（棉纤维制品的羊皮纸整理工艺）
acid fume 酸烟，酸雾
acid generative agent 酸发生剂
acid hydrolysis 酸性水解
acidic chlorination 酸性氯化《毛》
acidic pectinases 酸性果胶酶
acidification 加酸，酸化
acidifier 酸化剂；酸化器
acidifying 酸化
acidifying agent 酸化剂
acidimeter 酸［液］比重计，测酸仪，酸度计
acidimetric analysis 酸量滴定分析法
acidimetry 酸量滴定法
acidity 酸度；酸性
acid-leuco ［还原染料］隐色酸，酸隐色体
acid liberating salt 释酸盐
acid milling 酸缩绒
acid milling dyes 酸性耐缩绒染料
acid mordant dyes 酸性媒染染料
acid number 酸值
Acidol dyes 阿西多染料（酸性染料，商名，德国巴斯夫）
Acidol K dyes 阿西多 K 染料（以 1∶2 金属络合染料为基础，具有良好的配伍性、匀染性，商名，德国巴斯夫）
Acidol M dyes 阿西多 M 染料（双磺酸基 1∶2 金属络合染料，商名，德国巴斯夫）
acidolysis 酸解
acid peroxide bleach 酸性过氧化物漂白
acid-premetallized dyes 酸性金属络合染料
acid-proof 耐酸的
acid-proof alloy 耐酸合金
acid-proof fabric 耐酸织物
acid pump 酸泵，耐酸泵
acid radical 酸根，酸基
acid rain 酸雨
acid resist（＝acid resistance） 耐酸，酸防，抗酸
acid-resistant alloy 耐酸合金
acid resistant coating 耐酸涂层
acid-resistant linings 防酸内衬
acid resisting 耐酸的
acid resist varnish 耐酸清漆
acid rinsing 酸洗
acid salt 酸性盐
acid saponification 加酸皂化
acid saturation value 酸饱和值
acid saturator 浸酸槽
acid scouring 酸性精练
acid shock dyeing 酸性条件下快速染色
acid shock process 酸快速固色法
acid shrinking 酸缩
acid soap 酸性皂
acid splitting salts 裂酸盐，析酸盐，释酸盐
acid spotting 滴酸试验法（试验染色织物的耐酸色牢度）
acid stain 酸斑，酸渍
acid steaming 酸汽蒸
acid steeping 酸浸渍；酸退浆
acid traces 酸痕量

acid treatment 酸处理
acidulate 酸化
acid value 酸值,酸价
acid wash 酸洗;雪花洗(牛仔布)
ACIMIT 【德】意大利纺织机械制造商协会
ACIT 【法】法国纺织工业化学家协会
A-class goods 正品
ACM(advanced composite material) 先进复合材料
Acme burner 阿克梅[煤气]火口(不会发生回火的燃烧器)
Acme thread 阿克梅螺纹(一种英制梯形螺纹)
acorn[nut]starch 橡子粉,橡子淀粉,浆栎果淀粉
acoustextile 隔音纺织品
acoustic absorbent 吸音材料
acoustical absorption 吸音
acoustic baffling 隔音挡板
acoustic damping 声阻尼(材料)
acoustic fatigue 声致疲劳
acoustic felt 吸音毡
acoustic filter 消声器
acoustic insulation 隔音,隔声,声绝缘
acoustic intensity 声强[度]
acoustic meter 声级仪,噪声仪,比声计
acoustic pressure 声压
acoustic resilient layer 吸音弹性层(汽车用纺织品)
acoustic signal 声音信号
acoustic transducer 声换能器(用于烘房排风湿度监控仪)
acoustic treatment 隔音处理
acoustic warning 声音警告
Acquajet water blade 水喷射式刮刀(洗布机清洗装置)

Acrafix FH 阿克拉菲克司 FH(阿克拉明 F 法用的涂料交链剂,商名,德国拜耳)
Acrafix ML 阿克拉菲克司 ML(涂料印花交链剂,含低甲醛,商名,德国拜耳)
Acramin FWR powder 阿克拉明 FWR 粉(涂料印花黏合剂,商名,德国拜耳)
Acramin LC 阿克拉明 LC(涂料印花黏合剂,耐日晒老化、耐干洗,商名,德国拜耳)
Acramin pigment colours 阿克拉明涂料(印染用涂料,商名,德国拜耳)
Acramin SCN 阿克拉明 SCN(涂料印花黏合剂,耐日晒老化,可用于防拔染印花,商名,德国拜耳)
Acramoll W paste 阿克拉摩 W 浆(涂料印花黏合剂,商名,德国拜耳)
Acrapon A paste 阿克拉邦 A 浆(非离子型增稠剂,油水相乳化液,商名,德国拜耳)
acridine dyes 吖啶染料
acridone dyestuff 吖啶酮染料
acrolein 丙烯醛
acrolein polymer 丙烯醛类聚合物
acrolein resin 丙烯醛树脂
acrylamide〈AAid,AM〉丙烯酰胺
acrylamide dyes 丙烯酰胺染料
acrylate polymer 丙烯酸酯聚合物
acrylate sizes and finishes 聚丙烯酸酯浆料和整理剂
acrylic 聚丙烯腈系纤维(统称,由85%以上丙烯腈组成的纤维);丙烯酸[类];丙烯酸系衍生物
acrylic acid〈AA〉丙烯酸
acrylic acid derivatives 丙烯酸衍生物

acrylic amide〈AM〉 丙烯酰胺
acrylic binder 聚丙烯酸黏合剂
acrylic blanket 腈纶毯
acrylic boa 腈纶人造毛皮
acrylic-carboxylic acid complex 丙烯酸—羧酸复合物
acrylic ester emulsion 聚丙烯酸酯乳液
acrylic fibre 聚丙烯腈系纤维(聚合物中含丙烯腈不低于85％)
acrylic latex 聚丙烯酸胶乳
acrylic nitrile 聚丙烯腈
acrylic resin 聚丙烯酸[系]树脂
acrylic size 聚丙烯酸浆料
acrylonitrile 丙烯腈
acrylonitrile-butadiene rubber roll 丁腈橡胶辊(丙烯腈与丁二烯共聚物制成的橡胶辊)
acrylonitrile-butadiene-styrene〈ABS〉 丙烯腈—丁二烯—苯乙烯共聚物
acrylonitrile-styrene〈AS〉 丙烯腈—苯乙烯共聚物
ACS(American Chemical Society) 美国化学学会
A. C. static eliminator 交流静电消除器
A. C. synchroning system 交流同步系统《电》
ACT(after-clarification tank) 后澄清槽《环保》
actification 再生作用,复活作用
Actigard 安宁茄,防螨整理剂[商名,瑞士科莱恩]
actinic degradation 光降解[作用]
actinic light 光化灯
actinic resistance 耐光化性
actinic stability 光化稳定性
actinity 光化性

action backing 合成次底布(地毯专业词汇)
action limit 操作极限
action plan 行动计划
activable tracer 活化示踪剂
activate 活化;激化
activated adsorption 活化吸附作用
activated aeration process 活性曝气法《环保》
activated aluminum sulfate catalyst 活性硫酸铝催化剂
activated biofilter 活性生物滤池《环保》
activated carbon 活性炭
activated carbon filter 活性炭过滤器
activated charcoal 活性炭
activated clay 活性[黏]土
activated complex 活性络合物,活化络合物;活化复体,活化配合物
activated diffusion 活性扩散
activated fibre 活性纤维
activated molecule 活化分子
activated silica gel 活化硅胶
activated sludge 活性污泥
activated sludge bulking 活性污泥膨胀
activated sludge process 活性污泥法《环保》
activated state 活化状态
activating agent 活性剂,活化剂
activation 活化[作用],活性[化],激化
activation energy 活化能
activation energy of diffusion 扩散活化能
activation tank 活化槽《环保》
activator 活化剂,活性剂
active and fresh finish 防臭防菌整理
active bleach 活性漂白
active carbon 活性炭

active carbon adsorption and biochemical regeneration 活性炭吸附和生物化学再生法《环保》
active carbon adsorption process 活性炭吸附法《环保》
active carbon chemical treatment 活性炭化学处理（污水处理）
active carbon filter 活性炭过滤器《环保》
active cation site 活性阳离子位置
active center 活性中心；有效中心；活动中心
active charcoal 活性炭
active chlorine 活性氯，有效氯
active chlorine losses 活性氯损耗
active clay 活性黏土
active component 有效成分
active content 活性成分含量
active dressing 积极修饰（起绒辊）
active dyes 活性染料
active earth 活性土（漂白土）
active hydrogen 活性氢
active ingredient 有效成分
active level concentration 作用浓度；有效标准浓度
active oxygen 活性氧
active oxygen method 活性氧法（评定油脂稳定性的加速氧化试验法）
active protection system 积极防护系统
active site 活性部位
active sportswear 运动服式服装，宽松式运动衣
active standard 现行标准
active substance 活性物质（如酶、维生素等）
active volume 有效体积
active wear 运动服，功能性服装

activity 活度
activity coefficient 活度系数，活化系数
activity of dye 染料活度（有效染料浓度）
activity sampling 快速取样
actual weight 实际重量
actual yarn count 实际纱线线密度（纱支）
actuating mechanism 执行机构；促动器
actuating member 执行元件
actuating motor 伺服电动机
actuation 开动，传动
actuator 调节器；执行元件；传动装置；油压发动机（总称）；自控执行机构《电》
actuator drive 伺服电机和传动装置
ACU（acid cellulase unit） 酸性纤维素酶活力单位
acute dermal toxicity 急性皮肤毒性
acute inhalation toxicity 急性吸入毒性
acute oral toxicity 急性口服毒性
acute toxicity 急毒性，急性毒性《环保》
ACV（alkali centrifuge value） 碱离心值（织物树脂整理的特征参数）
acyclic 无环的；非同期(性)的
acyclic compound 无环化合物
acylable 可酰化的
acylamide 酰胺
acylaminoanthraquinone 酰氨基蒽醌
acylate 酰化［产物］
acylation 酰化作用
acyl cellulose 酰基纤维素
acyl peroxide 过氧化酰基
Adams-Nickerson colour difference equation 亚当斯—尼克尔森色差公式
Adams-Nickerson system〈ANS〉 亚当斯—尼克尔森［色差］系统
adaptability 适应性
adapter 适配器，接合器；［转换］接头转

接器;衔接器;附件,附加器;拾音器,拾波器

adapter connector 接头,连接器

adaptive clustering 自动分组(纺织品色光的自动分组软件)

adaptive colour shift 色适应位移(色度学用语)

adaptive microorganism 抗药性细菌

Adasil HS 安大矽 HS(非离子有机硅弹性体微乳液,适用于各类织物高档整理,赋予织物舒适、丰满的手感,商名,德国科宁)

ADC(analog-digital conversion) 模拟数字转换《计算机》

ADC(analogue to digital converter) 模拟数字转换《计算机》

ADC counts 模拟数字转换计数《计算机》

Adcon metering system 阿德康[加碱]自动计量系统(商名,德国制)

added touch 附属品(特别指装饰用品)

added value 附加价值

addendum 补遗;附录

adding and copying machine [花筒]连续曝光机

addition 附加,添加;附加物;加入(量);加成

addition agent 添加剂;加成剂;配合助剂

addition agent finish 添加剂整理

additional accelerator 辅助促进剂

additional appliance 附加设备

additional colours 调色[量],追加色

additional device 附加装置

additional equipment 附属设备

additional parts 附件

additional polymerization 加成聚合作用

additional regulations 补充规定

addition complex 加成络合物

addition highwet pickup 添加法高湿吸液

addition polymerization 加成聚合反应

addition process 加成工艺

addition product 附加物,加合物,加成[产]物

addition test [染料]添加测试

additive 附加的,加成的;添加剂,助剂

additive air 补充空气(目的在于改变混合可燃气体的浓度)

additive antifelt finishing 加法防毡缩整理

additive colour mixing (= additive colour mixture) 加色混色

additive colour process 加色法《测色》

additive complementary colour (= additive complementary light) 加法补色

additive dressing 多层组合服装

additive finish 添加剂整理;加法整理

additive finishing 加法整理

additive in resin finishing 树脂整理添加剂

additive level 添加剂总含量

additive mixture [of colours] 加法混色

additive mixture of colour stimuli 加法混色

additive pickup 添加吸液

additive primaries 加色法原色

additive process 加色法[混合]

additive process of antifelting finish 加法防毡缩整理工艺

additive property 加成性,加和性

additive rate 加成率

additive three primaries 加法三原色

additivity 加和性;相加性

additivity rule 相加规则,相加定律

add-on 加重率;加入量;加成;吸液量;给

液量
add-on size 上浆率,上浆量
add-on value 附加价值
adduct 加合物,加成物
adduct polymerization 加合聚合作用
adenosine triphosphate〈ATP〉三磷酸腺苷
adhere 黏附,黏着,黏合
adhered water 吸附水
adherence(=adherency) 黏附,黏着,黏合
adherent 黏合的,黏附的;黏合物,黏附体
adhering fibre 黏合纤维
adhesion 黏合,黏附;黏合力
adhesion agent 黏合剂
adhesion machine for screen printing fabric 筛网印花织物的贴布机
adhesion paste 黏着剂
adhesion preventives 防黏剂
adhesion promoter 增黏剂
adhesion strength 黏合强力,黏附强力
adhesion tension 黏合张力
adhesion test 黏合力测试,黏着试验
adhesive 胶黏剂,黏合剂;胶黏的
adhesive applicator 上胶装置,加黏着剂装置(圆网或平网印花时在印花导带上上胶,使织物黏贴在上面)
adhesive backed 黏合剂涂底
adhesive backing 黏合底布
adhesive band 黏合带
adhesive-bonded cloque 黏合泡泡纱
adhesive-bonded fabric 黏合织物,黏合法非织造物
adhesive-bonded knop textile 黏合法结子布
adhesive bonded nonwoven fabric 化学黏合法非织造布
adhesive-bonded pile carpets 黏合绒头地毯
adhesive-bonded pile coverings 黏合绒头地毯
adhesive-bonded textile 黏合纺织品
adhesive bonding 黏合,黏合法(非织造织物生产工艺)
adhesive bond strength tester 黏接强力试验机
adhesive capacity 黏合能力;黏合度
adhesive coat 黏接剂涂布,黏接涂层
adhesive fabric 黏合布
adhesive force 黏合力,黏着力;附着力
adhesive glue 黏合胶
adhesive label 胶黏标签
adhesive laminating machine 黏合层压机
adhesive oil 黏性油(机油)
adhesive plaster 橡皮膏
adhesive powder 粉状黏合剂
adhesive pretreatment 黏合预处理
adhesive property 黏合性
adhesive roll 黏合辊
adhesives for printing blankets 印花导带胶黏剂
adhesive spreader 胶着剂涂布器
adhesive strength 黏合强力
adhesive strike-through 胶黏剂渗透(台板印疵)
adhesive substance 黏着物;附着物
adhesive tape 胶黏带;橡皮膏带
adhesive tension 黏合张力
adhesive test 黏合试验,胶着试验
adhesivity 黏着性;黏附能力
adhibit 粘贴
adiabat 绝热线
adiabatic 绝热的,不传热的
adiabatic change 绝热变化[过程]

adiabatic drying 绝热干燥
adiathermic 绝热的
adiathermic property 保暖性
adipic acid 己二酸
adipic acid fibres 聚己二酸系纤维
Adire 阿代尔（非洲靛染防染印花产品）
adjacent colour 邻近色
adjacent double bond 邻近双键
adjacent fabric ［染色牢度试验中测试沾色用的］贴衬织物
adjective dyes 媒染染料
adjust 调整，校正，调节
adjustable 可调节的，可调整的，可移动的
adjustable bracket 调节架
adjustable spanner 活络扳手
adjustable v-belt 活络三角带
adjusted exhaust value〈AEV〉校正的吸尽提浸值（吸尽提浸值和染浴浓度的乘积）
adjuster 调整装置，调节器
adjusting hoop 调节箍
adjusting nut 调节螺母
adjusting roller 调节辊
adjusting screen 调节圆网（使对花准确）
adjusting screw 调节螺丝
adjusting wheel 调节手轮
adjustment 调整，调节
adjustment notch 调整标记；安装标记
adjuvant 助剂；助手
ADL（acceptable defect level） 允许疵点标准
Adler finish 艾德勒整理
ADMI（American Dye Manufacturers Institute） 美国染料制造商协会
administrator 管理员，行政官员

admiralty cloth 海军呢
admissible bending 容许弯曲度
admissible error 容许误差
admissible residual unbalance 容许残留不平衡量
admissible run-out 容许跳动量
admission valve 进入阀，进气阀
admixture 混合，混合物
adornment 装饰；装饰品
A. D. R.（American Dyestuff Reporter）《美国染料报道》（月刊）
Adron 105 安得隆 105（精纺毛纱的浆纱剂，含表面活性剂、抗静电剂、防腐剂的复配溶液，可提高单经单纬纱织造效率，商名，英国 PPT）
adsorbability 吸附能力
adsorbate 被吸附物
adsorbed water 吸着水
adsorbent 吸附剂；吸附体
adsorption 吸附作用
adsorption agent 吸附剂
adsorption analysis 吸附分析
adsorption by hydrogen bond 氢键吸附
adsorption catalysis 吸附催化作用
adsorption chromatography 吸附色谱法；吸附色层分离法
adsorption displacing activity 吸附取代（防止表面活性剂对纤维的吸附）
adsorption equilibrium 吸附平衡
adsorption film 吸附膜
adsorption indicator 吸附指示剂
adsorption isotherm 吸附等温线
adsorption layer 吸附层
adsorption potential 吸附势，吸附位
adsorption theory of dyeing 染色吸附理论
adsorptive capacity 吸附能力；吸附量

adulteration 掺杂
advanced biosystems〈ABS〉先进生化系统
advanced colour management capability 先进颜色管理能力
advanced composite material〈ACM〉先进复合材料
Advanced Denim Dyeing Process 先进的牛仔布染色工艺（具有高效、灵活、环保特性，瑞士科莱恩）
advanced multiaxial technology〈AMT〉先进的多轴向技术
Advanced Performance Finish〈APF〉功能整理[巴斯夫(BASF)公司多功能整理技术]
advanced treatment 深度处理
advanced waste treatment〈AWT〉深度废水处理《环保》
advanced waste water treatment 深度废水处理《环保》
advancing bow 中凸弓纬（织疵）
advancing colour 近似色，连感色
advancing contact angle 前进接触角
adverse effect 逆效果
AECC【西】西班牙质量控制协会
AEQCT【西】西班牙纺织化学家和染色家协会
aerated lagoon 曝气塘，氧化塘《环保》
aeration 充气，充氧；吹风；换气；曝气《环保》
aeration basin 曝气池《环保》
aerial condenser 空气冷凝器
aero-accelerator 加速曝气池《环保》
aerobacteria 好氧细菌，好气细菌
aerobic 需氧的，好气的
aerobic bacteria 好氧细菌，好气细菌
aerobic biodegradation 曝气生物降解《环保》
aerobic biological treatment 好氧生物处理，曝气生物处理《环保》
aerobic degradation 需氧降解，暴气降解
aerobic digestion 需氧消化
aerobic ensemble 健身服
aerobic fashion 吸氧健身运动服
aerobic fermentation 需氧发酵
aerobic micro-organisms 好氧型微生物
aerobic organism 需氧有机体，好气有机体
aerobic respiration 需氧呼吸
aerobic retting 好氧菌浸渍法
aerobioscope 空中微生物检测器
aerodromometer 气流速度表
Aero-Dryer 气流烘燥机，一种吸入式圆网烘燥设备[英国贝帝斯(Bates)]
aerodynamic brake spinning〈ABS〉气动控制纺纱
aerodynamic dyeing systems 气动染色系统
aerodynamic web formation 气流成网
aerogel 气凝胶
aerograph 喷纱染色器，喷枪（染色用）
aerometer 量气计，气体比重计
aeronautical cloth 航空用织物，航空气球用织物
aeroplane fabric 飞机翼布
aerosol 气溶胶；气悬体；烟雾剂
aerosol dyeing 气溶胶染色（由空气、染料、水蒸气组成的染色体系）
aerosol filter 气溶胶滤网
aerosol resistance 抗烟雾
aeruginous 铜绿色的
AES(atomic emission spectrometry) 原子发射光谱
aesthetic feeling(＝aestheticism) 美感，审

美感
aesthetic judgement 审美能力
aesthetic measure 灵感度;审美感
aesthetic property 美感性,美观性
aesthetics 美学;美感性,美观性
AEV(adjusted exhaust value) 校正的吸尽提浸值(吸尽提浸值和染浴浓度的乘积)
affinity 亲和力;亲合性
affinity chromatography 亲和层析
affinity factor 亲和力系数
affinity modification 改变亲和力
affinity modifier 亲和力改变剂
affinity of dyeing 染色亲和力
affinity of dyestuff 染料亲和力
affix 附加,贴上
Afflammit 阿法拉民耐洗阻燃剂[商名,英国索尔(THOR)]
Afghan blanket 阿富汗毛毯
Afghan carpet 阿富汗地毯
AFM(atomic force microscope) 原子力显微镜(测定纳米材料的显微镜)
AFNOR 【法】法国标准化协会
A-frame A形架(常指大卷装卷布机的机架)
African design 非洲风格图案
African look 非洲式样
African printing 非洲风格印花
African stripe 非洲条纹布(粗阔彩色条纹)
Afrida lac cloth 阿夫里达虫胶布(印度产,用虫胶染成深红色)
Afridi 阿夫里迪蜡防花布(东南亚产)
AFTC 【法】法国纺织品控制协会
after backwashing (精梳后的)复洗《毛》
after bleaching 后漂白

after boarding 后定形《针》
after brightening bath 后增艳浴;后增艳处理
after burner 补燃器(例如加装在热定形拉幅机后,可以达到净化废气、节约能源的目的)
after burner system 后燃烧炉方式《环保》
after burn time 续燃时间
after-care 售后服务
after chlorinate 后氯化,补充氯化
after chrome dyes 后铬媒染料
afterchrome mordant 后铬媒媒染
afterchrome [dyeing] process 后铬媒染色工艺
afterchroming 后铬媒[处理];铬盐后处理
after-clarification tank〈ACT〉 后澄清槽《环保》
afterclearing 后清洗
aftercoagulation 后凝固
aftercondensation 后缩合[作用]
after copper dyes 铜盐后处理染料
after coppering 铜盐后处理
after coppering test 铜盐后处理测试
after fibrillation 后原纤化
after flame 完全燃烧,烧尽;残焰,续燃(阻燃试验用语)
after flaming time 阴燃时间
after glow 余辉,阴燃
afterhardening 后硬化
afterheat 剩余热量,余热
afterimage 余像;余感色;残留影像
after pinning 后上针
after pinning belt 后上针皮带
afterprint washing machine 织物印花后水洗设备

after processing 后加工
afterproduct 副产品
aftersale service 售后服务
after-saponification 后皂化
after scouring [缩绒]后洗涤,二次皂洗;后精练
after service [商品售出后一定时间内的]保修服务
after shrinkage 后收缩
aftersoaping 后皂洗
aftersoaping agent 后皂洗剂
aftersteaming 后蒸化
afterstretching 后拉伸
after-tension 后张力
after treated direct dyes 经处理的直接染料
after treating agent 后处理剂
after treatment 后处理
after treatment of sulfur dyes 硫化染料的后处理
after treatment with bleaching powder 漂白粉后处理
after treatment with chrome salt 铬盐后处理
after treatment with copper sulfate 硫酸铜后处理
after treatment with formalin 甲醛后处理
after treatment with metallic salt 金属盐后处理
after treatment with oxidized substance 氧化物后处理
after treatment with peroxide 过氧化物后处理
after trestment with soap 皂洗后处理
afterwash 后水洗(指织物经印花或染色后的蒸洗处理)

afterwashing treatment 洗净后的处理
afterwaxing 后上蜡
afterwelt 袜口下[加固]段
AFU(AATCC Fading Unit) AATCC[制定的]光照褪色单位
against-scale 逆鳞片
against-scale direction 逆鳞片方向
against the skin 与皮肤接触
against to sample 与来样不符
agar-agar 琼脂,琼胶;石花菜,洋菜
agar plate method 琼脂平面培养法
agate 玛瑙棕[色](深棕色)
agate green 灰湖绿[色]
aged 老化的
aged black 精元,氧化元,苯胺黑,阿尼林黑
aged film 老化[的]薄膜
aged top 熟化毛条
age-hardening 时效硬化,经时硬化
ageing 老化,陈化;蒸化[储存],熟化(常温常湿堆置定形处理)
ageing characteristics of mucilages 胶浆的老化性能(美国印花技术用语)
ageing fastness 抗老化牢度
ageing of textile material 纺织品老化
ageing resistance 抗老化[性能],耐老化性
ageing room 蒸化室,蒸房;给湿房
ageing stability 老化稳定性
ageing test 老化试验
ageing time 老化时间
age-inhibiting additive 防老化[添加]剂,阻老化[添加]剂
agenda 操作规程
agent 药剂;代理商
ager 老化机;热蒸机;蒸化机

age resister 防老化剂
agglomerate 结块,凝聚;凝聚物;凝集物
agglomeration 凝聚[作用];附聚[作用];烧结[作用]
agglutination 凝聚,凝集
agglutination precipitation method 凝聚沉降法《环保》
aggregate 集合器;联动装置
aggregated particle 凝集粒子
aggregated plant 集料厂,石料加工厂
aggregate tension 总张力
aggregate unit 联合机组;联动装置
aggregation 聚集[作用],凝聚[作用]
aggregation effect 凝聚效果
aggregation of dye 染料的凝聚
aggregation precipitation 混凝沉淀《环保》
aggressive enzyme 攻击型酶
aggressive tack 干黏性
agile manufacturing 敏捷生产;灵活生产
aging 老化,陈化;蒸化[储存],熟化(常温常湿搁置定形处理)
agitation 搅拌,摇动
agitator 搅拌器,搅拌装置
agreement 协议
Agreement on Textile and Clothing〈ATC〉[世界]纺织及成衣协议
agricultural and forestry textile 农业与森林用纺织品
agricultural fabric 农业用布
AIC【法】国际色彩协会(法国)
A. I. C.(American Institute of Chemists) 美国化学工作者协会
AICE(American Industrial Health Council) 美国工业保健委员会
AICE(American Institute of Chemical Engineers) 美国化学工程师协会

AICE(American Institute of Consulting Engineers) 美国顾问工程师协会
AICTC【意】意大利纺织化学家和染色家协会
aid 辅助设备;辅助手段;工具
aids 助剂
AIHC(American Industrial Health Council) 美国工业保健委员会
air admission valve 进气阀,真空破坏阀
air-air heat exchanger 气—气热交换器
air bag 充气[安全]袋(汽车防冲撞安全用具)
airbag fabrics 充气袋织物
air ballooning unit 吹气装置(筒状针织物水洗后加工用)
airballoon system [圆筒针织物]气袋系统
air bellows 气袋(轧车用气动加压装置)
air blast dryer 气流式烘干机
air bleeder 放气装置,放气阀(用于放出液压系统中的空气)
air blower 鼓风机
air break(=air breaker) 空气断路器,空气开关
airbrush 喷笔,气喷
airbrush addition 喷笔附加(制印花图稿)
air brushing 气喷,气刷(一种喷色方法)
airbrush jet 气喷喷头,气刷喷头(喷刷印花用)
air brush printing 喷雾印花
air brush/valve jet 气刷/阀喷(喷墨印花机的一种喷射装置)
air bubble viscometer 气泡黏度计
air-bulked [textured]yarn 膨体[变形]纱
air chamber 风室,空气室
air circulation 空气循环,空气环流
air circulation regulator 空气循环调节器

air clamp 气夹(用气动方式夹紧自动网印机印花导带的夹子)
air cleaner 空气过滤器,空气净化器,空气滤清器
air cock 气旋塞
air compressor 空气压缩机
air condenser 空气冷凝器
air conditioner 空气调节器,温湿度调节器
air conditioning 空[气]调[节],温湿度调节
air conditioning apparatus 温湿度调节装置
air-conditioning of fabrics 织物温度调节
air-conditioning unit 空气调节装置
air conduit 风管,风道
air connector 气动接头
air contaminant 空气污染物
air contamination 空气污染
air content 含气量
air-cooled finish 凉爽整理(使织物纱线间保持空隙,以便穿着时有舒适感)
air cooler 空气冷却器,空气冷却管
air cooling 空气冷却
air-cured 晾晒于空气中的;晾干的
air-curing 空气晾干,用空气处理
air current 气流
air current drying machine 气流式烘燥机
air curtain 空气幕,风帘
air cushion 气垫
air cushion drying machine 气垫式烘燥机(一种非接触式烘燥设备)
air cushion squeeze 气垫式挤压
air cushion squeezer 气垫式轧车
air cushion stenter (＝air cushion tenter frame) 气垫式拉幅机

air cylinder 气缸
air damper 风门,空气挡板;空气减震器
air detection [含]空气检测
air distributor 空气分配器
air doctor coating 空气刮刀涂层
air draft 排风,通风;气流
air dried sample 风干试样
airdry 风干,晾干
airdry moisture regain 标准回潮率
air-dry weight [纺材]湿重(指标准回潮时的重量)
airduct 风管,风道;导气管;风筒
air duster 气流除尘器;气流吸尘器
air dye technology 气流染色技术
air-escape cock 空气阀,泄气阀
air evacuation 排气
air evacuation performance [空气]排空性能,排气性能
air evacuation valve 放气阀,排气阀
air exchange times 换气次数
air exhaust 排风,排气
air exhauster 抽气机,排气风机
air extraction system 空气抽吸系统
air extractor 抽气机,抽风机
air fabric 透气性织物
air filled squeeze roller 充气轧辊
air filter 空气过滤器,滤网
air filtration 空气过滤
Air Finish tumbling machine 一种转笼式整理机
airflow 气流,风量
air flow chamber 气室(风箱)
airflow choking plate 控制风门,调节风门
airflow dryer 热风烘燥机
airflow dyeing machine 气流染色机
airflow edge spreader 气流剥边器(非接触

式剥边装置,适用于细薄机织物和针织物)
airflow instrument 气流仪
airflow meter 气流计,气流仪
Air-flow squeegee 气袋式刮刀(荷兰 Stork 公司研制的用于圆网印花的一种新型刮刀,刀片后面有一气袋,具有刮浆均匀、渗透性好、节约浆料等优点)
air flue 气道;烟道
air free ager 不含空气的蒸化机,还原蒸化机
air freight 空运;空运费
air-fuel ratio 空气与燃料混合比(指重量混合比)
air gap [空]气隙
air gas 风煤气(含空气的煤气);含空气可燃气体
air gauge 空压计
air governer 空气调节器
air gun 空气枪
air gun rotary spraying unit 旋转式自动喷枪
air heater 空气加热器,热风器
air-heat set 热空气定形
air humidification 空气加湿
air humidifier 空气湿润器,空气增湿器
air-impermeable 不透气的
airing 透风,空气氧化;晾干;通风
airing arrangement 透风架,透风装置
airing frame 透风架
airing roller 透风滚筒
air injection beam dyeing 喷气经轴染色
air injection dyeing 空气喷射染色,气流染色
air inlet 进风口
air intake 进气口,进风口

air jet 喷气,空气喷射;喷气嘴
air jet bulking 喷气膨化
air jet ejector 空气喷射器
airjet type washing machine 喷气式水洗机(水洗槽内配有压缩空气喷射管,在洗液中产生振荡作用,以提高水洗效果)
air knife 空气刮浆刀;悬空刮刀(涂层机用)
air-laid system 气流成网法
air-lay drying machine 无接触热风烘燥机,气垫式烘燥机
air leakage 漏气
air lock 气封;气阻(管路中如有空气侵入,气泡易聚集在管路的曲折部分,引起管路不畅通);气闸,气阀
air loss 风耗,气耗
air-lubricated thrust bearing 空气润滑止推轴承
air manifold 空气总管,总风箱
air mattress 气垫
air metering unit 空气流量计
Airmint 中空聚酯纤维(商名,日本仓敷)
air moistener 空气湿润器,空气增湿器
air monitor 大气污染监测器
air nozzle 喷气嘴,喷风口
air-operated controller 气动调节器
air-operated valve 气动阀
Airo 1000 softening machine 爱罗 1000 型柔软整理机(用于各种纤维织物的柔软、烘干、水洗,使整理后织物达到蓬松、柔软等多种效果,商名,意大利白卡拉尼)
air outlet 出气口,出风口
air oven 烘箱
air oxidation 空气氧化
air oxidation machine 空气氧化机

air pad batch 气囊式轧卷《染》
air-pad system ［圆筒针织物］空气浸轧系统
air passage 空气通道；悬化，透风
air permeability 透气性
air permeability tester ［织物］透气性试验仪
air permeabillity test 透气性测试
air permeable waterproofing 透气性防水整理
air perviousness 透气性
air pilot valve 空气先导阀
air pipe 空气管道，风管
airplane luggage cloth 航空包布
air plant 通风装置；压缩空气站
air pocket 气包，气袋
air pollution 空气污染
Air Pollution Control Association〈APCA〉空气污染控制协会（美国）
Air Pollution Control Law 空气污染控制法
air pollution index〈API〉空气污染指数
air preheater 空气预热器
air-press mandrel system 空气加压［花筒］芯子系统《印》
air pressure 空气压
air pressure brake 压缩空气制动器
air pressure gauge 气压计，压力表
air pressure system 气压系统
air-proof 不透气的
air pump 气泵
air purge 空气吹扫
air purity respirator 空气净化呼吸器
air quenching chamber 空气骤冷室
air reduction valve 空气减压阀
air regenerating device 换气设备，空气循环设备
air retention 空气保持性
air screen 空气滤尘网，空气幕
air scrubber 空气洗涤器
air seal 气封（烘房等进出布口的一种封口措施）
air servo system 气动伺服系统
airship envelope fabric 飞艇（或飞船）气囊织物
airship fabric 气球织物，飞艇织物
air silent compressor 无噪声空气压缩机
Air Soft dyeing machine 一种气流染色机（英国制）
air space 空气层；空隙
air spray 气喷
air-steam circulating machine 空气蒸汽循环染色机（染袜子专用染色机）
air stream 气流
air/stream dyeing machine 用空气/水流系统操作的染色机
Air Stream dyeing machine 一种气流染色机（商名，德国第斯）
air stream turbulence 气流湍流
air structure 充气式结构
air-supply line 进风管，进风道
air supply slot outlet 条缝形送风口
air supported structure 充气结构
air suspension drier(= air suspension dryer) 无接触热风烘燥机，气垫式烘燥机
air switch 空气开关，空气断路器
air technology 空气技术（与空气有关技术，如制冷、除尘等）
air temperature probe 气温探头
airtex 网眼夏服料
air textured yarn〈ATY〉喷气［法］变形丝
air-tight 气密的，气［密］封的；不透气的

air-tight door 气封门
air-tight seal 空气密封
air trapper 捕气器
air tucking 空气打褶,空气起裥
air turbine 空气涡轮机
air twisting 气流加捻
air uncurler for knit fabric 针织物喷气式剥边装置
air velocity 风速
air vent 气孔,排气口;空气阀
air washer 空气洗涤室;空气洗涤器
air-water heat exchanger 气—水热交换器
airy fabric 透气性织物
AIS 【法】国际蚕丝协会(法国),国际肥皂和洗涤剂工业协会(法国)
aisle 通道(指机器与机器之间的走道)
AITIT 【法】国际印染工作者协会(法国)
Aizen Cathilon CD dyes 爱染卡磁隆 CD 染料(酸改性聚酯纤维专用阳离子染料,商名,日本保土谷)
Aizen Cathilon dyes 爱染卡磁隆染料(阳离子染料,商名,日本保土谷)
Aizen PP dyes 爱染 PP 染料(聚丙烯纤维用染料,商名,日本保土谷)
Ala 丙氨酸,氨基丙酸
alanine 丙氨酸,氨基丙酸(α-氨基丙酸)
alarm 警报;警报器
alarm and indicating equipment 报警指示装置
alarm gauge 气压报警器
alarm history 报警历史
alarm signal 报警信号
alarm thermometer 报警温度计
alarm unit 报警装置
Albegal A 阿白加 A(二性化聚乙二醇醚衍生物,羊毛染色匀染剂,商名,瑞士制)
Albegal B 阿白加 B(Lanasol 染料配套匀染剂,商名,瑞士制)
albumen (=albumin) 蛋白;白蛋白
albumin fibre 白蛋白纤维
albumin mordant dyeing 蛋白媒染
albumin paste 蛋白浆
albumin printing 蛋白印花
Alcalase 一种蛋白酶(有效清除丝胶,代替砂洗,商名,丹麦诺维信)
Alcamine CWS 阿卡明 CWS(季胺化合物,适用于各种纤维的柔软剂和抗静电剂,商名,瑞士制)
Alcian dyes 爱尔新染料(暂溶性酞菁染料,商名,英国制)
alcoholates 醇化物
alcohol blast burner 酒精喷灯
alcohol extractable matter 醇类提取物
alcoholic printing ink 醇性印墨(转移印花用)
alcohol/petrol-repellency 防酒精/石油性
alcohols 醇类
alcohol thermometer 酒精温度计
alcohol-water disperse dyes 醇—水可分散型染料
alcoholysis 醇解
Alcolube NSI 阿可溜 NSI(脂肪酸铵和有机硅柔软剂混合物,润滑柔软剂,商名,瑞士制)
Alcopol 020 阿可波 020(含二辛基磺化琥珀酸钠,阴离子润湿剂,商名,瑞士制)
Alcopol O conc. 60 阿可波 O 浓缩 60(含二辛硫代琥珀酸钠,阴离子润湿剂,商名,英国制)
Alcoprint PTF 阿可印 PTF(涂料印花用合成增稠剂,聚丙烯酸系,商名,瑞士

制)

Alcoprint PTP 阿可印 PTP(环保型印花用合成增稠剂,商名,瑞士制)

Alcoprint PT-RV 阿可印 PT-RV(环保型印花用合成增稠剂,商名,瑞士制)

Alcoprint RND 阿可印 RND(活性染料印花用合成增稠剂,耐电解质,商名,瑞士制)

alcosol 醇溶胶

aldehydes 醛类

aldol condensation 醛醇缩合

Aldrin 爱氏剂,氯甲桥萘(毒性大的防蛀剂,禁用)

alerting signal 报警信号

alert level 警戒基准《环保》

ALG (alginate) 海藻酸盐;海藻纤维

algicidal finishes 防海藻整理(防止纺织品上长海藻)

algin 海藻胶;海藻酸,藻朊[酸]

alginate ⟨ALG⟩ 海藻酸盐;海藻纤维

alginate fibre 藻酸纤维,海藻纤维

alginate thickener (= alginate thickenings) 海藻酸盐浆料;海藻酸盐增稠剂

alginate yarn 藻酸纤维纱,海藻纤维纱

alginic acid 海藻酸

Algol dyes 亚士林染料(还原染料,商名,德国制)

align 调整;定中心;校直

aligner 调中心工具(检查变形、变位的仪器总称)

aligning 校直;排成一行

aligning capacitor 调准电容器

alignment [直线]对准,校直;排成一行;准线;排顺,校准(指将机器的各部对准至正确的相互关系的位置,也指校准成一直线)

alignment mark 定位记号,对准记号

alignment pin 定位销,对中销

aliquot 等分[试样]

alive polymer 活性聚合物

alizarin 茜素(1,2-二羟基蒽醌)

alizarin dyes 茜素染料(媒染染料)

alizarine 茜素(1,2-二羟基蒽醌)

alizarine lake 茜素色淀

alizarin oil 茜素油;土耳其红油;磺化油

alizarin reds 茜草红布(用原色和茜草红色经纱及原色纬纱织成的条子布,西非商名)

alizarin synthesis 茜素合成

Aljaba printing machine 阿尔捷巴印花机(一种早期的圆网印花机)

Alkaflo 活性染料代用碱(无机盐强碱溶液,碱缓冲剂,用于活性染料染色,能控制染料水解,商名,美国先邦)

alkali 碱,强碱

alkali absorber 吸碱装置(丝光机用)

alkali-acid desizing 碱酸退浆

alkali binding agent 抗碱剂,碱中和剂

alkali blue 碱性蓝(酸性染料,可用作 pH 指示剂)

alkali boiling 碱煮

alkali boil-off 碱煮练

alkali buffer 碱缓冲

alkali cellulose 碱纤维素

alkali centrifuge value ⟨ACV⟩ 碱离心值(织物树脂整理的特征参数)

alkali cleaning 碱洗净;碱洗

alkali consumption 碱耗

alkali contracted 碱缩

alkali corrosion 碱性腐蚀

alkali damage 碱斑,碱损

alkali degradation 碱降解

alkali desizing 碱退浆
alkali deweighting finish(=alkali deweighting finishing) 碱减量整理(又称涤纶仿真丝绸整理)
alkali dialyser(=alkali dialyzer) 碱液渗析器(碱液回收装置)
alkali donor 碱给予体
alkali earth metal 碱土金属
alkali fastness 耐碱坚牢度
alkaliferous 含碱的
alkali fusion 碱熔融
alkalifying 碱化
alkali hydrolysis 碱水解
alkaliless desizing-scouring-bleaching process 无碱退煮漂工艺(绿色工艺)
alkalimeter 碳酸定量器(碳酸盐中二氧化碳定量测定装置)
alkalimetry 碱量滴定法
alkaline bath 碱性浴
alkaline carbonate 碱性碳酸盐
alkaline catalyst 碱性催化剂
alkaline degradation 碱降解
alkaline detergent 碱性净洗剂
alkaline disperse dyes 碱性分散染料(用于碱性浴染色)
alkaline milling 碱性缩绒
alkaline polymerization 碱性聚合
alkaline polysulphide 碱性多硫化物
alkaline scouring 碱性精练
alkaline sensitivity 碱敏[感]性,碱敏现象
alkalinity 碱度;碱性
alkali pectinases 碱性果胶酶
alkali peeling 碱减量,碱剥皮(使涤纶纤维变细的化学处理方法)
alkali peeling finish 碱剥皮整理(聚酯纤维碱减量工艺)
alkali polyphosphate 碱性多磷酸盐
alkali proof 耐碱
alkali reaction 碱性反应
alkali reclaim 碱液回收
alkali reserve(=alkali resist) 碱防染
alkali resistance 抗碱性,耐碱性
alkali-resist scouring agent 耐碱精练剂
alkali-sensitive link 碱敏键
alkali shock method [活性染料]碱快速固色法
alkali shrinkage single jersery 碱缩单面针织物
alkali side 碱性侧
alkali soap 碱性皂
alkali solubility 碱溶度
alkali solubility number [纤维素]碱溶性指数
alkali-soluble cotton 碱溶棉
alkali spotting 滴碱试验法(试验染色织物的耐碱色牢度)
alkali stain 碱斑;碱渍
alkali strike 碱性瞬染
alkali substitute 碱代替物;代用碱(用于活性染料染色)
alkali treatment 碱处理
alkali washing 碱清洗,碱洗
alkali yellowing 碱[致]泛黄(指羊毛)
alkalization 碱化作用;碱化
alkalized starch 碱淀粉
alkaloid 生物碱
alkane sulfonate 链烷磺酸盐;烷基磺酸盐
alkanolamine 烷基醇胺
alkanolamine soap 烷基醇胺皂
alkoxy 烷氧基

alkoxyalkyl sulfate 烷氧基烷基硫酸盐，烷氧基烷基硫酸酯
alkoxyl 烷氧基
alkoxylated alcohol 烷氧基化醇
alkoxysilane 烷氧硅烷
alkyd resin 醇酸树脂
alkyl alkoxy silicon ester 烷基烷氧基硅酯（硅防水整理剂）
alkylamide 烷基酰胺
alkylamine 烷基胺
alkylamine surfactant 烷基胺表面活性剂
alkylamino carboxylate surfactant 烷基胺基羧酸酯表面活性剂
alkylamino sulfate surfactant 烷基胺基硫酸酯表面活性剂
alkylamino sulfonate surfactant 烷基胺基磺酸酯表面活性剂
alkyl ammonium chloride 烷基氯化铵（阳离子型表面活性剂）
alkyl aryl compound 烷基芳基化合物
alkyl aryl polyglycol ether 烷基芳基聚乙二醇醚
alkyl aryl sulfonate 烷基芳基磺酸酯（阴离子表面活性剂）
alkylation 烷基取代，烷化
alkylation agent 烷化剂（硫化染料以烷化剂处理代替红矾氧化，可增加湿牢度，提高鲜艳度）
alkylation of cellulose 纤维素烷基化［工艺］（棉纤维上的羟基进行烷基化的变性处理）
alkyl benzene sulphonate〈ABS〉 烷基苯磺酸盐
alkylbenzolsulfonate 烷基苯磺酸盐
alkylcellulose 烷基纤维素

alkylhalosilane 烷基卤化硅烷（疏水整理剂）
alkyl ketene dimer 烷基乙烯酮二聚物（纤维素反应型的耐久防水柔软剂）
alkyl-methacrylate 烷基—甲基丙烯酸酯
alkylmorpholine surfactant 烷化吗啉（表面活性剂）
alkyl-naphthalene sulfonate〈＝alkyl naphthalene sulphonate〉烷基萘磺酸盐（聚酯针织物染色分散剂）
alkylphenol 烷基［苯］酚
alkylphenol ethoxylates〈APEO〉烷基酚聚氧乙烯醚（毒性较大的非离子表面活性剂，又名烷基苯酚聚乙二醇醚）
alkylphenol polyglycol ether 烷基苯酚聚乙二醇醚（非离子型活性剂，即烷基酚聚氧乙烯醚）
alkyl phosphate surfactant 烷基磷酸酯表面活性剂
N-alkyl polyamide N-烷基聚酰胺
alkylpolyglycol ether 烷基聚乙二醇醚（渗透剂）
alkyl polysaccharide glycoside〈APG〉烷基多糖苷（一种毒性低、易生物降解的"绿色"表面活性剂）
alkylpolysiloxane 烷基聚硅氧烷（消泡剂、拒水剂和柔软剂）
alkyl pyridinium chloride 烷基氯化吡啶［季铵］盐（防水剂）
alkyl pyridinium salt 烷基吡啶［季铵］盐（阳离子型活性型防水剂、柔软剂）
alkyl sulfate 烷基硫酸酯
alkyl sulfonate 烷基磺酸盐（阴离子型活性剂、净洗剂）
alkyl sulphonic acid 烷基磺酸

alkyl trimethyl ammonium bromide 烷基三甲基溴化铵
allaqueous system 全水性系统(非溶剂系统)
allaround pleated skirt 百褶裙
all cotton 全棉,纯棉
all cotton stretch yarn 全棉弹力纱
allen wrench 内六角扳手
allergens 过敏源
allergic 过敏性的
allergic dyestuffs 过敏性染料
allergic reactions(=allergic response) 过敏性反应
allergies 变态反应;过敏症
allergy of dyestuffs 染料的过敏性
alley 通道(指机器与机器之间的走道)
all fashion 全成形
alligator cloth 鳄鱼皮[式]漆布
all-in method 全料[印染]法
all-in printing process 全料印花法
all-in vat printing process 还原染料全料印花法
allocation 分配,布置;定位
allochroic(=allochromatic) 变色的;非本色的,带假色的
alloisomerism 立体异构[现象]
allopen screen 全开孔筛网
allosteric enzyme 变构酶
allotrope 同素异形体
all-over 满地,全幅;满地花纹
all-over colour 满地着色
all-over design 满地花纹图案,满地印花图案
all-over effect 满地花纹效应
all-over print 满地印花
all-over screens 满地印花筛网

allowable emission limit 排放容许浓度《环保》
allowable error (机件尺寸的)容许误差
allowable load 容许负载
allowable stress 容许应力
allowable value 容许值
allowance 公差(零件在加工制造时,容许的最大尺寸和最小尺寸的差值);[雕刻]分线留量
allowance error 容许误差
allowance for finish 精加工余量
allowance parts list 备件供应明细表
alloy 合金
alloying 掺混;熔合;共聚作用
alloy steel 合金钢
all-purpose 通用的
all-purpose detergent 通用洗涤剂
all-purpose soap 通用肥皂
all-round fastness 全面坚牢度
all round fibre 全能纤维(适合于各种用途)
all-round pattern 满地图案,满花
all round sewing 全面缝
all silk goods 全真丝织物(织物中没有蚕丝以外的纤维,但允许含有增重剂等化学品)
all-skin fibre 全皮层纤维
all skin rayon fibre 全皮层黏胶纤维
all steel clothing 全金属针布
all weather coat 风雨衣
all weather fabrics 全天候衣着;全天候织物
all weather garment 全天候服装
all wool 全毛,纯毛
all wool cloth 全毛织物
all wool hose 全羊毛袜

all wool yarn 全羊毛纱,纯羊毛纱
Allwörd's reaction 阿尔渥反应(羊毛浸于卤素水溶液中产生表皮囊状溶胀现象)
allyl alcohol 烯丙醇
allyl aldehyde 烯丙醛
allyl compound 烯丙基化合物
allylic resin 烯丙基树脂
allyl thermosetting polymer 烯丙基热固性聚合物
almond[brown] 杏仁棕[色](浅灰棕色)
almond green 杏叶绿[色](浅灰绿色)
aloe fibre 芦荟纤维
aloha shirt 夏威夷衫,香港衫
alongside printing with different sortiment of dyes 不同类别染料共同印花
Alpaca 阿尔帕卡织物;阿尔帕卡毛,羊驼毛;[碳化]阿尔帕卡再生毛
Alpaca wool 阿尔帕卡毛,羊驼毛
alpha 希腊字母α,可表示化学基团的位置
alpha-chlorohydrin[e] α-氯乙醇
alpha-fibre α-纤维(α-纤维素含量在94%以上的纤维)
alpha-keratin α-角蛋白
alpha-naphthol α-萘酚
alpha-olefinsulphonate α-烯(烃)磺酸盐(乳化剂)
alpha-rays α-射线
alterant 变色剂
alterant dyes 变色染料
alteration 变色;[服装]修改(服装出售时为了适合顾客体形而修改);变化,改变
alternate open washer 交流冲洗式平洗机(运转时织物在瓦楞式夹层中冲洗)
alternate twisted yarn 交替捻丝
alternating 交变,交替
alternating liquor circulation 浴液交替循环
alternative current〈A.C.〉交流[电]
alternative current motor 交流电动机
alternator 交流发电机
altitude adjustment 高度调整
alum 明矾
alumina 氧化铝,矾土
alumina-ceramic 氧化铝陶瓷,矾土陶瓷
alumina colloidal 胶态氧化铝(针织物防起球剂)
alumina gel 氧化铝凝胶
alumina mordant 铝媒染剂
alumina mordant dyeing 铝媒染色
alumina soap 铝皂(拒水整理用金属皂)
alumina white 矾土白
aluminium 铝
aluminium acetate 醋酸铝
aluminium acetate water proofing 醋酸铝防水
aluminium alloy 铝合金
aluminium ammonium sulphate 硫酸铝铵
aluminium and iron complex dyes 铝铁络合染料
aluminium bronze powders for printing 印花用铝青铜粉末
aluminium chlorate 氯酸铝
aluminium chlorhydroxide 碱式氯化铝(树脂整理催化剂、凝聚剂)
aluminium chloride 氯化铝
aluminium-containing polymer 含铝聚合

物
aluminium foil 铝箔
aluminium foil printing 铝箔印花
aluminium formate 甲酸铝
aluminium hydroxide 氢氧化铝
aluminium hydroxy chloride 碱式氯化铝（作为制防水剂铝皂的原料，也可作织物防滑剂）
aluminium mordant 铝媒染剂
aluminium oxide fibre 氧化铝纤维
aluminium potassium sulphate 硫酸铝钾，钾矾
aluminium silicate fibre 硅酸铝纤维
aluminium silicate refractory fibre 硅酸铝高温耐火纤维
aluminium soap water proofing 铝皂防水
aluminium stearate 硬脂酸铝
aluminium sulphate 硫酸铝
aluminium sulphate acetate 醋酸硫酸铝（茜素红染丝绸用媒染剂及防水剂）
aluminium thiocyanate 硫氰酸铝
aluminium waterproof 铝防水
aluminium yarns 铝皮，金银皮；铝线；金银线
aluminized 涂铝的，喷铝的
aluminosilicates 铝硅酸盐
aluminum【美】= aluminium
alum mordant 明矾媒染剂
alundum 氧化铝，铝氧粉，刚铝石
AM(acrylic amide, acrylamide) 丙烯酰胺
amalgamate 混合；合并；汞齐化
amalgamation 汞齐化[作用]
amalgam electrode 汞齐电极
amaranth colour 苋菜红色素（合成食品红色素）
amaranth purple 深青莲，紫罗兰
amber [yellow] 琥珀黄[色]（暗黄色）
ambient air 周围空气，环境空气
ambient conditions 环境状态，大气条件
Ambiente printing process 一种印花方法（局部着色，没有固定图案的自然花型，芬兰 Ambiente 公司开发）
ambient sensor 气温传感器
ambient temperature 室温，环境温度
ambiguity 模糊度（不清晰度，测色用语）
AM(aminized) cotton 胺化棉（通过胺化作用制取的变性棉纤维）
AM(animalized) cotton 动物质化棉，羊毛化棉
amendment 修正，更改
American Academy of Sciences〈AAS〉美国科学院
American Apparel & Footwear Association〈AAFA〉美国服饰与鞋类协会
American Apparel Manufacturers Association〈AAMA〉美国服装制造商协会
American Association for Textile Technology〈AATT〉美国纺织工艺协会
American Association of Textile Chemists and Colourists〈AATCC〉美国纺织化学家和染色家协会
American Chemical Society〈ACS〉美国化学学会
American Dye Manufacturers Institute〈ADMI〉美国染料制造商协会
American Dyestuff Reporter〈A. D. R.〉《美国染料报道》（月刊）
American Industrial Health Council〈AICE〉美国工业保健委员会

American Institute of Chemical Engineers〈AICE〉美国化学工程师协会
American Institute of Chemists〈A. I. C.〉美国化学工作者协会
American Institute of Consulting Engineers〈AICE〉美国顾问工程师协会
American mode 美国流行款式
American National Standards Institute〈ANSI〉美国国家标准研究所
American Patent〈AP〉美国专利
American Public Health Association〈APHA〉美国公共卫生协会
American Retail Federation〈ARF〉美国零售业联合会
American Society for Testing and Materials〈ASTM〉美国材料试验学会
American Society of Mechanical Engineers〈ASME〉美国机械工程师学会
American standard 美国标准
American Standard Association〈ASA〉美国标准协会(从 1966 年起改为 USASI)
American standard of Testing Materials〈ASTM〉美国材料试验标准
American Textile Export Co.〈AMTEC〉美国纺织品出口公司
American Textile Machinery Association〈ATMA〉美国[国际]纺织机械协会
American Textile Machinery Exhibition-International〈ATMA-I〉美国国际纺织机械展览会
American Textile Manufactures Institute〈ATMI〉美国纺织制品研究所
American wool grading system 美国羊毛分等制
Ames test 艾梅士试验《环保》

amianthus(=amiantus) 高级石棉,石绒
Amicor fibre 抗菌纤维(英国制)
amicron 超微粒,次微[胶]粒(直径小于 10^{-7} 厘米)
amidases 酰胺酶(裂解 C—N 键的酶)
amidated cotton 酰胺化棉
amidation 酰胺化[作用]
amide 酰胺;氨化物
amide resin 酰胺树脂
amide type solvent 酰胺型溶剂
amido bond 酰胺键
amido colours 酰胺染料
amido group [酰]氨基
amidolysis 酰胺解
amido-methylol reactant 羟甲基酰胺反应剂
amidosulfonic acid 氨基磺酸(酸性染料染深色时的助染剂)
amination 胺化[作用]
amine antioxidant 胺类抗氧剂,胺类防老[化]剂
amine-carboxyl end-balance 胺—羧端基平衡
amine cellulose 氨基纤维素
amine end group 末端氨基
amine odor 胺臭
amine soap 胺皂
amine surfactant 胺类表面活性剂
aminized〈AM〉cotton 胺化棉(通过胺化作用制取的变性棉纤维)
aminoacetic acid 氨基醋酸,甘氨酸
amino acid 氨基酸
amino acid derivatives 氨基酸衍生物
amino acid residues 氨基酸残基
amino alcohol 氨基醇
aminoalkoxysilane 氨基烷氧硅烷

amino amide 氨基酰胺
amino-azobenzene 氨基偶氮苯(染料的主要中间体)
amino azobenzene dyes 氨基偶氮苯染料
aminocaprolactam 己内酰胺
amino carboxylic acid 氨基羧酸
amino carboxylic acid condensation compound 氨基羧酸缩合物
amino carboxylic acid ester〈ACAE〉氨基羧酸酯
amino ethyl alcohol 氨基乙醇
aminoethylated cotton 氨乙基化棉(棉纤维经 2-氨乙基硫酸和浓烧碱液处理而成,是碱性溶解棉的一种)
amino ethyl cellulose 氨基乙基化纤维素
amino extra groups 碱性氨基酸
aminoformaldehyde resin 氨基甲醛树脂
aminofunctional silicone 功能性氨基有机硅
amino group 氨基;胺
aminoketone dyes 氨基酮染料
aminolysis 氨解
amino phenol 氨基苯酚
aminoplastics(＝aminoplasts) 氨基塑料,氨基树脂
aminopolycarboxylic acid 氨基多元羧酸
amino resin 氨基树脂(含有大量氨基氮的聚合物总称)
amino silane 氨基硅烷(用作硅防水剂)
aminosiloxane 氨基硅氧烷,氨基硅油
amino sulfonic acid 氨基磺酸
aminotriacetic acid 氨基三乙酸
aminotriazine 氨基三嗪(即三聚氰胺)
amino trimethyl carboxylic acid 氨三甲基羧酸
amiray ramie 苎麻[纤维](菲律宾名称)

ammeter 安培计,电流计,电流表
ammonia 氨[水]
ammonia complex 氨络合物
ammonia converter 氨转化器
ammonia cure 氨熏(阻燃整理后处理方法)
ammonia-dry-steam process 液氨干汽蒸法
ammonia finishing 液氨整理
ammonia gas 氨气
ammonia mercerizing 液氨丝光
ammonia test [for dyeing][染色织物的]氨水试剂(用于检测毛棉交织产品上的染料)
Ammoniator 氨熏机(商名)
ammonia-treatment chamber 氨处理室(液氨整理设备之一)
ammonia-water process 液氨水洗法,氨丝光工艺
ammonical nitrogen 氨氮《环保》
ammonical silver nitrate solution 硝酸银氨溶液(用于检测氧化与水解纤维素)
ammonio-cupric sulphate 硫酸铜铵,氨合硫酸铜
ammonium 铵
ammonium acetate 醋酸铵
ammonium acid fluoride 氟氢化铵(酸化剂,除锈剂,消毒剂)
ammonium alginate 海藻酸铵
ammonium alum 铵矾
ammonium carbonate 碳酸铵
ammonium chloride 氯化铵
ammonium chloride test 氯化铵试验
ammonium chromate 铬酸铵(印花过程中作媒介染料媒染剂)

ammonium compound quaternary 季铵化合物
ammonium ferric sulphate 硫酸铁铵
ammonium ferrous sulphate 硫酸亚铁铵
ammonium formate 甲酸铵
ammonium hydroxide 氢氧化铵
ammonium linoleate 亚油酸铵
ammonium persulphate 过硫酸铵(氧化剂,漂白剂)
ammonium phosphate 磷酸铵
ammonium polyacrylate 聚丙烯酸铵(浆料)
ammonium polysulphide 多硫化铵
ammonium p-toluene sulphonate (ATS) 对位甲苯磺酸铵(涂料印花催化剂)
ammonium soap 铵皂
ammonium sodium hydrogenphosphate 磷酸氢铵钠
ammonium stearate 硬脂酸铵
ammonium sulfamate 氨基磺酸铵(阻燃剂)
ammonium sulfate 硫酸铵
ammonium thiocyanate 硫氰酸铵
ammonium thioglycolate 巯基乙酸铵
ammonium vanadate 钒酸铵
ammonolysis 氨解[作用]
amorphism 无定形现象;非晶态
amorphous 非晶的,无定形的
amorphous band 非晶区谱带
amorphous cellulose 非晶态纤维素,无定形纤维素
amorphous content 非晶[态]区含量,无定形区含量
amorphous crystalline 无定形结晶
amorphous orientation 非晶取向
amorphous polymer 非晶型高分子

amorphous region 无定形区,非晶区
amorphous structure 无定形结构
amortization 缓冲,阻尼
ampere 安[培](电流单位)
amperemeter 安培计,电流计,电流表
amperometric titration 电流滴定法
amphibilic compound 双亲化合物(亲水又亲油)
amphibious craft 水陆两用船
amphipathic 亲水、疏水两重性的,双亲的
amphipathic adsorption 双亲吸附作用
amphipathic property 亲水亲油双亲性的
amphiphilic 亲水、疏水两重性的,双亲的
amphiprotic 两性的(既显示酸的性质,又显示碱的性质)
amphiprotic solvent 两性溶剂
amphivalent crosslinking agent 双性交联剂
ampholyte 两性电解质,两性物
ampholyte ion 两性电解质离子
ampholytics (= ampholytic surface active agent) 两性表面活性剂
amphoteric 两性[兼备]的(既显示酸的性质,又显示碱的性质)
amphoteric character 两性特征
amphoteric electrolyte 两性电解质
amphoteric fatty compound 两性脂肪族化合物
amphoteric ion exchange fibre 两性离子交换纤维
amphoteric surfactant 两性表面活性剂
amplifier 放大器;增音器;放大镜
Amsler type abrasion tester 阿姆斯拉型耐磨试验仪
AMT (advanced multiaxial technology) 先进的多轴向技术

AMTEC(American Textiles Export Co.)
美国纺织品出口公司
amunzen 阿蒙增绉纹呢(原是毛织品,现用棉或黏纤织造)
AMW(average molecular weight) 平均分子量
amyl acetate 醋酸戊酯
amyl alcohol 戊醇
amylase 淀粉酶
α-amylase α-淀粉酶
amylic 淀粉的
amylodextrin[e] 淀粉糊精,可溶性淀粉
amylogen 可溶性淀粉
amylolysis 淀粉分解
amylolytic activity 淀粉分解力
amylolytic enzymes 分解淀粉酶
amylopectase 支链淀粉酶
amylopectin[e] 支链淀粉,胶淀粉
amylopsin 胰淀粉酶
amylose 直链淀粉,糖淀粉
anaerobe(=anaerobia) 厌氧微生物,厌氧菌
anaerobic-aerobic treatment 厌氧需氧处理《环保》
anaerobic bacteria 厌氧细菌《环保》
anaerobic biodegradation 厌氧性生物降解
anaerobic biological treatment 无氧生物处理
anaerobic condition 厌气条件,厌氧条件
anaerobic digestion 厌氧消化
anaerobic fermentation 厌氧发酵,无氧发酵
anaerobic treatment 厌氧处理《环保》
anaerobic waste treatment 废水厌氧处理《环保》
analog data 模拟数据

analog digital control 模拟数字控制
analog-digital conversion〈ADC〉 模拟数字转换《计算机》
analog dosing 模拟量加料
analog element 模拟元件
analog input 模拟输入
analog module 模拟组件
analogous colours 类似色(如蓝光绿色和绿光蓝色)
analogue computer 模拟计算机
analogue display 模拟显示
analogue output 模拟输出
analogue to digital converter〈ADC〉 模拟数字转换《计算机》
analogy colour harmony 类比配色
analyser 分析器,分析机;检偏振器
analysis of blends 混纺织物分析
analysis of dyestuffs 染料分析
analysis of fabric 织物分析
analytical balance 分析天平
analytical column 分析柱
analytical error 分析误差
analytical methodology 分析方法论
analyticity 解析性
analyzer 分析器,分析机;检偏振器
anaphe silk 非洲阿纳菲野蚕丝
anatase 锐钛矿(二氧化钛)
Anatolian carpets 安那托利亚小毛毯(小亚细亚地区用的大几何图案毛毯)
anchor 支撑点,固定器
anchor agitator [小型移动式]锚式搅拌器
anchor bar 紧固杆,制动杆
anchor bolt 地脚螺栓
anchor coating 固定涂层(绒类织物固定绒毛的方法)
anchoring power [织物组织]固定能力

anchor method 锚法(测定助剂润湿性的方法)
anchor mixer [小型移动式]锚式搅拌机
anchor pin 固定销,连接销
anchor screw 地脚螺栓
anchor shackle 锚钩环(吊装装置)
anchor stirrer [小型移动式]锚式搅拌器
ancient madder 古茜草[印花],仿茜草[印花]
ancillary attachment 辅助附件,辅助附加装置
ancillary detergent 助洗剂
ancillary equipment 辅助设备,附加装置
ancillary system 辅助系统
ancillary tools 辅助工具
anemometer 风速计
angle bar 角钢,角铁
angle blade 角刮刀
angle brace 角撑
angle bracket 角撑架,角形托座
angle drive 转角传动
angle frame defect 印框疵(平版印花印框角压力过重)
angle gear 斜交轴伞齿轮
angle iron 角铁
angle lever 角形杠杆
angle of incidence 入射角
angle of internal friction 内摩擦角
angle of recovery 回复角;[织物]折皱回复角
angle of reflection 反射角
angle of refraction 折射角
angle of rotation 旋光度
angle plate 角板(有角度的金属板,作夹住和支托工件用)
angora 安哥拉山羊毛(也称马海毛);安哥拉兔毛;棉经马海毛纬平纹交织物;斜纹大衣呢;安哥拉兔毛[混纺]纱;安哥拉披肩
angora cashmere 马海毛斜纹薄织物
angora cloth 马海毛交织呢(棉经、马海毛纬)
Angora goat 安哥拉山羊
angora mohair 安哥拉马海毛
angora rabbit hair 安哥拉兔毛
angora sweater 兔毛衫
angora union 马海毛交织呢
angora yarn 安哥拉兔毛[混纺]纱
Angström (=angstrom) 埃(长度单位,等于 10^{-10} 米)
angular belting 皮带转角装置
angular contact bearing 斜座轴承
angular drainage 斜通道排水
angular flammability test 倾斜式阻燃试验
angular velocity 角速度
anhydrating agent 脱水剂
anhydride 无水物,酐
β-anhydroglucose β-脱水葡萄糖,β-失水葡萄糖
anhydroglucose unit 葡萄糖残基
anhydrohexitol 失水己糖醇
anhydrohexose 脱水己糖
anhydrous 无水的
anhydrous phosphate 无水磷酸盐
anilide 苯胺
aniline 阿尼林油,苯胺
aniline ager 苯胺黑蒸化机,精元蒸化机
aniline black 精元,氧化元,苯胺黑,阿尼林黑
aniline black ager 苯胺黑蒸化机,精元蒸化机
aniline black dyeing 精元染色

aniline black fastness property 苯胺黑牢度性能
aniline black range 精元染色联合机,氧化元染色联合机
aniline blue 苯胺蓝(碱性染料)
aniline brown 苯胺棕(碱性染料)
aniline colorimetric method 苯胺比色法
aniline dyes 苯胺染料
aniline hydrochloride 苯胺盐酸盐(阿尼林盐的主要成分)
aniline pinking 苯胺性泛红
aniline purple 苯胺紫
aniline resist printing 苯胺黑[地色]防染印花,精元[地色]防染印花
aniline salt 阿尼林盐,苯胺盐(通常指苯胺盐酸盐)
anilism 苯胺中毒
animal colloids 动物胶体
animal dye 动物染料(天然染料的一种,如胭脂红)
animal fibre 动物纤维
animal glue 动物胶
animal hair 动物毛,兽毛
animal hair-like fibre 仿毛纤维
animalization [纤维]动物质化(使纤维的染色性与羊毛相似)
animalized (AM) cotton 动物质化棉,羊毛化棉
animalized cellulose fibre 动物质化纤维素纤维,羊毛化纤维素纤维
animalized fibre 动物质化纤维,羊毛化纤维(经化学处理后的棉或人造纤维,染色性与羊毛相似)
animalized viscose rayon fibre 动物质化黏胶纤维
animalizing 动物质化处理,动物质化变性
animal protein fibre 动物蛋白质纤维
animal wax 动物蜡
anion 阴离子
anion activator(＝anion active agent) 阴离子活化剂
anion additive 负离子添加剂
anion complex 阴离子络合物
anion emission textiles 释放负离子纺织品
anion exchange fibre 阴离子交换纤维
anion exchange resin 阴离子交换树脂
anion FIR technique 负离子远红外技术
anionic 阴离子的
anionic catalyst 阴离子催化剂
anionic catalytic polymerization 阴离子催化聚合[作用]
anionic-cationic compound 两性离子化合物(电子中和化合物)
anionic dyes 阴离子染料
anionic exchanger 阴离子交换器
anionic polymerization 阴离子聚合[作用]
anionic polypropylene fibre 负离子丙纶纤维
anionic retardant 阴离子缓染剂
anionics(＝anionic surfactant) 阴离子剂,阴离子[型]表面活性剂
anionic site 阴离子位置
anionic softener 阴离子柔软剂
anion minerol 负离子矿物
anisomorphic 异构的;非同构的
anisotropic composite material 各向异性复合材料
anisotropic swelling 各向异性溶胀
anisotropy 各向异性[现象]
anklet 翻口短袜,套袜
ANLAB colour-difference equation ANLAB

色差方程
ANN（artificial nerve networks）人工神经网络（用以研究面料与辅料间织物行为关系）
annato（＝annatto）胭脂树红（浅橙色）；橙色植物染料
anneal 退火，热处理
annealed wool [定温]热处理的羊毛
annealed zone 缓冷区；热处理区；重结晶区
annealing [羊毛等]热处理，加速熟化；退火
annealing oven 热处理炉
annealing temperature 热处理温度
annealing treatment 热处理
annelation 增环反应，稠合作用（邻近化学环的结合）
Ann. Rep.（Annual Report）年报，年鉴
annual overhaul 年度检修
Annual Report〈Ann. Rep.〉年报，年鉴
annular ball bearing 径向球轴承
annular gear 内齿轮
annular seal 密封环
annular squeegee holder [圆网上胶用]环状刮刀夹
annunciator 指示器
anode 阳极（半导体器件）；正极（电子学）；板极（电子管）
anode glow 阳极辉光
anodic oxidation 阳极氧化
anomaloscope 色觉反常测验镜，色盲检测镜
anomalous colour vision 异常色觉
anomalous fading 反常[日晒]褪色（例如某些染料的耐日晒牢度随颜色的加深反而下降）

anomalous trichromatism 色弱,反常三原色性
ANS(Adams-Nickerson system) 亚当斯—尼克尔森[色差]系统
ANSI(American National Standards Institute) 美国国家标准研究所
antacid 抑酸剂
antagonism 对抗[作用],对抗性
antagonistic effect 对抗效果
antelope finish 仿羚羊皮整理
anthanthrone 蒽缔蒽酮，二苯并芘二酮
antherea pernyi [中国种]柞蚕
antherea silk [印度]柞蚕丝
anthracene 蒽
anthranilamide〈ATA〉邻氨基苯甲酰胺
anthranol 蒽酚
anthrapyridone 蒽吡啶酮
anthrapyrimidine 蒽素嘧啶
anthraquinone 蒽醌
anthraquinone dyes 蒽醌类染料
anthraquinone imine 蒽醌亚胺
anthraquinone-N-hydrazine 蒽醌-N-肼
anthraquinone vat dyes 蒽醌还原染料
Anthrasole 蒽台素染料（可溶性还原染料，商名，德国制）
anthrasol resist printing 溶蒽素防染印花
anthrasol-thermosol development〈ATE〉process 溶蒽素热熔显色法
anthrenus 各种地毯甲虫（羊毛制品上的蛀虫）
anthrenus beetle protection 地毯甲虫防护
anthrimide 蒽醌亚胺
anthrone 蒽酮
anthrone dyes 蒽酮染料
anthropometry 人体测量学
antiacid 抑酸剂

antiactivator 阻活化剂,活化阻止剂
antiadhesive 抗黏剂
antiager 抗老[化]剂,防老[化]剂
antiallergen 抗过敏
antiallergic treatment 抗过敏处理(合纤针织内衣抗过敏整理)
antibackstaining agent 防再沾污剂
antibacterial agent 抗菌剂
antibacterial alginate fibre 抗菌海藻纤维
antibacterial alginate/gelatin blended fibre 抗菌海藻酸明胶共混纤维(医用纺织纤维)
antibacterial fibre 抗菌纤维
antibacterial finish 抗菌整理,抑菌整理
antibacterial property 抗菌性
antibacterial textile 抗菌纺织品
antibactirial and purifying finish 抗菌防臭整理
antibagging finish 防拱胀整理(尤指肘膝处)
antibiosis 抗菌性
antibiotic agent 抗菌剂
antibiotic auxiliary 抗菌助剂
antibiotic fabric 抗菌织物
antibiotic finish 抗菌整理
antiblast technology 抗冲击技术
antiblaze 抗火焰剂,阻燃剂
Antiblaze 19T 恩的勃拉时 19T(涤纶阻燃剂,商名,美国制)
antiblockage device 防堵塞装置
antiblock agent(=anti-blocking agent) 防黏剂;抗阻塞剂
antibody 抗体
anticatalyser(=anticatalyst,anticatalyzer) 反催化剂,抗催化剂(双氧水稳定剂)
antichlor 脱氯剂

antichloration(=antichlorination) 脱氯,去氯
anticling finishes 抗贴身整理(对化纤织物的,近似抗静电整理)
anticlockwise 逆时针转
anticoagulant 阻凝剂,凝固防止剂
anticoagulation 阻凝[作用],抗凝[作用]
anticockle 抗皱
anticockle treatment 防皱整理(羊毛针织物)
anticonfiguration 反型,反式构型
anticontaminator 防污染剂
anticorrosion paint 防腐涂料,防腐漆
anticorrosive 防腐的;防锈的
anticorrosive agent 防腐蚀剂
anticorrosiveness 防腐性
anticorrosive oil 防锈油
anti-counterfeiting 防伪造,抗假冒
anticracking agent 抗裂解剂,抗裂剂
anticreaming agent 防稠剂
anticrease 防皱,抗皱[的]
anticrease finish 防皱整理
anticreasing agent 防皱剂
anticrocking agent 摩擦[脱色]牢度改进剂
anticurl 防卷边
anticurl agent 防卷边剂
anticurl finish 防卷边整理
antidarkening 防黯剂
antidazzling illumination 防耀眼照明
antidepinning device 防脱针装置(用于针板拉幅机,防止织物在拉幅运行时从针板上脱开)
antideteriorant 防坏剂,防老剂
antidiazosulphonate 反重氮磺酸盐
antidiazotate 反重氮盐,稳定重氮盐,反

偶氮酸盐,异重氮酸盐
antidusting 抗尘[作用];抗尘的
antielectrolyte 抗电解质
antielectromagnetic radiation fabric 防电磁辐射织物
antienzyme 抗酶
antiexplosion hatches 防爆天窗(烘干机用)
antiexposure suit 抗浸服(用于海上航行时防海水浸渍)
antifatigue 抗疲劳的,耐疲劳的
antifelting agent 防毡缩剂
antifelting treatment 防毡缩处理
antiferment 防酵剂,抗发酵剂
antifibrillation 抗原纤化[处理]
antiflaming[finish] 阻燃[整理],防焰[整理]
antiflaming agent 阻燃剂
antifloating agent 防浮剂
antiflocculating agent 抗絮凝剂
antiflushing agent 抗渗化剂(印花工艺用)
antifoam 消泡沫,防泡沫
antifoamer(=antifoaming agent) 防[泡]沫剂,脱泡剂,消泡剂
antifog agent 防雾剂(用于保持玻璃透明)
antifogging film 防雾薄膜
antifouling 防污,防塞
antifouling agent 防污剂;防腐剂
antifreezing 防冻
antifreezing agent 防冻剂
antifriction bearing 减摩轴承,滚动轴承
antifrosting agent 防起霜剂,防霜花剂《染》《整》
antifrosting process 防霜花处理,防局部脱色处理
antifrother(=antifrothing agent) 防泡沫剂,消泡剂
antifume finish 抗烟[气]整理;抗酸气整理
antifungal agent 抗真菌剂
antifuzzing and antipilling finish 防起毛起球整理
antifuzzing finish 防起毛整理
antifuzzing finish agent 防起毛整理剂
antigas clothes 防毒衣
antigasfading agent 烟熏褪色防止剂
antigelling agent 抗胶凝剂
antiglare colour 反炫光色泽
antiglossing pigment 消光涂料
antigreying agents 防起霜剂
anti-G suit 抗荷服(用于特技飞行)
antihemorrhagic fibre 止血纤维
antihole-melt finishing 抗熔孔整理
antihypothermia bag 抗低温[睡]袋
antiicing agent 防冻剂
antiincrustant 防水垢剂
antiinfection apparel 防感染服
antiknock 减震,消震
antiladdering[丝袜]防抽丝[整理]
antilivering 防浮色剂
antilubricant 防滑剂
antimatting finish 防退光整理
antimelting finishing 抗熔融整理
antimicrobial agent 抗微生物剂,抗菌剂
antimicrobial fibre 抗微生物纤维
antimicrobial finish 抗微生物整理
antimicrobic fibre 抗微生物纤维
antimicrobiotic finish 抗菌整理
antimicrobiotics 抗菌剂
antimigrant(=antimigrating agent) 防泳

移剂
antimildew agent 防霉剂,抗霉剂
antimilling 抗缩绒加工;羊毛防缩加工
antimist cloth 透明织物;防霉织物
antimony electrode 锑电极
antimony oxide-halogen-latex 氧化锑—卤素—乳胶(阻燃整理剂)
antimony potassium tartrate 吐酒石;酒石酸氧锑钾
antimosquito finish 驱蚊整理
antimoth 防蛀
antimycotic 抗霉菌的
antimycotic agent 抗霉菌剂
antimycotic finish 抗霉菌整理
antiodour finish 防臭整理,防气味整理
antioxidant 抗氧剂,防老[化]剂
antioxidant additive 抗氧添加剂
antiozonidate 防臭氧剂
antiperspirants 防汗剂,除臭剂
antiperspiration finish 防汗渍整理
antipicking finish [呢面]清洁整理
antipill fibre 抗起球纤维
antipilling agent 防起球剂
antipilling finishing 防起球整理
antipollution 防污染,去污染
antiprecipitant 抗沉淀剂
antiputrefactive 防腐的
antique bronze 古铜色(暗黄棕色)
antique finish 仿古整理
antique finished carpet 仿古[整理]地毯
antique rug 仿古地毯
antiradiation analysis 抗辐射分析
antiradioactive contamination agent 抗放射性污染用剂
antiradioactive contamination garment 抗放射性污染服装

antiradioactive contamination technology 抗放射性污染技术

antiradioactivity 抗放射性
antiredeposition agent 防再污染剂(抗再沉淀剂)
antireflecting coating 防反射涂层
antirelease spring 防松弹簧
antirheumatic 抗风湿的
antirheumatic fibres 防风湿[症]纤维(聚丙烯腈、氯化聚氯乙烯纤维等,具有低导热、高电位差等性能)
antirotting agent 防腐剂
antirot treatment 防腐处理,防腐整理;防菌整理
antirumble 消声器
antirust agent 防锈剂
antisag 防垂荡,防松垂;防伸长变形
antiscumming agent 消浮垢剂,脱浮沫剂
antisepsis 防腐作用
antiseptic [agent] 防腐剂,抗菌剂
antiseptic cotton 防腐棉,消毒棉
antiseptic dressing 绷带,外科包扎布
antiseptic finish 防腐整理
antiseptic gauze 消毒纱布
antiseptics 防腐剂,抗菌剂
antisetting agent 抗定形剂(羊毛染整工艺用)
antisetting mechanism 抗定形机理《毛》
antishrink 防缩
antishrink finish 防缩整理
antishrink setting 防缩定形
antishrink treatment 防缩处理;预缩处理
antiskid power belt 防滑传动带
antislip finish 防滑整理
antislip finishing agent 防[经纬纱]滑移整理剂

antismell finish 防臭整理
antisnag finish 防钩丝整理;防起毛整理,防擦毛整理
antisnagging agent 防钩丝整理剂
antisnarl device 防扭结装置
antisoil 防污,抗污
antisoil agent 防污剂
antisoiling finish 防污整理
antisoil-redeposition property 防尘污再沾着性
antisplash 抗泼水的(防水整理性能)
antisqueak 消声器,减声器
antistain finish 防污整理
antistaining agent 防沾污剂
antistaining detergent 防沾污净洗剂
antistat 抗静电剂;静电消除器
antistatic 抗静电
antistatic agent 抗静电剂
antistatic device 抗静电装置
antistatic effect 抗静电能力,抗静电效应
antistatic fibre 抗静电纤维
antistatic finish 抗静电整理
antistatic in dry cleaning 干洗用抗静电剂(防止干洗剂着火爆炸)
antistatic installation 抗静电设备;离子化装置
antistatic property 抗静电性
antistatic spray 抗静电喷雾
antistatic test 抗静电试验
antistatic treatment 抗静电处理
antistick chemicals 抗黏化学品
antistick film wrapping 薄膜包复防黏
antisticking agent 防黏剂,抗黏剂
antisunburn 防日晒
antiswelling agent 防膨化剂,防溶胀剂
antitack agent 防黏剂

antitarnish agent 防锈剂;防黯剂
antithixotropy 反触变性,抗触变性
antitickle finish 防痒整理
antitoxic 防毒的,抗毒的
antitoxic filtration material 防毒过滤材料
antiultraviolet finishing 抗紫外线整理
antiultraviolet-ray 抗紫外线
antiultraviolet-ray finishing 抗紫外线整理
antiultraviolet-ray finishing agent 抗紫外线整理剂
antiultraviolet-ray propylene filament 抗紫外线丙纶长丝
anti-UV agent 抗紫外线整理剂
antiwicking agent 抗渗化剂,防渗化剂《印》
antiwrap device 防卷缠装置
antiwrap guide 防卷缠装置
antiyellowing 防泛黄[整理]
antiyellowing finish 防泛黄整理
AOS (apparent opening size) 表观孔径
AOU (apparent oxygen utilization) 表观耗氧量
AOX (adsorbable organic halogen) 可吸附的有机卤素《环保》
AOX load 可吸附的有机卤素负荷,可吸附的有机卤素含量《环保》
AOX value 可吸附的有机卤化物值《环保》
AP (American Patent) 美国专利
APCA (Air Pollution Control Association) 空气污染控制协会(美国)
APD (automatic powder dispenser) 粉剂自动分配系统
APEO (alkylphenol ethoxylates) 烷基酚聚氧乙烯醚(毒性较大的非离子表面活性剂,又名烷基苯酚聚乙二醇醚)

APEO free 不含烷基酚聚氧乙烯醚的
Apermeter ［织物］透气性测定仪（商名）
aperture 孔径；光圈；开口；隙缝
aperture colour 小孔色；非相关色，光孔色
APF（Advanced Performance Finish） 功能整理［巴斯夫（BASF）公司多功能整理技术］
APG（alkyl polysaccharide glycoside） 烷基多糖苷（一种毒性低、易生物降解的"绿色"表面活性剂）
APHA（American Public Health Association） 美国公共卫生协会
API（air pollution index） 空气污染指数
APO（tris-aziridinyl phosphine oxide） 三乙烯亚氨基膦化氧（棉织物阻燃整理用）
apochromatic microscope objective 复消色差显微镜（使不同波长光同时成像）
apocynum 罗布麻
APO flame retardent APO（氧化膦）阻燃剂
Apollotex 阿波罗特克斯微波染色设备（商名，日本市金公司）
Apollotex Eletron Reactor 阿波罗特克斯微波染色机（新颖的连续染色整理设备，商名，日本市金公司）
APO resin APO树脂（三乙烯亚氨基膦化氧树脂）
aposematic colouration 警戒色
APO-THPC flame-retardant process 三乙烯亚氨基膦化氧—四羟甲基鳞化氯阻燃处理法
APP（atactic polypropylene） 无规［立构］聚丙烯
APP（atmospheric pressure plasma） 常压等离子体
apparatine 碱化淀粉浆（经烧碱处理并中和使黏性增强的马铃薯淀粉浆）
apparatus 装置，仪器，设备
apparatus analyse 仪器分析
apparatus error 仪器误差
apparel 服装（统称）
apparel fabric 衣料
apparel textiles 服装纺织品
apparel wool 衣料用羊毛
apparel yarn 衣料用纱
apparent activation energy 表观活化能
apparent constant 表观常数
apparent density 表观密度
apparent detergency 表观去污力（污布洗涤前后以反射率的改变所表示的去污力）
apparent energy of activation 表观活化能
apparent fluidity 表观流度
apparent length 表观长度
apparent melt viscosity 表观熔体黏度
apparent opening size〈AOS〉 表观孔径
apparent oxygen utilization〈AOU〉 表观耗氧量
apparent power 表观功率
apparent saturation value 表观饱和值
apparent shear viscosity 表观剪切黏度
apparent strength 表观强度
apparent thickness 表观厚度
apparent viscosity 表观黏度
apparent volume 表观容积，表观体积
apparent weight 表观重量
appearance 外观
appearance inspection 外观检查
appearance of dyeing 染色的外观
appearance of grounding ［印花］露底
appearance quality 外观质量
appearance rating 外观等级

appearance retention 外观保持性,外观稳定性
appearance retention test 外观保持性试验
appearance standard 外观标准,外观标样
appending label 商标标签
appendix cloth ［色牢度检验用］贴衬布
applegreen 苹果绿［色］
appliance 器具,用具
applicant 报验者,报验单位
applicating roller 给液辊;上胶辊
application diversity 应用多样性
application printing 直接印花(实用印花工艺,如直接辊筒印花、手工模板印花、筛网印花)
applicator nozzle 给液喷嘴(泡沫整理机的给液装置)
applicator roller 给液辊
applicator wheel 上胶轮(针织物拉幅机的机件,布边上胶用)
applied printing 直接印花(实用印花工艺,如直接辊筒印花、手工模板印花、筛网印花)
applique printing ［发泡式］立体印花
apposition dyeing 同浴染色;表面染色(只染纤维周围不及内芯)
appret 织物上浆整理
approved sample 封样,验收样(作对比用的实物标样)
apricot (=apricot buff) 杏黄［色］
apricot cream 米黄色乳剂,杏色乳剂
apricot skin 桃皮绒
apron 皮圈;皮板输送带;挡板圈裙
apron band 输送带
apron dryer 帘式烘燥机
apron fabrics 输送带织物
aprotic solvent 非质子溶剂;非酸碱溶剂

APS (automatic presetting system) 自动预调系统
AQL (acceptable quality level) 合格质量标准,验收合格标准
aqua 水(拉丁语);溶液;液体
aqua-blast washing machine 喷水水洗机(在水洗槽液面上安装两个高压喷射洗涤装置,向织物喷射高压水流,喷口对面装有能承受高压的辊筒)
aqua fortis 浓硝酸
aqua green 水绿［色］(浅黄绿色)
AquaJet Spunlace system 射流喷网系统［商名,德国福来司拿(Fleissner)］
Aqualuft 阿夸卢弗特液流染色机(低浴比染色机,商名,美国)
Aquamarble 高吸湿放湿聚酯短纤维(商名,日本东洋纺)
aquamarine 海蓝［色］,宝石蓝［色］(浅绿蓝色)
aquaplaning 溅浆(印疵)
aqua pura 蒸馏水,纯水
aqua regia 王水(浓硝酸与浓硫酸的混合)
Aquaroll mangle 阿夸罗浮压辊轧车(一种高效、均压轧车,上轧辊为橡胶辊,下轧辊为浮在水面上的小直径贝纶辊,利用水压力沿全长压向轧辊,使上下轧辊间获得较均匀的线压力)
Aquaroll water mangle 阿夸罗轧水轧车(德国基土特公司制造的一种三辊高效轧水设备,由橡胶支承辊、贝纶中轧辊及游动轧辊组成)
AquaTex technique 水刺整理工艺,射流喷网技术［商品,德国福来司拿(Fleissner)］
aquatic 水产的;水化的,含水的
aquatic fungi 水生真菌

Aquatran process 阿夸曲兰法（一种湿法转移印花法）
aqua vitae 酒精
Aquazym 淀粉酶（低、中温型,20～90℃,适用于轧卷退浆,商名,丹麦诺维信）
aqueous 水的,含水的;水状的;多水的
aqueous adhesive 水性胶黏剂
aqueous based synthetic thickeners 水相合成增稠剂
aqueous dispersion 分散水相,水分散体,水分散液
aqueous dyeing 水相染色
aqueous emulsion 水包油型乳状液;水乳胶体
aqueous extract 含水提取物;水提（出）物
aqueous polyurethane 水性聚氨脂
aqueous preparations 水相制剂
aqueous printing ink 水性印墨（转移印花用）
AR (aromide) 芳香族聚酰胺
arabesque 橘棕[色]
Arabian blue 阿拉伯蓝[色]（铁青色,深色）
Arabic gum 阿拉伯树胶,金合欢胶
arachin fibre 花生蛋白纤维
araldite 环氧类树脂
aralkyl 芳烷基（由一芳基取代的烷基）
aralkyl sulphonate 芳烷基磺酸盐
aramid 芳香族聚酰胺
aramid fibre 芳香族聚酰胺纤维
ARBE【德】德国服装工业研究集团
arbor 心轴;刀杆
arc chute 弧形槽
arc cutting 电弧切割
archil 苔色素（地衣植物,天然植物染料,呈红紫色）

architectural textile 建筑用纺织品
architecture of textile 织物结构
arch steamer 圆顶形汽蒸箱
arc lamp 弧光灯,炭光灯
arctic cap 防寒帽
arc welding 电弧焊
ardometer 辐射高温计,光测高温计,表面高温计
area bonding 面黏合
area burning rate 面积燃烧速度（阻燃试验用语）
area felting shrinkage 面积毡缩
area shrinkage 面积收缩率（一般用于羊毛防毡缩试验）
area weight measuring and control equipment 织物单位重量测定及控制装置（通过测定织物密度来探测织物单位重量,将测定值与预定值偏差指示出来,并利用内装式的调节器使织物单位重量保持在给定范围内）
areometer 液体比重计
areometry 液体比重测定法
ARF (American Retail Federation) 美国零售业联合会
Arg.（arginine）精氨酸,2-氨基-5-胍基戊酸
argent 银[白]色
Argentine merino 阿根廷美利奴毛
Argentine wool 阿根廷羊毛
argentometric titration 银量滴定法
argentometry 银盐定量,银量滴定
Arghan 菠萝叶纤维（属凤梨科,制织物和网等用,商名,英国制）
argilla 陶土,泥土,铝氧土,高岭土
arginine〈Arg.〉精氨酸,2-氨基-5-胍基戊酸

argon-arc welding 氩弧焊
argon plasma 氩等离子体
argon plasma surface treating 氩等离子体表面处理
argyle design 菱形图案,菱形花纹
Aridye pigment colours 阿里代涂料(印花用涂料,商名,美国制)
Aridye system 水/油相涂料印花系统(商名,美国制)
Arkofix 爱科菲系列免烫整理树脂(改良型 DMDHEU,低甲醛(小于20毫克/千克,商名,瑞士科莱恩)
arm 辐;臂,柄;杆
armature 电枢《电》;[继电器]衔铁
armature-boosted synchronizing system 电枢补偿同步系统《电》
armor 盔甲,保护层
armored cable 铠装电缆
armored fabric 防护[用]织物(石棉、玻璃纤维、棉或其他纤维织物,经硝化纤维素、橡胶或合成树脂涂层或浸渍处理)
armored thermometer 带套温度计,铠装温度计
armorial design 防护性图案,伪装图案
armor plate 防护板
armoured woven fine cloth 金属细布(用于过滤)
army blanket 军用毛毯
army brown 军服棕[色]
army cloth 军服织物
army duck 军用帆布
army look 军服式样
aroma finishing 芳香整理
aroma finishing agent 香味整理剂
aroma microcapsule finishing 微胶囊香味整理

aroma therapy finish 芳香治疗整理
aromatic amines 芳香胺类[染料]
aromatic based polymer 芳香聚合物
aromatic diamine 芳族二胺
aromatic dicarboxylic acid 芳族二羧酸
aromatic nylon fibre 芳族锦纶纤维,芳族聚酰胺纤维
aromatic polyamide-imide fibre 芳族聚酰胺—亚胺纤维(耐高温阻燃纤维)
aromatic polyimide fibre 卡美儿(Kermel)芳族聚酰亚胺纤维
aromatics 芳香剂;除臭剂
aromatic sulphonates 芳族磺酸盐
aromide〈AR〉芳香族聚酰胺
arrangement 排列,布置;装置,设备
arrester 制动器,制止器;避雷器;过压保险丝;锁定装置
arresting pawl 制动棘爪
Arrhenius equation 阿伦尼乌斯方程式
arrowroot 竹竿粉,木薯粉
ARS(automatic registration system)[圆网印花机]自动对花系统
arsenic 砷
arsenic sulphide 五硫化二砷
arsenic trace 痕量砷
arsenometry 亚砷酸滴定法
art applique 外饰艺术
artefact 人工制品
artefact pollution 人工污染
arterial replacement 人造血管,动脉的代替品
art flower 人造花(用作服饰)
artichoke green 洋蓟绿[色]
article mark 商品标志
article number 品号
article specific design 商品特殊设计

artificial blood vessel 人造血管
artificial cotton 人造棉
artificial dialysis membrane 人工透析膜
artificial down 人造羽绒
artificial draft 强制通风
artificial dyestuffs 人造染料
artificial enzyme 人造酶
artificial fibre 再生纤维
artificial flower 人造花
artificial flower finish 人造花卉整理(一种硬挺整理方法)
artificial fur 人造裘皮
artificial grass 人造草地
artificial implant 人造植入物
artificial intelligence 人工智能
artificial leather 人造革
artificial leather carrier webs 人造革基布
artificial light 人造光源
artificially soiled fabric 人造污垢织物(用于测洗涤效果)
artificial mohair 人造马海毛
artificial moire 人造波纹,人造云纹
artificial nerve networks〈ANN〉人工神经网络(用以研究面料与辅料间织物行为关系)
artificial neutral networks 人工神经网络
artificial organ 人造[内脏]器官
artificial perception 人工识别
artificial perspiration 人工汗渍
artificial silk 人造丝,人丝(即黏胶纤维丝)
artificial soil 人造污垢
artificial suede 人造麂皮
artificial sweaty liquid 人造汗渍液(测试耐汗渍牢度用)
artificial ultramarine 人造群青;红光鲜蓝色
artificial wool 人造[羊]毛
artillery twill 斜纹马裤呢
art imlique 内蕴艺术
artistic design 美术设计
artistic dyeing and finishing 艺术染整
artistic style 艺术风格
artistic tapestry 工艺美术壁毯
art silk 刺绣丝线
art square 工艺美术地毯(双面提花,两端有穗)
art-textile 艺术纺织品
art ticking 印花床垫;印花枕头布;印花床单
artware 工艺品
aryl 芳基
arylide 芳基化物
As 砷
AS【德】专利说明书
AS(acrylonitrile-styrene) 丙烯腈—苯乙烯共聚物
AS(asbestos fibre) 石棉纤维
AS(Australian Standard) 澳大利亚标准
ASA(American Standard Association) 美国标准协会(从 1966 年起改为 USASI)
Asahiguard AG-710/AG-480 一种氟碳聚合物(拒水拒油整理剂,商名,日本旭硝子)
asbestos 石棉
asbestos-ceramic fibre composite 石棉—陶瓷纤维复合材料
asbestos cloth 石棉布,石棉织物
asbestos cord 石棉绳(可作热风拉幅机烘房门板的绝热材料)
asbestos fibre〈AS〉石棉纤维
asbestos substitute 石棉替代品

ASBL【法】比利时纺织研究中心
ascending chromatography 上行色谱[法]，上行色层分离[法]
ascending design 上行设计（地毯图案）
ascending development 上行展开[法]
ascending ratio of strength 强力上升率
asche test 静电消除测试
ASD【法】瑞士文献资料协会
ASDC（Associate of the Society of Dyer & Colourists）染色工作者学会会员（英国）
ASENET【法】瑞士干洗和织物整理企业协会
asepsis 无菌；无毒
aseptic 无菌的
aseptic cotton 消毒棉，防腐棉
Asgard 一种耐火织物（商名，英国制）
ash 烟灰色；灰分
ash alkalinity 灰碱度
ash content 灰分含量
ashen 灰[色]的；苍白的
ash free cellulose 无灰[分]纤维素
ash grey 灰色
ash separator 飞灰分离器
Asiatic gum 亚洲树胶
aslant padder 斜轧车
asm（assembly）装配；组合，套；组件，装配件
ASME（American Society of Mechanical Engineers）美国机械工程师学会
ASP（asparagic acid, aspartic acid）天冬氨酸
asparagus green 浅豆绿[色]，龙须菜绿[色]
aspartic acid（= asparagic acid）〈ASP〉天冬氨酸

aspect 样子，外观，外表；方面；状况；观点
aspect ratio [纤维]直径比；长宽比
aspergillus microorganisms 曲霉微生物
asperity 粗糙度
asphalt waterproofing 沥青防水
aspiration 吸尘[作用]；吸气，吸入
aspiration psychrometer 吸气式[干湿球]湿度计
aspirator 吸气器，抽风机；自动清洁器
aspirator combined with dust collector 吸尘器
ASQ【德】德国统计质量控制研究学会
AS resin 丙烯腈—苯乙烯树脂
assay 化验，分析；检定
assemblies 衣片
assembling department 装配车间
assembling diagram（= assembling drawing）装配图
assembling pin 定位销
assembly〈asm〉装配；组合，套；组件，装配件
assembly line 装配线，流水作业装配线
assess（= assessment）评定，估价
assessment of singeing 烧毛评级
assessments of fastness 染色牢度评级
assimilated carbon 吸附炭（用于废水处理）
assimilation effect 同化效应
assistants 助剂
assistor 辅助机械装置
associate 缔合物
associated running 联合运行
Associate of British Chemical Manufacturers〈ABCM〉英国化学制造商协会
Associate of the Society of Dyer & Colourists〈ASDC〉染色工作者学会会员（英

国）
Associate of the Textile Institute〈ATI〉 纺织学院联合会（英国）
association 缔合,结合
association colloid 缔合胶体
association degree 缔合度
association heat 缔合热
association of surfactants 表面活性剂的缔合
association polymer 缔合聚合物
associative ability 缔合能力
assorted colours 杂色
assorted fibre 混杂纤维
assorting 拣选,分级,分类
assortment 品种；机器组合；分类,分等,分级；配棉成分
assortment for patterns and colours 配花配色
assortment for pieces 拼件
assured processability factor 安全操作系数
ASTI 【德】瑞士纺织整理工业协会
ASTM（American Society for Testing and Materials）美国材料试验学会
ASTM（American Standard of Testing Materials）美国材料试验标准
astrakhan 俄国羔皮；俄国羔羊毛（俄罗斯、伊朗和亚洲产）；充羔皮织物；经缎毛圈组织《针》
Astralplush 起绒整理剂（高分子复配物，用于灯芯绒、平绒，具有丰厚手感,亲水和抗静电,商名,美国先邦）
Astrazon dyes 阿斯屈拉松染料（腈纶用阳离子染料,商名,德国制）
astringent 收敛剂
astroquartz fabric 宇航石英织物（可耐1100℃高温）

Asulak E 961 一种印花罩印浆（氧化物混合的丙烯酸类树脂,适用于各种织物,手感柔软,薄膜弹性大,深地色上罩印效果好,商名,意大利米洛）
Asuperle FNC 一种氟碳聚合物（拒水拒油整理剂,商名,意大利米洛）
asymmetrical design 不对称花纹
asymmetrical monomer 不对称单体
asymmetry 不对称
asymptotic 渐近线的
asynchronism 异步性《电》
asynchronous motor 异步电动机,感应电动机
At 砹（化学元素符号）
AT（automatic test）自动测试
ATA（anthranilamide）邻氨基苯甲酰胺
atactic polymer 无规立构聚合物
atactic polypropylene〈APP〉 无规［立构］聚丙烯
ATC（Agreement on Textile and Clothing）［世界］纺织及成衣协议
ATE（anthrasol-thermosol development process 溶蒽素热熔显色法
ATF 【德】德国纺织组织研究学会
athletic clothing 运动衣
athletic hose 运动袜
athletic shirt 运动男衫
athletic shoes 运动鞋
ATI（Associate of the Textile Institute）纺织学院联合会（英国）
Atlas weather-o-meter 阿特拉斯耐气候色牢度仪
atm.（=atmos.,atmosphere）大气；大气压
ATMA（American Textile Machinery Association）美国［国际］纺织机械协会

ATME-I (American Textile Machinery Exhibition-International) 美国国际纺织机械展览会
ATMI (American Textile Manufactures Institute) 美国纺织制品研究所
atmmometer 蒸发计，汽化计
atmosphere〈atm.，atmos.〉大气；大气压
atmosphere standard 大气标准
atmospheric beck [针织品]常压[绳状]染色机
atmospheric contaminants 大气污染物
atmospheric control 空气控制，温湿度控制
atmospheric degradation resistance 耐气候降解性
atmospheric drop 常温排染《染》《整》
atmospheric dryer 大气干燥机，常温干燥机
atmospheric dyeable polyester fibre 常温可染聚酯纤维
atmospheric dyeing 常压染色
atmospheric fading 大气褪色；烟气褪色
atmospheric fumes 大气烟雾
atmospheric ionizer 大气离子发生器
atmospheric overflow dyeing machine 常压溢流染色机
atmospheric piece dyeing machine 常压匹染机
atmospheric pollutant 大气污染物（指大气中的二氧化硫、二氧化氮等污染物）
atmospheric pressure 大气压[力]
atmospheric pressure plasma〈APP〉常压等离子体
atmospheric quartz tube heater 常压式石英管加热器
atmospheric relief valve 排空阀，空气阀

atmospheric steamer 常压汽蒸箱
atmospheric temperature 常温
atmospheric winch 常温绞盘染机
atoleine(=atolin) 液体石腊
atomic absorption spectrometry〈AAS〉原子吸收光谱测定法
atomic absorption spectrophotometry 原子吸收分光光度测定法
atomic absorption spectroscopy〈AAS〉原子吸收光谱测定法
atomic bond 共价键
atomic configuration 原子构型
atomic emission spectrometry〈AES〉原子发射光谱
atomic force microscope〈AFM〉原子力显微镜（测定纳米材料的显微镜）
atomic linkage 原子键合
atomic mass 原子质量
atomic number 原子序数
atomic polarizability 原子极化性
atomic polarization 原子极化[作用]
atomic theory 原子理论
atomic weight〈At.Wt〉原子量
atomisation(=atomization) 喷雾作用，雾化
atomize 雾化，喷雾；粉化
atomized lubrication 喷雾润滑
atomizer 喷雾器，喷[水]嘴
atomizing 喷雾
atomizing adhesion 喷雾黏合
atomizing sprinkler 喷雾器
ATP (adenosine triphosphate) 三磷酸腺苷
ATS (ammonium-p-toluene sulphate) 对位甲苯磺酸铵（涂料印花用催化剂）
attached cushion [地毯]附加衬垫
attaching 缝合，钉合；贴；缚

attaching sleeves 附着袖套
attachment 附件,附属物;设备,装置
attachment screw 装配螺钉;紧固螺钉
attack 冲击;侵蚀,腐蚀;影响
attapulgite 凹凸棒石活性白土
attapulgite clay 凹凸棒石黏土(用于染料的脱色)
attegenus 毛皮甲虫(羊毛制品上的蛀虫)
attemperation 温度调节
attemperator 温度调节器;降温器
attendance 维护,保养;机台看管
attention time 看[机]台时间
attenuant 稀释剂
attractive force 吸引力(水分子)
attractiveness 诱目性,吸引性
attrition resistance 抗磨耗性
attritor 磨碎机,研磨机
ATV【德】德国废水处理技术协会
At. Wt(atomic weight) 原子量
ATY(air textured yarn) 喷气[法]变形丝
aubergine 茄皮紫色(乌紫色)
audibility 可听度;可听到
audible warning 音频警告
Au Portique【法】法国标准化协会[织物]可燃性标准
aurora[orange] 朝霞橘色(红橙色)
auroral 玫瑰红的;光亮的
Australian Commonwealth Engineering Standards Association〈ACESA〉澳大利亚联邦工程标准协会
Australian wool 澳大利亚羊毛
Australian Wool Commission〈AWC〉澳大利亚羊毛委员会
Australian Wool Corporation〈AWC〉澳大利亚羊毛协会
Australian Wool Innovation Limited〈AWI〉澳大利亚羊毛发展有限公司
Australian Wool Research Commission〈AWRC〉澳大利亚羊毛研究委员会
Australian Wool Testing Authority〈AWTA〉澳大利亚羊毛检验局
auto 自动装置,自动机械,自动器
auto-acceleration 自动加速,自动促进
auto batcher 自动卷布机
autochromatic plate 彩色板
autochrome process[媒染染料]同浴铬媒染色法
autoclave 高压釜,煮布锅
autoclave-setting 高压釜定形,蒸压器定形
autoclave-style beck 高压釜式[绳状]染色机
auto coding 自动编码
autocondensation 自动缩合
autocontrol 自动控制,自动调整
autoconvection 自动对流
autocorrection 自动校正
autocut-out 自动断路器
autocut-winder 自动剪卷联合机
autodigestion 自体消化,自溶[作用]
auto-dilution system 自动稀释系统
auto feed 自动喂给,自动送料
auto-fit 自适应
autofluorescence 自发荧光
Autofoam 全自动泡沫整理装置[商名,瑞士德塔(Datacolor)]
autogenous decomposition 自动分解,自生分解
autogenous welding 气焊
autographic recorder 自动记录器
autohesion 自黏作用;自黏力
autoignition 自燃;自动点火

autoignition temperature 自燃温度
autoignition time 自燃时间
autojetwet 自动喷湿系统(用于污水处理)
autojig 自动卷染机
autolevel controller 液面自动控制器
autolysis 自体分解[作用];[细胞]自溶作用
automat 自动装置;自动控制器
automate 使[工厂]自动化
automated dye weighing of dispensing system 自动化染料称重分配系统
automated factory 自动作业工厂
automated graphic technology 自动化图示技术《计算机》
automated imagery processing 自动成像处理《计算机》
automated imformation management system 自动信息处理系统《计算机》
automated line 自动线
automated packaging lines 自动化包装生产线
automated production 自动化生产
automated storage 自动化储存[物品]
automatic add 自动添加,自动补充
automatic addition control〈AAC〉自动加入控制
automatic air filter 自动滤气器
automatic air switch 自动空气开关
automatically operated valve 自动阀
automatically progressive alkali metering system 自动渐次加碱计量系统(用于活性染料吸尽染色)
automatic arc welder 自动弧焊机
automatic batch change device 布匹自动切断及打卷装置

automatic batcher 自动卷布机
automatic batch mixing 自动配料混合
automatic batch weighing 自动分批称量
automatic bundling machine 自动打包机
automatic burette 自动滴定管
automatic centralized lubrication system 自动集中润滑系统
automatic checkout and recording equipment 自动测试与记录装置
automatic choke 自动阻气门
automatic cloth feeder 自动喂布装置
automatic cloth guider 自动导布装置;自动吸边器
automatic coating machine 自动涂胶机(涂感光胶用)
automatic colour change 自动换色
automatic colour control 自动色泽控制
automatic colour drawing system 自动描稿系统
automatic colour matching system 自动测色配色系统
automatic combustion control〈ACC〉自动燃烧控制,燃烧自控系统
automatic compressed air guider 压缩空气自动导布器
automatic concentration control 自动浓度控制
automatic continuous laundering 自动化连续洗烫机
automatic control 自动控制
automatic cutting-off and measuring machine 自动量剪机
automatic developing processor 自动显影机
automatic dispensers 自动配液器,自动配料器
automatic dispose 自动分配;自动排列

automatic dissolving station 自动溶解装置
automatic dosing system 自动加料系统
automatic dosing & weighing system 自动计量称重系统
automatic dye dosing station 染料自动配料系统
automatic dye jig 自动卷染机,自动染缸
automatic edge control 布边自动控制［装置］
automatic edge cutter（针织物拉幅机的）自动切边器
automatic electric oven 自动电热烘箱
automatic end counting device 自动计道装置(卷染机的一种附属装置,计算卷道数)
automatic fabric inspecting system 自动验布系统
automatic fabric slitting apparatus 自动切条装置
automatic fast cleaning unit［轧槽］自动快速清洗装置
automatic feed 自动喂料
automatic filter spectrophotometer 自动滤色片分光光度计
automatic fire control system 自动防火系统
automatic flame cut-off arrangement 自动停火装置
automatic flame detector 自动火焰检测器
automatic flat-bed screen printing machine 自动平板筛网印花机
automatic flat screen printing carriage 自动平网印花走车
automatic flat screen printing machine 自动平网印花机
automatic gas suspension device 煤气自停装置(机器发生故障时能自动切断煤气燃烧器的煤气供应)
automatic horizontal cloth opening device 自动卧式开幅装置
automatic jigger 自动卷染机
automatic kier decatizing machine 自动罐蒸机(毛织物整理设备,卷绕、汽蒸、抽冷、出布等均自动控制)
automatic labelling unit 自动贴商标装置
automatic laser inspection and control equipment 激光自动验布装置
automatic length indicator 自动计长器
automatic level control 自动液面控制
automatic lighter 自动点火器
automatic lint screen belt 自动除棉绒筛网带(磨毛机除绒装置)
automatic liquid level controller 自动液面控制器(用于轧车的轧槽)
automatic loader 自动装料机
automatic lubrication installation 自动润滑装置
automatic lubrication pump 自动加油泵
automatic measuring and plaiting machine 自动码布折布机
automatic metering system 自动计量系统
automatic mode 自动模式
automatic monitoring system 自动监控系统,自动检测系统
automatic oiler 自动加油器
automatic operation 自动操作
automatic order process 自动化订货方法
automatic palletizing machine 自动化货盘装运机
automatic pattern repeat system 自动花回调节系统(自动对花系统)
automatic piler 自动堆布器

automatic pleating machine 电子自动打褶机
automatic plotting 自动绘图
automatic powder dispenser〈APD〉粉剂自动分配系统
automatic preselector 自动预选器(用计算机的自动控制器)
automatic presetting system〈APS〉自动预调系统
automatic press 自动烫衣机；自动印花机(印T恤衫用)
automatic pressure control 自动压力控制
automatic production 自动化生产
automatic production line 自动生产线
automatic product processing 自动化产品加工
automatic program control 自动程序控制
automatic programming 程序自动化,自动编程
automatic raising fillet sharpening device 起绒钢丝针布自动磨尖装置
automatic recorder 自动记录仪
automatic recording detector 自动记录检测器
automatic registration system〈ARS〉[圆网印花机]自动对花系统
automatic regulating system 自动调节系统
automatic remote control 自动遥控
automatic repeat adjustment system 自动花回调节系统(自动对花系统)
automatic rolling machine 自动卷布机
automatic rope piler 自动绳状堆布机
automatic scale 自动秤
automatic scanning 自动扫描
automatic screen coating machine 自动筛网涂胶机(有圆网和平网两种,在网上均匀地涂上感光胶)
automatic screen printing carriage 自动筛网印花走车
automatic screen printing machine 自动筛网印花机
automatic seam passage device 线缝通过自控装置
automatic sequence controlled computer 自动程序控制计算机
automatic signal 自动信号
automatic solvent dry cleaning machine 自动溶剂干洗机
automatic spectrogram 自动频谱仪
automatic stop motion 自停装置
automatic synchronizer 自动同步机
automatic temperature control 温度自动控制
automatic temperature-time control 自动温度—时间控制
automatic test〈AT〉自动测试
automatic thermoregulator 温度自动调节器
automatic thermostat 自动恒温器
automatic thin-layer spreader 自动薄层[膜]涂布器
automatic timer 自动定时器；自动程序装置
automatic titration 自动滴定
automatic transport 自动化运输[物品]
automatic voltage regulator〈AVR〉电压自动调节器；稳压器
automatic weft straightener 自动整纬器
automatic weighing with isotopes 同位素自动称重
automatic weight control [织物]重量自控
automatic weld 自动焊接

automatic wrapping machine 自动包装机
automating printing process 自动化印花方法
automation 自动化;自动[装置];自动操作
automation of transport 运输自动化
automation package 自动化软件包
automation technology 自动控制技术
automation textile 纺织工程自动化
automatization 自动化
automobile safety belt 汽车安全带
automotive fabric 汽车用织物
automotive seat cover 汽车座位套
automotive upholstery 汽车用装饰布
autooxidation 自[动]氧化
autopilot 自动引燃器
autoplastics 自体成形
autopolymer 自聚物
autopolymerization 自聚合
autopurification 自净作用
autoradiograph 自动辐射照相
autoregulator 自耦调压器
autoscutch 自动开幅机
autosetter 自动定形机
autoswitch 自动开关
Autotop dye 汽车装饰织物用高耐日晒牢度分散染料(商名,瑞士制)
autotracing system 自动描图系统(包括花样分色、修正、晒图设备)
autotransformer 自耦变压器
autovoltage regulator 自耦调压器
autowash 自动洗涤机
autozero control 自动调零控制
auxiliaries 助剂;辅助设备
auxiliary agent 助剂
auxiliary chain 副链

auxiliary device 辅助装置
auxiliary drawing 辅助图
auxiliary drive 辅助传动
auxiliary equipment 辅助设备,外部设备
auxiliary mechanism 辅助机构
auxiliary shaft 副轴
auxochrome(＝auxochromic group) 助色团
availability 有效使用率,利用率;可用性;有效性
available chlorine 有效氯
available component 有效成分
available energy 有效能量
available equipment 现有设备
available oxygen 有效氧
available time 有效工作时间,开机时间
average 平均值;平均;平均的
average concentration 平均浓度
average degree of polymerization 平均聚合度
average gradient 平均梯度
average molecular weight〈AMW〉 平均分子量
average orientation 平均取向度,平均定向度
average pile density 平均绒毛密度
average sewage 普通污水,一般污水
average specific heat 平均比热
average value 平均值
average viscosity 平均黏度
aviation and space industry 航空航天工业
avivage 后整理;柔软整理;丝鸣整理(法文术语)
avivage agent 后整理助剂,柔软整理剂,增艳剂,发亮剂
avivan 含有机溶剂的磺化植物油(润湿剂,平滑剂,柔软剂)

Avivan SFC 亚伟环 SFC(脂肪酸酰胺缩合物,纤维素纤维与合成纤维用润滑剂、柔软剂,商名,瑞士制)
Avivan SO 亚伟环 SO(有机硅乳液,聚乙烯混合物,各种纤维用柔软剂,可与树脂整理剂同浴使用,商名,瑞士制)
avocado 橄榄绿[色]
avoidable noise 可除噪声
Avolan IL 阿复兰 IL(1∶2 金属络合染料匀染剂,无泡,商名,德国拜耳)
Avolan IS 阿复兰 IS(涤毛混纺织物染色的分散剂、匀染剂,商名,德国拜耳)
Avolan IW 阿复兰 IW(1∶2 金属络合染料的匀染剂,腈纶及其混纺织物染色的分散匀染剂,商名,德国拜耳)
Avolan SCN 150 阿复兰 SCN 150(酸性染料高效低泡匀染剂,商名,德国拜耳)
Avolan UL 75 阿复兰 UL 75(金属络合染料用高效低泡匀染剂,商名,德国拜耳)
avometer 万用表
AVR(automatic voltage regulator) 电压自动调节器;稳压器
AWA 【德】德国水研究学会
AWC(Australian Wool Commission) 澳大利亚羊毛委员会
AWC(Australian Wool Corporation) 澳大利亚羊毛协会
A-weighted sound level A 声级(一种噪声值)《环保》
AWI(Australian Wool Innovation Limited) 澳大利亚羊毛发展有限公司
awning curtain 遮阳帷幔
awnings 篷帐布;帆布;椅布
awning stripes 条子篷帐布

awning textiles 遮阳纺织品
AWRC(Australian Wool Research Commission) 澳大利亚羊毛研究委员会
AWT(advanced waste treatment) 深度废水处理《环保》
AWTA(Australian Wool Testing Authority) 澳大利亚羊毛检验局
axial bearing 轴向轴承,止推轴承
axial cam 轴向凸轮
axial drive batcher 轴驱动卷布机
axial dyeing process 轴流染色方法
axial fan 轴流风机
axial flexible core 轴向伸缩筒管
axial flow 轴流
axial flow blower 轴流风机
axial flow fan 轴流风扇
axial flow pump 轴流泵
axial motion 轴向运动
axial orientation 轴向取向
axial pump 轴流泵
axial seal 轴封
axial thrust bearing 轴向止推轴承
axioflexible cylindrical dye tube 轴向伸缩染色筒管
axis 轴,轴线
axle 轴,车轴
axle box 轴箱
axle cap 轴帽
axle housing 轴套
axle load 轴荷载
axle neck 轴颈
Axminster carpet 阿克斯明斯特地毯
axonometrical drawing 轴测图
aza compounds 氮杂环化合物
azelon 人造蛋白质纤维
azeotrope 共沸混合物

azeotropic boiling point 共沸点
azeotropic copolymerization 共沸共聚作用
azetidine 吖丁啶，氮杂环丁烷
azide 叠氮酸盐
azidine dyes 叠氮染料
azidosulfonyl dyes 叠氮硫酰染料
azine dyes 吖嗪染料，氮[杂]苯染料
azirdine 氮丙啶，乙烯亚胺
aziridine compound 氮丙啶化合物
azobenzene 偶氮苯
azobisformamide〈ABFA〉偶氮二甲酰胺（发泡剂）
azo compound 偶氮化合物
azo dyes 偶氮染料
azoflavine 酸性偶氮黄
azofuchsine 酸性偶氮红
azoic base 不溶性偶氮染料色基
azoic coupling component 不溶性偶氮染料偶合组分（即色酚）
azoic developing bath [不溶性偶氮染料]显色浴
azoic diazo component 不溶性偶氮染料重氮组分
azoic disperse dye 偶氮分散染料
azoic dyeing 不溶性偶氮染料染色，冰染[染]料染色
azoic dye printing 不溶性偶氮染料印花
azoic dyes 不溶性偶氮染料，冰[染]染料
azoic printing compositions 不溶性偶氮染料印花组分（指色酚与稳定的重氮化合物的混合体）
azoics 不溶性偶氮染料
azoic salt 不溶性偶氮盐
azomethine dyestuff 偶氮甲碱染料，甲亚胺染料
azo phthalocyanine 偶氮酞菁
azo pigment 偶氮颜料
azoreductase 偶氮还原酶（用于染料降解）
azo-sulphite fading 偶氮—亚硫酸盐褪色
azo-sulphite formation 偶氮—亚硫酸盐形成物（某些偶氮染料染色物在二氧化硫气体中形成的偶氮键亚硫酸盐加成物，产生染色风印）
azotic acid 硝酸
azotize 氮化
azotometry 氮素计，定氮仪，测氮计；氮量分析法，氮滴定法
azo violet 偶氮紫
azoxybenzene 氧化偶氮苯
Aztec print 阿兹泰克民族花样（墨西哥印第安人制作的色彩鲜明的几何花样）
azure 中蓝[色]，天蓝[色]
Azure blue （钴、钾碱、硅石制成的）蓝色颜料
azurine 天青色苯胺染料
azurite 蓝色颜料
azurite blue 石青蓝[色]（灰绿蓝色）

B

BA(bioavailability) （药物等的）生物有效性，生物有效率，生物利用度
BA(butyl acrylate) 丙烯酸丁酯
baby clothing 婴儿衣物
baby lamb 羔羊毛
baby pink 淡粉红
baby textiles 婴儿纺织品
back and face effect [染色]织物的正反面效果
back backing 背面涂层
back brush [地毯]背面刷毛
back chroming 后铬媒处理
backcirculation 循环回流
back cloth 印花衬布
backcoating 背面涂层；背面上胶；底布涂层
back coating and drying machine for imitation fur fabric 人造毛皮背面涂层烘燥机
back draft 回风
back dyeing 后媒法染色（染色后再作铬媒处理）
backfilled fabric 背面重浆织物，单面重浆布
back filling 背面上浆，单面上浆；背纬，里纬
backfilling finish 背面上浆整理
back filling machine 单面上浆机
backfinishing [织物]背面整理
back fire 逆火，回火
back flow 逆流，回流

backgreige seam impression 印花衬布接缝痕（印疵）
back grey 印花衬布
back grey crease mark 印花衬布折痕
back grey mark 印花衬布缝头印
backgrey seam 印花衬布缝头
backgrey seam impression 印花衬布接缝痕（滚筒印花疵）
backgrey seam mark 印花衬布缝头印
backgrey washing machine 印花衬布水洗机
background and reverberant noise 外来杂音和回声[噪声]
background colour 地色
background roller 满地[印花]滚筒
backing 衬垫，底座，轴瓦；[织物]背面上胶；衬里；底布
backing cloth 底布；[印花]衬布
backing compound 上胶剂
backing construction 底布（或衬布）结构
backing fabric 底布；[印花]衬布
backing fabrics [簇绒地毯]背衬织物（包括底布和衬布）
backing layer 底布层
backing machine [地毯]背面上胶机
backing roller 衬托滚筒《印》
backing-up screw 止动螺钉
back kiss roll 反面接触舔液辊
backlash 齿隙，侧向间隙（两个啮合齿轮或两个机件之间的游隙或空隙）
back lighting 后照射光

backlining 反面贴布,反面衬布
backlit 后照射光
back-lit flexible face fabric 背亮塑面[灯箱广告用]织物(双轴向高强涤纶涂层织物,用于做灯箱广告)
back-lit situation 背光位置,逆光位置
back-lit textiles 背亮[灯箱广告用]纺织品(双轴向高强涤纶涂层织物,用于做灯箱广告)
back nut 锁紧螺母
back-off 补偿
back pressure 反压力
back-printing 背面[渗透]印花
back purge system 反冲洗装置
backsizing 单面上浆,背面上浆
backskin 仿鞣皮织物,仿翻皮织物
back-staining 返沾色;返沾污
back starching 单面上浆,背面上浆
back starch machine 单面上浆机
backstay 后撑条,背撑
back-stroke [印花刮刀]返回行程
backtanning 单宁后处理
back tenter 小型拉幅机(印花前整幅用)
back-thickening 后增稠
back thickening process [印浆]后增稠法
back titrating 后滴定,反滴定
back-to-back 反面对反面相叠
back to back knitted fabric 双面针织物
back to face 反面对正面相叠
back-to-face shading 正反面色差
back-travelling 反向巡行《机》
backup power source 备用电源
back view 后视图;背景
back washing equipment 循环洗涤设备,回洗设备
back washing machine 复洗机《毛》

BA cotton 用 BAP(三溴甲烷一磷酸三丙烯酯)改性的阻燃棉
bacteria 细菌
bacteria amylase 细菌淀粉酶
bacteria bed 生物滤床
bacteria degradation 微生物降解
bacteria destruction finish 抗菌整理
bacteria diastase [细菌]淀粉糖化酶
bacteria inhibitor 抑菌剂
bacterial desizing agent 细菌退浆剂
bacterial inhibition 抑菌性
bacterial metabolites 细菌新陈代谢
bacterial optical density 〈BOD,B.O.D.〉细菌光密度
bacterial stain 细菌染色,细菌着色
bacteria-proofing 防菌整理
bacteria repellancy 抗菌性
bacteria resistant 抗菌的
bacteria toxicity 细菌毒性
bactericidal action 杀菌作用
bactericidal finish 灭菌整理
bactericide 杀菌剂
bacteriostat 抑菌剂
bacteriostatic agent 抑菌剂
bacteriostatic fibre 抑菌性纤维
bacteriostatic finish 抑菌整理,卫生整理
Bactosol AP 酵素 AP(特殊细菌酶,用于消除过氧化氢、氯和过硼酸钠,商名,瑞士科莱恩)
Bactosol CA 酵素 CA(精选纤维素酶,作用 pH 范围为 5~6,商名,瑞士科莱恩)
Bactosol SI 酵素 SI(精选蛋白酶,对真丝有特效,作用 pH 范围为 7~9.5,商名,瑞士科莱恩)
Bactosol TK 酵素 TK(淀粉酶,用于冷酵

退浆法,商名,瑞士科莱恩)
Bactosol WO 酵素 WO(精选蛋白酶,对羊毛有特效,商名,瑞士科莱恩)
bad odour 臭味
bad register 对花不准《印》
bad whites 白色不白
bad work 疵品,疵点
baffle 挡板,缓冲板,挡圈
baffle plate [绳状染色机的]多孔隔板;折流板
baffler 阻尼器;挡板;隔音板
baffle ring 紧圈
Bafixan dyes 巴弗散染料(分散染料,用于转移印花,商名,德国巴斯夫)
bag cloth 袋用织物
bag degumming 装袋精练
bagging height 拱胀高度(衣服局部拱胀)
bagging load 拱胀负荷(衣服局部拱胀)
bagging machine 装袋机
bagging method(＝bagging method of dyeing) 装袋染色法
bag packing machine 装袋机
bag test of water proofing 吊水式防水性测试法,水兜法试验
baked print 已焙烘的印花布
baker [高温]焙烘机
baking [高温]焙烘,焙固
baking machine 焙烘机
baking oven [高温]焙烘箱
baking process 烘焙法
baking stove [高温]焙烘机
balance 秤,天平;平衡,均衡
balance bond theory 价键理论
balanced-pressure rotameter 压力补偿式转子流量计
balanced roller 张力调节辊

balanced vane pump 平衡叶轮泵
balance gear 补偿器;差动传动;平衡齿轮
balance lever 平衡杆,平衡杠杆
balance pan 天平盘
balance pivot 杠杆支点
balance potential 中点电位(如测试还原染料氧化还原中点时的电位值)
balance rheometer 平衡流变仪
balance sett 平衡[的]经纬密度
balance valve 平衡阀(一种使开启和关闭方向两边液体相等的液压阀门)
balance weight 平衡重锤,平衡锤
balance wheel 飞轮;摆轮
balancing tank 平衡槽
balancing transformer 平衡变压器
bale 包;捆;件
bale breaker 开包机
bale dyeing 纤维装包染色
bale mark 包装唛头,包装标志
bale press 打包机
baling 打包
baling machine(＝baling press) 打包机
ball 纱球,线球;绒线球,团绒;毛球,毛团;滚珠
ball and socket coupling 球窝连接器
ball bearing 滚珠轴承
ball burst apparatus 钢球式顶破强力试验仪
ball burst testing 钢球式顶破强力试验
ball check valve 球形止回阀(附有球阀和阀座的阀门,液体只能单方向通过)
ball collar thrust bearing [环形]滚珠止推轴承
ball coupling 球铰连接器
ball cup 球形环,球窝
ball float type level controller 浮球式液面

控制装置
ball float type steam trap 浮球式疏水器
ball hardness 布氏硬度;钢球硬度
balling up 起球(织疵)
ballistic fabrics 防弹织物,防弹布
ballistic impact 子弹撞击
ballistic limit 防弹射极限
ballistic protection 防冲击;防弹
ballistic-resistant fabric 防弹织物;耐冲击织物
ballistic test 冲击强力试验
ballistic testing machine 冲击式强力试验机
ballistic vest 防弹背心
ball-joints 球节
balljoint screw 万向球铰螺杆
ball mill 球磨机
ballmill refiner 球磨精研机
balloon 气圈,气囊
balloon fabric 气球织物
ballooning 缩纬整理
ballooning device 吹气装置
balloon shape 气圈形态
ball points 圆珠
ballproof clothes 防弹衣
ball race 球轴承座圈
ball screw 滚珠丝杆(用于拉幅机上针部分的导轨处,具有动作灵敏、耐磨性好、功率消耗少等优点)
ball top 毛球
ball valve 球阀
ball varnish 球蜡(干蜡)《雕刻》
ball viscometer 落球式黏度计
ball warp 球经
ball warp dyeing 球经染色(牛仔布经纱染色法)

ball warp dyeing machine 球经染色机(牛仔布经纱染色机)
ball winding machine 绕球机,团绒机
ball woollen knitting yarn 团绒
balsam 香油,香脂;蟹绿[色]
balsam green 青灰[色]
bamboo 竹黄[色](浅灰黄色)
bamboo charcoal fibre 竹炭纤维
bamboo charcoal textiles 竹炭纺织品
bamboo clip 竹夹(竹制简易夹子)
bamboo fibre 竹纤维
bamboo pulp fibre 竹浆纤维
bamboo viscose fibre 竹材黏胶纤维
banana 香蕉黄[色]
Bancare finish 班卡尔整理(棉织物非树脂性化学整理,商名)
Bancora shrink-proof process 班科勒防缩工艺(由界面聚合沉积的锦纶对羊毛进行防缩)
Bancroft process 班克罗夫特阻燃整理法(使用尿素和磷酸,美国商名)
band 绳;带;衣带;光[谱]带;频带
band absorbance 频带吸收率;光谱带吸收率
bandage 绷带
bandage cloth 绷带布
bandage gauze 纱布绷带
bandana 班丹纳印花绸,班丹纳印花布(印度制);印花大手帕
Bandana dyeing 班丹纳染色法(印度手工扎染法)
bandana print 班丹纳印花
bandanna 班丹纳印花绸,班丹纳印花布(印度制);印花大手帕
band conveyer 输送带,皮带运输机
band coupling 带形连接器,平接连接器

band fulling machine 带式缩呢机
band mark 扎绞印
band pulley 带轮,绳轮;带动滑车
band spectrum 带状谱
band spring 板簧
Banflam flame retardant finishing 本弗来姆阻燃整理
bang-thro prints 压透印花
banks of steam pipes 多排蒸汽管
banner 标语,旗帜
BAP(bromoform allyl phosphate) 三溴甲烷烯丙基磷酸酯
bar 杆,棒,条,巴(气体压力单位,等于100千帕斯卡)
bar code 条形码
bar code reader 条形码读取器
Barcol hardness 巴氏硬度
bare conductor 裸线《电》
bare-tube coil heat exchanger 光板盘管式热交换器
bare weight 皮重
Bariod rotary viscometer 巴列奥德旋转式黏度计
barium 钡
barium carbonate 碳酸钡
barium compound 钡的化合物
barium hydroxide 氢氧化钡
barium number 钡值(丝光纱、布的丝光度测定值)
barium sulphate 硫酸钡
bark 鼠灰[色]
bark cloth 树皮布(用树皮纤维织制)
bark crepe 人造丝树皮绉
barking drum 筒式剥皮机
barley starch 大麦淀粉,大麦淀粉浆
bar marking in screen printing 网印条痕

barmarks 棒痕(长环蒸化疵点)
Barnsley finish 巴恩斯利整理(厚亚麻斜纹布上轻浆和轧光整理方法,英国名称)
barometer 气压计,气压表
barometric condenser 气压冷凝器
barometric pressure 大气压
baro-thermo-hygrograph 气压—温度—湿度仪
Barotor[machine] 巴罗托染色机(高温高压平幅松式染色机)
barre coverage 遮盖纬向条痕
barre effect 纬向条痕,纬档
barre fabric 纬向色条织物(或针织物)
barre-free fabric 无条痕织物
barre mark 纬向条痕,纬档(织疵);横条(针织物疵点)
barrier 安全保持器;壁垒,阻挡层;位叠;栅栏;隔板
barrier property 抗渗性,阻隔性能(抵抗空气、氧气、水蒸气、油脂、酸、碱以及普通溶剂渗透的性能)
barriness 条花,纬向色档
barring 条痕,条花(织疵);起棱
barroches 未漂细布(东南亚制)
barry dyeing 色纬,纬色档(织疵)
bar stay 圆钢支撑;撑杆,撑条
barye 微巴(压强单位)
baryta 氧化钡,重土
barytes 重晶石,重土石,天然硫酸钡
BAS(basic adjustment station) 基本调节站
Basacryl dyes 贝西克尔染料(阳离子染料,染腈纶,商名,德国巴斯夫)
base 基;碱;基线,底线;底座
base bearing 主轴承,底轴承,基轴承

base burn 基材燃烧(阻燃试验用语)
base cloth 底布,基布,地布
base coat(=base coating) 底涂;打底涂布
base [fabric] colour [织物]地色(转移印花之前织物所染的色泽)
base electrode 基极电极
base exchange 阳荷性离子交换
base fabric 底布
base framework 底座
base length of pressure zone [刮刀]受压区基本长度
baseless felt 无基布毡
baseline 基线
baseline potential 基准电位,基准电势
base load 基本负荷
basematerial [carpet] 基材[地毯],地毯首层背底
base plate 座板;底座
base point ⟨BP⟩ 基点,原点
base printing [冰染料]色基印花
BASF 【德】巴登苯胺苏打工厂,巴斯夫公司(德国)
basic adjustment station ⟨BAS⟩ 基本调节站
basic aluminium acetate 碱式醋酸铝
1/3 basic aluminium chlorhydroxide 1/3 碱式氯化铝
5/6 basic aluminium chlorhydroxide 5/6 碱式氯化铝
basic aluminium chloride 碱式氯化铝
basic catalyst 碱性触媒
basic chromium chloride 碱式氯化铬
basic colour dip dyeing 碱性染料浸染
basic colour discharge 碱性染料着色拔染印花[法]
basic colour dyeing 碱性染料染色

basic colour mordant 碱性染料媒染
basic colour resist 碱性染料防染
basic control of digital analogue output 数字模拟输出基本控制
basic design 基础图案
basic-dyeable fibre 碱性染料可染型纤维(例如改性聚酯纤维)
basic dyeing 碱性染料染色
basic dye lake 碱性色淀
basic dyes 碱性染料,盐基染料
basic fabrics 基本织物,基本衣料
basic finish 基本整理,一般整理(使坯布成为可售商品的整理)
basic frequency 主频[率]
basic hole system 基孔制(孔的极限尺寸为一定,与不同极限尺寸的轴配合,以得到各种配合性质)
basicity 碱度;碱性
basic material 主料,基本材料;碱性材料
basic menu 基本菜单《计算机》
basic metabolism 基础代谢《环保》
basic pattern 基础(底组织)花纹,基础(底组织)图案
basic plane 基准平面
basic shade 主地色
basic shaft system 基轴制(轴的极限尺寸为一定,与不同极限尺寸的孔配合,以得到各种配合性质)
basic soap 碱性皂
basic stimulus 基础刺激[值]
basic style 碱性染料媒染印花法(先印单宁酸金属盐媒染剂,再用碱性染料染色)
Basilen dyes 巴杰兰染料(活性染料,商名,德国巴斯夫)
basket centrifuge 篮式离心[过滤]机,筐

式离心机

Basofil 巴索菲尔新型阻燃纤维（以密胺树脂为原料，商名，德国巴斯夫）

Basolan 2448 巴索兰 2448（过一硫酸盐，羊毛防缩用氧化剂，商名，德国巴斯夫）

Basolan AS-A 巴索兰 AS-A（羊毛染色用抗定形剂，过氧化氢激活物，商名，德国巴斯夫）

Basolan AS-B 巴索兰 AS-B（羊毛染色用抗定形剂，商名，德国巴斯夫）

Basolan AS process 巴索兰 AS 毛纤维抗定形工艺（德国巴斯夫）

Basolan DC 巴索兰 DC（二氯异氰尿酸盐，羊毛防缩剂，商名，德国巴斯夫）

Basolan F 巴索兰 F（阳离子化合物，羊毛丝绸提高湿牢度整理剂，羊毛防缩剂，商名，德国巴斯夫）

Basolan MW Micro 巴索兰 MW 微乳（高浓度氨基聚硅氧烷，羊毛防缩及柔软剂，商名，德国巴斯夫）

Basolan SW 巴索兰 SW（溶于异丙醇的聚乙醚型反应性预聚物，羊毛防缩及柔软剂，商名，德国巴斯夫）

basophilic 亲碱的；碱性[染料]易染的

bassora gum 黄蓍胶，刺槐树胶

bast-cotton blended fibre 麻棉混纺纤维

bastose 木质纤维素

bast soap 精练废液

batch 一批；分批[法]；布卷；（染色后）卷布

batch arm [卷染机上]卷布轴承臂

batch bleaching 分批漂白

batch box 单卷汽蒸箱（轧卷练漂或染色用）

batch-cart 布卷推车

batch charging 间歇加料

batch cold process [印染]冷堆法

batch controller 分批控制器

batch [preparation] correction 配料修改

batch cure 分批烘焙

batch cycle 卷布次数

batch digester 间歇[式]蒸煮锅，间歇[式]蒸煮器

batch drying 分批烘干

batch dyeing 分批染色；间歇式染色；分卷染色

batched jute 油麻（经软麻加乳处理后的黄麻）

batched roll 布卷

batched roll delivery 成卷落布

batcher 卷布机；计量给料器；进料量斗

batch filtration 间歇式过滤

batch finishing 分批整理

batching 卷布

batching arrangement 卷布装置

batching crease 打卷皱

batching drum 卷布滚筒

batching machine 卷布机

batching mechanism 卷布装置

batching off 退卷，展开，退绕，放轴

batching-off machine 放轴机，退卷机

batching on 打卷，卷轴

batching-on machine 卷轴机

batching-on roller（＝batching roller） 卷布辊

batching tension control 卷布张力控制

batching time 打卷时间

batching-up 堆置；折叠；卷取

batch number 批号

batch-on 上卷

batch process 间歇法，分批[生产]法

batch processing 成批处理,成批加工;分批处理

batch production 分批生产,间歇生产

batch roller 卷布辊

batch scouring 分批煮练,分批精练

batch steamer 间歇式蒸化机

batch system 分批法

batch to batch reproducibility 批间重现性

batch to batch variation 批与批之间的差异

batch type scouring machine 间歇式煮练机

batchwise 分批的,不连续的

batchwise dyeing 卷染;分批染色,间歇染色

batchwise dyeing machine 间歇式染色机

batchwise process 间歇式加工法,间歇式工艺

bath 浴,染浴;槽

bath exhaustion 染液吸尽

bath gown [长]浴衣

bath guide roller 浴中导布辊

bathing suit 浴衣

bath mat 浴室小地毯

bathochromatic effect 深色效应,向红效应(染料的最大吸收波长向长波方向移动,即向深色移动的效应)《测色》

bathochromatic group 红移基团,深色基团

bathochromic 深色效应,红移效应

bathochromic shift 向红移(染料的最大吸收波长向长波方向移动)《测色》

bath oiling 油浴法

Bathotex FS-H 增深剂(商名,日本明成)

bathotonic 自动交联作用的

bath ratio 浴比

bath robe 浴衣,晨衣

bath rug 浴室地毯

bath towel 浴巾

Batik(=Batik dyeing, Batik printing) 巴蒂克印花法,蜡防印花法,爪哇蜡脂防染印花法

Batik prints 巴蒂克印花布,蜡防[莎笼]花布

Batik sarong 巴蒂克印花布,巴蒂克莎笼,蜡防[莎笼]印花布(印尼产)

batiste 法国上等细亚麻布[宽81.28厘米(32英寸),长13.72米(15码),重0.4536千克(1磅)];细薄织物(包括精梳丝光棉布或涤棉混纺织物、提花黏胶织物、细支毛织物等)

batt 毛网(针织地毯的纤维絮片)

battery driven truck (=battery operated truck) 电瓶车

Batticks 巴蒂克印花布,蜡防[莎笼]花布

battik 蜡染法;蜡染花布

battle dress 防火防弹服,战地服装

batt-on-base felt 有基布毡,絮材—布基复合毡

BAT value 工作环境的生理容许值(职业接触人群的生物材料中某一化学物质或其代谢物的最高容许量)

baulk finish 生坯轻缩[呢]整理

Baumé 波美(液体相对密度单位)

Baumé degree 波美度

Baumé hydrometer 波美计

Baumé tester 波美计

bauxite 铝矾土

bay 赤褐色的,栗色的

bayadere stripe print 巴亚德印花横条布

Baylan NT 拜兰NT(羊毛低温染色助剂,商名,德国拜耳)

Baylase 生物精练用碱性果胶酶（商名，德国拜耳）
bay oil 月桂叶油，月桂油
bayonet lamp socket 卡口灯座
bayonet ring 卡口环（用于圆网印花机上圆网头的放置）
bayonet socket bulb 卡口灯泡
Baypret USV 拜普力 USV（抗皱免烫整理剂，手感改进剂，变性氨基酸乙酯的水制剂，不含甲醛，适用于低甲醛树脂整理中的添加剂，商名，德国拜耳）
BCC(British Colour Council) 英国颜料染料委员会
BCF(bulk continuous filament yarn) 膨体长丝
BCRA(British Ceramic Research Association) 英国陶瓷研究协会
BCRA tiles 英国陶瓷研究协会标准色板
BCS(blanket cleaning system) 印花导带洗涤系统
BDA machine for carpet printing BDA 地毯平网印花机
BDI(benzene diisocyanate) 对苯二异氰酸酯
beach 海滩[色]（淡橄榄灰色）
beach cloth 海滨薄呢（夏服用，全棉或棉毛交织）
beach coat 海滨服
beach towel 沙滩巾
beach umbrella 太阳伞
beach wear 海滨服（披在游泳衣外）
bead 玻璃砂
beading 起球
bead weld 珠焊；狭的焊缝
beaker 烧杯
beaker flask 锥形烧瓶

beam 织轴；经轴；梁，横杆；机身最大宽度；波束；射束；光束；射线电子束
beam alignment system 光束校直系统
beam bleaching 经轴漂白
beam drying machine 经轴烘燥机
beam dyeing 经轴染色
beam dyeing machine（＝ beam dyer） 经轴染色机
beaming-off 退轴，倒轴
beam jigger 经轴卷染机
beam pivot 杠杆支点
bearcloth 熊皮大衣；粗毛大衣呢，粗绒大衣呢
bearding [地毯]绒头发毛
bearer 支座，承垫，托架
bearer plate 座板，支承板
bearing 轴承，轴耳；支座；方位
bearing anchor 轴瓦固定螺钉
bearing bar 承受杆；支撑钢筋
bearing block 轴承座
bearing brass 黄铜轴套
bearing bronze 青铜轴套
bearing cage 轴承罩
bearing cap 轴承盖
bearing capacity 承载能力，容许载负
bearing case 轴承壳
bearing housings 轴承座，轴承箱
bearing plate 垫板，承重板
bearing sleeve 轴承套筒
bearing support 轴承座
bearskin 熊毛皮；起绒棉布；毛皮帽；粗厚羊毛大衣呢
beat 拍；打浆
beater 打浆机；打手；搅拌器
beater additive 打浆添加剂
beater roller 打布辊

beater washing machine 耙式洗涤机(羊毛染色后用)
beating machine 打绒机
Beaumé 波美(液体相对密度单位)
beaver 海狸棕[色](暗棕色)
beaver [cloth] 海狸呢,水獭呢(粗纺大衣呢);海狸绒布(双层双面)
beaver finish 海狸呢整理(顺毛起绒)
beck [染]槽,[染]缸(如染纱长槽,绞盘染色机的盛染液部分)
Beckamine 贝卡明[合成树脂](尿素与甲醛的预缩物,纤维素纤维防皱防缩整理剂,商名,美国制)
beck dyeing 绳状染色
Beckmann rearrangement 贝克曼重排反应
beck rub mark 染槽擦痕
beck-type cloth accumulator 浸没式存布箱
Becoflex compartment 贝科弗莱克斯水洗槽(采用回形穿布形式的高效平洗机,瑞士贝宁根)
bed 底板;床,台
bed clothes 床上用品(指被、褥等)
bedcover 床罩
bedding 床上用品;被褥
bedding and clothing 被服
bed knife 底刀,固定刀
bed linen 床用织物(用亚麻、棉或混纺材料制成)
bed mat 床褥;床单织物(凸纹布)
bed plate 底板,底座,托板
bed rug 床边地毯
bed sheet 床单,被单布;床罩布
bedsheeting 床单细布
bed shield 床罩织物
bedside mat 床边地毯
bedspread fabric 床罩布,床单布

beef tallow 牛油(浆料用)
Beer's law 比尔定律
beeswax 蜂蜡,黄蜡
beetle 甲虫
beetle calender 叠层轧光机,捶布用轧光机
beetle colour 黝绿色,甲虫色
beetle finish 捶布整理
beetle lustring 捶布增光
beetling 捶布,打布(织物整理工艺之一,可使织物增光,手感柔软)
beetling calender 捶打轧光机
beetling machine 捶布机
beetroot purple 甜菜根紫[色]
before washing 〈B.W.〉水洗前
begin colour 本色,天然色
behavior 【美】=behaviour
behaviour 特性,性能,性状,行为
behavioural science 行为科学
beige 本色的,未漂白的;米黄色,浅灰黄色;坯布,原色哗叽;混合线呢(交织或混纺);薄斜纹呢
Bekk's viscometer 贝克黏度计
Belfasin 44 贝尔法新 44(脂肪酸酰胺,各种纤维柔软剂,具抗静电性,商名,德国科宁)
Belfasin LUB 贝尔法新 LUB(脂肪衍生物,纱线、纤维润滑剂,商名,德国科宁)
Belfasin TVE 贝尔法新 TVE(脂肪衍生物,纤维润滑剂,防止缝纫针亮,商名,德国科宁)
Belfast finishing 贝尔法斯特整理(棉织物用 3-氯-1,2-环氧丙烷交联剂处理,使其具有防皱性能,商名,美国)
bell 圆锥体状,钟形罩;漏斗

bell crank 直角［形］曲柄,直角［形］杠杆
Bellflame 贝尔弗莱姆阻燃性黏胶纤维（商名,日本制）
bell form star steamer 钟式星形蒸箱（蒸箱内保持一定的正压,多用于丝绸织物的汽蒸）
Bellmanized finish 贝尔曼整理（轻薄棉布挺爽耐洗整理,商名）
bellow cylinder 气筒
bellows 波纹管,伸缩软管；风箱；洗耳球
bellows pressure 风箱压力
bellows seal 波形密封,伸缩筒式密封
bellows sealed control valve 波纹管密封控制阀
bellows steam trap 波纹管式疏水器（一种压力平衡式的排水阻汽装置）
bellows valve 波纹管阀
bell-shaped depinner 钟形脱针器（置于针板拉幅机出布区）
bell-shaped float steam trap 钟罩形浮子式疏水器（重力式排水阻汽装置）
Bel-O-Fast finish 贝尔法斯特整理（在棉织物上用环氧氯丙烷作碱性交联剂的多步整理工艺,美国制）
below par 低于［生丝］原重
Belsoft 200 贝尔索夫特200（脂肪酸酰胺,柔软剂,双氧水稳定剂,商名,德国科宁）
belt 皮带,带,输送带
belt brushing machine ［绒类织物］带状刷毛机
belt clamp 皮带扣
belt conveyer 运输带,带式输送机,皮带运输机
belt drive(=belt driving) 皮带传动
belt drying machine 带式烘干机

belt expander 皮带扩幅器
belt fabric 胶带［用］织物
belt fastener 皮带扣
belt filter 带式过滤器
belt fork 皮带叉
belt guider 皮带导布器
belting 传动带装置；皮带传动
belt lacing 皮带接头,皮带扣
beltpresses 带式压滤机《环保》
belt printer 带式印花机（用于针织品印花）
belt printing system 带式印花系统（用皮带转动台板印花）
belt pulley 皮带轮
belt shifting apparatus 皮带滑动装置
belt spreader 皮带扩幅装置（用于筒状针织物）
belt stretcher (= belt stretching machine) 皮带拉幅机,带式拉幅机
belt tensioner 皮带松紧调整器（用来调整风机、水泵等驱动皮带的松紧度）
belt uncurler 皮带剥边器（用于针织物热定形机）
beltwash station 带式洗涤站《环保》
bench work 钳工作业
bend bar expander 弯辊扩幅器,扩幅弯辊
bending 弯曲［度］；偏移［差］；弯头
bending angle 弯曲角
bending fatigue 弯曲疲劳
bending frequency 弯曲振动频率
bending handrail 拐弯扶杆
bending length 弯曲长度
bending recovery 弯曲回复
bending relaxation rigidity 弯曲松弛刚度
bending resistance 抗弯性
bending rigidity 抗弯刚度,弯曲刚度

bending stiffness 抗挠劲度,抗弯硬挺度
bending strain 弯曲应变
bending strength 抗弯强度
bending test 弯曲试验
bend strength 弯曲强力,抗弯强度
Ben-extracta 本氏浸轧机（一种高效平幅水洗机,商名,德国制）
bengaline 罗缎（横棱纹织物）；绨（黏胶纤维与棉交织物）
bent expander roller 扩幅弯辊
benthic plant 水底植物,深水植物
bentonite 皂土；膨润土（蒙脱土）；中性硅酸铝
bent plate 弧形板
bent stretching roller 扩幅弯辊
bent tube 弯管
bent tube boiler 弯水管锅炉
benzal chloride 二氯甲苯
benzaldehyde 苯甲醛
benzal green 孔雀绿（碱性三苯甲烷染料）
benzalkonium chloride 烷基二甲基苄基氯化铵（阳离子净洗剂,杀菌剂）
benzanthrene 苯并蒽
benzanthrone 苯并蒽酮
benzene 苯
benzene diisocyanate〈BDI〉 对苯二异氰酸酯
benzenediol 苯二酚
benzene ring 苯环
benzene sulfonyl hydrazide 苯磺酰肼
benzene test 苯试验《毛》
benzene-toluene-xylene〈BTX〉 苯—甲苯—二甲苯（有毒溶剂）
benzene-toluene-xylene recovery 苯—甲苯—二甲苯的回收
benzenoid 苯环型的

benzidine 联苯胺,对二氨基联苯
benzidine dyes 联苯胺染料（一类致癌染料）
benzimidazole derivative 苯并咪唑衍生物（聚丙烯腈纤维耐亚氯酸盐漂白浴的荧光增白剂）
benzimidazoline 苯并咪唑啉
benzin(＝benzine) 汽油,挥发油,石油精
benzine soap 汽油皂
benzo copper dyestuff 苯并铜染料,苯并坚牢铜盐染料
benzodifuranone disperse dye 苯并二呋喃酮[结构]分散染料
benzofuran 苯并呋喃
benzoic acid 苯甲酸,安息香酸
Benzoid-Brücke effect 贝楚德—朴尔克效应《测色》
benzol 粗苯,苯
benzophenone 苯酮
benzopurpurine 苯并红紫
benzoquinone 苯醌
benzothiazole 苯并噻唑
benzotriazole 苯并三唑
benzoxazinone 苯并噁嗪酮
benzoxazole 苯并噁唑
benzoyl 苯酰,苯甲酰
benzoylated 苯[甲]酰化的
benzoylated cotton 苯酰化棉
benzoylation 苯[甲]酰化作用
benzoyl peroxide 过氧化苯甲酰
benzyl 苯甲基；苄基
benzyl alcohol 苯甲醇；苄醇
benzylating agents 苄化剂
benzylation 苯[基]化作用
benzyl chloride 苄基氯（聚酰胺纤维用直接染料染色的促进剂）

benzyltrimethylammonium chloride 苄基三甲基氯化铵
berberine 小檗碱(天然染料)
berber-style carpets 本色地毯
beret 贝雷帽
bergmeal 硅藻土
Berlin blue 柏林蓝[色](也称普鲁士蓝)
beryl 海绿[色];绿柱石
beryl blue 水蓝[色]
beryllium compound 铍化合物
Beryl printing 贝里尔印花(18世纪模版印花)
BESA(British Engineering Standards Association) 英国工程标准协会
bestLEN 一种激光雕刻机(商名,荷兰斯托克)
BET(Brunauer-Emmett-Teller) 布鲁诺尔—埃米特—特勒,由三位科学家名字的首字母组成,表示比表面积的测试方法
beta 希腊字母β,可表示化学基团的位置
BET adsorption BET 多分子层吸附
BET adsorption isotherm BET 吸附等温线
beta gauge β射线密度测量计
beta-glucosidase β-葡萄糖苷酶(纤维素酶的成分)
beta-glucosidic linkage β-葡萄糖苷键
betaine 甜菜碱;三甲胺乙内酯(分散剂,稳定剂,用于士林染料印花)
betaine surfactant 甜菜碱类表面活性剂(分散剂,稳定剂,还原染料印花用)
beta-keratin β-角蛋白
beta meter β射线测定仪
beta-naphthol 2-萘酚,β-萘酚
beta ray β射线
beta-ray weightometer β射线织物[含湿]称重仪
beta spectrometer β射线光谱仪
BET equation BET 方程
Bethamin 1166 conc. 贝赛明 1166 浓(脂肪酸缩合物,非离子,涤纶低聚物粘移剂,用于减少染色中生成的低聚物,商名,德国波美)
BET isotherm BET 等温线
bevel gear 伞[形]齿轮,圆锥齿轮
bevel weld 斜角焊
BHES(bis-β-hydroxyethyl sulfone) β-双羟乙基砜
BHSO(British Health and Safety Organization) 英国卫生与安全组织
BHSR(British Health and Safety Regulation) 英国卫生与安全管理条例
BHT(bright,high tenacity) 有光泽和高强力[的]
bias 偏斜
bias current 偏流
bias filling(=bias weft) 纬斜,纬不正(织疵)
biaxial coating fabric 双轴向涂层织物
biaxial fabric 双轴向织物
biaxially oriented material 双轴取向材料
biaxial oriented polypropylene〈BOPP〉 双轴向拉伸聚丙烯
bib cock 弯嘴旋塞
bibliography 书目提要,文献目录;文献学,目录学
bicarbonate 碳酸氢盐
bichromate 重铬酸盐
bichromate after treatment 重铬酸盐后处理
Bicoflex padding mangle 比科弗莱克斯调压式均匀轧车(商名,德国克莱韦弗)

Bicoflex roll 比科弗莱克斯轧辊（采用气压以补偿挠度的均匀轧辊，多用于染色轧车，德国克莱韦弗）
bicoloured 双色的
bi-component 双组分
bicomponent complex film 双组分复合薄膜
bicomponent fibre 双组分纤维
bicomponent filament 双组分长丝
bicomponent structure ［纤维的］双组分结构
bicomponent yarn 双组分变形纱
biconstituent filament[yarn] 双组分长丝
bidimensional 二维的，二元的
bi-directional twin-blade cutter 双面刃的两面割刀装置
biege 原色哔叽
biflux 汽—汽热交换器
bifunctional catalyst 双官能团催化剂
bifunctional compound 双官能团化合物
bifunctional monomer 双官能团单体
bifunctional oligomer 双官能团低聚物
bifunctional reactive dye 双官能团活性染料
bifunctional reagent 双官能试剂
big batching installation 大卷装装置
big batch system 大卷装
big batch trolley 大卷装小车
big batch winder 大卷装卷布机
biguanide 双胍；缩二胍
bi-ion active agent 两性离子表面活性剂
BIL(built-in lubrication) 滑爽
bilateral fibres 双组分纤维
bi-lateral screen drive 圆网的双侧传动（可减少圆网的扭转变形）
bilateral structure of fibre 纤维的双边结构
bilayers 双分子层
BIL crease-resist process 滑爽抗皱处理（棉织物改性处理）
bimetallic-strip thermostat 双金属条恒温器
bimetallic switch 双金属开关
bimetallic temperature sensor 双金属热敏元件
bimetallic thermometer 双金属温度计
bimetal pulse generator 双金属脉冲发生器（用于报警装置）
bimetal relay 双金属继电器
bimetal steam trap 双金属片式疏水器（热静力式排水阻汽装置）
bimetal thermometer 双金属温度计
bin 堆布池
binary colours 双色
binary deflection jet 二元偏转连续喷射（喷墨印花术语）
bind 黏合；束缚；使坚固；凝固
binder 黏合剂
binder efficiency 黏合效应；组合效应
binder film 黏合剂薄膜
binder for nonwovens 非织造布黏合剂
binder for pigment printing 涂料印花用黏合剂
binder prints on polyester 聚酯纤维织物的涂料印花
binding agent 黏合剂；载色剂；接合剂
binding by full bath impregnation 全浴浸渍法黏合
binding by impression 用轧压法黏合
binding energy 结合能，黏合能
binding force 黏合力，结合力
binding muslin 书面布

binding power 黏合力,结合力
bind material 黏合料
binocular 双筒的,双目[的];双目镜
binocular magnifier 双筒放大镜
binocular microscope 双目显微镜,双筒显微镜
binpiler 堆布池甩布机
bio-abrasion 生物侵蚀
bio-absorbable 生物可吸收的
bio-active 生物活性
bio-active fibre 生物活性纤维
bio-active monomer 生物活性单体
bio-analysis 生物分析[法]
bioassay 生物测定[法],生物鉴定
bio-availability〈BA〉 (药物等的)生物有效性,生物有效率,生物利用度
biobalance 生物平衡
biobarrier 生物屏蔽层,生物保护膜
biobarrier fabric 生物防护织物
biobattery 生物蓄电池
biocatalyst 生物催化剂
biocatalytic exhaust air cleaning 废气生化净化
biocell 生物电池
biocellulose 生物纤维素
biochemical-coagulation method 生化—混凝法《环保》
biochemical degradation 生化降解
biochemical energy 生化能
biochemically active 生化活性的
biochemical oxidation 生化氧化
biochemical oxygen demand〈BOD〉 生化需氧量,生化耗氧量
biochemical purification 生化净化
biochemicals 生化药品
biochemistry 生物化学

biocide 杀生[物]剂(对象包括细菌类、海产生物、蛀虫、白蚁或鼠类等一切有害生物)
bioclean 无菌的;无有害物体的
bioclean room 无菌室
biocolloids 生物胶体
biocompatibility 生物相容性
biocomposite 生物复合材料
bioconcentration 生物浓度
bioconversion 生物转化
biocopolymer 生物共聚物
biodegradability 生物降解性
biodegradable 生物可降解的
biodegradable detergent 可生物降解的洗涤剂
biodegradable sulfur dye 可生物降解的硫化染料
biodegradable surfactant 可生物降解的表面活性剂
biodegradation 生物降解
biodegradation of textile chemical 纺织化学品的生物降解
biodesizing 生物退浆
bio disc 生物转盘《环保》
bioecology 生物生态学
bio-electric amplifier 生物电放大器
bioelectronics 生物电子学
bioelement 生物元件
bioeliminability 生物消除性
bio-energy 生物能[源]
bio-engineered fibre 生物工程纤维
bioengineering 生物工程
bioenzyme 生物酶
bioenzyme finish 生物酶整理
biofibre 生物纤维
biofilm formation 生物膜的形成

bio filter 生物滤池,生物过滤器《环保》
biofiltration 生物过滤
biofinishing 生物整理
bioflocculation process 生物絮凝法
biofouling 生物污垢
biofuels 生物燃料
bioguard treatment 抗微生物处理
biogum 生物胶
bio-inertness 生物惰性
biological accumulation 生物浓缩,生物积累
biological agent 生物制剂
biological amplification 生物富集
biological and chemical protective clothing 生化防护服
biological basin 生物池
biological contact oxidation process 生物接触氧化法《环保》
biological decomposition 生物分解
biological deep shaft process 深井微生物处理法
biological degradability 生物降解度,生物降解性
biological depolymerization agent 生物解聚剂
biological detergent 生物洗涤剂(如酶洗涤剂)
biological eliminability 生物可消除性
biological equilibrium 生态平衡
biological film 生物膜
biological film process 生物膜法
biological filter 生物滤池
biological filter loading 生物滤池负荷
biological filter media 生物过滤介质
biological filtration process 生物过滤法
biological index of pollution〈BIP〉生物学污染指数
biological inertness 生物惰性
biologically active fibre 生物活性纤维
biological magnification 生物富集《环保》
biological membrane 生物膜
biological modification 生物改性
biological monitoring 生物监测
biological oxidation pond process 生物氧化塘法《环保》
biological oxidation process 生物氧化法
biological oxygen demand〈BOD〉生物需氧量,生物耗氧量
biological purification 生物净化
biological resistance 抗生物[侵蚀]性
biological self-purification 生物自净作用
biological sensor 生物传感器,能感知生物体并将其转换为相应电信号的传感器
biological sewage treatment 污水生物处理
biological stain 生物染色,生物着色
biological textile 生态纺织品
biological tissue 生理组织
biological tower 生物塔
biological treatment 生物处理
biological waste water treatment 生物废水处理
bioluminescence 生物[性]发光,生物荧光
biolysis 生物降解
biomass 生物体
biomaterial 生物材料
biomedical application 生物医学应用
biomedical fibre 生物医药纤维
biomedical material 生物医药材料
biomedical polymer 生物医药用聚合物
biomedicine 生物医学
biomembrane 生物膜
biomimetic 仿生的

biomimetic chemistry 仿生化学
biomimetic fibre 仿生纤维
biomimetic function 仿生功能
biomimetic materials 仿生材料
biomimetics 仿生[物]技术
bionics 仿生学
bionics-oriented garment designing 服装仿生设计
bionics research 仿生学研究
bio-osmosis 生物渗透
bio-oxidation 生物氧化[作用]
bioplasma 生物原生质
bio-polishing 生物光洁整理，生物抛光整理
biopolymer 生物聚合物（生物体内高分子化合物的总称，如核酸、蛋白质等）
biopolymerization 生物聚合
biopolysaccharide 生物多糖
bioprecipitation 生物沉淀作用
BioPrep 比奥普列普果胶酶（用于棉制品煮练工序，代替传统碱煮练，商名，丹麦制）
bio-preparation 酶精练；生物精练；生物前处理
biopromoter 生物促染剂
bioreactor 生物反应器
bio-reduction 生物减量整理
biorefractive 抗生物[降解]的
biorefractory nature 抗生物降解性
bioresistance 抗生物[降解]作用
bioresistant detergent 抗生物[降解]洗涤剂
biorheology 生物流变学
bioscience 生物科学
bio-scouring 生物煮练，生物精练
biosensors 生物传感器（用于环境分析）

Biosil 毕奥西尔抗菌织物（商名，日本东洋纺）
bioslime 生物污泥
Biosoftal N 一种透湿散热助剂（具有表面活性的改性聚氨酯，瑞士 Textilcolor 公司）
biosorption process 生物吸附法
biostatic finishes 抑制生物用整理剂，抑制生物用油剂
biosteel 生物钢（如蜘蛛丝）
bio-stoning 生物石磨整理
biosynthesis 生物合成
biosynthetic gums 生物合成胶
biotechnology 生物技术，生物工艺学
biotextiles 生物纺织品
biotic attack 生物腐蚀
biotic environment 生物环境
biotic index 生物指数
biotoxicology 生物体毒素学
biotreatment 生物处理
bio-washing 酵素洗，酶洗
BIP (biological index of pollution) 生物学污染指数
biphase 双相
biphenyl 联[二]苯
birch grey 白桦灰[色]
bird's-eye fabric 鸟眼花纹织物
bird's-eye pattern 鸟眼花纹
bird's-eye weave 鸟眼花纹组织
bi-reactive group 双活性基团
birefringence 双折射
birefringence method 双折射法
bis- （构词成分）双，二
bis-benzimidazole brightener 双苯并咪唑增艳剂，双苯并咪唑增白剂
2,2-bis-(p-chlorophenyl)-1,1-dichloroeth-

ane〈DDD〉 滴滴滴（对氯苯基二氯乙烷，国际禁用的杀虫剂）

bishrinkage yarn 双收缩变形丝《化纤》

bis-β-hydroxyethyl sulfone〈BHES〉 β-双羟乙基砜

bishydroxymethyl phosphine acid 二羟甲基次膦酸

bishydroxymethyl phosphine oxide 二羟甲基氧化膦

bisindolindigo 双吲哚靛蓝

Bismarck brown 苯胺棕（碱性染料）

bis-methoxymethyl urea 双甲氧甲基脲，双甲醚[基]脲（防皱防缩整理剂）

bismethoxymethyl uron 二甲氧基甲基乌龙，二甲醚[基]乌龙（棉织物的防缩防皱交联剂）

bis-methyl phosphine oxide 二甲基氧化膦

bismuth compound 铋化合物

bisphenol A〈BPA〉 双酚 A，2,2-二(4-羟基苯基)丙烷，二酚基丙烷（有毒化学品，在塑料制品及热感应纸等物品中均存在）

bisque 淡血牙色，藕荷色

bisulfate(=bisulphate) 硫酸氢盐，酸式硫酸盐

bisulfide(=bisulphide) 二硫化物

bisulfite(=bisulphite) 亚硫酸氢盐，酸式亚硫酸盐

bit 位，比特（二进位制）《计算机》

bitter almond oil 苦杏仁酸，苦杏仁油

biuret 缩二脲

bivalent 二价的

black 黑色；黑颜料；黑斑

black alkali 黑灰（不纯的碳酸钠，造纸业黑液回收的碱灰）

black and white halftone 黑白网点；浓淡点图，黑白半色调

black ash 黑灰（不纯的碳酸钠，造纸业黑液回收的碱灰）

blackbody 黑体（辐射吸收率与投射于其上的辐射波长和表面温度无关，并且恒等于1，是研究热辐射规律的理想辐射体）《测色》

black body locus 黑体轨迹

blackbody radiation 黑体辐射

black body temperature 黑体温度

Blackburn printers 英国布拉克本地方的平纹棉织物印花坯布

black denim 黑色牛仔布，黑色劳动布

blackening 发黑度

black fibre 碳纤维

Black fine line DFC 线条黑 DFC（用于细线条效果的黑色印花浆，适合湿罩湿工艺及汽蒸，不需焙烘，商名，意大利宁柏迪）

blackish 稍黑的

black light 不可见光

black mordant 黑色媒染剂

black-out cloth 遮光布

black-out coating 遮光涂层

black-out curtain 暗幕

black-out finishing 遮光整理

black sheet 薄钢板；薄板

blacksmithing 锻造

black spots 黑点（疵病）

black washer 黑皮垫圈（指普通垫圈，与光垫圈相对）

blade 刮刀片；叶片；[开关的]闸刀

blade coating 刮刀涂层

blade mark 刮刀印

blade spread device 刮刀式扩幅装置（用于弹性针织物）

blade squeegees 刀刃式刮刀,刀刃式刮浆器(圆网印花用)

blade wheel boiling off machine 叶轮式精练机(用于绢纺原料和清除杂质的机器)

blanc 漂布(法国名称)

blanched places 擦伤痕《丝》

Blancometer 光电白度仪(商名)

Blancophor 勃仑可夫荧光增白剂(商名,德国拜耳)

blank 坯布;毛坯;无色的;空白试验

blank assay 空白试验

blank bleach 空白漂白

blank card 未打孔纹板

blank dyeing 空白染色(不加染料的对比试验)

blanket 毯,毛毯,厚垫布;印花衬毯;造纸毛毯;滚筒包衬

blanket and cushions 毯子和垫子

blanket cleaning system〈BCS〉印花导带洗涤系统

blanket cloth 毛毯大衣呢

blanket cylinder 毛毯滚筒

blanket drying machine 衬毯烘干机

blanket felt 毛毯毡

blanket finishing 毛毯整理

blanket finishing machine 呢毯整理机

blanket guidance 导带引导

blanket heating 导带加热

blanket mark 衬布痕;胶毯痕《印》

blanket position measuring system〈BPMS〉印花导带定位自动测定系统(商名,荷兰斯托克)

blanket seam 橡皮衬布缝头压痕(滚筒印花疵)

blanket tension 导带张力

blanket washer 毛毯洗涤机;印花导带洗涤机

blanket washer with filter 带有过滤装置的印花导带洗涤机(圆网印花)

blanket washing machine 橡皮衬布洗涤机,印花导带洗涤机《印》

blank experiment 空白试验

Blankit 勃来开脱(含稳定剂的连二亚硫酸钠及增白剂的混合物,漂白剂,商名,德国巴斯夫)

blank print 空白印花

blank roller 光面滚筒(滚筒印花中不刻花纹的滚筒)

blank screen 空白筛网

blank test 空白试验;试空车

blank vat 空白还原浴(只含有烧碱和保险粉的还原溶液)

blaquets 印花机[用]毛毯

blast 鼓风,送风;鼓风装置;喷净装置;冲去[存储器内容]

blast fan 鼓风机

blast pipe 喷管,送风管

blast preheater 鼓风预热器

blast protection 防冲击

blast tube 鼓风管;喷管;送风管

blatic 湖绿[色]

bleach 漂白

bleachability 漂白率;可漂性

bleach activators 漂白活化剂

Bleach Clean up 氧漂生物净化法(商名,丹麦制)

bleach consumption 漂液消耗[量]

bleach croft 漂白车间

bleach decomposition 漂白分解

bleached goods 漂白布匹

bleached jeans 褪色型牛仔裤(靛蓝布漂

白,有穿旧变白感）
bleached linen 漂白亚麻布
bleached linter 漂白棉[短]绒
bleached pulp〈BP〉漂白浆粕
bleached yarns 漂白纱
bleacher 漂白坯布；漂白工厂；漂白工人
bleachery 漂白工厂；漂白间
bleaching 漂白[工艺]；油墨或油漆严重褪色
bleaching agent 漂白剂
bleaching apparatus 漂白设备
bleaching assistant 漂白助剂
bleaching bath 漂白浴
bleaching by winch beck 漂白用绞盘槽
bleaching cistern 淋漂箱
bleaching damage 漂白损伤
bleaching-dyeing combined process 漂白染色一步法
bleaching earth 漂白土
bleaching fastness 漂白[坚]牢度
bleaching-in-clear 清漂，无皂漂白
bleaching intensity 漂白强度（指氧化程度）
bleaching in the bale 毛包储存漂白（洗毛机木槽加入少量氧化剂，使净毛在储存期间渐被漂白）
bleaching lime 漂白粉
bleaching machine 漂白机
bleaching-out 褪色的
bleaching powder 漂[白]粉
bleaching power 漂白能力
bleaching range 漂白联合机
bleaching range for tubular knit 筒状针织物的漂白机组
bleaching steamer 漂白蒸箱
bleaching white wools 羊毛漂白

bleaching with peracetic acid 过醋酸漂白
bleaching with sulphur 用硫黄漂白
Bleach-O-Matic 前处理化学液自动测定与控制仪（商名，荷兰布鲁曼）
blebbing（= blebby） 满地印花轮廓不清（印疵）
bleb rate 发泡率
bleed 放气；放液；渗出；冒油；渗色
bleed allowance 色浆渗化分线容许公差
bleed design 渗散印花花样
bleeder style printing 渗散印花，渗化印花
bleeding 渗化，渗色，流散，化开，色晕（印疵）
bleeding and staining 渗化和沾色（印疵）
bleeding by screen 拖版（平网印花疵）
bleeding of additive 添加剂的渗出现象
bleeding style 渗散印花（利用渗化使花纹轮廓模糊的印花方法）
bleed resistance 抗渗化性
bleed testing 渗色试验（染色牢度试验）
blemish 瑕疵；污点
blendability 可混合性，可掺合性
blend dyeing 混纺染色
blended-colour fabric 混色布，混色织物
blended dyes 混合染料（用于混纺织物染色）
blended fabric 混纺织物
blender 混合器，搅拌机
blend fabric 混纺织物
blending 混合；混纺；混色；毛皮染色
blending attachment 混合装置
blind 窗帘；百叶窗；挡板；失光；闭塞的
blind dyeing 盲染，盲目染色
blinding 失光（由于湿加工而引起的光泽损失）；堵塞
blind roller （可以卷成筒状的）窗帘布；

遮阳卷帘

blind seam 暗缝

blind spot [视网膜上]盲点；视线盲视的地方

blip culture 瞬间文化

blister 气泡；气孔，砂眼

blister cloth 泡泡织物；泡泡呢

blister fabric 泡泡纱；凸纹[浮线]织物

block 手工印花的雕花木板；滑块，滑车；熨衣板，定形板；信息块《计算机》

block copolymer 接枝共聚物；嵌段共聚物

block diagram 方框图（一种常用的电气图）

blocked system 闭塞系统；端基封闭法

block holder 裁剪样板架

blocking 布层粘连（未干透的印花布打卷后被印浆粘住）；阻染[现象]；浆块，蜡块；接枝；塞网；封端，保护；堵塞

blocking agent 抑制剂（用于阻染）；封闭剂

blocking effect 阻塞效应；封闭效应；抑制作用（抑制染料上染）

blocking of dye site 封闭上染位置

blocking problems 堵网问题

block pattern 服装样板

block polymerization 嵌段[共]聚合[作用]，整体聚合[作用]

block printing 手工模版印花

blood-cyanide value 血—氰化物值

blood red 血红[色]

bloom [毛织物]光泽整理；[丝和丝绒织物]绿灰色绒光；[染色布]表面平淡光泽；[地毯绒头]松开；浴面浮渣

blooming [锡盐]增艳（棉毛织物染色后用锡盐处理）；起霜，起晕；图像散乱；白化（透明膜表面发生白云现象）

bloomy lustre 鲜明光泽，丝绒[状]光泽

blossom [pink] 淡粉红[色]

blotch 满地花纹，印花色底；斑渍

blotch checks 印经蓝斑格子布

blotch coverage 满地罩印；满地印花面积

blotch ground 满地花纹，印花色底

blotch printing 满地印花，底色印花，单面印花

blotch roller 满地花纹滚筒，地色花筒

Blotch-Roll-Flash-Age technic 刮浆快蒸印花法

blotchy 斑污，斑渍

blotchy dyeing 斑渍染色

blotter action 吸墨水作用

blotting 浸透，渗出；沾吸作用，吸墨纸效应（应用于轧吸技术，以达到低给液的目的）

blotting action 吸墨水作用

blouse fabric 衬衫织物

blow cleaner 吹拭装置

blowdown 吹除，送风；排污；排气；放料，卸料

blowdown stream 排出液流，排出水流

blower 鼓风机；低压空气压缩机；（缝纫机的）推布送料器

blow finish 蒸呢

blowing 鼓风；转鼓汽蒸；蒸呢；喷吹；漏气

blowing agent 发泡剂，生气剂，起泡剂

blowing and cool air exhausting machine 蒸呢抽冷机

blowing machine 蒸呢机

blowing pipe 风管

blowing roller 蒸呢辊

blowing stereo-printing process 发泡立体印花工艺

blow-in pipe 进气管,冲气管
blow lamp 喷灯
blow-off pipe 排气管;排水管;泄泥管
blow ratio 充气率,发泡率(一定体积的液体发泡前的重量与同样体积泡沫重量之比)
blow tank 受料槽;出料槽
blow valve 安全阀;送风阀;排空阀
blue ashes 深灰蓝[色],潮蓝[色]
blue brighting 上蓝增白
blue brushing 刷蓝(在绒类织物上用蓖麻油和蓝色染料增艳的方法)
blue deficiency 蓝光不足
blue denim 蓝色牛仔布
blue "eating" dye combination 引起蓝色染料催化褪色的拼色(主要问题在还原染料)
blue gas 水煤气
blueing 上蓝(对漂白过程中带有黄色光的白纱或织物加一些蓝色染料)
blueing agents 上蓝剂
blueing tint 上蓝
blueish 带蓝色的
blue jean 蓝斜纹
blue print 蓝图
blue purple 品蓝[色]
bluer [色光]偏蓝[色]
blue reflectance 蓝光反射率
blue scale [耐日晒牢度]蓝色标准
blue shift effect 蓝移效应(向蓝色光谱带位移),浅色效应《测色》
blue stone 蓝石;青石(一种天然硫酸铜)
blue verdigris 碱式醋酸铜
blue vitriol 胆矾,蓝矾,硫酸铜
bluewhite finish 上蓝整理
blue wool fabric 蓝色毛织物标样(评定耐光色牢度用)
blue wool light fastness standard 耐光色牢度用蓝色毛织物标样
blue wool standards 蓝色毛织物标样(评定耐光色牢度用)
bluff 齿;爪;钩
bluing 上蓝(对漂白过程中带有黄色光的白纱或织物加一些蓝色染料)
bluish 带蓝色的
blur 污渍;[织物]剪毛不良(疵点名)
blurring [图像]模糊,混乱
blurry [花纹]模糊不清
blush 光亮;有光处理
BMA(butyl methacrylate) 甲基丙烯酸丁酯
BMC(bulk molding compound) 本体浇铸化合物
BMV(brush-sueding machine version) 刷绒[仿麂皮]起绒机形式
board 板,木板;盘,配电盘
boarder 定形机
boarding 热定形,定形[工艺](指合纤袜类等针织物的热定形)
boarding defect 电压纸板印花疵
boarding machine 定形机(专用于合纤袜类等针织物)
boarding press 热压定形机,热压定形
boardness 织物硬挺度
boardy 粗硬织物;硬挺度;手感粗硬,手感板硬
bobbin 筒管,筒子;绕线管《电》
bobbin aftertreatment machine 筒子纱后处理机
bobbin dyeing machine 筒子纱染色机
bobbin rack 筒管架,筒子架
bobbin shifting machine 筒子纱换芯机

BOD (biochemical oxygen demand) 生化需氧量,生化耗氧量
BOD (=B.O.D., bacterial optical density) 细菌光密度
BOD (=B.O.D., biological oxygen demand) 生物需氧量,生物耗氧量
BOD loading 生化需氧量负荷,BOD 负荷
BOD ultimate 最终生化需氧量
body armor 人体装甲
bodying agent 充实剂
body-scanner system 人体扫描系统
body-scanning 人体扫描(用计算机进行人体设计的制衣程序)
body setting machine 衣服成形机
body supporter textile 护身用纺织品
boiled 精练的,煮练的,脱胶的
boiled lawns 精练细亚麻平布
boiled linen 脱胶亚麻布;精练亚麻布
boiled linen yarn 精练亚麻纱
boiled-off liquor 精练[丝绸]残液,丝绸脱胶脚水
boiled-out cotton 煮练过的棉织物;煮练过的棉纱
boiled [off] silk 熟丝,脱胶蚕丝,精练蚕丝;熟绸,精练绸缎
boiled yarn 煮练[过的]纱
boiler 锅炉
boiler capacity 锅炉容量,锅炉蒸发量
boiler efficiency 锅炉效率
boiler evapourative section 锅炉蒸发区
boiler feedwater 锅炉进水
boiler flue 锅炉烟道
boiler horsepower 锅炉功率
boiler house 锅炉间
boiler output 锅炉蒸发量,锅炉出力
boiler plant 锅炉装置;锅炉房

boiler rating 锅炉额定蒸发量,锅炉额定出力
boiler scale 锅炉水垢
boiler water 锅炉水
boilfast colour 耐洗牢度高的染料
boiling 沸腾;沸煮;煮呢
boiling kier 煮布锅
boiling machine 煮呢机
boiling off 煮练,精练;脱胶
boiling-off liquor 练液
boiling-off loss 练耗,练减量
boiling-off test 练减检验,除胶检验《丝》
boiling-out 煮练,精练
boiling pan 煮釜;调浆锅
boiling-pan for thickeners 煮浆锅
boiling point 〈B.P., b.p.〉 沸点
boiling point elevation 沸点升高
boiling reduction in weight 煮沸减量率
boiling shrinkage 沸水收缩性
boiling wildness 练丝起毛
boil-off gum content 沸溶法含胶量
boil-off shrinkage 精练收缩[率],退浆收缩[率]
boil-out test 沸煮试验
boil-over 沸溢
boilup rate 蒸发量
Bolaform electrolytes 波拉型电解质(双季铵盐,可作为染腈纶的新型缓染剂)
Bolaform surfactant 波拉型表面活性剂(新型端基亲水性基团染色助剂,可作为洗涤剂等)
bold corduroy 粗条灯芯绒
bolometer [电阻]辐射热测定器
bolster fabric 承垫织物
bolt 螺栓,栓
bolt coupling 螺栓联轴节

bolting cloth 筛绢,筛绢织物;筛网
bolting silk 筛绢
Bombay Textile Research Association〈BTRA〉孟买纺织研究协会(印度)
bond 键;结合;黏合;黏合体
bond angle 键角
bond dissociation energy 键离解能
bond distance 键长
bonded 化合的;结合的;黏合的
bonded carpet 黏合地毯;黏绒地毯
bonded chemicals [纺织纤维]黏合用化学品
bonded cushion [地毯]黏合衬垫
bonded dyes 键合染料(活性染料)
bonded fabric 黏合织物(非织造织物)
bonded knitted fabric 黏合针织物
bonded rubber cushion 黏合橡胶衬垫
bond energy 键合能,键能
bonding 黏合[工艺];结合,联结,黏合
bonding agent 黏合剂
bonding effect 结合效应,黏合效应
bonding fabric 黏合衬;黏合织物
bonding finish 胶合整理(将两层材料借黏合剂黏合在一起,或将泡沫胶乳熔融涂抹在基布上进行热胶合)
bonding force 黏合力
bonding nonwoven 黏合法非织造布
bonding of secondary backing [地毯]第二层底布的黏合
bonding press 黏压机(用于领衬布黏合)
bonding strength 结合强度,黏合强度
bond length 键长
bond stability 键合稳定性,键稳定性
bond strength 黏合强力,结合强力
bond-weakening 键削弱,键减弱
bone colour 骨色(近乎白色)

bone dry 绝对干燥,透骨干燥
bone glue 骨胶
bonnet 阀帽,盖,机罩
bookbinding leather 书面皮革
book cloth 书面布
boom 浮栅栏
booster 升压器,增压器;增效剂
booster bath 增效浴,辅助浴
booster box [还原染料固着]增效箱,还原汽蒸箱
booster compressor 升压压缩机,增压压气机
booster pump 升压泵,增压泵
booster solvent 助溶剂
boost motor 助进电动机;加速器
boot 保护套,防潮套;罩
BOPP(biaxial oriented polypropylene) 双轴向拉伸聚丙烯
Bopp nickel screens 波普镍网(印花用镍网,瑞士波普公司)
boracic acid 硼酸
borate 硼酸盐
borax 硼砂;月石
bordeaux [葡萄]酒红[色](暗紫红色)
bore 孔,内径;镗,钻孔
bore of cylinder 印花滚筒芯孔
boric acid 硼酸
boric acid-iodine system 硼酸—碘法(检验织物上 PVA 浆料)
boring 扩孔,镗孔
boring machine 镗床
borohydride 氢硼化物
boron fibre 硼纤维
borosilicate 硼硅酸盐
boss 轮壳,外壳;轴套;套筒
Bosselé 博塞莱整理(拷花整理,法国名

称)
botanical print 植物图案印花花样
botany fabric 博坦尼精纺细毛织物[用16.67tex(60公支)以上羊毛制成]
both-side printing 双面印花
bottle green 瓶绿[色](带蓝光深绿色)
bottle neck 瓶颈领;关键,障碍,生产薄弱环节
bottom [织物的]地,布地;地色;底部;残渣
bottom cloth 衬垫织物,地织物
bottom dyeing 地色染色
bottom fabric 厚重织物
bottoming 打底;漂染前处理;[花纸]油墨涂布
bottoming denim 牛仔布(先染硫化染料后套染靛蓝)
bottom line 底线,盈亏线
bottom plate 底板
bottom roller 下罗拉,下排导辊
bottom view 仰视图
bouclé 珠皮呢(仿羔皮呢);结子线织物
bouclé carpet 毛圈式地毯
bouclé yarn 结子花式线,毛圈花式线
bouncing 回跳,弹起
boundary 界面;边界
boundary effect 界面效应
boundary friction 界面摩擦
boundary layer 界面层,附面层
boundary layer theory 界面层理论;附面层理论
boundary lubrication 界面润滑
boundary reflection 界面反射;边缘反射
boundary science 边缘科学
boundary surface 分界面
boundary tension 界面张力

bound seam 滚边缝接
bound water 结合水
bourette 结子织物;绵绸
bourette silk 䌷丝
bow 弓纬,纬弧(织疵)
bow correction rolls 整弓纬辊
bow distortion 弓纬,纬弧(织疵)
bowed expander 弧形扩幅辊,扩幅弯辊
bowed roller 弯辊;弧形辊
bowed weft 弓纬,纬弧(织疵)
bow expander roller 弯辊扩幅机,扩幅弯辊
bowing 弓纬(织疵)
bowking [棉布]石灰水煮练
bowl 轧辊;槽(用于织物湿加工的开口容器);碗;(油水分离机的)转鼓;[离心机]转筒;浮筒;滚筒
bowl deflection 轧辊挠度
bowl deflection meter 轧辊挠度计
bowl grinding machine 滚筒研磨机
bowl paper 轧辊纸(如羊毛纸等)
bow roller 弯辊,弧形辊
bow straightener 弓纬校直器;弯辊整纬器
bow straightening roller 整纬辊
bow strap 缓冲皮带
bow weft 弓纬,纬弧(织疵)
box dyeing 染槽染色(绳状染色)
box pilling tester 箱式起球试验仪
box spanner 套筒扳手
box wheel 对花齿轮(滚筒印花机的对花装置)
box wrench 套筒扳手
BP(base point) 基点,原点
BP(bleached pulp) 漂白浆粕
B.P.(＝b.p.,boiling point) 沸点

BP(B-power) 阳极电源,乙电源
BP(British Patent) 英国专利
BPA(bisphenol A) 双酚 A,2,2-二(4-羟基苯基)丙烷,二酚基丙烷(有毒化学品,在塑料制品及热感应纸等物品中均存在)
BPMS (blanket position measuring system) 印花导带定位自动测定系统(商名,荷兰斯托克)
B-power〈BP〉 阳极电源,乙电源
brace 支柱,撑臂;拉条;撑
bracing 拉条,撑条
bracing plate 撑板
bracing wire 支撑钢丝
bracket 托架,托座;支架,支座;偶撑;括号,方括号;摇框;固定夹
Bradford Dyers Association-carpet prin-ting machine 布雷德福染色协会地毯印花机(真空吸色地毯网印机)
braided fabric 编织物
braided textile carpet 编结地毯
braiding technology 编织技术
brake 制动器,刹车,闸;闸式测功器;制动,阻滞;止退器
brake arm 刹车臂
brake band 制动带(附有制动衬面的带状金属箍)
brake block 制动块,闸块,制动器闸瓦
brake chain 制动链,闸链
brake coupling 制动离合器
braked batches 带有制动装置的卷布器
brake disc 制动片,制动盘
brake fluid 刹车油
brake handle 制动手柄
brake horsepower 制动马力
brake lever 制动杆

brake motor 带有制动装置的电动机
brake roller 制动辊
brake shoe 制动块,闸瓦
brake spring 制动弹簧,闸簧
brake strap 制动带
brake wheel 刹车盘;制动轮
braking force 制动力
branch 分支;转移
branch circuit 分支电路
branched aliphatic hydrocarbon 支链脂肪烃
branched chain 支链
branched polymer 支链型高分子
branch pipe 支管
brand 牌子,商标;品种
brand name 品牌
brand name policy 品牌政策
brands [毛包]标志;[唛头印记]沾污毛
brass 黄铜色;黄铜;黄铜轴承衬
brass beetle shade 黄铜甲虫色
brass bush 黄铜轴承衬
brattice 转动帘
brattice conveyer 帘子运输带
brattice drying machine 转帘式烘干机
brattice machine 洗线机《毛》
brattice scouring machine 转帘式精练机
brazen yellow 黄铜色
braze welding 铜焊
Brazil wax 巴西蜡
brazing 铜焊
break 绒面裂纹;毛皮露底
breakage 破损,断裂;断头率,破损疵点
breakage by separation 破裂,破碎,断裂;断面,断口
break compound 预洗剂;头道洗涤用剂
breakdown 故障,损坏;击穿《电》

breakdown maintenance 停工维修,事故维修

breakdown voltage 击穿电压

breaker 电流断路器

breaking capacity [开关的]遮断功率,断流容量《电》

breaking-down test 破坏[性]试验

breaking elongation (= breaking extension) 断裂伸长,断裂延伸

breaking-in period 试车时期,试产时期

breaking length 断裂长度

breaking load 断裂负荷

breaking machine 揉布机

breaking strength 断裂强力

breaking stress 断裂应力

breaking tenacity 断裂强度,抗断韧度,断裂强力

breaking tension 断裂张力

breaking unit 断电单元,断流单元(机器保护措施)

breakmark 灰点(绸缎疵点)

break-open 烧穿(阻燃试验)

break through 突破;猛涨

breast roller 胸辊

breathability 透气性

breathable coating 透气性涂层

breathable fabric 透气织物

breathable multi-micro porous coated fabrics 透气多微孔涂层织物

breathable multi-micro porous laminated film fabrics 透气多微孔层压膜织物

breathable waterproof uniforms 透气防水服

breathe 使[织物]透气

breather hole 通气孔;透气孔

breathing coating 透气性涂层

breathing mask 口罩,呼吸面具

breeding-damaging substances 危害生育物质

breen 褐绿[色]

Breviol DE 布列维尔 DE(烷基聚乙二醇醚,羊毛、锦纶等染色匀染剂,商名,德国科宁)

Breviol SCN 布列维尔 SCN(胺缩合物,羊毛用匀染剂,商名,德国科宁)

Breviol WSM 布列维尔 WSM(蛋白质水解物,羊毛保护剂,商名,德国科宁)

brick-colour 砖[瓦]色;砖红色

brick red 砖红[色]

brick wall design 砖墙组织图案《印》

bridge bond 桥键

bridge formation 桥结[现象],架桥结构

bridge-linking mechanism 架桥机理

bridging 搭接,架桥

bridging flocculation 架桥凝聚

bridging of insulator 绝缘体的搭桥(圆网制造)

bridging-oxygen 桥氧

bright 高度光泽;色泽鲜艳;有光《化纤》

bright and dark method 亮暗法(耐日光牢度测试法,灯光时亮时暗地进行照射)

bright blue 中湖蓝[色]

bright colour 鲜艳色,鲜明色

brightened goods 增白织物

brightener 增白剂,增艳剂

brightening 增艳处理,增白处理

brightening acid 增亮酸(丝整理用)

brightening agent 增白剂,增艳剂

bright fibre 有光纤维

bright, high tenacity〈BHT〉 有光和高强力[的]
brightness 光亮度,光泽度,鲜明度
brightness axis 亮度轴《测色》
brightness contrast 亮度对比
bright rayon 有光人造丝
bright washer 光垫圈
brilliance(=brilliancy) 鲜艳,光泽;灿烂;辉度,亮度
brilliant 加光整理;有光内衣绸(丝棉交织)
brilliant blue 酞青[色],孔雀蓝[色]
brilliant shade 鲜艳色
brilliant yellow 亮黄(pH指示剂)
brine 盐水
Brinell hardness〈H_B,H_b,H. Br.〉 布氏硬度,压球硬度
bristle 鬃毛,刚毛
bristle effect 戗毛效应
bristle hair 死毛,戗毛
bristle roller 鬃毛辊(起绒机用)
British Ceramic Research Association〈BCRA〉 英国陶瓷研究协会
British Colour Council〈BCC〉 英国颜料染料委员会
British Engineering Standards Association〈BESA〉 英国工程标准协会
British gum 印染胶
British Health and Safety Organization〈BHSO〉 英国卫生与安全组织
British Health and Safety Regulation〈BHSR〉 英国卫生与安全管理条例
British Patent〈BP〉 英国专利
British Standard〈BS,B. S.〉 英国[工业]标准
British Standards Institution〈BSI,B. S. I.〉 英国标准学会
British Textile Machinery Association〈BTMA〉 英国纺织机械协会
British Textile Technology Group〈BTTG〉 英国纺织技术集团
British thermal unit〈BTU,B. T. U.,Btu,B. t. u.〉 英热单位(热量单位,等于1055.056焦耳)
British Wool Confederation〈BWC〉 英国羊毛联合会
British Wool Marketing Board〈BWMB〉 英国羊毛销售管理部,英国羊毛贸易部
brittle 脆[性]的;易碎的,易损坏的
brittle point 脆化点
brittle resistance 防脆性
brittle temperature 脆化温度(材料低于此温度是脆性的)
broadcasting fabric 喇叭布
broad cloth 阔幅布;细平布;绒面呢;阔幅厚黑呢
broad drawing equalizing machine 宽幅拉伸扩幅机
broad fabrics 阔幅布,阔幅织物
broadloom carpet 宽幅地毯
broad-spectrum microbiocidal agents 广谱杀菌剂
broad woven goods 宽幅织物
brocade 锦缎,花缎;凹凸花纹
broken crease 破边
broken line 虚线;折线;断线(印疵)
broken-open perspective view 剖开透视图
broken screen 花型不整,断花;印花筛网破裂
broken selvedge 断边,破边,裂边(织疵)

brom[o]- （词头）表示"溴的""含溴的""溴代的"

bromate discharge 溴化盐拔染

bromic acid 溴酸

bromide 溴化物

brominated phosphate 溴化磷酸盐（阻燃剂）

brominated phosphite 溴化亚磷酸盐（阻燃剂）

bromination 溴化作用

bromination process 溴化［防缩］处理（毛织物整理）

bromine 溴

bromine compound 溴的化合物

bromite 亚溴酸盐

bromite desizing 亚溴酸盐退浆

α-bromoacryloylamino group α-溴代丙烯酰氨基团（毛用活性染料中的一种活性基团）

bromoform allyl phosphate〈BAP〉 三溴甲烷烯丙基磷酸酯

bromo indigo 溴靛蓝（染料）

bromthymol blue 溴百里酚蓝（酸碱测定用指示剂）

Bronsted acid 布伦斯特酸，质子酸（能给出一个质子的物质）《试》

Bronsted base 布伦斯特碱（能接受一个质子的物质）《试》

bronze 青铜色，古铜色；青铜

bronze blue 深蓝［色］；蓝色颜料

bronzed fabric 古铜色装饰织物

bronze green 灰橄榄绿［色］；绿色颜料

bronze mist 秋香色

bronze powder 铜粉

bronze red 铜红［色］（棕橘色）；红色颜料

bronze yellow 杏黄［色］

bronziness 铜光（染疵）

bronzing 铜翳（由于染料集中在织物表面而形成的似金属发亮现象）；铜光浮色（硫化元染疵）；金铜斑（丝绸染疵）

bronzy dyeing 青铜色泽（染疵）

Brookfield viscometer 布鲁克菲尔德回转式黏度计（测印花糊料黏度用）

brooming 帚化

brown 棕［色］，褐［色］

Brownian diffusion 布朗扩散

Brownian motion 布朗运动

browning 变褐色（染疵）

browning proofing 防变褐色处理

brownish 带棕色的

brown linen 本色亚麻布

brown paper 牛皮纸

brown souring ［布匹煮练后］第一道吃酸工序

browse 浏览

Brückner roller printing machine 勃鲁克纳滚筒印花机（单承压辊斜列式）

bruise roller 压浆滚筒（系滚筒印花时装在色浆滚筒后面的一种光面滚筒，可使大面积印花均匀并消除露地）

Brunauer-Emmett-Teller〈BET〉 布鲁诺尔—埃米特—特勒，由三位科学家名字的首字母组成，表示比表面积的测试方法

brush 刷；电刷

brush damping 毛刷给湿

brush damping machine 转刷给湿机

brush-dewing machine 毛织品给湿机，毛刷辊给湿机，转刷给湿机

brushed fabric（＝**brushed goods**） 起绒织

物;拉绒织物
brushed pile 拉绒绒毛
brush furnisher 毛刷给浆滚筒(滚筒印花机的给浆装置)
brushing 刷布;刷呢;刷毛;刷绒
brushing defect 刷柳(丝绒整理疵)
brushing machine 刷毛机;刷布机
brushing machine for flannelette 绒布刷绒机
brushing printing 刷印(刷帚花版印花)
brushing unit 刷毛箱
brush patterning unit 印花装置;刷花装置
brush pile finishing 起绒整理
brush pilling tester [拉绒织物]起球试验仪
brush roller 毛刷罗拉,毛刷辊
brush sieving colour printing 刷滤法印花
brush stencil printing 镂空模版印花
brush-sueding machine version (BMV) 刷绒[仿麂皮]起绒机形式
brush-sueding method 刷毛[仿麂皮]起绒整理方法
brush-sueding roller 刷毛[仿麂皮]起绒滚筒
brush washer 毛刷清洗器
BS (=B.S.,British Standard) 英国[工业]标准
BSI (=B.S.I.,British Standard Institute) 英国标准学会
BTCA(butane tetracarboxylic acid) 1,2,3,4-丁烷四羧酸(无醛防皱整理剂)
BTMA (British Textile Machinery Association) 英国纺织机械协会
BTRA(Bombay Textile Research Association) 孟买纺织研究协会(印度)
BTTG (British Textile Technology Group) 英国纺织技术集团
BTU (=B.T.U.,Btu,B.t.u.,British thermal unit) 英热单位(热量单位,等于1055.056焦耳)
BTX (benzene-toluene-xylene) 苯—甲苯—二甲苯(有毒溶剂)
BTX content [有害溶剂]苯—甲苯—二甲苯的含量
BTX recovery 苯—甲苯—二甲苯的回收
bubble coating 泡沫涂层
bubble colouration process 泡沫染色法
bubble jet 喷泡式(热喷泡式,喷墨印花方式)
bubble jet printing 喷泡(气泡)式喷墨印花
bubble printing 气泡印花
bubble process 气泡[染整]工艺
bubbling tube 吹泡管(吹泡式液面检测器组件)
bucket 水斗,吊桶,勺子
bucket steam trap 水斗式疏水器(机械式的排水阻汽装置)
bucking kier 煮练釜
buckled selvedge [布]边不齐
buckling 扣紧
buck souring 温酸洗工艺(在亚麻布漂后直接进行)
bud green 浅草绿[色],鲜绿[色],春绿[色]
buffer 缓冲;缓冲剂;缓冲器;保险杠
buffer agent 缓冲剂
buffering action 缓冲作用
buffering agent 缓冲剂

buffering zone 缓冲区
buffer salts 缓冲盐(用于亚氯酸钠漂白或其他染整工艺)
buffer solution 缓冲溶液
buffer spring 缓冲弹簧
buffer stop 缓冲挡板;缓冲停车
buffer value 缓冲值
buffing [仿麂皮]磨毛工艺;抛光
buffing fabric 磨光用织物
buffing finishing 磨光整理
buffing machine 磨光机
buffing wheel 抛光轮
buff-muff coupling 刚性联轴器,套筒联轴节
bug repellent fabric 防虫织物
builder 增效剂;增色剂;助洗剂;辅助助剂
building textiles 建筑用纺织品
building up properties 提升性(染深性能)
build up 提升(染深性能);花筒积色;污垢积聚;沉积,沉积物
build-up agent 增深剂
build-up finish 增浓整理,增深整理
built detergent 复配洗涤剂(添加各种助洗剂的洗涤剂)
built-in antistatic property 内藏抗静电性能
built-in lubrication 〈BIL〉滑爽
built-in motor drive 内装式电动机传动
built-in type 内装式,嵌入式
built soap 组合肥皂,复配肥皂(肥皂中含有一种或多种碱性洗剂,含干皂量不低于50%)
built-up member 装配件
built-up shaft 组合轴

built-up type 组合式
built-up welding 堆焊
bulb 灯泡;球管
bulging effect 起拱效应(测试裤子膝盖部位)
bulk 松密度;大量;胀量
bulk cleaning 大量干洗(装饰织物的廉价干洗法,价格依重量计算)
bulk compressibility 体积压缩性
bulk continuous filament yarn 〈BCF〉膨体长丝
bulk degradation 本体降解
bulk development 蓬松显现[处理]
bulked yarn 膨化变形纱,膨体纱
bulk effect 体积效应;膨化效应
bulk fibre 散纤维
bulk grafting 本体接枝
bulkiness 蓬松性,蓬松度
bulking 蓬松化;污泥膨胀《环保》
bulking capacity 膨化能力
bulking effect 膨化效应
bulking intensity 蓬松度
bulking liquor 补给液,添加液《染整》
bulking property (膨体变形的)蓬松性
bulking sludge 膨胀污泥《环保》
bulking wool 膨体羊毛
bulk material 松散充填材料
bulk molding compound 〈BMC〉本体浇铸化合物
bulk polymer 本体聚合物
bulk polymerization 本体聚合
bulk sample 大样
bulk trials 大样试验
bulk uniformity 蓬松均匀度
bulk viscosity 本体黏度
bulk volume 毛体积

bulk weight 散重,总重
bulk working 大批量加工,大批量生产
bulk yarn 膨体纱
bulky character 丰满[感]特性,蓬松特性
bulky drying machine 大型烘燥机
bulky finish 膨体整理
bulky finish of acrylic knitting 腈纶绒线蓬松整理
bulky hand 丰满手感
bulky property 蓬松性;弹性
bulky yarn 蓬松纱(纤维本身具有蓬松性,如空心丝、异形丝、卷曲丝等所组成的复丝或短纤纱)
bulldog wrench 管子扳手,管子钳
bullet-proof cloth 防弹织物
bullet-proof clothes 防弹衣
bullet-proof garment 防弹服
bullet-proof vest 防弹背心
bumper 缓冲器,减震器;保险杆;消音器
bumper bar 保险杆
bumper milling machine 锤式缩呢机
bump grey 印花衬布
bumpiness [织物]表观不平整性
bumps 印花[机]衬布(英国术语)
bump seam 衬布缝头横档痕(印疵)
bunching 起皱;收缩
bunch yarn 雪花线,竹节花线
bunchy yarn 竹节纱(纱疵)
Bundesman testing 邦迪斯曼防水[度]试验法
bundle 捆,扎,束,小包
bundle dyeing 绞纱染色
bundle system 打包装置
bundle yarn 小包绞纱[重2.27千克(5磅)或4.54千克(10磅)];小包亚麻绞纱[每包重4.54千克(10磅)]
bundling 打小包;集束,成束
bundling press 小包机
Bunsen burner 本生灯(一种煤气灯)
bunter 撞头,触杆
Bunte salt 本特盐(过硫酸烷基酯,用于羊毛防缩)
Bunte salt dyes 本特盐染料
buoyancy bag 浮力袋
bur [羊毛]草籽,草刺;结节(呢坯纱疵);毛口(如花筒毛口,指花筒表面受钢芯轧纹而出现的疵点)
burette 滴定管
burgundy 深酒红[色]
"buried end" patterning effect "绒毛埋入法"提花效果
burin [手工]雕刻刀
burl dyed cloth 斑点着色的毛织物(染后显现植物性疵点,经着色涂盖,使色泽一致)
burl dyeing [毛织物]纤维素杂质低温套色;[棉毛交织物]棉经低温套色
burler 修呢工,验布工
burling 修补,修呢(修去毛织物的纱头和结头等);去除麻结;去除毛结
burling machine 验布机
burl mark 修补痕(织疵)
burned by oil 用油燃烧
burner [烧毛机]火口;燃烧器,喷烧器;喷灯
burner bank 喷烧器组合
burner cooling system [烧毛机]火口冷却装置
burner exhaust duct 燃烧室排气管道
burner ports 燃烧器[火口]

burner ribbon 燃烧器条板
burner shifter ［烧毛机］火口调幅装置
burnetising(＝burnetizing) 氯化锌防腐整理(用于帆布和绳索)
burning behaviour 燃烧性能
burning off discharge 烂花拔印
burning off mordant 烂花媒染
burning proof 防燃,防灼,抗燃
burning rate 燃烧率
burning resistance 耐燃性
burning retardant 阻燃剂
burning temperature 燃烧温度
burning test 燃烧试验,灼烧试验
burnished gold 浅秋香色,草黄［色］
burnisher 磨光辊,抛光辊;磨光器,抛光器
burnishing ［花筒］磨光;［花筒］抛光;擦光;光泽
burnishing brush 磨光［刷］辊,抛光辊;抛光刷
burnishing roller 抛光辊;刷光辊
burn-off rate 融蚀率
burn-out printing 烂花工艺,烂花印花
burn-out style 烂花印花法
burnt discharge 烂花
burnt gas fumes 燃烧烟气
burn-through time 烧穿时间
burnt lime 烧石灰,煅石灰,生石灰,氧化钙
burnt-out discharge 烂花拔印
burnt-out fabric 烂花织物
burnt-out lace 烂花花边
burnt-out mordant 烂花媒染
burnt-out print 烂花花布;烂花
burnt-out printing 烂花工艺,烂花印花
burnt-out process 烂花工艺

burnt-out style 烂花印花法
burr ［羊毛］草籽,草刺;结节(呢坯纱疵);毛口(如花筒毛口,指花筒受钢芯轧纹而出现的疵点)
burr dyeing 带草籽染色
burr removal 去除草籽(羊毛炭化)
burst fibre 裂膜纤维
burst film yarn 裂膜丝
bursting ［织物］顶破
bursting strength ［织物］顶破强力,顶破强度
bursting strength tester ［织物］顶破强力试验机
bursting test ［织物］顶破强力试验;［织物］破裂试验
burst pressure 爆裂压力;顶破压力
burst safeguard 爆裂安全装置(置于热辊的一端,当其夹层压力大于规定值时会破裂)
bus 总线,母线《计算机》
bus-bar 汇流条,母线(在配电装置中,用来连接电源与各馈电支路,接受和分配电能的中间导体)
Buser type automatic screen printing machine 布赛尔型自动筛网印花机(商名,瑞士制)
bush(＝bushing) 轴套,衬套,轴瓦;套管
bush sleeve 套筒
busy print 繁花图案印花(以散乱线条为主的中到大面积印花)
butadiene 丁二烯(通常指1,3-丁二烯)
butadiene diepoxide 丁二烯二环氧化合物
butadiene rubber 丁二烯橡胶(一种合成橡胶)

butadiene-rubber roll 丁二烯橡胶辊
butane tetracarboxylic acid 〈BTCA〉1,2,3,4-丁烷四羧酸(无醛防皱整理剂)
butanol 丁醇
butcher 结子粗绸(平纹或斜纹不规则组合);平粗织物
butene 丁烯
butter cord 灯芯布
butterfly blue 蝴蝶蓝[色]
butterfly nut 蝶形螺母
butterfly screw 元宝螺钉,蝶形螺钉
butterfly seam 蝶形缝迹
butterfly throttle 蝶形节气门(可根据不同要求调节混合气的供应量)
butterfly valve 蝶形阀(一种节流阀)
butterfly yellow 蝴蝶黄[色](浅绿黄色)
butterscotch 杏黄[色]
butter yellow 奶油黄[色]
butt joint 对接
button switch 按钮开关
button weld 点焊
butt seam 对接[接合];[织物]对接缝合,平缝
butt welding 对头焊接
n-butyl acetate 丙烯酸正丁酯
butyl acrylate〈BA〉丙烯酸丁酯
butyl alcohol 丁醇
butyl benzoate 苯[甲]酸丁酯(载体及纤维素醚溶剂,增塑剂)
butyl benzyl phthalate 邻苯二甲酸丁苄酯(可用作地毯底层的增塑剂)
butyl coated fabric 丁基涂层织物
butylene glycol 丁二醇
butyl glycol 丁基乙二醇
butyl methacrylate〈BMA〉甲基丙烯酸丁酯
butyl naphthalene sulfonate 萘磺酸丁酯
butyl oleate 油酸丁酯(可用作增塑剂、溶剂、润滑剂、抛光剂)
butyric acid 丁酸
buyer 客户,买主
buyer's market 买方市场
buyer's material 客供原料,来料
buzzer 离心脱水机;砂轮;蜂音器
B. W. (before washing) 水洗前
BWC (British Wool Confederation) 英国羊毛联合会
BWMB (British Wool Marketing Board) 英国羊毛销售管理部,英国羊毛贸易部
by-mordant 辅助媒染剂
by-pass 旁路,分路;旁通管,间道管;支路,侧道;回油活门
bypass butterfly valve 旁路蝶形阀
bypass flow control 旁通流控制
by-pass valve 旁通阀
by-path 旁路,旁通
by-product 副产品
by-product level 副产品等级
byte 字节(二位制),位组《计算机》

C

CA(cellulose acetate) 纤维素醋酸酯,醋酸纤维素酯
CA(Chemical Abstracts) 《化学文摘》(美国刊物)
CA(citric acid) 柠檬酸
cabbage red 甘兰红(天然染料)
cabinet 箱,柜,室;小房
cabinet dry 风干,晾干;热空气干燥法
cabinet dryer 箱式烘干机
cabinet steamer 箱式汽蒸机
cable 粗索,缆;电缆;软轴
cable cords 宽条灯芯绒
cable-spun technology 缆型纺纱技术
cable thread 三股缝纫线
CAC(computer aided colouring) 计算机辅助测配色
cachmerette 仿开司米斜纹绒(丝或棉经,羊毛纬,背面拉绒);女式斜纹呢(绢丝经,粗纺毛纱纬)
cachmire shawl 羊绒披巾
cachou extract 儿茶精(媒染剂,含单宁约40%)
CAD(computer aided design) 计算机辅助设计
CADC(colour analysis display computer) 颜色分析显示计算机
CADD(computer aided design and drafting) 计算机辅助设计与制图
cade oil 杜松油
cadet 紫灰[色]
cadet cloth 军校制服呢(粗纺,紧密,重缩绒)
cadet gray 浅蟹灰[色],蓝灰[色]
cadmium 镉
cadmium compound 镉化合物
cadmium ethylene diamine hydroxide 乙二胺氢氧化镉
cadmium orange 血牙色,海螺红[色]
cadmium red 镉红(无机颜料)
cadmium sulphide 硫化镉
cadmium yellow 镉黄(硫化镉)
cadoxam 镉铵溶液
cadoxen retention value 氢氧化镉乙二胺溶液中停留值(测定树脂整理交联度用)
cadoxen solution 氢氧化镉乙二胺溶液(用于测定纤维上萤光增白剂;用于测定棉与化纤混纺织物中棉纤维含量)
CAD technology 计算机辅助设计技术
CAE(computer aided engineering) 计算机辅助工程管理
caesious 青灰色的
CAGD(computer aided garment design) 计算机辅助服装设计
cage 脱水笼;盒,罩;[滚动轴承的]保持架
cage boiling off machine 笼式脱胶机(用于绢纺原料的脱胶、除杂)
cage dryer 笼式烘燥机
cageless bearing 无保持架轴承,开口轴承

cage-like fluorine-carbon polymer 笼状结构氟碳聚合物
cake 滤饼,滤渣,丝饼
cake bleaching 丝饼漂白
cake dyeing 丝饼染色
cake dyeing machine 丝饼染色机
cake net 丝饼网
cake washing machine 丝饼水洗机
caking 黏结,凝结
cal(calorie) 卡[路里]（热量单位）
calami blue 菖蒲蓝[色]
calamine violet 水锌紫[色]
calandria 排管式;[蒸发设备的]加热件,加热体
calandria heat exchanger 排管式热交换器
calcareous 石灰的,含钙的
calcination 煅烧
calcined 煅烧过的
calcined pigment 煅烧颜料
calcite 方解石
calcium acetate test 醋酸钙试验（测定纤维素羧基用）
calcium alginate 海藻酸钙
calcium alginate fibre 海藻酸钙纤维
calcium chloride 氯化钙
calcium compound 钙化合物
calcium hardness 钙[质]硬度
calcium hydroxide 氢氧化钙
calcium hypochlorite 次氯酸钙;漂白粉
calcium oleate 油酸钙,钙皂
calcium oxide 氧化钙
calcium oxychloride 次氯酸钙;漂白粉
calcium reactions 钙反应
calcium soap 钙皂
calcium stain 钙斑

calcium stearate 硬脂酸钙
Calcobond dyes 含羟甲基胺基团的活性染料（用酸性催化剂固着,商名,美国制）
calculator 计算机,计算装置,计算器
Caledon SF dyes 克力登 SF 染料（经特别处理的蒽醌类还原染料,用于连续染色及树脂整理一步法,商名,德国巴斯夫）
Caledon yellow paper 克力登黄试纸（测试还原浴中保险粉用量）
calender 轧光;轧光机
calender bonded 热轧黏合（非织造布）
calender bowl 轧光滚筒,轧水滚筒
calender coating 轧光涂层法
calendered cloth 轧光布
calendered film 压延薄膜
calendered heat setting process〈CHS〉轧光热定形工艺
calendered moire 波纹轧光
calendered nonwoven fabric 热轧法非织造布
calender felt 轧光毡
calender finish 轧光整理
calendering 轧光,压光
calendering crease 轧光绉
calendering machine 轧光机
calendering spots 轧光损伤
calender man 轧光工
calender mark 轧光印
calender press 辊压机
calender printing 滚筒印花
calender produced point coating 轧平点涂层（粉点涂后轧扁,用于起绒织物）
calender roller 轧光滚筒
calender spreading 轧辊[涂层]黏合法

calender-type transfer printing machine 热辊式转移印花机
Calgon 卡尔康(以六聚偏磷酸钠为主要成分的软水剂,商名,德国拜耳)
caliatour wood 卡利埃图尔红木染料(不溶性染料,东南亚产)
calibrated 刻度的,分度的;校正的
calibrated microscope 定标显微镜
calibrate monitor 标定监测器
calibration 校准,测定;标定,分度
calibration data for computer colour matching 电脑配色基础数据
calibration graph 校准图
calibration paper 校准纸(在耐光牢度试验时,用于光源调整或测定标准褪色时间的一种特制试验纸)
calibration standard 校准用基准,检定用基准,检校用标准
calico 印花棉布;漂白,平[纹]布(比细棉布重,英国名称)
Calico Printer's Association〈CPA〉 棉布印花工作者协会(英国)
calico printing 棉布印花
calico strainer 细布过滤装置
caliper [织物]厚度
calipers【美】= callipers
call 调入,引入
callatome 显微镜切片器
calligraphic print 手绘图案印花,流畅线条花纹图案印花
callipers 卡钳,卡尺;圆规
caloee fibre 南美苎麻
calomel electrode 甘汞电极
calorie〈cal〉 卡[路里](热量单位)
calorific intensity [发]热强度
calorific potential 热势,潜热

calorific power 产热率,发热量
calorific radiation 热辐射
calorific value 热值
calorifier 发热器,加热器;煮水器
calorimeter 量热器,测烫器
calorimetric analysis 热量分析
calorimetric pyrometer 量热高温计
calorimetry 量热[法],量热学
calpreta fabric 防缩有光绉织物
cam 凸轮,偏心轮;桃盘
CAM(computer aided manufacture) 计算机辅助生产
camaieu effect [印花]同色深浅效应
camber 弯度,曲度;曲面
cambered bowl 中高轧辊,凸面轧辊
cambered cylinder 弯辊,弯辊筒
cam bowl 凸轮滚子
cambric 细薄布,细纺(平纹薄型棉或亚麻织物)
cambric finish 细薄布整理(经烧毛、轧光及上轻浆)
cam drive 凸轮传动
camel 驼色(暗棕色)
cameleon 织物闪色效应(在同一梭口中织入两根或三根色泽不同的纬纱,或用各色斑点纱织成而产生的视觉效应)
cameleon fibre 变色纤维,闪色纤维
camel hair 骆驼毛,驼绒毛
camel suede 仿驼毛呢棉布,仿麂皮绒
camelteen 仿驼毛薄呢(精纺毛纱制)
cameo brown 豆红[色],豆沙色
cameo printing 凸纹滚筒印花;凸版印刷
cameo printing machine 凸纹滚筒印花机

camera [linen] 粗松亚麻织物(漂白或不漂白,漂白者通常染浅色,一般为黄色,法国制)

cam follower 凸轮随动件(随同凸轮动作并传至推杆)

CAMOS (computer aided machineshop operating system) 计算机辅助车间操作系统

camouflage 伪装,掩饰,保护色

camouflage colouring 伪装染色

camouflage fabric 伪装[用]织物

camouflage look 迷彩服式样,伪装型

camouflage net 伪装网,防空网

camouflage pants 迷彩裤

camouflage pattern 伪装迷彩

camouflage pattern unit 迷彩服

camouflage print 伪装印花布,迷彩印花布

camouflage printing 迷彩印花,伪装印花

camouflage shorts 伪装短裤

camouflage sport suit 迷彩运动服

camouflage utility cap 伪装便帽

cam pack 凸轮部件,凸轮组件

campact dyeing system 快速吸尽染色法,快速紧凑染色法(英国ICI公司开发的分散染料吸尽染色法)

camphor 樟脑,α-莰酮

camping 野营服

camping tentage(= camping tent cloth) 篷帐布

cam shaft 凸轮轴

camwood 坎伍德红木染料(一种不溶性红木染料)

can 容器;圆筒;烘筒,烘缸

CAN(controller area network) 控制器局域网《计算机》

Canadian Association of Textile Colourists and Chemists〈CATCC〉 加拿大纺织染色家与化学家协会

Canadian Textiles Institute〈CTI〉 加拿大纺织学会

Canad. Text. J.(Canadian Textile Journal)《加拿大纺织杂志》(月刊)

canal 管道

canal blue 浅湖蓝[色],浅竹青[色]

canal dryer 管道式烘干机

canary yellow 金丝雀黄[色](嫩黄色)

CAN-bus 控制器局域网总线技术《计算机》

cancel 撤销

cancerogenic dyestuff 致癌性染料

cancerogenic substance (= cancerogenous substance) 致癌物质

cancerogeny of dyestuff 染料的致癌性

can decatizing machine 圆筒蒸呢机

candle [旧]烛光单位(=1.0067坎)

candle power 烛光

candry 圆筒烘燥,烘筒烘燥

can dryer 烘筒烘燥机

can dyeing 滚筒染色

cane sugar 蔗糖

canister 滤毒罐(防毒面具用)

cannele cord 花式灯芯绒

cannetille 金银线(刺绣用);金银线花边;经棱条织物(用于服装及装饰)

canopy 顶罩,顶盖,天棚

cantilever 悬臂;交叉支架

cantilever beam 悬梁

cantilever method 悬臂式测定法,刚软度测定法《试》(织物刚性试验)

Canton crepe 重双绉;广绫(做拷绸用,

中国制)
canton finish 棉布平光整理
Canton satin 广绫
canvas 粗帆布;篷帐;十字布;刺绣用网形粗布;油画布
canvas cloth 稀薄多股平纹毛织物;紧捻平纹棉织物
canvas disc wetting test 帆布片润湿试验
canvas duck 粗帆布(亚麻短纤维纱织成)
Canvas Products Association International〈CPAI〉 国际帆布制品协会
caoutchouc 橡胶,生橡胶
cap 盖,罩,帽
CAP(computer aided pattern) 计算机辅助图案设计
capacitance 电容,容量
capacitor 电容器
capacitor-start motor 电容启动电动机
capacity 容量;电容;能力;能量
capacity production 生产能力
capacity rating 额定功率;设备能力
CAPC(computer aided pattern creation) 计算机辅助图案创作
CAPD(computer aided pattern design) 计算机辅助图案设计
capillarimeter 毛细检液器,毛细管测液器
capillarity 毛细作用,毛细[现象]
capillarity of fibre 纤维的毛细管性
capillary 毛细作用的,毛细现象的
capillary action 毛细管作用
capillary active 毛细有效的
capillary activity 毛细管活性
capillary analysis 毛细管分析
capillary bore 毛细管孔

capillary characteristic 毛细管特性
capillary condensation 毛细管凝缩
capillary cross section effect 毛细管截面效应
capillary dehydration 毛细脱水
capillary effect 毛细管效应
capillary electrophoresis〈CE〉 毛细管电泳法
capillary elevation 毛细上升
capillary flow 毛细流动
capillary flow measuring method 毛细管黏度测定法
capillary force 毛细作用力
capillary gel electrophoresis〈CGE〉 毛细管凝胶电泳法
capillary melting point 毛细管法熔点
capillary meniscus 毛细管弯月面
capillary pressure 毛细管压力
capillary rheometer 毛细管流变仪
capillary rise method 毛细管上升法(测表面张力)
capillary seepage 毛细渗透
capillary transport 毛细管输送
capillary tube 毛细管
capillary-type penetration 毛细管式渗透[作用]
capillary visco[si]meter 毛细管黏度计
capillary water 毛细水,渗透水
capillator 毛细管比色计
capillometer 毛细试验仪
capital repair 大修
caplastometer 黏度计
CAPP(computer aided production planning) 计算机辅助生产计划
capped nut [盖]螺母
capped pipe 封头管子

capping group 封端基团
Capri 卡普里蓝[色](绿蓝色)
capric acid 癸酸
caproic acid 己酸
caprolactam 己内酰胺
caprolactam polymerization 己内酰胺聚合[作用]
Capron 卡普纶(锦纶 6,商名)
Capron hosiery 卡普纶丝袜(也称锦纶丝袜)
capryl alcohol 辛醇,亚羊脂醇(消泡剂)
caprylic acid 辛酸
capsanthin colour 辣椒红色素(食用天然植物色素)
capsicum red 辣椒红[色]
capstan 绞盘,绞车;曳引机;主动轮
capstan nut 带孔螺母,转塔螺母
capstan roller 拖引滚筒
capsule 胶囊;小气泡
capture [尘粒]捕获
capucine orange 金丝雀黄[色](嫩黄色)
caracule 羔皮;仿羔皮;充羔皮织物
caramel 淡褐[色],酱[色],黄棕[色]
caramel brown 焦糖棕[色]
carbamate 氨基甲酸酯;氨基甲酸盐
carbamate reactant 氨基甲酸酯反应物
carbamic acid 氨基甲酸
carbamide 脲,尿素
carbamide peroxide 过氧化脲
carbamide resin 聚脲树脂,碳酰胺树脂
carbamoyl 氨基甲酰基
carbamoylation 氨基甲酰化作用
carbamoylethylated cotton (＝carbamyl-ethylated cotton) 氨基甲酰乙基化棉(棉纤维经丙烯酰胺和淡烧碱液处理而成)

carbamyl 氨基甲酰基
carbanilate group 苯氨基甲酸酯基
carbanilation 苯氨基甲酸化
carbazole 咔唑;9-氮杂芴
carbazole dyes 咔唑(结构)染料
carbazole reaction 咔唑反应(用咔唑试剂测定尿素系树脂和糖类的呈色反应)
carbazole-sulfuric acid reaction 咔唑—硫酸反应
carbenium ions 碳正离子
carbide brushes 碳化物刷(用于刷绒)
carbide fibre [金属]碳化物纤维
carbinol 甲醇(一般用于复合词中)
carbitol 卡必醇(俗名),二甘醇一乙醚(溶剂,乳化剂,柔软剂)
carboard stencil 纸板刻花模板
carbocyanine dyes 羰花青染料
carbocyclic compound 碳环化合物
carbocyclic ring 碳环
carbohydrate 碳水化合物,糖类
carbohydrazole 碳酸肼(甲醛清洗剂)
Carbolan dyes 卡普仑染料(酸性耐缩绒染料,商名,原英卜内门)
carbolic acid 石炭酸,苯酚(俗称)
carbomoyl sulfonate group 氨基甲酰磺酸酯基团
carbonaceous 含碳的,碳质的
carbonaceous material 炭素材料
carbon arc 碳弧
carbon arc lamp 碳弧灯
carbonate 碳酸盐
carbon bisulfide (＝carbon bisulphide) 二硫化碳
carbon black 炭黑,墨灰,黑烟末
carbon black fibre 炭黑纤维

carbon bush 石墨轴衬,碳轴衬
carbon-carbon crosslink 碳—碳交联;碳—碳交联键
carbon chain compound 碳链化合物
carbon chain fibre 碳链纤维
carbon chain polymer 碳链聚合物
carbon dioxide 二氧化碳
carbon disulfide (= carbon disulphide) 二硫化碳
carbon electric arc 碳电弧
carbon electrode 碳电极
carbon fibre〈c-fibre, CF〉碳[素]纤维
carbon fibre composite 碳纤维复合材料
carbon fibre laminates 碳纤维层压材料
carbon fibre plastic composite 碳纤维塑料复合材料
carbon fibre reinforced plastic 碳纤维增强塑料
carbon fibre reinforced polymer〈CFRP〉碳纤维增强复合材料
carbon fibre reinforced polymer composite 碳纤维增强聚合物复合材料
carbon fibre reinforced thermoplastic 碳纤维增强热塑性塑料
carbon filament 碳素长丝
carbonic acid 碳酸
carbonium colours 碳鎓染料
carbonium ion 碳鎓离子,碳正离子
carbonization(= carbonisation) 炭化[作用];碳化[作用]
carbonized wool 炭化羊毛
carbonizer 炭化机
carbonizing(= carbonising) 炭化
carbonizing agents and assistants 炭化剂和辅助剂
carbonizing chamber 炭化室
carbonizing process 炭化过程;碳化过程
carbonizing treatment 炭化处理
carbon monoxide 一氧化碳
carbon nanotubes 碳纳米管
carbon oxychloride 碳酰氯(即光气)
carbon paper 碳素纸
carbon paper marker 复写纸划样;用复写纸作标志(如检查轧辊压力)
carbon residue 残留炭
carbon steel 碳钢
carbon steel fibre 碳钢纤维
carbon tetrachloride 四氯化碳,四氯甲烷
carbon tissue process 碳素纸印象法(照相凹版雕刻)
carbonylation 羰基化作用
carbonyl diamine 碳酰二胺,尿素(俗称)
carbonyl dyes 羰基染料
carbonyl group 羰基
carbonyl value 羰基值
Carbopol 846 卡博普尔846(涂料印花用合成增稠剂,粉状固体,商名,美国古立德)
carborundum 碳化硅,金钢砂
carborundum fillet 金刚砂带
carborundum raising 金刚砂起绒
carboxy 羧基
carboxy-haemoglobin value 羧基血红蛋白值
carboxyl 羧基
carboxylate surfactant 羧酸盐(或羧酸酯)型表面活性剂
carboxylation 羧化作用
carboxyl content 羧基含量
carboxyl end group 羧端基
carboxylic acid 羧酸

carboxylic acid type cation exchange fibre membrane 羧酸型阳离子交换纤维膜
carboxymethylated cotton 羧甲基化棉
carboxymethylation 羧甲基化
carboxymethyl cellulose 羧甲基纤维素
carboxymethyl cellulose thickening 羧甲基纤维素糊
carboxymethyl cotton 羧甲基化棉
carboxymethyl hydroxyethyl cellulose 〈CMHEC〉羧甲基羟乙基纤维素
carboxymethyl locust bean thickening 羧甲基皂荚胶糊
carboxymethyl starch〈C. M. S.〉羧甲基淀粉
carboxymethyl starch thickening 羧甲基淀粉糊
carburation 汽化;渗碳
carburator 汽化器;[煤气及空气]混合器;渗碳器
carbureting pilot 汽化式引燃器,空气煤气混合式引燃器
carburettor(=carburetter) 汽化器(使空气和汽油形成可燃混合气)
carcinogen 致癌物,诱癌剂
carcinogenic activity 癌变活性
carcinogenic dyestuff 致癌染料
carcinogenicity 致癌性
cardan 平衡环;万向节;万向接头;活节连接器
cardan joint 万向接头,万向节(由一个十字轴和两个轴叉组成)
cardan shaft 万向节轴(装有万向节的轴,如传动轴等)
cardboard core (卷布匹用的)硬纸板;卷芯板
cardboard sleeve 纸粕筒管(供卷布用)
cardboard tube 刚纸筒管;纸粕筒管
card clothing 钢丝针布(用于起毛机等设备)
carded felt 毛毡
card fillet 钢丝针布
cardinal cloth 主教呢(红色)
cardinal red 深红[色],鲜红[色]
cardinal stimuli 基准刺激
card mounting machine 针布包绕机
card programming 卡片程序控制
card raising machine 钢丝起绒机,钢丝起毛机,钢丝拉绒机
card reader 读卡器,卡片输入机
card test 卡片试验(测润湿效率)
card winding machine 针布包绕机
card wire 针布梳针
card wire point 针布梳针头
care claim 使用要求
care claim category 表明使用类别
care instruction 使用说明
care label 使用须知标签
care of fabrics 织物的维护(包括洗涤、熨烫、晒晾、刷、防蛀、去污、储藏等)
carlonized 卡罗内兹工艺(纤维通过浸渍形成保护层,具有防水、可洗、防污、防磨、防静电等作用)
carmine 胭脂红[色],洋红[色](鲜红色);红色天然颜料
carmine rose 桃红[色],喜蛋红[色],品红[色]
carminic acid 胭脂红酸
carmoisine 红色酸性染料;红色食用染料
carnation 淡红[色],肉色
carnauba wax 巴西棕榈蜡,卡瑙巴蜡
Caroat 卡罗盐(过一硫酸氢钾,可用作

羊毛氧化防缩剂）
carob-bean 角豆树豆,稻子豆
carob seed gum（＝carob gum）角豆树胶（印花糊料）
Caro's acid 卡罗氏酸,过一硫酸（可用作羊毛氧化防缩剂）
caroset 法兰绒（法国制法兰绒的通称）
carousel hosiery boarder 旋转式袜子定形机
carpet 地毯,毡毯
carpet ageing 地毯老化
Carpet and Rug Institute, USA〔CRI〕 美国地毯协会
carpet back coating 地毯背面涂层
carpet backing 地毯背衬
carpet back singeing 地毯底背烧毛
carpet base material 地毯底布
carpet beetle 地毯甲虫
carpet binding machine 地毯黏合机
carpet cleaning 地毯清洁
carpet comfort 地毯舒适度
carpet contaminant 地毯污染测试
carpet continuous dyeing plant 地毯连续染色设备
carpet cushion 地毯衬垫
carpet design 地毯花纹,地毯设计
carpet drying machine 地毯烘燥机
carpet durability test 地毯耐用性试验
carpet dust & water extractor 地毯除尘吸湿机
carpet dye bock 地毯染色槽
carpet dyeing machine 地毯染色机
carpet finishing 地毯整理
carpet frostiness 地毯露地（疵点）
carpeting 地面用品
carpet jet machine 地毯喷射染色机

carpet knitting machine 地毯针织机
carpet latexing machine 地毯上乳胶机
carpet lining 地毯衬垫
carpet moth 地毯蛀虫
carpet over flow machine 地毯溢流染色机
carpet pile 地毯绒毛,地毯绒头
carpet pile penetration 地毯绒头渗透
carpet prewashing & bulking unit 地毯预洗及膨化机
carpet printing 地毯印花
carpet printing machine 地毯印花机
carpet print transfer calender 地毯转移印花热轧辊
carpet rolling and measuring machine 地毯卷量机
carpet shampooing procedure 地毯清洁法
carpet shearing 地毯剪毛
carpet shearing machine 地毯剪毛机
carpet singeing 地毯烧毛
carpet slitting and trimming machine 地毯剪切修整机
carpet soiling 地毯沾污
carpet spray printing 地毯喷射印花
carpet square 组合地毯,小方地毯
carpet squeeze roller 地毯轧水辊
carpet steamer 地毯蒸化机
carpet surface 地毯表面
carpet sweeper 地毯清扫器
carpet tile 组合地毯,小方地毯
carpet transfer printing 地毯转移印花
carpet transport roller 毯子输送辊筒
carpet tufted 簇绒地毯
carpet underlay 地毯衬垫
carpet walk test 地毯的行走测试

carpet washer 地毯水洗机
carpet wear tester 地毯耐磨试验机
carpet weaving 地毯织造
carpet wool 地毯用羊毛
carrageen(= carragheen) 角叉菜,鹿角菜(红藻类,用以印花浆料)
carrageenate(= carragheenate) 角叉菜胶,鹿角菜胶(耐硬水、耐高温、耐酸和碱的印花浆料)
carragheenin 角叉菜胶
carriage 小电车;[绢网印花]行车;三角座支架;车;台架,托架
carriage wrench 套筒扳手
carrier 染色载体;载气(气相色谱);载液(液相色谱);托架;悬挂架,支持物
carrier bar [布环]挂棒(用于长环蒸化机、短环烘燥机等设备,既在箱内循环运行,又能自转)
carrier driving shaft 过桥传动轴
carrier dyeing 载体染色,载体染色法
carrier fabric 载体织物;介质织物
carrier gear 过桥齿轮
carrier layer 载体层
carrier pulley 过桥皮带盘
carrier removal 载体去除
carrier rollers 过桥辊(防止滑动)
carrier spot 载体染斑(染疵)
carrier wheel 过桥齿轮
carrot orange 胡萝卜橘黄[色]
carrot red 胡萝卜红[色](橙色)
carrotting [毛皮]毡合预处理;兔毛预缩处理
carrotting agent 预缩处理剂
Carr's melton 卡氏麦尔登呢(高级麦尔登呢,英国制)
carrying capacity 运送[机器]的容量

carrying rope 运送绳索(机器上运送布用的绳索);传送带
carry-off heat 散热,排热
carthamin 红花素(红色植物染料)
cartridge filter 过滤筒,筒式过滤器,管形过滤器
cartridge heating element 筒形加热器,筒式加热元件
cartridge magnetic tape 盒式磁带
car upholstery 汽车内装饰织物
carve 雕刻
carved pile fabric 浮雕绒头织物,剪花绒头织物
CAS(Chemical Abstracts Service) 化学文摘社(美国)
CAS(Chinese Academy of Sciences) 中国科学院
CAS(colour accreditation scheme) 色彩认证计划
cascade 级,级联;小瀑布;串联
cascade alternator washer 阶式振荡水洗机,级联涡流水洗机(一种高效水洗机,主要由两块凹凸波纹板组成,在板壁间喷射强力的液流,被洗织物通过时,借助于涡流提高水洗效果)
cascade control 串级调节,级联控制
cascade control system 双闭环调速系统,级联控制系统
cascade device 冲液装置
cascade dispensing unit 冲流式[印浆]调配单元
cascade dripcooler 水淋式冷却器
cascade dyeing 冲流染色
cascade dyeing machine 冲流染色机
cascade flow 阶式流动,瀑布式流动
cascade heat exchanger 阶式热交换器

cascadelock steamer 冲流式(瀑布式)封口的蒸箱,瀑布式封口的蒸箱

cascades and suction boxes 冲吸碱液装置(用于丝光机)

cascade sealing method 喷流水封法

cascade stripe 阶式条纹

cascade washer 冲流式[服用品]洗涤机,冲碱装置(用于丝光机)

case 箱,柜;盒,壳,罩

cased-muff coupling 刚性联轴器,套管联轴节

case hardening 表面淬火

casein 酪蛋白,酪素

casein fibre 酪蛋白纤维

casein glue 酪蛋白胶

casein plastic 酪蛋白塑料

casein protein fibre 酪蛋白质纤维

casein rayon 酪蛋白人造丝

casein wool 酪蛋白人造毛

case number 实例号

cashew resin 漆树树脂

Cashgora 开司哥拉羊毛

cashmere 开司米,开丝米山羊绒,羊绒;开司米织物,[山]羊绒织物;精纺毛针织物;松软striation棉织物

cashmere fancy suiting 羊绒花呢

cashmere finishing 仿开司米整理

cashmere hair 开司米,山羊毛

cashmere hosiery yarn 开司米针织纱线(精纺毛纱制)

cashmere knit goods 羊绒针织物;开司米针织物

cashmerette 仿开司米斜纹绒(丝或棉经,羊毛纬,背面拉绒);女式斜纹呢(绢丝经,粗纺毛纱纬)

Cashmilan 开司米纶(聚丙烯腈短纤维,商名,日本制)

cash on delivery〈COD,C.O.D.〉货到付款

cashoo 儿茶,阿仙药(红色植物染料)

casing 罩壳,外壳,套

casing seal 套管密封

CASS(cyclic activated slude system) 循环式活性污泥系统(污水处理)《环保》

cassava starch 木薯淀粉

cassel yellow 氯化铅黄(黄色颜料)

cassette 盒,箱,卡

cassette holder [圆网]印花头托座

cast coating 热辊层压法(利用热辊压层焙烘使树脂与织物胶合)

castellated nut 槽顶螺母;带眼螺母

castellated shaft 花键轴

caster 小脚轮(装在小车底端,可以朝任何方向推动,印染厂堆布车常用此轮)

cast foam 浇铸法泡沫材料

castile soap 上等丝光皂,橄榄油皂

casting 铸件;铸造

casting article 铸件

cast iron 铸铁,生铁

cast iron flanges [烘筒]铸铁闷头

cast iron slit gas burner 狭缝式铸铁火口

castor 海狸呢,厚呢;仿麂皮处理的羊皮;海狸毛棕[色](棕灰色);海狸皮;小脚轮

castor oil 蓖麻油

cast steel 铸钢

casual inspection 不定期检查,临时检查

CAT(computer aided testing) 计算机辅助测试

catabolism 分解代谢
catalase 过氧化氢酶(用于水解过氧化氢)
catalogue 目录,产品样本
catalysis 催化[作用],触媒[作用]
catalyst 催化剂
catalyst abrasion 催化剂磨蚀
catalyst carrier 催化剂载体
catalyst in deferred cure 后烘焙催化剂(促进交联)
Catalyst KR 催化剂KR(含活性氯化镁64%,商名,美国制)
catalyst regeneration 催化剂再生[作用]
catalyst selectivity 催化剂选择性
catalysts for resin finishing 树脂整理催化剂
Catalyst X-4 催化剂X-4(含硝酸锌活性成分25%,商名,美国制)
catalytic action 催化作用
catalytic activity 催化活性;催化活度
catalytic afterburner 催化后燃室
catalytic cashmere scouring 催化洗绒法
catalytic chamber 催化室
catalytic combustion 催化燃烧
catalytic combustion process of industrial organic gases 有机废气催化燃烧法
catalytic composite 复合催化剂
catalytic cooxidant 辅助氧化催化剂
catalytic cracking 催化裂解[法]
catalytic current 催化电流
catalytic damage [纤维的]催化损伤
catalytic decomposition 催化分解
catalytic degradation 催化解聚,催化降解
catalytic dehydrogenation 催化脱氢
catalytic exhaust purifier 废气催化净化器
catalytic fading 催化褪色(拼色时一种染料受另一种染料的影响而使耐光牢度下降)
catalytic gas infrared heating system 接触式气体红外线加热系统
catalytic incineration 催化焚烧
catalytic oxidation 催化氧化
catalytic poison 催化剂中毒
catalytic reaction 催化反应
catalytic tendering 催化脆损
catalytic vapour phase polymerization 催化气相聚合
catalyzed polymerization 催化聚合[作用]
cataphoresis 电泳
CATCC (Canadian Association of Textile Colourists and Chemists) 加拿大纺织染色家与化学家协会
catch 掣子,挡器;[门的]门扣
catch lever 挡杆,掣子杆
catch mechanism 捕捉[钩住]机构
catch pawl 挡爪
catch plate 挡板;刹车板
catch spring 挡簧
catch tray 收集盘
catchword 关键字(目录索引用)
CATD (computer aided textile design) 计算机辅助纺织品设计
catechin 儿茶素(茶叶中含有的植物染料色素)
catechu 儿茶(印度棕色植物染料);儿茶棕[色](棕色)
catechu dyeing 儿茶染色法(儿茶丹宁作为植物染料)
catgut 羊肠线

cathode 阴极，负极
cathode excitation 阴极激发
cathode ray tube〈CRT〉 阴极射线管《计算机》
catholyte 阴极电解液
cation 阳离子，正离子
cation active high-molecular electrolyte 阳离子活性高分子电解质
cation exchange capacity 阳离子交换能力
cation exchanger 阳离子交换器
cation exchange resin 阳离子交换树脂
cationic 阳荷电的，阳离子性的
cationic aftertreatment agent 阳离子后处理剂
cationic aftertreatment resin 阳离子后处理树脂（染色固色剂）
cationic brightener 阳离子增白剂
cationic catalyst 阳离子催化剂
cationic conditioning agent 阳离子调节剂
cationic detergent 阳离子洗涤剂
cationic dyeable polyamide 阳离子可染聚酰胺纤维
cationic dyeable polyester〈CDP〉 阳离子可染涤纶，阳离子可染聚酯纤维
cationic dyeable variants 阳离子染料可染型变性聚合物
cationic dye in dispersed form 阳离子分散型染料
cationic dyes 阳离子染料
cationic dyes for all purpose 普通型阳离子染料
cationic emulsifier 阳离子型乳化剂
cationic fibre-reactive dyes 阳离子型活性染料
cationic finishes 阳离子整理剂
cationic isothiuronium dye salt 阳离子异硫脲染料盐（在弱碱中水解成硫羟染料，可染羊毛）
cationic modification 阳离子改性
cationic modification technology 阳离子改性技术
cationic polymerization 阳离子［催化］聚合［作用］
cationic pretreatment 阳离子化预处理
cationic protein derivatives 阳离子蛋白质衍生物
cationic reactive dyes 阳离子型活性染料
cationic resin 阳离子树脂
cationic retarding agent 阳离子阻染剂
cationics 阳离子表面活性剂
cationic sites 阳离子［结合］位置
cationic softener 阳离子型柔软剂
cationic starch 阳离子淀粉
cationic surface active agent〈CSAA〉 阳离子表面活性剂
cationic surfactant 阳离子表面活性剂
cationic urea resin 阳离子性尿素树脂
cationizing agent 阳离子化助剂
cation transfer system 阳离子转移染色法（用于毛腈混纺同浴染色）
cation transport 阳离子转移
CATP（computer aided textile printing） 计算机辅助纺织品印花
Cat's eye 夜间闪光织物（商名）
cat whiskers 猫须状（牛仔裤的一种整理效果）
CAUS（Colour Association of the United States） 美国色彩协会
cause and effect diagram 因果图

causterising 烂花工艺,烂花印花
causterized pattern 烂花印花花样
causterized starch thickening 碱化淀粉糊
caustic 苛性碱;腐蚀剂
causticaire index 氢氧化钠法;棉成熟度指数
causticaire method 氢氧化钠法(测棉纤维成熟度)
causticaire scale 氢氧化钠法棉成熟度指数
causticaire value 氢氧化钠法棉成熟度值
caustic discharge 碱性拔染
caustic embrittlement 碱性脆损
causticity 苛性,腐蚀性
causticization 苛化作用
causticizing 苛[性]化(氢氧化钠处理)
causticizing number 苛化指数
causticizing treatment 苛化处理
causticizing treatment of viscose fabric 黏胶织物的苛化处理
caustic J-box 碱煮练用 J 形箱
caustic leaching 碱浸提
caustic lime 生石灰,苛性石灰
caustic lye cooling arrangement 烧碱液冷却装置
caustic make-up pump 碱液补充泵
caustic over-printing 烧碱罩印(用于泡泡纱印花)
caustic potash 苛性钾,氢氧化钾
caustic printing process 烧碱印花法(产生泡泡纱效果)
caustic recovery plant 烧碱回收装置
caustic recuperator 去碱汽蒸箱(用于丝光机)
caustic reduction 碱减量
caustic reduction process 碱减量工艺(聚酯纤维碱剥皮工艺)
caustic resistance 耐碱性
caustic saturator 浸碱槽
caustic scouring 碱精练
caustic shock process 烧碱快速固色法
caustic soda 烧碱,苛性钠,氢氧化钠
caustic soda crepe 烧碱[液印]泡泡纱
caustic soda printing 烧碱印花(产生泡泡纱效果)
caustic soda recovery unit 烧碱液回收装置
caustic soda softening 苛性钠软水法
caustic stage 烧碱煮练阶段
caustic suction unit 吸碱装置(用于丝光机)
caustic treatment 碱处理
caustic wash tower 碱洗塔
caustification 苛化作用
caution label 警告标签
CAV(critical application value) 临界施加值
Cavalite dyes 卡伐莱特染料(活性染料,含乙烯砜活性基团,商名,美国制)
cavernous 多孔的
cavitation 空化[作用],空穴[作用](由于总压力的减小而在流体中形成局部空腔或气泡);气穴
cavity 空腔,凹处;空穴
cavity formation 空腔形成,空穴成形
cavity structure 中空结构(纸或纤维)
cavity vat 浆桶夹套煮浆桶,给浆桶
cavity wall 双层壁;水汀夹板;空心墙
CB(chemical/biological) 化学/生物的
CBA(chemical blowing agent) 化学发泡

剂

CBH(cellobiohydrolase) 纤维素水解酶

CBR(continuous bleaching range) 连续漂白联合机

CBWT(Confederation of British Wool Textiles) 英国毛纺织联合会

cc(＝c. c., cubic centimetre) 立方厘米

CCM(＝C. C. M., ccm, computer colour matching) [电子]计算机配色[法]

CCMS(colour control & management system) 色彩控制管理系统

CCM software 计算机配色软件

CCol(chartered colourist) [英国染色工作者学会的]注册印染工作者

CCPB technology(controlled cold pad batch technology) 可控冷轧堆染色技术

CCR(combined cooling and rinsing) 洗涤和冷却同步进行（高温高压染色法）

CCS(computer colour searching) 计算机色样检索

CCS(computer colour selector) 计算机选色装置

CCU(combined cellulase units) 混合纤维素；酶活力单位

CDC(colour design console) 颜色花样图案控制台

CDC(column development chromatography) 柱展开色谱法

CDP(cationic dyeable polyester) 阳离子可染涤纶，阳离子可染聚酯纤维

CD-ROM(compact disk-read only memory) 只读光盘存储器《计算机》

CDS(computerized dispenser system) 计算机化分配器系统，自动液体与粉末分配系统

CE(capillary electrophoresis) 毛细管电泳法

ceiling 天花板,顶板；最高限额,上限

ceiling coil 顶盘管（装于蒸箱顶部的盘管,使其加热,以防滴水）

ceiling outlet 天花板出风口

ceiling temperature 最高温度；顶板温度

celadon gray 灰绿[色]

celadon green 青瓷绿[色]（浅灰绿色）

celandine green 秋香色,茶绿[色]

celestial blue 天空蓝[色]

cell 微泡；细胞；电池；空气囊；小室

cellaring 地窖给湿,窖藏

cellar moistening 地窖给湿

cell cavity 细胞腔

cell dryer 热板烘燥机（织物从两层夹板之间通过,常用作中间烘燥设备）

Cellestren dyes 泽莱斯特伦染料（分散染料,相对分子质量较大,在有棉膨化剂的情况下,可使涤纶和棉染成一致色泽,商名,德国巴斯夫）

cell membrane 细胞膜

cell membrane complex〈CMC〉细胞膜复合材料

cell membrane composite 细胞膜复合材料

cellobiase 纤维二糖酶

cellobiohydrolase〈CBH〉纤维素水解酶

cellobiose 纤维二糖

celloidin 火棉,火棉胶,纯硝酸纤维素

cellolose reactant type 纤维素反应型

Cellolube Supra 赛璐路贝苏普拉柔软剂（聚硅氧烷聚乙烯复合物,适用于各种纤维,阳离子性,商名,美国先邦）

cellophane 赛璐玢,纤维素薄膜,黏胶薄

膜,玻璃纸(俗称)
cellophane coated fabrics 赛璐玢涂层织物
cellophane yarn 赛璐玢薄膜纱,玻璃纸纱
cellosolve 2-乙氧基乙醇,乙二醇乙醚,赛洛索尔弗溶剂(印染用溶纤剂和工业溶剂,商名,美国制)
cellucotton 棉絮,棉团
cellular 细胞的,蜂窝状的
cellular fibre 多孔纤维,泡沫纤维
cellular photocatalyst fibre 多孔性光触媒纤维
cellular plastics 泡沫塑料
cellular structure 多孔性结构
cellulase 纤维素酶
cellulose 纤维素
α-cellulose α-纤维素,甲种纤维素
β-cellulose β-纤维素,乙种纤维素
γ-cellulose γ-纤维素,丙种纤维素
cellulose Ⅰ 纤维素Ⅰ(天然纤维素,如棉、麻等)
cellulose Ⅱ 纤维素Ⅱ(水合纤维素,丝光纤维素)
celluloseⅢ 纤维素Ⅲ(纤维素氨化物的分解物)
celluloseⅣ 纤维素Ⅳ(在200℃以上分解的碱纤维素或纤维素黄酸钠)
cellulose acetate〈CA〉纤维素醋酸酯,醋酸纤维素酯
cellulose acetate dyes 醋酯纤维染料
cellulose acetate fibre 醋酯纤维,醋酸纤维素纤维
cellulose acetate film 纤维素醋酯薄膜,醋酸纤维素薄膜
cellulose base fibre 纤维素纤维

cellulose blends 纤维素纤维混合制品
cellulose bonded non-wovens 纤维素黏合法非织造布
cellulose carbamate fibre 纤维素氨基甲酸酯纤维
cellulose combustion 纤维素燃烧
cellulose degradation 纤维素降解
cellulose degree of polymerization 纤维素聚合度
cellulose derivative 纤维素衍生物
cellulose diacetate 纤维素二醋酸酯,二醋酸纤维素
cellulose digester 纤维素蒸[煮]球
cellulose dyes 纤维素纤维用染料
cellulose ester 纤维素酯
cellulose ester bonding 纤维素酯黏合
cellulose estimation 纤维素测定
cellulose ether 纤维素醚
cellulose fibre 纤维素纤维
cellulose fibril 纤维素原纤维;纤维素微丝
cellulose film 赛璐玢,纤维素薄膜,黏胶薄膜,玻璃纸(俗称)
cellulose filter aid 纤维素助滤剂
cellulose glycolate 羧甲基纤维素,纤维素乙醇酸醚
cellulose graft 接枝纤维素
cellulose grafting 纤维素接枝
cellulose hydrate 水合纤维素
cellulose image print 纤维素阴影印花(用纤维素作浆的白色涂料印花)
cellulose intergrowth 纤维素共生物
cellulose micelle 纤维素微胞
cellulose modal〈CMD〉莫代尔纤维(国际代码)
cellulose modifier 纤维素纤维改性剂

cellulose nitrate 硝酸纤维素,纤维素硝酸酯
cellulose number 纤维素值,铜值
cellulose phosphate 磷酸纤维素
cellulose reactant 纤维素反应剂
cellulose surface grafting 纤维素表面接枝
cellulose swelling 纤维素膨化
cellulose triacetate〈CT，CTA〉 纤维素三醋酸酯,三醋酸纤维素
cellulose triacetate fibre 三醋酯纤维,三醋酸纤维
cellulose value 纤维素值,铜值
cellulose varnish ［硝化］纤维素清漆
cellulose viscose〈CV〉 黏胶纤维
cellulose xanthate（= cellulose xanthogenate） 纤维素黄原酸酯钠
cellulosic composite fibre 纤维素复合纤维
cellulosic man-made fibre 纤维素系化学纤维（以纤维素为原料的化学纤维）
cellulosics 纤维素制品
Cellusoft L 纤柔酶 L（纤维素酶,用于光洁柔软整理,商名,丹麦诺维信）
Cellusoft PLUS L 纤柔酶 PLUS L（纤维素酶,用于光洁柔软整理,商名,丹麦诺维信）
cell wall ［纤维］细胞壁
cell wall thickness ［纤维］胞壁厚度
celsius degree 摄氏温度
celsius temperature 摄氏温度
celsius thermometer 摄氏温度计
CEMATEX【法】欧洲纺织机械制造商委员会
cement 水泥；黏合剂；胶接剂；黏合,胶接

cementation 黏结,胶结；表面硬化；渗碳［法］；渗入处理；渗金属法；置换沉淀
cement grey 水泥灰色
cementing medium 胶结剂,黏结剂
cement matrix 水泥浆,水泥胶结料
cement steel 渗碳钢,表面硬化钢
CEN【法】欧洲标准化委员会
cendre 灰色［的］
center【美】= centre
centesimal balance 百分天平
centi-（构词成分）表示"厘、百分之一"
centigrade 摄氏温度
centigrade degree 摄氏温度；百分温度
centigrade scale 摄氏标度,百分温标
centigrade thermometer 摄氏温度计
centimeter 厘米
centipoise〈cP〉 厘泊(黏度单位)
centistokes 厘斯(动力黏度单位)
central axis plane 中心轴面
central cabinet 中心控制室
central centralized lubrication 中央集中润滑
central control 集中控制
central cylinder 印花大锡林（辊筒印花机的承压大辊筒）
central cylinder rotary screen printing machine 辐射式圆网印花机（几个圆网围绕一个承压大辊筒排列）
central inspection station 中心试验室；中心检验室
centralized control 集中控制
centralized lubrication device 集中润滑装置
centralized monitoring system 集中监控系统,集中监视系统
central line 中心线

central mixing and dissolving station 中央化料站
central processing unit〈CPU〉 中央处理器,中央处理单元
central spindle 中心轴
central supporting bowl 中固轧辊,中支轧辊
centre 中心,中央;集中,居中
centre cutting device 剖幅装置
centre drive batcher 中心驱动卷布机
centreing device 对中装置
centre line 中心线
centre spindle 中心轴
centre-to-centre distance 中心距
centre to selvedge variation 边中色差
centre winder 中央卷布机
centrifugae mercerizing machine 离心式丝光机(用于针织品,包含直辊轧碱、成卷隐形、离心去碱、洗涤等过程,商名,德国制)
centrifugal 离心的
centrifugal analysis 离心分析
centrifugal atomiser 离心式喷雾器
centrifugal basket 离心机转筒,离心机吊篮
centrifugal beam hydro extractor 经轴离心脱水机
centrifugal blower 离心式鼓风机
centrifugal cake bleaching 离心式丝饼漂白
centrifugal chromatography 离心色谱[法]
centrifugal clutch 离心式离合器
centrifugal damping machine 离心给湿机
centrifugal dehydrator 离心脱水机

centrifugal disk atomiser 离心式圆盘雾化器
centrifugal drier 离心式脱水机;离心干燥机
centrifugal dust collector 离心式除尘器
centrifugal dyeing machine 离心式染色机
centrifugal extractor 离心萃取器;离心脱水机
centrifugal fan 离心式风扇
centrifugal force 离心力
centrifugal friction clutch 离心式摩擦离合器
centrifugalization 离心分离[作用]
centrifugally cast stainless steel roll 离心浇注不锈钢辊
centrifugal precipitator 离心分离机
centrifugal pump 离心泵
centrifugal separator 离心分离机
centrifugal washing 离心洗涤
centrifuge 离心机,离心脱水机
centrifuge for yarn packages 筒子纱离心脱水机
centrifuge method 离心方法
centrifuge swelling 离心容胀率
centrifuging 离心脱水[工艺],离心[作用]
centr[e]ing 定中心
centr[e]ing adjustment 对准中心;中心调整;合轴调整
centr[e]ing apparatus 中心整位器(使织物居中进入拉幅机等加工设备)
centr[e]ing mark 布铗痕;钩破洞
CEO(chief executive officer) 业务最高负责人,首席执行官
CEP(chemical engineering products) 化

学工程产品
ceramic 陶瓷的
ceramic acid-resistant pipe 耐酸陶管
ceramic brushes 陶瓷刷(用于刷绒)
ceramic coating 陶瓷涂层
ceramic element 陶瓷元件
ceramic fibre 硅酸盐纤维,陶瓷纤维
ceramic finishing 陶瓷整理
ceramic infrared heater 陶质红外加热器
ceramic matrix composite〈CMC〉 陶瓷基复合材料
ceramic-metal composite 陶瓷与金属复合材料
ceramic plate 陶瓷板(用于红外线燃烧器)
ceramic powder finish 陶瓷粉整理
ceramics 陶瓷
ceramic tile 陶瓷贴片,陶瓷瓦
Ceranine HCL 丝柔软 HCL(阳离子脂肪酸缩合物,柔软剂,商名,瑞士科莱恩)
Ceranine HCS 丝柔软 HCS(阳离子柔软剂,酸稳定性好,商名,瑞士科莱恩)
Ceranine RW 丝柔软 RW(毛巾用柔软剂,具再润湿性,商名,瑞士科莱恩)
Ceranine SIL 丝柔软 SIL(含硅阳离子柔软剂,商名,瑞士科莱恩)
ceré 上蜡色布(法国名称)
cerecloth 蜡布,漆布
cerement [cloth] 蜡布,漆布
ceric sulfate 硫酸高铈
cerise 樱桃红[色](淡红色)
cerium compound 铈化合物
Cerofil 76 切罗菲尔 76(蜡和聚硅氧烷乳胶,用以改善棉及合成纤维混纺缝线的滑动性能,商名,意大利制)

Cerol EWL 西罗尔 EWL(改良型热固树脂,防水剂,商名,瑞士科莱恩)
Cerol WB 西罗尔 WB(脂肪酸衍生物的吡啶盐,阳离子防水剂、柔软剂、润滑剂,商名,瑞士科莱恩)
Cerol ZN 西罗尔 ZN(硅酸锆石蜡分散剂,防水剂,商名,瑞士科莱恩)
certificate 证明书,合格证
certificated quality level 质量保证,质量认证
certificate of approval 合格证
certificate of conformity 合格证
certificate of delivery 交货证书
certification 检定,验证;证书,鉴定书
certification of proof 检验证书
certified variety 鉴定品种,鉴定项目
certified washable or dry cleanable 可洗或可干洗的合格标记(国际织物防护协会的合格印记)
cerulean 天蓝[色](淡蓝色)
cerulean blue 蓝色;蓝色颜料
cesium compound 铯化合物
cesium hydroxide 氢氧化铯
cesspipe 污水管
cetine 鲸蜡
cetyl alcohol 十六烷基醇
cetyl betaine 鲸蜡基甜菜碱(具有润湿、匀染、分散、柔软、净洗作用)
cetyl dimethyl benzylammonium chloride 十六烷基二甲基苄基氯化铵(季铵盐杀菌剂、乳化剂、润湿剂)
cetyl pyridinium chloride 十六烷基吡啶氯化物(防水剂,柔软剂)
cetyl sodium sulphate 十六烷基硫酸钠
cetyl trimethyl ammonium bromide〈CTAB〉 十六烷基三甲基溴化铵(阳

离子柔软剂）
CF（carbon fibre）碳［素］纤维
CF（clean factory）清洁工厂
CF（comfort factor）舒适度
CF applicator（chemical foam applicator）化学泡沫装置
c-fibre（carbon fibre）碳［素］纤维
cfm（cubic feet per minute）立方英尺每分钟（排气量单位）
CFRP（carbon fibre reinforced polymer）碳纤维增强复合材料
CFS（chemical foam system）化学泡沫系统
CFS（continuous flushing system）连续冲洗系统
CFS（continuous feeding system）连续加料系统
CFS value（cycle felting severity value）循环毡缩指数
CFT（chemical foam technology）化学泡沫技术
CF value（comfort factor value）舒适指数
CG（computer graphics）计算机图形学
CGE（capillary gel electrophoresis）毛细管凝胶电泳法
CG/IP（computer graphic and image processing）计算机图形/图像处理技术
C.G.S. system 厘米·克·秒单位制
ch.（＝chem.，chemical）化学的
ch.（＝chem.，chemistry）化学
chadding 涂敷，覆面
chafe mark 擦伤痕（疵点）
chafing resistance 耐磨性
chagreen 仿皮革布
chain 链，链条；连锁，链接；链式

chain alignment 链排列
chain belt 链带
chain block 链滑车,起重葫芦
chain change-over device 布铗链换向装置
chain cleaner 链条清洁器（喷射蒸汽以清洁拉幅链条）
chain drive 链传动
chain-dyeing 链状染色（用于不耐张力织物的绳状染色,也用于绞纱串成链状染色）
chain-extending agent 链段增长剂
chain guide rail 链条导轨
chain hoist 链式起重机
chain length 链长（分子链）
chainless mercerizing range 弯辊丝光机,无链丝光机
chainless padless mercerizing range 直辊丝光机
chain link numbers 链节数（分子链）
chain mercerising range（＝chain mercerizer）布铗丝光机
chain molecule 链型分子
chain rack wheel 链条轮
chain rail 链条导轨
chain reaction 连锁反应；链式反应
chain regulating agent 链长调节剂
chain return wheel 转向链条盘
chain saw 链锯
chain segment mobility 链段运动（聚合物）
chain sprocket 链轮
chain stenter 链式拉幅机
chain stitch sewing machine 链式线缝缝纫机
chain stud 链条凸头

chain tension adjuster 链张力调整器
chain tensioner 链条松紧调整器
chain tenter 链式拉幅机
chain termination 链终止[作用]
chain terminator 链终止剂
chain tightening device 链条收紧装置
chain track 链条轨道(拉幅机)
chain track guide 链条轨道导板
chain traject 链轨
chain transmission 链条传动
chain wheel 链轮;链条盘
chaisel (=chainsel) 细亚麻布
chaising calender 叠层轧光机
chalk 白垩;粉笔;浅莲灰[色]
chalk finish 白垩粉整理(丝或人造丝织物的无光及加重整理)
chalkiness 白条痕,白色擦痕[疵];失光;光泽黯淡发白(印疵)
chalking 钙质处理;白垩处理;失光,擦白[疵]
chalk mark 细白条痕;白垩斑(染疵)
chalky 白垩色
chamber 室;箱;盒
chamber drier(= chamber dryer) 箱式烘燥机;烘燥间,烘房
chamber drying 烘房烘燥
chambray 青年布(中支纱平纹交织,棉或涤棉混纺织物)
chameleon 三色调效应(由单色经丝和两种不同色的纬丝交织得到的效应)
chameleon fibre 变色纤维(含有光敏性染料,能随光强的变化而改变色泽)
chameleon paint 示温漆,温度指示漆(变色漆)
chameleon printing 变色印花
chamfer 倒棱,倒角;沟,槽,凹线

chamfering 倒棱,倒角
chamois leather 麂皮革
chamois suede 仿麂皮棉织物
changeable colour 可变色
changeable effect 闪光效应
changeable mechanism 可变机构
changeant 闪光效应;闪光织物(经纬异色,使织物在不同角度反映出不同色泽)
changeant dyeing 闪光染色(使交织物的经纬染成不同色泽)
change colour ways 换色位《印》
change control desk 变换控制台
change design 换花型
change gear 变换齿轮,变速齿轮,变向轮,挂轮
change gearbox 变速齿轮箱
change in colour 变色
change in shade 变色
change of stroke 行程变化;冲程变化
change-over 转换;转换开关
change-over clips/needles 铗/针的转换(拉幅机上针铗两用机构)
change-over time 调换[品种]时间
change-pole motor 变极式电动机(一种双速电动机)
changer 变换器,换流器《电》
change-speed gearbox 变速器,变速箱
change-speed motor 分级调速式电动机,多速电动机
change time 更换品种停台时间
change wheel 变换齿轮,变速齿轮;变轮
channel 槽钢;槽,渠;通道;[电缆]管道
channel gulley 集污槽
channeling 沟流《染》,导水沟

char 炭,炭化
character 特性,性质,特征;字母,符号
characteristic 特性,性能;特性曲线;规格;鉴定
characteristic absorption band 特征吸收谱带
characteristic curve 特性曲线
characteristic equation 特性方程式
characteristic fading curve 褪色特征曲线
characteristics of nano-material 纳米材料的特性
characteristic test 特性试验
characteristic value 特征值
characterization 特征鉴定;表征[法];品质鉴定
characterize 表征;说明;鉴定
charcoal black 炭黑
charcoal grey 炭灰色
charge 加料;充电;充气;电荷;负载;费用
charge neutralization 电荷中和[作用]
charger 充电设备
charging 增重《丝》;加料;充电
chariot 小车(指车间内物料搬运器具)
char length 烧焦长度,炭化长度(织物防燃试验项目)
charring (=charing) 炭化[作用]
chart 图,图表;表,一览表
chartered colourist〈CCol〉[英国染色工作者学会的]注册印染工作者
chartreuse green 卡尔特绿[色](黄绿色)
chartreuse yellow 卡尔特黄[色](浅绿黄色)
chasing [布匹]叠层轧光

chasing calender 叠层轧光机
chassis [轧染机]轧液槽;印花浆盘
chatoyant [织物]闪光效应(法国名称)
chatter 印花跳刀疵(滚筒印花疵);深浅条(网印疵)
check 格纹;检验,控制,校验
check-back fabric 格背双层织物
check band 缓冲绳,制动绳
check effect 格子花纹
checker 检验器,校验装置;检验员;格子花,方格图案
checkerwork 彩格图案(不同色交替排列成正方形)
check gingham 格子布
checking for zero 零位调整
check lever 止回杆
check list 检查表,检查目录
check motion 防冲装置
check nut 防松螺母,锁紧螺母
check out 检查,检验;调整,调试
check out system 检查系统
check out test set〈COTS〉 检[验]测[试]设备
check pattern 格子花纹
check pawl 棘爪,止回棘爪
check point 检测点,基准点
check ring 止动环,卡环
checks 各色格子织物
check sample 试验标准布样
check screw 压紧螺钉,止动螺钉,定位螺钉
check shirting 格子衬衫布
check spring 止动弹簧,复原弹簧,止回弹簧
check strap 缓冲皮带
check valve 单向阀,止回阀,逆止阀;检

验开关
check voile 缎条巴里纱
check-weigh system 检测—称重系统
cheese 筒子纱；扁柱形筒子（通常直径大于高度）
cheese colour 干酪色，奶酪色
cheese density 筒子纱卷绕密度
cheese dryer 筒子纱烘干机
cheese dyeing 筒子纱染色
cheese dyeing machine 筒子纱染色机
cheese mercerizer 筒子纱丝光机
cheese winder 筒子纱卷绕机
chelate 螯合的；螯合物
chelate compound 螯[形化]合物
chelate fibre 螯合纤维
chelate indicator 螯合指示剂
chelate polymer 螯合聚合物
chelate resist printing 螯合防染印花
chelate ring 螯形环
chelating agent 螯合剂
chelating dyes 螯合染料
chelation 螯合作用
chelation value 螯合值（测螯合剂效率）
chelatometric titration 螯合滴定[法]；络合滴定[法]
chem.(＝ch., chemical) 化学的
chem.(＝ch., chemistry) 化学
chemi- (构词成分)表示"化学的"
chemic 漂液（次氯酸钙或次氯酸钠溶液）
chemical〈ch., chem.〉化学的
Chemical Abstracts〈CA〉《化学文摘》（美国刊物）
Chemical Abstracts Service〈CAS〉化学文摘社（美国）
chemical acitivity 化学活度

chemical affinity 化学亲合性，化学亲和性
chemical agent protective clothing 防化[学作用]服装
chemical analysis 化学分析
chemical attack 化学侵蚀
chemical/biological〈CB〉化学/生物的
chemical-biological protective garment 防生化服装
chemical blowing agent〈CBA〉化学发泡剂
chemical bond 化学键，化学结合
chemical bonding 化学黏合工艺
chemical carcinogen 化学致癌物
chemical cellulose 化学纯纤维素
chemical change 化学变化
chemical cistern 漂液槽
chemical composition 化学组成
chemical conditioner [污泥]脱水助剂
chemical constitution 化学构造
chemical cracking 化学裂解
chemical creep 化学蠕变
chemical crimp 化学处理[所得的]卷曲
chemical crosslink 化学交联
chemical decatizing 化学蒸呢
chemical degradation 化学裂解
chemical degumming 化学脱胶
chemical detergent 化学洗涤剂
chemical detoxification of pesticide 杀虫剂的化学解毒（用于农业工作服）
chemical embossed floor coverings 化学轧花地毯
chemical embroidery 化学烂花
chemical engineering 化学工程
chemical engineering products〈CEP〉化学工程产品

chemical environmental resistance 耐化学性；耐化学环境性
chemical equation 化学方程式
chemical equilibrium 化学平衡
chemical equivalent 化学当量
chemical etching 化学侵蚀
chemical etching machine ［花筒］化学腐蚀机
chemical fibre 化学纤维
chemical fibre fabric 化学纤维织物
chemical finishing 化学整理
chemical fixation 化学固着
chemical foam applicator〈CF applicator〉化学泡沫装置
chemical foamer and feed system 化学泡沫发生和喂给系统
chemical foam system〈CFS〉化学泡沫系统
chemical foam technology 化学泡沫技术
chemical formula 化学式
chemical grafting 化学接枝
chemical identification of fibres 纤维的化学鉴定
chemical index 化学指标
Chemical Industry Institute of Toxicology〈CIIT〉化学工业毒物研究所
chemical inertness 化学惰性
chemical inert support 化学惰性载体
chemical inhibitor 化学抑制剂；抗氧剂
chemical initiation 化学引发（接枝聚合）
chemical leather 合成革，人造革
chemically-bonded water 化学结合水
chemically modified cotton 化学变性棉
chemically modified fibre 化学变性纤维
chemically non-corrodible 耐化学腐蚀剂的
chemically pure〈CP, C.P.〉化学纯
chemically reactive dyes 活性染料，反应性染料
Chemical Manufacture Association〈CMA〉化学品制造商协会（美国）
chemical metering 化学品计量
chemical metering station 化学计量站
chemical modification 化学变性［作用］
chemical monitoring 化学监测
chemical nickel plating 化学镀镍（用于金属或纺织品上）
chemical oxygen consumption〈COC〉化学耗氧量
chemical oxygen demand〈COD, C.O.D.〉化学需氧量
chemical pad 化学浸轧
chemical padder 化学品浸轧轧车
chemical permanent set 化学持久定形
chemical potential 化学位
chemical processing 化学加工，化学处理
chemical promoter 化学促进剂
chemical proofing 耐药品性，耐化学药品性
chemical property 化学性质
chemical protection〈CP〉化学防护
chemical protective clothing〈CPC〉防化服
chemical protective helmet 防化头盔
chemical pulp〈CP, C.P.〉化学浆粕，化学浆
chemical quilting 化学绗缝
chemical reaction 化学反应
chemical recovery 化学回收
chemical recovery system 化学回收装置
chemical relaxation 化学松弛

chemical resistance 化学稳定性,耐化学药品性
chemicals 化学品
chemical scission 化学断键,化学断链
chemical setting 化学定形
chemical setting agent [羊毛]化学定形剂
chemical shift 化学位移
chemical shrink 化学收缩,化学预缩
chemical sizing method 化学上浆法
chemical stability 化学稳定性
chemical staple fibre 化学短纤维
chemical stress relaxation 化学应力松弛
chemical stretch 化学弹力整理（一般指松式丝光整理）
Chemical Substances Control Act〈CSCA〉化学物质控制法规
chemical suits 防化服装
chemical tendering 化学脆化,化学脆损
chemical testing 化学测试法
chemical texturing 化学变形法
chemical thermodynamics 化学热力学
chemical transformation 化学转变,化学变换
chemical vapour coagulation method〈CVC method〉化学蒸发凝聚法（制纳米微粒的方法）
chemical warfare〈CW〉化学战争
Chemical Weapons Convention〈CWC〉化学武器公约（禁止有害化学品的输出）
chemic cistern 漂液槽
chemick 漂液（次氯酸钙或次氯酸钠溶液）
chemiluminescence 化学发光
chemisorption 化学吸附

chemisorptive fibre 化学吸着性纤维
chemist 化学家,化学师
chemistry〈ch.，chem.〉化学
chemo- （构词成分）表示"化学的"
chemosensor 化学传感器
chemosynthesis 化学合成
chemothermal 化学热
chem pad 化学浸轧
chemset process 化学定形工艺
Cheney method 切尼染色法（加入苯酚使聚酰胺纤维膨化的染色法,商名,美国）
Chenille carpet 雪尼尔地毯,绳绒地毯
chequer 格子花,方格图案
cherry [red] 樱[桃]红[色]
chessboard-like 棋盘花纹式样的,方格花纹的
chessboard pattern 棋盘格（电子雕刻网点排列的一种方式）
chest 柜,室,箱
chest drying machine 夹板烘燥机
chesting 布卷式轧光工艺
chesting calender 布卷式轧光机
chest interlining 胸衬
chestnut [brown] 栗[壳]棕[色]
chestnut black 栗子黑[色],栗[壳]黑[色]
cheviot finish 切维奥特整理（能使粗纺毛织物织纹清晰）
chica red 契卡红（中美土著用的红色染料）
chicken yellow 鸡黄[色]
chief engineer 总工程师
chief executive officer〈CEO〉业务最高负责人,首席执行官
chief value of cotton〈CVC〉以棉为主的

涤棉混纺纺织品(棉比例50%以上,俗称倒比例)
chiffon 雪纺绸,薄纱,薄绢,薄绸
chiffon crepe 雪纺绉绸(似乔其绉,较柔软)
chiffonized 丝绒柔光整理
chiffon velours 雪纺丝绒(软薄丝绒)
chili 红棕[色],铁锈色
chilled bowl 铸铁辊
chiller 冷冻装置
chilling 冷却;冷凝
chimney 烟囱,烟道管;尘塔
chimney damper 烟囱挡板,烟道挡板
chimney draft 自然拔风,烟囱排风
chimney drain 烟囱式排水
China blue 中国蓝(商名),酸性水溶青,溶性蓝
china clay 陶土,瓷土
China crepe 中国绉
china finish 陶土整理,重浆轧光整理
China grass 苎麻
China hemp 中国大麻
China International Textile Machine Exhibition〈CITME〉 中国国际纺织机械展览会
China jute 茼麻,青麻,芙蓉麻,天津麻
chinchilla 灰鼠呢,珠皮呢(羊毛粗大衣呢);银灰色
chinchilla cloth finishing 灰鼠呢整理,珠皮呢整理(厚呢经缩绒、起绒后用灰鼠呢面整理机将呢面茸毛搓成小珠形)
chinchilla machine 珠皮呢面整理机,灰鼠呢面整理机
chine printing 经纱印花
chine ribbon 印经缎带

Chinese Academy of Sciences〈CAS〉 中国科学院
Chinese gown 旗袍;长衫
Chinese knot 中国结
Chinese linen 夏布
Chinese stencil print 中国蓝印花布
chinese white 白岭土,陶土,瓷土
chine silk 印色丝(绞丝印花或经丝印花)
chine velvet 印经丝绒
chine yarn 印花纱线
chinlon 锦纶
chino fabric 丝光斜纹棉织品
chints(=chintz) 摩擦轧光印花棉布(一般印有华丽和大花型图案)
chintz calender 摩擦轧光机
chintz finish 摩擦轧光整理
chips 碎片,片屑;切片
chisel 凿(刀);凿子
chitin 甲壳质(织物的防雨剂、硬挺剂及丝网印花的浆料和黏合剂)
chitin fibre 甲壳质纤维
chitopoly 甲壳素黏胶复合抗菌纤维
chitosan 脱乙酰的甲壳质,壳聚糖
chitosan fibre 壳胺纤维,壳聚糖纤维
chloral 氯醛,三氯乙醛
chloral hydrate 水合氯醛,水合三氯乙醛
chloramine 氯胺
chloramine-T 氯胺 T,对甲苯磺酰氯胺钠(漂白剂,氧化剂)
chloranil 氯醌,四氯代苯对醌,四氯苯醌
chloranisidine 氯代茴香胺,氯代甲氧苯胺
Chlorantine dye 克罗兰丁染料(直接耐

晒染料,商名,德国制)
chlorate 氯酸盐
chlorate discharge 氯酸盐拔染
chlorate-steam process 氯酸盐汽蒸工艺
Chlorazol fast dyes 克罗腊素尔耐晒染料(直接耐晒染料,商名,英国制)
chlorendic acid 氯菌酸(阻燃剂,增塑剂,防霉剂)
chlorendic anhydride 氯菌酸酐(阻燃剂,固化剂)
chloric acid 氯酸
chloride 氯化物
chloride of lime 漂白粉
chloride of soda 次氯酸钠
chlorinated compound 含氯化合物,氯化化合物
chlorinated hydrocarbon 氯代烃
chlorinated hydroxy diphenyl ether 氯化羟基二苯醚(杀菌剂)
chlorinated lime 漂白粉
chlorinated paraffin 氯化石蜡
chlorinated polyethylene〈CPE〉 氯化聚乙烯
chlorinated solvent 氯化溶剂
chlorinated starch 氯化淀粉
chlorinated water 含氯的水
chlorinated wool 氯化防缩羊毛
chlorination 氯化[作用]
chlorination acid 氯酸
chlorination of wool 羊毛的氯化防缩处理
chlorination/resin finish 氯化/树脂整理
chlorination rubber 氯化橡胶
chlorination shrink proofing 氯化防缩整理
chlorination treatment 氯化处理(可提高羊毛的染色性及防缩性)
chlorine 氯
chlorine bleach 氯漂[白]
chlorine bleaching agent 含氯漂白剂
chlorine bleach liquor 氯漂[白]液,次氯酸钠溶液
chlorine compound 氯化合物
chlorine cylinder 液氯瓶
chlorine damage 氯损
chlorine demand 需氯量
chlorine dioxide 二氧化氯
chlorine donor 氯给予体(供氯物质)
chlorine fastness 耐氯漂牢度
chlorine-Hercosett treatment 氯化—赫科塞特防缩处理(毛织物经氯化后再用赫科塞特树脂整理)
chlorine peroxide bleach 氯及双氧水漂白,氯氧双漂
chlorine peroxide bleaching range 氯氧双漂机组
chlorine/resin system 氯化/树脂法(羊毛防毡缩处理)
chlorine resistance 耐氯性
chlorine resist finish 抗氯[树脂]整理
chlorine-resist resin 耐氯树脂
chlorine retention 留氯作用,吸氯作用
chlorine retention and yellowing 吸氯泛黄
chlorine retention damage 吸氯破损,残留氯破损
chlorite 亚氯酸钠
chlorite activating agent 亚氯酸钠活化剂
chlorite-bisulphite redox system 亚氯酸盐—亚硫酸氢盐氧化还原[引发]体系

chlorite bleach 亚氯酸钠漂白
chlorite bleaching range 亚氯酸钠漂白机组
chloroacetic acid 氯代醋酸,氯代醋酸(制造羧甲基纤维素的原料)
chloroacetylamino dyes 氯乙酰胺[型]染料(活性染料)
chlorobenzene 氯苯
chloro cyanuric acid 氯氰尿酸
chlorodifluoropyrimidine group 二氟一氯嘧啶基团(活性染料的一种活性基团)
chlorofibre〈CLF〉含氯纤维
chlorofluoro carbons 氯氟碳
chloroform 氯仿,三氯甲烷
chlorohydrin[e] 氯乙醇
chloroisocyanuric acid 氯代异氰脲酸
chloroorganic dyeing carrier 含氯有机染色载体
chlorophenol red 氯酚红
chlorophyll 叶绿素
chloroprene 氯丁二烯
chloroprene rubber roll 氯丁橡胶辊
chloropropanol 氯丙醇
chloropyrimidine dyes 氯嘧啶[型]染料;氯间二氮[杂]苯[型]染料(活性染料)
chlorosulfonated polyethylene〈CSPE〉氯磺化聚乙烯
chlorosulfonic acid 氯磺酸
chlorosulphonyl isocyanate 氯磺酰异氰酸盐
chlorotriazinyl dye 氯三嗪基染料(活性染料的一类)
chlorous acid 亚氯酸
chlorzyme process 氯化酶工艺

CHM(Colour Harmony Manual)色彩调和手册,配色协调手册(美国)
chock 制动器;塞块;轧辊轴承;垫木
chocolate[brown]巧克力[棕]色
choice of mesh for screen printing fabrics 筛网印花织物网目的选择
choke 阻气门;阻塞;抑制;[管的]闭塞部分
cholesterin(=cholesterol)胆甾醇,异辛甾烯醇(脂肪族单元醇,为羊毛脂主要成分)
chopped fibre 碎段纤维
chopping machine 切片机;打印机
CHP(combined heat & power)热电联产
chroma 彩度
chroma contrast 彩度对比
chroma scale 彩度标度
chroma-scan 色度扫描
chromat- (构词成分)表示"颜色""染色质"
chromate discharge 铬酸盐拔染
chromate mordant 铬酸盐媒染剂(两份铬酸钾或钠,四份硫酸铵)
chromate printing 铬酸盐法印花(可溶性还原染料印花方法之一)
chromate[dyeing]process [媒染染料]同浴铬媒染色法
chromatic 颜色的,色彩的
chromatic aberration 色[相]差
chromatic adaptation 色适应性
chromatic characteristic 色彩特性
chromatic circle 色环
chromatic colour 有彩色
chromatic difference 色差
chromatic dispersion 色散[现象]

chromatic dyeing 袖筐绞纱染色(与袖笼绞纱染色相仿,具有绞纱染色和卷装染色两者优点的染色法)
chromaticity 色度;色品;染色性
chromaticity chart 色品图(指国际照明委员会色品图)
chromaticity coordinate 色品坐标,色度坐标
chromaticity diagram 色品图,色度图
chromaticity spacing 色度间距
chromaticness 色品度
chromatic resolving power 色分辨能力
chromatics 比色法,颜色学
chromatic spectrum 彩色光谱
chromatic transference scale 彩色沾色样卡
chromatic value 色彩值,色度值
chromatism 色[相]差
chromatograph 色谱法,色层分析法
chromatography-mass spectroscopic analysis 色谱—质谱联用分析
chrome 铬;重铬酸钾,重铬酸钠(媒染剂)
chrome complex dyes 含铬染料,铬络合染料
chrome dyeing process 铬媒染色法
chrome dyes 铬媒染料
chrome gelation 铬明胶
chrome green 铬绿(无机颜料)
chrome immobilization 铬固定
chrome limits 铬的限量
chrome mobilization 铬转移
chrome mordant 铬媒剂
chrome mordant azo dye 铬媒染偶氮染料
chrome mordanting 铬媒染

chrome mordant process 预铬媒染法,先铬媒染法
chrome orange 铬橙(无机颜料)
chrome photo printing 铬影印
chrome plating 镀铬
chrome plating apparatus 花筒镀铬设备,镀铬设备
chrome plating installation [花筒]镀铬设备
chrome printing 铬媒染料印花,媒染染料印花
chrome release 铬的释放
chrome residue 残留铬
chrome-topped 铬后处理
chrometrope 铬变素(酸性染料)
chrome yellow 铬黄(无机颜料)
chromic acid 铬酸
chromic anhydride 铬酐,三氧化铬
chromiferous dyes 含铬络合染料
chrominance 彩矢量
chroming 铬媒处理;镀铬(花筒)
chromium 铬
chromium fluoride 氟化铬(媒染剂)
chromium plating 镀铬
chromium reaction 铬反应
chromiun salt 铬盐
chromo- (构词成分)表示"铬""颜色""色素"
chromogen 色原体(含发色团但无助色团的无色有机化合物)
Chromojet 地毯喷射印花装置(商名,奥地利齐默)
chromojet printing of carpets 地毯喷射印花
chromolithography 彩色平版印刷术;彩色石印术

chromometer 比色计
chromophore 发色团,生色团,生色基
chromophore theory 发色团理论
Chromo Tex jet printing machine 克罗马塔克斯喷墨印花机(商名,奥地利齐默)
Chromotronic jet-spray screenless system 电子控制无筛网喷射印花法(商名,奥地利齐默)
Chromotronic machine 电子控制无筛网喷射印花机(奥地利齐默)
Chromotronic microjet printing machine 电子控制微喷射印花机(每排细微喷嘴只印一种颜色,由电子计算机控制,奥地利齐默)
chromotropic acid 铬变酸,1,8-二羟萘-3,6-二磺酸(用于树脂整理织物上游离甲醛的测定)《试》
chromotropic acid method 铬变酸法(游离甲醛测试方法)
chromotropism(=chromotropy) 异色异构[现象]
chromous acid 亚铬酸
chromoxane green 铬绿
chronic oral toxicity 慢性口服毒性
chronic toxicity 慢性毒性《环保》
chronograph 计时器
chronological sequence 顺时[间]顺序
chronometer 精密计时器,航行表
chronometry 测时学
chrysanthemum 高粱红(天然色素)
chryso - (构词成分)表示"黄色""金黄色""金"
chrysoidine 碱性橘橙染料
chrysophenine 直接冻黄
CHS (calendered heat setting process) 轧光热定形工艺
chthode ray tube 阴极射线管
chuck 钻轧头;卡盘;[电磁]吸盘
Chunri 琼里扎染法;印度扎染棉平布
chute 斜管,斜槽;[降落]伞
chute fabrics 降落伞织物(包括布水雷、布炸弹等用的织物)

chute washing 斜槽清洗
CI (=C. I., Colour Index) 《染料索引》(书名);颜色指数,比色指数
Cibacet dyes 汽巴缓脱染料(染醋酯纤维及聚酰胺纤维用的分散染料,商名,瑞士制)
Cibacron C dyes 汽巴克隆 C 染料(一氟均三嗪乙烯砜双活性基团染料,商名,瑞士制)
Cibacron F dyes 汽巴克隆 F 染料(一氟均三嗪型活性染料,商名,瑞士制)
Cibacron FN dyes 汽巴克隆 FN 染料(一氯均三嗪乙烯砜双活性基团新型活性染料,商名,瑞士制)
Cibacron HW dyes 汽巴克隆 HW 染料(适用于高、中温染色的活性染料,商名,瑞士制)
Cibacron LS dyes 汽巴克隆 LS 染料(含两个一氟均三嗪活性基团染料,适用于低盐染色,商名,瑞士制)
Cibacron P-T/P dyes 汽巴克隆 P-T/P 染料(低亲和性活性染料,适用于轧染及印花,商名,瑞士制)
Cibacron S 汽巴克隆 S(高提升性活性染料,由2~3个发色体和2~3个活性基团组成)
Cibafast N-2 汽巴法司特 N-2(耐日晒牢度改进剂,适用于锦纶,商名,瑞士制)

Cibafast W 汽巴法司特 W（耐日晒牢度改进剂，适用于羊毛，亦可降低羊毛损伤，商名，瑞士制）

Cibafluid C 汽巴染浴宝 C（聚醚类共聚物，适用于各种纤维，防止织物擦伤、磨损及折皱，商名，瑞士制）

CID【法】国际表面活性剂委员会

CIE【法】国际照明委员会

CIE colour designation 国际照明委员会表色法

CIE LAB 国际照明委员会 LAB 色差公式（国际色差测算公式，CIE 代表国际照明委员会，L 代表亮度，A 代表红—绿轴坐标数字，B 代表黄—蓝轴坐标数字）

CIE LAB colour difference equation 国际照明委员会 LAB 系统色差公式

CIE 1976 $L^*a^*b^*$ colour difference formula 国际照明委员会 $1976L^*a^*b^*$ 色差公式

CIE LAB colour space 国际照明委员会 LAB 系统色空间

CIE 1976 $L^*a^*b^*$ colour space 国际照明委员会 $1976L^*a^*b^*$ 色空间

CIE LAB system 国际照明委员会 LAB 颜色体系

CIE LUV 国际照明委员会 LUV 色差公式

CIE 1976 $L^*u^*v^*$ colour difference formula 国际照明委员会 $1976L^*u^*v^*$ 色差公式

CIE 1976 $L^*u^*v^*$ colour space 国际照明委员会 $1976L^*u^*v^*$ 色空间

CIE LUV system 国际照明委员会 LUV 颜色体系

CIE spectral tri-stimulus values 国际照明委员会光谱三色刺激值

CIE 1931 standard colourimetric observer 国际照明委员会 1931 标准色度观察者

CIE 1931 standard colourimetric system 国际照明委员会 1931 标准色度系统

CIE standard colourimetric system 国际照明委员会标准色度系统

CIE 1964 standard colourimetric system 国际照明委员会 1964 标准色度系统

CIE standard illuminant A 国际照明委员会标准照明体 A（钨丝灯光）

CIE standard illuminant C 国际照明委员会标准照明体 C（平均昼光）

CIE standard illuminants 国际照明委员会标准照明体

CIE standard illuminating and viewing conditions 国际照明委员会标准照明和观察条件

CIE standard light sources（=CIE standard sources） 国际照明委员会标准光源

CIE 1964 supplementary standard colourimetric observer 国际照明委员会 1964 补充标准色度观察者

CIE 1964 supplementary standard colourimetric system 国际照明委员会 1964 补充标准色度系统

CIE-whiteness formula 国际照明委员会白度公式

CIF（cost, insurance and freight） 到岸价格，包括成本、保险费及运费的价格

cigarette test 可燃性测试

CIIT（Chemical Industry Institute of Toxicology） 化学工业毒物研究所

CIJ（continuous ink-jet） 连续喷墨[法]

CIJ printing 连续喷墨印花
cilana 永久绉纹效应(泡泡效应)
Cilander type rope washing machine 雪兰达型绳状水洗机(采用多管逆流,德国门泽尔)
CIM(computer intergrated manufacturing) 计算机集成制造
cinnamic acid 肉桂酸,苯基-α-丙烯酸
Cirasol PI 锡腊苏PI(防污整理剂,商名,英国卜内门)
circlip [开口]簧环
circuit 电路,回路
circuit breaker 断路器《电》
circuit diagram 电路图
circuitry 电路图,电路系统,电路
circular brush 圆形毛刷
circular brushing machine 圆形刷呢机;圆形刷毛机
circular chart 圆形图表
circular dichroism 圆形偏振光二色性
circular ended wrench 套筒扳手
circular hank washing machine 圆形绞纱水洗机
circular knife (针织物拉幅机的切边剖幅装置的)圆形刀片,(剪毛机的)螺旋刀
circular knit 圆筒形针织物
circular knit hosiery 圆筒袜
circular knit pile fabric 筒形起绒针织物
circular knit setting machine 圆筒形针织布定形机
circular nozzle 圆孔喷风管
circular polarization 圆形偏振光
circular pressure brush 上针转刷(针板拉幅机进布区的上针装置)
circular raising machine 回转起绒机,回转刮绒机,回转起毛机
circular screen printing machine 圆网印花机
circular singeing 筒形针织物烧毛
circular stenter 圆形拉幅机
circular stretching machine 圆形拉幅机
circular washing machine 圆形水洗机
circular waveguide 圆波导
circular web 圆筒形针织物
circulating air drying 循环热风烘燥
circulating bath 循环浴
circulating cistern 循环淋液槽
circulating fan 循环风机
circulating hot oil unit 热油循环装置
circulating liquor dyeing 染液循环染色
circulating liquor dyeing machine 染液循环染色机
circulating pipe 循环管
circulating pump 循环泵
circulating speed 循环速度
circulating type hank dyeing machine 循环式绞纱染色机;循环式绞丝染色机
circulation 循环[量],环流[量]
circulation compressor 循环压缩机
circulation tank 回流槽
circulator 循环泵;循环管
circulatory dryer 循环风烘干机
circulatory flow dryer 循环回流式烘燥器
circulatory flow pressure dryer 循环加压烘干机
circumferential distortion 周向变形
circumferential seam 环缝,包缝
ciré 上蜡色布(法国名称)
cire [仿漆皮]蜡光整理(丝织物经上蜡

热轧后产生高度光泽)
cire calendar 蜡光整理轧光机
cire effect 蜡光效果
cire finishing 蜡光整理,光泽处理
cire satin 蜡光缎
CIRFS【法】国际黏胶纤维和合成纤维委员会
cis-form 顺式
cis-isomer 顺式异构体
cistern 槽,水槽,储液槽,蓄水器
2-cistern open washer 双格平洗槽
cis-trans isomerism 顺反异构现象
CITME(China International Textile Machine Exhibition) 中国国际纺织机械展览会
citrate discharge 柠檬酸盐拔染
citric acid〈CA〉 柠檬酸
citron 香橼黄
citrone yellow 香橼黄;黄色颜料
city gas 城市煤气
Civilian Supply Administration〈CSA〉 民需供应局
civil textiles 民用纺织品
C. L.(confidence limit) 置信区间,置信[界]限,信任界限
clack valve 瓣阀,止回阀
claddings [机器]外包层
cladding yarn 外包纱,外缠纱(包缠在外层的纱或丝)
claded yarn 包芯纱,包缠纱
claim 专利范围;索赔
claim indemnity 索赔
clamming 轧制[刻模]钢芯
clamming machine 压模机(花筒雕刻用);钢芯雕刻机(印花辊筒雕刻设备之一),子母机

clamp 钳;夹,夹板
clamp coupling 纵向夹紧联轴节
clamper 夹持器;接线板
clamping bar 压板,夹杆
clamping device 夹紧装置
clamping jaw 夹口,钳口
clamping plate 压板
clamp nut 花螺母,压紧螺母
clams 压模机(花筒雕刻用);钢芯雕刻机(印花辊筒雕刻设备之一),子母机
clan green shade 蓝绿[色]
claret [red][葡萄]酒红[色](暗紫红色)
claret violet [葡萄]酒紫[色]
clarification temperature 澄清温度
clarifier 沉降分离装置
clarity 透明度,清晰度
Clarte [织物的]表面清洁整理(法国名称)
classer 分级员,分级工
classic denim 古典牛仔布
classification 分类,分级
class settling 分级沉降
claw clutch 爪式离合器,爪式接合器
claw coupling 爪式接合器,爪形联轴节
clawker 棘轮撑头
clay 黏土;泥土;白土;泥土色(浅棕色)
cleanability 抗垢力,除尘度
clean colour 纯洁色(指不带灰、暗、脏、垢的色泽)
clean content 净量
cleaner 除尘器,清洁器;清洁工;除杂机
clean factory〈CF〉 清洁工厂
clean fashion 清洁时尚;清洁式样(世界大纺织品批发商设定的生态标记)
clean finishing [毛织物]显纹整理

Clean Guard 一种新型除臭织物(日本制)
cleaning 洗涤,清除;除杂,除尘
cleaning activity 清洁活性(干洗剂指标)
cleaning agent 清洁剂,清洗剂
cleaning and shearing machine 清洗剪毛机
cleaning bath 清洗浴
cleaning cloths 清洁布
cleaning doctor 刮[浆]刀;清洁刮刀
cleaning foam 泡沫清洁剂
cleaning in situ 原地清洗
cleaning mask 清洁用防护面具
cleaning performance test 净洗性能试验
cleaning process 清洁生产,清洁工艺(无污染生产)
cleaning room clothing 净洗室工作服
cleaning trunk 除尘箱
cleanliness 纯净度,清洁度
cleanness 清洁[度]
clean print 轮廓光洁的印花花纹
clean production 清洁生产(无污染生产)
clean products 清洁产品(无污染产品)
clean room 洁净室
cleansing 清洗,清除
cleansing oil 清洁用油,冲洗油
Cleanstar 清洁星(一种起绒辊清洁装置,商名,德国 Xetma)
clean touch 清洁触感(近似抛光工艺)
clean-up [环境]净化
clean-up time [机器的]清洁时间
clear [不含颜料黏合剂的]涂料印花浆;稀释的合成增稠剂浆
clearance 间隙(指机体之间的空隙量)

clearance angle 留隙角
clearance fit 动配合,间隙配合
clearance gear 间隙离合齿轮
clear area 有效截面积
clear coat 透明涂层;洁净涂层;透明胶涂层
clear concentrate 浓缩浆(未加涂料色浆的增稠剂,美国印花技术用语)
clear-cut 轮廓清晰
clear-cut finish 剪净整理,光洁整理《毛》
clear finish 光洁整理;光面整理《毛》
clearing brush 清洁刷,清洁毛刷
clearing cloth 揩布,擦布
clearing gun 去油污枪
clearing property 洗净性,清除性
clear span structure 大跨度结构
clear white 纯白[色]
clear width 净宽
clear woollen finish 毛呢光面整理
cleavage 剥层;分裂,劈裂;断裂;解离[性]
cleavage of dyes 染料解聚
cleavage product 解聚产物
clematis 铁线莲紫[色](红紫色)
cleopatra 克娄巴特拉蓝[色](鲜蓝色)
clerical grey 深灰[色]
clevis bolt U 形螺栓
CLF (chlorofibre) 含氯纤维
cliche 手工印花模版
click 棘爪,掣子,撑头
climate garment 调温服
climatic chamber 空调室,空调箱,人工气候室
climatic conditioning cabinet 空[气]调[节]箱,恒温恒湿箱

climatic unit 空调机
climatized garment 调温服
climatized room 空调室
climatizer 空调室；烘燥室（圆网雕刻用）
climbing roller 上升辊
clinging property 紧身性，贴身性，缠合性
cling-test ［织物的］紧贴性试验；静电吸附试验
clip 布铗，百脚（俗称）；夹子，夹，接线柱；压紧；一季剪毛量
clip chain 布铗链条，百脚链
clip chain for width stretching 拉幅链
clip chain mercerizing range 布铗［链条］丝光联合机
clip defect 布铗疵点（布铗拉幅疵点）
clip expander 小拉幅装置（用于机械防缩联合机，布铗及导轨均比一般拉幅机小）
clip feeler 布铗触指
clip guide 布铗导轨
clip knife 布铗刀口
clipless mercerizing machine 无布铗丝光机，直辊丝光机
clipless mercerizing range 直辊丝光联合机
clip mark 布铗痕（布铗拉幅疵点）；钩破洞
clip mercerizing machine 布铗丝光机
clip opener 开铗器
clipper seam 夹扣缝
clip piece mercerizing machine 布铗丝光机
clipping 剪毛
clipping and carving ［地毯］剪花

clipping failure trip 脱铗自停（布铗拉幅机脱铗时自动停车）
clipping machine ［地毯］剪毛机
clip rail 布铗导轨
clip spots 布铗疵点，布铗斑
clip spring 夹箍弹簧，开口弹簧
clip sprocket 布铗链条盘
clip stenter(=clip tenter) 布铗拉幅机
clip tentering 布铗拉幅
clip tentering machine 布铗拉幅机
clip tester 布铗测试器
clip type dryer 布铗式烘干机
clip wheel 布铗链条盘
CLOC (concentration of limit of oxygen for combustion) 燃烧限氧指数
clock arrangement 测长装置
clock motion 测长装置；测长运动
clock printing machine 袜类印花机（指袜边印花）
clock pulse 时钟脉冲，定时脉冲
clock rate 时钟频率
clockwise rotation 顺时针转动
clockwise sequence 顺时针次序
clogging 阻塞，堵塞
clogging of rollers or screens 花筒嵌塞，绢网阻塞（筛网印花疵）
clogging of screen 塞网，堵眼（筛网印花疵）
cloky 泡泡组织织物，泡泡点纹
clone 克隆
cloqué 泡泡组织织物，泡泡点纹（法国名称）
cloqué effect 泡泡纱效应，泡泡隆起效应
cloqué fabric 凸纹［浮线］针织物
cloqué organdy 泡泡蝉翼绉（又称瑞士

蝉翼绉,经耐久坚挺硬化处理,部分透明,最初为瑞士生产)
clorafin 氯化石蜡(防霉、防水、防火剂)
close clearance 紧公差,小间隙
closed bearing 封闭式轴承
closed chamber 密封室
closed circuit 闭合电路
closed construction 密闭结构
closed coupling 固定联轴器;永久接合
closed kier 高压煮布锅;密闭煮布锅
closed loop 闭环《电》
closed loop control 闭环控制
closed loop driving system 闭环调速系统
closed loop feedback control 闭环反馈控制
closed loop system 闭环系统
closed seam 暗缝,包缝
closed-toe 无缝袜头,封闭袜头
closed vessel 密闭容器
closed washer 蒸洗槽,加盖平洗槽
close face finish 密实呢面整理(缩绒和起绒后,紧靠呢面剪毛,再加压,使呢面密实)
close fitting 紧贴
close-knit 细网眼针织物
closely bound goods 紧密织物
closely woven fabric 紧密织物(指相邻纱线间距离小于纱线直径)
close-mesh 细网眼
close-packed structure 紧密结构
close running fit 紧动配合,紧转配合
close stitch 密针脚
close texture [织物]紧密结构,紧密组织
close-up view 近视图,特写视图

close weave 紧密织物
close working fit 紧滑配合
close zippers 封住拉链洗涤(洗衣机用语)
closing 缝合;填空整理(在低级棉织物上加填充剂,再经轧光,使经纬间的空隙填满)
closing plate 封板,闷板(平洗槽的一零部件,当若干格平洗槽相连时,卸去此板,换上逆流管,以达到逐格倒流的效果)
closing thread 缝线
closure 密闭,封闭,闭锁;闭塞物
cloth 织物;布;呢绒
cloth abrasion tester 织物耐磨试验仪
cloth accumulator 容布箱,存布箱
cloth analysis 织物分析
cloth area weighing device 织物单位重量测定装置
cloth bag degumming 装袋精练法
cloth beam 卷布辊;经轴
cloth beam dyeing machine 经轴染色机
cloth bleaching 织物漂白
cloth breakdown [织物]断头(连续加工疵点)
cloth breaking machine 揉布机
cloth brusher 刷布机
cloth carbonisation 呢坯炭化
cloth clip 布铗,百脚(俗称)
cloth construction 织物结构
cloth constructor 织物花纹设计员,织物设计员
cloth container 容布箱
cloth content 容布量
cloth contraction 织物收缩;织物缩率
cloth conveyor system 送布系统

cloth cooling and conditioning machine 冷呢给湿机
cloth counter(＝cloth counting glass) 织物密度镜
cloth cover 布面丰满
cloth covered roll 包布辊(落布辊包覆织物,以增加摩擦,有利落布)
cloth crimp tester 织物皱缩试验仪
cloth cutting machine 切布机,裁衣机
cloth defects 织物疵点,织疵
cloth deflector 布匹转向器
cloth delivery device 出布装置
cloth discharge device 出布装置
cloth doffer 落布工
cloth doubler 合布机;层布贴合机
cloth drying machine 织物烘燥机,烘布机
cloth emerizing machine 金刚砂辊磨布机;磨毛机
cloth entrance slot 布入口
clothes 衣服,衣着
clothes dryer 干衣机,烘衣机
clothes for surgical utilization 外科用服装
clothes line 晒衣绳
clothes moths 衣服蛀虫
clothes wadding 衣服垫料
cloth examination 验布,织物检验
cloth examiner 验布工;织物验收员
cloth expander 扩幅装置,扩幅辊,布匹扩展器
cloth feeder(＝cloth feeding device) 喂布装置,进布装置
cloth feeding roller 进布辊
cloth feeler 探边器
cloth finish 呢面整理;人造丝织物起绒整理
cloth folder(＝cloth folding machine) 折布机,码布机
cloth front pressing 热熔贴面
cloth gripper 布铗
cloth guider 导布装置,导布器;吸边器
cloth guide rod 导布棒,挂布棒(用于长环蒸化机内,能自转)
cloth guide roller 导布辊
cloth guide tube 导布管
cloth guiding device 导布装置
cloth holder 布架
clothier 服装商;呢绒布匹商;裁缝
cloth impact tester 织物冲击试验仪
clothing 衣服,衣着
clothing comfort 服装舒适性
clothing design 服装设计
clothing felt 毛毡
clothing for cold weather 防寒服
clothing industry 衣着工业
clothing insulation 衣服绝缘性,衣服保暖性
clothing micro-climate monitor system 服装微气候监测系统
clothing ornament 服饰
clothing physiology 衣着生理学
clothing pressure 衣着压
clothing sanitation 衣服卫生
clothing technology 服装技术
clothing test 衣着试验
clothing textiles 衣着用纺织品
clothing toxins 衣橱毒素,衣服毒素(衣服上含有游离甲醛等毒素)
clothing value〈clo value〉克罗值(服装保温值),纺织品和服装隔热性的国际通用单位

clothing variable 衣着变数
cloth in rope form 绳状布，绳状织物
cloth insertion sheet 双面胶布
cloth inspecting machine 验布机
cloth inspecting table 验布台
cloth inspection 验布，织物检验
cloth laying machine 摊布机（台板平网印花时，将织物平铺在印花台板上）
cloth lifter 提布轮
cloth-lined wall paper 用布加固的墙纸
cloth looking machine 看布机，验布机
cloth make-up 纺织品整理
cloth mandrel 卷布辊
cloth measuring and cutting machine 量裁机
cloth mellowing machine 揉布机
cloth notcher 切布装置
cloth paper 压呢纸板
cloth pattern cutting machine 切布样机
cloth plaiting machine 码布机，折布机
cloth press 打布包机，织物打包机；压呢机
cloth pressing-on roller 轧布辊
cloth printed cretonnes 大花型印花装饰布
cloth proofing 布上涂胶
cloth puncture strength tester 织物穿孔强力试验仪
cloth quality 织物质量，织物品级
cloth raising machine 起绒机，拉绒机，刮绒机
cloth recipient 容布器
cloth repeated stress machine 织物反复应力试验机
cloth reversing device 翻布装置
cloth rewinder 卷布机

cloth roll 布卷；卷布辊
cloth roller 卷布辊，卷绸辊
cloth rolling machine 卷布机
cloth roll size detector 布卷尺寸检测器
cloth roll stands 布卷架
cloth roll-up 卷布
clothroom 布房间（整理布匹的车间）
cloth sample 布样
cloth sanding machine 织物磨毛机
cloth scouring 布匹煮练
cloth separator 隔布挡
cloth sewing 缝布
cloth shearing machine 剪毛机；刮布机
cloth slitting machine 切布机，剖幅机，切绷带机
cloth specimen 布样
cloth spreader 折布机，铺幅机《针》
cloth spreading machine 折布机
cloth stamping machine 织物打印机
cloth storage container 容布箱
cloth take-up 织物卷取
cloth temperature measuring device 织物温度测定装置
cloth temperature probe 布温探头
cloth tensioner 紧布器，紧布架
cloth thickness gauge 织物测厚规，织物厚度［测量］仪
cloth truck 堆布车
cloth turning device 织物翻面装置
cloth turning machine 翻布机（圆筒形针织坯布用）
cloth turn over device 织物翻面装置
cloth vacuum extractor 织物真空脱水机
cloth washing machine 织物水洗机，洗布机
cloth waste 碎布

cloth wearing tester(＝clothwear testing machine) 织物耐磨试验仪
cloth weight 织物重量
clothy appearance 布面线纹清晰
clothy feel 轻软手感(指轻浆织物的手感)
clothy hand 布质坚实、柔软而有身骨
clotted soap 皂块,块皂;凝块皂
clotting 烧结
cloudiness 云斑(疵点)
cloudiness dertermination 浊度测定法
clouding 云纹工艺(染绸时做成云纹斑点);污斑(刮浆不清疵病)
cloud pink 淡藕红色,淡妃色
cloud point 浊点
cloud point index 烛点指数
cloud point titration 浊点滴定法
cloud print 云斑(疵点)
cloud test 浊点试验
cloudy 云斑的;云斑(疵点)
cloudy goods 云斑织物
clo value (clothing value) 克罗值(服装保温值),纺织品和服装隔热性的国际通用单位
clove 红灰[色]
cluett process 一种机械预缩工艺
clutch 离合器,联轴器
clutch box 离合器箱
clutch cam 离合器凸轮
clutch control lever 离合器控制杆
clutch facing 离合器衬片
clutch fork 离合器叉
clutch handle 离合器手柄,快慢扳手
clutch plate 离合器盘片
clutch shaft 离合器轴
clutch shifter 离合器拨叉,分离叉

clutch shifter shoe 离合器滑瓦
clutch shoe 离合器闸瓦
CMA (Chemical Manufacture Association) 化学品制造商协会(美国)
CMA (conventional moisture allowance) 习惯容许含湿
CMC(cell membrane complex) 细胞膜复合材料
CMC (ceramic matrix composite) 陶瓷基复合材料
CMC(＝C. M. C., carboxymethyl cellulose) 羧甲基纤维素
CMC (Colour Measurement Committee) 颜色度量委员会
CMC colour difference formula 颜色度量委员会色差公式
CMD (cellulose modal) 莫代尔纤维(国际代码)
CMHEC (carboxymethyl hydroxyethyl cellulose) 羧甲基羟乙基纤维素
CMP (computer match prediction) 计算机配色预测
C. M. S. (carboxymethyl starch) 羧甲基淀粉
CMS(colour matching system) 配色系统
CMYK value 青、品红、黄、黑四原色值(喷墨印花用)
CNC (computerized numerical control) 计算机数字化控制
coacervation 凝聚
coacervation dyeing 胶体分相凝聚层染色法(主要用于毛织物的连续染色)
coach wrench 双开活动扳手
coagulant 凝集剂
coagulated polyurethane film 聚氨酯凝固膜(用于湿法涂层)

coagulating bath 凝固浴
coagulating coating line 凝固涂层生产线
coagulating coating process 凝固涂层法
coagulating point 凝固点,凝结点
coagulating sedimentation 混凝沉淀《环保》
coagulating thickeners 凝固增稠剂
coagulation 凝结,凝固
coagulation basin 混凝池
coagulation phenomena 凝固现象
coagulation plant 凝固设备
coagulation pressure floatation process 凝聚上浮法《环保》
coagulation process of polyurethane film 聚氨酯薄膜凝固法(湿法涂层)
coagulation value 凝固值
coagulative precipitation process 凝聚沉淀法《环保》
coagulum 凝结物,凝块(胶体溶液凝固后所得凝集块状物)
coaldust stained yarn 煤灰纱(疵点)
coal equivalent 标准煤
coalescence 聚结[作用]
coalescer 聚结器
coal gas 煤气
coalsteam ratio 锅炉煤汽比
coal tar colours(＝**coal tar dyes**) 煤焦油染料,合成染料
coalwater ratio 锅炉煤水比
coarse adjustment 粗调节
coarse cloth 粗织物
coarse counts 粗支[数],低支[数]
coarse emulsion 粗[滴]乳状液
coarse filter 粗滤器,初滤器
coarsely-dispersed system 粗分散系统

coarse pitch 粗节距
coarse sheeting 粗平布
coastal defences 海岸防御工事(土工布的应用)
coat 男式上衣,外套;妇女、孩童短大衣;涂层
coated backing 涂层基布
coated denim 牛仔涂层面料
coated fabric 涂层织物,上胶织物(如漆布)
coated gloves 涂塑手套
coated technical textiles 产业用涂层纺织品
coater 涂胶机;涂料器,涂层器;涂胶工
coating 上衣衣料;[织物]涂料;上胶;涂层;涂层工序
coating agent 涂层剂;涂胶剂
coating compound 涂层化合物
coating finish 涂层整理
coating finishing machine 涂层整理机(防水、防油、防污等特种整理设备)
coating knife 涂层刮刀
coating machine 涂层机;涂料器;(照相雕刻的)上胶机
coating modifier 涂层改进剂
coating polymers 涂层聚合物
coating property 被覆性
coating substrates 涂层基材
coating technology 涂层技术
coating yarn 涂塑线(用玻璃丝或金属丝作芯,外涂热塑性树脂)
coat thickness measuring instrument 涂层测厚仪(用放射性射线进行检测)
coaxial 同轴的,共轴的
cobalt 钴
cobalt blue 钴蓝[色](青光蓝色);蓝色

矿物颜料
cobalt complex dyes 含钴络合染料
cobalt green 钴绿[色]
cobaltiferous dyes 含钴络合染料
cobalt yellow 钴黄[色]
cobble 疵点再染;织物[染整]修补;再染色,织物回修
cobble black 回染黑色
cobbler 退修的染整织物
cobbling [染疵]织物回修
Cobral M/A&M/B 高宝 M/A、M/B(羊毛弹性整理剂,用于后整理,可代替氨纶,商名,意大利宁柏迪)
cobweb finish 蛛网整理
COC(chemical oxygen consumption） 化学耗氧量
cocarcinogen 辅致癌物[质],助致癌物[质]
cocatalyst 辅助催化剂,辅触媒
cochineal 胭脂虫红[色](大红色);天然红色染料
cock 旋塞,龙头
cockled effect 折裥效果
cockled fabric 起皱织物(疵点)
cockled selvedge 皱边(织疵)
cockling [织物的]皱面外观;织物起皱
cock spanner 旋塞扳手
cocktail coat(=cocktail dress, cocktail suit) 燕尾服
cock valve 旋塞阀
cocoa brown 可可棕[色](黄棕色)
cocoanut oil 椰子油
COD(=C. O. D., cash on delivery) 货到付款
COD(=C. O. D., chemical oxygen demand) 化学需氧量

COD(=C. O. D., cyclooctadiene） 环辛二烯
code 代码;编码;符号;规则;标准
code name 代码名
coder 编码器
code value 代码值
coding 编码
co-dyeing 共染,同染(染色时加入整理剂,使染品收到染色以外的效果,如增加柔软、耐光等)
coefficient 系数,率;折算率;程度
coefficient of diffusion 扩散系数
coefficient of dynamic friction 动摩擦系数
coefficient of elasticity 弹性系数
coefficient of expansion 膨胀系数
coefficient of extension 伸长系数
coefficient of friction 摩擦系数
coefficient of heat transfer 热转移系数
coefficient of inplane thermal expansion 面内热膨胀系数
coefficient of length variation 长度变异系数
coefficient of processing 加工系数
coefficient of thermal expansion〈CTE〉 热膨胀系数
coefficient of thixotropy 触变系数
coefficient of transfer 转移系数
coefieient of viscosity 黏度系数
coemulsifier 辅助乳化剂;辅助乳化器
coenzyme 辅酶
coeruleine 媒染棓酸绿
cogenerating heat and power(=cogeneration of power & heat) 热电联产,热电并供
cogged belt 齿形带,同步传动带

cogged belt drive 齿形带传动
cog-wheel coupling 齿形联轴节,离合联轴节
cohere 黏结,黏合
coherence(＝coherency) 相干性;内聚现象;内聚力;黏结
coherent 黏附的;黏结的
cohesible 可黏合的;可黏结的
cohesion 内聚[现象];内聚力;黏合
cohesion energy 内聚能
cohesive 黏的;内聚的
cohesive-energy density 内聚能密度;聚集能密度;黏结能密度
cohesive force 黏合力;内聚力
cohesiveness 黏结性;内聚性
cohesive setting 内聚定形
cohesive strength 黏结强度;内聚强度
Cohras prints 科勒斯蜡防印花布(东南亚有色棉花制)
cohydrolysis 共水解作用
COI(critical oxygen index) 临界氧指数
coil 盘管,蛇形管;线圈《电》
coiled radiator 盘管式散热器
coil heater 盘管加热器
coil heat exchanger 蛇形管热交换器,盘管热交换器
coiling 盘旋;卷曲;盘曲
coil pipe 盘管
coil spring 圈簧,盘簧
coil-tube boiler 盘管锅炉
colcothar 英国红[色],氧化铁红[色]
cold batch bleach 冷卷漂白
cold black finish 冷黑整理(一种抗紫外线技术)
cold calendering 冷轧光
cold colours 冷色(蓝、绿色等)

cold crack temperature 低温龟裂温度,冷裂温度
cold creep 低温蠕变
cold dwell curing 冷堆焙烘
cold dwell method(＝cold dwell process) 冷堆法
cold dyeing 低温染色
cold-embossing 冷轧纹[成形工艺]
cold fix method [活性染料]冷固色法
cold flame 低温燃烧
cold mercerizing 低温丝光法(处理温度在5℃以下)
cold mordanting 冷媒染
cold pad azoic dyeing 冷轧不溶性偶氮染料染色
cold pad-batch bleaching with hydrogen peroxide 双氧水冷轧堆漂白
cold pad-batch dyeing〈CPB dyeing〉 冷轧堆染色
cold pad-batch pretreatment 冷轧堆前处理
cold pad dyeing 冷轧堆染色法
cold pattern roller 冷轧花滚筒
cold pigmentation process [还原染料]悬浮体冷染法(用于针织品)
cold plasma 冷等离子体
cold pressing [呢绒]冷压
cold print-batch process 印花冷卷固色法
cold resistance 抗低温性能
cold rinse 室温漂洗;冷水冲洗
cold roll 冷轧;冷卷
cold setting 冷固化;常温硬化
cold start 冷态启动
cold stress 冷应力
cold system optical brightener 低温荧光

增白剂
cold type reactive dye 冷染型活性染料
cold water dispersible starch 冷水分散型淀粉
cold water shrinkage 冷水收缩率
cold water shrinking [呢绒]冷水预缩
cold water shrinking machine [呢绒]冷水预缩机
cold water soap 低温洗涤剂(可用凉水洗涤)
coleoptera 甲虫
Colid 可立特(苎麻、涤纶和黏胶丝混纺色织布)
collagen 胶原蛋白(骨胶原)
collapse 密实,致密(纤维结构由疏松状变成紧密状);崩溃;瘪缩
collapse rate [泡沫]破灭速率
collapse test 破坏性试验
collar 挡圈;轴环;套管
collar bearing 环形止推轴承
collar lining 领衬;轴环衬垫
collar nut 环形螺母;凸边螺帽
collar stiffener 衣领硬挺剂
collate 依序整理;对比,校对,检验
collecting agent 促集剂,捕集剂
collecting groove 聚凝槽
collecting system 收集系统
collecting well 集水井
collection 时装展览;服装精品;收集[样]品;采集
collection efficiency 聚集效率《环保》
collection printing 印花花样汇总
collection tray 集液槽
collector 积布器;集料器;捕收剂
colligative phenomena 综合现象
colligative property 综合性质

collimated beam 平行光束
collimating lens 准直透镜
collision 碰撞
collodion 胶棉,火棉胶(可溶性火棉的溶液)
collodion embedding 火棉胶包埋法(显微镜切片术)
collodion filter 火棉胶过滤器
colloid 胶体,胶质;胶态
colloidal alumina 胶态氧化铝
colloidal antimony oxide 胶态氧化锑
colloidal dispersion 胶态分散体
colloidal dyes 胶态染料(在染液中能以超显微聚集体形态存在)
colloidal electrolyte 胶态电解质
colloidal particle 胶粒,胶体微粒;胶态质点
colloidal particulate silica 二氧化硅胶态微粒
colloidal property 胶体性质
colloidal solution 胶体溶液
colloidal suspension 胶态悬浮体
colloid chemistry 胶体化学
colloid mill 胶体研磨机,胶体磨
colloid theory of dyeing 染色胶体理论
Colloresin process 科洛雷辛工艺(两相还原染料印花法)
collosol 溶胶
collotype 珂罗版(印刷制版技术)
colophony 松香,松脂
color 【美】= colour
colour 颜色;颜料;染料
colourability 上染性,染色性
colour abrasion 色磨损
colour accent 缀色增亮(用少量明亮的色彩克服图案的暗淡感)

colour accreditation scheme〈CAS〉色彩认证计划
colour acid 色酸
colour analyser 颜色分析仪
colour analysis display computer〈CADC〉颜色分析显示计算机
colour anion 色阴离子
colourant 着色剂,染色剂
colourant 色素;染料;颜料;着色剂
colourant layer 染色层
colourant-light stabilizer 色料光稳定剂
colourant mixture computer〈COMIC〉配色计算机
colourant paper 色素纸(亦称碳素纸,凹版制版法中用)
colour appearance system 色表系统
colour association 色的联想,色的联系
Colour Association of the United States〈CAUS〉美国色彩协会
colour assortment 色的组合
colouration 色彩使用,配色;染色;着色;彩色[度]
colour atlas 色度表,色卡本,色卡簿
colour axis 原色轴《测色》
colour balance 色平衡
colour bar 色档
colour base 色基;色素碱基(碱性染料的发色母体)
colour bleeding 渗色
colour blend yarn 混色纱
colour blindness 色盲
colour block 色轮,色印;墨印装置
colour bodies 发色体
colour boiling room 调色间,配色间
colour box [印花机的]色浆盘
colour build-up 染料积聚;染料色泽提升
colour cards 染色样本
colour cast [带有某种]色光
colour cation 色阳离子
colour change 变色
colour change effect 变色效应
colour change printing 变色印花
colour characteristics 色彩特性[曲线]
colour chart 色卡,彩色图表,比色图表
colour chip 色卡;色切片《测色》
colour chlorine fastness 耐氯色牢度
colour circle 色环
colour clarity 色泽清晰度
Colour Clear™ effluent treatment system 还原剂和催化剂双组分化学处理印染废水系统(商名,美国制)
colour coat 着色剂涂布;着色剂涂层
colour code 色标,色码
colour combinations 配色,拼色
colour comparator 比色仪
colour competence 色彩识别能力
colour computer 配色电子计算机
colour concentrate 浓色体,浓色物料;染料浓缩液
colour conditioning 色彩调节(指环境色调布置以减少疲劳,增加安全等)
colour confidence 色彩可靠性
colour consistency 印花色浆稠度;色泽一致性
colour constancy 色恒定性
colour container washing machine 浆桶洗涤机
colour content 纯色含量
colour contrast 色对比,色对比度
colour control 色样控制;[工场环境]色调管理

colour control & management system 〈CCMS〉色彩控制管理系统
colour coordination 色彩调和;颜色坐标
colour correction 颜色校正
colour couplers 彩色偶合剂(彩色照相术)
colour craft process 段染
colour data processing system 颜色数据处理系统(由测色仪、软件及个人计算机等组成)
colour density 色饱和度
colour depth 色浓度,色深度
colour depth of waste water 废水色度《环保》
colour designation 色的表示
colour design console〈CDC〉颜色花样图案控制台
colour detection 测色
colour detector 颜色探测器
colour developers 彩色显色剂(彩色照相术)
colour development 显色,发色
colour deviation 色差;色偏移
colour difference 色差(指肉眼可识别的色泽差异),色光不符
colour difference formula 色差公式
colour difference meter 色差计
colour discharge 着色拔染
colour dispensary 配色间,染料间
colour distortion 色彩失真
colour doctor 色浆刮刀,刮浆刀
colour dripping 滴浆(印疵)
coloured 有色的,着色的
coloured band 色谱带
coloured contaminant 有色污染物;有色杂质
coloured cotton 彩棉
coloured discharge printing 着色拔染印花
coloured fibre 有色纤维
coloured goods 色布,花布
coloured resist printing 着色防染印花
coloured spot 色斑,色花(疵点)
coloured stripe 花色条带;彩条
coloured-white effect 染色留白效应
coloured woven cloth 色织布
coloured woven goods 色织[织]物
colour effect 配色效应;色彩效应
colour embossing [凹凸花纹]着色轧印
colour equation 色[等价]方程式,色方程
Colour Eye 滤色片测色仪(商名,美国制)
colour fading 褪色
colour fastness 染色[坚]牢度
colour fastness figures 染色牢度等级
colour fastness grading 染色牢度评级
colour fastness number(= colour fastness rating) 染色牢度等级
colour fastness to acid boiling 耐酸煮色牢度
colour fastness to acid felting 耐酸毡缩色牢度
colour fastness to acid milling 耐酸缩绒色牢度
colour fastness to acid perspiration 耐酸性汗渍色牢度
colour fastness to acids 耐酸色牢度
colour fastness to acid spotting 耐酸滴色牢度
colour fastness to air 耐空气色牢度
colour fastness to air pollution 耐空气污染色牢度

colour fastness to alkaline milling 耐碱性缩绒色牢度,耐碱性毡缩色牢度

colour fastness to alkaline perspiration 耐碱性汗渍色牢度

colour fastness to alkalis 耐碱色牢度

colour fastness to alkali spotting 耐碱滴色牢度

colour fastness to aqueous agents 耐水剂色牢度

colour fastness to atmospheric contaminant 耐大气污染色牢度

colour fastness to atmospheric gases 耐空气色牢度

colour fastness to bleaching agents 耐漂白剂色牢度

colour fastness to bleaching with chlorine 耐氯漂色牢度

colour fastness to bleaching with hypochlorite 耐次氯酸盐漂白色牢度

colour fastness to bleaching with peroxide 耐过氧化物漂白色牢度

colour fastness to bleaching with sodium chlorite 耐亚氯酸钠漂白色牢度

colour fastness to bleeding 耐渗色牢度

colour fastness to blocking 耐布层粘连沾色牢度(未干透印花布打卷后被印浆粘住);耐布层粘连色牢度

colour fastness to blooming 耐晕色牢度

colour fastness to boiling 耐沸煮色牢度

colour fastness to boiling alkalis 耐碱煮色牢度

colour fastness to boiling off 耐脱胶色牢度,耐精练色牢度

colour fastness to boiling soap 耐皂煮色牢度

colour fastness to brightening 耐增艳色牢度,耐艳化色牢度;耐上油处理色牢度;耐增柔处理色牢度《毛》

colour fastness to brush-washing 耐刷洗色牢度

colour fastness to burnt gas fume 耐烟熏色牢度

colour fastness to carbonizing 耐炭化色牢度

colour fastness to carbonizing with aluminium chloride 耐氯化铝炭化色牢度《毛》

colour fastness to carbonizing with sulphuric acid 耐硫酸炭化色牢度

colour fastness to carrier 耐载体色牢度

colour fastness to cement 耐黏合色牢度

colour fastness to chafing 耐摩擦色牢度,耐磨损色牢度

colour fastness to chemicking 耐漂色牢度,耐氯漂色牢度,耐氯漂色牢度

colour fastness to chlorinated acid 耐氯酸色牢度

colour fastness to chlorinated water 耐含氯水色牢度

colour fastness to chlorination 耐氯化色牢度

colour fastness to chlorine 耐氯色牢度

colour fastness to chlorine bleaching 耐氯漂色牢度

colour fastness to chlorite bleaching 耐亚氯酸钠漂白色牢度

colour fastness to chromium salt in dye bath 耐染浴中铬盐色牢度

colour fastness to cold washing 耐冷水洗涤色牢度,耐冷洗色牢度

colour fastness to crabbing 耐煮呢色牢度,耐热水色牢度

colour fastness to crocking 耐摩擦色牢度

colour fastness to cross dyeing 耐交染色牢度

colour fastness to curing 耐焙烘色牢度，耐焙固色牢度，耐干热定形色牢度

colour fastness to daylight 耐天然光色牢度

colour fastness to decatizing 耐蒸煮色牢度

colour fastness to degumming 耐脱胶色牢度

colour fastness to discharge 耐拔染色牢度

colour fastness to dry cleaning 耐干洗色牢度

colour fastness to dryheat 耐干热色牢度

colour fastness to fadeometer exposure 耐褪色试验仪暴晒色牢度

colour fastness to finishing operations 耐整理工艺色牢度，耐整理色牢度

colour fastness to formaldehyde 耐甲醛色牢度

colour fastness to fulling 耐缩绒色牢度

colour fastness to fume fading 耐烟熏色牢度

colour fastness to gas fumes 耐烟气色牢度

colour fastness to hand washing 耐手洗色牢度

colour fastness to heat 耐热色牢度

colour fastness to heat setting 耐干热定形色牢度；耐焙固色牢度

colour fastness to home washing 耐家洗色牢度

colour fastness to hot pressing 耐热压色牢度

colour fastness to hot water 耐热水色牢度

colour fastness to hypochlorite bleaching 耐次氯酸盐漂白色牢度

colour fastness to iron and copper in dye bath 耐染浴中铁和铜色牢度

colour fastness to ironing 耐熨烫色牢度

colour fastness to kier boiling 耐煮布色牢度

colour fastness to laundering 耐机械洗涤色牢度

colour fastness to light 耐光色牢度

colour fastness to light and washing 耐光及耐洗色牢度

colour fastness to light and weathering 耐光和气候色牢度

colour fastness to light through glass 耐透过玻璃的光色牢度

colour fastness to lime 耐石灰色牢度

colour fastness to lime soap 耐碱皂色牢度，耐石灰皂洗色牢度

colour fastness to marking off 耐搭色牢度；耐压烫色牢度，耐熨烫色牢度；耐涂层热烫搭色牢度

colour fastness to mercerizing 耐丝光色牢度

colour fastness to metals in dye-bath 耐染浴中金属色牢度

colour fastness to migration 耐泳移色牢度

colour fastness to milling 耐缩绒色牢度

colour fastness to nitrogen oxides 耐氧化氮色牢度

colour fastness to normal use 耐［一般使］用色牢度

colour fastness to oil 耐油色牢度
colour fastness to organic solvent 耐有机溶剂色牢度
colour fastness to overprinting 耐罩印色牢度
colour fastness to ozone in atmosphere 耐大气中臭氧色牢度
colour fastness to perchloroethylene 耐全氯乙烯色牢度
colour fastness to peroxide bleaching 耐过氧化物漂白色牢度
colour fastness to peroxide washing 耐过氧化物水洗色牢度
colour fastness to perspiration 耐汗渍色牢度
colour fastness to planking 耐酸缩绒色牢度
colour fastness to pleating and fixation in dry heat 耐干热褶裥、定形色牢度
colour fastness to potting 耐沸水色牢度
colour fastness to preliminary light 耐初期日晒牢度
colour fastness to pressing 耐压烫色牢度
colour fastness to processing 耐加工色牢度
colour fastness to rain 耐雨水色牢度
colour fastness to reduction for aniline black 耐苯胺黑还原色牢度
colour fastness to rubbing 耐摩擦色牢度
colour fastness to rubbing off 耐擦净色牢度,耐摩擦色牢度;[涂层]耐磨色牢度
colour fastness to rubbing with organic solvent 耐有机溶剂中摩擦牢度
colour fastness to saliva 耐唾液色牢度
colour fastness to salt water 耐盐水色牢度,耐海水色牢度
colour fastness to scouring 耐煮练色牢度
colour fastness to sea water 耐海水色牢度
colour fastness to shampooing 耐洗发剂色牢度,[地毯]耐洗剂色牢度
colour fastness to sizing 耐上浆色牢度
colour fastness to soaping 耐皂洗色牢度
colour fastness to soda boiling 耐碱煮色牢度
colour fastness to sodium hypochlorite 耐次氯酸钠色牢度
colour fastness to solvents 耐溶剂色牢度
colour fastness to solvent spotting 耐溶剂液滴色牢度
colour fastness to spotting 耐液滴色牢度
colour fastness to steaming 耐汽蒸色牢度
colour fastness to steam pleating 耐汽蒸褶裥色牢度
colour fastness to storage 耐储存色牢度
colour fastness to stoving 耐硫熏色牢度
colour fastness to sublimation 耐升华色牢度
colour fastness to sulphurous acid 耐亚硫酸色牢度
colour fastness to sunlight 耐日晒色牢度
colour fastness to swimming bath water 耐游泳池水色牢度
colour fastness to tetrachloroethylene 耐四氯乙烯色牢度
colour fastness to topping 耐套染色牢度,耐交染色牢度
colour fastness to trichloroethylene 耐三

氯乙烯色牢度
colour fastness to vulcanization 耐硫化色牢度
colour fastness to washing 耐洗色牢度
colour fastness to water 耐水浸色牢度
colour fastness to water drops 耐水滴色牢度
colour fastness to water impregation 耐水浸色牢度
colour fastness to water spotting 耐水滴色牢度
colour fastness to weathering 耐气候色牢度
colour fastness to wet processing 耐湿加工色牢度
colour fastness to wool dyeing 耐羊毛染色色牢度
colour fatigue 色疲劳《测色》
colour fibre migration 毛毯透色(毛毯疵点)
colour filter 滤色器,滤色镜
colour fixing 固色处理
colour fixing agent 固色剂
colour fly 飞花混入(织疵)
colour formulation 配色
colourful 颜色丰富的,多色的
colour furnisher 印花给浆滚筒
colour furnishing roller 印花给浆滚筒
colour gamut [全]色域
colour-glittering fabrics 闪色织物
colour-glittering printing 闪色印花,闪烁印花
colour gradation 色调层次
colour grade(=colour grading)［孟塞尔制］色彩分级
colour grinding machine 染料研磨机

colour halo 色晕
colour harmony 色［和］谐,色彩调和,配色协调
Colour Harmony Manual〈CHM〉 色彩调和手册,配色协调手册(美国)
colourimeter 比色计,色度(差)计
colourimeter tube 比色管
colourimetrical analysis 比色分析［法］
colourimetric detector tube 比色法检测管
colourimetric estimation 比色定量
colourimetric method 比色法
colourimetric purity 测色纯度;色度纯度
colourimetric shift 测色的色移位;色度位移
colourimetric specification 色度规格
colourimetric test 比色试验
colourimetry 比色法;色度学,测色学
colour in artificial light 人造光源色彩
Colour Index〈CI、C. I.〉《染料索引》(书名);颜色指数,比色指数
colour index number 染料索引号数
colour in fashion textile 纺织品流行色
colouring 着色
colouring material(= colouring matter) 着色剂;染料;色素
colouring power 着色力,染色力
colour inlay effect 凹纹着色效应
colour intense dispersion 染透分散作用,着色分散作用
colour intensity 颜色强度
colourist 印染工作者;配色技师
colour kitchen 配色间
colour knife 色浆刮刀,刮浆刀
colour lacking uniformity 色不匀,色缺

乏均匀

colour lake 色淀
colourless 无色的
colourless dyestuff 无色染料(如荧光增白剂)
colour lift-off [筛印]色转移
colour location 颜色定位
colour management capability 色彩管理能力
colour management system〈CMS〉 自动颜色管理系统
colour manipulation 色彩管理
colour masking 色彩蒙版修正
colour matching 比色;仿色;[照来样]配色
colour matching cabinet 测配色用标准光源箱
colour matching computer 配色电子计算机
colour matching fluorescent lamp 配比色用荧光灯
colour matching function 配色函数
colour matching system〈CMS〉 配色系统
colour measurement 测色
Colour Measurement Committee〈CMC〉颜色度量委员会
colour measuring 测色
colour measuring instruments 测色仪
colour measuring system 测色系统
colour migration on drying 烘干中染料泳移
colour mixer 混色仪,混色器
colour mixing machine 调浆机(调色设备)
colour mixing room 调色间,配色间

colour mixtures 混色
colour mixture triangle 拼色三角形
colour modifier 色的修饰语
colour monitor 彩色图案监控器
colour name 色名
colour notation 表色法
colour number 色数(表示颜色浓淡的数值)
colour on colour 重色配色法(彩色物体上再上颜色)
colour original 彩色原稿
colour out 脱浆(印疵)
colour overlapping 搭色
colour palette 调色板,色谱
colour pan 调色浆锅,打浆锅
colour paste 色浆
colour paste for pigment printing 涂料印花用色浆
colour paste strainer 印浆过滤器,滤浆机
colour patch 色渍(染疵)
colour pencil 测温笔(随温度变色)
colour perception 色知觉
colour picture processing 色图像处理
colour planning 色彩计划,色彩设计
colour position calculation 色位[印浆]计算(通过计算机计算印花残浆组分达到回用的科学方法)
colour printer 彩色打印机
colour printing with different type of dyes in one paste 各种染料同浆印花
colour producing reagent 显色试剂
colour profile 颜色分布(图像处理)
colour proofing system 着色试样系统
colour properties 颜色性质
colour quality control 颜色质量控制

colour quality control system 颜色品质控制系统
colour range of indicator 指示剂变色范围
colour reaction 显色反应
colour recognition 颜色识别
colour rendering 演色性;显色性
colourrendering index 显色(性)指数;演色性指数(照明光源与标准日光的指数对比)
colour rendition 色重演作用,彩色再现
colour repairing agent 修色剂
colour reproduction 色复现,色再现[作用]
colour reproductivity 色的再现性
colour resist[printing] 着色防染印花
colour resist over printing 防印印花
colour retention 保色性
colour reversal film 彩色反转胶片
colour roller 印花滚筒
colour rotation 色顺(滚筒印花时雕刻花筒的排列顺序)
Colour Rules [海门丁格]颜色标尺
colour safe dyeing system 可靠的染色系统;保险的染色系统
colour sample 色样
colour sampling device 打样装置
colour saturation 彩度;饱和度
colour scale 颜色标准级别,颜色分度
colour scales Lab 颜色标度 Lab(L＝亮度,a＝绿轴线,b＝黄蓝轴线)
colour scanner 彩色扫描机
colour scheme 配色方案
colour screen 滤色镜
colour sector analysis 颜色分类分析
colour sensation 色感觉

colour sensitivity 感色灵敏度
colour sensitizer 彩色感光剂
colour separate photograph 分色照相
colour separation 色分离;分色
colour separation negative 分色负片(胶片)
colour separation system 分色系统
colour seperation and film make 分色制版《印》
colour shade 色光,色调,色泽
colour shade difference 色光偏差
colour shifting 色移动;色转移
colour shop 调色间,配色间
colour sieving unit 染料溶液过滤器,滤[色]浆器
colour simulator 色彩模拟器
colour smear 拖浆,拖毛(印疵)
colour soiling 沾色
colour solid 色立体(色点立体表示法)
colour space 色空间
colour specification 色的表示法,颜色表示
colour speck 色斑
colour splashes 拖浆,溅浆
colour spray process 喷色工艺
colour spun dyes 纺前着色染料
colour stability 色素稳定性
colour staining 色斑,色渍
colour standards 色样标准
colour stimulus 色刺激
colour stimulus function 色刺激函数
colour strainer 滤[色]浆器
colour streak 色条
colour strength 颜色深度;颜色力份,颜色强度
colour strength of dyes 染料强度,染料

力份
colour stripper 剥色剂
Colour Strip^TM 染料脱色系统(用于染色织物的剥色处理,商名,美国制)
colour sublimation 染料升华
colour system 表色系统
colour temperature 色温[度]
colour terminology 颜色术语
colour tolerances 色公差;颜色宽容度
colour tone 色调
colour trailing 拖色
colour transfer belt 色转移带(用于毛条印花)
colour transference 色转移
colour tree 色树(孟塞尔色立体)
colour triangle 原色三角形
colour trough 印浆槽,色浆盘
colour value 色值(用不同染料染得同样色泽,所费价值之比);给色量,得色量
colour video screen 彩色影像荧光屏
colour vision 色视觉
colour volume 色浆用量(印花用语)
colour wash fastness 耐洗色牢度
colourways [配色]色位
colourway simulator 配色色位重现装置;花样重现装置
colour weave 色织
colour working 调色
colour-woven 色织的
colour-woven-like printing process 仿色织印花工艺
colour[ed] yarn fabric 色织物
colour yield 得色量,给色量
column 柱,架;塔;列;[表格]栏
column adsorption chromatography 柱吸附色谱法;柱吸附色层分离法
column chromatography 柱型色层分离法
column development chromatography 〈CDC〉柱展开色谱法
column type biofilter 塔式生物滤池《环保》
column type metering stations 塔式计量站
colza oil 菜籽油
Combi Jigger 康比高温高压卷染机(全部计算机控制,商名,意大利制)
combinability 配伍性,相容性
combinable dyestuff 配伍性染料
combination 化合[作用];混合;联合;连衫裤
combination bearing 组合轴承
combination bleach 联合漂白
combination chain 针板布铗两用链
combination desize/dyeing 退浆、染色联合工艺
combination desize/scour process 退浆精练一浴法
combination dyeing 拼色染色(用一种以上的染料染色)
combination dyes 拼混染料
combination fabric 混合织物;混纺织物
combination milling 联合缩呢工艺
combination of tones 色调的组合
combination process 联合工艺
combination shades 拼色色泽;拼色样卡
combination starter 综合启动器《电》
combination steamer [织物精练用]组合式汽蒸机,复合式汽蒸箱
combination type milling machine 联合缩绒机(精练与缩绒同时进行)

combined brush 组合式刷绒

combined brush and sueding 组合式刷绒和[仿麂皮]起绒

combined brush, sueding and emerizing 组合式刷绒、[仿麂皮]起绒和金钢砂起绒

combined cellulase units〈CCU〉混合纤维素;酶活力单位

combined chlorine-peroxide bleach 氯氧联合漂白

combined cooling and rinsing〈CCR〉洗涤和冷却同步进行(高温高压染色法)

combined dyeing and centrifugal equipment 染色兼离心设备

combined dyeing and finishing 染色整理一浴法

combined fabric 复合织物

combined filament yarn 混纤丝,混合长丝纱(不同品种的长丝均一混合后所成的复丝)

combined formaldehyde 结合甲醛

combined heat & power〈CHP〉热电联产

combined hot and cold pressing machine 冷热压呢联合机

combined open-width and rope washing range 平幅绳状联合洗涤机

combined package 拼件成包

combined pieces and combined colours 拼匹调色,拼色拼件

combined pin and clip chain 针板布铗两用链

combined pin and clip stenter 针板布铗两用拉幅机

combined pretreatment 短流程前处理

combined process of antifelting finish 混合法防缩整理

combined raising and emerizing 组合式起绒和金钢砂起绒

combined raising and shearing 组合式起绒和剪绒

combined resistance 复合耐力

combined roller and conveyer steamer 导辊及履带复合汽蒸箱(蒸箱上部为由导辊组成的加热区,下部为由平板履带组成的反应区)

combined scouring and milling machine 洗呢缩呢联合机

combined sewage 混合污水,合流污水

combined stage bleaching 联合漂白法,联漂

combined vacuum saturator and steaming box 真空浸渍及汽蒸联合机

combined waste water treatment 复合废水处理

combining 层压,胶合

Combi Soft 康比多功能柔软剂(意大利制)

Combi-Squeegee 康比混合刮刀,康比混合刮浆器

Combi-Washer 康比复合式洗涤机(绳状平幅联合洗涤机,商名,意大利美齐拉)

combustibility 可燃性

combustible 可燃的,易燃的

combustible fibre 可燃纤维

combustible mixture 可燃混合气(空气和燃料按一定的可以燃烧的比例混合)

combustible waste 可燃烧性废弃物《环保》

combustion 燃烧
combustion air fan 助燃风机
combustion behaviour 燃烧特性
combustion drying 燃烧烘干
combustion efficiency 燃烧效率
combustion index 燃烧指数
combustor 燃烧室
Come Clean 克姆克林防污织物(商名)
come off 褪[红]色
comformability 舒适性；顺应性
comformity [色泽]一致性
comfort 穿着舒适；盖被(美国名称)
comfortable 舒适的；盖被(美国名称)
comfortable finishing 舒适整理
comfortable fit 舒适合身
comfort durability [服用]舒适耐久性
comfort fabrics 舒适性织物
comfort factor〈CF〉舒适度
comfort factor value〈CF value〉舒适指数
comfort finish 舒适整理
comfort stretch apparel 舒适弹力服装
comfort stretch fabric 舒适弹力织物
comfort wear 穿着舒适
COMIC(colourant mixture computer) 配色计算机
command 指令；控制
command code 指令码，操作码
commander switch 主令开关
commercial continuous gravure printing machine 商用印纸连续凹版印花机，商用连续照相凹版印刷机(可用于印转移印花纸)
commercial fastness properties 商业[通用]色牢度
commercial fibre 商品纤维

commercial finish 工业整理，商品化整理
commercialized dyes 商品化染料
commercial mass 商业重量
commercial print 广告花样印花(常印在圆领针织短袖衫上)，商用印花
commercial regain 商业用回潮率
commercial residual shrinkage 商业公定残余缩水率
commercial size 工业规模，商品化规模
commercial standard〈CS〉商业标准
commercial trade mark 商标
commercial transfer printing paper 商品转印纸
commercial use 商业用途
commingling 掺混，掺和
comminution 磨碎
commission 调试
commission baking 代客加工烘焙
commission dyeing 代客染色
commission dyer 代客加工染色厂
commission finisher 代客加工染整厂
commissioning 投产，运转，正式使用，委托加工
commission printing factory 代客加工印花厂
commission wash 代客加工洗涤
commodity 商品
commodity value 商品价值
common bevel gear 直齿伞形齿轮
common impression cylinder 普通承压辊
common ion effect 共同离子效应
common parts 通用零件
common salt 常用盐(氯化钠、硫酸钠)，食盐

common stack 总风道
Commonwealth Scientific and Industrial Research Organization〈CSIRO〉 联邦科学与工业研究组织（澳大利亚）
communication 通信
communication error 通信错误
commutativity 可交换性；可互换性
commutator 整流器；换向器
commutator-less motor 无换向器电动机，无刷电动机
commutator motor 三相交流整流子电动机，整流子电动机
compact 紧凑的；小型的，小巧，袖珍的；密致的；结实的
compact cloth 紧密织物（指相邻纱线间距离小于纱线直径）
compact coating 致密涂层，坚固涂层
compact disk-read only memory〈CD-ROM〉 只读光盘存储器《计算机》
compact dyeing system 紧凑染色法，快速吸尽染色法（缩短染色时间的染色法）
compact handle 坚实手感
compacting ［织物］热压预缩，压缩加工，紧凑加工
compacting calender 热压预缩机（针织品用）
compacting drying condition 致密干燥条件（使纤维密实）
compacting machine 挤压式预缩机《针》
compacting method 挤压预缩法
compaction resistance 压实阻力
compactness 紧密性，紧密度，密实度，充实度
compactor 挤压式预缩机（用于加工针织物）

compact package dyeing machine 紧凑型筒子纱染色机
compact-type jet dyeing machine 筒式喷射染色机［机器外形呈圆筒状，浴比小，为1∶(4～7)］
company standard 企业标准，公司标准
comparable equalibrium exhaustion 相对平衡吸尽
comparative clothing 比较衣着学
comparative dyeing 对比染色
comparative test 比较试验，对比试验
comparator 比长仪，比较仪
compartment 分隔间；格；箱；室
compartment drying machine 分段烘干机
compatibility 相容性，可混用性，配伍性
compatibility index 相容性指数
compatibility of dyes 染料配伍性
compatibility value〈CV〉 相容性值
compatibilizer 配伍剂
compatible 可配伍的，相容的；适合的
compatible dyes 配伍［性］染料；可混用的染料
compatible index 配伍指数
compatible shrinkage 相容收缩（指黏合织物的面布和底布或服装的面料、衬料以及辅料具有相等的收缩程度）
compensated induction motor 补偿感应电动机
compensated repulsion motor 补偿排斥式电动机
compensating creel 补偿式轴架
compensating device 补偿机构
compensating fabric feed 进布补偿器
compensating lattice 补偿帘子
compensating roller 调整辊，调节辊，张

力辊
compensating unit 补偿装置,松紧架
compensation 补偿;补偿器
compensation gear 补偿装置;差动齿轮
compensation inlet 补风口
compensator 松紧架(织物张力调节装置),张力辊,升降辊;补偿器
compensator-controlled D. C. drive motor 补偿器控制的直流电动机
compensator dancing roller 松紧调节辊,张力辊(织物张力调节装置)
compensatory motion 补偿运动
competing dyeing(＝competitive dyeing) 竞染
complement 余,补;配套;定员,编制人数
complementary chromaticity coordinates 补色色度坐标《测色》
complementary chromaticity diagram 补色色度图标《测色》
complementary colour 互补色,补色
complementary dominant wavelength 补色主波长
complementary error 附加误差
complementary wavelength 补色波长
complete combustion 完全燃烧
complete design 完全图案
completely washable fabric 可充分洗涤的织物(用 71.7℃ 热水机洗而不过分缩水或变形变色)
complete treatment 完全处理(如污水处理,包括凝结、沉淀、过滤及消毒)
complex 络合的;络合物;络合基
complex builder 络合剂;络合组成
complex catalyst 络合催化剂,综合催化剂

complex coacervation 复凝聚法
complex colour harmony 复式配色,不调和的配色方法(不同于常规配色)
complex compound 络合物
complex dye 复合染料
complex enzyme 复合酶
complex fibre 复合纤维
complex film 复合薄膜
complex formation 络合物构成(多价螯合作用)
complex-forming ability 络合能力
compleximetry 络合滴定
complex indicator 络合指示剂
complexing 络合化
complexing agent 络合[物形成]剂
complex ion 配离子,络离子
complex pollution 复合污染
complex salt 络盐
compliance 顺应性,符合度
component 元件,构件;组成部分;成分
component assembly 零件装配
composite 复合材料
composite bonding 复合键合(静电键合与共价键合)
composite bonding theory 复合[材料]黏合理论
composite buoyancy materials 复合浮力材料
composite dye 拼合染料,复合染料
composite enzyme 复合酶
composite fibre 复合纤维
composite micro fibre 复合超细纤维
composite specimen 组合试样
composite yarn 包芯纱,复合纱
composition 组成,构成;构图(花纹设计)

composition analysis 成分分析,组成分析
composition cloth 防水帆布;多层复合布
composition doctor 小刀,铜[刮]刀(用以刮除印花机花筒上的棉纱、棉絮、棉屑等带浆杂质)
composition fibre 复合纤维
composting 堆肥《环保》
compound bearing 组合轴承
compound cloth 多层织物
compound colour 混合色;调和色
compound dyeing 拼色染色(用混合色染料染色)
compound enzyme 复合酶
compound fabric 复合织物,多层织物
compound indicator 复合指示剂
compounding 配料
compounding ingredient 配料成分;掺合剂
compounding ratio 配料比
compound milling machine 双滚筒缩绒机
compound motor 复激电动机
compound shade 复合色,拼色
compound-wound D.C. motor 复励式直流电动机
comprehensive starter 综合起动器
comprehensive utilization 综合利用
compress 压缩;打包机
compressed air 压缩空气
compressed air hydro-extracting 加压空气脱水
compressed air valve 压缩空气阀
compressed cotton bowl 棉纤维轧辊,花衣辊筒
compressed elastic recovery 压缩弹性回复《试》
compressed linen bowl 麻纤维轧辊
compressed paper bowl 纸粕轧辊
compresser 压缩机
compressibility 压缩性;压缩系数
compressibility ratio 压缩率《试》
compression 压缩
compressional resilience 压缩回弹[性]
compressional resistance 耐压性
compression and recovery test 压缩和回复试验
compression bandage 压缩性绷带
compression deformation 压缩变形
compression elasticity 压缩弹性
compression fatigue tester 压缩疲劳试验仪
compressionmeter 压缩试验仪
compression ratio 压缩比
compression resilience 压缩回弹[性]
compression roll 压[液]辊
compression spring 压缩弹簧,压力弹簧
compression stiffness 压缩刚性
compression strength 抗压强度,耐压强度
compression stress relaxation 压缩应力松弛
compression test[ing] 耐压试验
compression type cooler 压缩型冷却器
compressive buckling 加压弯曲
compressive elasticity 压缩弹性[率]
compressive force 压缩力
compressive shrinkage 机械预缩;机械预缩整理
compressive shrinkage machine 机械预缩机

compressive shrinkage range 机械预缩机组，机械防缩联合机

compressive strength 抗压强度，耐压强度

compressive stress 压缩应力

compressive stuffing 压缩填充

compressor 压缩机

compulsory operation 强制运行

computer [电子]计算机，计算装置

computer abort 计算机的异常中止

computer aided colouring〈CAC〉 计算机辅助测配色

computer aided design and drafting〈CADD〉 计算机辅助设计与制图

computer aided design〈CAD〉 计算机辅助设计

computer aided engineering〈CAE〉 计算机辅助工程管理

computer aided garment design〈CAGD〉 计算机辅助服装设计

computer aided machineshop operating system〈CAMOS〉 计算机辅助车间操作系统

computer aided manufacture〈CAM〉 计算机辅助生产

computer aided pattern〈CAP〉 计算机辅助图案设计

computer aided pattern creation〈CAPC〉 计算机辅助图案创作

computer aided pattern design〈CAPD〉 计算机辅助图案设计

computer aided production〈CAP〉 计算机辅助生产

computer aided production planning〈CAPP〉 计算机辅助生产计划

computer aided testing〈CAT〉 计算机辅助测试

computer aided textile design〈CATD〉 计算机辅助纺织品设计

computer aided textile printing〈CATP〉 计算机辅助纺织品印花

computer code 计算机代码

computer colour matching〈CCM，C. C. M.，ccm〉 [电子]计算机配色[法]

computer colour matching system 计算机配色系统

computer colour searching〈CCS〉 计算机色样检索

computer colour selector〈CCS〉 计算机选色装置

computer colour separating 计算机分色描稿

computer control [电子]计算机监控

computer controlled continuous chemical metering 计算机控制化学品连续计量

computer controlled dyestuff injection system 计算机控制染料注射装置

computer data processing 计算机数据处理

computer graphic and image processing〈CG/IP〉 计算机图形/图像处理技术

computer graphics〈CG〉 计算机图形学

computer intergrated manufacturing〈CIM〉 计算机集成制造

computerized batch process 计算机化间歇式生产工艺

computerized colour matching 计算机配色

computerized colour measure system（=**computerized colour measuring system**）计算机化测色系统

computerized colour separation system 计算机化分色系统
computerized control 计算机控制
computerized dispenser system〈CDS〉计算机化分配器系统,自动液体与粉末分配系统
computerized dyeing 计算机化染色工艺
computerized engraving equipment 计算机控制的雕刻设备
computerized fabric designing 计算机化织物设计
computerized high-resolution scanner 计算机化高分辨能力的扫描器
computerized jet-printing machine 计算机喷墨印花机,数码喷墨印花机
computerized monitoring system 计算机监测系统
computerized numerical control〈CNC〉计算机数字化控制
computerized process control 计算机化工艺控制
computerized screen printing carriage 计算机控制筛网印花走车
computerized searching system 计算机检索系统
computerized system 计算机系统
computerized weighing, labelling and packing system 计算机化称重、贴商标及包装系统
computer language 计算机语言
computer match prediction〈CMP〉计算机配色预测
computer memory 计算机存储器
computer micro meter 计算机测微仪
computer point 计算机布点(圆网孔排列方式)
computer point screen 计算机布点的圆网(用于点布涂层技术)
computer processor 计算机信息处理机
computer program 计算机程序
computer programming 计算机程序设计
computer simulation 计算机模拟
computer simulation for dyeing process 计算机模拟染色过程
computer technology 计算机技术
computer tomography〈CT〉计算机 X 射线断层照相术
computer virus 计算机病毒
computer vision system 计算机视觉系统(测定纤维色泽用)
concave 凹面;凹的
concave roll 中凹辊(用于凹凸辊整纬器校正弓纬疵点)
concave surface 凹曲面
concave weld 凹焊缝
Concavex roller 凹凸辊(一种整纬装置,可整 S 形弓纬等疵点,德国 Krantz)
concentrate 浓缩,蒸浓;浓缩物
concentrated 浓的,浓缩的;集中的
concentrated load 集中载荷
concentrating solar power〈CSP〉聚焦型太阳能热发电
concentration 浓度;浓缩
concentration cell 浓差电池
concentration coefficient of diffusion 扩散[的]浓度系数
concentration coefficient of sedimentation 沉积[的]浓度系数
concentration dependence 浓度依存性
concentration diffusion 浓差扩散

concentration distribution 浓度分布
concentration fluctuation 浓度涨落
concentration gradient 浓度梯度
concentration of limit of oxygen for combustion〈CLOC〉 燃烧限氧指数
concentration potential 浓差电势
concentration profile 浓度分布曲线
concentrator 浓缩器
concentric adjustable bearing 同心可调轴承
concentric cylinder measuring system 同心圆筒测定法（黏度测定）
concentricity 同心度《机》
concept of sustainable development 可持续发展概念
conch shell 深藕红[色]
concordance 一致，协调；索引；便览
concrete grey 水泥灰[色]
concurrent process 顺流式工序；顺流加工法
condensable 可缩合的，可凝聚的
condensate 冷凝水
condensate pump 冷凝泵
condensate return 冷却水回收
condensate water 冷凝水
condensating point 雾点
condensation 冷凝；凝聚；缩合
condensation copolymerization 共缩聚作用
condensation heat 凝结热，冷凝热
condensation heater 凝结加热器
condensation polymer 缩[合]聚[合]物
condensation polymerisation 缩聚作用
condensation reaction 缩合反应
condensation resin 缩合树脂
condensation zone 凝聚区

condensed film 凝聚膜
condensed naphthalene sulfonate 凝聚萘磺酸盐
condensed phase 凝聚相，凝结相
condensed phosphate 缩合磷酸盐
condensed sulfur dyes 缩聚染料，硫化缩聚染料
condensed water 冷凝水
condense dyes 缩聚染料
condenser 冷凝器；电容器
condenser motor 电容式电动机
condenser plate 冷凝器金属板
condenser-spun 废纺；粗梳毛纺
condensing action 凝聚作用
condensing agent 缩合剂；冷凝剂
condensing steam 凝结蒸汽，冷凝蒸汽
condition 含湿量；调湿；条件
conditional matches 条件等色
conditioned 调湿的；调节的
conditioned atmosphere 公定温湿度
conditioned blast air 空调鼓风，调温调湿送风
conditioned elongation 调湿伸长
conditioned sludge feed 调节污泥喂入《环保》
conditioned weight 公量，公定回潮率重量
conditioner 空气调节器；调湿器
conditioning 调湿；调温；调整；修整
conditioning apparatus 给湿装置
conditioning fibre and textile 调温纤维和纺织品
conditioning function 调温功能
conditioning garment 调温服
conditioning machine 调节温湿度机；给湿机

conditioning of calender bowls 调节轧光辊
conditioning oven 烘箱,调温箱
conditioning rack 给湿架;调节架
conditioning tentering machine 给湿拉幅机
conditioning test 公量检验,水分检验
conditioning test oven 水分检验烘箱
conduct 传导;导管,套管
conductance 电导;热导;传导
conductance of insulating materials 绝缘材料导电性
conduction 传导;电导,导电性
conduction metry 电导[定量]学;导热率测定;导电率测定
conductive bonding agent 传导性黏着剂
conductive fabric 导电织物
conductive fibre 导电纤维
conductive heating 传导热
conductive latex 导电乳胶
conductive textile 传导纺织品(如具有去静电、抗热等效果)
conductivity 传导性,传导率;导电率
conductivity agent 导电添加剂
conductivity coefficient 传导系数,导热系数;导电系数
conductivity measurement 导电率测定
conductometric analysis 电导[定量]分析
conductometric titration 电导[定量]滴定[法]
conductometry 导电率分析
conductor 导体,导线;导条器
conduit 导管;导线管;风道;水道
cone 锥形筒子,锥形轮
cone and disc viscometer 锥盘式黏度计

cone and plate viscosimeter 锥板式黏度计
cone belt 锥轮皮带,铁炮皮带
cone calorimeter 锥形量热计
cone clutch 锥形离合器
cone compensating device 锥形补偿装置
cone coupling 锥形联轴节
cone dyeing 筒子染色
cone dyeing machine 筒子染色机
cone expander 锥形扩幅器
cone key 锥尾销
cone mounting [印花辊]锥形装置
cone plate measuring system 锥板式测定法(黏度测定)
cone pulley 锥轮,铁炮皮带盘
cone regulating device 铁炮调节装置,锥轮调节装置
cone test of water proofing 测防水性的锥形漏斗试验
cone winding 筒子卷绕机
Confederation of British Wool Textiles 〈CBWT〉英国毛纺织联合会
confetti dot 彩色点子图案
confidence level 置信水平《数》
confidence limit 〈C. L.〉置信区间,置信[界]限,信任界限
configuration 构形,形状,结构;组合,配置
configuration file 配置文件《计算机》
configuration of holes [圆网]孔的结构形态
confinement of foam 泡沫的范围(泡沫加工中泡沫粒径大小)
confirmatory reagent 确认试剂
confirm no 不确认
confirm yes 确认

conformability 顺应性;适合性;贴身性
conformation 构象;构型;适应,一致性
conformity certification 合格证书
conformity in shade 色泽符合
congeal 冻凝,凝固,冻结
congealing starch 冻凝淀粉
congelation 冻凝
conglutination 黏附作用
Congo red 刚果红(直接红色染料)
Congo red damage test 刚果红[纤维]损伤测定
Congo red test paper 刚果红试纸
conical expander 锥形扩布器
conical head rivet 圆锥铆钉
conical opening roller 锥形扩幅辊
conical seal 锥形封口
conical winding 锥形卷绕机
coniferaldehyde 松柏醛(彩棉成分)
coning oil and winding lubricants 络纱油和络筒润滑剂
conjugated chain 共轭链
conjugated compound 共轭化合物
conjugated double bond 共轭双键
conjugated fibre 复合纤维
conjugated linkage 共轭键
conjugated micro fibre 共轭型超细纤维;共轭型微纤维
conjugated structure 共轭结构
conjugated yarn 共轭丝,复合丝
connecting arm 连杆,连接臂
connecting clamp 连接夹
connecting gear 过桥齿轮
connecting link 连杆
connecting pipe 连接管,管接头
connecting rod 连杆
connection diagram 相互连接图(一套设备中不同单元之间的全部连接图)
connector 连接管,连接器,插接件
consequent-poles motor 交替磁极式电动机(有绕组极与无绕组极交替);变极式双速电动机
conservation of energy 能量守恒,能量不灭;能量节约,节能
conservative 防腐剂;防腐的
conservative colour 保守色
consistence 稠度,稠性;一致性
consistency 稠度,浓度;一致性;连续性
consistency curve 稠度曲线
consistency index 稠度指数
consistency property 相容性质
consistometer 稠度计,浓度计
console 控制台,仪表板
console control panel 落地式控制仪表盘
console panel 操纵板,控制盘
console switch 操纵开关
consolidation 凝固,硬化[作用];[污泥的]浓缩
consolidation shrinkage [织物]自然紧缩;强化收缩《针》
consolute temperature 临界共溶温度
conspicuity 能见度
constant 常数;恒量;恒定,不变
constant air conditioning 恒温恒湿
constantan 康铜(一种镍铜合金)
constant displacement pump 均匀送料泵,计量泵
constant drying condition 恒干燥,恒干燥条件
constant drying rate 等速烘燥
constant head test 恒水头试验
constant horsepower characteristic [电动机]恒定马力特性

constant pitch screw 等[螺]距螺杆
constant speed drive 恒速传动
constant speed motor 定速电动机
constant temperature and humidity 恒温恒湿,定温定湿
constant temperature dyeing 恒温染色
constant temperature method 恒温法
constant temperature process 恒温工艺
constituent 组分,成分
constitutional formula 结构式
construction 结构;作图,设计
construction fabrics 建筑用织物
constructive viscosity 结构黏度
constructive viscosity index 结构黏度指数
consulting 咨询,顾问
consulting engineer 顾问工程师
consumable parts 易耗件
consumer 消费者
Consumer Product Safety Commission 〈CPSC〉消费制品安全委员会(美国)
consumption 消费;消耗,消耗量;费用
contact 接触
contact adhesive 接触法胶黏剂
contact angle 抱合角,接触角
contact area 接触面
contact biological filter 接触生物滤池《环保》
contact combination 触点组
contact damage 接触损伤
contact drying 接触烘燥(如烘筒烘燥)
contact feeler 接触式探针
contact-free dryer 无接触烘燥机
contact-free measurement 非接触式测量
contact heat setter 接触式热定形机
contact heat thermosol unit 接触式热熔染色装置
contacting sedimentation tank 接触沉降槽《环保》
contact irritation 接触刺激
contact ledge 挡板(安全装置)
contactless 无接触
contactless dry 无接触烘干
contactless dryer 无接触烘干机
contactless method 无接触方法
contactless pickup 无触点传感器
contactor 接触器;触点
contact oxidation method 接触氧化法
contact pressure 接触压力
contact printing process 接触式印花工艺
contact probe 接触式探头
contact sensitization 接触过敏
contact singeing 接触式烧毛
contact singeing machine 接触式烧毛机
contact switch 接触开关
contact times 接触次数(筒子纱染色的染浴循环次数)
contact warmth 接触温度
contact winder 接触式卷布机
container 储布器;容器,槽,储藏器;盛袜器
contaminant 污染物,沾污物
contaminated surface 污染表面
contamination 沾染,污染
contamination accident 污染事件,污染事故
contamination precipitation 杂质的沉淀
contents 内容;含量
contexture 织物;结构;交织
N-contg. fibre (N-containing fibre) 含氮纤维

continuous automatic titrimeter 自动连续滴定仪

continuous baling press 连续打包机

continuous batching machine 连续卷布机

continuous bath 续染浴；连续浴；老脚水回用；连缸；套缸

continuous beam dyeing machine 连续经轴染色机

continuous beater 连续打浆机

continuous belt ion exchange system 连续带式离子交换装置

continuous bleaching range〈CBR〉连续漂白联合机

continuous bleaching system 连续式漂白法

continuous boil-off machine 连续煮练机

continuous boil-out and bleach 连续煮漂

continuous centrifuge hydroextractor 连续离心脱水机

continuous control 连续控制

continuous crabbing machine 连续煮呢机

continuous cure 连续焙烘

continuous curing machine 连续焙烘机

continuous decatizing 连续煮呢

continuous decatizing machine 连续蒸呢机

continuous denier reduction range 连续减量加工设备，连续式［聚酯纤维］纤度降低加工设备

continuous drop production 连续液滴喷射式［印花］（喷墨印花的一种方法）

continuous dry finishing line 连续干式整理生产线《毛》

continuous drying machine with a perfo-rated drum 连续多孔圆筒烘燥机

continuous dyeing range 连续染色联合机

continuous feed and bleed operation ［超过滤］连续喂入流出方法《环保》

continuous feeding system〈CFS〉连续加料系统

continuous filament yarn 复丝，长丝纱

continuous finishing range 连续整理联合机

continuous flushing system〈CFS〉连续冲洗系统

continuous high temperature pressure steamer 高温高压连续蒸化机

continuous ink-jet〈CIJ〉连续喷墨［法］

continuous kier 连续加工煮布锅

continuous knit mercerizing machine 针织物连续丝光机

continuous load 连续负载

continuously controlled processing technique 连续控制加工技术

continuously variable crown roller system〈CVC voller system〉连续可变锥度轧辊系统（德国蒙福茨）

continuous monitoring 连续监测

continuous open-width creping machine 连续平幅起绉机

continuous open-width washing range 连续平幅水洗联合机

continuous operation 连续操作；连续运转

continuous peroxide bleach 连续过氧化物漂白

continuous phase 连续相

continuous photoprinting machine 连续照相印花机

continuous pressure steamer 连续高压蒸化机
continuous process 连续化工序
continuous rapid dyeing equipment 连续快速染色设备
continuous relaxation machine 连续松弛机
continuous relaxing washer 连续松弛水洗机
continuous ribbon dyeing range 连续带子染色机
continuous rope washing machine 连续绳状水洗机
continuous scouring 连续煮练；连续洗涤
continuous scouring machine 连续精练机
continuous semidecatizing machine 连续半蒸呢机，连续开式蒸呢机
continuous shrinking and bulking range 连续收缩膨化联合机
continuous shrinking and drying range 连续收缩烘燥联合机
continuous sludge discharge machine 连续排泥式油水分离机《毛》
continuous solvent washing machine 连续溶剂洗涤机
continuous spectrum 连续[光]谱
continuous steamer 连续式蒸化机
continuous steaming and shrinking machine 连续汽蒸预缩机（筒状针织物的整理设备）
continuous stream〈CS〉连续液流（喷射式印花机的一种喷墨方式）
continuous tensionless openwidth washing range 连续无张力平洗机
continuous tone 连续[浓淡]色调

continuous top dyeing 毛条连续染色
continuous tow and top dyeing machine 连续毛条染色机
continuous transfer printing machine 连续转移印花机
continuous unit-pattern 连续纹样
continuous vat 连续还原浴
continuous vat dyeing 还原染料连续染色
continuous xerogel process 连续干燥凝胶法（合成纤维凝胶纺丝法）
continuous yarn printing machine 连续纱线印花机
Contipress 连续呢毯压烫机
contour 外形，轮廓；略图；恒值
contoured roller 多角滚筒，波形辊
contract carpet 配套地毯，订货地毯
contraction 收缩；收缩量；短缩；缩度
contraction percentage of area 断面收缩率
contraction temperature 收缩温度
contractive colour 收缩色
contractometer 收缩性试验仪
contracurrent(= contra flow) 逆流；对流
contrast 对照；对比[度]；反差
contrast colour 反衬色；对照色；对比色
contrast dyeing 差别化染色
contrast dyes 反差染料；差别染料
contrast effect （交织物染色的）闪色效应，双色效应
contrast gloss 对比光泽，反衬光泽
contrast grade 反差度，对比度
contrast harmony 对比的调和（配色用语）
contrast index 反差指数
contrast minus 负反差

contrast of hue 色相对比
contrast of tone 色调对比
contrast plus 正反差
contributing length 有效长度
control board 仪表板
control box 控制箱,操纵箱
control box stand 控制箱安装架
control butterflies 蝶形控制阀
control button 操纵按钮,控制按钮
control by concentration 浓度控制《环保》
control by immutable weight 总量控制《环保》
control cabinet 控制柜
control cable 控制电缆
control chart 控制图
control console 控制台,操纵台
control damper 调节挡板
control desk 控制台
control distributor [染色]调节分配器
control finger 接触指杆(用于电气或杠杆式吸边器等装置)
control flap 控制风门
control heater 调节加热器
control instrument 控制仪器
controllability 可控性
controllable technology of nano material 纳米材料的可控制备技术
controlled affinity 受控亲和力
controlled cold pad batch technology〈CCPB technology〉可控冷轧堆染色技术
controlled colouration 受控染色,受控着色
controlled compressive shrinking range 可控机械预缩机组

controlled crown rolls system 可控凸面式均匀轧辊装置
controlled extraction process 控制提取法(溶剂防缩法)
controlled humidity room 恒湿室
controlled member 被控件,控制对象
controlled plant 被控设备
controlled shrinkage annealing 控制收缩热处理
controlled-sudser (= controlled sudsing detergent) 控泡洗涤剂,低泡洗涤剂
controlled system with teleservice facility 远程服务设施控制系统
controlled variable 受控变量
controller 控制器,调节器;管理员
controller area network〈CAN〉控制器局域网《计算机》
controller pilot 调整器控制活门
control level 控制水准,控制水平
control lever 操纵杆,控制杆
control limit 控制界限
controlling mechanism 开关机构,控制机构
control loop 控制线路圈
control mechanism 控制机构,开关机构
control memory 控制储存器
control message 控制信息,控制信号
control module 控制组件
control monitor 监控器
control of cloth tension 织物张力控制
control of liquid level 液面控制
control over amount of colour applied 色浆应用量控制
control panel 操纵台,控制盘,控制屏
control relay 控制继电器
control sample 对照样品

controls and instrumentation 控制仪器
control sensor unit 控制传感装置
control signal 控制信号
control standard 控制标准
control station 控制站;主站
control switch 控制开关
control system 控制系统;管理制度
control technology 控制技术
control test 控制[性]试验,对照试验
convection 对流
convection coefficient 对流换热系数
convection dryer 对流烘燥机(如热风烘燥机)
convection drying 对流烘燥,对流烘干
convection drying process 对流烘燥法
convection heat dissipation system 对流散热装置(用于热定形机的冷却区)
convection heating 对流加热
convection heating system 对流加热体系
convection [drying] oven 对流烘箱
convective diffusion 对流扩散
convective heat 对流热
convective null-heat-balance method 对流热平衡消除法
convector 对流放热器
conventional art 传统技术,常规技术
conventional design 传统纹样
conventional energy source 常规能源
conventional method 常规方法
conventional moisture allowance〈CMA〉习惯容许含湿
conventional printing 直接印花
conventional screen 普通[结构]圆网
conventional type 普通型号,常规类型
conventional washing machine 普通平洗机

convention moisture regain 公定回潮率
convergence(＝convergency) 收敛;会聚[度];聚焦
convergent lens 聚光镜
convergent synthesis 收敛合成法
conversational print 风俗画印花布(用于领带及衬衫)
conversion 印染加工(印染术语,即把毛坯加工为成品);转化,转换;转化率;换算
conversion chart 换算表
conversion effect 转色效应(利用两种不同性质染料叠印产生双色效应);转化效应
conversion effect printing 转化效应印花
conversion factor 换算系数
conversion printing 转化印花(两种不同性质染料叠印,产生多色效果)
conversion rate 转化率
converted fabrics 加工织物(经过染整加工的织物)
converted starch 糊精,转化淀粉
converted ticking 漂白印花被套布
converter 转化器;换流器;染整厂;布匹加工的批发商(美国名称)
convertible plasticizer 可转化的增塑剂
convertible top 可拆卸顶盖
converting 印染加工
convex 凸面;凸的
convex expander roller 弯辊扩幅器,扩幅弯辊
convex roll 中凸辊(用于凹凸辊整纬器校正弓纬疵点)
convex weld 凸焊缝
conveyer 运输带;运输机,输送设备
conveyer dryer 履带式烘燥机;导带式烘

燥机
conveyer type scouring and bleaching machine 履带式练漂机
conveyer type washer 履带式水洗机（针织物连续水洗设备）
conveyor 运输带；运输机，输送设备
conveyor belt [印布] 输送导带（用于圆网印花后的烘干机）
conveyor belting 运输带，传送带
conveyor belt shrinking machine 履带式预缩机
conveyor belt washer 输送 [布] 导带的洗涤器（用于圆网印花后的烘干机）
conveyor dwell unit 履带堆置设备
conveyor lattice 输送帘子
conveyor pressing system 履带式烫衣机
conveyor [belt] steamer 履带式汽蒸机
conveyor system 传送带 [流水] 作业法
conveyor tip [烘干机] 导带触头
cook 蒸煮；煮
cooker 蒸煮釜，蒸煮锅
cooker for thickeners 煮糊锅
cooking kettle 煮浆桶
coolant 冷却剂
cool colour 冷色（给人以寒冷感觉的颜色）
cool-down method 降温 [后加碱] 法（活性染料染色方法）
cool-down rate 冷却率
Cool Dry fabric 吸湿排汗功能性织物（商名，中国制）
cooler 冷却器，冷却装置，致冷装置
cool feeling 凉爽感
cooling agent 冷冻剂，冷却剂
cooling and conditioning machine 冷却给湿机

cooling cans（＝**cooling cylinder**） 冷却滚筒，冷水辊
cooling drum 冷水辊，冷却辊
cooling fins 散热片
cooling medium 冷却介质，冷却媒质，冷却剂
cooling method 冷却法
cooling ribs [电动机上的] 冷却肋片
cooling towers 冷却塔
cool iron 低温熨烫
cool-keeping fabrics 凉爽织物
Coolmax 柯梦丝（六角凹槽形合成纤维，优良透湿性，用于制运动衣，休闲服，商名，美国杜邦）
Coolmax fabrics 柯梦丝面料
Coolnice 导湿涤纶（商名，中国制）
Coolplus 吸湿排汗纤维（商名）
Coomassie dyes 科麦西染料（酸性染料，商名，英国卜内门）
coordinate 配套（花样）；坐标 [系]；配位的；配价的；并列的
coordinate bond 配价键
coordinated ionic polymerization 配位离子聚合作用
coordination 配位；配合；协调
coordination catalyst 配位催化剂
coordination compound 配位化合物
coordination number 配位数
cooxidant 辅助氧化剂
copal gum 硬树脂，柯巴脂（硬树胶）
copal varnish 柯巴上光油（用于透明底色和上光）
cop dyeing 管纱染色
cop dyeing machine 管纱染色机
co-PET 共聚酯，共溶性聚酯纤维
co-pigment 辅颜料（对天然染料有增色

co-plasticizer 辅增塑剂
copolyamide 共聚酰胺(热熔胶)
copolycondensate 共缩聚物
copolycondensation 共缩聚[作用]
copolyester 共聚酯(热熔胶)
copolymer 共聚体,共聚物
copolymer fibre 共聚物纤维
copolymerization 共聚合[作用]
copolymer of acrylic ester 丙烯酸酯共聚物
copper 铜
copper[red] 铜红[色](棕橘色)
copper aftertreatment 铜后处理
copper ammonia fibre 铜氨纤维
copperas 绿矾,水绿矾,硫酸亚铁(作媒染剂,真丝加重用)
copperas black 铜盐精元
copper[aniline] black [铜盐]精元,[铜盐]阿尼林元
copper bowl 铜[轧]辊(表面轧有花纹,用作凹凸轧花辊)
copper complex dyes 铜盐络合染料
copper compound 铜化合物
copper green 铜绿[色]
coppering [花筒]镀铜
coppering installation [花筒]镀铜设备
copper-iodine anti-oxidant system 铜—碘抗氧剂体系
copper-ion method 铜离子法
copper-lined 衬铜的
copper mordant 铜媒染剂
copper naphthenate 环烷酸铜(防霉剂,防腐剂)
copper number 铜值
copper pentachlorophenate 五氯代苯酚铜(防霉剂,防腐剂)
copper phthalocyanine 铜酞菁[蓝]
copper plate printing 铜版印花
copper plate singeing machine 铜版烧毛机
copper-8-quinolinolate 8-羟基喹啉铜(杀霉菌剂)
copper rayon 铜氨人造丝
copper[printing]roller 印花铜滚筒,花筒
copper rust 铜锈色
copper stencil printing 铜刻版印花
copper sulphate 硫酸铜
copper treatment 铜盐后处理
copper vitriol 硫酸铜,蓝矾,胆矾
Coprantex BNL 铜盐后处理固色剂BNL(不污染环境,商名,瑞士汽巴)
coprecipitation 共沉淀
coproduct 副产品
copy 复写,复制;复制品
copying 复印
copying and adding machine 拷贝机(用于感光雕刻制板)
copy machine 拷贝机,复印机
copy printing 转移印花
copyright 版权
copyrighted 版权所有
coral 珊瑚色
corbeau 乌鸦黑[色]
cord 绳,索,棱,凸纹;灯芯绒;帘子线;双丝绸
cord carpet 凸纹地毯,条纹地毯
corded velveteen 灯芯绒[织物],条绒[织物]
corded voile 经向凸条巴里纱
cord effect 经向凸纹效应

cord fabric 轮胎布,帘子布
cord lane 条绒类织物
Cordura 考杜拉高强锦纶(商名,美国杜邦)
corduroy 灯芯绒[织物],条绒[织物]
corduroy and velvet finishing 灯芯绒毛绒整理
corduroy cutting blade 灯芯绒割绒刀
corduroy cutting machine 灯芯绒割绒机
core 绳,线;芯子;磁心
coreactant 共反应物
coreaction 共反应
core competence 核心能力
core crosslinking 芯层交联
core crosslinking treatment 透心交联处理
core mercerising machine 透芯丝光机
core mercerization 透芯丝光;纤[维]芯[子]丝光
core-sheath compound fibre 皮芯型复合纤维
core shell 核壳型
core shell polyacrylate emulsion 核壳型丙烯酸酯乳液
core-shell structure 核—壳结构(黏合剂制造中形成的结构)
core-skin fibre 皮芯纤维
core-spun spandex 包芯氨纶丝,包芯氨纶纱
core-spun stretch yarn 包芯弹力丝(芯为弹力长丝,外包短纤维)
core-spun yarn〈C.S.Y.〉 嵌芯纱;包芯纱
core temperature 中心温度
core thread 包芯线
corinth 酱红[色]
corinth pink 桃红[色]
cork 软木;浅棕[色],豆沙色

cork method 木塞法(切纤维截面用)
cornering 转向性能
corner seal 角封
corner stay 角撑
corner weld 角焊
corn fibre 玉米纤维
cornisilk 淡黄[色];佛手黄[色]
corn protein fibre 玉米蛋白纤维
corn starch 玉米淀粉
corona 电晕
corona discharge 电晕放电(等离子体术语);电量放电(一种等离子体技术)
corona discharge plasma 电晕放电等离子体
coronizing [玻璃布]高温处理(起柔软、抗皱、易着色等作用)
corporatewash 批量洗涤
corr.(corrugated) 波纹的,折皱的
corrected effective temperature 修正有效温度
corrected temperature〈Tc〉 标正温度
correction 修正,改正,校准;修正值
correction coefficient 校正系数,修正系数
correction coefficient of temperature and humidity 温湿度修正系数
correction factor 校正因数,修正系数
correlated colour temperature 相关色温
correlation coefficient 相关系数
correspondence 对应,相当,一致
corrode 腐蚀,侵蚀
corrosion 腐蚀,侵蚀
corrosion inhibitor 防腐蚀剂
corrosion prevention by galvanic polarization 电流极化抗腐蚀处理
corrosion prevention passivation 钝化抗腐蚀作用

corrosion-proof 防腐蚀的
corrosion protection agent 防腐蚀剂
corrosion resistance 耐腐蚀性
corrosion-resistant material 防腐蚀材料
corrosion resisting 耐蚀的,耐腐的
corrosion stability 耐腐蚀性
corrosion test 腐蚀试验
corrosive agent 腐蚀剂
corrosive substance 腐蚀物品
corrugate 起波纹;波[纹]状的
corrugated〈corr.〉 波纹的,折皱的
corrugated fin 波形散热片
corrugated pipe 波纹管
corrugated plate 波形板
corrugated round bar 圆竹节钢
corrugated square bar 方竹节钢
corrugation 波纹度,波曲度;沟槽
cortex 皮质
cortical cell 外皮细胞(羊毛)
cortical fibre 韧皮纤维,树皮纤维
cortical layer 外皮层
corundum 刚玉,金刚砂
cosecant 余割《数》
cosine 余弦《数》
cosmetic effects 芳香效果
cosolubilization 共增溶解[作用]
cosolubilizer 共增溶解剂
cosolvent 潜溶剂
cospinning fibre 共纺纤维(也称混抽纤维)
cospun yarn 共纺丝,混纤丝
cosse green 豆荚绿[色](鲜黄绿色)
cost 成本;费用;价格
cost analysis 成本分析
cost comparison 成本比较
cost effectiveness 成本效率

cost, insurance and freight〈CIF〉 到岸价格,包括成本、保险费及运费的价格
costive 不透水织物,不透气织物
cost saving 成本节约
costume 服装;戏装;化装服
costume cambric 杂色低档轧光平布
cosurfactant 辅表面活性剂(和表面活性剂一起使用)
COTS (check out test set) 检[验]测[试]设备
cottage form star steamer 立式星形蒸箱
cottage steamer 小型立式蒸箱(用于印花织物的间歇式高压汽蒸)
cotter 栓,开尾销;楔
cotter bolt 带销螺栓,带开尾销螺栓;地脚螺栓
cotter joint 销接合
cotter pin 开尾销,开口销
Cottestren dyes 科特斯特伦染料(染聚酯纤维/纤维素纤维的分散/还原混合染料,商名,德国制)
cotton 棉花;棉布,棉织品;棉纱;棉线;棉的
cotton and wool mixtures 棉毛交织物,棉毛混纺织物
Cotton-Art Process 柯登安特法(一种棉布湿转移印花法,丹麦制)
cotton bleaching 棉漂白
cotton bowl 棉纤维轧辊,花衣滚筒
cotton broadcloth 宽幅棉布
cotton chintz 擦光印花棉布
Cottonclarin 88 柯登克拉林88(阴离子表面活性剂混合物,高碱稳定性,用于前处理,商名,德国科宁)
cotton condenser spinning 废纺
cotton covered wire 纱包线《电》

cotton covering 纱包绝缘《电》
cotton crepe 棉绉纱,绉布
cotton dewaxing 原棉脱蜡
cotton fabric 棉织物
cotton fibre 棉纤维
cotton flannel 棉法兰绒,绒布
cotton hosiery 棉纱线袜
cotton interlock singlets 棉毛衫
cotton interlock trousers 棉毛裤
cotton lead chromate 棉铬酸铅处理(先用醋酸铅再用重铬酸钠或钾处理)
cotton-like fibre 仿棉纤维,具有棉纤维凉爽、透湿等优良性能的合成纤维
cotton-like finishing 仿棉整理
cotton mill 棉纺织厂
cotton piece goods 棉布
cotton polyester blend 棉和聚酯[纤维]混纺,涤棉混纺
cotton poplin 棉府绸
cotton press cloth 打包棉布
cotton print 印花棉布
cotton roll 棉花滚筒
cotton seed fragments 棉籽壳,棉籽皮,棉籽散片
cotton shell 棉籽壳,籽屑
cotton souring 棉布酸处理(用弱酸溶液中和残余碱性)
cotton spandex knit fabrics 棉/氨纶针织物
cotton spandex woven fabrics 棉/氨纶机织物
cotton textiles 棉纺织品
cotton type fibre 棉型纤维
cotton velveteen 平绒,棉绒,纬绒
cotton wadding 棉絮,棉胎
cotton waste 废棉;回花;回丝

cotton wax 棉蜡
cottony 似棉的,起毛的,柔软的
cottony finish 仿棉整理
couleur de rose 【法】玫瑰色,粉红色
coulomb 库仑(电量单位)
coulomb force 库仑力
coulombic interaction 电荷相互作用
coulombic repulsion 电荷排斥力
coulometric analysis 电量分析
coulometric titration 电量滴定
coumarone-indene resin 苯并呋喃—茚树脂
count 支数;号数;织物经纬密度;计数
counter 计数器
counter acting open-width washing machine 逆流平洗机
counteraction 反作用
counter-bearing 平衡轴承
counter clockwise rotation 逆时针转动
counter-current 对流,逆流;反向电流
counter-current airing 对流通风,逆流通风
counter-current distribution method 逆流分析法
counter-current drying system 逆流烘燥系统
counter-current flow 反向流动,逆流
counter-current process 逆流加工法
counter-current washing system 逆流洗涤系统
counter diffusion 相互扩散
counter flow 逆流,反向流动
counter flow dryer 逆流式烘燥机
counter flow principle 逆流原理
counter flow process 逆流法
counter flow valve 逆流阀

counter force 反作用力
counter ion 抗衡离子;平衡离子;反电离子
counter pattern 对称花纹
counter pile roller 逆针起绒辊,逆针起毛辊
counter pressure 反压力
counter seal ring 反封环(泵中部件)
counter shaft 副轴(皮带传动系统中主动轴与被动轴之间的中间轴)
counter spring 缓冲弹簧,平衡弹簧
counter-sunk head bolt 埋头螺栓
counter-sunk head rivet 埋头铆钉
counter-test 反向试验
counter weight 平衡重锤
counter-weighted dancer roller 重锤式松紧调节辊
couple 力偶;连接
coupled dyes 显色染料,偶合染料
coupling [偶氮染料的]偶合;联轴节
coupling agent 偶联剂,偶合剂
coupling bath 偶合浴,显色浴
coupling component 偶合组分
coupling connection 联轴节连接
coupling constant 偶合常数
coupling dyes 偶合染料
coupling flange 连接法兰
coupling lever 连接杆,离合器杆,联轴杆
coupling pin 连接销钉
coupling process 偶合工艺
coupling reaction 偶合反应
coursewise 横向
Courtex M 一种抗菌聚丙烯腈纤维(商名,英国制)
Courtrai finish 考特赖整理(爱尔兰亚麻布不加光粗整理)

coutinous flushing system 连续冲洗系统
covalency 共价
covalent bond 共价键
covalent chemical bond 共价化学键
covalent link 共价键
covalent reaction 共价反应
cover 盖,罩;包覆层;涂层;布面丰满;密满
coverage [织物的]被覆性,覆盖性
coverage coating 覆盖涂层
coverage power 覆盖能力,被覆能力
coverall [衣裤相连的]工作服
cover guard 盖板,压板
covering cambered bowl 中高轧辊,凸面轧辊(轧辊的包覆层车削成中间部分的直径大于两端的直径,以补偿轧辊的挠度)
covering power 遮盖能力,盖染力
covering property 盖染性,遮盖性,被覆性,覆盖性
covering yarn 包芯纱
cover plate 罩板,盖板
cover printing [全面]罩印
cover roller 罩印花筒(花筒上刻有精细的图案,如细点、细线、方格等,用于罩印在另一花布上,形成一种柔和的图案地色)
cover sheet 铁皮盖
coverspun yarn 包芯纱,包绕纱
coverstock 包覆料,包布(卫生巾、尿布等)
cover with access hole 人孔盖
cover with hand hole 手孔盖
cover with lamp hole 灯孔盖
cowboy suit 牛仔装
cozy 保温套

cP (centipoise) 厘泊(黏度单位)
CP(chemical protection) 化学防护
CP(=C.P., chemically pure) 化学纯
CP(=C.P., chemical pulp) 化学浆粕,化学浆
CPA (Calico Printer's Association) 棉布印花工作者协会(英国)
CPAI(Canvas Products Association International) 国际帆布制品协会
CPB dyeing(cold pad-batch dyeing) 冷轧堆染色
CPC(chemical protective clothing) 防化服
CPE(chlorinated polyethylene) 氯化聚乙烯
CPSC (Consumer Product Safety Commission) 消费制品安全委员会(美国)
CPTA (cyclopentane tetracarboxylic acid) 环戊烷四羧酸
CPU (central processing unit) 中央处理器,中央处理单元
crab [缩小机的]蟹钳部分;起重绞车,起重机
crabbed fabric 已煮呢
crabbing 煮呢
crabbing machine 煮呢机
crabbing roller 煮呢辊
crack 裂化;裂缝,裂痕;稀弄(织疵);松档(织疵)
crack dyeing 裂纹防染(一种使用特殊防染浆的防染法)
cracked screen 筛网疵,筛网色迹,破损筛网
cracked selvedge 破边(疵点)
cracking pattern 裂纹图形
crackle finish 裂纹整理;龟裂效应
crackling effect 裂纹效应,龟裂效应,冰花效应;皱纹[涂层]整理
crack of printing paste 印浆的开裂
crack printing 裂纹防染(一种使用特殊防染浆的防染法)
cradle pantograph machine 摇架式缩小雕刻机
cramming machine [凹纹]轧模机(印花雕刻用)
cramoisy(=cramoisie) 深红色的;红布
cranberry 酸果蔓红[色](暗红色)
crane 起重机,吊车
crank 弯轴
crank driving 曲柄传动
crank fulling mill 曲柄缩绒机
crank pin 曲柄销,联杆销
crank shaft 弯轴
crap 废料
crape cloth 绉织物
craping defect 皱疵;皱纹不匀
craping machine 起绉机
crash 粗布(纱线粗细不匀,布面粗糙)
crash finishing 折绉加工
crash towel 粗毛巾
crating 装箱(指机器用木条或木板装箱待运)
Crava-clean 克赖伐清洁处理(防污、耐久压烫、抗静电一浴法处理,烘焙前后均可用,商名)
Cravenette 克赖文内特防水处理(商名);防雨卡,防水织物,雨衣料
Crawford-Pickering warp printing machine 克劳福德—毕克林经纱印花机(应用于地毯印花)
crawl gear 爬行齿轮,爬行档
crawling speed running 慢速运行,寸行(印染设备专用名词)

crawl speed 慢车速度,慢速
crayon discharge 蜡笔描绘拔染(用于丝绸拔染)
crayon dyeing 色笔描绘染色(用直接染料、阿拉伯树胶和膨润土制成的色笔,在湿布上描绘图案,再进行汽蒸着色)
craze 裂纹;开裂
crazing 龟裂,裂开;裂纹化
CRC process CRC 热还原苛化处理法(棉布退浆、煮练、丝光一步法,瑞士科莱恩)
cream 奶黄[色]
cream bleaching 不完全漂白(亚麻漂白达到的奶白程度)
creamed linen yarn 漂白亚麻纱
creaming 乳化;乳状分层
cream of latex 胶乳
cream of tartar 酒石酸氢钾
creamy white 奶白色
creasability 折皱性
crease 折印,折痕,皱纹;褶裥
crease abrasion test 折皱耐磨试验
crease acceptance 允许折皱度
crease and shrink resistant finish 防皱防缩整理
crease angle 折皱角
crease angle method 折皱度试验法(织物折皱弹性试验)
crease edge-wear 折边磨损
crease elasticity 折皱弹性
crease folding 皱折折叠,折皱
crease formation 折皱形成
crease inhibitor 防皱整理剂
crease in putting on 贴布折皱印(印花坯布在台板上贴布不平)
crease marks 折印,皱痕(染疵)

crease pressed-in 熨烫折痕;熨烫裤线
creaseproof 不皱的,防皱的
crease recovery 折皱回复性,弹性回复度,回能性
crease recovery angle 折皱回复角
crease recovery tester 折皱回复测定器,折皱弹性测定仪
crease resistance 抗皱性,防皱性
crease resistance after washing 水洗后的抗皱性
crease-resistant agent 防皱整理剂
crease resistant fabric 防皱织物
crease resistant finish〈CRF〉 防皱整理
crease resist finish 防皱整理
crease resist finishing agent 抗皱整理剂
crease retention 折缝耐久性;褶裥保持性
crease sensitivity 易折皱性[能],折皱灵敏性
crease setter [呢裤]褶印定形器
crease sharpness 褶裥清晰度
crease-shedding 抗皱,除皱[处理]
crease-shrinkage 皱缩
creasing machine 折幅机
creasing power 折皱力
creasing property 折皱性
creative image processing 创造性的图像处理
creepage speed 爬行速度,寸行速度
creeping device 寸动装置
Creora brand 新型氨纶丝品牌(商名,韩国晓星)
Creora H-100D 一种黑色氨纶(商名,韩国晓星)
crepe-apparatus 起绉装置
crepe (=crêpe) 绉丝;绉线;绉布;绉绸;绉呢;起绉组织

crepe de chine 双绉（丝织物）；中国绉纱（棉织物，法国名称）
crepe effect 绉纹效应，泡泡效应
crepe embossing 绉纹拷花
crepe fabric 绉织物
crepe finish 织物绉缩整理
crepe georgette 乔其纱，乔其绉
crepe jersey 乔赛绉（类似针织的直条绉纹丝织物）
crepe plisse 泡泡纱（在棉布上印浓碱，产生条状或点状绉缩）
crepe printing 绉缩印花；印制泡泡纱[工艺]
crepe satin plain 素绉缎
crepe-weavelike wrinkle 绉纱型皱纹
creping 起皱整理（强捻织物用水或碱液等处理，使织物表面形成皱纹）
crepon finishing 绉缩整理
crepon georgette 乔其绉；顺纡乔其
crepon printed effects 印花绉效应
cresol 甲酚
cresol-formaldehyde resin 甲酚—甲醛树脂
cresol purple 甲酚紫
cresol red 甲酚红，邻甲酚磺酞
Crestfil 克雷斯特菲尔（仿麂皮织物，商名，日本制）
crest value 峰值
cresylic acid 甲苯基酸（俗称），混合甲苯酚（渗透剂，防霉剂，丝光助剂）
CRF（crease resistant finish） 防皱整理
CRI（Carpet and Rug Institute, USA） 美国地毯协会
crimp 皱缩；织缩；卷曲；波纹弯边
crimp angle 卷曲角度
crimp appear 卷曲显现

crimp contraction 皱缩率，卷缩率
crimp development 卷曲展现
crimp durability 卷曲持久性
crimp effect 卷曲效应
crimp elasticity 卷曲弹性
crimp fabric 绉纹织物，泡泡纱
crimp fixing 卷曲的固定
crimpness 卷曲度
crimp percent 卷曲率
crimp-proof finish 防皱整理
crimp ratio 卷曲率
crimp recovery 卷曲回复率
crimp removal 卷曲的去除
crimp retention 卷曲保持
crimp retraction 卷曲收缩率
crimp roll 展幅辊（全幅沟槽辊，用于防止织物幅面折皱）
crimps 绉纹织物
crimpset 卷曲定形
crimp stability 卷曲稳定性
crimp style 皱缩式样《印》
crimp style printing 皱缩印花，泡泡纱印花
crimp tendency 卷曲性
crimson 玫红[色]，绯红[色]
crinkle [织物]绉纹；条子泡泡纱
crinkle cloth 泡泡绉
crinkled crepe 泡泡绉
crinkle fabric 绉条织物
crinkle finish 皱缩整理；泡泡纱整理
crinkle resistant 防皱的
crinkle-retention 绉缩保持性
crinkling 使卷曲，使折皱
crinkly cloth 绉纹织物
Criscoat P-1120 paste 克列斯特柯特 P-1120 浆（溶剂型丙烯酸酯涂层剂，商

名,日本油墨)
crisp 挺爽手感;英国优质细亚麻布
crisper [仿毛皮]卷绒整理机
crisp handle 挺爽手感
crispness 挺爽性
crispwhite 纯白色的
crispy handle 挺爽手感
Crisvon NX 克列斯封 NX(溶剂型涂层剂的交链剂,异氰酸酯类,商名,日本油墨)
criterion 标准,规范;准则;尺度
critical 临界的,极限的;转折[点]的,关键性的;临界[值]
critical add-on value 临界附加值
critical application value〈CAV〉临界施加值
critical components 关键元件
critical concentration 临界浓度
critical condition 临界条件
critical consolute temperature 临界共溶温度
critical constant 临界常数
critical defect 严重疵点;致命疵点
critical dissolving temperature 临界溶解温度
critical dissolving time 临界溶解时间
critical fabric tension 织物临界张力(染整设备的导布辊能承受,且不会产生永久性弯曲的最大的织物张力)
critical fault 临界故障
critical fibre length 纤维临界长度
critical heat 转化潜热,临界热
critical ignition time 临界着火时间
critical interfacial tension 临界界面张力
critical length 断裂长度
critical micelle concentration 临界胶束浓度
critical moisture content 临界含水量
critical operation 关键操作
critical oxygen determinator 临界氧浓度测试仪(耐燃性试验)
critical oxygen index〈COI〉临界氧指数
critical point 临界点
critical potential 临界电势,临界电位;临界[位]势能
critical pressing 临界压烫(后烘焙前的熨烫平挺工作)
critical pressure 临界压力
critical region 临界区域,判别区域
critical setting 精密调节
critical shade 难染色调
critical solution temperature 临界溶解温度
critical state 临界状态
critical surface tension〈CST〉临界表面张力
critical surface tension effect〈CSTE〉临界表面张力效果
critical temperature of wetting 临界润湿温度
critical temperature〈Tc〉临界温度
critical velocity 临界速度
croceate 藏红花色的
crock fastness 耐摩擦[色]牢度
crocking 摩擦脱色
crock-meter 耐摩擦[色]牢度测定器
crock resistance [染色]耐摩擦性能
crock test 耐摩擦[色]牢度试验
cropping [织物]剪毛;剪毛整理落布
cropping cylinder [地毯剪毛机的]剪毛辊
cropping machine 剪毛机,剪呢机
croquis (图案)草图
cross 十字头,交叉;四通[管]

cross air blasting 侧吹风,横向吹风
cross baffle 横隔板
cross beam 横梁
cross belt 交叉皮带
cross brushing machine 横向刷毛机
cross contamination 交叉污染,交互沾染
cross cut 横向切割
cross cutting machine 横向剪毛机,纬向剪绒机
cross dyed cloth 交染织物
cross dye effect 交染效应
cross dyeing 交染(混纺或交织物分别用不同染料染得不同色泽)
crossed belt 交叉皮带
cross flow heat exchanger 交叉流式热交换器(流体在其中相互垂直地流动)
cross-flow membrane filtration 横流膜过滤《环保》
cross head 十字头;T字顶架
cross line pattern 十字格《雕刻》
cross line screen 十字线网版
cross link(＝cross linkage) 交联
crosslinkable plasticizer 可交联增塑剂
cross link density 交联[点]密度
cross linked polymer 交联聚合物
cross linked spherocolloids 交联球状胶体
cross linked starch 交联淀粉(一种变性淀粉)
cross linker 交联剂
cross linking 交联
cross linking agent 交联剂
Cross linking agent EH 交联剂 EH(涂料印花用,商名)
cross linking degree 交联程度
cross-printing 交叉印花(染料与助剂交叉叠印的效果)

cross rail 横轨,横档
cross raising 纬向起绒
cross rod 拉杆,连杆;横杆
cross section 截面,横断面
cross section of fibre 纤维截面
cross shaft 横轴
cross shearing machine 横向剪毛机,纬向剪绒机
cross-staining 交染沾色(指色织或混纺染色中的相互沾色)
crosswise colour difference 横向色差
crosswise contraction 横向收缩
crosswise fold 横向折叠
crotonic acid 巴豆酸,丁烯酸
crow black 乌黑[色]
crown bowl 中高轧辊,凸面轧辊
crown gear 差动器侧面伞齿轮
crow's feet 鸡爪印,皱纹印(疵点)
Croydon finish 克劳伊登整理(整理后可使棉布纹路清晰)
CRT(cathode ray tube) 阴极射线管《计算机》
CRT display 阴极射线管式显示器《计算机》
crude dye 原染料
crude oil 原油
crumpling 起皱(织疵)
crumpling resistance 抗皱性
crunch 摩擦响声
crushed berry 赤豆红[色]
crushed foam 挤压泡沫(将能发泡的化合物固着于织物,挤压后使其起泡)
crushed foam coatings 挤压泡沫涂层
crushed imitation leather 拷花人造革
crushed plush 拷花长毛绒,压花长毛绒
crushed velvet 拷花丝绒,压花丝绒

crush effect 压浆作用《印》
crusher 光板压浆辊,压辊;(砂轮)非金刚石整修器,砂轮刀
crush finish 折皱整理
crushing 压浆作用(由于后面的印花花筒轧压而造成印花浓度、亮度等减弱)
crushing machine 粉碎机
crushing-release 挤压—松释[作用]
crushing unit [泡沫]挤压单元
crush mark 压印;皱痕;揉皱痕
crushproofing 防皱[整理]
crushproof pile 抗压绒头
crush resistance 抗皱性;抗压回复性
crush-resistant 立绒抗压树脂处理(用合成树脂处理,使绒头不易倒伏)
crush roll 压浆辊;挤压滚筒(用于泡沫染整)
crush roller 光板压浆滚筒
crust 水垢,硬垢
cryogen 冷冻剂,制冷剂
cryogenics 低温学;低温实验法
cryogenic temperature 低温温度(通常低于-100℃)
cryoscope 冰点测定器
cryoscopic method 凝固点下降法《试》
crystal〈XTAL, X-tal〉 晶体;结晶的
crystal growth 晶体生长,单晶生长
crystal gum 结晶树脂,结晶胶
crystal gum thickening 结晶胶糊
crystalline 结晶的,晶状的
crystalline-amorphous transition 晶态—非晶态转变
crystalline and amorphous ratio 晶区—非晶区比
crystalline melting point 晶体熔点(以 T_m 表示)

crystalline modification 晶态改变,晶体变型
crystalline orientation 结晶取向
crystalline particle 结晶粒子
crystalline polymer 结晶[性]聚合物
crystalline region 结晶区
crystallinity 结晶度;结晶性
crystallite 微晶
crystallization 结晶[作用],晶化
crystallization exotherm 结晶放热
crystallization kinetics 结晶动力学
crystallization morphology 结晶形态学
crystallize 结晶
crystallized soda 晶碱
crystal perfection 结晶完整性
crystal size 晶粒大小
crystal violet 结晶紫(红光紫色碱性染料)
crystal wash 水晶洗(牛仔布用)
CS(commercial standard) 商业标准
CS(continuous stream) 连续液流(喷射式印花机的一种喷墨方式)
CSA(Civilian Supply Administration) 民需供应局
CSAA(cationic surface active agent) 阳离子表面活性剂
CSCA(Chemical Substances Control Act) 化学物质控制法规
CSIRO(Commonwealth Scientific and Industrial Research Organization) 联邦科学与工业研究组织(澳大利亚)
CSIRO magnate-salt process 澳大利亚联邦科学与工业研究组织高锰酸钾—食盐防缩处理
CSIRO-set 西罗整理(澳大利亚联邦科学与工业研究组织对毛织物进行褶裥整

理的方法）

CSIRO-set process 澳大利亚联邦科学与工业研究组织的定形法《毛》

CSP(concentrating solar power) 聚焦型太阳能热发电

CSPE(chlorosulfonated polyethylene) 氯磺化聚乙烯

cSt 厘斯（运动黏度单位）

CST(critical surface tension) 临界表面张力

CSTE(critical surface tension effect) 临界表面张力效果

C. S. Y. (core-spun yarn) 嵌芯纱；包芯纱

CT(computer tomography) 计算机 X 射线断层照相术

CT(=CTA，cellulose triacetate) 纤维素三醋酸酯；三醋酯纤维素

CTAB(cetyl trimethyl ammonium bromide) 十六烷基三甲基溴化铵（阳离子柔软剂）

CTE(coefficient of thermal expansion) 热膨胀系数

CTI(Canadian Textiles Institute) 加拿大纺织学会

Cubex machine 库贝克斯洗涤试验仪（测试毛织物的毡缩度）

Cubex tester 库贝克斯缩水率试验仪

cubic centimetre⟨cc，c. c.⟩ 立方厘米

cubic feet per minute⟨cfm⟩ 立方英尺每分钟（排气量单位）

Cubitex 3-D printing system 寇别特克斯 3D 印花系统（具有立体凹凸效果的印花方法，用近红外线加工，日本开发）

cudbear 苔色素（紫红植物染料）；地衣红

cuddling blanket 睡毯

cuff 袖口；裤子翻边

cuff accents 袖口、袴口翻边特色（牛仔服）

cuff wear 折边磨损（袖口、袴口磨损）

cuit silk 全脱胶丝，熟丝，全练丝

cumulative 累加的，累积的

cumulative time 累积时间

cuoxam 铜氨液；氢氧化铜氨溶液

cup 杯，帽；圈杯；油杯

cup head pin 圆头销

cuprammonium 铜氨

cuprammonium fibre 铜氨纤维

cuprammonium fluidity 铜氨流动度

cuprammonium hydroxide 铜氨液；氢氧化铜氨溶液

cuprammonium process 铜氨法（制造铜氨人造丝的方法）

cuprammonium rayon 铜氨丝，铜氨人造丝

cuprammonium solution 铜氨溶液

cuprammonium viscosity test 铜氨［液］黏度试验

cuprammonium woven fabric 铜氨纤维机织物

cuprate silk 铜氨人造丝

cuprene fibre 铜氨纤维

cupric acetate 醋酸铜

cupric sulphate 硫酸铜

cupri-ethylene diamine solution 铜乙二胺溶液

cupriferous dyes 含铜［络合］染料

cuprous chloride 氯化亚铜

cup seaming machine 包缝机

cup spring washer 杯形弹簧垫圈

cup weight 杯重（泡沫整理技术术语，即发泡率的倒数）

curative 固化剂；硬化剂；熟化剂

curcurmin colour 姜黄色素(天然植物染料,食用色素)
cure 烘焙;硫化;固化;塑化
cure box 高温烘焙箱
cure condition 烘焙条件
cured fabric 烘焙织物
cured resin 固化树脂
curer 焙固机,烘焙箱
curing [高温]焙烘,焙固
curing agent 固化剂;硫化剂;熟化剂
curing catalyst 焙烘催化剂
curing cycle 烘焙循环
curing degree 固化度;硫化度;熟化度
curing machine 烘焙机,焙固机
curing oven 烘焙箱
curing speed 硫化速度;固化速度
curing temperature 焙烘温度
curing time 焙烘时间
curl 毛圈;卷曲;[织物]卷边
curled edge 卷边,翻边(织疵)
curled pile 卷曲绒头
curled selvedge 卷边,翻边(织疵)
curl effect 卷曲效应
curliness 卷曲度
curling elasticity 卷曲弹性
curl plate 弧形板
curl tendency 卷曲倾向
curl texture [纤维]卷曲结构
current 电流;液流;流
current-carrying components 带电元件
current collector 集电器(平网印花行车和起重行车电力牵引系统用来与架空接触导线或导轨相接触的装置)
current conductor 电流导线
current distribution 电流分配,电流分布
current efficiency 电流效率

current fashion 流行式样
current feedback 电流反馈
current meter 电流表;流速仪
current regulator 电流调节器
current repair 小修(指设备的经常性的现场修理)
current settings 当前设置
current shade 流行色泽
current transformer 电流互感器,变流器
curtain 帷幔,幕,窗帘
curtain boom 幕式拦油栅
curtain fabric 窗帘织物
curtain gauze 窗帘纱
curtain net 网眼窗帘
curvature 曲率《数》;弯曲;弓纬,弓形布面(织疵)
curvature radius 曲率半径《数》
curved bar expander 弯辊扩幅器,扩幅弯辊
Curved Blade Applicator 弯刀给液机(一种低给液装置,由给液泵、输液泵及工艺控制装置等组成,商名,美国)
curved cylinder expander 弯辊开幅器
curved displacement 弓纬,纬弧
curved expander roll 弯辊扩幅器,扩幅弯辊
curve disk 弧形盘,钩盘
curved pile [地毯]剪花绒头
curved plate 弧形板
curved restraining shoe 弧形制动器,弧形制动闸
curved roller 弯辊
curved straightener 弯辊整纬器
curved straightening roller 整纬弯辊
cushion 垫子,衬垫
cushion back carpet 衬垫毯

cushioning effect 缓冲作用,减震作用;垫层作用
customary suction pump 常用抽吸泵
custom-engineering [服装]定制工程
customization [服装]定制,定做
custom-made 定制的,定做的
custom-tailor 定制,定做
custom tufted 手工簇绒
cut across [疵布]开匹,开剪
cut and sewn 裁剪与缝制
cut and try method 试探法,逐次逼近法
cut away drawing 剖视图
cutch 儿茶(印度棕色植物染料);儿茶棕[色](棕色)
cut clear 乳化稀释浆,合成稀释浆(印花用)
cuticle 表皮,外皮
cuticular layer (羊毛的)表皮层
cuticular ridge (羊毛的)表皮层隆脊
cuticular scales 表皮鳞片
cutin 角[皮]质(植物)
cutinase 角质酶(用于去除棉纤维中角质素等)
cut length 切断长度,切段长度
cut listing 破边(织疵)
cut loop pile 圈割绒
cut mark 分匹印,墨印
cut marks 搭色(色浆黏度不良造成的疵点)
cut-off selvedge extractor 集边器
cut-off switch 断路开关
cut-off valve 断流阀
cut-open view 剖视图
cut-out 停车;切断,断开;排汽阀
cut out method [浆印]制版法,刻纸板法
cut piece 衣片

cut pile 割绒
cut pile fabric 割绒织物
cut selvedge 破边(织疵)
cutter 切刀;切断机
cutter roller 切割辊
cut through [疵布]开匹,开剪
cutting [呢绒]剪毛;拔染处理;切断[电路]《电》
cutting agent 稀释剂
cutting and finishing machine 割绒整理机(加工丝绒和灯芯绒)
cutting blade 割刀刀片
cutting disc 圆刀
cutting edge engraving 锐边雕刻
cutting machine 剪毛机;裁剪机;割绒机
cutting machine for velvets 割绒机
cutting motif fabric 剪花织物
cutting-off 切断
cutting oil 切削油
cutting resistance 抗切断
cutting tools 刀具
cuttle 折叠
cuttling machine [织物]对折机
cuttling up nozzle [织物松式染整设备的]折堆喷嘴
CV(cellulose viscose) 黏胶纤维
CV(compatability value) 相容性值
CVC(chief value of cotton) 以棉为主的涤棉混纺纺织品(棉比例50%以上,俗称倒比例)
CVC method(chemical vapour coagulation method) 化学蒸发凝聚法(制纳米微粒的方法)
CVC roller system(continuously variable crown roller system) 连续可变锥度轧辊系统(德国蒙福茨)

CW(chemical warfare) 化学战争
CWC(Chemical Weapons Convention) 化学武器公约(禁止有害化学品的输出)
cyan 青色,蓝绿色
cyan- (构词成分)表示"深蓝""青色""氰""氰基"
cyanamide 氨基氰;氨腈
cyanate 氰酸盐
cyan-blue 深蓝[色],天蓝[色],湖蓝[色],翠蓝[色],晴空蓝[色]
cyanidation 氰化
cyanide 氰化物
cyanine blue 深蓝[色];深紫蓝[色]
cyanine dye 花青染料(蓝色染料)
cyano- (构词成分)表示"深蓝""青色""氰""氰基"
cyano compound 氰基化合物
cyanoethylated cotton 氰乙基化棉
cyanoethylation 氰乙基化[作用](棉或黏胶纤维经丙烯腈变性处理,使之耐磨损、耐高温、防菌、防霉、增加强力、增强吸色力等)
cyanoguanidine 双氰胺,二聚氨基氰,氰基胍
cyanuric acid 氰尿酸,三聚氰酸
cyanuric chloride 氰尿酰氯,三聚氰[酰]氯,三聚氯氰
cycle 循环、周期
cycle felting severity value〈CFS value〉 循环毡缩指数
cycle index 循环[完成]次数
cycle test 反复试验
cycle time 周期时间,循环时间
cycle time controller 循环时间控制器(用于分批染色)
cycle triphosphazene based flame retardants 环状磷腈阻燃剂
cyclic 环状的,环式的;环化的
cyclic activated sludge system〈CASS〉 循环式活性污泥系统(污水处理)《环保》
cyclic compound 环状化合物
cyclic ethylene urea 环亚乙基脲
cyclic ethylene urea-formaldehyde 环亚乙基脲甲醛(抗皱整理用)
cyclic oligosaccharides 环状糊精(环状低聚糖,可用作整理剂)
cyclic train 周转轮系(用于煮糊搅拌器、整纬装置、滚筒印花的电动对花装置等)
cycling suit 自行车服
cycling wear 自行车服
cyclisation (=cyclization) 环化作用,成环作用
cyclo- (词头)表示"圆的""轮转的""环的""环状的"
cyclodextrins 环状糊精(环状低聚糖,可用作整理剂)
cyclohexane 环己烷
cyclohexanone 环己酮
cyclohexanone resins 环己酮树脂
cycloid 摆线
cycloidal gear 摆线齿轮(齿廓曲线为摆线的齿轮)
cycloidal gear reducer 摆线减速器
cycloidal planetary reducer 行星摆线减速器
cycloidal pump 摆旋泵
cyclone dust extractor 旋风除尘器
cyclooctadiene〈COD,C.O.D.〉 环辛二烯
cyclopentance tetracarboxylic acid〈CPTA〉 环戊烷四羧酸
cyclopropane 环丙烷

Cycloprothrin 环普罗士林（较安全的羊毛防蛀剂，类似于除虫菊酯）
cylinder 滚筒；烘筒；[提花机]花筒；印花铜辊
cylinder brushing machine 圆筒刷毛机
cylinder drying 滚筒烘燥
cylinder drying machine 圆筒烘燥机
cylinder ironing 滚筒式熨烫机
cylinder mill 花筒轧纹印模，钢芯
cylinder milling machine 压辊缩呢机，压辊毡缩机
cylinder press 滚筒式压呢机
cylinder printing 滚筒印花
cylinder printing machine 滚筒印花机
cylinder shrinking machine 圆筒缩绒机
cylinder singeing machine 圆筒烧毛机（使用灼热的铸铁辊筒）
cylinder sizing machine 圆筒式浆纱机
cylinder slasher 圆筒型浆纱机

cylinder teaseling machine 滚筒式起绒机
cylindrical beater 滚筒打手
cylindrical beck 圆筒式绳状染色机
cylindrically wound package [织物]筒状卷装
cylindrical pin 圆栓销
cylindrical press 滚筒式压呢机
cylindrical spur gear 圆柱正齿轮
cylindrical stainless steel housing 圆形不锈钢箱体
cypress [green] 柏树绿[色]（暗黄绿色）
cysteine 半胱氨酸；巯基丙氨酸
cystin bridge 胱氨酸桥基
cystin cross-link 胱氨酸交联
cystine 胱氨酸；双巯丙氨酸
cystine linkage [羊毛分子结构式中的]胱氨酸键，硫联
cyto chemistry 细胞化学

D

D55 国际照明委员会标准照明体 D55（平均自然昼光,色温 5503K）
D65 国际照明委员会标准照明体 D65（平均自然昼光,色温 6504K）
D75 国际照明委员会标准照明体 D75（平均自然昼光,色温 7504K）
DAB(diazo aminobenzene) 重氮氨基苯（发泡剂）
dabbing print 压印（用染料印工业纺织品）
dacian cloth 拷花布,压花布,浮雕布
Dacron 大可纶（聚酯长丝和短纤维,商名,美国制）
DAD(draft addendum) ［国际标准］补充草案
DAED(diacetyl ethylene diamine) 二乙酰乙二胺（双氧水活化剂原料）
DAF(direct air flow) 直接［空］气流
daily sheet 日程表
dairy salt 食盐
Dalanol EMP 迪兰匀染剂（氧乙烯缩聚物,用于羊毛、锦纶的染色匀染,商名,英国 PPT）
daltonism 色盲（尤指红绿色盲）
dam 坝
DAM(draft amendment) ［国际标准］修正草案
damage 危害;故障,损坏
damage by friction 擦伤
damage coefficient 损伤系数
damaged and kept〈D and K〉 认赔（指印染厂商承认次货的赔偿）
damaged area 损毁面积（阻燃试验）
damaged fabric 有损伤的织物
damaged length 损毁长度（阻燃试验）
damage factor 损伤因子
damage of singing 烧毛损伤
damage tolerance 耐损伤性
damas chine 印经花缎
damask finish 仿锦缎整理
damask printing 印花仿锦缎
damask satin 缎纹织锦
damask swiss organdy 仿花缎印花织物（透明底纹上用涂料或一般印花作出不透明的花纹）
damask velour 织锦丝绒
damas lisere 金银线花缎
damasque 印花皱缩效应
dam board 挡板,挡浆板
dammar gum 达马树脂,达马胶
damnification 损伤
damp 潮湿;打湿;阻抑,衰减
damp air 湿空气
damp centrifuged 离心去湿
damp cure 湿焙烘
damp dry 半干,带潮
damped oscillation 阻尼摆动
dampening 给湿;使衰减
damper 缓冲器,减震器,阻尼器;挡板;调节风门,气闸
damper control 挡板调节,挡板控制
damper regulator 气闸调节器

damping 给湿；喷雾；阻尼，减震
damping capacity 阻尼容量，衰减容量；阻抑量；调湿量
damping conditioning 调湿，调节含潮
damping effect 阻尼效应，减震作用，阻抑效应
damping machine 给湿机，喷雾机
dampness 潮湿；湿度
damp pressing 汽蒸熨烫
damp-proof 防潮的
damp proofing 防湿
damp proof package cloth 防湿包装布
damp-proof wall 防湿墙
damp rag 湿布（熨烫服装时用）
damp setting 湿定形
damp storage closet 给湿室
damson 暗紫色
dance roller（＝dancer roll, dancing roller）松紧调节辊，张力调节辊，升降[导]辊
dancing roller accummulator 升降辊储布器，升降辊松紧架（可储存一定量的织物，以调节其张力）
dancing roller compensator 升降辊松紧架
Dancool 但库耳涤棉混纺耐久定形织物（商名）
D and D(direct and developed dyeing)［拔染印花用］直接染料和后处理染色工艺
D and D dyes(direct and developed dyes) 重氮显色直接染料
D and K(damaged and kept) 认赔（指印染厂商承认次货的赔偿）
Danfit process 丹菲特防缩整理（用于针织物的防缩整理）
dangerous substance 危险物质
Daniel Kochlin process 丹尼尔科赫林媒染拔染法（应用氧化拔染剂的媒染染料拔染方法）
DAP(diallyl phthalate) 邻苯二甲酸二烯丙酯
DAP(diammonium phosphate) 磷酸氢二铵
daphne cobalt 钴蓝
dark adaptation 暗[视]适应
dark and white spots 深浅细点
dark blue 深蓝[色]
dark blue purple 藏青[色]
dark brown 深棕[色]
dark clear series 暗澄清色《测色》
dark colour 深色，暗色；暗色调
dark-colouring agent 增深剂
dark dyeing polyamide 深染型聚酰胺；异染型聚酰胺（差异染色）
darkening 加深，使暗
darkening agent 色加深剂，色变暗剂，增深剂
dark field 暗[视]场
dark-field condenser 暗视场集光器
dark field electron microscopy 暗视场电子显微术
dark-field image 暗视场图像
dark grey 深灰[色]
dark-ground illumination 暗视野照明
dark ground illuminator 暗视野照明器
dark heavy pattern 深色花纹
dark indigo 靛蓝
darkness 暗度
dark olive green 灰茶绿[色]
dark rose taupe 古铜色
dark selvedge 深边（印疵）
dark shade 暗色泽
dark solution titration 深色溶液滴定

dark stripiness 黑条(染疵)
darner 织补机;织补工
darning 织补
darning needle 织补[用]针
darning stitch 织补针迹
darts and steels 刀线(滚筒印花疵,由于刮浆刀刀刃损坏而造成的往复状线痕);嵌花筒铁屑
dash board 仪表板,控制板
dasher 桨式搅拌器,叶片式搅拌器
Dash Line liquor circulating dyeing machine 达什莱恩液流染色机(一种染色设备,多用于针织物的染色,具有温差小、染色均匀、浴比小等优点,日本大岛机械)
Dash Line range 达什莱恩联合机(一种常温常压连续加工线,可用于煮练、漂白、松弛及印花后水洗等,具有浴比小、不起皱、张力低等优点,日本大岛机械)
dashpot 缓冲筒减震器
data 数据,资料;信息(datum 的复数)
data acquisition systems 数据采集系统
data bank 资料库;数据资料情报中心
data-base 数据库
Datacolor 瑞士电子测配色技术公司
data handling 数据处理
data-in (计算机的)输入数据
data logger 数据记录器,数字式记录仪
data logging system 数据记录系统
data management system 数据收集管理系统《计算机》
datamation 数据自动处理《计算机》
data-out (计算机的)输出数据
data package 数据包(用于设备上的数据储存)

data processing〈DP〉数据处理
data processor 数据处理机;数据处理部件,数据处理程序
data reduction 数据简化
data retrieval 数据检索
dating 确定年代
dative bond 配[价]键
datum 数据,资料;信息(复数为 data)
datum line 基准线
datum temperature 基准温度
dauber 上蜡辊
daub printing 涂抹印花
daul beam measurement 双光束测量
Davis-Gibson filter〈DG filter〉 台维斯—吉布森滤光器(即 DG 滤光器)
davit-mounting 吊装
dawn grey 黎明灰[色]
Day-glow 日光荧光染料(商名)
daylight 日光
daylight booth 日光灯柜
daylight colour 日光色(指朝北光线下的颜色)
daylight fluorescent lamp 日光荧光灯
daylight fluorescent pigment 日光荧光涂料
daylight lamp 日光灯
daylight test 日光试验,耐日光牢度试验
day shift 日班
day time clothes 白昼服
day wear 白昼服
dazzling 光彩炫目的
dB(=DB. ,db. ,db,decibel) 分贝(音强单位,电平单位)
DB(dry bulb) [温度计的]干球
DBD(dielectric barrier discharge) [常压]介电屏蔽放电,电介质屏蔽放电

DBP(dibutyl phthalate) ［邻］苯二［甲］酸二丁酯(增塑剂)

DBS(dodecyl benzene sulfonate) 十二烷基苯磺酸酯

DC (= D. C. , dc, , d. c, d-c, direct current) 直流［电］

DC 109 道康宁 109(溶剂型有机硅聚合物,用于羊毛防缩,商名,美国道康宁)

DC 544 道康宁 544(有机硅消泡剂,高温安定性好,商名,美国道康宁)

DC 1111 道康宁 1111(羟基二甲基硅氧烷聚合物,整理剂,商名,美国道康宁)

DC 5700 道康宁 5700(有机硅抗菌剂,耐久性、安全性好,商名,美国道康宁)

DC BY22-856D 道康宁 BY22-856D(有机硅,羊毛用柔软防缩剂,商名,美国道康宁)

DCCA(dichloroisocyanuric acid) 二氯异氰尿酸(羊毛氯化剂)

DC control circuit 直流控制回路

DC distributing equipment 直流配电装置

DC insulation resistance 直流绝缘电阻

DCM (dichloromethane) 二氯甲烷

DC motor 直流电动机

DC multi-motor drive 多单元直流［电动机］传动(指联合机中各单元机均备有一个电动机,进行同步传动)

DC Q2-8586 道康宁 Q2-8586(氨基有机硅乳液,亲水性柔软剂,耐剪切,商名,美国道康宁)

DCRA (dry crease recovery angle) 干态折皱回复角

DCRO(Dyers' and Cleaners' Research Organization) 染色与干洗业研究组织(英国)

DC source 直流电源

DC T4-0118A 道康宁 T4-0118A(耐久性柔软防水整理剂,商名,美国道康宁)

DC tacho-generator 直流测速发电机

DCT dye(dichlorotriazine dyes) 二氯均三嗪染料(含二氯均三嗪活性基团的染料)

DC transformer 直流互感器

DD (differential dyeing) 差异染色(指混纺、交织物或具有不同上染性能的同类纤维织物在一浴中染得不同色泽)

DD(draft data) 数据草案

DDC(direct digital control) 直接数字控制

DDD［2,2-bis(*p*-chlorophenyl)-1,1-dichloroethane］ 滴滴滴,2,2-双(对氯苯基)-1,1-二氯乙烷(国际禁用的杀虫剂)

DDD (dilute, disperse and decontaminate) 稀释、分散与去污(废物去除系统)

DDI (dye distribution index) 染料分散指数

DDM (direct dyebath monitoring) 直接染浴监控

DDS(drug delivery system) 药物传送系统

DDT(dichlorodiphenyl trichloroethane) 滴滴涕,双对氯苯基三氯乙烷(国际禁用杀虫剂)

de- (词头)表示"脱""去""除""解""减"

deacetylated chitin 脱乙酰甲壳质

deacetylation 脱乙酰基［作用］

deacidification 脱酸［作用］

deacidizing 去酸,脱酸；还原,脱氧

deactivation 减活化作用,失活作用

deactivator 去活化剂

dead abutment 固定支座

dead appearance 黯淡外观

dead axle 静轴,不转轴

dead centre 死心,死点,止点

dead colour 死色,呆色,萎色
dead dyeing 无光泽染色
deadening agent 黯淡剂,消光剂
deadening fabric 隔音布
dead fibre 死纤维
dead fold 死[褶]裥
dead halt 完全停机(不能恢复正常运转的停机)
dead knife 固定刀,底刀
deadline 限期,截止时间;安全界限
deadlock 停顿,停止作用,闭锁
dead match 与来样完全符合(指染色)
dead oil 厚润滑油,重油
dead plate 底刀
dead set screw pressing 永久定形螺旋压烫
dead spots 死角
dead steam 废汽
dead time 停滞时间;空载时间;失效时间
dead weight 自重,静负荷,静重
dead weighting 静重加压
deaerate 除去空气,除去气体
deaerated biological treatment 脱气生化处理
deaerated water 脱氧水
de-aerating agent 脱泡剂;去空气剂
deaeration 脱泡[作用]
deaerator 脱泡器,脱泡桶;脱泡剂;除氧器,除气器
deafener 消声器,减音器
deageing 返生,反熟化(指羊毛)
deairing 去空气
de-airing agent 去空气剂
dealkylation 脱烷基[作用],脱烃[作用]
deamination 脱氨基[作用],脱氨基化
de-and-re-chrome 剥铬再镀铬

deannealed wool 消除[定温]热处理的羊毛
deaquation 脱水[作用]
Debaca range 旦巴卡整理联合机(一种涤纶织物碱减量整理设备,织物整理后光泽如丝,手感丰满,改善吸湿性,意大利斯伯罗托—利马)
debatching 退卷
debatching machine 退卷机
debonding 松解[工艺]
debris 碎片,碎屑
debubbling 脱泡,消泡
debug 排除故障
Debye factor 德拜因数
dec.(decimal) 十进的;小数的
dec.(declination) 倾斜;偏差;磁偏
dec.(=decomp.,decomposition) 分解
Deca-BDE(decabromodiphenyl ether) 十溴联苯醚(有毒阻燃剂)
decabromodiphenyl ether〈Deca-BDE〉十溴联苯醚(有毒阻燃剂)
decabromo diphenyl oxide 十溴代[二]苯醚;十溴代二苯基氧(阻燃剂)
decade 十的;十进制的
decade counter 十进制计数器
decalcification of water 水的软化
decalcified 脱钙的
decalcifier unit 自动脱钙单元(软化水装置)
decal printing 移画印花法,转移印花
decarboxylate 脱羧基;脱羧物
decarboxylation 脱羧基[作用]
decating 蒸呢,蒸煮
decating blanket 蒸呢包布
decating faults 蒸呢印渍
decating knit goods 蒸煮的针织品

decating mark [蒸呢]皱痕;搭头印(疵点)
decationize 除去阳离子
decatize(=decatizing) 蒸呢
decatizer(=decator) 蒸呢机
decatizing fastness 汽蒸牢度
decatizing machine 蒸呢机(毛织物整理设备)
decatizing wrapper 蒸呢包布
decay 衰变;分解;腐烂
decay constant 衰变常数
decay curve 衰变曲线
decay of water-repellency 防水效果的衰减
decay stress 衰变应力
decay time 衰减时间
decelerating flow 减速流动
decelerating plastic flow 减速塑性流动
deceleration 减速
deceleration rate 减速速度
decelerator 减速器,减速装置
decelerometer 减速计
dechlorinating agent 脱氯剂(用以脱除漂白纤维上的剩余氯)
dechlorination 脱氯[作用]
dechroming 去铬,剥铬
dechroming installation [花筒]脱(剥)铬设备
decibel〈dB,DB.,db.,db〉分贝(音强单位,电平单位)
decimal〈dec.〉十进的;小数的
decitex〈dt,dtex〉分特克斯,分特(线密度单位,为1特克斯的1/10)
decker 脱水机;调料机;层
declination〈dec.〉倾斜;偏差;磁偏
declutch bearing 离合器[操纵]轴承
decolor【美】=decolour

decolour 褪色;脱色,去色
decolourant 脱色剂
decolouration(=decolouring) 脱色
decolouring assistant 脱色助剂
decolouring by oxidation 氧化脱色
decolouriser 脱色剂
decolourization 脱色[作用]
decolourize 剥色
decolourizer 脱色剂
decolourizing 脱色,剥色
decolourizing agent 去色剂,脱色剂,褪色剂
decolourizing carbon 脱色碳
decompose 分解
decomposition〈dec.,decomp.〉分解
decompression 减压[作用];降压
decompression valve 泄压阀
decontamination 净化,纯化
deconvolutions count 解捻值(用以表示棉纤维在丝光过程中的膨化程度)
decorating colour 点缀色
decoration (衣服的)装饰;装潢;边饰
decorative design 装饰图案
decorative effect 装饰效应
decorative fabric 装饰[用]织物
decorticating machine 剥皮机
decrement 减量,递减
decrimping 退卷曲,使伸直,卷曲弛解
Decroline 德科林(甲醛次硫酸锌,拔染剂,商名,德国巴斯夫)
decrystallization 非晶化
decrystallized cotton 消晶化棉,非晶化棉
Dedeco steamer 代代科[星形]蒸化机(商名,德国制)
deduce 演绎,推演
dedusted dyestuff 不扬尘染料

dedusting 去尘
dedusting agent 除尘剂
dedusting property 去尘性能
deemulsification 浮浊澄清[作用];反乳化[作用];消乳化[作用]
deemulsifier(=demulsifying agent) 消乳化剂,反乳化剂,破乳化剂,浮浊[液]澄清剂
de-energize 切断电源,失去能量
deep colour 浓色,深色
deep colour effect 深色效应,向红效应
deep-dye [地毯的]透染,透印
deep-dyeable fibre 可深染[的]纤维
deep dye carpet printing process 地毯深染印花加工法(地毯面向下,多色印版向上顶压地毯表面而产生印花)
deep dyeing 染深,深染(如纤维经变性后能吸收更多染料)
deep-dyeing assistant 增深剂
deep dyeing promoter 增深剂
deep-dyeing variant 深染性变异体《化纤》
deep dye process 透染工艺(地毯)
deep engraving roller 刻度较深的花筒
deepening agent 增深剂
deepening colour 深色,浓色
deep etching 深蚀《雕刻》
deep freezing 深度冷冻
deep gray 深灰[色]
deep immersion trough 深浸槽
deepness of colour 色泽深度
deep pile 长毛绒;仿毛皮织物(统称)
deep pile fabric 长毛绒织物
deep red brown 酱红[色]
deep shade 深色,浓色
deep shaft aeration process 深坑曝气法《环保》

deep sky blue 深天蓝[色]
deep-textured fabric 有立体感的织物
deep water 深海,远洋
deep well pumps 深井泵
deerskin 鹿皮;(揩拭用)麂皮
default 默认,缺省《计算机》
default settings 默认设定值《计算机》
defecation 澄清[作用]
defecator 澄清槽,滤清器
defect 疵点;故障;损伤
defect by knotting 结子疵
defect dyeing or printing to crease of under cloth 衬布折皱引起的染疵或印疵
defect elimination during dying 染疵消除
defect in finishing 加工疵点;整理疵点
defective 有疵点的,有缺陷的
defective area 疵损面积
defective colour vision [色]视觉缺陷,异常色觉
defective goods 疵布
defective selvedge 边不良,毛边
defect printing to uneven point of under cloth 衬布不整齐引起的印疵
defect rate 疵品率
defects in grey goods 坯布疵品率
defects of the equipment 设备缺陷
defects on fabric surface 织物表面点
defense apparel 防护服
deferrable 易牵拉的,能延迟的
deferred cure(=deferred curing) 后焙烘,延迟焙烘(制成服装后的焙烘)
deferring relay 延时继电器
deferrize 除铁
defibrillation 去除原纤化
defilade property 遮蔽性
defining aperture 辨色光孔

definition [花纹轮廓]清晰度;定义,解说
deflecting roller 转向导辊
deflection 偏转度,偏转;挠度,挠曲
deflection station [喷墨印花]偏转站
deflector 导风板,导向装置;偏转器
deflexion 偏转度,偏转;挠度,挠曲
defloculant(= deflocculating agent) 反絮凝剂,抗絮凝剂
deflocculation 反絮凝[作用],抗絮凝[作用]
deflocculation agent 反絮凝剂,抗絮凝剂
defoam 去泡(沫),消泡
defoamant 消泡剂
defoamer(= defoaming agent) 消泡[沫]剂,去沫剂
deformability 形变性
deformation 变形,形变,走样
deformational resilience 变形回弹[性能]
deformation damage 变形损伤
deformation of squeeze rollers 压辊变形
deformation-recovery 变形回复
deformation resilience 变形回弹[性能]
deformation resistance 耐变形性,抗变形性
deformation set 永久形变,[形]变[固]定
defuzzing 脱绒毛,脱茸毛
deg.(degree) 度[数];程度
DEG(diethylene glycol) 二甘醇(助溶剂和吸湿剂)
degasification 脱气[作用]
degasser 除气剂;脱气器,排气器
degassing 放气,排气;脱泡;消[除]毒[气]
degenerate 退化;变质;衰退;简并
degenerated curve 退化曲线
de-glossing 消光

deglossing 消光
degradability 可降解性
degradable organic effluent components 可分解有机污水成分
degradation 降解;降等
degradation by bacteria 细菌降解[作用],菌解[作用]
degradation of cellulose 纤维素降解
degradation of main chain 主链裂解
degradation property 降解性;[纤维]脆化性
degradation-resistant fibre 抗降解纤维
degradation testing 老化试验
degradation with aging 老化降解
degradation with shearing 剪切降解
degradative process 降解处理[法]
degrade 由深到浅的色调
degraded cellulose 脆化纤维素;降解纤维素
degraded chitosan 降解壳聚糖
degrading 降解;降级[处理];降等
degreasant [羊毛]脱脂剂
degrease [羊毛]脱脂
degreased wool 脱脂羊毛
degreasing 脱脂
degreasing agent [羊毛]脱脂剂
degreasing machine 脱脂机
degreasing power 脱脂力
degree〈deg.〉 度[数];程度
degree celsius 摄氏度
degree centigrade 百分度;摄氏度
degree of acetalization 缩醛度
degree of acetylation 乙酰化度
degree of activity 活化度
degree of adhesion 黏合程度
degree of automation 自动化程度,自动操

作程度
degree of Baumé (= degree of Beaumé) 波美度(液体浓度单位)
degree of bulging 膨胀度,膨化度
degree of confidence 置信度
degree of conversion 转化度;转化率
degree of coverage 覆盖度
degree of crimp 卷曲度
degree of crosslinking 交联度
degree of crystallization 结晶[程]度
degree of cure 熟化[程]度;焙烘程度
degree of deacetylation 脱乙酰化度
degree of depolarization 去极化度;消偏振度
degree of desizing 退浆[程]度
degree of dispersion 分散度
degree of dissociation 解离度
degree of dyeing 染着度,上染度
degree of dyeing power 染着度
degree of exhaustion 吸色率;上色程度
degree of exhaution-time curve 上染速率曲线
degree of finish 光洁度,加工精度
degree of fixation 固定度,固着度
degree of flexibility 挠度,柔曲度
degree of flexural modulus 屈曲弹性度
degree of freedom ⟨DF, df⟩ 自由度
degree of grafting 接枝度
degree of ionization 电离度
degree of irregularity 不均匀程度,不匀率
degree of lateral order 侧序度
degree of lignification 木质化程度
degree of mercerization 丝光度
degree of migration 移染度;泳移度
degree of molecular orientation 分子取向度

degree of orientation 取向度,定向[程]度
degree of penetration 渗透度,透染度
degree of polarization 偏光度,极化[程]度
degree of polycondensation 缩聚[程]度
degree of polymerization ⟨DP⟩ 聚合度
degree of preswelling 预膨化度
degree of purification 净化度
degree of residual crimp 剩余卷曲度,残余卷曲度
degree of ripeness 熟成度
degree of saponification 皂化度
degree of saturation 饱和度
degree of setting 固着度,定形度
degree of shrinkage 收缩度
degree of slip 活络度
degree of substitution ⟨DS, D.S.⟩ 取代度,置换[程]度
degree of sulfonation 磺化度
degree of super heat 过热度
degree of swelling 泡胀度,膨胀度,溶胀度
degree of transparency 透明度
degree of Twaddell 特沃德尔度(液体密度单位)
degree of unsaturation 不饱和度
degree of water hardness 水硬度
degree of whiteness 白度
degree of xanthation 黄原酸酯化度,黄原化度
degree of yellowing 泛黄度
degummase 脱胶酶
degummed silk 熟丝,精练蚕丝;精练丝织物
degummed silk fabric 熟绸,精练绸缎
degummed silk yarn 精练丝
degumming 脱胶[工艺];精练[工艺]
degumming agent 脱胶剂,精练剂

degumming at low temperature 低温脱胶，低温精练
degumming barrel 精练桶《丝》
degumming bowl(＝degumming box) 练槽
degumming by fermentation 发酵脱胶
degumming loss 练减，练耗（丝绸精练后的减重）
degumming loss percentage 练减率，脱胶率
degumming machine 脱胶机
degumming of ramie 苎麻脱胶[工艺]
degumming pan 精练锅《丝》
degumming silk 脱胶蚕丝
degumming soap 脱胶肥皂
degumming with chemical agents and schapping 用化学剂腐化脱胶
dehairing 去除戗毛；[骆驼毛或山羊毛]分绒工艺；脱毛；[羊绒]分梳
dehumidification 减湿[作用]；除湿
dehumidifier 减湿器，干燥器
dehumidify 除湿
dehydrate 脱水；甩水
dehydrated weight 干重，脱水重量
dehydrating agent 脱水剂
dehydration 脱水[作用]
dehydration column 脱水塔
dehydration column reboiler 脱水塔再沸器
dehydrator 脱水器，除水器
dehydro-alanine 脱氢丙氨酸
dehydrochlorination 脱氯化氢
dehydrocyclization 脱氢环化
dehydrogenase 脱氢酶
dehydrogenation 脱氢[作用]
deice 除冰
deicing agent 防冰剂，除冰剂
deicing property 防冰性能，除冰性能
deionisation 消除电离作用；去离子作用

deionized distilled water 脱离子蒸馏水
deionized water 脱离子水；无离子水
D.E.K. 【德】德国染色坚牢度委员会
Dekol N 德科尔 N（提纯的亚硫酸盐纤维素废液，阴离子性，作分散剂、匀染剂、胶体保护剂用，商名，德国巴斯夫）
Dekol SN-S 德科尔 SN-S（聚丙烯酸酯，分散剂及保护胶体，浸染用染色助剂及皂洗助剂，商名，德国巴斯夫）
Del(delete) 删除；删除键《计算机》
delaminating 脱层，离层，分层
delay 延时，延迟
delayed action 延迟作用
delayed coagulation 缓凝[作用]
delayed cure 延迟焙烘
delayed cure finish 延迟焙烘整理（服装树脂整理焙烘工艺）
delayed deformation 缓弹性变形；延迟形变；迟缓性形变
delayed elastic recovery〈DER〉 缓弹性回复[率]
delayed recovery 延迟回复
delay-elastic deformation 缓弹性形变
delay element 延时元件
delay relay 延迟继电器
delete〈Del〉 删除；删除键《计算机》
deleterious 有毒的；有害的
deleterious effect 有害作用，有害影响
delft blue 东方蓝[色]（暗蓝色）
delicate colour 娇嫩色泽，柔和色泽
delicate fabrics 宜轻洗的织物；易损织物；轻薄织物，易变形织物
delignification 脱除木质素[作用]
deliming 脱钙，除钙；脱灰
delint 剥除短绒
delinting 剥棉籽绒

deliquate 冲淡,稀释
deliquescence 潮解
deliquescent 潮解的,吸湿的;潮解剂,吸湿剂
deliquescent agent 潮解剂,吸湿剂
deliquescent effect 吸湿性,潮解性
deliquium 潮解物
delivery 输出;出布
delivery by packed 随机供应
delivery end 出布端;输出端《电》
delivery end framing box [拉幅机]车尾箱
delivery hose 输出软管
delivery plaiter 落布架
delivery rate 输出率,输送率;给料速度,输料速度
delivery roller 落布辊,出布辊
delivery seal 出布处封口
delivery tube 导布管(喷射染色机中的导布装置)
delphinium blue 翠雀蓝[色]
delta connection 三角联接《电》
deluster 【美】=delustre
delustrant(=delusterant) 消光剂
delustrant content 消光成分
delustre 退光,消光,使无光;消光剂,除光剂,退光剂
delustred drying 烘燥失光,消光烘燥
delustred fibre 消光纤维,无光纤维
delustred grade 消光度
delustred printing 消光印花(指白涂料印花)
delustred rayon yarn 无光人造丝
delustre printing 消光白色印花
delustring 除光,退光;消光[工艺]
delustring agent 消光剂
deluxe [fabric] 高光泽电光整理布(常用于直贡、里子布等)
demineralization 脱矿质[作用];软化[作用]
demineralized water 软水
demineralizing plant 软水处理站;脱盐装置
demister 除雾器;除沫器
demitint 晕色(介于深、浅色之间)
demodulator 解调器《计算机》
demudding device 排污泥装置
demulcent 缓和的,减轻刺激的;软化剂
demulsification 破乳化[作用]
demulsifier (=demulsifying agent) 破乳化剂
demulsifying resistant 抗破乳剂
den (=dens., denier) 旦,旦尼尔(衡量纱线粗细的单位)
den. (=dens., density) 密度;比重;浓度
denaturant 变性剂
denaturated alcohol 变性酒精
denaturation 变性[作用]
denature 变性,使……变性
denatured guar gum thickening 变性瓜耳豆胶糊
dendrimer 树状大分子
dendrimer for water-proof finish 用于防水整理的树状大分子
denier 〈den, dens.〉 旦,旦尼尔(衡量纱线粗细的单位)
denier irregularity 线密度不匀率
denier number 〈Nd〉 旦数
DeniLite 丹丽来(漆酶,用于靛蓝织物漂白,具有不同于常规氯漂的独特整理效果,商名,丹麦诺维信)
denim 粗斜棉布,劳动布,牛仔布
Denimax 丹尼麦克斯(洗牛仔布用纤维

素酶,商名,丹麦诺维信)
Denimax Acid SBX 丹尼麦克斯酸性SBX(丹粒酶,即酸性纤维素酶,用于牛仔布处理,商名,丹麦诺维信)
Denimax BT 丹尼麦克斯BT(丹粒酶,即中性纤维素酶,用于牛仔布处理,商名,丹麦诺维信)
denim finishing range 牛仔布后整理生产线
denim garment dyeing 牛仔布成衣染色
denim look 劳动布风格,牛仔布风格
denim pants 粗斜纹布裤,劳动布裤,牛仔裤
denims 劳动布工作服,牛仔裤
denim suit 劳动布工作服,牛仔服
Denisol 丹尼素(成衣染色的涂料,可产生石磨效果,商名,意大利宁柏迪)
denitrated collodion 脱硝火棉胶
denitration 脱硝,去硝,除硝
denitrification 脱氮[作用]
dens. (=den, denier) 旦,旦尼尔(衡量纱线粗细的单位)
dens. (=den., density) 密度;比重;浓度
dense pattern 密纹图案
dense texture 紧密结构[织物],紧密组织
densification 增浓[作用],稠化[作用];压实
densimeter 密度计;比重计
densitometer 密度计;比重计;光密度计
density 〈den., dens.〉 密度;比重;浓度
density bottle 密度[测定]瓶
density-gradient centrifugal analysis 密度梯度离心分析
density of fabric 织物密度
density variation 密度变化;密度不匀性
denterium 氘,重氢

deodorant 脱臭剂,除臭剂
deodoration 除臭;除气味
deodor function 消臭功能
deodorisation 脱臭[作用]
deodorizer 脱臭剂,除臭剂
deodorizing carbon 脱臭炭
deodorizing finish 除臭整理
de-oiling 脱油,去油
deoxidation 脱氧
deoxygenation 脱氧[作用]
deoxyribonucleic acid 〈DNA〉 脱氧核糖核酸
DEP (diglycidyl ether of polyethyleneglycol) 聚乙二醇二缩水甘油醚
depainting 去漆,除漆;去漆渍(去除羊毛上标记颜色)
Department of Scientific and Industrial Research 〈DSIR〉 科学与工业研究局(英国)
Department of Transportation 〈DOT〉 运输部
departure 偏移,偏离
dependability 可靠性
dependence 相关;依赖
dependent variable 应变量《数》
dephenolizing 脱酚
depiction 雕刻图案;描绘
depigmentation 防止色素沉淀;脱色素[作用]
depilation 脱毛
depilatory 脱毛剂
depinning protection 防脱针装置
depinning roll 脱针辊(置于针板拉幅机出布区)
depletion 消耗量;用尽;倒空
depolymerization 解聚[作用]

depolymerizator 解聚[合]器,解聚[合]机
depolymerizing agent 解聚剂
deposit 沉积;沉积物
depositing bleach 堆置漂白
depositing plasma 沉积性等离子体(表面改性的主要工艺)
deposition 沉淀[作用];沉淀物
deposition potential 析出电位,析出势能
depreciation 折旧;减值;损耗
depress 降低,抑制;按下
depressant 抑制剂
depression 降温,降压;凹部
depression of freezing point 冰点下降
depressurizing 降压,减压
depth 深度;色泽浓度
depth of foam treatment 泡沫处理深度
depth of immersion 浸入深度
depth of shade 色度,色泽深度,色深
depuration 纯化,净化,提纯
DER (delayed elastic recovery) 缓弹性回复[率]
deresination 脱树脂[作用]
derivation 公式推导;偏差
derivatives of anthraquinone with fused ring structure 带有稠环结构的蒽醌衍生物(蒽醌染料结构)
derma 真皮,皮肤
dermal toxicity 皮肤毒性
dermatology 皮肤病学
dermestidae 幼蛀虫
dermestid beetles 蛀蚀毛皮的幼甲虫
desalinization 脱盐[作用];[海水的]淡化
desalinization of brackish water 咸水脱盐作用
desalt 脱盐
desalted water 脱盐水

desamination [羊毛]脱氨基作用
descaling 消除鳞片(使产生防缩、防毡并的效果)
description 说明,说明书;品名
descumming 脱泡沫,撇浮质
desetting 去除定形[作用]
desiccant 干燥剂
desiccantor 干燥器
desiccating agent 干燥剂
desiccation 干燥[作用];脱水
desiccator 干燥器
design 花纹;设计;图案
design area 花型范围;花型面积
designation 标示;名称;代号
designation of nominal width 公称宽度的标记
design brushing finish 刷花整理
design brushing machine 刷花机(织物干整理设备)
design details input 花纹详图输入
design digitalization 图案数字化
design flocking 图案植绒
design graph 花纹图样,设计图样
design interpretation 图案的表达
design limitation 图案限制
design motif 花样单元
design number 花号,印花图案号
design potential 图案优势
design processing [花样]设计处理
design register [印花]对花
design room 设计室,打样室
design sample 花样样品
design screen 设计图案显示荧光屏
design studios 图案摄影室
design style 设计风格
design test 鉴定试验;设计试验

design transport system〈DTS〉 花样[图案]输送系统
desizability 退浆能力,退浆性
desized cotton 退浆棉布
desize washer 退浆水洗机
desizing 退浆[工艺]
desizing agent 退浆剂
desizing by ferment 纤维素酶退浆
desizing cistern 退浆槽
desizing machine 退浆机
desizing mangle 退浆轧车
desizing percentage 退浆率
desizing with enzyme 纤维素酶退浆
desk 实验台
desk calculator(＝desk computer, desk top) 台式计算机
desmolipase 不溶性脂肪酶
desolvation 去溶剂化[作用]
desolventizing 脱溶剂
desorber 解吸器
desorption 解吸[作用](清除吸附气体)
desorption of moisture 脱湿[作用],减湿[作用]
desoxyindigo 脱氧靛蓝
despecking 去斑点
destarch 脱浆,退浆
destarching 退浆[工艺],退浆处理
destatic 除静电
destaticizer 除静电剂
destruction 破坏;断裂
destruction test (＝destructive method) 破坏性试验法
desulfonation 脱磺酸基[作用]
desulfurizing 脱硫,去硫
desulphuration(＝desulphurization) 脱硫[作用]

desuperheater [过热蒸汽]减温器
deswelling 消膨胀,退溶胀[作用]
deswollen state of cellulose 纤维素纤维的消溶胀状态
DETA(diethylenetriamine) 二亚乙基三胺（染色助剂,胶乳稳定剂,阻氧化剂,加硫促进剂）
detachable column 单立柱
detachable part 可装拆件
detackifier 脱黏剂
detail 详图;细节,详情;零件,部件
detail [图案]清晰度
detail reference 零件图号
details and outlines of printing 印花清晰度与轮廓
detail views 详图
detectability 检测能力,探测能力
detecting element 传感元件
detection 检验,探测;检波
detection agent 检定剂;检测剂
detection and assessment of photo-chromism 光致变色的检验与评定
detection of metals 金属检测
detectivity 探测率,检测能力
detector 检测器,探测器;检波器;探索装置
β-detector β-射线检测仪（用于测定织物密度和重量）
detector head 探测[器]头;指示器;检波头
detector knife 检测刀（剪毛机上用）
detent 棘爪制动器,锁销
detention period 停留时间
detention tank 滞留[沉淀]槽
detent spring 棘爪簧
detergency 洗涤性;洗净力,去污力

detergency builder 助洗剂
detergency efficiency 去污效率
detergency promoter 助洗剂
detergency test 脱垢试验,去污试验
detergent 洗涤剂,净洗剂
detergent analysis 洗涤剂分析
detergent auxiliary 洗涤助剂,去垢助剂
detergent base 洗涤剂基料,去垢剂基料
detergent booster 去污增效剂
detergent index 洗涤指数
detergent residues 洗涤剂残余
detergents and wetting agents 洗涤剂和润湿剂
detergent sanitizer 净洗消毒剂
detergent scouring 洗涤剂煮练
deterioration 劣化[作用];降解[作用];变质,变坏;退化[作用]
determination 测定,鉴定;规定,定义
determine〈detm.〉 确定;测定;决定
determining molecular weight 分子量测定
detersive 洗净剂,清洁剂
detm.(determine) 确定;测定;决定
DETO(Dyes Environmental and Toxicological Organization) 染料环境和毒物学组织
detonation 爆破
detoxicate 解毒
detrition 磨耗;耗损
detwister 退捻(将绳状织物退捻,以便开幅)
detwister-squeezer 退捻轧水机
detwisting and opening machine 退捻开幅机
Deutsche Wool Institute〈DWI〉 德国羊毛研究所
devapouration 止汽化[作用]

developed area 展开面积
developed colours 显色染料
developed dyeing 显色染色[工艺](指直接重氮染料及不溶性偶氮染料的显色工艺)
developed projection 展开投影图
developer 显色剂
Developer AN 显色剂 AN(稳定的乙萘酚钠,商名,德国制)
developing 显色[工艺];显影
developing agent 显色剂
developing and soaping range 显色皂洗联合机
developing bath 显色浴,显色液
developing dyeing method 显色染色法
developing dyes 显色染料
developing liquor 显色液
developing machine 显影机(照相雕刻的一种附属设备)
developing steamer 显色蒸箱
developing trough 显色槽
development 发展;研制;显色;显影(照相技术术语)
deviation 偏差,差异;偏移
deviation in shade 色差
device 设备,装置;元件;方法,手段
devolatilization 排气,除挥发物
devorant style 烂花印花法
devore effect 烂花效应
devore style 烂花印花法
devulcanization 脱硫[作用]
dewatering 脱水[作用];脱水[工艺]
dewaxed cotton 脱脂棉
dewaxing 脱蜡,去蜡
dew dropping [蒸箱顶部]滴水(蒸化疵点)

deweighting 减量处理,碱减量处理
deweighting loss percentage 减量率
deweight printing 减量印花
dewing 给湿,喷雾
dewing machine 喷雾机,给湿机
dew point〈d. p.〉 露点
dew point hygrometer 露点湿度计
dew point method 露点测湿法
Dew Print machine 给湿印花机(一种湿法转移印花机,英国制)
Dew Print process 给湿印花法(一种连续湿转移印花法,英国开发)
dextrin[e] 糊精
dextrinization 糊精化[作用]
dextrinization activity 糊精化活力
dextrin thickening 糊精糊
dextrorotatory 右旋的(有机化合物结构)
dextrose 右旋糖,葡萄糖
DF(=df, degree of freedom) 自由度
3D fabric(three-dimensional fabric) 三维[结构]织物
DFE(differential frictional effect)[羊毛]摩擦差异效应
DG filter(Davis-Gibson filter) 台维斯—吉布森滤光器(即 DG 滤光器)
DG finish(durable glaze finish) 耐久性光泽整理,耐久上光整理
DHDMEU (4,5-dihydroxy-1,3-dimethyl ethylene urea) 4,5-二羟基-1,3-二甲基乙烯脲(无甲醛交联整理剂)
DHDMI (4,5-dihydroxy-1,3-dimethyl-2-imidazolinone) 二羟基二甲基咪唑啉酮(无甲醛交联树脂)
DHEU(dihydroxy ethylene urea) 二羟基亚乙基脲
dia.(=diag., diagram) 图,图表,图解,示意图;电路图
diacetal 二乙缩醛
diacetyl ethylene diamine〈DAED〉 二乙酰乙二胺(双氧水活化剂原料)
diacid(dibasic acid) 二酸,二元酸;二盐基酸
diacidic base 二元碱
Dia-Film 一种透湿防水的热塑性聚氨酯薄膜的品牌
diag.(=dia., diagram) 图,图表,图解,示意图;电路图
diagnosis [设备]诊断
diagonal arranging bowls 交叉轧辊
diagonal brushing machine 斜交式刷绒机;斜向刷布机
diagonal displacement 对角线形纬斜(织疵)
diagonal pattern 斜纹图案,斜纹花型,斜交式图案
diagonal repeat adjustment 花位对角调节
diagonal stay 对角拉撑,斜撑条
diagonal straightening roller 对角整纬辊(校正对角线纬斜)
diagonal strut 斜撑
diagonal weft 对角纬斜,斜线形纬斜
diagram〈dia., diag.〉 图,图表,图解,示意图;电路图
diagrammatic sketch 示意图,草图
diagrammatic view 图示,简图
Diagum 达伊胶(植物种子胶衍生物,印花糊料,商名,德国大美)
dial [刻]度盘;转盘
dialdehyde 二醛
dialdehyde chitosan 双醛壳聚糖
dialdehyde cotton fabric 双醛基棉织品
dialdehyde cotton fibre 双醛基棉纤维(用

于改性)
dialdehydes 双醛类(用于耐久压烫整理)
dialdehyde starch 二醛淀粉
dial gauge 千分表,[度盘式]指示表
dialkyl benzene 二烷基苯
diallyl phthalate〈DAP〉 邻苯二甲酸二烯丙酯
dial thermometer 圆刻度盘温度计
dialysis 渗析,透析
dialysis membrane 渗析膜,透析膜
diamagnetic 抗磁的
diameter 直径
diameter pitch 径节
diaminostilbene 二氨基芪(荧光增白剂的主结构)
diaminostilbene disulfonic acid〈D. S. D. acid〉 二氨基芪二磺酸(一种荧光增白剂原料)
Diamira dye 大爱米拉染料(活性染料,含乙烯砜基团,商名,日本三菱)
diammonium phosphate〈DAP〉 磷酸氢二铵
diamond graver 金刚石刻刀
diamond holder lever [缩小机]针架
diamond pattern 菱形图案
diamond pigment 钻石涂料
diamond point 金刚钻针《雕刻》
diamond printing 钻石印花(具有钻石光的印花方法)
diamond stylus 金刚钻针《雕刻》
dianisidine blue 联茴香胺蓝,直接偶氮蓝
Dianix AM 大爱尼克司 AM(一种高耐光牢度分散染料,商名,德国制)
Dianix dyes 大爱尼克司染料(分散染料,商名,德国制)
Dianix HLA 大爱尼克司 HLA(高耐光牢度分散染料,商名,德国制)
dianthrone 二蒽酮
diaper 尿布;彩色格子斜纹棉布;细亚麻毛巾布;菱形花纹织物
diaphaneity 透明度
diaphanometer 透明度计
diaphragm 隔板;薄膜,膜片,隔膜;光阑
diaphragm burst test 薄膜顶破试验
diaphragm cylinder 薄膜气缸
diaphragm for explosion proof 防爆膜
diaphragm gauge 膜盒式压力计,薄膜式压力计
diaphragm loading device 薄膜加压装置
diaphragm pump 隔膜泵
diaphragm steam trap 隔膜式疏水器(一种热静力式的排水阻汽装置)
diaphragm valve 隔膜阀
Diaplex 大爱普莱克斯(透气、防水新型聚氨酯,日本三菱重工)
diapositive 透明正片,反底片,幻灯片
diapositive film 透明正片(胶片)
diarylmethane dyestuff 二芳基甲烷染料
diastase(=diastatic enzyme) 淀粉糖化酶,淀粉酶制剂(退浆剂)
diathermic oil 热载油
diatomaceous earth 硅藻土
diatomaceous earth filter 硅藻土过滤《环保》
diatomite 硅藻土
diatomite adsorption treatment 硅藻土吸附法《环保》
diazo aminobenzene〈DAB〉 重氮氨基苯(发泡剂)
diazo component 重氮组分
diazo compound 重氮化合物
diazo-coupling 重氮偶合

diazo dyes 偶氮染料,重氮染料
diazonium salt 重氮盐
diazo titration 重氮滴定
diazotization 重氮化;重氮化反应
diazotize colour 重氮化染料
diazotized 重氮化的
diazotized and developed dyes 重氮显色染料
diazotized base 重氮化色基
diazotizing 重氮化;重氮化反应
diazotype 重氮印象法;重氮化型[照相材料]
dibasic acid〈diacid〉 二酸,二元酸;二盐基酸
dibutyl phthalate〈DBP〉 [邻]苯二[甲]酸二丁酯(增塑剂)
dicarboxylic acid 二羧酸
o-dichlorobenzene 邻二氯苯(染色载体)
dichlorobutadiene 二氯丁二烯(合成橡胶)
dichlorocyanuric acid 二氯氰酸
dichlorodiphenyl trichloromethane〈DDT〉 滴滴涕,双对氯苯基三氯乙烷(国际禁用杀虫剂)
dichloroisocyanuric acid〈DCCA〉 二氯异氰尿酸(羊毛氯化剂)
dichloromethane〈DCM〉 二氯甲烷
dichlorophen 二氯芬,二羟二氯二苯甲烷(防蛀、防腐、防霉及耐气候的整理剂)
dichloropropanol 二氯丙醇(棉织物湿防皱树脂整理剂)
dichloropyrimidine dye 二氯嘧啶染料(含二氯嘧啶活性基团的活性染料)
dichloroquinoxaline dye 二氯喹噁啉染料(含二氯喹噁啉活性基团的活性染料)
dichlorotriazine dye〈DCT dye〉 二氯均三嗪染料(含二氯均三嗪活性基团的活性染料)
dichroic 二色性的
dichroic dyestuff 二色性染料
dichroic ratio 二[向]色性比
dichroism 二色性,二向色性
dichromate process 重铬酸盐法(可溶性还原染料染色法之一)
dichromatic 二色性的
dichromatism 二色性
Dicrylan WK 特奇连 WK(改性聚硅氧烷,非离子型,改善手感、抗皱、抗起毛起球耐久整理剂,商名,瑞士制)

dictionary〈dict.〉 字典,辞典
dicyandiamide 双氰胺,二聚氨基氰;氰基胍
dicyandiamide formaldehyde resin 双氰胺甲醛树脂(纤维素织物阻燃整理剂)
dicyanodiamide 双氰胺,二聚氨基氰;氰基胍
di-diol bond 双二醇键
die [刻纹]铜模,冲模,硬模,阴模;型板,压出板
die and mill engraving 钢芯雕刻
die-away test 衰减试验(洗涤剂活性物生物降解率的测定方法)
die block 模块,模板
die cast 压铸,模铸
die-casting method 压铸法
die coating 模压涂层,挤压涂层
die cutter 冲压裁剪机,钢芯雕刻机《印》
Dieldrin 狄氏剂(氧桥氯甲桥萘,防蛀等用,毒性大,国外已禁用)
dielectric 电介质;绝缘材料
dielectric barrier discharge〈DBD〉 [常压]介电屏蔽放电,电介质屏蔽放电
dielectric breakdown test 绝缘击穿试验,

电介质击穿试验

dielectric breakdown voltage 绝缘击穿电压,电介质击穿电压

dielectric constant 介电常数,电容率

dielectric dryer 高频烘燥机

dielectric drying 高频烘燥

dielectric heating 绝缘加热

dielectric high-frequency drying 电介质高频烘燥

dielectric hysteresis 电介质滞后现象

dielectric loss 绝缘损失

dielectric point 等电点

dielectric property 介电性,不导电性

dielectrics 绝缘材料,电介质

die milling machine 钢模刻纹机

die pantograph machine 钢模缩小机(印花雕刻用)

diepoxide 二环氧化物

diesel engine 内燃机,柴油机

diesel fuel 柴油机燃料

diethanolamide surfactant 二乙醇酰胺表面活性剂(浸透剂,起泡剂)

diethanolamine 二乙醇胺

diethanolamine soap 二乙醇胺皂

diethylamine 二乙胺(选择性溶剂,杀虫剂,聚化作用抑止剂)

diethylene glycol〈DEG〉 二甘醇(助溶剂和吸湿剂)

diethylenetriamine〈DETA〉 二亚乙基三胺(染色助剂,胶乳稳定剂,阻氧化剂,加硫促进剂)

diethylenetriamine pentaacetic acid〈DTPA〉 二亚乙基三胺五醋酸(有机螯合剂)

diethylenetriamine pentaacetic acid pentasodium salt 二亚乙基三胺五醋酸五钠(双氧水漂白浴的金属络合物形成剂)

diethylenetriamine pentamethylene phosphonic acid〈DTPP〉 二亚乙基三胺五亚甲基膦酸(生物降解性较优)

diethylenglycol ester surfactant 二甘醇酯型表面活性剂(乳化分散剂,纤维用油剂,染色助剂)

diethyl ether [二]乙醚

diethyl phthalate 酞酸二乙酯,邻苯二甲酸二乙酯(聚酯纤维染色载体)

diethyl tartrate 酒石酸二乙酯

difference〈dif.,diff.〉 差异,区别

difference of shade under artificial light 非自然光下的色差

different colours of both sides 双面异色

different fineness 异细度

differential 差速器,差动装置;微分《数》

differential block 差动滑轮

differential brake 差动闸

differential calorimetry 差示量热法(试样和某一参考材料反应热、吸收热、水解热等的测量和比较方法)

differential control 差动控制

differential diffusion coefficient 微分扩散系数

differential distribution 差示分布

differential dyed fibre 异染纤维

differential dyeing〈DD〉 差异染色(指混纺、交织物或具有不同上染性能的同类纤维织物在一浴中染得不同色泽)

differential dyeing fibre 差异染色性纤维(具有不同染色特性的变性同类纤维)

differential dyeing method 差异染色鉴别法

differential dyeing test 差异染色试验法

differential feed 差动喂布;差动送纱

differential fibre 差别化纤维,改性纤维

differential friction [羊毛顺逆鳞片]摩擦差异
differential frictional effect〈DFE〉[羊毛]摩擦差异效应
differential gauge 微分计;微分气压计
differential gear 差动齿轮;补偿器
differential-geared device 差动齿轮装置
differential gearing system 差动轮系
differential manometer 差示压力计
differential motion 差动;差动装置,差微装置
differential pressure 压差,压力降
differential pressure indicator 压差指示器
differential pressure switch 压差开关
differential pump 差动泵(往复泵的一种)
differential refractometer 差示折光计,差示折射仪
differential relay 差动继电器
differential scanning calorimeter〈DSC〉差示扫描量热计
differential scanning calorimetry〈DSC〉差示扫描量热法
differential sensitivity method 差示敏感度法
differential shrinkage 差示收缩量;差异收缩量
differential speed 差微速度
differential switch 差动开关
differential taking-up mechanism 差微式卷取装置
differential thermal analysis〈DTA〉差热分析
differential thermo-gravimetric analysis 差示热重分析法
differential thermo-mechanical analysis 差示热机械分析

differential thermometer 差示温度计
differential transformer 差示变压器,差动变压器
differential vapour pressure thermometer 差示蒸汽压温度计
differential viscosity 微分黏度
differential weft straightener 差动式整纬器
differential winder 差速卷布机
differentiator 微分电路,差动电路;差示器,差示装置
different shade on edge 布边色差
diffracted grating 衍射光栅
diffraction 衍射,绕射
diffraction diagram 衍射图
diffraction grating 衍射光栅
diffraction pattern 衍射图样
diffraction spectrum 衍射光谱
diffractometer 衍射计,衍射仪
diffractormetry 衍射法
diffuse adsorption 扩散吸着
diffused illumination 漫散照明,扩散照明
diffused light 扩散光
diffuser 散气管(活性污泥处理)《环保》;扩散体,扩散器;喷雾器;浸提器;洗料器;扩散剂
diffuse reflectance 漫反射度,漫反射
diffuse reflection 扩散反射
diffuse reflection factor 扩散反射率
diffuse transmittance 漫射透光率,漫射透射比
diffusibility 扩散能力
diffusible resin 扩散性树脂
diffusion 扩散[作用];渗滤;漫射
diffusional adsorption layer 扩散吸附层
diffusional boundary layer 扩散边界层

diffusional double electric layer 扩散双电层
diffusion cell 扩散池
diffusion coefficient 扩散系数
diffusion constant 扩散常数
diffusion current 扩散电流
diffusion effect 扩散效应
diffusion equation 扩散方程式
diffusion in non-steady state 非稳态扩散
diffusion in steady state 稳态扩散
diffusion layer 扩散层
diffusion mechanisms in dyeing 染色扩散机理
diffusion of dyes 染料的扩散
diffusion potential 扩散电位
diffusion rate 扩散速度；扩散率
diffusion step 扩散阶段
diffusive reflection 漫反射
diffusivity 扩散系数；扩散率；漫射率
difluorochloropyrimidine 二氟一氯嘧啶（活性染料的一种活性基团）
difunctional 双官能团［的］；有两种功能的
difunctional comonomer 二官能［团］共聚单体
difunctional epoxy compound 双官能环氧化合物
digallic acid 单宁，二棓酸，单宁酸，鞣酸
digested sludge 消化污泥
digester 浸煮器，蒸煮器，蒸煮锅；消化池
digester cycle 蒸煮周期
digestion 消化
digestion tank 消化池
digestive gas 消化气体
digestive treatment 消化处理《环保》
digestor 浸煮器，蒸煮器，蒸煮锅；消化池

digit 数［字］；数［字］位；位数
digital 数字的，计数的；数字式
digital-analog converter 数字模拟转换器
digital camera 数码照相机
digital computer 数字计算机
digital controlled 数字式控制
digital densitometer 数字密度计
digital display 数字显示
digital electronic computer 电子数字计算机
digital encoder 数字编码器
digital frequency converter 数字式变频器
digital image process technology 数码图像处理技术
digital indication 数字显示
digital indicator 数字显示装置
digital ink jet printing 数码喷墨印花
digital integrator 数字式积分器
digital inverter 数字式变频器
digitalization 数码作用，数码技术
digital jet printing machine 数码喷墨印花机
digital measurement technology 数字测量技术
digital moisture indicator 数字式湿度显示器
digital picture library 数码图形库《计算机》
digital plotter 数码绘图机《计算机》
digital printing 数码印花（运用计算机完成的印花技术，如喷墨印花）
digital printing systems〈DPS〉 数码印花系统
digital print-out （计算机的）数字印出
digital process control computer 数字式工艺控制计算机

digital read out 数字读出
digital servomotors 数码伺服电动机
digital shrink computer 数字式缩率计算机
digital softwinder 数码松式卷布器(张力小)
digital speed indicator 数字式测速仪
digital static printing 数码静电印花
digital stencil-less printing 数码无版印花
digital switch 数字式开关
digital technology 数码技术
digital TEMP-RH metre 数字式温湿度计
digital textile 数码纺织,数码纺织品
digital textile sample printer 数码印花布打样机
digital thermometer 数字式温度计
digital tone 数字色调(由数值电路调整的色调)
digital versatile disc〈DVD〉 数字多功能光盘《计算机》
digital video disc〈DVD〉 数字化视频光盘《计算机》
digital web guiding 数字式织物引导器
digital weighing device 数字式秤重装置
digital width indicator 门幅数字显示器(用于拉幅机)
digital xerographic printing technique 数码静电印花技术
digital yield indicator 数字式产量指示器
digitize 数字化
digitizer 数字转换器
diglycidyl ester 甘油二酸酯
diglycidyl ether of 1,4-butandiol 1,4-丁二醇二缩水甘油醚
diglycidyl ether of polyethyleneglycol〈DEP〉 聚乙二醇二缩水甘油醚

diglycol 二甘醇(助溶剂和吸湿剂)
diglycol ester 二乙二醇酯,二甘醇酯
diglycol ester surfactant 二甘醇酯表面活性剂
diglycol stearate 硬脂酸二甘醇酯(润滑剂和柔软剂)
dihydric alcohol 二元醇
2,6-dihydroxyanthraquinone 2,6-双羟基蒽醌(熔态金属还原染料染色的导氢剂)
4,5-dihydroxy-1,3-dimethyl ethylene urea〈DHDMEU〉 4,5-二羟基-1,3-二甲基乙烯脲(无甲醛交联整理剂)
4,5-dihydroxy-1,3-dimethyl-2-imidazolinone〈DHDMI〉 二羟基二甲基咪唑啉酮(无甲醛交联树脂)
dihydroxy ethylene urea〈DHEU〉 二羟基亚乙基脲
dihydroxy tartaric acid 二羟基酒石酸
diisocyanate 二异氰酸酯,二异氰酸盐
dil.(dilute) 稀释
dil.(dissolve) 溶解
Dilasoft KLP 德纳素 KLP(亲水性柔软剂,商名,瑞士科莱恩)
Dilasoft RN 德纳素 RN(亲水性柔软剂,商名,瑞士科莱恩)
Dilasoft TF 德纳素 TF(亲水性柔软剂,商名,瑞士科莱恩)
dilatability 膨胀性,延伸性
dilatancy 滞胀,胀流性,触稠性,摇稠现象(流变特性,流体或胶体在搅拌中变得黏稠而在静置时仍复原的现象)
dilatant 滞胀的,胀流的,触稠性的;膨胀的,扩张的
dilatant flow 胀流型流动
dilatant flow behavior 膨胀流动性

dilatate 膨胀
dilatation 膨胀[作用]
Dilatin 狄拉廷(聚酯纤维染色载体,商名,瑞士科莱恩)
dilation 膨胀[作用]
dilatometer 膨胀计
dilatometric thermometer 膨胀温度计
dilatometry 膨胀测量法
diluent 稀释剂
diluent limit 稀释限界
dilute〈dil.〉稀释
diluter 稀释剂
dilution 冲淡,稀释
dilution, disperse and decontaminate〈DDD〉稀释、分散与去污(废物去除系统)
dilution ratio 稀释比
dilution viscometer 稀释黏度计
dilution wash 稀释清洗
dim 微暗的;使暗淡
dim colour 暗色
dim effect 朦胧[印花]效果
dim effect printing 迷彩印花;暗淡印花;朦胧印花
Dimensa mercerizing machine 狄蒙沙拉幅直辊丝光机(商名,德国贝宁格)
dimension 尺寸;因次;量纲
dimensional analysis 量纲分析,因次分析
dimensional change 尺寸变化(织物在染整加工过程中产生的长度、面积或形状的变化)
dimensional change tester 尺寸变化试验仪
dimensional measurement 尺寸测定
dimensional metrology 尺寸测量法
dimensional recovery 尺寸回复
dimensional restorability 形状复原性,尺寸复原性

dimensional stability 形稳性,尺寸稳定性
dimensional stability tester 尺寸稳定试验仪
dimensional tolerance 尺寸公差
dimension line 尺寸线
dimension scale 尺寸比例
3-dimentional effect 三维效应,立体效应
3-dimentional fabric〈3D fabric〉三维织物
dimer 二聚物
dimerization 二聚反应
dimethyl acetyl amide〈DMAA〉二甲基乙酰胺
N,N-dimethyl-azetidinium chloride〈DMAC〉二甲基—氯化吖丁啶(棉纤维改性剂)
dimethyl dichlorosilane 二甲基二氯硅烷(耐洗的防水剂和柔软剂)
dimethyl formamide〈DMF〉二甲基甲酰胺(鉴别活性染料及合成纤维的试剂)
dimethylfumarate〈DMF〉富马酸二甲酯(一种防霉剂,此产品因能引起接触性皮炎,欧盟已禁止使用)
dimethylmethylol urea resin 二甲羟甲基脲树脂
dimethylol acetone 二羟甲基丙酮(防缩整理剂)
dimethylol adipic acid amide 二羟甲基己二酰二胺(耐洗的织物防缩防皱活性树脂)
dimethylol butanediol diurethane 二羟甲基二氨基甲酸丁二酯(防缩防皱交联剂)
dimethylol carbamate 二羟甲基甲氨酸酯
dimethylol dicarbamate 二羟甲基二氨基甲酸酯(防缩防皱交联剂)
dimethylol dihydroxy ethylene urea〈DMDHEU〉二羟甲基二羟基亚乙基脲(一

种耐皂洗、氯漂的棉织物树脂整理剂）

dimetholethylene urea〈DMEU〉二羟甲基亚乙基脲（防缩防皱树脂整理剂）

dimethylol-formamide 二羟甲基甲酰胺（棉织物的防缩防皱交联剂）

dimethylol hydroxy propylene urea〈DMHPU〉二羟甲基羟基亚丙基脲

dimethylol propylene urea〈DMPU〉二羟甲基亚丙基脲（耐洗的棉织物防缩防皱树脂整理剂）

dimethylolpropylene urea 二羟甲基亚丙基脲

dimethylol triazone precondensate 二羟甲基三嗪酮初缩体

dimethylol urea〈DMU〉二羟甲基脲（一种脲醛树脂，棉或黏胶织物的防缩防皱树脂整理剂）

dimethylol urea resin〈DMUR〉二羟甲基脲树脂

dimethyl sulphone 二甲砜

dimethyl sulphoxide〈DMSO〉二甲[基]亚砜

dimethylurea and glyoxal〈DMUG〉二甲基脲和乙二醛（树脂整理剂）

dimmer 遮光器；调光器；减光器

dim style 朦胧式样

DIN【德】德国工业标准

DIN colourimetric system DIN 测色系统

dinyl 联[二]苯（热载体或聚酯纤维的染色载体）

dioctyl phthalate〈D.O.P.〉邻苯二[甲]酸二辛酯（增塑剂）

dioctyl sodium sulfosuccinate 二辛酯琥珀酸磺酸钠（阴离子表面活性剂）

dioctyl sulfosuccinate 二辛基磺基琥珀酸酯

diode 二极管

diol 二醇

diol bond 二醇键

dioxane 二噁烷（溶剂）

dioxin 二噁英（毒性极大的持久性有机污染物）

dioxolone 二噁茂酮，二氧杂茂酮

dip 浸渍

dip bleaching 浸漂

dip bonding 浸渍黏合[法]

dip-coating 浸涂

dip composition 浸渍合剂，浸渍组分

dip-drain technique 浸渍—流干法；浸渍—排液法

dip-dyed hosiery 浸染针织物；浸染袜类

dip-dyeing 浸染

dip dyeing machine 浸染机

dip dyeing with oxidation colour 氧化染料浸染

diphehyl amine 二苯基胺

diphenyl 联[二]苯（热载体或聚酯纤维的染色载体）

diphenyl black 联苯黑，二苯黑

diphenyl black dip dyeing 联苯黑浸染

diphenyl ether [二]苯醚

diphenyl heating kettle 联苯加热釜

diphenyl methane 二苯甲烷

diphenyl methane diisocyanate 二苯甲烷二异氰酸盐（或酯，聚脲烷树脂，地毯用黏合剂）

4,4′-diphenyl methane diisocyanate〈MDI〉4,4′-二苯甲烷二异氰酸酯

diphenyl methane dye 二苯甲烷染料

diphenyl oxide [二]苯醚（热载体）

diphenyl steam boiler 联苯蒸汽[锅]炉

dip-hydro 浸渍—脱水（毛衫湿处理法）

dip-hydro technique 水溶液浸渍整理工艺
diphyl 联[二]苯(热载体或聚酯纤维的染色载体)
dipolar ion 偶极离子
dipole molecule 偶极分子
dipole moment 偶极矩
dipolymer 二聚物
dip padding 浸轧工艺
dipped fabrics 浸染织物,浸渍织物
dipped hose 浸染袜子
dip pickup 浸渍附着量
dipping 浸染
dipping process 浸渍过程
dipping vat 浸染槽
dip-printing system 浸渍印花法(纱线印花)
dip-squeeze system 浸轧装置
dip starching 浸渍上浆
dip-steam process 浸渍—汽蒸工艺
dip-stick [浸]测深[度]标尺
dip test 浸渍试验
dip treating 浸渍处理
dip trough 浸渍槽
dip-tumble process 浸渍—转筒法(溶剂防缩用)
dip-tumbler 浸渍转筒
direct-acting pump 直接驱动泵,直接联动泵
direct air flow〈DAF〉 直接[空]气流
direct-air-flow system 直接气流系统
direct and developed dyeing〈D and D〉 [拔染印花用]直接染料和后处理染色工艺
direct and developed dyes〈D and D dyes〉 重氮显色直接染料
direct azo dyes 直接偶氮染料

direct calorimetry 直接热量测定法
direct coating 直接涂层,直接上胶
direct colour 直接染料
direct-contact heat exchanger 直接接触式热交换器
direct cotton dye 直接染棉染料,染棉直接染料
direct coupling gear 直接偶联齿轮;联轴节传动
direct current〈DC,D. c. ,dc, d. c. ,d-c〉 直流[电]
direct current amplifier 直流放大器
direct current control voltage 直流控制电压
direct current generator 直流发电机
direct current mains 直流电源
direct current three-wire system 直流三线制《电》
direct digital control〈DDC〉 直接数字控制
direct digital dye-jet[textile]printing 直接数码喷射印花
direct digital printer 直接数码印花机
direct drive rotary printing machine 直接驱动式圆网印花机(荷兰斯托克制)
direct dyebath monitoring〈DDM〉 直接染浴监控
direct dyeing 直接染色
direct dyes 直接染料
direct dyes dip dyeing 直接染料浸染法
direct dyes discharge 直接染料拔染法
direct dyes for after-treatment 需要后处理的直接染料
direct dyes for metal salt after-treatment 用金属盐后处理的直接染料
direct dyes resist 直接染料防染法

direct engraving process 直接雕刻工艺
direct esterification 直接酯化[作用]
direct fast turquoise blue 直接耐晒翠蓝
direct-fired heater 直接火焰加热器
direct gas-heating system 煤气直接燃烧加热系统
direct gravure coating 凹印直接涂层
direct heat dryer 直接加热烘燥机
direct-heating gas burner 直接加热煤气灯,直接加热煤气燃烧器
directing roller 导[向]辊
directional frictional effect 顺摩擦效应
directional friction effect of wool 羊毛定向摩擦效应
directionality 方向性
directional valve 方向阀,换向阀
directional vanes 导向叶轮
direction of twill 斜纹方向
direct printing 直接印花
direct printing style 直接印花法
direct printing with mordant dyes 媒染染料直接印花
direct printing with substantive colour 直接染料直接印花
direct reflectance 正反射率
direct reflection 正反射(镜面反射)
direct-reflection factor 正反射率
direct sizing 直接上浆
direct steam 直接蒸汽,活蒸汽,一次蒸汽
direct style 直接印花法
direct to-garment〈DTG〉 直接制衣
direct to-garment printing 直接制衣喷墨印花
direct transmission 正透过,正透射
direct-transmission factor 正透过率
direct transmittance 正透过率

direct weighting 直接加压
Diresul RDT dyes 特丽素 RDT 染料(环保型硫化染料,含硫极低,无不良气味,商名,瑞士科莱恩)
dirt 污垢
dirt and oil 油污
dirt and oil patch 油污斑
dirt catcher 滤尘器,除尘器
dirt collector 吸尘器
dirt factor 污垢因素
dirt removal 去除污垢
dirt removing power 去污能力
dirt repellent finish 拒污整理,防污整理
dirt resistance 抗污性
dirt retardant 阻垢剂
dirt stains 污渍
dirt test 沾污试验
dirty colour 沾污的色浆(滚筒印花中前后色浆传色造成)
dirtying 沾污;沾色
DIS(Draft International Standard) 国际标准草案
disadvantage 疵点;不方便;缺点
disaggregation 离解作用
disassembly 拆卸,分解
disassembly and dismantling [机器]拆卸
disassociation 解离[作用]
disazo dyes 双偶氮染料,双重氮染料
disazo dyestuff 二重氮化染料,双重氮染料
disc 圆盘,圆片
disc brake 盘式制动器,圆盘刹车
disc clutch 圆盘离合器
disc comparator 圆盘比色计
disc feed 圆盘送料
discharge 拔染;放出;放电;放流量

Discharge 2000 拔印素 2000（涂料拔染系统，无需使用蒸汽而达到拔染效果，商名，意大利宁柏迪）
dischargeability 拔染性
dischargeable 可拔染的
dischargeable dyeings 可拔染的染地，可染的地色
dischargeable with difficulty 难拔染
discharge accelerating agent ［印花］助拔剂
discharge agent 拔染剂，咬白剂，拔白剂
discharge amount standard 排出量标准
discharge capacity 排放量，流量；排水能力
discharge design 拔印花样
discharge device 出布装置
discharge edges ［拔染］色晕；拔染渗边（印疵）
discharge fan 排气风扇
discharge grounds 拔染地色
discharge paste 拔染浆，拔染印浆
discharge pressure （气压机的）排放压力
discharge printing 拔染印花
discharge prints 拔染印花织物
discharge process 拔染工艺
discharge-resistant dyes 防拔染［用］染料；防拔染［着色］染料
discharge-resist printing 防拔染印花
discharges with mordant dyes 媒染染料拔染
discharge test 拔染试验
discharge valve 放汽阀，排水阀；排液阀；闸流管《电》
discharge winch 出布绞盘
discharging 拔染；蚕丝精练；放电
discharging auxiliaries 拔染助剂
discharging of preparing agent 处理剂拔染（指媒染染料的处理剂）
discharging of tannin mordant 单宁媒染拔染
disc knife 圆片刀
discolour 脱色
discolouration 褪色；脱色；变色
discoloured wool 尿污毛，变色毛，褪色毛
discolouring 脱色，变色
discolourization 褪色，脱色，变色
discomfort 不舒适感
discomfort index 不舒适指数
discomfort threshold 不舒适界限
disconnector 断路器，切断开关
discontinuous dyeing machine 间歇式染色机
discontinuous open-width washing machine 间歇式平幅水洗机
discontinuous process 不连续工艺，间歇法
discrepancy 不符合；差异
discrete relaxation 不连续松弛，分立的松弛
discrimination 鉴别，识别
discrimination threshold 识别界限
discriminator 鉴频器；鉴别器，甄别器
disc steam trap 圆盘式疏水器
disc tension device 圆盘张力装置
disc valve 圆盘阀
disengagement 分离，解开，脱离
disengaging bar 关闭杆
disfigured design 对花不准（印疵）
disharmony 不调和
dished drum end 碟形汽包封头
dished head 碟形封头，凸封头
disincrustant 防水锈剂，消除锅垢剂
disinfectant 清毒剂
disinfectant cleaning 消毒剂清洗

disinfection 消毒,灭菌
disintegrating machine 粉碎机,分解机
disk 圆盘,圆片
disk brake pad 盘式制动磨擦块
diskette 塑料磁盘,软盘
disk file 磁盘文件
disk operating system〈DOS〉 磁盘操作系统《计算机》
disk test 圆盘试验(测染料分散性用)
dislocation 位错,变位,位移;结晶转移
dismantle 拆卸,拆除
dismantling 拆卸
disodium hydrogen phosphate 磷酸氢二钠
disorder 无序,紊乱
disordered region 无序区
disordered structure 无规结构
disorientation 消定向[作用],解取向[作用]
disoxidation 减氧[作用];还原[作用]
dispatcher 配电器
dispenser 分配器
dispensing and measuring device 调配计[量]测[定]装置(染液)
dispensing and weighing station 称料调配站(染料)
dispergator 解胶剂,胶溶剂
dispersable vat dyes 分散型还原染料
dispersancy 分散力
dispersant 分散剂
disperse 分散
disperse-acid composite dyes 分散—酸性组合染料
dispersed acetate dyes 分散性醋酯纤维染料;醋酯纤维用分散染料
dispersed aeration process 分散曝气法《环保》

disperse-diazo composite dyes 分散—偶氮组合染料
disperse disazo dyes 分散重氮染料;双偶氮分散染料
dispersed light 色散光
dispersed phase 分散[内]相
dispersed vat dyes 分散型还原染料(微粒型)
disperse dye printing 分散染料印花
disperse dyes 分散染料
disperse dye sensitivity 分散染料的[化学]敏感性
disperse dyes E type E型分散染料
disperse dyes for acetate rayon 醋酯纤维用分散染料
disperse dyes for polyester fibre 涤纶用分散染料
disperse dyes for PP fibre 丙纶用分散染料
disperse dyes for T/C fibre 涤/棉用分散染料
disperse dyes for transfer printing 转移印花用分散染料
disperse dye solubility 分散染料的溶解性
disperse dyes SE type SE分散型染料
disperse dyes S type S型分散染料
disperse metallic dyes 分散金属络合染料
disperse-neutral composite dyes 分散—中性组合染料
disperse PC dyes 分散[双酯结构]染料
disperse premetallized dyes 分散金属络合染料
disperser 分散器;分散剂
disperse-reactive composite dyes 分散—活性组合染料
Disperser WA 分散剂WA

disperse-solubilized vat composite dyes 分散—可溶性还原组合染料

disperse system 分散[体]系(含有分散质与分散介质的液体)

disperse thickeners 分散型增稠剂

disperse-vat composite dyes 分散—还原组合染料

dispersibility 分散性;分散能力

dispersing 分散

dispersing action 分散作用

dispersing agent 分散剂

dispersing auxiliary 分散助剂

dispersing power 分散能力

dispersing reactive dyes 分散型活性染料

dispersion 分散体,分散相;分散[作用];色散

dispersion adhesives 分散型胶黏剂

dispersion coefficient 分散系数

dispersion degree 分散度

dispersion dot coating 分散点涂层

dispersion force 分散力,弥散力

dispersion for coating 涂层用分散体系

dispersion for finishing 整理用分散体系

dispersion index 分散指数;色散指数

dispersion medium 分散介质,分散媒质

dispersion of light 光的分散

dispersion rate 离散率;分散率

dispersion stability 分散稳定性

dispersion test unit [染料]分散性试验仪

dispersivity 分散性

dispersoid 分散胶体

Dispersol dyes 地司潘素染料(分散染料,商名,德国巴斯夫)

Dispersol PC dyes 地司潘素 PC 染料(含双羧酸酯分散染料,用于碱性拔染,商名,德国巴斯夫)

Dispersol SF dyes 地司潘素 SF 染料(含苯二呋喃结构分散染料,坚牢度和提升力良好,商名,德国巴斯夫)

Dispersol XF dyes 地司潘素 XF 染料(含偶氮杂环和部分双酯基团分散染料,易清洗,湿牢度好,商名,德国巴斯夫)

displacement 位移;排量(指液压泵或液压电动机理论上每转输出或需要输入的液体体积)

displacement and rotation 置换和旋转(计算机分色术语)

displacement chromatography 置换色谱

displacement consistency 置换浓度

displacement development 置换显影

displacement device 排液器(置于液槽中央,以减少储液量,节约染化料)

displacement diagram 位移图

displacement method 替换法

displacement printing 置换印花;防染印花

displacement pump 往复泵,活塞泵,柱塞泵,排代泵

displacement resist printing 置换防染印花

displacement washing 置换水洗

displacer 排液器《机》,置换剂《化》

displacing carriage 移动运载车

displacing element 排液器,排液元件

display 显示,显像;显示器;展览品(商品)

disposable fabric 用即弃织物

disposable laminates 用即弃层压物

disposable textiles 废弃纺织品

disposal 处理,整理,配置

disposal ecology 处理生态学

disproportionation 歧化作用;不相称

disrepair 失修

disruptive voltage 击穿电压

dissimilation 异化，异化作用
dissipation 散逸[作用]，分散[作用]；耗散
dissipation of heat 热逸散
dissipative muffler 消声套管，消声器（指一种衬有吸声材料的导管，当气体通过它时能吸收声能）
dissociated ammonia 解离氨
dissociation 解离[作用]
dissociation constant 解离常数，电离常数
dissociation energy 解离能
dissociation temperature 解离温度
dissolution 溶解[作用]
dissolution agent 溶剂
dissolution assistant 溶解助剂
dissolution heat 溶解热
dissolvability 溶[解]度；[可]溶性
dissolve ⟨dil.⟩ 溶解
dissolved-air floatation [加压]气浮法
dissolved organic carbon ⟨DOC⟩ 溶解有机碳《环保》
dissolved oxygen ⟨DO⟩ 溶解氧
dissolver 溶解机
dissolving agent 溶解剂
dissolving machine 溶解机
dissolving power 溶解力
dissolving salt 溶解盐
Dissolving salt B 溶解盐 B（苄基氨基苯磺酸钠，还原染料印花，助溶剂，商名，德国、法国制）
dissonant colours 不协调色；异质染料
dissymmetric molecule 不对称分子
dissymmetry 非对称[现象]
distance sensor 遥测传感器
distant reading thermometer 远距离读数温度计
distant thermometer 遥测温度计
distilling still 蒸馏釜
distillation 蒸馏
distillation apparatus 蒸馏装置
distillation column 蒸馏柱，蒸馏塔
distillation diffusion 蒸馏扩散
distillation range 蒸馏区间，沸腾范围
distillation reactor 蒸馏反应器
distillation under reduced pressure 减压蒸馏
distilled water 蒸馏水
distillery 蒸馏室
distorted selvedge 布边歪斜
distortion 扭变,畸变；歪边(织疵)；(信号等)失真
distortion resistance 抗歪曲，抗歪斜
distributing dyestuffs 匀染染料
distributing property 匀染性能
distributing roller 匀浆辊
distribution 分配；分布
distribution board 配电盘，配电板
distribution coefficient 分配系数
distribution curve 分布曲线
distribution manifold [糊料]分配管
distribution of molecular weight 分子量分布
distribution of polymerization degree 聚合度分布
distribution of sizes 粒度分布
distribution ratio 分配比
distribution switchboard 配电盘，配电板
distribution temperature 分布温度
distributor 分配器；配水器；配电器
distributor pipe 配液管
disturbance 扰动，干扰；故障；激波
disturbance flow 湍流；干扰流动

disulfide-bis-ethylsulfone 二硫双乙基砜（新活性染料中间体）
disulfide bond 二硫键
disulfide bridge 二硫[化]桥键
disulfide interchange 二硫交换反应
disulphide bond〈SS-bond〉 二硫[化]键
dithio compounds 二硫代化合物
dithionite 连二亚硫酸盐（保险粉的化学名称）
dityrosine 双酪氨酸
divalent 二价的
divariant system 二变系，双变量系统
divergence(＝divergency) 发散[度]；偏差
divergent rectangular fabric flowing tube 扩散状矩形导布管
divergent synthesis 发散合成法
diversification 多样化
diversion 转换，转向
divided bearing 对开轴承
divider bar 隔布档，分布辊（分隔绳状织物的辊棒）
dividing head 分度头
diving suit 潜水服
divinyl sulfone compound 二乙烯砜化合物
divinyl sulfone〈DVS〉 双乙烯砜（极活泼的反应性化合物）
division 间隔；部，段，片
divisional colour 分界色
division plate 隔板
division sheet 隔板
DMAA(dimethyl acetyl amide) 二甲基乙酰胺
DMAC(N, N-dimethyl-azetidinium chloride) 二甲基—氯化吖丁啶（棉纤维改性剂）
DMDHEU(dimethylol dihydroxy ethylene urea) 二羟甲基二羟基亚乙基脲（一种耐皂洗、氯漂的棉织物树脂整理剂）
DMEU(dimethylolethylene urea) 二羟甲基亚乙基脲（防缩防皱树脂整理剂）
DMF(dimethyl formamide) 二甲基甲酰胺（鉴别活性染料及合成纤维的试剂）
DMF（dimethylfumarate） 富马酸二甲酯（一种防霉剂,此产品因能引起接触性皮炎,欧盟已禁止使用）
DMHPU（dimethylol hydroxy propylene urea） 二羟甲基羟基亚丙基脲
DMPU(dimethylol propylene urea) 二羟甲基亚丙基脲（耐洗的棉织物防缩防皱树脂整理剂）
DMSO(dimethyl sulphoxide) 二甲[基]亚砜
DMU(dimethylol urea) 二羟甲基脲（一种脲醛树脂,棉或黏胶织物的防缩防皱树脂整理剂）
DMUG(dimethylurea and glyoxal) 二甲基脲和乙二醛（树脂整理剂）
DMUR(dimethylol urea resin) 二羟甲基脲树脂
DNA（deoxyribonucleic acid） 脱氧核糖核酸
DO(dissolved oxygen) 溶解氧
DO(drawing office) 制图室,设计室
dobby cloth 多臂提花织物
DOC(dissolved organic carbon) 溶解有机碳《环保》
doctor 印花刮刀
doctor blade [印花]刮刀；卷曲刀
doctor blade holder 刮刀支架
doctor blade washing machine 刮刀清洗机
doctor knife [印花]刮刀
doctor knife grinding machine 磨刮刀机

doctor lifting 跳刀,浮刀(印疵)
doctor mark 刀线,拖刀,刮刀条花,刀疵(印疵)
doctor polishing machine 磨刮刀机
doctor-roll 涂胶量控制辊;匀浆辊
doctor roll washing machine 刮刀清洗机
doctor sharpening stone 磨刀石
doctor shears 刮刀钢夹板
doctor stain 拖刀[污迹],刀线
doctor steps 刮刀架
doctor streaks 刀线,拖刀,刮刀条花,刀疵(印疵)
doctor traverse in roller printing 滚筒印花中刮刀往复运动
documentation 文献资料工作;记录;文件
DOD(drop on demand) 按需滴液(喷墨印花术语)
dodecyl benzene 十二烷基苯(以单烷基苯为主的工业混合物,净洗剂的中间体)
dodecyl benzene sulfonate〈DBS〉十二烷基苯磺酸酯
dodecyl salicylate 水杨酸十二酯(紫外线吸收剂)
dodecyl sulfate 十二烷基硫酸酯
dodecyl trimethyl ammonium chloride 十二[烷]基三甲基氯化铵(阳离子表面活性剂)
doe 浅棕灰[色],驼灰[色]
doeskin 驼丝锦(紧密缎纹毛织物);羚羊皮;仿麂皮;仿麂皮织物
doeskin finish 仿麂皮整理
dog 止块,挡块,轧头,夹头
dog bolt 活节螺栓,轧头螺栓
dog catch 掣子,夹子;擒纵器
dog clutch 爪形离合器
dog coupling 爪形联轴节

dogging test 追踪试验
doll head 进汽轴头
dolly 洗呢机;捶打洗涤机《针》,短皂洗机
dolly dyestuff 家用染料
dolly roller 洗呢机导辊
dolly scouring machine 捶打精练机
dome 拱顶;汽室
domestic 国产的,自造的;家庭的
domestic art 手艺
domestic laundering 家庭洗涤法(国际标准化组织规定的耐洗试验法)
domestic laundry soap 家用洗涤皂
domestics 家用纺织物;衬衫料(美国名称)
domestic textiles 内销纺织品;家用纺织品
domestic washing machine 家用洗衣机
domestic waste 家庭废弃物
dominant harmony 主色彩调和(一种具有主色的协调设计)
dominant shade 主色[调]
dominant wavelength 主波长《测色》
donkey boiler 辅助锅炉
donkey stitching machine 坯布缝头机,简易缝头机
Donnan equilibrium 道南[膜]平衡(以半透膜分隔的非扩散质与扩散质的溶液之间的静电电位差存在的平衡状态)
Donnan membrane equilibrium 道南膜平衡
Donnan pipette 道南吸管
Donnan's distribution 道南分布
Donnan's potential 道南电位
Donnan's theory 道南理论
donor 给予体
donor-acceptor chromogen 给予—接受电

子发色体

D. O. P. (dioctyl phthalate) 邻苯二[甲]酸二辛酯(增塑剂)

dope 涂布漆胶(用以增进不透水、不透气等性能);飞机翼涂料

doped fabric 上过涂料的飞机翼布

dope-dyed 纺液着色[的],纺丝原液着色[的],纺前着色[的]《化纤》

dope-dyed fibre 纺液着色纤维《化纤》

dope dyeing 原液着色

Dorlastan 多拉斯坦(聚氨基甲酸乙酯纤维即氨纶,商名,德国拜耳)

Dorlastan CC 多拉斯坦 CC(氨纶弹力丝,商名,德国制)

dorsal view 背视图

dorsetteen 毛经丝纬织物

DOS(disk operating system) 磁盘操作系统《计算机》

dosage 剂量;配料量

dosage bunker 配料槽;剂量槽

dose 剂量;用量

dose level 剂量水平

dose rate 剂量加入速度

dosimeter 测剂量器,剂量仪器

dosing 剂量;配料量

dosing chamber 投配室

dosing device 计量装置

dosing feeder 计量进料装置

dosing machine 计量机

dosing of chemicals 化学品计量

dosing pump 计量泵

dosing system 计量系统;加料系统

dot 点;点纹;网点

DOT(Department of Transportation) 运输部

dot coating 点布涂层,点纹涂层

dote 腐烂;腐败物

dot gain 点增益[印墨印到织物上发生扩散渗色]

dot matrix 点阵

dot-screen 点纹筛网

dotted 点子的,点纹的,有斑点的

dotted line 点线,虚线

double acting brushing machine 双动式刷呢机;双向刷毛[绒]机

double-acting raising machine 顺逆交替起绒机,双动式起绒机;双动式起毛机

double-acting shrinking machine 双动式预缩整理机

double and twist denim 双色纬纱劳动布,双色纬纱粗斜纹布

double backing 第二层底布,双层底布(地毯)

double-back tape 双面黏合带

double-ball bearing 双列滚珠轴承

double batch reaction chamber 叠卷式汽蒸反应箱

double batch system 叠卷式[连续]练漂联合机

double batch system continous scouring and bleaching range 叠卷式连续练漂联合机(织物平幅汽蒸练漂设备)

double batch type 叠卷式

double-bath dyeing 两浴法染色

double beam spectrophotometer 双光束型分光光度计

double bed rotary press 双座式滚筒压呢机

double belt 双层皮带

double blanket 双幅毛毯

double bleaching 两次漂白法

double bleaching with sodium chlorite and

hydrogen peroxide 亚氯酸钠双氧水双漂法

double bleaching with sodium hypochlorite and hydrogen peroxide 次氯酸钠双氧水双漂法

double bond 双键

double cantilever beam 双悬臂梁

double carpet 双层地毯

double cased copper pan 夹套铜质煮浆锅

double chemical lace 立体[双面]烂花花边

double clip 双层布铗（两个布铗座一上一下固定在一起，可同时运送两层织物）

double clip mercerizing machine 双层布铗丝光机

double cloth 双层织物

double cloth run 回形穿布[运行]，织物双层运行

double cloth way 回形穿布

double coating 二次涂层

double-coil spring 双圈弹簧

double-coloured twist yarn 双色花线

double column cylinder drying machine 双柱烘筒烘燥机

double concentric jets 双股同心射流（由中心圆形射流和彼此分隔而又同轴的环状射流组成）

double coupling 双联轴节

double crossway system 回形穿布法

double cure method 二次焙烘法

double cutter shearing machine 双刀辊剪毛机

double decating 双轴蒸呢

double deck 双层

double-deck motor 双鼠笼感应电动机

double diaphragm pump 双膜片泵

double-dip-single-nip padder 两浸一轧车

double dip-sizing 两次浸渍上浆

double dot coating 双点涂层

double drum drier 双转鼓干燥机，双滚筒干燥机

doubled yarn 合股纱线

double dyed（＝double dyeing） 两次染色（指交织物）

double-effect evaporator 双效蒸发器

double electric layer 双电层

double face coating 双面涂层

double face cutting machine 双面剪毛机

double faced 双面织物；双面的

double-faced flocking 双面植绒

double face printing 双面印花

double face transfer printing press 双面转移印花压烫机

double glue-down 二次涂胶

double green 甲基绿

double-jacketed cylinder 夹套热辊

double jersey 双面针织物，双面乔赛

double jet singeing burner 双喷嘴烧毛火口

double kier 双层煮布锅

double knit 双面针织物

double-layer ager 双层蒸化机

double-layer chainless mercerizing range 双层直辊丝光机

double-layer clip chain mercerizing range 双层布铗丝光机

double layer clip mercerizing machine 双层布铗丝光机

double-layer mercerizing machine 双层丝光机

double-layer stenter 双层拉幅机

double-layer suppression 双电子层（电偶

极子层）

double loop washing compartments 双环式洗涤分隔槽

double mercerized finish 全丝光；双丝光（指纱线与织物均进行丝光）

double-milled 两次缩绒

double milled fabric 双缩绒织物

double milling machine 双辊缩呢机

double-nip-double-dip type dye padder 两浸两轧型轧染机

double oil of vitriol〈D.O.V.〉双硫酸，浓硫酸（商名）

double padder 两道轧染机

double passage 回形穿布；双头进布

double passage clipless mercerizing machine 双幅直辊丝光机

double pass spectrometer 双光路分光光度计

double-pass stenter 往复式拉幅机

double pile 双面绒头织物

double pile velveteen 双层丝绒，双层立绒，双层天鹅绒

double pipe 套管

double plaiting device 双落布装置

double plush 双面长毛绒；针织绒布

double-ply 双股；双层

double point coating 双点涂层（先粉点，后浆点）

double-pole switch 双刀开关

double printing 第二次印花（罩印）；双面印花（布的正反面相同或不同花纹印花）；叠印（底纹上再印花）；叠印疵

double protein blended yarn 双蛋白纤维混纺纱线

double protein fibre 双蛋白纤维

double resist 二次防染

double roll box 双道辊浆槽（浆槽中有两对浸浆辊）

double-row ball bearing 双列球轴承

double-row self-aligning bearing 双排自调轴承

double-row spherical ball bearing 双列球面球轴承（常用于烘筒）

double salt 复盐

double screen process 双筛网法

double selvedge 卷边，翻边（织疵）

double sheet dyeing ［经轴］双片染色

double shrinkproof fabric 双防缩织物

double side coating 双面涂层

double-sided 双面的

double-sided plush 双面绒，双面长毛绒

double-sided transfer paper 双面转移印花纸

double-side singeing 双面烧毛

double sizing 双面涂胶；双面上浆

double sole 加固［的］袜底，夹底

double spanner 双头扳手

double spiral steamer 双层螺旋形蒸箱

double spot 双点［黏合衬］

double-spot coating 双点涂层（每个涂层点有一个熔点较低的芯和熔点较高的壳，有利于加强黏合）

double squeegee system 双刮刀系统

double squirrel cage induction motor 双鼠笼感应电动机

double stack cylinder drying machine 双柱烘筒烘燥机

double stick hank dyeing machine 双棒绞纱染色机

double-table transfer printing press 双面转移印花压烫机

double texture fabric 双层胶布

double thread steamer 回形穿布蒸箱
double thread-up ［平洗槽］回形穿布
double thread washer 回形穿布水洗机
double thrust bearing 双向止推轴承
double tier dry cylinder 双排烘筒
double toe 加固袜头
double tone colour 双色调色（深浅同色效果）
double velvet 双层丝绒
double-wall pan 夹层煮锅
double waxing 双面上蜡
double-web cloth infeed 双幅进布
double width 双幅
double width ageing machine 双幅蒸化机
double width clipless mercerizing machine 双幅直辊丝光机
double width washing machine 双幅平洗机
double-woven blankets 双层织造［的］毯子
double woven fabric 双层织物
double-woven pile fabric 双层织造［的］长毛绒织物
double yarn 合股纱
doubling and lapping machine 对折卷板机
doubling and plaiting machine 对折折布机
doubling and rolling machine 对折卷布机；对折机
doubling-over test 折叠试验
doupion silk 双宫绸，疙瘩绸
D. O. V. (double oil of vitriol) 双硫酸，浓硫酸（商名）
dove 鸽灰［色］（浅灰色）
dove colour 淡红灰［色］，暖灰［色］
dove gray 紫灰［色］
dowel pin 定位销，暗销，合缝销
down 羽绒（如鸭绒、鹅绒）；［有色纱线］纵向平视外观

down cloth 鸭绒织物，水鸟绒毛织物
down fall 崩溃
down garments 羽绒服装
downgraded 降等级的
down hair 绒毛，茸毛
down-like 仿羽绒
download 下载《计算机》
downpipe 水落管
down-proof fabric 防羽绒织物
down proof finish 防羽绒整理
downproofness test 防羽绒刺出性试验
downproofs 防羽绒刺出性织物
down quilt 羽绒被
down stream 顺流
downstream operation 下道工序，后道工序
down stroke 下行冲程；下行式
downtake 降液管；下导管；运输带
down-time 停台时间，停机时间
down time for colour change 换色停机时间
down time hours 下机时间，下机时数
downward force 向下作用力
down wear 羽绒服
Dow textile solvent system 道氏树脂防缩法（用溶于氯化溶剂中的树脂处理毛织物及其混纺织物，商名，美国制）
Dowtherm 道生［热载体］（联苯及联苯醚的混合物）
Dowtherm heating 道生加热（使用道生热载体进行加热）
Dowtherm service facilities 道生热载体供应装置
Dow XLA 一种新型弹性纤维（商名，美国陶氏化学公司）
dozen(doz.) 一打，十二个

DP 4168 DP 4168防水易去污剂(含氟丙烯酸共聚物,对各种织物特别是高档毛料、针织面料具有防水易去污能力,商名,英国PPT)
DP(data processing) 数据处理
DP(degree of polymerization) 聚合度
DP (Deutsche Patent) 德国专利
d. p. (dew point) 露点
DP(durable press) 耐久压烫,耐久定形
DP finishing(durable press finishing) 耐久压烫整理
dpi(drop per inch) 点/英寸;分辨率(印花精细程度)
DP rating(durable press rating) 耐久压烫等级
3D print (three dimensional print) 三维立体印刷;三维立体打印
3D printing (three dimensional printing) 三维立体印花
DPS(digital printing system) 数码印花系统
drabness 无光度,消光度
drab printing fabric 拉毛印花布
draft 通风;气流;草图
draft addendum〈DAD〉 [国际标准]补充草案
draft amendment〈DAM〉 [国际标准]修正草案
draft data〈DD〉 数据草案
drafted pattern 意匠图;花纹设计图
draft gauge 通风计
drafting board 制图板;制图桌
drafting paper 绘图纸
Draft International Standard〈DIS〉 国际标准草案
drag 卷绕张力;阻力;拖浆(印疵)

drag coefficient 阻力系数
drag down [印花机]刮刀口上的积垢
Drage viscometer 德雷奇黏度计(旋转式)
drag roller 拖布辊
drain 排水管;排放口
drainage 排水;排污水;排水设备;下水道
drainage fabric 排水滤布
drainage index 排水指数
drainage tester 滤水性试验器
drain board 淌水板
drain cock 排水旋塞
drainer 排水器;储浆池;滴干板
draining at high temperature 高温排液
draining table 沥干台,滴干台
drain mark 水纹,水渍,水印(染疵)
drain off 去水,排水,脱水
drain pipe 排泄管,排水管
drain plug 排水塞
drain valve 排泄阀;疏水阀
drapability [织物]悬垂性
drapability coefficient 悬垂系数
drape [织物]悬垂性;窗帘;褶裥;自然皱
drape coefficient 悬垂系数
draped skirt 自然皱裙;垂饰裙
drapeometer [织物]悬垂性试验仪
drapery 毛织物;装饰[用]织物;匹头织物(布匹、呢绒的总称);悬挂织物;帷幕
drapery fabric 装饰[用]织物,装饰布
drape test 悬垂性试验
drape tester [织物]悬垂性试验仪
draping 悬垂性
draping property 悬垂性[能]
draping quality 悬垂性能
draught 通风,排气
draught chamber 通风室,排气间

draught gauge 通风计
draught mark 风印,风渍
draught principle 通风原理
draughtsman's paper 绘图纸
drawability 可拉伸性
drawback 缺陷;障碍
draw-etching 写真雕刻,写真蚀刻
drawing 制图,绘图;图纸
drawing desk 描稿桌
drawing foil 描稿片
drawing godet roller 拉伸导丝辊
drawing-in roller 喂给辊,喂入罗拉
drawing instrument 绘图仪器
drawing number 图纸编号
drawing office〈DO〉 制图室,设计室
drawing pen 制图笔
drawing roller 拖布辊
drawing screens 可拉出的过滤网
draw-in roller 进布辊
drawn piece 纬斜,弓斜
drawn pile finish 顺绒整理
drawn sample 抽样
draw-off roll 牵拉辊;导布器
draw-off roller 牵引辊,送出罗拉,引出辊,出布辊
draw-out roller 出布辊
draw roller 拖动辊,拖布辊
draw textured yarn〈DTY〉 拉伸变形丝
dress 服装(尤指外衣)
dress down 便服,简装
dressed fabric 上浆整理的织物;起绒圈针织物
dresser 上浆机;缩绒机
dresser roll 上浆辊,轧浆辊
dress face finish 重缩绒整理(不露底纹)
dressing 修饰整理(起绒辊用)

dressing brush 上浆刷
dressing gauze 绷带纱布
dressing material 绷扎材料
dressing roll 上浆辊,轧浆辊
dressing roller 修饰辊(起绒机上使用)
dressing trough 浆槽
drier 干燥机,干燥器,烘缸;干燥剂,催干剂,速干剂
drill 中、厚斜纹织物
drill type burner 钻孔型火口
Drimafon dyes 黛玛芳染料(黛棉丽染料和福隆染料的混合染料,商名,瑞士科莱恩)
Drimafon ECO process 黛玛芳 ECO 法(涤棉混纺织物碱性一浴染色法,瑞士科莱恩)
Drimafon K process 黛玛芳 K 法(涤棉混纺织物染色法,活性/分散染料一浴二步法,瑞士科莱恩)
Drimafon XN process 黛玛芳 XN 法(涤棉混纺织物染色法,分散/活性染料一浴二步法,瑞士科莱恩)
Drimagen ER/E2R 黛棉匀 ER/E2R(芳香族聚醚磺化物,活性染料染色匀染剂,商名,瑞士科莱恩)
Drimalan dyes 黛毛丽染料(毛用活性染料,弱酸染色,适用于羊毛、氯化羊毛、真丝、锦纶等染色,商名,瑞士科莱恩)
Drimarene CDG dyes 黛棉丽冷溶颗粒活性染料(商名,瑞士科莱恩)
Drimarene CL-C dyes 黛棉丽 CL-C 染料(高色牢度活性染料,适用于连续染色和冷轧堆染色,商名,瑞士科莱恩)
Drimarene CL dyes 黛棉丽 CL 染料(含一氯均三嗪和乙烯砜双活性基团,属中温型活性染料,商名,瑞士科莱恩)

Drimarene HF dyes 黛棉丽 HF 染料（含二氟嘧啶和乙烯砜双活性基团，高固着率，商名，瑞士科莱恩）

Drimarene K/R dyes 黛棉丽 K/R 染料（含二氟一氯嘧啶活性基团，反应性高，固色温度 40～60℃，适合于染色和印花，商名，瑞士科莱恩）

Drimarene P dyes 黛棉丽 P 染料（含一氯均三嗪活性基团，属中等反应性活性染料，适用于印花、连续染色，商名，瑞士科莱恩）

Drimarene S dyes 黛棉丽 S 染料（含乙烯砜活性基团，属中等反应性活性染料，固色温度 60℃，适用于浸染和连续染色，商名，瑞士科莱恩）

Drimarene X/XN dyes 黛棉丽 X/XN 染料（三氯嘧啶和改良均三嗪类活性基团，适用于高温浸染固色，商名，瑞士科莱恩）

drip catching tray 集水盘（横导辊水洗机用）

drip cooler 水淋冷却器

drip dry(＝drip-drying) 滴干

drip-dry garment 滴干服装

drip leg 集液包

dripping point 滴落点

drip printing 滴印法（荷兰开发，地毯印花用）

drip-proof motor 防滴式电动机

drip vessel 滴油槽，滴油器

Dri-sol process 德里索防缩法（用磺酰氯进行羊毛氯化防缩法）

drive 传动；驱动机构

drive aggregates 传动联动装置

drive cam 导凸轮，导动凸轮

drive fit 压配合，紧配合（轴颈比轴孔稍大，装配时借助压力将轴颈压入轴孔）

drive gear 主动齿轮；传动机构

driven element 驱动子；激励单元

driven gear 从动齿轮

driven shaft 从动轴，被动轴

driven wheel 从动轮

drive-pipe 套管，竖管

drive pulley 主动皮带轮，传动轮

driver 驱动器；发动机；激励器

driver roller 传动辊，主动辊

drive shaft 主动轴，传动轴

driving mechanism 传动机构；传动机理

driving shaft 主动轴，传动轴

driving wheel 主动轮

drop 落下，掉下；滴，水滴；降低，下降

drop cutters 液滴切刀（用于地毯多流染色机）

drop formation 水滴形成（机器内部顶上蒸汽冷凝成水滴）

droplet 微滴，小滴

drop lubrication 滴液润滑

drop method 液滴法，滴注法（测定表面张力）

drop oiler 液滴加油器

drop on demand〈DOD〉 按需滴液（喷墨印花术语）

drop penetration test 水滴渗透试验（测防水性）

dropper 升降辊

drop per inch〈dpi〉 点/英寸；分辨率（印花精细程度）

dropping bottle 滴瓶《试》

dropping mercury electrode 滴下水银电极

dropping system 滴料系统（打样用）

drop pipette 滴管

drop-proof motor 防滴式电动机

drop-proof system 防滴水系统(蒸化机顶部有盘香管加热,以免蒸汽冷凝滴水)
drop separator 油[水]滴分离器
drop test 点滴试验;冲击试验;落锤试验
drop time 滴下时间
drug delivery system〈DDS〉药物传送系统
drugget 印花地毯,粗毛地毯;棉毛混纺地毯;粗而耐用、作地毯用的织物;地毯罩布
drug store 染料间(美国俗称)
drum 圆筒,鼓轮
drum brake 鼓式制动器
drum cleaning machine 滚筒干洗机,滚筒式洗涤机
drum drier 转筒烘燥机,鼓式干燥器
drum dryer 烘筒烘燥机,圆网烘燥机;转筒烘燥机,鼓式干燥器
drum dyeing 转筒染色(染袜或成衣)
drum dyeing centrifuging machine 圆筒染色离心机
drum dyeing machine 滚筒染色机,圆筒染色机
drum end 汽包封头
drum partitions 滚筒[内部]分隔结构
drum perforations 滚筒内孔
drum printing 转筒印花(地毯经纱印花)
drum printing machine 滚筒印花机
drum pumps 桶泵
drum reactor 鼓式反应器
drum scourer 转鼓精练机(袜子等细纤度织物用)
drum stenter(=drum tenter) 圆筒拉幅烘燥机《针》
drum type washer 转鼓式洗涤机
drum washing machine 转鼓式洗涤机
dry 干燥,烘干

dry and wet bulb temperature 干湿球温度
dry bearing 无油轴承
dry beetling 干打绉
dry bleaching 干漂白
dry blowing [呢绒]干蒸
dry brushing 干刷绒,干刷毛
dry bulb〈DB〉[温度计的]干球
dry bulb temperature 干球温度
dry bulb thermometer 干球温度计
dry can 烘筒,烘缸
dry carbonization 干式炭化法
dry carbonizing 干炭化
dry cell 干电池
dry chemical 干化学品
dry chemicking 干漂
dry cleanability 耐干洗性
dry cleaner 干洗机
dry cleaning 干洗(织物用有机溶剂净洗)
dry cleaning agents 干洗剂
dry-cleaning antifoams 干洗消泡剂
dry cleaning fastness 耐干洗牢度
dry cleaning intensifier 干洗增效剂
dry cleaning machine 干洗机
dry-cleaning machine 干洗机
dry cleaning regulations 干洗规范
dry-cleaning resistant finish 耐干洗整理
dry-cleaning shrinkage 干洗收缩
dry cleaning solvent 干洗溶剂
dry coating 干法涂层
dry compound cleaner [地毯]混合干洗法
dry content 固体含量,含固量
dry crease recovery 干折皱回复
dry crease recovery angle〈DCRA〉干态折皱回复角
dry creasing-Lintrak 林曲莱克法干褶裥处理

dry crocking 耐干摩擦[色]牢度

dry crosslinking 干态交联,干交联[反应] (纤维浸轧交联剂后在干燥不膨化状态时焙烘)

dry cure 干焙烘

dry decating(=dry decatising) 干蒸呢,干蒸

dry decatizer 干蒸呢机

dry decatizing machine 干蒸呢机,蒸呢机

dry distillation 干蒸馏

dry dyeing 非水相染色(指溶剂染色等)

dry elongation 干伸长

dryer 干燥机,干燥器,烘缸;干燥剂,催干剂,速干剂

dryer felt 烘缸毛毯

dryer wagon 干燥[机]料车(干燥机用)

dry filling 干布化学[浸轧]处理

dry filtration 干法过滤

dry finishing 干整理(织物在干燥状态下整理的总称,包括轧光、起绒、拉幅、修补、剪毛、电压、干蒸呢等工序)

Dry Fit 一种用超细合成纤维结构制成的运动服品牌,有好的透湿排汗性能(商品,美国)

dry foam [地毯]泡沫干燥剂

dry gas 干空气,干煤气

dry-gas method 干气法(用盐酸蒸汽的炭化法)

dry handle 滑爽手感

dry heat curing 干热焙烘

dry heat fixation 干热固着;干热定形

dry heat setting 干热定形

dry heat steam 干热蒸汽

dry heat tenter setting 干热拉幅定形[工艺]

dry heat tranfer printing 干热转移印花

dry heat treatment method 干热处理法

dry ice 干冰,固态二氧化碳

drying 干燥,烘干

drying and brushing machine 烘刷机

drying and heat setting stenter 烘燥热定形拉幅机

drying and tentering machine 烘燥拉幅机

drying apron 烘燥用履带

drying by radiation 辐射烘燥

drying by spraying 喷雾烘燥

drying can 烘筒,烘缸(俗称)

drying capacity 烘燥能力

drying chamber 烘房,烘燥室,干燥箱,烘箱

drying chest 烘燥夹板(俗称蒸汽夹板,夹板内通入蒸汽,织物在两层夹板之间通过,进行不接触烘燥)

drying compartment 烘房

drying cottage 热风烘燥室

drying cylinder 烘筒,烘缸(俗称)

drying in frozen 冷冻干燥

drying in the open 自然干燥

drying machine 烘燥机,烘干机

drying of wool fabrics 羊毛织物烘干

drying oil 干性油

drying on cans 烘筒烘燥

drying on frame 拉幅机烘燥

drying oven 烘房,烘箱

drying rate 烘燥率,烘率

drying retarder 干燥抑制剂

drying shrinkage 干燥收缩[率]

drying stenter 烘燥拉幅机

drying stiffness 干燥硬度

drying stove 烘箱,烘燥炉

drying systems 烘干系统

drying technology 烘干技术

drying temperature 烘干温度
drying tenter 干式拉幅机
drying time 烘干时间
drying tumbler 转鼓式烘燥机;烘燥鼓,烘燥转筒
drying unit 烘燥装置
dry ironing 干熨烫
dry lamination 干法叠层
dry mercerizing 干丝光(适用于针织物,先浸轧碱液,然后在气垫式拉幅机中烘燥)
dry mercerizing machine 烘干法丝光机
dry napping 干式起绒
dryness 干度(干蒸汽成分的重量,即一千克湿蒸汽中干蒸汽的相对重量)
dry nipping process 干轧加工
dry-packing 干填充
dry powder bonding 干粉黏合
dry raised fabric 干起绒织物
dry raising 干式起绒
dry relaxation 干式松弛
dry relaxing 干松弛
dry reserve dyeing 干防染
dry rubbing 耐干摩擦[色]牢度
dry rub fastness 干摩擦牢度
dry saturated steam 干饱和蒸汽
dry setter 干热定形机
dry setting 干热定形
dry setting machine 干热定形机
dry sieving 干筛
dry soil repellency 抗干污性
dry soil resistance 抗干污性
dry soil test 干污试验(防污试验法)
dry stain removal 干法去渍,干法去垢
dry steam 干蒸汽(即干饱和蒸汽)
dry steaming 干蒸

dry strength 干强力
dry tack 干黏合性
dry-taping machine 干法上浆机
dry tenacity 干强度
dry tenacity and elongation 干强伸度
dry tensile strength 干[断裂]强力
dry touch 滑爽感
dry-touch and skin friendly fabric 干爽舒适织物
dry transfer 干烫转移[鉴定法](针织业中用热熨斗烫压商标等标志,以鉴定其真伪),热转移
dry transfer printing 干热转移印花
dry transfer printing machine 干热转移印花机
dry tumble 干式转筒整理
dry weight 干重,干燥重量
dry wrinkle recovery 干折皱回复
DS(=D.S., degree of substitution) 取代度,置换[程]度
DSC(differential scanning calorimeter) 差示扫描量热计
DSC(differential scanning calorimetry) 差示扫描量热法
D.S.D. acid(diaminostilbene disulfonic acid) 二氨基芪二磺酸(一种荧光增白剂原料)
DSIR(Department of Scientific and Industrial Research) 科学与工业研究局(英国)
dt(=dtex, decitex) 分特[克斯](线密度单位,为1特克斯的1/10)
DTA(differential thermal analysis) 差热分析
dtex(=dt, decitex) 分特[克斯](线密度单位,为1特克斯的1/10)

DTG(direct to-garment) 直接制衣
DTPA(diethylenetriamine pentaacetic acid) 二亚乙基三胺五醋酸(有机螯合剂)
DTPP (diethylenetriamine pentamethylene phosphonic acid) 二亚乙基三胺五甲基膦酸(生物降解性较优)
DTS(design transport system) 花样[图案]输送系统
DTY(draw textured yarn) 拉伸变形丝
dual burner 复式燃烧器(可用两种燃料)
dual catalyst 双功能催化剂
dual circuit 对偶电路
dual circuit temperature control device 对偶电路的温度控制器
dual coating system 双重浸胶体系,双涂层体系
dual control system 复式控制系统
dual drive 双重传动
dual firing (用两种燃料)混合加热
dual fuel burner 可用两种燃料的燃烧器
dual function 双重作用
dual mode machine 双重模式机器
dual printer 复式印花机(圆网与滚筒复式印花设备)
dual process 有双重作用的复合加工过程
dual ratio reduction 双减速比机构
dual-servo motor drive system 双伺服电机传动体系
dual system 双重系统,对偶系统
dual top and bottom steam boxes 复式上下汽蒸箱
dual wave length thin layer chromatogram scanner 双波长薄层色谱扫描仪
Dubosqq colourimeter 杜氏比色计
duck 帆布;粗布
duct 导管;通道,甬道

ductility 延展性;韧性;可塑性
ductwork 管道系统
dull 色光暗淡,无光,消光
dull appearance 无光外观
dull colour 暗淡色,灰暗色
dull finish 消光整理;平光柔软整理
dull habotai 无光纺
dulling agent 消光剂,除光剂
dulling effect 暗晦效应,灰黯现象,无光效应
dull-lustered rayon yarn 无光人造丝[纱]
dullness 无光度,消光度
dumper 印染前验布工
dumping [垃圾]倾倒《环保》
dump plate 翻转板
dump steam 废气
dun 暗褐色[的],鼠灰(灰褐)色[的]
dune yellow 沙丘黄[色](淡黄褐色)
dunging 牛粪后处理(媒染后用牛粪或磷酸钠和磷酸钙处理,以固着或去除剩余媒染剂)
dunking roller 浸渍辊
dunnage bag 包装袋
Du Nouy surface tension balance 杜氏表面张力计
duplet bond 双键,偶键
duplex 二重的,双向的,双的;双联的;复式的《机》
duplex coating 双面涂层
duplex cylinder dryer 双面烘筒烘燥机
duplex driving 复式传动,双速传动
duplexed system 双套系统
duplex engraving 双面印花滚筒雕刻
duplex fabric in different colours 双面异色布
duplex fibre [皮芯型]复合纤维

duplex flat table pantograph engraving machine 双面印花滚筒平台缩小雕刻机(同时雕刻一对正反花纹的滚筒)
duplexing 双,双重,双向
duplex milling machine 双辊式缩呢机
duplex printing 双面印花,双面复合印花
duplex printing machine 双面印花机
duplex prints 双面[复合]印花织物
duplex pump 双缸泵
duplex raising machine 双面起毛机
duplex reciprocating pump 双筒往复泵,复式往复泵
duplex rotory screen printing machine 双面圆网印花机
duplex screen printing machine 双面筛网印花机
duplex shreiner[ing] calender 双层电光机
duplicate part 备件
duplicate printing 双面复合印花,双面印花
duplicate printing machine 双面印花机
duplicate sample 存档样本,留底样本
duplicate tests 平行试验,双试验
duplicating 复制,复印
duplicating paper 复写纸
duprene(=duprene rubber) 氯丁橡胶
Durabeau 杜腊堡整理(提高针织品光泽、柔软、不变形、防脱散、防水等性能,商名)
durability 耐久性,耐用性
durability coefficient 耐用系数(耐磨次数与强力损失之比)
durability of wear 耐穿性
durability test 耐用性试验
durability under repeated washing 耐重复洗涤性

durable anti-bacterial finishing 耐久抗菌整理
durable antistatic finish 持久性抗静电整理
durable crease 耐久褶裥;耐久折痕
durable finish 耐久性整理(耐水洗、皂洗、干洗的整理)
durable flame retardant finishing 耐久阻燃整理
durable glaze finish〈DG finish〉 耐久性光泽整理,耐久上光整理
durable mechanical effect 耐久机械处理效应(织物经树脂浸轧,再经缎光机、擦光机、凹凸轧花机或折叠机等压轧,最后经高温焙烘而获得耐久的效应)
durable pleating 耐久褶裥
durable press〈DP〉 耐久压烫,耐久定形
durable press finishing〈DP finishing〉 耐久压烫整理
durable press plastic replicas 耐久压烫塑料复制模,立体塑料复制模(用于对照评级)
durable press rating〈DP rating〉 耐久压烫等级
durable setting 耐久性定形
durable softening and water-repellent finish 耐久柔软拒水整理
durable soil-release 耐久去污
Duralized process 纯棉液氨处理工艺(液氨与预缩加工的组合,商名,美国)
duration 持续时间;持久;延续性
duration of afterflow 阴燃期(阻燃试验用语)
duration of coagulation 凝结期间;凝固时间
duration of flame 有焰燃烧期

Durene 杜兰丝光棉线(商名,美国制)
duromer 硬质体,刚性体
durometer 硬度计
duroplasts 硬塑料
dusky green blue 深青灰[色]
dust 淡褐[色];灰尘,粉末
dust absorption 吸尘性
dust accumulation 尘埃累积
dust bag 集尘袋
dust box 集尘箱
dust catcher 除尘器,集尘器
dust chamber 滤尘器;尘室
dust chimney 尘塔
dust coat 防尘衣
dust collecting fan 除尘风机
dust collection 集尘,除尘
dust collection effect 集尘效应
dust collector 集尘器
dust colour 灰暗色
dust concentration 粉尘浓度
dust diminisher 减尘剂(空调系统的空气洗涤器中使尘粒带电,被电极吸附)
duster 防尘衣
dust exhaust device 除尘装置
dust explosion 粉尘爆炸
dust explosion potential 尘埃爆炸潜能
dust extraction unit 吸尘装置
dustfall 粉尘降下
dust filtration 滤尘
dust free 无尘的
dust ignition 粉尖点火
dusting machine 除尘机
dusting test 粉尘试验
dust inhibitors 防尘助剂;粉尘抑制剂
dustladen air 含尘空气
dust-laying agent 消尘剂

dustless dyes 无尘染料
dust level 粉尘浓度
dust like 粉状的;尘状的
dust/lint eliminator 灰尘/短绒消除器(印花用)
dust/lint removal 灰尘/短绒去除器
dust loading 粉尘含量,粉尘浓度
dust measurement 粉尘测量
dust problems 粉尘问题
dustproof 防尘
dust proofing agent 防尘剂
dust proof mask 防尘罩
dust-regulation limit 灰尘限定范围
dust release 去尘
dust removal unit 去尘装置
dust resistant fabric 防尘织物
dust resistant finish 防尘整理
dust scrubber 洗净式集尘装置《环保》
dust-tight [密封]尘埃不入的;防尘的
dusty grey 土灰[色]
dusty jade green 浅灰绿[色],豆绿[色]
dusty wrapper [家具等的]防尘套
dutch blue 浅紫蓝[色]
duty 工作,工作制度;功率;生产率;负荷,负载
duty cycle 工作状态循环;暂载率
DVD(digital versatile disc) 数字多功能光盘《计算机》
DVD(digital video disc) 数字化视频光盘《计算机》
DVS(divinyl sulfone) 双乙烯砜(极活泼的反应性化合物)
dwell 静止,停留,机器运转中有规则小停顿
dwell chamber [暂停]堆置室
dwelling compartment 堆置槽,堆置室

dwelling time control 定形时间控制,滞留时间控制
dwell process 堆置法(染整加工方法)
dwell time 停留时间(指织物在练漂、汽蒸、松弛等设备中的加工处理时间),停顿时间
dwell zone 停留时间
DWI(Deutsche Wool Institute) 德国羊毛研究所
dyads 补色配色
Dybln dyes 迪布尔恩染料(分散染料的一种,在有棉膨化剂的存在下,可使涤棉混纺纤维染成一致色泽,商名,美国制)
dyeability 可染性
dye absorption spectra method 染料吸收光谱法(测临界胶束浓度)
dye adsorption 染料吸附[作用]
dye affinity 染料亲和力
dye aggregate 染料集合体,染料聚集体
dye aggregation 染料集合,染料聚集
dye applicator slot 给染液缝隙
dye bags (针织品的)染色用网袋
dye bath 染浴,染液
dye bath additive 染浴添加物
dye bath lubricant 染浴润滑剂
dye bath monitoring and control 染浴监控
dye bath preparation 染浴配制
dye bath recycles 染浴循环数
dye bath reuse 染浴再利用
dye beam 染色经轴,染色布轴
dye beck 染槽;绳状染色机
dye blocking agent 阻染剂
dye boarder (袜子的)染色定形[联合]机
dye box 染槽
dye carrier 染色载体;导染剂

dye/chemical dispensing 染料/化学品配制
dye circle 染色循环[次数];染色周期
dye classification 染料分类
dye cleavage 染料裂解
dye compatibility 染料相容性
dye concentration 染料浓度
dyed cotton 染色棉布
dye deepening agents 染色增深剂
dye depth 染色深度
dye developing machine 显色机
dyed fibre tracers 染色示踪纤维
dyed goods [染]色布
dyed in the piece 匹染
dyed in the wool 原毛染色
dyed in the yarn 纱线染色
dye disaggregation 染料解聚作用
dye dispersant 染料分散剂
dye dispersion 染料分散性
dye-dissolving agent 染料助溶剂
dye distribution index(DDI) 染料分散指数
dye distributor 染料配液槽,染料分配器
dyed oxford 染色牛津布
dyed shirtings 染色平布,染色细布
dyed style 媒染剂印花后染色法(一种早期的印花方法)
dyed style printing 媒染剂印花后染色法(一种早期的印花方法)
dye dusting test apparatus 染料粉尘试验仪(美国 AATCC)
dyed yarn 染色纱
dye estimation 染料测定
dye exhaustion 染料吸尽率,上色率
dye fastness 染色[坚]牢度
dye fastness test 染色牢度试验

dye-fibre bond 染料纤维结合键
dye fixing 染色固色
dye-fixing agent 染料固着剂,固色剂
dye-fixing percentage 固色率
dye fleck 染色斑点
dye for acetate fibre 醋酯纤维用染料
dye for cellulose fibre 纤维素纤维用染料
dye for cotton 棉用染料
dye formula 染色处方
dye for nylon 锦纶用染料
dye for polyamide fibre 聚酰胺纤维用染料
dye for polypropylene fibre 丙纶染料
dyehouse 染色间,染色工场,染坊
dyehouse communication centre 染料控制中心
dye hydrolysates 染料水解产物
dyeing 染色,染色工艺,染色工程
dyeing ability 着色力,染色亲和力,上染性能
dyeing acid 染色酸,染色用酸剂(包括染色时释酸的化学品,用于调节染色pH)
dyeing alkali 染色碱,染色[印花]用碱剂(包括染色或印花时释碱的化学品)
dyeing and finishing industry 染整工业
dyeing and finishing plant 染整厂
dyeing and finishing wastes 染整厂废物,印染污水
dyeing and finishing waste water 染整废水
dyeing and scouring of raw silk with weighting 生丝练染并增重
dyeing apparatus under atomspheric pressure 常温常压染色设备
dyeing assistants(＝dyeing auxiliaries) 染色助剂

dyeing behaviour 染色性能
dyeing by ultrasonics 超声波染色
dyeing capacity 着色力,上染能力
dyeing carrier 染色载体;导染剂
dyeing coefficient 染色系数
dyeing concentration 染色浓度
dyeing crease mark 染色皱印
dyeing cycle 染色周期,染色循环
dyeing defect 色花(染疵)
dyeing depth 染色深度
dyeing equilibrium 染色平衡
dyeing equipment 染色设备
dyeing factory 染色工厂
dyeing fault 染疵
dyeing formula 染色处方,染色配方
dyeing formulation 染浴组成
dyeing in boiling state 沸态染色
dyeing in jigger 染缸染色,卷染机染色
dyeing in loop 环状染色
dyeing in open width 平幅染色
dyeing in rope form 绳状染色
dyeing in supercritical carbon dioxide medium 超临界二氧化碳介质染色
dyeing in width 平幅染色
dyeing jigger 卷染机
dyeing kettle 染色锅
dyeing kinetics 染色动力学
dyeing lubricant 染色润滑剂,染色抗皱剂(染色时防止折皱疵)
dyeing machine 染色机
dyeing machine of packing system 充填式染色机,筒装式染色机
dyeing management system 染色管理系统
dyeing mechanism 染色机理
dyeing method 染色方法
dyeing method by hydrosulfite vatting 次亚

硫酸盐还原染色法
dyeing mill 染色工厂
dyeing padder 染色轧车
dyeing pan 染色锅
dyeing power 染色力
dyeing process accompanying scouring 伴随精练的染色法,练染合并法
dyeing program [me] 染色程序
dyeing property 染色性能
dyeing range 染色装置,染色联合机
dyeing rate 染色率,上染速度
dyeing recipe prediction system 染色处方预测系统
dyeing-sizing machine 染色上浆联合机,浆染联合机
dyeing spech 染渍,色斑
dyeing spot 染色色点
dyeing steamer 染色蒸箱
dyeing tank 染槽
dyeing theory 染色理论
dyeing transition temperature 染色转移温度,染色转变温度
dyeing vessel 染缸
dyeing with boiled silk 精练丝染色
dyeing with liquid ammonia 液氨染色
dyeing with weight 增重染色法《丝》
dye in residual bath 残浴中染色
dye intermediates 染料中间体
dye jig 卷染机,染缸(俗称)
dye jigger 卷染机,染缸
dye kitchen 染液配料间;调[色]浆间;染料[称料]间
dye laser 有色激光
dye leveller 匀染剂
dye levelling 匀染的
dye levelness 匀染率;匀染法

dye liquor 染液
dye-liquor jets 染液喷嘴
dye lot 染色批量,染批
dye mark 染渍(染疵)
dye migration 染料泳移
dye mobility 染料迁移性
dye net 染色用网袋(染袜用)
dye-ometer 染色计
dyeometer 上色率测定仪(连续测定染浴浓度)
dye on the beam 卷染;经轴染色
dye padder 染色轧车
dye pan 染锅,染槽
dye penetrate test 染色渗透检验
dye penetration 染料渗透
dyer 染色工作者,染色工,染色技师
dye receptive properties 染料吸取性
dye receptive site 染料吸取位置
dye recipe 染色处方
dye recovery bath 拖缸(利用染缸残余染料套染回收节约染料的方法)
dye release characteristic 染料释出特性
dye resistant 拒染的,难染的
dye retardant 缓染剂
Dyers' and Cleaners' Research Organization 〈DCRO〉 染色与干洗业研究组织(英国)
dyer's bleach 半漂
dyer's greenweed 植物染料
dyers' package 染色用筒子
dyer's pitch 第一次染色打样处方
dyer's woad 菘蓝(植物靛青染料)
dyes 染料
dye saturation value 染料饱和值
dye selection 染料选择
Dyes Enviromental and Toxicological Organ-

ization〈DETO〉 染料环境和毒物学组织

dyes for copper-salt after-treatment 用铜盐后处理的染料

dyes for dope dyeing 纺前着色用染料,纺丝原液着色染料

dyes for formaldehyde-after-treatment 用甲醛后处理的染料

dyes for half-wool 半毛染色用染料

dyes for leather 皮革染料

dyes for polyester fibre 涤纶用染料

dyes for transfer printing 转移印花用染料

dyes for vinylon 维[尼]纶用染料

dyes for viscose rayon 黏胶丝用染料

dye site 染座,受色位置,上染位置(纤维中能和染料分子结合的官能基团位置)

dye-sizing 染色上浆

Dyes of MAK classes Ⅲ A1 and Ⅲ A2 MAK Ⅲ A1 和 MAK Ⅲ A2 组中的染料(德国公布禁用的致癌染料种类)

dye solubilizer 染料溶剂;染料助溶剂

dye solution 染料溶液

dye speck 染渍,色斑

dye spot 飘色疵点(染料粉末飞散而造成的色点);色点,色斑

dyespring 钢丝筒管,弹簧筒管(卷装染色用)

dye springs [染色筒子]弹簧管

dyes substantive assistants 亲染料性助剂,染料亲合性助剂

dyes substantive levelling agents 染料亲合性匀染剂

dyestain 染渍

dye-stick [染纱槽中]挂纱圆棒

dye streak 色柳,染色条花(疵点)

dye strength 染料力份,染料浓度

dye strike 初期染着

dye strike rate 染色上色率,染料瞬染率

dyestuff evaluation and analysis 染料评定与分析

dyestuff/fibre equilibrium 染料/纤维平衡

dyestuff grinding machine 染料研磨机

dyestuff identification on fibre 纤维上染料鉴定

dyestuff nomenclature in three parts 染料三段命名法

dyestuffs 染料

dye-substantive assistant 亲染料性助剂

dye substantivity 染料直接性

dye transfer 染料迁移

dye-transfer system 染液转移法(地毯印花用)

dye tumbler 转笼式染色机

dye turntable 染料转台(调配染料用)

dye under pressure 高压染色

dye uniformity 染色均一性

dye uptake 上染率,上色率,吸色率

dye-variant fibres 差异染色性纤维(具有不同染色特性的变性同类纤维)

dye vat 染槽

dye-weave polychromatic dyeing machine 彩色织花型染色机(染液经斜板流到织物上,经轧车形成花纹)

dye-weave system 染液喷射花型法(地毯印花用)

dyeweed 植物染料

dye winch 绳状染色机

dyewood 木本植物染料(含染料的植物,如苏木、巴西木、红木等)

dyewood extract 木本植物染料萃取物,植物染料提取物

dyeworks 染色工厂
dye yield 得色量
Dylan DC 迪兰 DC(二氯异氰尿酸钠,用于羊毛表面鳞片氯化处理,使毛制品具机可洗防毡缩效果,商名,英国 PPT)
Dylan GRB batch process 迪兰羊毛防缩间歇法工艺(英国 PPT)
Dylanize process 迪兰[毛织物]防缩法(商名,英国 PPT)
Dylan process 迪兰[毛织物]防缩法(商名,英国 PPT)
Dylan Salt 迪兰盐(过一硫酸盐,羊毛防缩剂,商名,英国 PPT)
Dylan Ultrasoft CW 迪兰超柔软剂 CW(硅酮类聚合物,羊毛防缩剂柔软剂,商名,英国 PPT)
dynamic 动态的;动力的
dynamic abrasion 动态磨损
dynamic absorption 动态吸收
dynamic absorption test 动态吸收试验
dynamic adhesive property 动态黏着性
dynamical balance testing instrument 动平衡试验仪
dynamic balance(＝dynamic balancing) 动态平衡
dynamic behaviour 动态性状
dynamic bioassay test 动态生物鉴定试验
dynamic boundary layer 动力边界层
dynamic buckling 动态变形,动态弯曲
dynamic characteristic 动态特性
dynamic deformation 动态变形
dynamic detector head 动态检测头
dynamic drape 动态悬垂性
dynamic elasticity 动态弹性
dynamic elestic modulus 动态弹性系数

dynamic equilibrium 动平衡,动态平衡
dynamic fatigue 动态疲劳
dynamic fatigue resistance 耐动态疲劳性
dynamic foam height 动态泡沫高度
dynamic formula 动态式
dynamic friction 动摩擦
dynamic immersion absorption test 动态浸渍吸收试验(测织物防水性)
dynamic isomerism 动态异构[现象]
dynamic load 动态负荷
dynamic measurement 动态测定
dynamic modulus 动态模量
dynamic pressure 动压力
dynamic recovery 动态回复
dynamic relaxation 动态松弛[作用]
dynamic resilience 动态回弹
dynamic response 动态响应
dynamic response test 动态特性试验
dynamic rigidity 动态刚性
dynamics 动力学
dynamic state 动态
dynamic test 动态试验
dynamic viscoelastic 动态黏弹性
dynamic viscoelastometer 动态黏弹仪
dynamic viscosity 动力黏度(即绝对黏度,单位为泊)
dynamic water absorption 动态吸水
dynamite 炸药
dynamited silk 增重丝绸
dynamiting 增重工艺《丝》
dynamo 发电机
dynamo- (词头)表示"力""动力"
dynamometer 测力计;功率计
dynamomotor(＝dynamotor) 电动发电机
dynamo oil 电动机油
Dyna pump 代纳泵(一种循环泵)

dyne 达因(力的单位)

Dytran inks 迪特兰油墨染料(转移印花用的油墨型分散染料,商名,瑞典制)

E

EA (ethyl acrylate) 丙烯酸乙酯
early failure 过早损坏
early flax 石棉
ear protection 耳罩
earth colour 土色(深棕色或茶色)
earth colours 矿物颜料
earth connection 接地
earth fault 接地故障《机》
earth-free 不[需]接地的
earthing 接地《电》
earthing resistance 接地电阻
earth moving 土方工程
ease 舒适；拨开
ease of care 免烫，随便穿
ease of extinguishability 易熄灭性
ease of ignition 易[点]燃性
easily biodegradable auxiliary 易生物降解助剂
easily biodegradable surfactant 易生物降解表面活性剂
easily dischargeable 易拔去的《印》
easing gear 松动装置(安全阀内装置)
eastern carpet 东方地毯
Eastman 伊斯脱曼(美国染料公司)
Eastman dyes 伊斯脱曼染料(分散染料，商名，美国柯达)
Eastobrite 伊斯托布赖特荧光增白剂(商名，美国制)
Eastone dyes 伊思通染料(分散染料，商名，美国制)
easy care 免烫，随便穿

easy care catalyst 免烫催化剂
easycare finish 免烫整理，洗可穿整理
easy care performance 免烫性能
easy care property 免烫性能，随便穿性能
easy care resin 免烫树脂
easy cationic dyeable polyester〈ECDP〉[常压]阳离子染料易染型涤纶
easy clear 易洗除浮色
easy disperse dyeable polyester〈EDDP〉分散染料易染聚酯纤维
easy dyeable fibre 易染纤维
easy foaming property 易起泡性
Easy Living 伊锡利文(由大可纶、棉和莱克拉弹性纤维制成的具有超防水性的弹性织物，商名，美国制)
easy slide fit 轻滑配合
easy-to-move 容易操作
easy worn out parts 易损件
easyworn parts 易损件
EB (electric beam) 电子束
ebonite 硬橡胶，胶木
ebony 深蟹青[色]，暗蓝[色]，乌木黑[色]
ebony brown 茶褐[色]，乌木棕[色]
ebullated bed 流化床，沸腾床
ebullience washing 煮沸洗涤
ebulliometer (＝ebullioscope) 沸点升高计
ebullioscopy 沸点升高测定法
ebullition 沸腾
EC (European Community) 欧洲共同体，欧共体

eccentric 偏心的；偏心轮
eccentric adjuster 偏心调节装置
eccentric adjustment 凸轮调整，偏心调整
eccentric cam 偏心凸轮
eccentric disc(＝eccentric disk) 偏心盘
eccentric drive 偏心轮传动
eccentric engraving 偏心轮雕刻法
eccentricity 偏心距；偏心率
eccentric pin 偏心销
eccentric shaft 偏心轴
eccentric wheel 偏心轮
ECD(electron capture detector) 电子俘获检测仪
ECDP(easy cationic dyeable polyester) ［常压］阳离子染料易染型涤纶
EC eco audit 欧共体生态审核
EC eco label 欧共体生态标志(生态标签)
ECF(electrochemical fluorination) 电化学氟化法（制造含氟整理剂的技术）
ECFC(European Continental Fastness Convention) 欧洲大陆染色坚牢度公约
EC guide line 欧共体准则
ECHA(European Chemicals Agency) 欧盟化学品管理局
echelette 红外光栅
Echolux effect 埃可乐特殊轧纹效应(电光机的钢辊上刻出按图案花样交叉排列的斜纹线，使织物轧成明暗不同如绸缎光泽的花纹)
eclipse 失色，掩暗；遮蔽
ECM(European Common Market) 欧洲经济共同体
ECO(Environmental Control Organization) 环境控制组织(美国)
eco-activity 生态活动
ecobalance 生态平衡

ecobiotic 生态生物的
eco bleach 生态漂白新工艺
eco-car 环保汽车
ecocide 生态破坏
eco circle ［纤维］循环利用
eco-diaper 生态尿布(可被生物降解)
eco-dyestuff 生态染料(无有害物质，生产时三废少)
eco-efficiency 生态效益
Eco-Flash process 易固发色工艺(经济型高温固色工艺，德国 BTM 公司)
Eco-Flash unit 易固发色单元机(经济型高温固色机，德国 BTM 公司)
Eco-Foam Unit 生态泡沫系统（无需使用尿素的方法，商名，荷兰斯托克）
eco-friendly auxiliary 生态友好助剂
eco-friendly dyes 生态友好染料
eco-friendly process 生态友好工艺，环境友好工艺
Eco-Guide 生态指南，环保指南
ecolabelling of textiles 纺织品生态标签
eco label requirement 生态标签需求
ecolabels 生态标签
Ecological and Toxicological Association of Dyes and Organic Pigments Manufacturers〈ETAD〉 染料及有机颜料制造商生态和毒理学协会
ecological auxiliary 生态助剂，环保型助剂
ecological balance 生态平衡
ecological constants 生态常数
ecological dyestuffs 生态染料，环保型染料
ecological edict 生态法令
ecological equilibrium 生态平衡
ecological evaluation 生态学评价
ecological fabric 生态织物

ecological fashion 生态时装；自然式服装
ecological fibre 生态纤维，无[环境]污染纤维
ecological finish(= ecological finishing) 生态整理
ecological harmlessness 生态无害性
ecologically advanced 生态改良的
ecological problems 生态问题
ecological requirements 生态需求
ecological sustainability 生态可持续性
ecological textile 生态纺织品
ecological viewpoint 生态观点
ecology 生态学
ecology look 生态风貌
economical kiss roll padding 用面轧辊的经济轧液法
economic benefits 经济效益
economic boiler 经济型锅炉
economic calculation 经济核算
economic contract 经济合同
economizer 节煤器；废气预热器
economy price 经济实惠的价格
economy size 经济尺寸
economy trough 经济轧槽
economy type 实用型；经济型
Econtrol MXL process 纯棉织物免烫整理新工艺（商名，德国）
E Control process E型控制工艺（活性染料涤/棉织物连续轧染新工艺，高效、节能、低污染，商名，德国 Monfort，Zeneca 联合开发）
eco physiology 生态生理学
eco-plan 生态计划
Eco-steam process [活性染料]短流程湿蒸工艺
Eco Swat processing 节约水和时间的染色工艺
eco system 生态系统
Eco-tex [欧共体]纺织品生态标签；生态服装；生态纺织品
eco-textiles 生态纺织品，环保纺织品
ecotoxicity 生态毒性
ecotoxicological testing 生态毒性试验
ecotoxicology 生态毒性学
Eco-Vision 生态眼光，环保眼光
ECR(electron cyclotron resonance) 电子回旋加速器共振（产生微波设备）
ecrase 压花效应，绉纹效应（指纺织品和皮革制品）
ECR plasma 微波回旋共振等离子体
ecru 米黄[色]；淡灰褐[色]；本色的；未漂白的
ecruing [生丝]半脱胶法
ecru method [生丝]半脱胶法
ecru silk [yarn] 半脱胶丝
ecru silk cloth 本色生丝织物
ecru yarns 本色纱
ECU(endo cellulase units) 内切纤维素酶活力单位
EDA(electron donor-acceptor) 电子供受体
EDANA(European Disposables and Non-Wovens Association) 欧洲耗材及非织造布协会
EDCC(Environmental Detection and Control Center) 环境监测中心
EDDP(easy disperse dyeable polyester) 分散染料易染聚酯纤维
eddy current 涡流
eddy current clutch 转差离合器，涡流离合器
eddy current magnetic couplings 涡流磁联轴器

eddy flow 涡流
edge 刀口；布边；边饰
edge abrasion 布边磨损
edge binding 粘边
edge crease offsetting device 吸边器
edge-crimp process 布边起绉法
edge curl (= edge curling) 卷边（印疵）
edge cutter 切边机
edge cutting 切边
edge cutting device 切边装置（装于针织物拉幅机的出布区，切去毛边，达到规定的门幅要求）
edgedecurler 剥边器（将针织物卷边剥开展平）
edge extracting device 集边装置
edge feeler 探边器
edge finish ［毛巾］布边整理
edge fusing 热熔成边
edge gathering device 集边装置
edge glueing device 布边上浆装置
edge guide 吸边器；导布器
edge guiding sensors 吸边传感器
edge gumming 浆边
edge gumming device 布边上浆装置
edge marks 布边色差，边色不匀，边印（圆编针织物平摊染色时造成的疵点）
edge-on-edge 缝料相叠
edge or fold abrasion resistance 耐折边磨
edge pick stitching machine 边缝缝合机
edge roll ［针织物的］卷边
edge runner 轮碾机，研磨机
edge scanning arrangement 布边扫描检测装置
edge seam 边缝，缝边
edge sensor 探边器
edge slack 荷叶边，宽急边（织疵）

edge spreader 剥边器
edgestitcher 缝边器
edge straightener 布边伸直器
edge-to-edge ［布］边到［布］边
edge-to-edge printing 边到边印花，全幅印花（针织品）
edge trimming 整边，剪边，修边
edge trimming device 切边装置
edge trim remover 集边装置
edgetrol 布边控制装置
edge uncurler 剥边器
edge wear 坏边
edge weave 边组织
edge zone heater 边区加热器
edging lace 镶边花边
edging tape 包边带，滚边带
EDMA (ethylene dimethacrylate) 二甲基丙烯酸次乙酯
EDP system (electronic data processing system) 电子数据处理系统
EDTA (ethylene diamine tetra acetic acid) 乙二胺四乙酸（有机螯合剂）
eductor 喷射器；排泄器
eductor condenser 喷射式冷凝器
eductor pump 喷射泵
edulcoration 纯化，除杂
EEC (environment effluent concentration) 环保污物浓度
EEC (European Economic Community) 欧洲经济共同体
EEE system (efficiency, economy, ecology system) 效率、经济、生态体系（染整工艺三要素）
EF (extra fine) 超精度，极细
eff. (effective) 有效的
eff. (efficiency) 效率，功效

effect 效应,效果;花纹
effect colour of printing 花色印花(增加着色效果的印花,如金银粉印花,闪烁印花等)
effect dyeing 花色效应染色(如间隔染色、交染、蜡防染色等)
effect finishes 效果整理
effective〈eff.〉有效的
effective bandwidth 有效波长带宽
effective bond length 有效键长
effective chlorine 有效氯
effective dipole moment 等效偶极矩,有效偶极矩
effective dye concentration 有效染料浓度
effective half life〈EHL〉实际半衰期
effective horse power〈EHP〉有效马力
effective insulation ratio 有效隔热比
effective opening size〈EOS〉有效孔径
effective pile thickness 有效绒毛厚度
effective porosity 有效孔隙度(纤维或织物的)有效疏松度
effective temperature 实效温度
effective transmission 有效透射;有效传动
effective value 实效值
effective wave length〈EWL〉有效波长
effect of antielectricity 抗电效果
effect of double print [印花]叠色效应
effect of metal dyebaths 金属染槽效果
effect of order 有序效应
effect on aquatic life 对水生物的影响《环保》
effect side [织物]正面
effect threads 花色线
effect yarn 花色线,(花式线的)装饰纱
effervescence 泡腾,起泡[沫]
efficiency〈eff.〉效率,功效

efficiency curve 有效曲线
efficiency, economy, ecology system〈EEE system〉效率、经济、生态体系(染整工艺三要素)
efficiency of fan 风机效率
efficiency of pump 泵效率
efficient fossil power generation 高效火力发电
efflorescence 风化;粉化
effluent 污水,废水;出水,出流
effluent disposal 污染处理,污水处理
effluent gas analysis 废气分析
effluent purification 污水净化
effluent quality standard 排放质量标准
effluent quality standard for dyeing and printing industry 印染工业污水排放标准
effluent toxicity reduction 降低污水毒性
effluent treatment 污水处理,废水处理
efflux 射流
efflux method 流出法(测黏度)
effuser 扩散管,喷管
effuxion 射流;流出,排出;喷出;渗出物
EFTA(European Free Trade Association)欧洲自由贸易联合会
EFTC(European Free Trade Committee)欧洲自由贸易委员会
eg(=e. g., exempli gratia) 例如
EG(endo glucanase) 内切葡聚糖酶
EG(ethylene glycol) 乙二醇
egg albumin fibre 卵清蛋白纤维
eggshell finish 暗光整理(织物经过一对表面刻有细小高低花纹的滚筒轧压而成)
egg yellow 蛋黄[色]
EGU(endo glucanase unit) 内切葡聚糖酶

活力单位(纤维素酶)
Egyptian blue 埃及蓝[色]
EH(enviromental hormone) 环境荷尔蒙(对人类健康和生态环境危害很大的化学物质)
EHA(electric hemming apparatus) 电动卷边装置;缝线防偏装置
EHL(effective half life) 实际半衰期
EHP(effective horse power) 有效马力
Ehrlich's reagent 埃尔利希试剂(0.3%对二甲氨基苯甲醛的稀盐酸溶液)
Eicken psychrometer 艾氏湿度计(瑞士制)
eight colours printing machine 八色印花机
ejector 喷射器
ejector jet pump 喷射泵,注射泵
Ekofast 埃科法斯特连续加压蒸呢机(商名,英国制)
Ekofast continuous pressure decatizing machine 埃科法斯特连续压力蒸呢机(商名,英国制)
Ekofast machine 埃科法斯特连续罐蒸机(供连续蒸呢、烘呢、定形和压呢用)
EL(Environment Labels) 环境标签(国际环境 ISO 14000 标准)
Elaine 伊兰油酸(羊毛润滑剂,煮练助剂,商名)
Elanit washing machine 伊赖尼铁水洗机,用于弹性针织物的平幅水洗(商名,德国制)
Elanstane 伊赖斯坦(聚氨基甲酸乙酯弹性纤维,欧洲对氨纶的统称)
elasticator 弹性剂(增加纤维弹性的添加剂)
elastic bearing 弹性轴承
elastic calender bowl 弹性轧辊

elastic constant 弹性常数
elastic cord 松紧线,松紧绳(有橡皮芯的丝线或棉线)
elastic corduroy 弹力灯芯绒
elastic coupling 弹性联轴节
elastic curve 弹性曲线
elastic deformation 弹性变形
elastic distortion 弹性畸变
elastic elongation 弹性伸长
elastic extension 弹性伸长率
elastic fabric 弹性织物
elastic filament 弹力丝
elastic finish 弹性整理
elastic finishing agent 弹性整理剂
elastic finish sheeting 弹性整理稀薄平布
elastic fore-effect 弹性前效
elasticity 弹性
elasticity of compression 压缩弹性
elasticity of elongation 伸长弹性[率]
elasticity of flexure 弯曲弹性
elasticity of shape 形态弹性
elasticity of torsion 扭转弹性,扭曲弹性
elasticity of volume 体积弹性
elasticity testing 弹性试验
elasticized fabric 弹性织物
elasticizer 弹性剂(增加纤维弹性的添加剂)
elastic limit 弹性极限,弹性限度
elastic liquid 弹性液体
elastic modulus 弹性模量(即杨氏模量)
elastic nylon 弹力锦纶
elastic packing 弹性填料,弹性密封
elastic performance 弹性
elastic reaction 弹性作用,弹性反应
elastic recovery 弹性回复
elastic recovery angle 弹性回复角

elastic recovery of curl 卷曲弹性回复
elastic recovery percentage of elongation 拉伸弹性回复率
elastic region 弹性界限
elastic resilience 回弹性
elastic seam 弹性缝
elastic sewing thread 橡筋包覆缝纫线
elastic sock 弹力短袜
elastic stocking 弹力长筒袜
elastic strain 弹性应变
elastic wrinkle resistant fabric 弹性抗皱织物
elastic yarn 弹力丝,橡筋线
elastomer 弹性体
elastomer gel 弹性凝胶
elastomeric 弹性体[的]
elastomeric base pad 弹性底垫
elastomeric fibre 弹性纤维（聚氨基甲酸酯弹性纤维的统称）
elastomeric film 弹性薄膜
elastomeric laminates 弹性层压物
elastomeric pad method 弹性垫试验法
elastomeric radiation-curable binders 可辐射固化的弹性体黏合剂
elbow 弯头；弯管,肘管
elbow bend 管子弯头
elbow joint 弯头结合
elbow union 弯头套管
elderberry 莲灰[色]
elec. (electrical) 电的
elec. (electricity) 电；电学
electrical〈elec.〉电的
electrical air floatation 电气浮法《环保》
electrical arc welding 电弧焊
electrical cabinet 电气柜
electrical capacity-type moisture meter 电容式测湿仪
electrical diagram 电气图
electrical effect 电气效果
electrical equipment noise 电器噪声
electrical erosion 电腐蚀
electrical finish [织物]电绝缘整理
electrical humidity measurement 电子湿度测定
electrical insulator 电绝缘体
electrical interaction 电气相互作用
electrical layout diagram 电气安装图
electrical leakage resistance 抗电气泄漏，漏电阻抗
electrically-conductive additives 导电添加剂
electrically-conductive adhesives 导电粘接剂
electrically conductive fibre 导电纤维
electrically-conductive products 导电产品
electrically heated cylinder 电加热滚筒
electrically heated flat bed press 电热平板压烫机（间歇式转移印花设备）
electrically heated roller 电热辊
electrically [heated] infrared predryer 电热红外线预烘机
electrically transplanting 静电植绒工艺
electrical moisture meter 电气测湿计
electrical provision 电气设备
electrical pulse 电脉冲
electrical repulsive force 电气反斥力
electrical resistivity 电抗值；电阻率
electrical signal 电信号
electrical slip 电动机转差
electrical stop motion 电气自停装置
electrical strip heater 电热丝式加热器
electrical surface phenomenon 界面带电现

electric arc welding 电弧焊
electric axis 电轴
electric beam curing methed 电子束焙烘法
electric blanket 电热毯
electric brake 电闸,电刹车
electric charge 电荷
electric charge density 电荷密度
electric circuit diagram 电气原理图
electric cloth cutter 电裁衣刀
electric comsumption 电力消耗
electric conductivity 导电性;电导率
electric contact thermometer 电接点温度计
electric controller 电力控制器
electric control valve 电动调节阀
electric coupling 电磁联轴器;电耦合
electric current 电流
electric dipole 电偶极子
electric discharge 放电
electric discharge ozone generator 放电臭氧发生器
electric double layer 双电层
electric-driven bowl 电动辊,主动辊
electric eye 电眼,光电池
electric glow discharge method 辉光放电（低温等离子体技术）
electric heat drying 电热烘燥
electric heating 电加热
electric heating blanket 电热毛毯
electric heating hydraulic press 电热式水压压呢机
electric heating plate 电热板
electric heating press 电热压呢机
electric hemming apparatus〈EHA〉电动卷边装置;缝线防偏装置

electric hoist [block] 电动葫芦
electric-hydraulic fabric infeed device 电动液压进布装置
electric ignition 电火花点火,电气点火
electric insulation 电绝缘性
electric interlock 电气联锁;电接合,电连接
electric iron 电熨斗
electricity〈elec.〉电;电学
electricity generation by steam pressure difference 蒸汽压差发电
electric jacket 电热套
electric mains 输电干线
electric oiler 电动加油器
electric-osmose 电渗[透]
electric polisher 电热抛光机,烫光机
electric potential 电位
electric power 电力;电功率
electric press 电压呢机,电烫衣机
electric pump 电动泵
electric resistance 电阻
electric schematic diagram 电气原理图
electric sealing machine 电气熔缝机,高频熔缝机
electric sewing machine 电动缝纫机
electric shoe 电热靴（毛毡预缩机重要部件）
electric singeing 电气烧毛
electric singeing machine 电热板烧毛机
electric smog〈E-smog〉电子雾
electric smog protection 电子雾的防护
electric spark 电火花
electric steel 电钢（合金钢）
electric tape cloth 黄蜡布,电线包布
electric transmission 输电
electric washing machine 电动洗涤机

electric welding 电焊
electric wiring diagram 电气线路图
electrifiable 可起电的,可带电的
electrifying 毛绒电光[整理]
electroanalysis 电分析
electrobalance 电天平,电光天平(俗称);电平衡
electro blowing 电喷法(纳米纤维形成技术的一种)
electrocapillarity 电毛细[管]现象
electrochemical analysis 电化学分析
electrochemical equivalent 电化学当量
electrochemical fluorination〈ECF〉 电化学氟化法(制造含氟整理剂的技术)
electro chemical measurement 电化学测定
electrochemical reduction dyeing equipment〈ERDE〉用电化学还原的染色设备(可避免化学品还原所产生的污染)
electrochemistry 电化学
electrocoating 电涂;静电植绒
electrocolour method of carpet printing 电子喷印法地毯印花
electroconductive resin 导电[性]树脂
electro-conductive textile 导电纺织品
electroconsole unit 电气控制台
electrocorrosion 电腐蚀
electrocratic 电稳的
electrocurtain 电屏,帷幕式电子加速器
electrode 电极;焊条
electrode boiler 电极锅炉
electrode holder 焊条钳
electrodeposition coating 电沉积涂层
electrode potential 电极电位
electrodialyzer 电渗析器
electro etching 电解刻蚀
electrofixer 电热固色器(用于还原染料两相法印花后的固色);电定形器(预缩机)
electrofixer rapid ager 电热固色快速蒸化机
electroflotation method 电解浮选法;电解浮渣法
electro-functional dyes 电功能性染料
electrofying 电光整理(用于长绒针织物)
electrographic analysis 电子(X光)图像分析
electrogravimetric analysis 电重量分析
electrogravimetry 电重量分析法
electrohydraulic load 电动液压加载
electrokinetic potential ζ电势;ζ电位;动电势;动电位
electroluminescence 电致发光,场致发光
electrolyser 电解池,电解槽
electrolysis 电解[作用]
electrolyte 电解质
electrolyte free solution 无电解质溶液
electrolyte resistance 抗电解质
electrolyte sensitive thickener 电解质敏感的增稠剂
electrolyte sensitivity 电解质敏感性
electrolytic 电解的,电解质的
electrolytic agent 电解剂
electrolytic analysis 电解分析
electrolytic bleach(= electrolytic bleaching)电解漂白
electrolytic bleaching apparatus 电解漂白设备
electrolytic cell 电解[电]池
electrolytic corrosion 电解腐蚀
electrolytic cuprous ion method 电解亚铜离子染色法(用于腈纶染色)
electrolytic dissociation 电离[作用]

electrolytic etching 电解刻蚀
electrolytic etching technique 电解刻蚀技术
electrolytic fluoration 电解氟化
electrolytic purification of waste water 废水电解法处理
electrolytic scouring 电解精练
electrolytic solution 电解溶液
electrolytic water proofing method 电解防水法
electromagnet 电磁铁
electromagnetic〈EM〉电磁的
electromagnetic brake 电磁闸
electromagnetic button 电磁式按钮
electromagnetic cloth guiders 电动吸边器
electromagnetic clutch 电磁离合器
electromagnetic concentration meter 电磁浓度计
electromagnetic control 电磁控制
electromagnetic device 电磁装置
electromagnetic disc brake 电磁刹车盘
electromagnetic disturbing shield 电磁干扰屏蔽
electromagnetic flowmeter 电磁流量计
electromagnetic gear 电磁传动,电磁联轴节
electromagnetic impulse 电磁脉冲
electromagnetic induction 电磁感应
electromagnetic interaction 电磁相互作用
electromagnetic radiation 电磁辐射
electromagnetic radiation shielding 电磁辐射屏蔽
electromagnetic sensor 电磁传感器
electromagnetic separation 电磁分离
electromagnetic valve 电磁阀
electromagnetic wave 电磁波

electromagnetic wave absorbing material 电磁波吸收材料
electromagnetic wave shielding 电磁波屏蔽
electro magnetism 电磁;电磁学
electromechanical controlled weft straightener 机电控制整纬器
electromechanical device 机电设备
electromechanical edge feeler 电动探边器
electro-mechanic integration 电机一体化
electrometer 电位计
electrometric analysis 电位分析
electrometric method 电测法,量电法
electrometric pH indicator pH 电测指示仪
electrometric titration 电势滴定
electromobile 电动车,电瓶车
electromotive force〈EMF, E. M. F., emf, e. m. f.〉电动势
electromoulding 电铸
electron acceptability 电子接受能力
electron-accepting agent 电子接受剂
electron accepting group 电子接受基团
electron acceptor 电子接受剂,电子接受体
electron affinity 电子亲势
electronation 增电子[作用](即还原作用)
electron attractive group 吸电子基团
electron attractivity (= electron attractor) 电子吸引体
electron beam 电子束
electron beam fixation 电子束固着
electron beam fixation dyes 电子束固着染料
electron beam radiation 电子束辐射
electron camera 电子摄像机
electron capture detector〈ECD〉电子俘获

检测仪
electron cloud density 电子云密度
electron curtain 电子帘
electron cyclotron resonance〈ECR〉电子回旋加速器共振(产生微波设备)
electron-deficiency 缺电子
electron-deficient 缺电子的
electron diffraction 电子衍射[法]
electron diffraction diagram 电子衍射图
electron donability 电子给予性
electron-donating agent 电子给予剂
electron donative group 电子给予基团,供电子基团
electron donor 电子给予体
electron donor-acceptor〈EDA〉电子供受体
electronegative 负电性的,阴电性的
electronegativity 负电性,阴电性
electron emission spectrometry 电子发射光谱测定法
electron emission spectroscopy 电子发射光谱
electronically controlled engraving head 电子雕刻头
electronically monitored and controlled dispensing equipment 电子监控[染料]调配设备
electronic balance 电子秤
electronic board 电子板
electronic circuitry 电子线路
electronic cloud density 电子云密度
electronic colour matching device 电子配色装置
electronic colour scanner 电子彩色扫描器
electronic colour separating device 电子分色装置

electronic component 电子元件
electronic computer(= electronic computing machine)电子计算机
electronic conduction 电子导电
electronic control console 电子控制台
electronic counting 电子计数
electronic cycle programming and control 电子周期程序编制和控制
electronic data processing system〈EDP system〉电子数据处理系统
electronic data reading, storing and transfer system 电子数据读数、存储和传送系统
electronic device for fault diagnosis 电子瑕疵分析装置;电子故障分析装置
electronic diagnosis 电子鉴别诊断;电子识别
electronic-electromagnetic system 电子—电磁系统
electronic emission 电子发射
electronic engraving 电子雕刻[法]
electronic engraving machine 电子雕刻机
electronic feeler 电子探测器
electronic guiding 电子导布
electronic instrument 电子仪器
electronic length measuring apparatus〈ELMA〉电子长度测试仪
electronic length measuring unit 电子长度计量装置
electronic manometer 电子压力计
electronic microscope 电子显微镜
electronic moisture content controller 电子湿度控制器
electronic moisture meter 电子测湿计
electronic photoengraving machine 光电雕刻机

electronic polarization 电子极化作用
electronic precision scales 精密电子秤
electronic printing and colour scanning system 电子印花分色系统
electronic program[me] control system 电子程序控制系统
electronic pulses 电子脉冲
electronic scanner 电子扫描器
electronic seam detector 电子缝头探测器（装于轧光机、起绒机等设备上，探测织物的线缝，及时将信号传至自控装置，使辊筒稍许松开，以免轧坏）
electronic setting temperature 电子[控制]定形温度
electronic signal 电子信号
electronic spectrum 电子光谱
electronic speed probe 电子测速器
electronic stop motion 电子自停装置
electronic temperature sensor 电子测温器
electronic tuning and control equipment 电子调控装置
electronic video recorder 电子录像器
electron magnetic resonance spectroscopy 电磁共振光谱学
electron micrograph 电子显微镜相片
electron microscope 电子显微镜
electron microscope photograph 电子显微照相
electron microscopy 电子显微术
electron optic[al] 电子光学的
electron optics 电子光学
electron pair bond 共价键
electron probe 电子探针，电子探头
electron repelling group 排斥电子基团
electron shell 电子[壳]层
electron spectrometer 电子分光仪

electron spectroscopy for chemical analysis〈ESCA〉电子能谱分析，化学分析用电子光谱
electron spin resonance〈ESR〉电子自旋共振
electron spin resonance spectrometry 电子自旋共振分光测定法
electrontimer 电子定时控制器
electron transfer 电子移动
electron transfer spectrum 电子移动谱
electron tube 电子管
electron volt 电子伏特
electron withdrawing group 吸电子基团
electro-optical application 电光应用
electrooptic effect 电光效应
electro-osmosis 电渗[透]
electropeter 整流器，转换器
electrophilic 亲电[子]的，吸电[子]的
electrophilic reaction 亲电反应
electrophoresis 电泳[法]
electrophoretic deposition 电泳沉积
electrophotography 电子照相术
electrophotometer 电子光度计(测火焰用)
electroplated roll 电镀辊
electroplating 电镀法，电镀
electropneumatic control 电动气动控制
electro-pneumatic piston 电控气动活塞
electropolishing 电抛光
electropositive 正电性的，阳电性的
electropsychrometer 电气湿度计
electropyrometer 电阻[测]高温计
electroreduction 电解还原
electro-rheological fluids 电流变体
electroscope 验电器
electroslag welding 电渣焊
electrospray lubricator 电动喷雾式加油器

electrostatic 静电的
electrostatic adsorption 静电吸附［作用］
electrostatic agent 静电［消除］剂
electrostatically spraying 静电喷射
electrostatic attraction 静电吸引
electrostatic bond 离子键
electrostatic capacity 静电容量
electrostatic charge 静电荷
electrostatic decay half-life 静电半衰期
electrostatic deposition 静电沉降（过滤机理）
electrostatic filter 静电滤尘器
electrostatic flocking 静电植绒
electrostatic flock printing 静电植绒印花
electrostatic heat 静电热
electrostatic high tension machine 静电高压发生机
electrostatic inkjet 静电式按需喷墨（印花术语）
electrostatic interaction 静电相互作用
electrostatic precipitator 静电除尘器；静电沉淀器
electrostatic printing 静电印花
electrostatic printing machine 静电印花机
electrostatic propensity of carpets 地毯的静电倾向《试》
electrostatic unit〈ESU, E. S. U.〉 静电单位
electrostylus 电唱针，电记录针，电描画针
electrotimer 定时继电器
electrovalence 电［性］价，电化价
electrovalent bond 离子键
electrovolumetric analysis 电容量分析
electroweight analysis 电重量分析
electrowelding 电焊
electrozero control 电动调零控制

electrozero raising machine 电调零点起毛机
Elemendorf testing method of tearing strength 埃尔门多夫撕破强力测试法
element 元素；单体，单元；元件；零件；电池
elementary 元素的；基础的；初等的
elementary analysis 元素分析
elementary colour 原色
elementary time 工序时间，操作时间
element[ary] filament （复丝中的）单丝，单纤维
elevated float dryer 架空式悬浮烘燥机
elevated tenter frame 架空式拉幅机
elevation 上升，升高；正视图；纵剖图
elevation of boiling point 沸点上升
elevation of temperature 升温
elevator 升降机；电梯
eliminability 可消除性
β-elimination β-消除［作用］（乙烯砜型活性染料化学反应之一）
elimination reaction 消除反应
elimination test 淘汰试验，排除试验［法］
eliminator ［空调用］空气净化器；静电消除器
eliminostatic 静电消除
ellagitannic acid 鞣花单宁酸
ellipse 椭圆
ELMA (electronic length measuring apparatus) 电子长度测试仪
Elmendorf method 埃尔门多夫织物撕破强力试验法
Elmendorf tear tester 埃尔门多夫撕破强力试验仪
Elmillimess 电动测微仪
elongation 伸长；伸长率

elongation at break 断裂伸长率
elongation at specified load 定负荷伸长率
elongation percentage 伸长率
elongation strength testing 强力伸长试验
eluant 展开剂(色层法);洗提液
eluate 洗出液
elysian overcoating 拷花大衣呢
EM(electromagnetic) 电磁的
EMA(ethyl methacrylate) 甲基丙烯酸乙酯
E-mail 电子邮件
embedding 包埋法(显微镜切片术)
embedding medium 包埋介质,镶嵌介质
embedment 埋置[法];包埋,埋封
embo printing [凹凸花纹]着色轧印
emboss 凹凸轧花,浮雕印花,拷花,压花
embossed backing 压花底布
embossed calico 压花布,压花棉平布
embossed cloth 拷花布,压花布,凹凸花纹布,浮雕布
embossed crepe 轧纹泡泡纱
embossed effect 压花效应,拷花效应,凹凸花纹效应,浮雕效应
embossed finish 拷花整理
embossed foam 凹凸花纹泡沫底板(用于地毯)
embossed georgette velvet 拷花乔其丝绒
embossed groove 轧花沟槽
embossed overcoating 拷花大衣呢
embossed pattern 凹凸纹花纹,浮雕花纹
embossed plush 拷花长毛绒
embossed velvet 拷花丝绒;拷花棉绒
emboss finish 浮雕轧花整理;凹凸轧花整理;拷花整理
embossing 凹凸轧花,浮雕印花,拷花,压花;轧花整理,拷花整理

embossing bonding (非织造织物的)轧花黏合法
embossing calender 凹凸轧花机,拷花轧压机,浮雕轧压机
embossing fault 轧纹疵
embossing felt 轧花毛毯
embossing machine 凹凸轧花机,压花机,拷花机
embossing roller 轧花滚筒,拷花滚筒,压花滚筒,浮雕滚筒
emboss mark 辊压痕(整理疵)
embrittlement 发脆,脆化,脆变
embrittlement temperature 脆化温度
embrittlement time 脆化时间
embroidery effect 绣花效应
embroidery thread 绣花线
EMC(Engineering Manpower Commission) 工程人力委员会(美国)
emerald 绿宝石绿[色]
emerald green 翡翠绿[色];绿色颜料
emeraldine 苯胺绿[色](苯胺黑氧化过程中的色泽)
emergency control 事故调节;紧急控制器
emergency electricity generators 紧急用发电机
emergency light 事故信号灯
emergency operation 紧急情况,突然事件
emergency shut-down 紧急停车,安全停车
emergency staircase 安全梯
emergency stop 紧急停运,紧急停车
emergency stop bottons 紧急停车钮
emerized fabric 金刚砂起绒织物,仿麂皮起绒织物
emerizing machine 金刚砂起绒机
emerizing machine version〈EMV〉 金刚砂起绒机形式(磨毛机)

emerizing method 金刚砂起绒工艺,充麂皮整理
emerizing roll 金刚砂磨辊
emery 金刚砂
emery beam 金刚砂卷布辊
emery cloth [金刚]砂布
emery-coated beam(=emery-covered roller) 金刚砂卷布辊,金刚砂导布辊;磨辊
emery disc 金钢砂盘
emery fillet [金刚]砂带
emery grinder 金刚砂磨辊
emery machine 金刚砂起绒机,磨毛机
emery paper 金刚砂纸
emery raising machine 金钢砂起绒机
emery roller 金刚砂辊,磨辊
emery wheel [金刚]砂轮
emerzing machine version 金刚砂起绒机形式(磨毛机)
EMF(=E. M. F., emf, e. m. f., electromotive force) 电动势
Emila rotary viscometer 埃米拉旋转式黏度计
emission 发射;散发;放射性
emission factor 发射系数,发射因子
emission of volatile components [纺织品上]挥发性成分的散发
emission register 排放记录器;发射记录仪
emission spectral analysis 发射光谱分析
emission spectrum 发射光谱
emissions potential 潜在排放量
emissivity 发射系数,比辐射;发射率
emit 放射(光、热等)
emit light 辐射光
emittance 放射,发射,辐射
emitted energy 发射能量

emitter 发射极,发射体
emitting surface 辐射面(如金属红外加热器的红外线辐射表面)
Emmon's fire triangle 依氏燃烧三角[原理]
emollient 软化剂;润滑剂;润肤剂
empire tube 绝缘套管
empirical curve 经验曲线
empirical equation 经验方程,经验公式
empirical formula 经验公式
empty 贫色(色度学用语)
empty dyeing 染浅淡色
emptying 排放,排空
EMS(Environment Management Systems) 环境监测系统(国际环境 ISO 14000 标准)
emulate 仿真,仿效
emulator 仿真器;仿真程序
emulgator 乳化剂
emulsator 乳化器
emulsibility 乳化性
emulsifiable oil 乳化油
emulsification 乳化[作用]
emulsified oil 乳化油
emulsified solvent 乳化溶剂
emulsified textile lubricants 乳化润滑剂
emulsifier 乳化剂;乳化器
emulsifier of the oil-in-water type 水包油型乳化剂
Emulsifier VA 爱莫尔雪菲尔 VA(聚乙二醇醚类的混合物,非离子型,乳化增效剂,可增加油/水相乳液的黏度,商名,德国拜耳)
emulsify 乳化
emulsifying agent 乳化剂
emulsifying machine 乳化机,乳化器

emulsifying power 乳化能力
emulsion 乳浊液，乳胶，乳剂
emulsion binder 乳化黏合剂
emulsion breaker 破乳液剂，破乳[化]剂
emulsion copolymerization 乳液共聚作用
emulsion dyeing 乳化液染色(合纤载体染色法之一)
emulsion finishing 乳化整理
emulsion pad process 乳化液轧染
emulsion paste 乳化[印花]浆
emulsion polymer 乳液聚合物
emulsion polymerisation (= emulsion polymerization) 乳液聚合
emulsion polymerization type resin 乳化聚合型树脂
emulsion printing 乳化浆印花
emulsion screen 乳化法制[圆]网
emulsion speed 感光[胶]速度
emulsion stabilization 乳液稳定性
emulsion stabilizer 乳液稳定剂
emulsion system 乳化系统，乳化法
emulsion thickener 乳化糊；乳液增稠剂；印花乳浆
emulsion triangle 乳液三角(三相三角)
emulsion type polyacrylic ester copolymer 乳液型聚丙烯酸酯共聚物
emulsion wax 乳化蜡
emulsoid 乳胶[体]
Emulsur M 埃苗瑟 M(非离子型合成增稠剂，聚乙二醇醚类化合物，适用于涂料印花乳化浆的增稠，商名，日本松井色素)
Emulsur V 埃苗瑟 V(非离子型合成增稠剂，聚乙二醇醚衍生物，适用于涂料印花乳化浆的增稠，商名，日本松井色素)

EMV (emerzing machine version) 金刚砂起绒机形式(磨毛机)
EN (European Norm) 欧洲标准
enamel 搪瓷，釉
enamel blue 珐琅蓝[色](紫蓝色)
enameling duck 涂层帆布(作油布、帆布鞋、书面布、人造皮革等用)
enamelled cloth 漆[皮]布
enamelled wire 漆包线
enamel paint 瓷漆
enamel printing 漆印印花
Enc. (encyclop[a]edia) 百科全书
encapsulate 包胶的，用胶囊包起来的
encapsulated aluminum and nickel powder 胶囊化铝镍粉(用于涂层)
encapsulated dyes 胶囊化染料
encapsulated stannous chloride 胶囊化氯化亚锡(防拔染剂)
encapsulating film 包覆膜，囊包膜
encapsulation 密封；封装；形成囊状包[现象]；包囊[作用]
enclosed batching chamber 封闭式卷布室
enclosed blowing machine 密闭式蒸绒机
enclosed boiling kier 封闭式精练罐，高压精练罐
enclosed dyeing machine 封闭式染色机
enclosed hood 封闭罩子
enclosed jigger 封闭式卷染机
enclosed open-width washer 加盖平洗槽，蒸洗槽
enclosed scale thermometer 内标温度计
enclosed type 封闭式
enclosed type hood 封闭式安全罩
enclosed type motor 封闭式电动机
enclosure 外壳，箱体；附件；箱，罩
encoder 编码器

encrustation 结垢;结皮
encyclop[a]edia〈Enc.〉百科全书
end 端,封头;[卷染]道数;零头布;接头(连续加工中布匹缝头)
end bond 端键
end-capped 末端封闭
end capping polyurethane fixative [活性]封端聚氨酯固色剂
end cloth 头子布,接头布;导布
ended fabric 零头布
end effect 末端效应
end elevation 侧面图,侧视图
end fent 零头布,短码布;头子布
end float 轴向间隙
end frame 车尾机架
end group 端基
end-group concentration 端基浓度
end group distribution 端基分布
end group titration 端基滴定[法]
ending [染色]头梢色差,两端色差
ending marks 结头痕
end initiation 端基引发
end item 成品检验,末道工序;最后项目
end item examination 成品检验,最终项目检验
endless belt 环[状]带,循环皮带
endless blanket 环形橡胶导带(滚筒或圆网印花机用的无接缝橡胶毯)
endless chain 循环链
endless conveyor 循环传送带
endless fabric 环[带]形织物
endless felt 循环毛毡
endless flexible steel band 挠曲性的环形钢带
endless screw 蜗杆
endless weaving 环形织造

end mark 梢印(盖在每匹或每段织物两端的印章)
endo activity 内切活性
endo cellulase 内切型纤维素酶
endo cellulase units〈ECU〉内切纤维素酶活力单位
endocrine 内分泌的,激素的
endocuticle 内表皮层,内皮层
endo glucanase〈EG〉内切葡聚糖酶
endo glucanase unit〈EGU〉内切葡聚糖酶活力单位(纤维素酶)
endosmosis 内渗[现象]
endosperm 胚乳
endotherm 吸热
endothermal transitions 吸热转变
endothermic 吸热的
endothermic reaction 吸热反应
end plate 端板,端盖
end point 终点《试》;[使用]寿命
end point correction 终点校正《试》
end product 成品,最终产品
end-ring [圆网]闷头
end-ring glue [圆网]闷头胶
end-ring glueing device [圆网]闷头胶黏装置
end-ring glueing machine [圆网]闷头上涂胶机
end-ring glue remover [圆网]闷头脱胶剂
end-rings for rotary screen 圆网端环,圆网闷头
end sewing [坯布]缝头(原布间将坯布缝头连接,以利连续加工)
ends per inch〈epi〉每英寸经纱根数;经密
end stamping 布端打印
end stitching [坯布]缝头

endurance 耐久性,耐用性;耐磨性
endurance bending test 耐弯曲疲劳试验
endurance crack 疲劳断裂,疲劳裂纹
endurance limit 疲劳极限
endurance test 疲劳度试验,耐久性试验
end-use 产品用途,最终用途
end use field 使用范围
end-use test 使用试验
end use textile product 最终用途的纺织制品
end view 端视图
energetic balance 能量平衡
energize 激发,加强
energy balance 能量平衡[结算]
energy barrier 能量屏障
energy carriers 载能体
energy conservation 能量节约;能量守恒
energy consumption 耗能
energy cost 能量成本
energy dissipation 能量消散
energy efficiency 能源效率
energy exchange 能量交换
energy flow diagram 能流图
energy level 能[量]级,能量水平
energy loss 能量损耗
energy of adhesion 黏合能
energy of deformation 应变能
energy of retraction 弹性回复能
energy recovery 能量回收
energy recuperation device 能量回收装置
energy retrieval system 能源回收系统
energy savings 节能
energy storage fabric 储能织物
energy-to-break 断裂能
eng.(engine) 发动机,引擎
eng.(engineer) 工程师

eng.(engineering) 工程,工程学
Eng.(England) 英国
Eng.(English) 英语;英国的
eng.(engrave) 雕刻
engagement 啮合;接合
engine〈eng.〉 发动机,引擎
engineer〈eng.〉 工程师
engineered flexibility 工程伸缩性
engineered materials 工程材料
engineered textiles 工程纺织品
engineering〈eng.〉 工程,工程学
engineering atmospheric pressure 工程大气压(压力单位,等于1千克力/平方厘米❶)
engineering change 工艺更改
engineering fabric 工程织物
engineering fibre 工程纤维
Engineering Manpower Commission〈EMC〉 工程人力委员会(美国)
engineering of civil 土木工程
engineering of hydraulic & dam 水力和水坝工程
engineering of land reclamation & environment 土地开垦和环境工程
engineering of transport 运输工程
engineering plastics 工程塑料
engineering time 维护检修时间
engineers' cloth 工作服布(如劳动布、厚斜纹布)
engine oil 机油,机器润滑油
England〈Eng.〉 英国
Engler visco[si]meter 恩氏黏度计
Engler viscosity 恩氏黏度

---

❶ 1千克力/平方厘米≈98.1千帕斯卡。

English〈Eng.〉英语;英国的
English blowing 英式蒸呢法(顺逆向两次蒸呢)
English count 英制支数
English red 英国红,氧化铁红
English system 英制(长度、质量单位分别为英寸、磅)
English yarn count〈N_e〉英制纱线支数
engrained 深染的
engrave〈eng.〉雕刻
engraved copper roller 刻花铜辊;花筒(俗称)
engraved cylinder 刻花滚筒
engraved point (刻度盘的)铭记点,标记点
engraved roller 印花滚筒,花筒;刻纹滚筒(滚筒表面刻有微孔的滚筒)
engraved roller printing 雕刻滚筒印花
engraved roll pad 纹辊浸轧法(低给液法)
engraved screen 雕刻筛网
engraver 花筒雕刻工;雕刻刀,刻头
engraving 雕刻,刻纹
engraving chisel 雕刻刀,雕刻凿子
engraving depth 雕刻深度
engraving die 雕刻钢模
engraving hammer 雕刻锤
engraving head 雕刻头
engraving machine 雕刻机
engraving of embossing roller 压纹滚筒雕刻
engraving stylus 雕刻刀,雕刻针
engraving width 雕刻幅
engraving with wax jet machine 用喷蜡机[雕刻]制网
enhanced 增艳;明艳
enhancement 增强,提高

enl.(**enlarged**) 扩大的;放大的
enl.(**enlargement**) 扩大;放大
enlarged〈enl.〉扩大的;放大的
enlarged scale 倍尺
enlargement〈enl.〉扩大;放大
enlargement and reduction 放大和缩小(电脑图案处理)
enlarger 放大器
enlarging a design 放大花样
enlarging and reducing range 缩放范围
enlarging camera 放样机,放映机
enlarging machine 放样机,放映机
enleavage 拔染印花
enol 烯醇
enriching recovery 浓缩回收
ensanguined 血红色的
ent.(**entrance**) 入口,进口
entering arrangment 进布装置,进布架
entering device 进布装置
entering roller (＝entering rolls) 进布辊,喂入辊
enthalpy 焓,热函
enthalpy-entropy chart 焓熵图
enthalpy-entropy diagram 焓熵图
enthalpy of dissolution 溶解热函
enthalpy of dyeing 染色热函
enthalpy of hydration 水合热函
enthalpy of reaction 反应热函
enthalpy of swelling 膨胀热函
enthalpy of wetting 润湿热函
entire bleach 完全漂白
entrance〈ent.〉入口,进口
entrance keeper 进布挡边器
entrance slit 入口狭缝
entrapped air 陷捕空气,截留空气
entropy 熵,热荷

entropy of dyeing 染色熵
entropy of reaction 反应熵
entry 入口;进入,输入
entry end 进布端
entry end framing box [拉幅机]车头箱
entry frame(＝entry gantry) 进布机架
entry mouth of the steamer 蒸箱的进口
entry slot 进布口
entry zone 进布区
envelope 打包布,麻袋布
envenomation 表面毒化
environment 环境

environmental bearing 环境负荷
environmental biology 环境生物学
environmental capacity 环境容量
environmental chamber 人工环境室;人工气候室
environmental chemistry 环境化学
environmental communication 环境交流活动
environmental conscious process 环保意识型流程
environmental conservation 环境保护
environmental contamination 环境污染
environmental control chamber 人工气候室
Environmental Control Organization〈ECO〉环境控制组织(美国)
environmental criteria 环境标准;环境准则
environmental degradation 环境降解
Environmental Detection and Control Center〈EDCC〉环境监测中心
environmental disruption 环境破坏
environmental engineering 环境工程
environmental factor 环境因素

environmental-friendly process 环境友好[染整]工艺
environmental garment 环保服装
environmental hormone〈EH〉环境荷尔蒙(对人类健康和生态环境危害很大的化学物质)
environmental impact 环境影响
environmental impact assessment 环境影响评定
environmental legislation 环境法规
environmentally friendly finishing process 环境友好整理工艺
environmentally responsible product 无环境污染产品
environmentally safe 生态安全[产品]
environmental management 环境管理
environmental monitoring 环境监测
environmental pollution 环境污染,公害
environmental pollution control〈EPC〉环境污染控制
environmental protection 环境保护
Environmental Protection Agency〈EPA〉环境保护局(美国)
environmental quality pattern 环境质量模型
Environmental Quality Standard〈EQS〉环保质量标准
environmental resistance 环境阻力
environmental sciences 环境科学
environmental sensitive 环境敏感
environmental sensitive hydrogels 环境敏感水凝胶
environmental simulation lab 环境[模拟]试验室
environmental standard 环境保护标准,环境标准

environmental stress index 环境重点指标
environmental water quality standard 水质环境标准
environment effluent concentration〈EEC〉环保污物浓度
Environment Labels〈EL〉环境标鉴(国际环境 ISO 14000 标准)
Environment Management Systems〈EMS〉环境监督系统(国际环境 ISO 14000 标准)
Environment Protect Evaluate〈EPE〉环境保护实效评价(国际环境 ISO 14000 标准)
environment quality standard 环保质量标准
environment technology 环境技术
environment type auxiliary 环保型助剂
environment type dyestuff 环保型染料
enzymatic biofinish 酶生物整理
enzymatic bleach clean-up process (漂浴中)去除剩余双氧水的酶洗法
enzymatic defibrillation [用]酶去原纤化
enzymatic degradation 酶[催化]降解
enzymatic degumming 酶脱胶
enzymatic descaling [用]酶去羊毛鳞片
enzymatic desizing 酶退浆
enzymatic hydrolysis 酶[催化]水解
enzymatic removal 酶[催化]消除[法]
enzymatic rinsing system 酶型皂洗系统(活性染料皂洗剂,可提高耐湿摩擦牢度,且节能节水)
enzymatic shock 酶快速退浆
enzymatic stonewashing 酶石磨洗
enzymatic system 酶系
enzyme 酶
enzyme activity 酶活力

enzyme carrying agent 酶载体
enzyme conversion 用酶转化
enzyme preparation 酶制剂
enzyme reduction treatment 酶减量处理
enzyme saturator 浸酶槽
enzyme scouring 酶精练
enzyme steeping 酶浸渍[退浆]
enzyme stop 酶处理的消除
enzyme substrate complex 酶底物复合体
enzyme transglutaminase 反式谷氨酸转化酶
enzyme washing 酶洗
enzyme wool 经酶处理的羊毛
enzymic protein 酶类蛋白质
enzymology 酶学
enzymolysis 酶解[作用]
EO(ETO,ethylene oxide) 环氧乙烷;乙烯化氧
EOQC(European Organisation for Quality Control) 欧洲质量控制组织
EOS(effecive opening size) 有效孔径
eosine 朝红,伊红,酸性曙红,曙红钠
EOX(extractable organic halogen) 可萃取的有机卤化物
EP(extra pure) 特纯的
EPA(Environmental Protection Agency) 环境保护局(美国)
EPC(environmental pollution control) 环境污染控制
EPDM(ethylene-propylene diene monomer) 乙烯—丙烯二烯单体
EPE(Environment Protect Evaluate) 环境保护实效评价(国际环境 ISO 14000 标准)
epi(ends per inch) 每英寸经纱根数;经密
epichlorohydrin 3-氯-1,2-环氧丙烷;表氯

醇(俗称)
epicuticle 外层薄膜,外表皮;鳞片表层《毛》
epicyclic gear 周转齿轮
epicyclic gear train 周转轮系
epicyclic reduction gear unit 行星减速装置
epidermis 表皮
epipolic 荧光[性]的
epoxidation 环氧化作用
epoxide 环氧化物
epoxide resin 环氧树脂(线型树脂,棉织物防缩防皱剂)
epoxy crease resist finishing agent 环氧类防皱整理剂
epoxy propyl-trimethyl ammonium chloride 环氧丙基三甲基氯化铵
epoxy resin 环氧树脂(线型树脂,棉织物防缩防皱剂)
epoxy resin cure 环氧树脂固化
epoxysiloxane 环氧硅氧烷
Eppenbach homomixer 埃彭巴赫[涂料树脂]高速搅拌器
EP reagent (extra pure reagent) 一级试剂
Epsom salt 七水合硫酸镁,泻盐(可作阻燃剂及染色助剂)
Epton method 埃泼顿法(阴离子、阳离子活性剂定量分析法)
EQS(Environmental Quality Standard) 环保质量标准
equalization and neutralization lagoon 均合中和池《环保》
equalization pond [污水处理的]生物氧化塘,均合池
equalizer 平衡器,补偿器;拉幅机,平拉机
equalizer roll 扩幅辊

equalizing 匀染
equalizing acid dyes 匀染[性]酸性染料
equalizing device 平衡装置
equalizing effect 均衡效应,调整效应
equalizing frame 拉幅机,平拉机
equalizing squeegee 均浆刮刀(圆网印花或平网印花时,导带上胶用)
equalizing stenter 整幅拉幅机,平幅机
equalizing tank 调节池,均衡槽
equational box 差速装置
equation of linear regression 线性回归方程
equi-energy spectrum 等能光谱
equilibrium [反应]平衡,均衡;平均
equilibrium absorption 平衡吸收[作用]
equilibrium adsorption isotherm 平衡吸附等温线
equilibrium behaviour 平衡行为
equilibrium condition 平衡状态,平衡条件
equilibrium constant 平衡常数
equilibrium contact time 平衡接触时间
equilibrium degree of exhaustion 平衡吸尽度
equilibrium dyeing 平衡染色
equilibrium exhaustion 平衡吸尽
equilibrium liquid 平衡液
equilibrium melting point 平衡熔点
equilibrium modulus 平衡模量
equilibrium moisture content 平衡含湿率
equilibrium moisture regain 平衡回潮率
equilibrium relative humidity〈ERH〉 平衡态相对湿度
equilibrium state 平衡状态
equilibrium swelling 平衡溶胀
equilibrium viscosity 平衡黏度
equilibrium water content 平衡含水量
equipment 设备,装备,装置

equipment appurtenance 设备附件
equipment compatibility 设备兼容性;设备互换性;设备相容性
equipment control 设备管理
equipment failure 设备故障,设备失效
equipment intact rate 设备完好率
equipment repair 设备检修
equipment replacement 设备更新
equipment running rate 设备运转率
equipment utilizational rate 设备利用率
equivalence 等价(化合价相等)
equivalence point 当量点,等效点
equivalent 当量,等效
equivalent adsorption 等效吸附
equivalent circuit diagram 等效电路图
equivalent conductivity 当量导电率,等效导电率
equivalent evapouration 蒸发当量
equivalent focal length 当量焦距,等效焦距
equivalent steam setting temperature 等效蒸汽定形温度
equivalent weight 当量
eradicator 去毛装置;去污装置
erase 擦除,删除
erasing knife 刮刀
ERDE (electrochemical reduction dyeing equipment) 用电化学还原的染色设备(代替化学品还原所产生的污染)
erecting shop 装配车间
erection 架设,安装
erection drawing 安装图
erect pile 直立绒头,立绒
ergonomic control systems 人类工程控制系统,人机工程控制系统
ergonomics 人类工程学,人机工程学,工效学
Ergonomics of clothing 服装工效学(研究人体、服装、环境之间相互关系的科学)
ergosterol 麦角固(甾)醇
ERH (equilibrium relative humidity) 平衡态相对湿度
Erifon process 埃利砜[防火整理]法(纤维素织物以钛、锑化合物处理的防火整理法,商名,美国制)
Eriochrome dyes 艳丽华媒介染料(用于羊毛染色、毛条印花,商名,瑞士制)
Erional RF 伊利那尔 RF(防染、匀染剂,可改善毛锦混纺及锦棉混纺染色的匀一色效果,商名,瑞士制)
Erionyl A dyes 伊利尼尔 A 型染料(弱酸性染料,适用于锦纶及羊毛染色,上染率高、移染性和条花覆盖性良好,商名,瑞士制)
Eriopon CRN 伊利朋 CRN(清洗剂,强力洗涤及皂煮用,特别适用于高反差棉色织物,商名,瑞士制)
Erlenmeyer flask 锥形烧瓶,爱伦美氏瓶
erosion 侵蚀,腐蚀,磨蚀,刻蚀
erosion method 腐蚀法,侵蚀法,刻蚀度(激光雕刻术语)
error 谬误;错误;误差
error-diffusion screening 误差扩散网点(喷印术语)
error-free regulating system 无误差调节系统
error message 错误信息
error rate 误差率
error system 误差检测系统
erythrosine 赤藓红(四碘荧光素)
ES(ethyl sulfone) 二乙基砜

ESCA(electron spectroscopy for chemical analysis) 电子能谱分析,化学分析用电子光谱
escapechute 救生降落伞
escape opening 排气孔
escape pipe 排气管
escaping gas 逸出的气体
E-smog(electric smog) 电子雾
especially〈esp., especc.〉特别,尤其
Esqual T-150 埃斯阔尔 T-150(涤纶匀染剂,特殊非离子、阴离子表面活性剂,在涤纶织物快速染色中能防止起色斑,避免筒子纱内外色差,商名,大祥)
ESR(electron spin resonance) 电子自旋共振
essence 香精;香料;本质,实体
essential oil [香]精油
essential value 酯化值
Esser system loose fibre dyeing machine 埃塞散纤维染色机
est.(estimate) 估计,估价;预算
est.(estimated) 估计的
ester 酯
esterase 酯酶
ester cellulose 纤维素酯
ester gum 酯化胶
esterification 酯化
esterification oil 酯化油
esterification waterproofing 酯化防水整理法(纤维素用酸性氯化物或烯酮、异氰酸盐等酯化得到防水性)
esterified starch 酯化淀粉
ester interchange 酯交换
ester linkage 酯键[合]
ester number 酯[化]值
ester oil 酯油(以酯组成的合成油)

ester suponification 酯皂化
ester value 酯[化]值
ester wax 酯化蜡
esthetic design carpet 工艺美术地毯
esthetics 美学;美观性,美感性(纺织品的手感、颜色、光泽、式样、结构等通过目视手触,使人感受的特性)
estimate〈est.〉估计,估价;预算
estimated〈est.〉估计的
estimation 评价,评定;测定
estrogen 雌激素
ESU(＝E. S. U., electrostatic unit) 静电单位
ETAD(Ecological and Toxicological Association of Dyes and Organic Pigments Manufactures) 染料及有机颜料制造商生态和毒理学协会
etamine glace 印花丝毛薄织物
etchant 浸蚀剂;蚀刻剂
etched-out fabric 烂花织物
etched screen 蚀刻筛网
etching [花筒]腐蚀;烂花[工艺];蚀刻法
etching apparatus 花筒腐蚀设备
etching discharge 烂花拔染法
etching solution 腐蚀溶液
etching trough 花筒腐蚀槽
etching varnish 蚀刻清漆
ethane 乙烷
ethanol 乙醇
ethanolamine 乙醇胺,2-羟基乙胺,氨基乙醇
ethene 乙烯
ethenoxy unit 氧乙烯单位(即环氧乙烯单位)
ether 醚,乙醚
ethereal blue 醚蓝[色],纯蓝[色]

ether extraction 醚萃取
etherification 醚化[作用]
etherification of cellulose 纤维素的醚化
etherification uniformity 醚化均匀度
etherification waterproofing 醚化防水整理法(一种活性防水剂与纤维素结合成醚,有良好的耐水性和耐久性)
etherified cotton 醚化棉
etherified starch 醚化淀粉
etherifying agent 醚化剂
ether linkage 醚键
ether link content 醚键含量
ether sulfates 醚硫酸盐
ether-type dimethylol urea 醚型二羟甲基脲,醚化二羟甲基脲(防皱整理剂)
ethnic pattern 民族图案
ethoxylated 乙氧基[化]
ethoxylated ester 羟乙基化酯
ethoxylated fatty acid 羟乙基化脂肪酸
ethoxylation 乙氧基化作用
ethoxyline resin 羟乙基苯胺树脂
ethyl acetate 乙酸乙酯
ethyl acrylate〈EA〉丙烯酸乙酯
ethylamine 乙胺,氨基乙烷
ethyl cellulose 乙基纤维素
ethylene 乙烯
ethylene diamine 乙二胺
ethylene diamine tetra acetic acid〈EDTA〉乙二胺四乙酸(有机螯合剂)
ethylene dimethacrylate〈EDMA〉二甲基丙烯酸次乙酯
ethylene glycol 乙二醇
ethylene glycol ether 乙二醇醚(硝化纤维素的溶剂)
ethylene glycol monostearate 硬脂酸乙二醇酯(遮光剂、乳化剂、稳定剂)

ethylene imine 乙烯亚胺
ethyleneimine isocyanate 乙烯亚胺异氰酸盐(防水整理剂)
ethylene-maleic anhydride 乙烯—马来酸酐
ethylene-maleic anhydride copolymer 乙烯—马来酸酐共聚物(可用作液态净洗剂、纺织物浆料、分散染料染色的缓冲剂)
ethylene oxide〈EO,ETO〉环氧乙烷;乙烯化氧
ethylene oxide condensate emulsifier 环氧乙烷缩合乳化剂
ethylene-propylene diene monomer〈EPDM〉乙烯—丙烯二烯单体
ethylene terephthalate 对苯二甲酸乙二[醇]酯
ethylene urea 环亚乙基脲,乙烯脲
ethylene urea formalehyde resin 乙烯脲甲醛树脂
ethylene urea triazine precondensate 乙烯脲三嗪预缩物(棉织物的耐氯防缩防皱剂)
ethylene vinyl acetate(EVA) 乙烯基醋酸乙烯酯
ethylene-vinyl acetate copolymer 乙烯—醋酸乙烯共聚物
ethyl methacrylate〈EMA〉甲基丙烯酸乙酯
ethylsulfonamide 乙磺酰胺
ethyl sulfone〈ES〉二乙基砜
ethyl triazone resin 乙基三嗪酮树脂(棉织物防缩防皱和耐氯整理剂)
etiolation 褪色
ETO(=ED,ethylene oxide) 环氧乙烷;乙烯化氧

Eton blue 伊顿蓝

ETSA(European Textile Services Association) 欧洲纺织品服务协会

EU(European Union) 欧洲联盟,欧盟

eucalptus 浅绿灰[色]

Eulan finish 优兰防蛀整理

Eulan SPA 优兰 SPA(多用途防虫、防蛀剂,合成除虫菊酯,适用于羊毛及其混纺织物,商名,德国拜耳)

Eulan U33 优兰 U33(国外较早的防蛀剂,商名,德国拜耳)

Eulysin S 优力新 S(pH 调节剂,脂肪族二羧酸混合物,用于染色印花工艺,商名,德国巴斯夫)

Eulysin WP 优力新 WP(pH 调节剂,低升华性有机酯,用于羊毛、锦纶染色,商名,德国巴斯夫)

eurhodine colouring matters 二胺吩嗪染料

eurhodiole colouring matters 二羟吩嗪染料

European Chemicals Agency〈ECHA〉欧盟化学品管理局

European clothes 西装

European Colour Fastness Establishment 欧洲色牢度组织

European Common Market〈ECM〉欧洲经济共同体

European Community〈EC〉欧洲共同体,欧共体

European Continental Fastness Convention〈ECFC〉欧洲大陆染色坚牢度公约

European Disposables and Non-Wovens Association〈EDANA〉欧洲耗材及非织造布协会

European Economic Community〈EEC〉欧洲经济共同体

European Environmental Label 欧洲环保标签

European Free Trade Association〈EFTA〉欧洲自由贸易联合会

European Free Trade Committee〈EFTC〉欧洲自由贸易委员会

European Norm〈EN〉欧洲标准

European Organisation for Quality Control〈EOQC〉欧洲质量控制组织

European Standard 欧洲标准

European Textile Services Association〈ETSA〉欧洲纺织品服务协会

European Union〈EU〉欧洲联盟,欧盟

European Union Eco-label 欧共同体生态标签

European Union Environmental Symbol 欧[洲联]盟环保标志

European Wet Cleaning Committee〈EWCC〉欧洲湿洗委员会

eutectic mixture 低共熔混合物

eutectic point 低共熔点

eutrophication 富营养作用《环保》

eutrophic water 富营养水,过营养水,过肥水《环保》

EVA(ethylene vinyl acetate) 乙烯基醋酸乙烯酯

evacuate 抽空;排气

evacuated chamber(=evacuated housing) 真空室

evacuated quartz tube heater 抽真空式石英管加热器

evacuated tube collector 真空管式[太阳能]收集器

evacuation 抽空,抽真空

evaluation 鉴定,评定;计算,求值

evaluation of dye 染料评价,染料鉴定

evaluation report 评价决策报告
evapourating basin（＝evapourating dish） 蒸发皿
evapourating surface 蒸发面,汽化面
evapouration 蒸发
evapouration capacity 蒸发量(常用来表示烘燥设备的能力,即一小时内汽化水分的质量,单位为千克/小时)
evapouration plant 蒸发设备
evapouration residue 蒸发残留
evapourative drying 蒸发式烘燥
evapourative heat loss 蒸发热损失
evapourator 蒸发器
evapourator coil 蒸发盘管
evapourimeter(＝evapourometer) 蒸发计
even dyeing 匀染
evening light 黄昏光
evenness 均匀,均匀度
evenness defect 均匀瑕疵;丝条斑
evenning effect 均匀作用
even-speed compensator 匀速补偿器
event alert 事故警告,报警(设备运转过程中事故的警告提示)
even touch 均匀触感(如剪毛绒)
ever crease 永皱的
ever-cut finish 烂花整理
ever fast 不褪色
Everglaze 耐久光泽整理(商名)
everglaze minicare 电光;摩擦轧光整理
evergreen 冬青绿[色]
everlasting cloth 永固缎纹织物(全毛或毛棉交织,鞋料等用)
ever pleat 耐久褶裥
evolution 析出,放出
EWCC（European Wet Cleaning Committee） 欧洲湿洗委员会

EWL（effective wave length） 有效波长
E-wool 生态防缩羊毛
E-wool treatment 生态防缩羊毛处理
exaltation 练浓《化》;精练;升高;加浓(色彩)
examination〈exam,examn.〉 检验,检查
examination certification 检验证书
examine 检验,检查
excellent definition 轮廓清晰(指印花)
excess air coefficient 过量空气系数
excess air control 过量空气控制
excess heat 余热
excessive equivalent adsorption 超等效吸附
excessive pressure 超压
excess scouring 过度精练
excess weight 过重,超重;多余重量
exchanger 交换器
exchange reaction 交换反应
exchange titration 置换滴定
excitation 激磁,励磁《电》
excitation failure 磁场失磁《电》
excitation purity 色纯度,[色]刺激纯度,[色]兴奋纯度
excitation region 激励波区
excited state 激发态
exciter 激发机,励磁机
exciting coil 激磁线圈,激磁绕组
exciting lamp 激励灯
exciting winding 激磁绕组,激磁线圈
exclusion chromatography 排阻色谱法,筛析色谱法
execute input/output〈XIO〉 执行输入/输出
execution drawing 施工图
execution time 执行时间;完成时间
executive component 执行元件,执行机构

executive system 操作系统；执行系统
exempli gratia〈eg, e. g.〉 例如
exergy 㶲；有效能；可用能
exhaust 排出；排除装置；排气管
exhaust air analysis 废气分析
exhaust air cleaning system 排气净化系统
exhaust air disposal 废气处理
exhaust air duct 排气管道
exhaust air heat recovery system 排气热回收系统
exhaust air pollution measurement 废气污染测定
exhaust canopy 排气顶罩
exhaust damper 排气风门
exhaust dyeing 浸染法；竭染法（染色后在同浴中再染色，以测试染料吸尽程度）
exhauster 排气风扇；排汽机；鼓风机；排粉［尘］机
exhaust fan 排气风扇
exhaust gas 尾气
exhaust gas cleaning 废气清洁
exhaust gas desulfurization 排烟脱硫
exhaust hood 排气罩
exhausting 抽空；排气
exhausting agent 尽染剂（促使染料吸尽的助剂）
exhaustion 尽染［作用］；上染率
exhaustion curve ［染色］吸尽曲线
exhaustion of bath 吸尽浴
exhaustion property 上染性能；吸尽性能；尽染性能
exhaustion rate 上染率
exhaustion rate coefficient 吸尽率系数
exhaust manifold 排气总管
exhaust muffler 排气消声器
exhaust pipe 出水管，排水管；排气管；废气管
exhaust resistance 排放阻力，流出阻力
exhausts 排气量
exhaust steam 废汽
exhaust thermosol process 吸尽热熔工艺
exhaust value 吸尽值
exhaust valve 排气阀，排放阀
exit 出口；排气口，输出，子程序的出口《计算机》
exit device 出布装置
exit end framing box ［热拉机］车尾箱
exit frame(＝exit gantry) 出布机架
exit slit 出射［狭］缝
exit slot 出布口
exo activity 外切活性
exo cellular cell 初生胞壁
exo cellulase 外切型纤维素酶
exocuticle 外角质层《毛》
exo-glucanase 外切葡聚糖酶（纤维素酶的成分）
exometer 荧光计
exothermic 放热的
exothermic catalyst 放热催化剂
exothermic element 放热元件
exothermic reaction 放热反应
exotic 异国情调
exotic colour 奇异颜色
exotic style 异国情调式样
expanded perlite 膨胀珍珠岩
expanded steam 扩容蒸汽
expander 扩幅装置，展幅辊；开幅器；［圆筒针织物的］扩张器
expander arrangement 扩幅器
expander bar 开幅板，扩幅棒
expander cap 膨胀套
expander roller 开幅辊

expander uncurler 扩幅器
expanding 扩幅,伸幅
expanding agent 发泡剂
expanding device for fabric 织物扩幅设备
expanding drum 开幅滚筒
expanding plate 扩幅板
expanding roller 开幅辊
expansibility 可扩张性;可膨胀性
expansion 膨胀
expansion bearing 活动支承(根据温度变化而有伸胀或收缩变化)
expansion bend 膨胀弯管
expansion compensator 膨胀补偿器
expansion joint 伸缩[接]缝;伸缩接头,涨缩接合
expansion ratio 充气率,发泡率,膨胀率
expansion tank 染液调整储桶(卷装染色设备);膨胀槽,扩容箱
expansive colour 膨胀色,奢华的颜色
expansivity [热]膨胀系数
experiment 试验,实验
experimental 试验性的,实验性的
experimental apparatus 实验设备
experimental correlogram 实验相关图
experimental design 实验设计
experimental dyeing 试验染色
experimental error 实验误差
experimental method 实验设计
experimental plant 实验工厂
expertise 专长,专门知识(专门技能);专家评价
explanation〈xpln.〉解释,说明
explanatory diagram 说明图(一种电气图,常见的有方框图、电路图等)
explanatory notes 说明,凡例
exploded view 展开图;部件分解图

exploitation 开发
explore 探索;浏览
explorer 浏览器《计》
explosion 爆炸
explosion diaphragm 防爆膜
explosion hatches [防]爆炸天窗(印花用干燥机的防爆设施)
explosion hazard 爆炸危险
explosion prevention system 防爆方法
explosion proof induction motor 防爆电动机
explosion suppression system 消除爆炸方法
explosive limit 爆炸极限
exponential distribution 指数分布
export capacity 出口量
export proportion 外销比重
export quantity 出口量
export rate 出口率
exposed film 已曝光的胶片
exposing all night 夜曝法
exposition(=exposure) 暴露;曝光
exposure chamber 曝气仓
exposure light test 曝光试验
exposure machine 曝光机(照相雕刻设备)
exposure meter 曝光表
exposure period 曝光期,曝晒期
exposure rack 曝光台
exposure test 曝光试验,暴晒试验
exposure time 曝光时间,暴晒时间
exposure to air 透风
exposure to sunlight 日光暴晒
expression 轧液率
exsiccant 干燥剂;干燥的
exsiccate 使干燥,弄干
exsiccation 干燥[作用]

exsiccator 干燥器
extended 冲淡,稀释(包括液体或固体); 延长,扩展
extended aeration 延时曝气
extended aeration process 延时曝气法《环保》
extended dyeing cycle 延长染色周期
extended linear macromolecule 伸直线型大分子
extended service 长期运行
extender 补充剂,增充剂;增量剂;稀释剂
extending agent 填充剂;填料
extensibility 伸长性,延伸性,延展性
extensional creep 拉伸蠕变,伸长蠕变
extensional deformation 拉伸形变
extensional motion 拉伸运动
extension at break 断裂伸长率
extension cycles 伸张次数(表示被测试织物的拉伸、回复性能)
extension measuring unit 伸长测定装置
extension percentage 伸长率
extension under given load 定负荷伸长
extension work 扩建工作
extensometer 伸长[测试]计
extent 程度;尺寸;范围
extent of devitrification 透明消失[程]度,失透[程]度
extent of polymerization 聚合[程]度
extent of reaction 反应[程]度
external antistatic agent(＝external antistatics) 外用抗静电剂
external cross-linking agent 外交联剂
external delustering [纤维]外部消光
external diameter 外径
external gauge 外径规
external gear 外齿轮,外啮合齿轮

external heated roller singeing machine 外热式圆筒烧毛机
external heat exchanger 机外热交换器
external indicator 外指示剂
external interference 外干涉
external multitubular heater [锅]外多管加热器(煮布锅)
external [coating] polymer [羊毛纤维]表面处理聚合物
external thread 外螺纹
external view 外视图
extinction 熄灭,消灭;消光;消失;衰减
extinction coefficient 消光系数
extinction value 消光度
extinguishability 可扑灭性
extinguisher 灭火器
extinguishing agent 灭火剂
extinguishing media 灭火介质
extracellular enzyme 细胞外酶
extra concentrated 〈extra conc.〉 特浓的(染料名称后的标记)
extract 萃取[物],抽出[物]
extractable organic halogen 〈EOX〉 可萃取的有机卤化物
extracted cloth 炭化呢
extracting 萃取法;炭化《毛》
extracting agent 提取剂,萃取剂
extracting machine 轧液机,轧车;绞榨机;提取机,萃取机
extraction 离心脱水;提取,萃取;炭化《毛》
extraction bath 萃取浴,抽提浴
extraction column 萃取塔,提取塔
extraction of a coacervate 凝聚物萃取[法]
extraction roller 提布辊
extraction scour/extraction bleach 萃取煮

练/萃取漂白
extraction solvent 萃取溶剂;抽提溶剂
extractor 脱水机;轧液机,轧车
extractor pad （圆筒形针织物的）轧液机
extract printing 拔染印花
extract style printing 天然染料型印花
extract ventilation unit 排气通风装置
extract wool 炭化再生毛
extra fine〈EF〉超精度,极细
extra fullness 特等丰满度
extra heavy size 特重浆,特厚浆
extra large〈XL〉特大号
extraneous matter 杂质
extra pale 特淡的,非常淡的
extra pure〈EP〉特纯的
extra pure reagent〈EP reagent〉一级试剂
extras ［设备］额外件
extra stretch 超弹力
extra wide cloth 超宽幅布
extreme dimension 极限尺寸
extreme value 极限值
extrication 放出,游离,摆脱
extrudability 挤出性;挤出能力
extrudate 挤出物

extrudation 挤出,压出
extrude 挤压;挤出,压出
extruded catalyst 挤压催化剂
extruded fabric 挤压织物（指塑料直接成布的非织造织物）
extruded latex 挤压胶乳
extruding machine 挤压机
extrusion 挤压
extrusion coater 挤压涂层机
extrusion coating 挤出涂覆,挤压涂层
extrusion coating process 挤压涂层法
extrusion lamination 薄膜挤压叠层法
extrusion press 挤压机
exudation 渗出［物］;流出［物］
ex-warehouse 仓库交货
eye bolt 环头螺栓（一端卷成环状,可作为吊环之用）
eye irritation 眼刺激性,眼的刺激
eyelet fabric 网眼布
eyepiece 目镜
eyepiece micrometer 目镜千分尺,目镜测微计
eye protection 眼罩
Eyring viscosity 艾林黏度

F

F(Fahrenheit) 华氏[温度]
FAA(Federal Aviation Administration) 联邦航空管理局(美国)
fabinet 天然纤维黏合人造革
fabric 织物,织品;布
3D fabric (three dimensional fabric) 三维织物
fabric absorption test 织物吸湿试验
fabric accumulator 储布器,容布箱,堆布装置
fabric air permeability test 织物透气性试验
fabric analysing glass 织物分析镜,织物密度镜
fabric analysis 织物分析
fabric appearance 织物外观
fabric appearance inspection 织物外观检验
fabric assurance by simple test (FAST) 织物风格简便测试
fabricated structure 装配式结构
fabrication 生产,加工,制作;装配;构造
fabric board 卷布板
fabric breathability 织物透气性
fabric bursting strength test(＝fabric bursting test) 织物顶破强力试验
fabric center supporting device 织物中央支承装置(置于针织物拉幅机进布区)
fabric cling-testing 织物黏附性试验
fabric clippings 碎布屑
fabric comfort(＝fabric comfortability) 织物穿着舒适性
fabric compacting 织物压缩加工
fabric composition 织物的组成(指纤维成分)
fabric construction 织物结构(指经纬密度、纱线线密度、织物单位面积质量、织物组织等)
fabric container 容布箱,储布器,存布器
fabric content 织物的纤维组分
fabric conveyer system 织物传送系统
fabric crimp 织物绉纹
fabric cutting machine 裁布机
fabric defects 织物疵点
fabric delivery 出布
fabric density 织物密度
fabric design 织物设计
fabric dwelling tube 储布管(管式喷射染色机的主要部件)
fabric distortion 织物变形
fabric doubling machine 织物对折机
fabric drum 卷布筒
fabric drying machine 织物烘燥机
fabric durability test 织物耐用性试验
fabric elasticity test 织物弹性试验
fabric engineering 织物工程设计
fabric extension 织物伸长
fabric external appearance inspection 织物外观检验
fabric fall 织物悬垂性
fabric fault detector 织物探疵器
fabric feed device 进布装置

fabric finish 织物整理
fabric finishing agent 织物整理剂
fabric finishing machine 织物整理机
fabric flamability 织物可燃性
fabric flow diagram 织物流程图,工艺流程图
fabric flow test 织物动态测试(染色)
fabric foam laminates 织物泡沫层压制品
fabric folding machine 折布机
fabric growth 织物增长
fabric guard 护料装置,拦布装置
fabric guide lattice 导布帘子
fabric guide 导布装置,导布器,吸边器
fabric guide roll 导布辊
fabric handle 织物手感,织制风格
fabric handling system 织物加工设备
fabric heat conductivity 织物传热性能
fabric heat conductivity test 织物传热性试验
fabric holding capacity 容布量
fabric hole [织物上]破洞
fabric illuminator [圆型针织机]照布灯
fabric impact test 织物冲击强力试验
fabric in rope form 绳状[加工]织物
fabric inspecting table 验布台
fabric inspection 验布,织物检验
fabric inspection machine 织物检验机
fabric introduction 穿布
fabric lifting device 织物提升装置
fabric light 织物照明
fabric load 坯布投入量
fabric loading port 织物装填口(间歇式染色设备的进布口)
fabric loop dryer 长环烘燥机,悬挂式烘燥机
fabric moistening plant 织物给湿装置

fabric of medium length fibre 中长纤维织物
fabric opening machine 筒状织物剖幅机
fabric packaging machine 布匹打包机
fabric passage 织物运行路线
fabric porosity 织物多孔性;织物孔隙度;织物紧密度
fabric position detector system〈FPD〉织物位置探测系统
fabric presser 压布装置
fabric property 织物性质
fabric rating standards 织物分级实物标样
Fabric Research Laboratory〈FRL,F. R. L.〉织物研究试验所(美国)
fabric resilience 织物回弹性
fabric roll 布卷
fabric rolling-up device 卷布装置
fabric rubbing test 织物摩擦试验
fabric run scheme 织物运行图
fabric sanding 织物[仿麂皮]磨毛起绒工艺
fabric serviceability 织物服用性能
fabric sewability tester 织物缝纫性能试验仪
fabric shaker tester 织物振动疲劳试验仪
fabric sheen 织物光泽
fabric shift tester 织物经纬纱线滑移性试验仪
fabric shock test 织物冲击强力试验
fabric shrinkage 织物缩水率,织物收缩率
fabric shrinkage test 织物收缩率试验,织物缩水试验
fabric shrinking machine 织物预缩机
fabric silket machine 织物丝光整理机
fabric simulation 织物仿真
fabric simulation CAD technology 织物仿

真计算机辅助设计技术
fabric slitting device 织物剖幅装置(筒状针织物的剖幅设备)
fabric softeners 织物柔软剂
fabric specification 织物规格
fabric speed indicator 布速指示器
fabric spreading machine 织物铺展机
fabric stabilization 织物幅宽控制
fabric stiffness test 织物硬挺度试验
fabric strength and elongation tester 织物强力与伸长试验仪
fabric strength test 织物强力试验
fabric stretch 织物弹性,织物伸长回弹性;织物伸长
fabric strighteners 织物整纬器
fabric structure 织物结构
fabric style 织物品种;织物风格
fabric style tester 织物风格测试仪
fabric substrate 被作用织物;被染整织物
fabric surface 织物表面
fabric take off(＝fabric take-up) 织物卷取
fabric take-up and batching device 织物卷取和打卷装置
fabric tear preventing 织物防裂
fabric temperature measuring and control device 织物温度监控装置
fabric temperature sensor 织物温度传感器,织物测温计
fabric tension control 织物张力控制
fabric tension measurement 织物张力测量
fabric transport 织物传送
fabric tube 圆筒形针织物;管状织物
fabric turning device 翻布机
fabric tyre 织物衬里轮胎
fabric visualization 织物形象化

fabric visualized simulation technology 织物形象化仿真技术
fabric washing machine 织物洗涤机
fabric weight 织物重量
fabric width 织物幅宽,织物门幅
fabric width measuring device 织物门幅测定装置
fabrography 网印技术
face [织物]正面,布面;表面,面;工作面
face cloth [揩]面[毛]巾;方巾;经面织物;光面女式呢(驼马绒毛制)
face coating 表面涂层
face contact 表面接触
face finish 绒面整理,呢面整理(包括缩绒、洗呢及起绒等工序)
face finished fabrics 正面整理织物
face finishing machinery 织物表面整理机械(包括起毛机、剪毛机、蒸呢机等)
face gear 平面齿轮
face goods 光洁毛织物
face-hardening 表面淬火
face plate 面板;(缝纫机的)台板;荧光屏
face seam 毯面拼接
face shield 防护面罩
face side (织物的)正面,布面
face side singeing 正面烧毛
face stamping 表面打印
face texture 布面结构
face to back shade deviation 染色正反面色差
face-to-back variation 正反面色差(染疵)
face to face bonded-pile carpet 双层割绒地毯
face-to-face carpet 双层地毯(可割成两个割绒地毯)
face towel 毛巾,面巾

face width 机宽
face yarn 面纱,正面纱线
facilities 设备,装置,工具
facing 罩色,留白沾色(滚筒印疵,俗称刮不清)
facing tool 车平面刀
facsimile engraving 传真雕刻(电子雕刻)
facsimile fabric 仿真织物
facsimile fibre 仿真纤维
facsimile printing 摹真印花
FACT(factor analysis chart technique) 作业因素分析图示法
factice 油膏(合成柔软剂,应用于橡胶和塑料)
factor 系数
factor analysis chart technique〈FACT〉 作业因素分析图示法
factor of rigidity 刚性系数
factor of safety〈f/s〉 安全系数
factory 工厂,工场
factory overhead 制造费用
factory transport 厂内运输
faddish 喜欢赶时髦的;一时流行的
fade 褪色
faded denim 褪色型劳动布,返旧牛仔布
faded look type 褪色外观型,陈旧外观型
fade-in 渐显(图像的逐渐显出,照相用语)
fadeless 不褪色的
fade-ometer 耐晒色牢度试验仪,褪色试验器
fade-out 渐隐(图像的逐渐消失,照相用语)
fade-out finishing 褪色处理,仿旧处理
fade-out jeans 褪色型牛仔裤
fade proof 防褪色

Fadex F 辉涤斯 F(紫外线吸收剂,苯系衍生物,能改善分散染料染涤纶的日晒牢度,商名,瑞士科莱恩)
fading 褪色,变色
fading by resin finish 树脂整理变色
fading lamp 耐晒牢度试验用灯
fading test 褪色性试验
fading unit 褪色色差单位
fad style 一时风行款式
Fahrenheit〈F〉 华氏[温度]
Fahrenheit's temperature scale 华氏温度刻度
fail-safe-control 防障控制,保安控制(具有自动防止故障特性的控制)
failure 故障,事故;失败,失效;断裂
failure criterion 破裂标准;故障判据
failure-free operation 正常运行
faint 暗淡的(指色)
fair 展览会;交易会;中等品
fair drying percentage 适干率
fake fur 人造毛皮
faller roller 松紧辊,张力调节辊,升降[导]辊
falling ball viscometer(=falling ball viscosimeter) 落球式黏度计
falling-drop method 滴下法(液体密度测定法之一)
falling-film evapourator 降膜式蒸发器
falling head test 变水头[渗透]试验
falling-pendulum apparatus 落锤式织物撕破强力测试仪
falling rate of drying 干燥速度渐减[阶段]
falling rate period of drying 降速烘燥阶段
falling-sphere damage test 落球[法]损伤测定(用落球法测定纤维素纤维的损

伤度)
falling sphere viscometer 落球式黏度计
fall-on printing 叠印
fall-on prints 叠印印花布
fall-on style 叠印法
fallow 淡棕色
false bottom 假底,花铁板
false colour 不坚牢色
false set 暂时定形
false twist-textured yarn 假捻变形纱
fan 风扇,风箱,通风机;叶片,翼
fancies 花式织物,花色货品
fancy 花式;时新织物
fancy blanket 花式毯,提花毯
fancy cutting 花纹剪绒
fancy cutting machine 花式剪毛机;花纹剪呢机
fancy dimity 变化麻纱
fancy-dyed fabric 花式染色织物
fancy fabric 变化组织[针]织物,花式[针]织物
fancy knitting yarn 花色绒线
fancy lace 花式花边织物
fancy overcoating woolen 花式大衣呢
fancy print 花式印花
fancy raising 花式起毛,局部起毛
fancy-raising flannelette 花色起毛绒
fancy shade 杂色
fancy shearing 花纹剪绒
fancy suiting 花呢;花式套头衣料
fancy woollens 粗纺花呢
fancy yarn 花式纱线
fan grille 风扇格栅
FAO(Food and Agricultural Organization) [联合国]粮农组织
F. A. R. (Federal Aviation Regulations) 联邦航空条例(美国)
farina 马铃薯淀粉
far infrared absorption spectrum 远红外吸收光谱
far infrared detector 远红外探测器
far infrared fibre and textile 远红外纤维及其纺织品
far infrared〈FIR〉 远红外,远红外线
far infrared heating element 远红外加热元件
far infrared molecular laser 远红外分子激光器
far infrared ray 远红外线
far infrared region 远红外区
fasciated yarn 包缠纱
fashion 时新式样;流行,时兴;样子;形式
fashionable 时髦的,非常讲究的
fashionable outwear 时尚外衣
fashion colour 流行色
fashion currency 时装流行
fashion cycle 流行周期
fashion design 时装设计
fashion designer 时装设计者
fashion dress 时装
fashion dynamics 时装动态
fashioned hosiery 成形针织品,成形袜
fashion life 时装流行寿命
fashion life cycle 时装流行周期
fashion message 时装信息
fashion mode 时新式样,流行
fashion model 时装模特
fashion performance 时装表演
fashion shade 流行色
fashion theme 流行倾向,款式动向
fast 坚牢的;快的
FAST(fabric assurance by simple test) 织

物风格简便测试

fast acting catalysts 快速催化剂

fast base(=fast colour base) [不溶性偶氮染料的]色基,显色基

fast colour dyeing 坚牢染色

fast colours 坚牢染料;坚牢色泽

fast colour salt [不溶性偶氮染料的]色盐,显色盐

fast coupling 硬性联轴节

fast degumming agent 快速精练剂;快速脱胶剂

fast dyed yarn 不褪色纱线

fast dyeing 坚牢染色

fast dyes 不褪色染料,坚牢染料

fast elastic deformation 急弹性变形

fast elasticity 急弹性

fastener 紧固零件;纽扣;揿钮

fastening nut 紧固螺母

fastening pin 安全销,保险销

fastening screw 紧固螺钉

fastness 牢度,坚牢度(常指染料、涂料、树脂、浆料、橡胶等附着物对纺织品的牢度)

fastness for dyeing 染色坚牢度

fastness grading 染色牢度评级

fastness improver 牢度增进剂

fastness of dye 染料的牢度

fastness promoter [染色]牢度增进剂

fastness rating 牢度等级

fastness standards 色牢度标准

fastness test [染色]牢度试验

fastness test to acid felting 耐酸缩绒坚牢度试验

fastness test to bleaching with hypochlorite 耐次氯酸钠漂白坚牢度试验

fastness test to burst gas fume 耐烟熏坚牢度试验

fastness test to hand washing 耐手洗坚牢度试验

fastness test to metals in the dye bath 耐染浴中金属坚牢度试验

fastness test to peroxide bleaching 耐双氧水漂白坚牢度试验

fastness to abrasion 耐磨损牢度

fastness to acids 耐酸牢度

fastness to acid spotting 耐淡酸渍牢度

fastness to air pollution 耐空气污染牢度

fastness to alkaline storage 带碱储存牢度

fastness to alkalis 耐碱牢度

fastness to alkali spotting 耐淡碱渍牢度

fastness to anti-crease processing 耐防皱加工牢度

fastness to artificial light 耐人造光牢度

fastness to atmospheric gases 耐大气牢度

fastness to bleaching 耐漂白牢度

fastness to boiling 耐沸煮牢度

fastness to brushing 耐刷洗牢度

fastness to carbonizing 耐炭化牢度

fastness to chemical washing 耐化学洗涤牢度(专指毛毯)

fastness to chlorinated water 耐氯水牢度

fastness to chlorination 耐氯化牢度

fastness to chlorine-bleaching 耐氯漂牢度

fastness to crocking 耐摩擦脱色牢度

fastness to crossdyeing 耐交染牢度

fastness to daylight 耐天然光牢度

fastness to decatizing 耐蒸煮牢度

fastness to degumming 耐脱胶牢度《丝》

fastness to dry chemicking 耐干漂牢度

fastness to dry-cleaning 耐干洗牢度

fastness to dryheat 耐干热牢度

fastness to dust 耐尘污牢度

fastness to gas fumes 耐烟气牢度，耐烟熏牢度

fastness to home washing 耐[家常]洗涤牢度

fastness to hot pressing 耐热压烫牢度

fastness to hot water 耐热水牢度

fastness to ironing 耐熨烫牢度

fastness to laundering 耐机洗牢度

fastness to mercerizing 耐丝光牢度

fastness to milling 耐缩绒牢度，耐毡合牢度

fastness to nitrogen oxides 耐氧化氮牢度

fastness to perspiration 耐汗渍牢度

fastness to planking 耐酸缩绒牢度

fastness to pleating 耐褶裥牢度

fastness to potting 耐沸水牢度

fastness to rain 耐雨淋牢度

fastness to rubbing 耐摩擦牢度

fastness to scrooping 耐丝鸣[处理]牢度

fastness to sea water 耐海水牢度

fastness to soaping 耐皂洗牢度

fastness to soda boiling 耐碱煮牢度

fastness to soil burial 耐土埋牢度

fastness to [organic] solvents 耐[有机]溶剂牢度

fastness to steaming 耐汽蒸牢度

fastness to stoving 耐二氧化硫气体牢度，耐硫熏牢度

fastness to sublimation 耐升华牢度

fastness to trubenising 耐胶合牢度

fastness to vulcanization 耐硫化牢度

fastness to washing 耐洗牢度

fastness to water 耐水浸牢度

fastness to water spotting 耐水渍牢度

fastness to wear 耐穿着牢度

fastness to weathering 耐气候牢度

fastness wash [提高]坚牢度的洗涤

fast operate relay 快动作继电器

fast print dyestuff 坚牢印花染料

Fastran transfer process 法斯曲兰转移印花法（酸性染料羊毛半湿转移印花法）

fast repeat 快速对花

fast salt [不溶性偶氮染料的]色盐，显色盐

fast setting 快速固色；快速定形

fast start 快速启动

FAST system 织物风格评价系统（澳大利亚）

fast wet 快湿

fat 脂肪，油脂

fat colours 油溶性染料

fat-free aromatic products 脱脂芳香族化合物

fatigue 疲劳

fatigue deterioration 疲劳降解，疲劳劣化

fatigue endurance limit 耐疲劳极限

fatigue failure 疲劳破坏，疲劳致损

fatigue lifetime 疲劳寿命

fatigue limit 疲劳极限

fatigue machine 疲劳度试验机

fatigue phenomenon 疲劳现象

fatigue proof 耐疲劳的

fatigue property 耐疲劳性

fatigue resistance 抗疲劳性

fatigue rupture 疲劳断损，疲劳破坏

fatigue strength 疲劳强度

fatigue testing 疲劳度试验

fatigue under flexing 挠曲疲劳

fatigue under scrubbing 摩擦疲劳

fatiguing load 疲劳负荷

fat-modified aminoplasts 脂肪改性氨基树脂

fatty acid 脂肪酸
fatty acid alkylolamine condensate 脂肪酸烷醇胺缩合物(非离子型表面活性剂)
fatty acid sulfate 脂肪酸硫酸酯(阴离子型表面活性剂,如红油或酯化油)
fatty alcohol 脂肪族醇
fatty alcohol sulfate 脂肪醇硫酸酯(阴离子型表面活性剂,用作羊毛及其他纤维的精练剂、染色助剂)
fatty alcohol sulphonates 脂肪醇磺酸盐
fatty amide derivative 脂肪酰胺衍生物
fatty anhydrides 脂肪酸酐
fatty glyceride 脂肪酸甘油酯
fatty ketone 脂肪酮
fatty matter content 油脂含量
fault 疵点;故障;失效;错误
fault analysis 故障分析
fault detection 疵点检测
fault diagnosis 故障诊断
fault finder 故障检测器
fault-free cloth 正品布,无疵点布
fault-free fabric 无疵点织物
faultless 无疵点的
faultless cloth 正品布,无疵点布
fault marker 疵点标记器
fault marking and recoding device 织疵标记及记录装置
fault message 故障信息
fault rate 次品率
fault tree analysis〈FTA〉事故树型分析
faulty shade 错色
faulty work 疵品;废品;疵病、疵点
fayence prints 靛青印花棉布
FB(=f.b., freight bill) 装货清单
FBA(fluorescent brighting agent, fluorescent brightening agent) 荧光增白剂

FBI(Federation of British Industry) 英国工业联合会
FC(franchise chain) 合同连锁店
FC(free control) 任意调节,自由调节
FC finish (fluorochemical finish) 氟化合物整理(提高拒油性)
FCT(fog chamber technique) 雾室技术(用于甲醛捕集)
FDA(fibre diameter analyzer) 纤维直径分析仪
FDA(Food and Drug Administration) 食品和药物管理局(美国)
FDTY(fully drawn texturing yarn) 全拉伸变形丝
FDY(fully drawn yarn) 全拉伸丝
feasibility study 可行性调查
feasible 可实行的,现实的
feather proofness test 防羽绒刺出性试验
feather quilt 羽绒被
feather yarn 羽毛纱
feature 特征,特点;要点
fecal sewage 生活污水
Federal Aviation Administration〈FAA〉联邦航空管理局(美国)
Federal Aviation Regulations〈F. A. R.〉联邦航空条例(美国)
Federal Board of health in Germany 德国医疗卫生联邦委员会
Federal Flammability Standard 联邦燃烧性标准(美国)
Federal Insecticide, Fungicide, and Rodenticide Act〈FIFRA〉联邦杀虫剂、杀菌剂、灭鼠剂法案
Federal Specifications〈FS〉联邦规格(美国)
Federal Standard〈Fed Std〉联邦标准(美

国)

Federal Trade Commission〈FTC〉 联邦贸易委员会(美国)

Federation of British Industry〈FBI〉 英国工业联合会

Fed Std(Federal Standard) 联邦标准(美国)

feed 给料;运送

feed arm 喂料臂,送布臂

feedback 反馈,回授,回输

feedback control 反馈控制器

feedback fields 反馈区

feedback loop 反馈回路

feedback system 反馈系统

feed/delivery speed ratio 喂入/输出速度比

feed end 喂入端

feeder 给料机;导纱器;喂给帘[子];喂给装置;输送器

feeder cable 馈电线

feedforward 前馈

feedforward control 前馈控制《电》

feed funnel 进料漏斗

feeding box 供料桶,给液桶

feeding liquor 补充液

feeding pipe 给液管

feeding ratio 补液比(轧槽初始浓度与补充液浓度之比)

feed pump 喂料泵,喂入泵

feed regulator 喂料调节器

feed roller 喂料辊;送网辊;送布辊

feed water 补给水

feedwater connection 给水管接头

feedwater makeup 补给水

feel [织物]手感

feeler 灵敏元件;探头;触指;塞尺,厚薄规,测隙规;仿形器

feeler gauge 触指规;检测规

feeler mechanism 检测机构

feeling [织物]手感

feeling finishing 风格整理

feeling motion 探测装置

feeling spindle 探针

feet〈ft.〉 英尺

Fehling's solution 费林溶液《试》

Felisol [欧洲纺织品]高色牢度标签

Fellow of the Society of Dyers and Colourists〈F. S. D. C.〉 染色工作者学会会员(英国)

felt 毛毡,毡;毡缩,毡化,缩呢,缩绒;[起毛拉绒的]起毡,塞毛

feltability 毡合性,缩绒性

felt blanket drying unit 呢毯烘燥装置

felt calender 呢毯轧光机,呢毯压烫机

felt cleaner 毛布洗涤器,毛毯洗涤器

felt deadener 吸音毡

felt dryer(=felt drying machine) 毛毯烘燥机,呢毯烘燥机

felted carpet 毛毡地毯,地毡

felted fabric 缩绒织物

felted flannel 双面绒棉毯

felted mattress [棉]毡垫

felted texture 毡制品

felted yarn 缩绒羊毛纱

felt fabric 毡合织物,毛毡,毡呢

felt floorcovering 毛毡地毯,地毡

felt-free wool 防毡缩羊毛

felt goods 毡制品

felting 毡化,毡合;缩绒,缩呢;毡;[起毛拉绒的]起毡,塞毛

felting ability 成毡性能,缩绒性

felting behaviour 缩绒性状,毡缩性状

felting dimensional change 缩绒尺寸变化
felting finish 毡化整理,缩绒整理;拉绒整理(棉织物的拉绒,刷光及重压)
felting machine 缩绒机,缩呢机
felting needle 制毡针
felting phenomenon 毡合现象
felting power 缩绒性,缩呢性,毡合性
felting process 毡合工艺,毡缩工艺,缩绒工程
felting propensity(=felting property) 毡合性,缩绒性,缩呢性
felting quality 缩绒质量;毡合性,缩绒性,缩呢性
felting shrinkage 毡合收缩率;缩绒缩率
felt-like structure 类似毡材结构
felt machine 呢毯机
feltmaking 制毡
felt pad 毡垫
felt press machine 毛毡压烫机
felt proofing 防毡缩[整理]
felt reinforcing 毡合增强剂
felt resistance 抗毡缩性
felt roller 毡辊
felt tarpaulin 防雨毡
female fitting 内螺纹
female flange 凹法兰
female spanner 套筒扳手
female thread 阴螺纹
fence boom 栅帘;撇油器
fencing 零头布
fender 防护板,护木
fent [滚筒印花]打样;小块布
Fenton reagent 芬顿试剂(双氧水与二价铁离子组成的氧化体系)
fents 零头布,短码布;坏布片
FEP(formaldehyde evolution potential) 甲醛逸出潜力
ferment 酵素;发酵
fermentation 发酵[作用]
fermentation degumming 发酵练丝法
fermentation scouring of vegetable fibre 植物纤维发酵精练
fermentation vat 发酵缸;发酵还原染料
ferment barrel(=ferment pot) 发酵桶
fern green 蕨叶绿[色]
ferric alum 铁[铵]矾;铁明矾
ferric chloride 三氯化铁
ferric tartarate 酒石酸铁
ferricyanide discharge 氰铁酸盐拔染
ferritic steel 普通钢
ferroprussiate paper 蓝图纸,晒图纸
ferrous acetate 醋酸亚铁
ferrous chloride 氯化亚铁
ferrous metal 黑色金属
ferrous sulfate 硫酸亚铁
ferrule 箍;套圈,环圈
fertilizer 肥料
FES(flame emission spectroscopy) 火焰发射光谱[学]
festoon 长环悬挂式
festoon ager 长环蒸化机,悬挂式蒸化机
festoon drying machine 长环烘燥机,悬挂式烘燥机
festooning faults 成环疵点(布环成形不良)
festooning pole [布环]挂棒
festoon scouring machine 悬挂式精练机
festoon steamer 长环蒸化机,悬挂式蒸化机
f-f(free formaldehyde) 游离甲醛
FFA(Flammability Fabrics Act) 易燃性织物法(美国)

F3-fibre (multi coaramide) 芳纶Ⅲ,多元共聚芳纶(商名,中国制)
FFT(foam finishing technology) 泡沫整理技术
FFT application system 泡沫整理给液装置(美国生产的一种新型低给液设备,给液率可低至 10%)
FFT applicator 泡沫整理技术用施加器
FI(flammability index) 易燃性指数
FIA(flow injection analysis) 喷流分析《试》
fiber 【美】=fibre
fibering 纤维化
fiberize 使成纤维,使纤维化
fibration 纤维形成;纤维化
fibre 纤维;纤维制品;刚纸
fibre analysis 纤维分析
fibre and yarn lubricant 纤维和纱线润滑剂
fibre-based material 纤维基材料
fibre bonded carpet 黏合地毯
fibre bonded cloth 胶合纤维布,非织造织物
fibre bonding device 纤维黏合设备
fibre bowl 纤维[轧]辊(如羊毛纸辊、棉纤维辊、纸粕轧辊等)
fibre coating technique 纤维涂层技术
fibre conditioner 纤维调湿器
fibre cushion [地毯]纤维衬垫
fibre diameter analysis 纤维直径分析
fibre diameter analyzer〈FDA〉 纤维直径分析仪
fibre dichroism 纤维二色性
fibre diffraction pattern 纤维衍射图样
fibre dyed fabric 色织物
fibre dyed yarn 色纺纱

fibre fill 纤维填塞物
fibrefill 纤维填塞物
fibre filled roll 纤维充填辊
fibre fine structure 纤维微细结构
fibre finish 纤维处理剂(润滑,抗静电)
fibre flock 纤维绒,纤维屑
fibre fly 纤维绒毛,飞花,飞毛
fibre forming 纤维成型
fibre gear 纤维层压板齿轮
fibre glass 玻璃丝,玻璃纤维
fibreglass fabric 玻璃纤维织物
fibre identification 纤维[的]鉴别
fibre identification reagent 纤维鉴别用试剂
fibre identification stain 纤维鉴定着色剂
fibre impact tester 纤维冲击强力试验仪
fibre lubrication 纤维润滑[作用]
fibrelysin 溶纤维剂
fibre melting point 纤维熔点
fibre microanalyzer〈FMA〉 纤维显微[图像]分析仪
fibre migration 纤维移动
fibre modification 纤维改性[作用]
fibre modifier 纤维变性剂,纤维改性剂
fibre morphology 纤维形态学;纤维形态
fibre-optic cable 光纤电缆
fibre optic material 光纤材料
fibre optics 纤维光学
fibre optics sensor〈FOS〉 光导纤维传感器
fibre orientation 纤维取向[度],纤维定向[度]
fibre or yarn dyed fabric 色织布
fibre packing density 纤维填充密度,纤维密集度
fibre pores 纤维毛孔
fibre porosity 纤维的多孔性

fibre property 纤维性质
fibre protection agent (=fibre protective agent) 纤维保护剂
fibre reactant 纤维交链剂;纤维反应剂（只与纤维起共价交链作用而本身不聚合）
fibre reactant type 纤维反应型
fibre reactive dyes 纤维反应性染料
fibre reactive finishes 纤维反应性整理剂
fibre reactivity 纤维反应性
fibre regeneration 纤维化再生
fibre reinforced composite 纤维增强复合材料
fibre reinforced concrete 纤维增强混凝土
fibre reinforced metal composite 纤维增强金属复合材料
fibre-reinforced plastics〈FRP〉纤维增强塑料
fibre-related biotechnology 纤维相关生物技术
fibre saturation value 纤维饱和值
fibre set process 织物定形整理
fibre structure 纤维结构
fibre substantive leveling agents 纤维亲和性匀染剂
fibre tendering 纤维损伤
fibre to fibre friction 纤维间摩擦
fibre washer 纤维垫圈
fibre whiteness 纤维白度
fibril 原纤维
fibrillar distribution 原纤分布
fibrillated yarn 裂膜纱,原纤化纤维
fibrillation 原纤状结构;原纤维化[作用],原纤化
fibrillation control 原纤化的控制
fibrillation control technology 原纤化控制技术、控制工艺
fibrillation index 原纤化指数（用以表示莱赛尔纤维的原纤化程度）
fibrillation of printing paste 印浆裂膜现象
fibrillation prevention 原纤化的防止
fibrillation property 原纤化性能
fibrillation stabilization 原纤化稳定
fibril structure 原纤结构
fibrinogen 纤维蛋白原《生化》
fibroin 丝素;丝朊;丝心蛋白
fibroin agent 丝素整理剂
fibroin coating 丝心蛋白涂覆（棉布或麻布涂丝心蛋白的碱液,烘干后有丝的光泽）
fibroin finishing 丝素整理（丝的防皱）
fibroin protector 丝素保护剂
fibrous 含有纤维的,纤维组成的;纤维状的
fibrous architecture 纤维状构造
fibrous carbon 纤维状碳,碳纤维
fibrous finish 带绒毛整理
fibrous heat insulator 纤维状保温材料,纤维状绝热材料
fibrous high-temperature-resistant material 耐高温纤维材料
fibrous ion changer 纤维状离子交换体
fibrous material 纤维状物质
fibrous structure 纤维状结构
ficelle 灰褐色的
fickle coloured 闪光[效应]
Fick's equation 菲克方程（纤维单位面积和染料扩散数量的关系）
Fick's Law of diffusion 菲克扩散定律（染料扩散）
FID(flame ionization detector) 火焰离子

化检测仪
fidelity 保真度,重现精度(印花图像)
field 现场;视野
Fielden hygrometer 菲尔登湿度计
field excitation 励磁《电》
field-regulated synchronizing system 调磁同步系统《电》
field regulation 磁场调整
field report 现场报告
field resistance 磁场电阻
field rheostat 磁场变阻器
field trial(＝**field test**) 现场试验
field winding 磁场绕组
fiery red 火红[色]
FIFRA (Federal Insecticide, Fungicide, and Rodenticide Act) 联邦杀虫剂、杀菌剂、灭鼠剂法案(美国)
fig. (**figure**) 花纹;数字;形状;图,插图,附图
fig brown 无花果棕[色]
figural motive 图案花型
figurative prints 绘画格调印花布
figure〈**fig.**〉花纹;数字;形状;图,插图,附图
figure colour 闪光效应
figured carpet 花纹地毯
figured-effect 花纹效应
figured plush fabric [提花]长毛绒织物
figured towel 提花毛巾
figure of merit〈**FOM**〉优值;灵敏值;品质指数
figuring 花纹,图案
filament〈**fil.**〉长丝;单纤维
filament acetate yarn 醋酯纤维纱线
filamentary 长丝的,纤维的
filament denier 纤维线密度(以旦为单位),单丝线密度(以旦为单位)
filament rayon yarn 人造丝;人造丝纱线
filament winding 绕丝
file 文件,文件存储器;外存储器,外存件《计算机》
filemot 枯叶色,黄褐色
filled bowl 层压轧辊(如羊毛纸辊、棉纤维辊、纸粕轧辊等)
filled cloth 上浆布,增重布
filled finish 上浆整理
filled soap 填充皂
filler 填料;垫板;加重剂
fillet 钢丝针布
fillet grinding device 磨钢丝针布装置
fillet mounting machine (＝**fillet winding machine**) 钢丝针布包绕机
filling agent of cloth 布料用填充剂
filling and stiffening agent 填充硬挺剂
filling band in shade 纬色档(织疵或印染疵)
filling bands 纬档,纬向条痕(织疵)
filling bar 横档,纬向条痕(织疵)
filling cord pique 横向灯芯布,横向凸条布
filling doctor 铲印浆刮刀
filling ikat 纬纱扎染布
filling mass 填料
filling of contours 充实轮廓(电脑图案处理)
filling product 添加物,填充料
filling pulled-in 边拖纬(织疵)
filling run out 缺纬(织疵)
filling skewness 纬[纱歪]斜度
filling snarl 纬[纱结]圈
filling straightener 正纬装置;扩幅装置
filling streaks 横档,纬向条花

filling-stretch woven fabrics 纬向拉伸织物
filling-wise 纬向
fill up 覆盖;罩染,套染
film 膜,表面膜;薄膜;(涂层)薄膜;照相胶片
film adhesive 薄膜黏合剂
film and fabric laminating 薄膜和织物层压
film application 薄膜涂覆
film-based fibre 薄膜纤维
film builder 成膜剂
film coating 薄膜涂层
film copying machine 胶片拷贝机(照相雕刻的附属设备)
film cutting table 胶片切割台
film drying cabinet 胶片烘干箱
film elasticity [浆]膜弹性(印花糊料)
film evaluation 胶片评级(起球的评级方法)
film extruder 薄膜挤出机
film extrusion 薄膜挤出
film fibre forming 薄膜成纤
film fibrillation 薄膜原纤化
film forming 成膜
film forming agent 成膜剂
film forming properties 成膜性能
film forming speed 成膜速度
film laminating 薄膜层压
film plotter 胶片绘图机《雕刻》
film printing 薄膜印花
film pump 薄膜泵
film puncher 胶片打孔机
film release system 脱膜体系(转移印花或涂层)
film release system transfer printing 脱膜法转移印花
film release transfer printing 脱膜法转移印花
film scanner 胶片扫描器
film screen printing 筛网印花,绢网印花
film slitting 薄膜分割(成纤)
film splitting 裂膜,薄膜分裂
film step and repeat machine 连晒机,连拍机
film strength 浆膜强度
film tape 薄膜带
film yarn 薄膜纱
filter〈filt.,Flt〉过滤器;滤纸;滤波器;筛选程序《计算机》;滤光器;滤色片
filterability 过滤性;过滤率
filter aid 助滤剂
filter cake 滤饼
filter cartridge 过滤筒
filter cloth 滤布
filter equipment 过滤设备;滤尘设备
filter factor 滤色镜系数
filtering flask 过滤瓶《试》
filtering out of imagination 筛出规律;图像筛出(计算机分色术语)
filter layer 过滤层
filter loading 生物滤池负荷
filter mats 过滤垫
filter medium 过滤介质
filter pack 过滤组合件
filter paper 滤纸
filter photometer 滤色光度计
filter pooling 过滤污水池
filter press 压滤机,压滤器
filter residue scale 滤渣级别(用以评定分散性染料的分散度)
filter saturation 过滤饱和度
filter sieve 过滤筛,过滤网

filtrate 滤[出]液
filtration 过滤,滤清
filtration adjuvant 过滤助剂
filtration fabric [过]滤布
filtration material 过滤材料
filtration plants 过滤设备
filtration textiles 过滤用纺织品
fin 散热片,翅片(金属的凸出薄片,用以增加散热面积)
final bleaching 最终漂白,末道漂白
final coat 外观涂层整理;最后一道涂层
final drying 最终干燥
final finish 后整理
final heat setting 后热定形
final inspection 出厂检验,出门检验
financial management 财务管理
finder 探测器,瞄准装置
fine 细粉的(染料名称后的标记);细的;优良的
fine adjustment 精密调节,微调
fine carpet 精细地毯
fine chemicals 精细化学品
fine coating 精细涂层
fine corduroy 细条灯芯绒
fine cover pattern 地色印花花纹,满地印花花纹
fine denier fibre 细旦纤维
fine denier yarn 细旦丝
fine dispersion 均匀分散
fine-draw 修布,织补
fine emulsion 细[滴]乳状液
fine fabric 细薄织物,精细织物
fine filter 精滤器
Fine Gum 法因胶(合成印花糊料,纤维素纤维衍生物,适用于合成纤维、棉、黏纤织物的印花,商名,日本第一工业制药)
fine laundering 轻洗(指洗涤丝绸、毛织品)
fine line [印花]细线条
fine measuring instrument 精密测试仪器
fine mesh 细网眼
fine mesh rotary screen 细目圆网
fine needle corduroy 细条灯芯绒
fineness 细度;纯度;光洁度
fine plain 细平布
fine powder 细粉
fine structure 微细结构
fine texture 微细组织
fine timing 精密调速(微调)
fine turbidity meter 精密浊度计
finger 测厚规;指针;突指,触指;导纱器
finger guard 指形导布器
finger roller [三指剥边器]螺旋辊
finish 整理[法];整理剂;精修
finish applicator 给油盘;给油装置
finish chlorine fastness 耐氯漂整理色牢度
finish decatizing 后整理蒸呢
finish decatizing machine 蒸呢后整理机
finished appearance 成品外观
finished carpet 地毯成品
finished count 成品[布]密度
finished dimension 整理后尺寸
finished fabric 成品布
finished product 成品
finished silk fabric 熟绸,练熟绸匹
finished size 成品尺寸
finished washer 抛光垫圈
finished weight 成品重量,整理后重量
finished width [of cloth] 成品整理后宽度,成品幅宽
finished yarn 加工纱线

finished yield 整理后重量

finisher 整理厂;(服装)整烫工;后蒸发器

finish-free knitted good 免整理针织物(无需后整理的针织物)

finishing 织物整理;后处理

finishing agent 整理剂

finishing allowance 精加工留量

finishing auxiliaries 整理助剂

finishing bar 整理档(整理疵)

finishing cotton 棉布整理

finishing dirts 整理污迹

finishing jobber 代加工整理厂

finishing machine 整理机

finishing mangle 整理轧车

finishing mixture [整理用]浆料

finishing of degummed silk fabric 脱胶丝织物整理

finishing of knitting wool 绒线整理

finishing of plush 长毛绒整理

finishing of wool textiles 毛织物整理

finishing of woven fabrics 机织物整理

finishing oven 整理烘房,长筒袜定形烘房

finishing paste 整理浆

finishing plant 整理工厂,染整工厂

finishing process 整理工序;染整加工

finishing range 整理机组

finishing room 成品间;整理间

finishing routine 整理工艺程序

finishing scouring 复练,后精练

finishing scutching 整理开幅

finishing spot 污斑,渍斑;整理污迹

finishing starch 整理用淀粉

finishing streak 整理折痕,整理条痕

finishing tenter 整理拉幅机

finishing width 整理门幅

finish inspection 整理检验

finish migration 整理剂泳移

finish oil 整理油剂

finish presser 整理熨烫工;整理熨烫机

finish pressing 熨烫整理(服装)

finish roll 给油辊

finish stains 整理污迹

Finisol SE 菲尼索 SE(溶剂型有机硅羊毛防毡缩剂,商名,德国波美)

Finisol WA 菲尼索 WA(溶剂型有机硅羊毛防毡缩剂,商名,德国波美)

finite dyebath 有限染浴

finite swelling 有限溶胀

finned clip 翅片式布铗(布铗的垂直面上有若干加强筋,类似热交换器的翅片,布铗与布边的接触面采用镂空形式,因而布边的烘燥、定形效果较好)

finned tube exchanger 翅片管换热器

fin-tube coil heat exchanger 翅状盘管式热交换器

FIR(far infra red) 远红外,远红外线

fire alarms 火警警报器

fire and explosion hazards 着火与爆炸危险

fire behaviour 着火性能(阻燃试验)

fire door 防火门

fire effluents 火灾废气

fire extinguisher 灭火机,灭火器

fire fighter uniform 消防服

fire fighting procedure 消防措施

Firegard 法耶加德阻燃整理剂(商名)

fire hazard classification 火灾分类

fire hose 水龙带,消防带

fire hydrant 消防栓,灭火龙头

fire load 火势**

fire load density 火势强度(阻燃试验)
fireman uniform 消防员制服
fire plug 消火栓
fire preventing coating 防火涂层
fire proof 防火的
fire proof boom 防火栅
fire proof fibre 防火纤维,不燃纤维
fire proof finish 防火整理
fire proofing agent 防火剂
fire proofness test 防火试验
fire protection garment 消防服
fire red 火红[色](深红橘色)
fire resistance 耐火性
fire resistance tester 耐火性试验机
fire resistant 抗火的,防火的
fire resistant fabric 抗火织物
fire resistant finish 抗火整理
fire retardancy 阻燃性
fire retardancy treatment 阻燃处理
fire retardant 阻燃剂;阻燃布
fire retardant anti-melt viscose fibre 阻燃抗熔融黏胶纤维
fire retardant fibre 阻燃纤维
fire retarding agent 阻燃剂
fire, smoke and smolder retardants 防明燃、烟燃和余燃的阻燃剂
fire test 燃烧试验
fire, water, weather, and mildew resistance〈FWWMR〉防火、拒水、耐气候和防霉[整理法]
fir green 冷杉绿[色],松针绿[色](浅暗绿色)
firing order 点火顺序
firing rate 燃烧强度,燃烧率
firing temperature〈F.T.〉着火点,着火温度

firing time test 燃烧速度试验
firm handle 厚实手感(指有身骨)
firmness 厚实,[织物]有身骨;坚实性
first aid procedure 急救措施
first backing [地毯]第一层底布
first bath 头缸,初染浴《染》
first break (来样的)初度变色
first coat 预涂层,底涂层
first drop of liquid 初见液
first grade 头等,高级
first law of thermodynamics 热力学第一定律
first print 第一次印花(指先印的地色)
first printed resist 先印[后染]防染印花
first quality 一等品,优质
first stage polymerization 初级聚合,前段聚合
first strike 初期瞬染
first washing 第一次洗涤
Fischer metering pump 菲舍计量泵
fish brine smell 咸鱼味
fishery commodity 渔业用品
fish eyes 鱼眼(印疵,由于色浆中有未溶解的胶粒而造成的未印出的小点),色点;小洞
fish odour 鱼腥臭(氨基树脂整理后因储存过久而产生)
fish scale pattern 鱼鳞花纹
fishtail burner 扇形火焰燃烧器
fish toxicity 鱼的毒性《环保》
fishy handle 手感顺滑
fishy odour 鱼腥臭(氨基树脂整理后因储存过久而产生)
fissility 分裂性
fissure 裂伤(织疵);裂隙,裂缝
fit 配合;装配;调整;[印花]对花准确

fitness 适应性,适用性
fitter 裁剪试样工;装配工;机工;钳工
fitting 符合;装配;零件;试衣;(服装鞋袜的)尺寸,尺码(英国名称)
fitting cross 对花十字记号
fitting room 试衣室
fitting-up [结构]部件成型,[结构]部件安装
fit tolerance 配合公差
five-roller stretcher 五辊拉伸机
five-tier short loop dryer 五层短环烘燥机
fix 固色;固着,固定;装配;(安装后)调整,整修
Fixapret CPN 服丝平 CPN(树脂整理剂,主成分为二羟甲基二羟基乙烯脲,商名,德国巴斯夫)
Fixapret FR-ECO 服丝平 FR-ECO(低甲醛树脂整理剂,商名,德国巴斯夫)
Fixapret PH 服丝平 PH(树脂整理剂,二羟甲基丙烯脲,商名,德国巴斯夫)
fixation 固色;固着,固定;定影;装配
fixation accelerant 固着加速剂
fixation agent 固色剂
fixation bath 固着浴;固化浴
fixation dynamics 固色动力学
fixation value 固着值
fixation yield 固色率,固色量,固着率
fixative 固色剂;固着剂;定影剂
fixative for direct colours 直接染料固色剂
fixed bearing 固定支承
fixed displacement motor 定量电动机(指排量不可调节的液压电动机)
fixed displacement pump 定排量泵(工作时每个循环的液体排量不变)
fixed film biological reactor 固定膜生物反应器

fixed frame tentering machine 定幅拉幅机
fixed phase 固定相
fixed plate singeing machine 固定式热板烧毛机
fixed resin content 树脂固着量
fixed return flow plate 固定回流板
fixed support 固定支座
fixed temperature and humidity 恒温恒湿
fixed tentering machine 定幅拉幅机
fixer 保全工,修车工;固着剂,固色剂;定影剂
fixing 固着
fixing agent 固着剂;固色剂;定影剂
fixing bath 固着浴;固色浴
fixing machine 固着机;固色机
fixings 附件;设备,装备
fixing salt 定影剂
fixing screw 定位螺钉,固定螺钉
fixing solution 固着液
fixing temperature 固着温度
fixing time 凝结时间,硬化时间,固化时间;定形时间
fixing type trough 固定式轧槽
Fixogene CD 菲克索琴 CD(直接染料湿牢度改善剂,合成阳离子性树脂,商名,英国卜内门)
fixture 夹具;固定量;附件,装置
fizz 漏气
flabby handle 松弛手感
flake fabric 植绒织物
flaking off 剥落(印花浆烘后被剥落)
flame 有焰燃烧(阻燃试验)
flame [red] 火红[色](深红橘色)
flame accelerator 促燃剂
flame back coating 火焰背面上胶
flame backing 火焰背面上胶,焰熔背面

上胶
flame-blowing (玻璃纤维的)火焰吹制
flame bonding 火焰黏合,焰熔黏合
flame-bonding 火焰胶合
flame breakthrough 烧穿
flame checking 阻燃整理(防止火焰蔓延的处理)
flame control 火焰调节,火焰控制
flame detector 火焰检测器
flame drying 火焰烘燥
flame electrode 火焰电极
flame emission spectroscopy〈FES〉 火焰发射光谱[学]
flame extinction 熄火,火焰熄灭
flame eye 火焰监视器
flame failure 熄火,灭火
flame-free finish 无焰整理
flame ignition 点火
flame ignitor 点火装置
flame intensity 火焰强度
flame ionization detector〈FID〉 火焰离子化检测仪
flame ion mass spectrometry 火焰离子质谱分析
flame laminating 火焰胶合法,火焰层压法
flame laminating machine 火焰层压机
flame lamination 火焰胶合,焰熔层压
flameless combustion 无焰燃烧
flame monitor 火焰监察器
flame photometric detector〈FPD〉 火焰光度检测器
flame photometry〈FP〉 火焰光度[测定]法
flame proof 防燃的
flameproof fabric 防燃织物,防火织物

flameproof fibre 防燃纤维
flame proofing agent 防燃剂,阻燃剂
flame proofing finish 防燃整理
flame proofing mechanism 阻燃机理
flame propagation rate 火焰蔓延速度
flame propagation test 火焰蔓延性试验
flame reaction 燃烧反应
flame repellent 拒燃的;拒燃剂
flame resistance〈FR〉 抗燃
flame resistant 抗燃的
flame resistant fibre 抗燃纤维
flame resistant resin 抗燃树脂,耐火树脂
flame retardant〈FR〉 阻燃的;阻燃剂
flame retardant additive 阻燃添加剂
flame retardant fabrics〈F. R. fabrics〉 阻燃织物
flame retardant finish 阻燃整理
flame retardant property 阻燃性
flame retardants〈FRs〉 阻燃剂
flame retardant variants 阻燃变性[纤维]品种
flame retard fibre 阻燃纤维
flame scarlet 大红[色],猩红[色],焰红[色]
flame sensor 火焰传感器
flame singeing 火焰烧毛
flame singeing machine 气体烧毛机
flame speed rate 火焰蔓延速度
flame spread 展焰性
flame test 焰色试验
flame thermocouple detector 火焰热电偶检测器
flame tip 焰舌
flame velocity 火焰速度
flame yarn 烧毛纱;双色纱;杂色纱
flaming mode 燃烧方式

flamingo 橘红色
flammability 可燃性,易燃性
Flammability Fabrics Act〈FFA〉 易燃性织物法(美国)
flammability hazard 可燃性危险;可燃性事故
flammability index〈FI〉 易燃性指数
flammability standard 可燃性标准
flammability temperature 着火温度
flammability test 可燃性试验,着火试验
flammability tester 可燃性测试仪
flammable 易燃的
flammable fabric 易燃性织物
flammable limits in air 空气中燃烧极限
flammable liquids 可燃性液体
flammable material 易燃材料
flange 凸缘,法兰,边盘
flange bearing 接盘式轴承,法兰轴承
flange coupling 凸缘联轴器
flanged edge 法兰边
flanged fitting 法兰连接件
flange pipe 法兰管,凸缘管
flange support 凸缘支架
flanging 折边,翻边,卷边(金属板材的一种加工方法)
flanging machine 折边机,外缘翻边机
flannel 法兰绒
flannelette 绒布
flannel finish [棉布]仿法兰绒整理;拉绒整理
flap 挡板,风门;片状物
flap valve 翻板式阀门,瓣阀
flaring exit 圆锥形出口,喇叭形出口
flash 闪光,闪烁;光泽,亮度;快速的,瞬时的
flash ageing 快速蒸化(用于两相法印花

flash ager 快速蒸化机(用于两相法印花工艺)
flash boiler 快速蒸发锅炉
flash cure 快速焙烘(成衣印花用)
flash deaerator 瞬间脱泡器,快速脱泡器
flash dryer 闪蒸烘燥器,快速干燥器
flash drying 急骤干燥,快速干燥
flash dyeing 快速染色
flasher 闪蒸器;闪光器
flash evapourating apparatus 扩容蒸浓设备(烧碱回收)
flash evapouration 扩容蒸发
flash fire 急骤燃烧,暴燃
flash fire propensity 闪火倾向(阻燃试验),火花倾向
flashing 闪光,闪光光泽
flashing off 急骤馏出,闪蒸
flashing reactor 闪蒸反应器;快速反应器
flash-off 速蒸,闪蒸
flash-off steam 闪蒸蒸汽(压力突降时由水形成的蒸汽)
flash point 闪点
flash ranging〈FR〉 光测,光测距离
flash steaming 快速蒸化(用于两相法印花)
flash temperature 闪燃温度,闪点
flask 烧瓶
flat abrasion 平面磨蚀
flat abrasion test 平磨试验
flat bed heat transfer printing press 平板转移印花压烫机
flat bed laminating machine 平板层压机
flat bed press 平板压烫机;平板热压机(间歇式转移印花用)
flat bed printing machine with rotary unit 带圆网装置的平网印花机(平圆网印

花机）

flat bed screen printing 平板[筛]网印[花]

flat bed screen printing machine 台板平网印花机

flat beetling 平幅捶布[整理]

flat belt 扁平传动带,扁平运输带

flat cable 扁形电缆

flat curing 平幅焙烘

flat die 平版钢芯,[刻纹]平钢模

flat dryer 卧式烘干机

flat embossing 拷花整理

flat embossing machine 拷花机,拷花整理机

flat fabric 平幅针织物

flat foam 泡沫塑料；背衬

flat head bolt 平头螺栓,平帽螺栓

flat iron 熨斗,烙铁；扁钢

flat key 平伏键

flat layer drier 平式铺层烘干机

flat-on view 俯视图

flat plate press 平板压烫机

flat plate pressing machine 平板压烫机

flat plate printing 平板印花

flat [screen] printing 平网印花

flat quartz plate heater 平板式石英加热器

flat screen 平板筛网,平网

flat screen printing machine 平网印花机

flat set 平整处理,平整定形

flat sheet transfer printing machine 平板衣片转移印花机

flat sheet transfer unit 平板转移印花机

flat spring 板簧,片簧

flat surface 布面平整；呢面平整

flat table pantograph engraving machine 平台缩小雕刻机

flattening [染色]平淡,浅薄

flattening & glazing calender 轧光机,摩擦轧光机

flatwork 家用织物（如手帕、餐巾、台布等）

flavanthrene(＝flavanthrone) 黄烷士林,黄蒽酮（阴丹士林黄 G 的化学名称）

flavanthrone paper 还原黄试纸,黄蒽酮试纸（用于试验还原浴中保险粉是否足够）

flavone 黄酮

Flavourzyme 酸性蛋白酶（商名,丹麦）

flaw 疵点；缺点；裂缝

flawless 无缺点的,无疵点的；无裂隙的

flax 亚麻；亚麻黄[色]（暗淡黄棕色）

flax-like fabric 仿麻织物

flax-like fibre 仿麻型纤维

flax linen 亚麻布

flax paper bowl 亚麻纸[轧]辊

fleck 斑点；染斑（染疵）

flecked yarn 霜花纱

flecky 斑点,染斑

fleece finish （毛毯的）起绒整理

fleece wash 洗套毛

fleecy fabric 起绒织物

fleecy printing 起绒印花

fleecy sweat pants 卫生裤

fleecy sweat shirts(＝fleecy sweat T-shirt) 卫生衫,棉绒衫

flesh blond 肉棕色

flex 弯曲,挠曲；花线,皮线

flex abrasion [屈]曲磨[损]；挠曲磨损性

flex abrasion tester 曲磨测试仪

flex cracking test （涂层的）耐屈挠龟裂强度试验,挠曲疲劳试验

flexibility 挠性,柔软性;适应性,机动性
flexibility resistance 耐折性,柔韧性
flexiblade coater 弹性刮刀式涂层机
flexible bearing 挠性轴承
flexible chain 挠性链
flexible dye tubes 弹性染色筒子纱管
flexible film laminating 薄膜层压
flexible manufacturing system (FMS) 柔性制造系统
flexible pipe 软管
flexible roller 挠性辊
flexible schedule 弹性工作时间表
flexible side group 柔性侧基
flexible time 弹性时间,灵活时间
flexing abrasion resistance 耐曲磨
flexing abrasion tester 耐曲磨试验仪
flexing elasticity 挠曲弹性
flexing fatigue 弯曲疲劳
flexing resistance 抗挠曲性
flexing stress 屈曲应力,挠曲应力
flexion 挠曲,弯曲
Flexitronic system 灵敏的织物张力调节系统(德国埃克斯梯玛)
flex life 挠曲寿命
flexnip system 高给液量浸轧系统
flexographic ink 橡胶版轮转印刷印墨
flexographic printing 橡胶版轮转印刷;橡胶版轮转印花;凸版印刷
flexographic printing ink 橡胶版轮转印刷印墨
flexometer 挠度计;曲率计;屈曲磨损试验器
flex resistance 抗弯性,抗挠性
flex rigidity 弯曲刚度,抗弯刚度
flexroll-sleeve 弹性辊套(尼龙套筒套在轧光辊外,可提高轧辊的回弹性和耐热性)

flex stiffness 弯曲刚度,抗弯刚度
flexural fatigue 弯曲疲劳,挠曲疲劳
flexural modulus 弯曲模量
flexural resilience 弯曲回弹性,挠曲回弹性;弯曲回弹能,挠曲回弹能
flexural resistance 抗弯性,抗挠性
flexural strain 挠曲应变,弯曲应变
flexural strength 抗弯强力,挠曲强度
flick 绒毛(指起绒织物的绒毛)
flicker 闪烁,闪光
flicker bar 轻打棒(用于刷毛机)
flinger 甩水圈,甩油圈
flint dry 高温烘燥
flip flop 触发器,触发电路
flip flop switch 快速开关,触发式开关
flip pallet 转动式印花台板(针织衣片印花)
float 浮子,浮筒
floatation 浮上分离,浮选
floatation method 浮上分离法《环保》
float compensation device 喂入补偿器
float compensator 浮动式补偿器(调节织物张力)
float coupling 浮动式联轴节,自由回转接头
float drier 无接触热风烘燥机
float dryer 气垫式烘燥机
float gauge 浮子式液面计
floating ball steam trap 浮球式疏水器
floating cloth 贴布不良,贴布浮起(台板印疵)
floating coater 浮刀涂层机
floating drum thermal transfer machine 浮筒加热式转移印花机
floating film dryer 无接触热风烘燥机,悬

空带式烘燥机
floating jet stenter 气垫式喷嘴热风拉幅机
floating knife coater 浮动刮刀涂布器
floating knife method 浮动刮刀加工法
floating [out] phenomenon 浮色现象(涂料印花)
floating roller 浮动辊
floating roll guide 悬浮导辊,张力调节辊
floating separation 浮升分离法
floating suspension nozzle method 喷嘴式热风悬浮烘燥法
floating-suspension stenter 气垫式拉幅机
floating system 浮上系统《环保》
floating type stretcher 无张力拉幅装置
floating web dryer 无接触式热风烘燥机,悬空带式烘燥机
float level indicator 浮标液位指示器
float-on-air dryer 无接触热风烘燥机,气垫式烘燥机
floats 悬浮染色织物
float-shearing 浮动式剪毛(剪毛程度较轻)
float switch 悬浮式开关,浮动开关
float type flowmeter 浮子式流量计
floc 絮凝物
flocculant 絮凝剂
flocculation 絮凝作用
flocculation value 絮凝值,凝固值
flocculator 絮凝器
flocculent 絮凝的,絮结的
flock (供植绒用的)短绒;絮状沉淀
flock binder 植绒黏合剂
flock bond strength 植绒黏合强力
flock carpet 植绒地毯
flock coating 植绒涂层

flock dot(=flock-dotting) 植绒点子,植绒花纹
flocked carpet 植绒地毯
flocked fabric(=flocked goods) 植绒织物
flocked pile 植绒
flocked sheeting 平绒
flocked twill 斜纹绒
flocked yarn 植绒纱
flocker 植绒机
flock finishing 植绒整理
flocking 植绒[工艺];静电植绒;密绒,短绒
flocking adhesive 植绒黏着剂
flocking agent 植绒剂
flocking machine 植绒机
flocking process 植绒法(非织造织物工艺)
flock powder 绒屑,短绒(植绒用)
flock printed handkerchief 植绒印花手帕
flock printed sheer 植绒印花薄绸
flock printing 植绒印花
flock radiator 静电植绒喷纤器,喷绒器
flock testing 植绒织物测试
flock transfer paper 转移植绒纸
flock transfer printing 植绒转移印花
flock working 植绒加工
flocky 毛绒绒的
flood bar 浆棒(印花调浆用)
flood ring 液流环(装在某些喷射染色机的喷嘴处,与节流阀相连并可调节液流循环时用或不用喷嘴)
flood stroke 溢浆初刮(第一次刮浆,使色浆均匀铺开,但未透过筛网)
flood stroke doctor 溢浆刮刀(将色浆从平网一端刮至另一端)
floorcovering 地毯

floor mounting hood 落地烟罩

floor performance property 地板性能,地面性能

floor plan 平面图

floor space 占地面积;车间面积,厂地面积

Flop effect 随角异色效应(从正视角到较大斜角观察涂层产生的颜色和亮度变化的效应)

floppy 下垂的,松弛的

floppy disc 软磁盘

floppy disc drive 软磁盘机

floral [design] 花卉图案,彩花式图案

floral motif 花卉花纹

floral pattern 花卉图案

floral print 花卉图案印花

flotation process 浮选法,气浮法

flour 面粉;粉状物质

flow 水流;气流;流量;流动

flowability 流动性

flow back 逆流,倒流

flow birefringence method 流动双折射法

flow chart 流程图

flow control additives 控制流动的[涂层]添加剂

flow control valve 流量控制阀

flow curves 流动曲线

flow diagram 流程图

flow dividing valve 分流阀(单一液流按一定比例自动分成两个支流的流量控制阀)

flow dyeing 渗透染色

flow-form polychromatic dyeing machine 液流型多色染色机(染液流在轧车的上滚筒上,使织物上染)

flow gauge 流量计

flow indicator 流量指示计

flow injection analysis〈FIA〉 喷流分析《试》

flow interceptor 截流器

flow measurement 流量测定

flow mechanism in dyeing 染液的流动机理

flow meter 流速计,流量计

flow moderators 流动缓和剂,流变性改进剂(糊料)

flow monitoring device 流量监控装置

flow of textile goods 纺织品的流动(包括运送、储存、定货等)

flow print(＝flow printing) 渗化印花(织物在潮湿状态下印花,花纹有模糊感)

flow property 流动性

flow rate [泵]流量

flow reactor 连续反应器

flow regulator 流量调节器

flow resistance 流阻

flow sheet [工艺]流程图,布置图

flow straightener 导流板,流体导向叶片

flow tank 沉淀池

flowthrough dryer 径流烘燥机

flow through quantity 流[过]量

flow washing process 流动水洗工艺

Flt(＝filt.,filter) 过滤器;滤波器;滤纸;筛选程序《计算机》;滤光片;滤色片

fluctuating flow 波动流动

fluctuation 升降,波动;变动;起伏

flue 烟道,尘道

flue dust collector 除尘器

flue gas analysis 烟道气分析

flue gas neutralization 烟道气中和

fluff 绒毛,蓬松物;起毛;变松;抖松(羽毛等)

fluff collector screen 绒毛滤网
fluff detector 毛羽检测计,绒毛检测器
fluff effect 绒毛效果(绒类织物)
fluffing 起毛
fluffy yarn 毛绒纱
fluid bed drying machine 流化床烘干机
fluid bed laboratory threebath dyeing machine 实验室流化床三浴染色机
fluid bed process 流化床法
fluid char adsorption process 活性炭流化吸附
fluid drive coupling 液力偶合器
fluid friction 流体摩擦
fluidic 流体的,液体的;射流的
fluidic oscillator 流体振荡器
fluidics 射流技术;流控技术
fluidized bed 流化床
fluidized bed boiler 流化床;锅炉
fluidized bed catalytic cracker 流化床催化裂化装置
fluidized bed dryer 流化床干燥器
fluidized bed dyeing 流化床染色
fluidized bed dyeing machine 流化床染色机
fluidized bed furnace 流化床加热炉
fluidized bed incinerator 流化床焚烧炉
fluidized bed process 流化床法
fluidized catalyst 流化催化剂
fluidmeter 流度计,黏度计
fluid proof clothing 防液流服
fluid resistant clothing 抗液流服
fluids 流体
fluid seal 液封(高温高压前处理设备中进、出布口的一种封口形式)
Fluidyer 地毯染色机(商名,德国寇斯脱)
fluoboric acid 氟硼酸

fluoranthene 荧蒽
fluorescein[e] 荧光素[黄]
fluorescence 荧光
fluorescence emmission 荧光发射
fluorescence-exciting ultraviolet region 发荧光的紫外光区
fluorescence microscopy 荧光显微镜术
fluorescence quenching 荧光猝火
fluorescence spectrum 荧光光谱
fluorescent bleach 荧光漂白;荧光增白
fluorescent bleaching agent (= fluoresent brightener) 荧光增白剂
fluorescent brighting(= fluorescent brightening) 荧光增白
fluorescent brighting agent (= fluorescent brightening agent)〈FBA〉荧光增白剂
fluorescent dyeing 荧光染色
fluorescent dyes 荧光染料
fluorescent fabrics 荧光织物
fluorescent illuminated magnifier 荧光照明放大镜
fluorescent lamp 荧光灯
fluorescent lighting 荧光照明
fluorescent pigment printing 荧光涂料印花
fluorescent pigments 荧光颜料,荧光涂料
fluorescent reactive dye 荧光活性染料
fluorescent salt 荧光盐
fluorescent screen 荧光屏
fluorescent spectrometry 荧光光谱法
fluorescent whitening 荧光增白
fluorescent whitening agent〈FWA〉荧光增白剂
fluorescer 荧光增白剂
fluoridation 氟化反应
fluoride 氟化物

fluorimeter 荧光[测定]计
fluorimetric analysis 荧光分析
fluorimetry 荧光测定法
fluorinated resins 含氟树脂
fluorinated surfactants 含氟表面活性剂
fluorine compound 含氟化合物
fluorocarbon 碳氟化合物
fluorocarbon coating 氟碳化合物涂层（防油防污涂层）
fluorocarbon compound 氟碳化合物
fluorocarbon fibre 碳氟纤维
fluorocarbon polymerization 氟碳聚合（有机氟防油树脂的合成工艺）
fluorocarbon silane 碳氟[基]硅烷
fluorochemical finish〈FC finish〉氟化合物整理（提高拒油性）
fluorochemical oil and water repellent finish 拒油、拒水氟化合物整理
fluorochemicals 含氟化学品
fluorogene group 荧光诱发基团
Fluorol 弗卢罗（荧光增白剂溶剂染料，商名，美国制）
fluorophore 荧光团
fluorophotometer 荧光光度计
fluororesin 氟树脂；荧光树脂
fluoro telomers 含氟调聚物（防水防油助剂的原料），氟碳化合物
fluorozirconate process 氟锆酸盐处理（羊毛阻燃）
fluorspar 萤石，氟石
flushability 可冲洗性
flushed away 冲刷掉
flushed printing 罩印
flusher cleaning attachment 冲洗[清洁]装置
flushing 渗开，渗化（印花疵）；冲洗；涂料

移相[工艺]
flushing colours 胶状颜料
flushing ring 冲洗环（喷射染色机加料喷射装置的一个组件）
flush resistance 抗渗化性
flute 沟槽，凹槽
fluted drum 开槽轮
fluted effect [织物轧压]沟纹线效应,电光效应
fluted nut 槽形螺母
fluted roller 沟槽辊
fluttering 抖动；波动，脉动
flux 通量；焊剂，焊药，助熔剂
fluxing temperature 熔化温度
fly-ash collector 除尘器
flying wear 飞行服
fly nut 蝶形螺母,翼形螺帽,元宝螺帽
fly wheel 飞轮
FMA(fibre microanalyzer) 纤维显微[图像]分析仪
FMC-2 colour difference formula FMC-2 色差公式
foam 泡沫；泡沫塑料
foamability 起泡性，发泡能力
foam application 泡沫染整，泡沫加工
foam application unit 泡沫给液装置
foam attenuated fabric 发泡挤压[非织造]织物
foam backed textile 泡沫塑料衬里纺织品
foam backing 泡沫塑料衬里
foam bonding 泡沫材料黏合,泡沫黏合
foam booster 发泡剂
foam breaker 泡沫抑制剂，消泡剂
foam coated fabric 泡沫塑料涂层织物
foam coating 泡沫涂层
foam crushing 泡沫压破（泡沫在加工中

被挤压破裂)
foam cushion method 泡沫垫层法(织物透湿性试验)
foam decay 泡沫衰减
foam decay half life period 泡沫半衰期
foam degumming 泡沫精练,泡沫脱胶
foam degumming apparatus 泡沫精练机,泡沫脱胶机
foam density 泡沫密度
foam dot coating 泡沫点涂层
foam drainage 泡沫排泄
foam dyeing 泡沫染色
foamed impregnation 发泡浸渍法(黏合法非织造物工艺)
foamed polystyrene 聚苯乙烯泡沫体
foamed polyurethane 聚氨酯泡沫体
foamed rubber 泡沫橡皮,泡沫橡胶
foamer 发泡器,泡沫发生器
foam extinguishers 消泡剂
foam finishing 泡沫整理
foam finishing technology〈FFT〉 泡沫整理技术
foam flame bonding 泡沫塑料热熔黏合(利用火焰)
foam generator 泡沫发生器(用压缩空气和发泡剂使整理液发泡,以供泡沫整理用)
foamicide 消泡剂
foam impregnation process 泡沫浸渍工艺
foam improver 增泡剂
foaming 起泡,发泡
foaming adhesion 泡沫黏合
foaming agent 发泡剂
foaming auxiliary agent 发泡助剂
foaming backing 发泡背涂
foaming behaviour 起泡性

foaming binder 发泡印花黏合剂
foaming booster 泡沫增效剂
foaming device 发泡装置
foaming equipment 泡沫涂层装置;发泡装置
foaming head 发泡头
foaming microcapsule 发泡微胶囊
foaming power 起泡[能]力
foaming printing 发泡印花
foaming process 泡沫法
foaming property 起泡性
foaming substance 起泡物质
foam inhibitor 泡沫抑制剂,消泡剂
foam laminated fabric goods 泡沫层压织物
foam laminated material 泡沫塑料层压材料(织物)
foam laminating 泡沫塑料层压黏合
foam laminating machine 泡沫塑料层压机
foam mercerization 泡沫丝光工艺
foam multicolour dyeing 泡沫多色染色
foam/nonwoven layer 泡沫/非织造布层(汽车用纺织品)
foam pad 泡沫垫
foam padder 泡沫轧车(用于地毯染色)
foam performance evaluation 泡沫性能评价
foam preventing agent 消泡剂,泡沫抑制剂
foam printing 泡沫印花
foam printing bonding 泡沫印花黏合
foam printing paste 泡沫印花浆
foam processing 泡沫加工
foam promoter 助起泡剂;发泡助剂
foam rubber 泡沫橡皮,泡沫橡胶
foam rubber backing 泡沫橡胶底衬

foam shampoo 起泡皂液
foam sizing 泡沫上浆
foam stability 泡沫稳定性
foam stabilizer 泡沫稳定剂
foam suppressor 泡沫抑制剂,消泡剂
foam-to-fabric process 泡沫成布法(非织造织物工艺)
foam underlay 泡沫地毯衬垫,泡沫垫底层
foam value 发泡值
foam volume 泡沫容积
FOB(＝F. O. B., fob, f. o. b., free on board) 船上交货;离岸价格
focal distance(＝focal length) 焦距
focal point 焦点
focus 焦点,聚焦;焦距
focusing 调焦,准焦
focusing camera 聚焦照相法;聚焦照相机
focusing mirror 聚焦镜
fog 雾气
fog chamber technique〈FCT〉 雾室技术(用于甲醛捕集)
fogging 起雾
fogging test 雾气试验
foil 薄片,箔
foil coating 箔片涂层,薄膜涂层
foil colourimeter 金属箔比色计
foil dress [金]箔装
foil insulation 箔绝热
foil printing [金]箔印花
foil printing machine 箔片印花机
foil separating machine 箔片剥离机
foil strips 箔带
foil type keyboard 薄片式键盘《计算机》
fold [布匹]折叠
foldability 可折叠性

fold abrasion 折边磨
fold abrasion resistance 折边抗磨性能
folded selvedge 卷边,翻边(织疵)
folder 折布工;折布机;织物折断强度试验器;折边器;[平幅]落布架
fold hanging type dryer 长环烘燥机,悬挂式烘燥机
folding 折布
folding endurance 耐折性;耐折度
folding fault 折印,折痕;折疵
folding machine 折布机,码布机
folding makeup 定长折叠
folding plate 卷板
folding test 折皱试验
fold marks 折皱印(织疵)
Foldo Rol Calender 圆筒针织物用轧光定形机(商名,美国制)
fold[ed] yarn [合]股线
follower 从动单元,随动件,从动件;跟踪器
follower motor 从动电动机
follow-up 跟踪[系统],随动[系统],伺服[系统]
FOM (figure of merit) 优值;灵敏值;品质因数
fondant prints 渗化型印花布
fondu printing 虹彩印花
food additive 食品添加剂
Food and Agricultural Organization〈FAO〉 [联合国]粮农组织
Food and Drug Administration〈FDA〉 食品和药物管理局(美国)
food dyes 食用染料
foot〈ft.〉 英尺
footing 底脚,基底
foot pedal 脚踏板

foot pound〈f. p.〉 英尺磅
footstep 踏脚板；支座；止推轴承
footstep bearing 托杯轴承；立轴承
foot valve 底阀
for.（foreign） 外来的；外国的
Foraperle 防油防污含氟树脂（商名，法国制）
forbidden dyes 禁用染料（致癌或对人体有害染料）
forced air blast 强制通风喷吹
forced circulation 强制循环
forced convection dryer 强制对流烘燥机
forced draft 压力通风，强制通风
forced draft oven 强制送风炉
forced draught 压力通风，强制通风
forced drying 加速[强]烘燥
forced feed lubrication 压油润滑法
forced liquor circulation 强制液流循环
forced lubrication 压送式润滑
forced oil lubrication 强制加油
forced vibration 强制振动
force fit 压配合，压入配合
force of cohesion 内聚力；黏合力
force of flocculation 絮凝力
forcing jack（＝forcing machine） 印花滚筒装拆机
forecast（＝forecasting） 预测；预报
forecooling 预冷却
foreign〈for.〉 外来的；外国的
foreign exchange〈F/X, fx〉 外汇
foreign fibres 异形纤维
foreign impurities 杂质，含杂
foreign matter 异物；杂质
foreign matters test 附着物检验，杂质检验
foreman 工长，领班

foreman dyer 染色工长，染色领班
foreman finisher 整理工长，整理领班
forerun 初馏物
forerunner 导布
foreshot 初馏物
forest 森林色（墨绿色）
forest green 森林绿[色]（暗黄绿色）
forewarmer 预热器
forget-me-not 勿忘我；灰蓝[色]，潮蓝[色]（植物色）
forked lever 叉形杆
fork follower 叉形随件
fork lift truck [装货]铲车
fork type washing machine 叉式洗涤机
formal concentration 克式量浓度《试》
formaldehyde 甲醛，蚁醛
formaldehyde-actived cold bleach 甲醛活化法冷漂
formaldehyde condensation product 甲醛缩合产品
formaldehyde evolution potential〈FEP〉 甲醛逸出潜力
formaldehyde free easy care finishing agent 无甲醛免烫整理剂
formaldehyde free finish 无甲醛整理
formaldehyde-free finishing agents 无甲醛整理剂
formaldehyde limits in air 大气中甲醛限定
formaldehyde monitor 甲醛监测器
formaldehyde odor 甲醛味
formaldehyde precondensate 甲醛预缩物
formaldehyde release 甲醛释放
formaldehyde scavenger（＝formaldehyde-scavenging agent） 甲醛捕集剂，[游离]甲醛清除剂

formaldehyde sodium hydrosulfite 甲醛化连二亚硫酸钠

formaldehyde sodium sulphoxylate 雕白粉;甲醛化次硫酸氢钠

formaldehyde treatment 甲醛处理

formaldehyde urea copolymer 脲醛共聚物

formaldehyde vapour phase treatment 甲醛气相整理

formal dress 礼服

formalin 福尔马林,甲醛水[溶液]

formalin after-treatment 甲醛后处理

formamide 甲酰胺

formamide-formaldehyde resin 甲酰胺甲醛树脂(棉织物的防缩防皱整理剂)

formamidine sulphinate 甲脒亚磺酸盐(碱煮还原剂)

formamidinesulphinic acid 甲脒亚磺酸(用于丝毛还原染料印花)

formanilide 甲酰替苯胺,苯甲酰胺

format 格式,形式

formate 甲酸盐;甲酸酯

formation ability 尺寸稳定性,形态稳定性

formatting 格式化,格式编排《计算机》

formazan dyes 甲䐶(结构)染料

formiate 甲酸盐;甲酸酯

formic acid 蚁酸,甲酸

formic acid colloid of methylolmelamine resin 羟甲基三聚氰胺蚁酸胶体树脂(棉织物防缩防皱整理剂及防菌剂)

formless 无定形的

form-memory polyurethane 形状记忆聚氨酯

Formotex 一种高湿模量黏胶纤维(商名,台湾化学纤维股份有限公司生产)

form stability shirts 形状稳定衬衣(免烫整理衬衣)

form stability suits 形状稳定服

formula 式,公式;[染整]工艺计划;[染整]处方

formula constitution 配方,处方

formula search 配方搜索

formulation 配方,处方;列出公式

formulation station 染液配料间

formula weight 式量

Fornax W 福尔纳克斯 W(抗滑移、抗起毛起球助剂,阳离子型,改性胶状聚硅酸,商名,瑞士制)

Foron AS dyes 富隆 AS 型染料(分散染料,耐日晒色牢度极佳,应用于汽车用布,商名,瑞士科莱恩)

Foron E dyes 富隆 E 型染料(分散染料,升华牢度低,应用于载体染色和条斑严重的物料,商名,瑞士科莱恩)

Foron RD dyes 富隆 RD 型染料(快速染色分散染料,提升高,可缩短染色时间,商名,瑞士科莱恩)

Foron S dyes 富隆 S 型染料(分散染料,升华牢度高,应用于高温染色、热熔染色和印花,商名,瑞士科莱恩)

Foron SE dyes 富隆 SE 型染料(分散染料,升华牢度中等,应用于各种涤纶染色,商名,瑞士科莱恩)

Foron S-WF dyes 富隆 S-WF 型染料(分散染料,湿牢度、耐热迁移性好,适用于涤纶超细纤维以及涤纶混纺产品的印染,商名,瑞士科莱恩)

Forosyn dyes 富隆新染料(分散、酸性混合染料,适用于涤毛混纺染色,商名,瑞士科莱恩)

forse plug 阳模《雕刻》

fortifier 增强剂,增白剂

fortifying fibre 增强纤维
Foryl FK-N 法雷尔 FK-N(非离子型,表面活性剂复配物,用于含高油脂材料及严重沾污纺织品的清洗,乳化、分散力极佳,商名,德国科宁)
for your information〈FYI, fyi〉供参考
FOS(fibre optics sensor) 光导纤维传感器
fossil 化石,化石的
fossil fuels 矿物燃料
foulard 轧液机,轧染机,打底机
foulard stability 轧车稳定性
foul gas 有害气体,秽气
fouling 结垢;结焦;污塞;结皮;污垢
foundation 座;基础;地基;底布
foundation bolt 地脚螺丝
foundation garment 妇女胸衣
foundation pattern 简单[单元]图案
foundation plan 地脚示意图
foundation plate 底板,基础板
four-bar linkage 四连杆机构
four colour printing 四色印花(三原色印花)
Fourier transform infrared spectroscopy〈FTIR〉傅里叶变换红外光谱
four-roller weft straightener 四辊整纬器
fourRs principle(reduction, recovery, reuse, recycle)[清洁生产]4R 原则(减少、回收、回用、循环)
four-tube jet dyeing machine 四管喷射染色机
fox 狐狸棕[色]
FOY(fully-oriented yarn) 全取向丝
FP(flame photometry) 火焰光度[测定]法
f. p.(foot pound) 英尺磅

f. p.(=FP, freezing point) 凝固点;冰点
F. P.(French Patent) 法国专利
FPD(fabric position detector system) 织物位置探测系统
FPD(flame photometric detector) 火焰光度检测器
FQCY (=Fqcy, frequency) 频率
FR(flame resistance) 抗燃
FR(flame retardant) 阻燃的;阻燃剂
FR(flash ranging) 光测,光测距离
fractal polyester filament 分形涤纶长丝
fraction 馏分;级分;碎片
fractional condensing unit 小型压缩冷凝机组
fractional distillation 分馏
fractional emission 分级放射
fractional hydrolysis 分级水解,分步水解
fractional motor 微型电动机
fractional precipitation 分级沉淀
fractional reflectivity 反射分率《测色》
fractional solubility of dyes 染料分级降解性
fractional solution 分级溶解
fractionate 分馏;分级
fractionated dyeing 分步上染
fractionating tower 精馏塔,分馏塔
fractionation 分级;分馏
fractography 断面显微镜观察
fracture 破裂,破碎,断裂;断面,断口
fragment 碎片
fragrance 香味
fragrance finishing 芳香整理
fragrant finishes 芳香整理剂
fraise 法国草莓红[色]
frame 架,座架;框;轧车;拉幅烘燥机
frame crane 龙门式起重机

frame degumming 框[式]丝脱胶,框[式]丝精练
frame dryer 框式烘燥机
frame drying 框式[拉幅]烘燥
frame head 机头,车头
frame impression 框子印
frame mark 绞盘擦痕(绳状染色);机架擦痕(拉幅机烘燥定形产生),框架印(印疵)
frame side [机架]墙板
frame side of squeezer 轧车座
framework 构架[工程];机架,框架
framing 拉幅;铺布;机架
franchise chain〈FC〉合同连锁店
Franklin process 卷装染色法(棉纱或其他纱线)
fray 擦伤,磨损(织疵)
fraying 纱线位移(整理疵)
fray proof 防磨损
free 游离的;自由的;松的;有空的;免除
free acid 游离酸
free alkali 游离[烧]碱
free area 自由区
free base 显色基(不溶性偶氮染料的重氮组分)
free charge 自由电荷
free control〈FC〉任意调节,自由调节
free convection 自由对流
freedom 间隙;(涂层间或两层间的)剥离
freedom fibre 自由纤维(新型弹性纤维)
free dye contamination 自由染料污染物(干洗溶剂中的杂质);色污染
free edge 毛边
free electron pairs 自由电子对
free energy of dyeing 染色自由能
free energy of reaction 反应自由能

free energy of swelling 膨润自由能
free fit 自由配合
free flowing powder 易流动粉末
free formaldehyde〈f-f〉游离甲醛
free formaldehyde scavenger 游离甲醛清除剂
free from motes 无斑点的(指染色)
free liquor 流动液量(绳状染色中除被织物吸去的液量外,用以维持流动的液量)
free of harmful substances 不含有害物质
free of tackiness 不黏着的
free on board〈FOB, F. O. B. ,fob,f. o. b.〉船上交货;离岸价格
free radical 自由基
free radical inhibitor 游离基防止剂
free radical initiators 自由基引发剂
free radical trap mechanism 游离基团捕集机理
free roller 张力调节辊
free running 空转
free running fit 轻转配合
free sedimentation 自由沉降
free settling 自由沉降,自然沉降
free shrinkage 自由收缩
free support 活动支座,自由支承
free surface 自由面
free surface moisture 表面水分,外水分
free-to-shrink 自由收缩
free volume model of diffusion 自由体积扩散模型
free water 游离水
free wheel 活轮,游滑轮;单向离合器
free-wheel clutch 活轮离合器(单向离合器)
free-wheel time 活轮时间(导辊打滑)

freeze-drying 冷冻干燥；冷冻升华（不稳定溶液在冷冻状态下真空蒸干）
freeze-thaw stability 凝冻—解冻稳定性（涂料黏合剂等化学品的性能）
freezing 凝冻
freezing mixture 冷冻混合物；冰冻剂，凝冻剂
freezing point〈FP, f. p.〉凝固点；冰点
freezing point depression 凝冻点（冰点）降低
freezing resistance 耐寒性，耐冻结性
freezing test 凝冻试验
freight bill〈FB, f. b.〉装货清单
French blue 法国蓝［色］（红光鲜蓝色）
French chalk 滑石，皂石
French finish suiting 无光上浆棉平布
French garn count 法制纱线支数（以 N_g 表示）
French lawn finish 法国细麻布式整理（坚实，上光，在有张力状况下干燥，保持硬挺）
French Patent〈F. P.〉法国专利
frequency〈FQCY, Fqcy〉频率
frequency changer 变频器
frequency controlled 变频控制
frequency converter 频率转换器
frequency drive 变频传动
frequency meter 频率测量仪
frequency multiplier 倍频器
frequency regulated 变频调节的《电》
fresh bath 新［配染］浴
freshly prepared 现配，现制
fresh water 淡水
Freundlich adsorption equation 弗罗因德利希吸附公式
Freundlich's adsorption isotherm 弗罗因德利希吸附等温线
F. R. fabrics（flame retardant fabrics）阻燃织物
friability 脆性；易碎性
friction 摩擦
frictional characterization 摩擦特性
frictional difference 摩擦差
frictional durability 摩擦耐久性
frictional force 摩擦力
frictional properties 摩擦性能
frictional resistance 耐摩擦性
frictional strength 摩擦强度
friction block 掣铁，摩擦块（制动用）
friction brake 摩擦闸，摩擦制动器
friction burning 摩擦烧
friction calender 摩擦轧光机（钢辊与纸粕辊速度不同，使布产生擦光效应）
friction calendering 摩擦轧光［工艺］
friction clutch 摩擦离合器
friction coefficient 摩擦系数
friction finish 摩擦轧光整理
friction gear 摩擦传动
frictioning 摩擦整理，摩擦加工
friction mark 擦毛，擦白（绸缎疵）
friction nip 摩擦轧点
friction pasting 摩擦制糊
friction roller 摩擦辊
friction starching machine 摩擦式上浆机（一种三辊面轧式上浆机，上辊表面速度比中辊快，能产生摩擦挤压作用，使织物手感丰满厚实，下辊为给浆辊）
friction starching mangle 摩擦上浆机
friction towel 擦背浴巾布
friction washer 摩擦垫圈（起止动作用）
fridge 冷藏库
friezing machine 起绒机，起毛机

frill machine 折裥机
fringe selvedge （织物的）毛边
fringe speed 线速度
fringe test 流苏染色试验（拆去部分经纱或纬纱，染色后进行对比）
frizzing 呢面卷结整理
frizzing machine 卷结机，起球机
FRL（＝F. R. L.，Fabric Research Laboratory）织物研究试验所（美国）
fronce 皱褶
front elevation 正面；正视图
front pantograph arrangement 前缩放尺（缩小机）
front view 前视图；正视[图]
frost appearance 霜白外观（染整疵）
frosted corduroy fabric 霜花灯芯绒
frosted glass plate 毛玻璃片
frosted yarn 霜花纱
frostiness 霜花
frosting 霜白疵，起霜花（织物经穿着后局部脱色发白）；霜白现象（厚织物印染疵）；[棉布表面]暗淡光泽（由轻度轧浆整理所致）；白雾（真丝绸精练病疵）
frosting effect 霜花效果
frosting finish 霜花整理；雪花整理
frosting phenomenon 消光现象
frost proof 防霜冻
frost wash 霜花洗；雪花洗
froth 泡沫，浮沫；起泡[沫]
froth backcoating 背面泡沫涂层
froth boiling 泡沫精练
froth boiling apparatus 泡沫精练机；泡沫脱胶机
froth degumming 泡沫脱胶，泡沫精练《丝》

froth dyeing 泡沫染色
frothed backing system 泡沫涂层系统
frothing 起泡
frothing agent 起泡剂
froth promoter 泡沫促进剂
froth stability 泡沫稳定性
froth stain 泡斑
FRP（fibre-reinforced plastics）纤维增强塑料
FRs（flame retardants）阻燃剂
fructose 左旋糖，果糖
fruit's colour 水果色彩
fruit stain 果汁污渍
f/s（factor of safety）安全系数
FS（Federal Specifications）联邦规格（美国）
FSA fibre 高吸水纤维（商名，英国考陶尔）
F. S. D. C.（Fellow of the Society of Dyers and Colourists）印染工作者学会会员（英国）
F. T.（firing temperature）着火点，着火温度
ft.（foot, feet）英尺
FTA（fault tree analysis）事故树型分析
FTC（Federal Trade Commission）联邦贸易委员会（美国）
F-test F 试验《数》
FTIR（Fourier transform infrared spectroscopy）傅里叶变换红外光谱
FU（fuse）保险丝，熔断器，熔丝
fuchsia pink 浅红莲[色]
fuchsia red 深红莲[色]，玫红[色]
fuchsin[e] 品红[色]；碱性品红
Fuchun habotai 富春纺，青春纺
fuel cell 油箱

fuel economizer 燃料节省器
fuel evapouration rate 燃料产汽率（蒸发量与燃料耗量之比）
fuel flexibility 燃料[改烧]适应性
fuel oil 燃油
fuel value 燃料热值
fugimeter 耐晒牢度试验仪
fugitive 易褪色的，不坚牢的
fugitive dyes 易褪色染料
fugitiveness 褪色性；不稳定性；易挥发性；短效性
fugitive tinting 易褪的着色（纺织材料标志用）
fugitive tinting colours 标志色，着色剂，指色剂
fugitometer 耐晒牢度试验仪
fukiye prints 喷印花布（日本用铜漏斗喷印）
fulcrum 支点；转动中心
fulcrum pin 支销；顶针（用于缩小机）
full automatic card controlled dyeing plant 全自动卡片控制染色机
full automatic microprocessor-based control 全自动微机控制
full bath treatment 大浴比处理（染色）
full bleaching 全漂白
full chintz 深浅红色地擦光印花布；全棉布
full colour （奥斯瓦尔德制的）纯[彩]色
full concentration liquor 饱和染液
full coverage print 满地罩印印花（不留白）
full coverage printing 满地印花
full decatizing 罐蒸；全蒸；加压蒸呢
full decatizing effect 永久性蒸呢效果
full decatizing machine 罐蒸机

full decator 密闭蒸呢机，加压蒸呢机
full development 充分发色
full dull 全无光，全消光
full dull yarn 全无光丝
full dyeing 深色染色
fulled fabric 缩绒织物
fuller 制毡工；漂洗工；缩绒工
fuller's earth 漂泥，漂土
full face finish 丰满呢面整理（缩绒起绒后剪毛加压，使呢面丰满）
full facial cover 全封闭面罩
full fashion 全成形[衣服]
full feel 丰满手感
full felting counter pile 全部反向起绒
full felting pile 全部正向起绒
full finish 双面整理
full frame type 整机式
full front fusing 全面熔合，全面黏合[领衬布]
full handle 丰满手感
fulling 缩绒，缩呢，毡合
fulling agent 缩绒剂
fulling fold 缩呢折痕（疵点）
fulling machine(＝fulling mill machine) 缩绒机，缩呢机
fulling roller 缩呢辊
fulling soap 缩呢肥皂，毡合肥皂
fulling stocks 槌式缩绒机
fulling trough 缩呢槽，毡合槽
full journal bearing 全围式轴承（轴瓦在360°范围内包围着轴颈的滑动轴承）
full lightly 适度缩绒；预缩（美国名称）
full mercerized finish 全丝光[整理]
fullness 丰满度，丰满性
fullness of shade 色泽丰满度
full package 满管，满载

full print 饱满深色印花
full radiation(＝full radiator) 全辐射体；黑体《测色》
full scale 实物大小；实际尺寸；满刻度；原样尺寸
full scale load 满[幅度]负荷
full scale production 最大规模生产
full scale test 大规模试验
full shade 饱和色(指纤维最高得色深度)
full size 足尺，原样大小
full strength print 饱满深色印花
full surface coating 全面涂层
full washable textiles 可机洗纺织品
full weight 全重，毛重
full white yarn 全漂白纱
full width 全幅，全幅
full width rolled on tube 全幅卷筒
full width scouring machine 平幅洗呢机；平幅洗布机
full width washing machine 平幅水洗机
fully-bleached linen 全漂亚麻布
fully-cured 全焙烘
fully drawn texturing yarn〈FDTY〉 全拉伸变形丝
fully drawn yarn〈FDY〉 全拉伸丝
fully-fashioned 全成型的
fully fashioned dyeing 全成型的染色(指成衣染色)
fully-flooded dyeing machine 全浸没式染色机
fully-flooded high temperature jet-dyeing machine 全浸没式高温喷射染色机
fully-oriented yarn〈FOY〉 全取向丝
fully-relaxed dimension 全松弛处理后尺寸
fully-shrunk 全预缩的

fulvous 茶色，暗黄色，黄褐色
fumaric acid 富马酸，反丁烯二酸
fume 烟气，烟雾
fume fading 烟气褪色
fume hood 通风罩
fume-proofing finish 防烟气整理
fumigant 烟熏剂
fuming nitric acid 发烟硝酸
fuming sulfuric acid 发烟硫酸
function 操作；作用；功能
functional 功能的；官能的
functional chemicals 功能性助剂
functional clothing 功能性衣着
functional composite 功能性复合材料
functional dyes 功能性染料
functional elements 功能元件，起作用的元件
functional end group 官能性端基
functional fabric 功能织物(具有防火、防雨、防蛀、防缩等性能)
functional finish 功能整理(指防雨、防火、防蛀、防缩、防皱等各种特种用途整理)
functional group 官能团
functional group analysis 官能团分析
functional integrity 功能完整性
functionalization 功能化作用
functional polymer 官能聚合物，功能聚合物(如离子交换树脂、螯合物树脂等)
functional properties 官能性，功能性
functional sportwear 功能性运动衣
functional test 功能性试验
functional textile 功能性纺织品
functional unit 操作部件
function finish 功能整理
fundamental chain 母链，主链

fundamental colour 地色；基础色，基本色《测色》
fundamental measure 基本尺寸
fundamental response curve 基本感度曲线《测色》
fungal amylase 霉菌淀粉酶
fungal degradation 真菌降解
fungicide 杀霉菌剂；杀真菌剂
fungicide finish 抗菌整理
fungistat 防霉的
fungistatic agent 抑［霉］菌剂
fungus 霉菌；真菌
fungus resistance 耐霉性，抗霉性，防霉性
funnel 烟囱；漏斗
fur 毛，绒毛；毛皮

furan 呋喃
furan resin 呋喃树脂
fur brush 毛刷
fur cloth 仿毛皮织物
fur dyeing 毛皮染色
fur fabric 仿毛皮织物；长毛绒织物
furfural（＝furfuraldehyde） 呋喃甲醛；糠醛
furfural resin 糠醛树脂
furfuran 呋喃
furfurol 糠醛
furfuryl alcohol 呋喃甲醇
fur imitation plush 仿毛皮长毛绒织物
fur-like fabric 仿毛皮织物
furnace black 炭黑
furnisher 给浆辊（用于滚筒印花机）
furnishing fabric 装饰［用］织物，家具布
furnishing prints 装饰印花布
furnishing properties 给色浆转移性（印花浆从给浆滚筒经花筒转移到布面的移动性）

furnishing roller 给浆辊
furniture fabric 家具［用］织物
furniture upholstery 家具装饰材料
furrow 皱纹，起皱；沾污，染污
fur shading 人造毛皮润色（使其色泽具有由深到浅的层次色调，类似真毛皮）
fuscous 淡褐灰［色］
fuse〈FU〉熔断器，保险丝，熔丝
fused collar fabric 热熔黏合领衬织物
fused interfacing 热熔衬
fused joint 熔融黏结（非织造布加工）
fused mass 熔融物
fused salt 熔盐
fused seam 压熔缝合
fused shirt collar 热熔黏合衬衫领
fuse fusible interlining 热压薄膜衬
fusibility 可熔性
fusible 易熔的；可熔化的
fusible interlining 热熔衬（一种黏合非织造衬里）
fusible seaming 热熔缝合
fusing 熔化，熔合
fusing interfacings 热熔贴面，熔合面
fusing machine 黏合熨烫机
fusing material 熔合材料
fusing press 压烫熔合机
fusing shrinkage 热熔收缩
fusion 热熔
fusion bonding 溶融黏合
fusion-bonding process 热熔胶合法
fusion lamination 热熔层压
fusion［contraction］shrinkage 热熔收缩
fusion technology 黏合工艺；熔合工艺
fustic 佛提树，黄颜木（天然植物黄染料）
futurology 未来学
fuze〈fz.〉熔丝，保险丝

fuzz 微毛,绒毛,绒毛;毛丝
fuzz ball 起球(织疵)
fuzziness 绒毛效应
fuzzing and pilling 起毛起球
fuzz prevention 防止起毛
fuzz resistance 抗起毛性
fuzz-snag performance 起毛钩丝性
fuzzy texture 绒毛表面
FWA(fluorescent whitening agent) 荧光增白剂
FWWMR(fire, water, weather, and mildew resistance) 防火、拒水、耐气候和防霉［整理法］
F/X(=fx, foreign exchange) 外汇
FYI(=fyi, for your information) 供参考
Fyrol 76 法罗尔76(乙烯磷酸盐低聚物,羊毛、棉、毛棉混纺织物用阻燃剂,商名,美国制)
Fyrol 76 finish 法罗尔76整理法(一种阻燃整理工艺)
fz.(fuze) 熔丝,保险丝

G

Ga 镓(化学元素)
gabardine(＝gabardeen) 轧别丁,华达呢
G-acid G-酸;2-萘酚-6,8-二磺酸
gadget 机件;配置;装置
gag 闭塞,封闭;压紧装置;压板
gage 表,计;量规,隔距片
gain value 增益值
gal.(gallon) 加仑(液量单位)
galactan 多缩半乳糖,半乳糖胶
galactomannan(＝galactomannoglycan)〈GM〉半乳甘露聚糖(多糖类糊料主成分)
galactose 半乳糖
galacturonic acid 半乳糖醛酸(海藻酸盐主成分)
galalith 酪朊塑料
gall 磨损,擦伤;瑕疵
gall extract 五倍子提取物,没食子提取物
gallic acid 棓酸,五倍子酸
Gallipoli oil 加利波利[橄榄]油(润滑羊毛用)
gallnut 棓子,五倍子,没食子
gallocyanin 棓花青,媒染棓酸青(作为碱性染料或媒染染料)
gallon〈gal.〉加仑(液量单位)
gallotannic acid 棓单宁酸,鞣酸
Gall-Riedel formula 盖尔—里旦尔公式(表示印染深度)
gall soaps 五倍子皂,没食子皂
galv.(galvanometer) [灵敏]电流计(用来测量微小电流和检查电路中有无电流通过的一种电流检测仪器);检流计
galvanic cell 电池
galvanization 镀锌
galvanize 镀锌;通电流
galvanized pipe 镀锌管子,白铁管(俗称)
galvano- (构词成分)表示"电的""电流的"
galvano direct design screen making process〈GDD〉电铸直接花样筛网制造法
galvanometer〈galv.〉[灵敏]电流计(用来测量微小电流和检查电路中有无电流通过的一种电流检测仪器);检流计
galvanoplastic screen 电铸法制成的花网
galvanoplastic screen process 电铸花网工艺
galvanoplastic screen process by laser engraving 激光雕刻[直接]电铸花网工艺
galvanoscope 验电流器
galvano[nickel]screens 电铸镍网
gambiered Canton silk 莨绸,拷绸(中国制)
gamboge 藤黄[树脂];橙黄色
gamma 希腊字母γ,可表示化学基团的位置
gamma-ray γ射线
gamsa 甘姆萨绸(醋酯丝经,黏胶丝绉线纬,绉面经背组织)
gamut 色域,色卡
Ganalok process 甘纳洛克树脂整理(用双

乙烯砜为交联剂,商名)
gang drive 集体传动,合组传动
ganging 同轴;机械连接
gantry ager 门式蒸化机,快速蒸化机,架桥式蒸化机
gantry crane 龙门起重机
gantry steamer 架桥式汽蒸箱
Ganz formula 甘茨白度公式
gap 间隙,间隔;缺口(齿轮间裂口);差距
gap coating 隙缝涂层法,孔隙涂层法(美国)
gap winding 间隔卷布
Garan finish 加兰整理(一种提高玻璃纤维布湿强度等性能的整理,商名,美国)
garden carpets 波斯地毯
garden furnishings 户外用布(如各种篷布、躺椅用帆布等)
gardenia 栀子(可提取染料的植物,含黄色素)
gardenia blue 栀子蓝色素(天然植物染料,食用色素)
gardenia colour 栀子黄色素(天然植物染料,食用色素)
gardenia yellow 栀子黄色素(天然植物染料,食用色素)
Gardinal 加丁努尔(阴离子净洗剂,商名,法国制)
Gardinol 加丁诺尔(磺化脂肪醇,润湿剂,羊毛缩绒剂)
gare 死毛,戗毛
garment 衣服(一般指长袍、外套)
garment brushing in dry cleaning pretreatment 成衣干洗预刷
garment component 服装组件

garment computer aided design〈GCAD〉 服装计算机辅助设计
garment construction 服装结构
garment dip process 成衣浸渍整理(成衣免烫整理)
garment dyeing 衣件染色,成衣染色
garment dyeing machine 衣件染色机,成衣染色机
garment dyeing shrinkage 成衣染色缩水率
garment fabrics innovation and design center 服装面料创新和设计中心
garment finishing 成衣整理,服装整理
garment industry 服装工业
garment interlining 服装内衬
garment leather 成衣用皮革
garment-making accessories 成衣用辅料
garment manufacture 服装制造
garment molding 服装熨烫定形
garment parts 衣片
garment physiology 服装生理特性
garment pieces 衣片
garment prebrushing 成衣预刷
garment press 烫衣板
garment printing technology 成衣印花技术
garment setting 服装定形[整理],成衣定形[整理]
garment shrinkage 服装收缩,成衣收缩
garment spray process 成衣喷洒整理(成衣免烫整理)
garment steamer 服装熨烫机
garment wear test 服装穿着试验
garment wet processing 服装湿加工,成衣湿加工
garnet [red] 石榴石红[色],酱红[色]
garnet brown 石榴石棕[色](红棕色)
gas absorbent 气体吸收剂

gas adsorption cell 气体吸收池
gas air mixer 煤气空气混合器
gas analysator 气体分析器
gas analysis 气体分析
gas analysis apparatus 气体分析器
gas black 气黑,气烟末
gas blanketing 气体覆盖
gas bleaching 气体漂白
gas blue printing 靛蓝印花
gas booster 煤气增压器
gas bubble 气泡
gas bubble method 气泡法
gas burner 煤气[喷]灯;煤气燃烧器
gas carburizing 气体渗碳
gas cavity cell 气孔池
gas cell fabric 气囊织物
gas chlorinating bath 气体氯化浴
gas chlorination [毛织物]气体氯化
gas chromatogram 气相色谱图
gas chromatography〈GC〉气相色谱[法];气体色层分离法
gas cleaning unit 气体净化装置
gas combustion 气体燃烧
gas conduit 烟道
gas constant 气体常数
gas cushion 气垫
gas cutting 气割
gas detector 气体检测器
gas discharge etching 气体放电蚀刻[法]
gas discharging lamps 气体发射灯
gaseous chlorination 气态氯化处理
gaseous diffusion process 气体扩散方法
gaseous effluent 气态排出物
gaseous gap [充]气[间]隙
gaseous injection dyeing 气态喷射式染色
gaseous plasma 气体等离子体
gaseous polymerization 气相聚合作用
gaseous quenching device 气体骤冷装置
gas fading 烟气褪色
gas fading apparatus 耐烟薰褪色牢度试验仪
gas fading inhibitor 烟气褪色抑制剂
gas fading proof finish 防止气体褪色整理,防烟退整理
gas family 石油气(如甲烷、乙烷、丙烷、丁烷等)
gas fastness 耐烟气色牢度
gas filtering equipment 滤气装置
gas filtering tower 滤气塔
gas fired boiler 燃气锅炉
gas fired infrared predryer 煤气红外预烘机
gas flame singeing machine 煤气[火焰]烧毛机,气体烧毛机
gas flow silencer 气流消声器
gas-flow transfer printing 气流法热转移印花
gas fume 烟气
gas fume fading 烟气褪色
gas fume fastness 耐烟熏色牢度
gas gel chromatography 气体凝胶色谱[法]
gas generator 煤气发生炉;气体发生器
gas heated infrared predryer 煤气加热红外预烘机
gas heating 煤气加热;气体加热
gas holder 气柜
gas hydrate 天然气水合物
gasification 气化
gasificator 气化器
gasifier 煤气发生器,气化器
gas impermeable 不透气的

gas infrared ray drying machine 煤气红外烘燥机
gas injection dyeing 气态喷射式染色
gasket 衬垫,垫圈,垫片,密封垫;填料
gas laser 气体激光器
gas laws 气体定律
gas liquid chromatography〈GLC〉气液色谱[分析]法
gas manometer 气体压力计
gas mask 防毒面具
gasohol 酒精汽油(由汽油与酒精混合而成)
gas oil ratio〈GOR〉油气比
gasol 液化石油气,气体油
gasoline(＝gasolene) 汽油
gasometer 气量计;煤气表
gas partition chromatography〈GPC〉气相分配色谱[法]
gas permeability 透气性
gas permeability test 透气性试验
gas phase 气相
gas phase chromatography analysis 气相色谱分析
gas phase crosslinking 气相交联(如热转移印花)
gas phase crosslinking process 气相交联法
gas phase fluorination 气相氟化作用
gas phase polymerization 气相聚合
gas phase processing 气相处理
gas phase separation 气相分离
gas phase xanthation 气相黄化
gas pilot 煤气引燃器
gas plate singeing 煤气铜板烧毛
gas plating method 气镀[金属薄膜]法
gas pocket 气袋;(塑料表面的)气窝
gas pressure spring 气压弹簧

gas producer 煤气发生炉
gas producing plant 煤气发生器
gas proofing 不透气
gas proofness 不透气性;气密性
gas proportioning damper 烟气比例调节挡板
gas propulsion 气流推进(气流染色中,气流推动织物运行于机器中进行染色)
gas protective clothing 防毒气服装
gas purger 气吹扫设备
gas radiant predryer 气体辐射热预烘机
gas sampler 气体取样器《环保》
gas sampling 气体取样
gas scrubber 气体洗涤器
gas seal 气封,惰性气保护
gassed end 烧毛[经]纱
gassed georgette 细特(高支)烧毛线绉布;充乔其纱
gassed yarn 烧毛纱
gas seepage 气体渗出
gas sensor 气体传感器
gas separator 气体分离器
gassing 烧毛
gas singeing 煤气烧毛,气体烧毛
gas singeing machine 气体烧毛机
gassing frame 烧毛机
gassing machine 烧毛机
gassing machine for yarn 纱线烧毛机
gassing off 放气
gas solid chromatography 气—固色层分析
gas sweetening 气体纯化
gas testing tube 气体检测管
gas thermometer 气体温度计
gas tight 气密;不透气的
gas tight casing 气密式外壳
gas tightness 气密性

Gaston County Micro Monitor 加斯顿微监控器(最新的控制系统,一名操作人员在中央控制台进行操纵,便可自动控制多达32台染色机,美国制)

gas trace analysis 气体痕量分析

gas tube 离子管;充气管;气体管线

gas turbine generator 燃气涡轮发电机

gas turbine in combined heat & power generation 热电联产汽轮机

gas washer 气体洗涤器

gas washing bottle 气体洗涤瓶

gate 门;门电路;栅极;控制极;选通

gate paddle agitator 框叶式搅拌器

gate turned-off thyristor 可关断可控硅

gate valve 闸阀,滑门阀;选通管

gather 皱裥;打褶裥

gatherer (= gathering attachment) 打裥器,打裥装置

GATT (General Agreement on Tariffs & Trade) 关税及贸易总协定

gatterwalking 缩呢机,缩绒机

gatti 菱形花布(东南亚制)

gaudivi 印花粗布(印度制)

gauffer 拷花,轧花;起皱;打皱褶

gauffer calender 凹凸纹轧花机,印纹轧压机

gauffered cloth (= gauffered fabric) 轧纹布,拷花布,轧花凸纹织物

gauffrage effect 轧花效应《整》

gauffré crepe [轧花]泡泡绉

gauffré(= gaufré) 拷花,轧花(法国名称)

gauffré ribbon 拷花带,轧花带

gauffré satin 轧花缎

gauffré silk 轧花绸

gauffré velvet 雕花天鹅绒

gauge 隔距;隔距片;评估,判断;测量;尺度,标准;测量仪器

gauge adjustment 隔距调整

gauge block 隔距板,块规

gauge glass 计液玻管;锅炉水位指示玻璃管

gauge mark 定位标记

gauge outfit 测量头,表头

gauge pressure 计示压力

gauss 高斯(磁场强度单位)

gauze 纱罗组织;纱罗织物;纱布

gauze cloth 纱罗织物,薄纱布

gauze curtain fabric 窗帘纱

gauzed stencil paper 网状型纸(手工印花用)

gauze fabric 纱罗织物;薄纱布;薄绢

gauze mark 筛网印痕

G. B.(Great Britain) 英国;大不列颠

GC(gas chromatography) 气相色谱[法];气体色层分离法

GCAD(garment computer aided design) 服装计算机辅助设计

g-cal(gram-calorie) 克卡路里

GCR(grey component replacement) 灰成分替代(印刷分色用术语)

Gd 钆(化学元素)

G/D(= g/d, grams per denier) 每旦纤维质量(用克表示);克每旦,克/旦

GDCh 【德】德国化学协会

GDD(galvano direct design screen making process) 电铸直接花样筛网制造法

GDME (glycol dimethyl ether) 乙二醇二甲基醚

GDP(glow discharge plasma) 辉光放电等离子体

Ge 锗(化学元素)

gear 齿轮;传动装置

gear box 齿轮箱
gear carriage 齿轮架
gear cluster 齿轮组
gear crimped yarn 齿轮卷曲法变形丝
gear drive 齿轮传动
gear end 机头,车头
gear guard 齿轮护板,齿轮罩
gear housing 齿轮箱
gearing 齿轮装置;传动装置
gearing diagram 传动图
gearing layout 传动图
gear motor 齿轮电动机;有减速器的电动机
gear pitch 齿轮节距
gear pump 齿轮泵
gear rack 齿条
gear ratio 齿轮速比
gear shift 变速,调档;变速机构
gear stud 齿轮芯轴
gear train [齿]轮系
gear-type metering pump 齿轮式计量泵
gearwheel pump 齿轮泵
gegenion 反离子
Geiger counter 盖革计数器
Geiger-Müller counter tube 盖革—弥勒计数管
gel 凝胶,冻胶;胶滞体
gel.(gelatinous) 胶凝的,凝胶的;胶状的
gelatification oven 胶凝烘箱
gelatin 明胶,水骨胶,动物胶,精制胶;胶[体],凝结体,胶质;全力丁,全力片
gelatinase 明胶酶
gelatination 胶凝作用;凝胶化
gelatine size 明胶浆
gelatin fibre 凝胶纤维
gelatinization 胶凝[作用];凝胶化[作用]

gelatinizing point of starch 淀粉的胶化温度
gelatinizing theory 胶凝理论(指羊毛缩绒的机理)
gelatinous〈gel.〉 胶凝的;凝胶的;胶状的
gelatinous fibre 凝胶纤维
gelatin temperature 胶凝温度
gel chromatography 凝胶色谱[法]
gel coherence 凝胶内聚力,凝胶凝集
gel deposition 凝胶沉积[作用]
gel dyeing 凝胶着色法
gel filtration 凝胶过滤[法]
gellant 胶凝剂
gelling 胶凝作用
gelling agent 胶凝剂
gelling machine 胶凝机
gelose 琼脂,琼胶,石花菜,洋菜
gel permeation chromatography〈GPC〉 凝胶渗透色谱法
gel point 凝胶点
gel swelling 凝胶溶胀
gel time 胶凝时间
gel-whitening 凝胶增白(合成纤维的一种增白方法)
gemmatein 埃蕈染料(从埃蕈中得来的天然色素)
genappe yarn 烧毛[精纺]毛纱
genapping 烧毛
gene 基因
gene modified cellulase 基因改性纤维素酶
gene-modified starch 基因改性淀粉
General Agreement on Tariffs & Trade〈GATT〉 关税及贸易总协定
general arrangement drawing 总装图
general assembly 总装
general cleaning 普通的洗涤

general colour rendering index 一般色再演指数
general layout 总平面图;总体布置
general overhaul 大检修
general precipitation 自然沉降
general purpose computer〈GPC〉通用计算机
general purpose relay 通用继电器
general purpose rubber 通用橡胶
General Services Administration〈GSA〉总务管理局(美国)
general view 全视图,总图,概略图
generating surface 蒸发受热面
generation 产生;发生
generator 编制程序;发电机;发生器;振荡器;发送器;传感器
gene research 基因研究
generic class 属类
generic name 通称,属称
gene technology 基因技术
genetically modified 基因改性的
genetic code 基因编码
genetic engineering 基因工程
geneticists 研究遗传技术的科学家
genetic material 基因物质,遗传物质(即DNA)
genetic regression〈GR〉遗传退化
gene-transferred technology 转基因技术
genome map 基因图
genomics research 基因学研究
gentian blue 龙胆蓝[色](浅紫蓝色)
gentian violet 龙胆紫(碱性紫色染料)
gentle washing program 缓和的洗涤程序
genuine part 正品配件(指原厂出品的配件)
genuine ultramarine 天然群青;艳蓝色

geochemistry 地球化学
geo-composite 土工复合材料
geomaterial 土工材料
geo-membrane 土工膜
geometrical optical illusion 几何学的错视
geometric design 几何图案
geometric formula 几何结构公式
geometric isomer 几何异构体
geometric pattern 几何图案
geometric porosity 多孔性结构
geometric properties 结构特性
geometric spot 几何图案点子
geometry of cloth 织物几何结构
geo-net 土工网
georgette 乔其纱
georgette brocade 织花乔其纱
georgette crepe 乔其绉;乔其纱
georgette velvet 乔其丝绒
georgette with printed velvet flower 烂花乔其绒
georgetton 棉乔其纱
geo-synthetic fabric 土工合成纤维织物
geo-synthetics 土工合成材料
geotechnical fabric 土工[技术]织物
geotextile 土建布,土工布,土工织物
geotextile composite 复合土工织物
geotextile-related products 土工织物相关产品
geothermal energy 地热能
Ger.(German) 德国的;德国人
geranium pink 妃色,绯红[色]
German〈Ger.〉德国的;德国人
German new legal regulation 德国新法规
German Patent〈GP〉德国专利
German physiological-contact ordinance 德国生理学接触法案(关于禁止用致癌

染料的法案）

German prints 德国花布（蓝地白花粗棉布）

German water hardness 德国制水硬度

germ-destroying finish 杀菌整理

germ free condition 无菌条件

germicidal agent 杀菌剂

germicidal finish 杀菌整理

germicide 杀菌剂

germifuga 抗菌剂

germ resistant 防腐剂，防菌剂；防腐的，杀菌的

germ resistant fabrics 抗菌织物

gerotor pump 摆线泵（齿形为摆线的内啮合齿轮泵）

Gesamt textile Research Council 纺织工业组织研究委员会（德国）

Gf 对鱼毒性指数（德国废水处理法中用）

GF(glass fibre) 玻璃纤维

GH 水硬度单位

ghatpot 绫

ghatto gum 茄多胶（阿拉伯树胶代用品，又称印度树胶、亚洲树胶）

GH fibre GH阻燃纤维（商名，日本制）

ghosting 虚像，虚印（转移印花疵）

ghosting effect 虚印作用

ghosting in transfer printing 转移印花虚印

ghost print 虚印

giant batch rolls 大卷装滚筒

giant batch trolley 大卷装小车

giant batch winder 大卷装卷布机

giant pressured jig 高压大卷装染机

gib and cotter 合楔

Gibbs adsorption equation 吉布斯吸附方程

Gibbs adsorption theory 吉布斯吸附理论

Gibbs-Marangoni effect 吉布斯—马拉高尼效应（泡沫膜弹性理论）

Gibbs phase rule 吉布斯相位定律

gib key 凹形槽

Gifu dyeing process for polyester 吉富染色法（用酚作膨化剂染色，日本）

gig 起毛刺果；起绒机；起绒，拉毛

gig barrel 起绒刺果滚筒

gigging 起绒，拉毛

gigging machine 起绒机，起毛机，刺果起绒机

gill 及耳（英美制液量单位，英制1及耳＝0.1421升；美制1及耳＝0.118升）

gill box 针梳机（制毛条的设备）

gilled heating pipe 翼片加热管

gimbal 万向接头

gimp box yarn 螺旋包芯纱（装饰用）

gingu 姜色（棕色）

gingham 方格色织布，条格平布

gingham print 仿色织条格印花

girder 大梁；撑柱，撑杆

glacé 光泽效应（由整理取得，来自法语）

glacé finish 光泽整理

glacés 有光织物（以棉经、马海毛纬制成）

glacé silk 闪光绸

glacé thread(= glacé yarn) 蜡光线；上光线

glacial 冰的

glacial acetic acid 冰醋酸

glacial acetic acid solubility test 冰醋酸溶解性试验

glacial acetic acid test for dye class identification 冰醋酸法测染料分类

glacial dyes 冰染料,纳夫妥染料
gland 压盖,密封压盖,填料盖
gland packing 压盖填料,压盖密封,轴封装置
gland steam condenser 轴封凝汽器
glare 眩目;眩光;显眼
glarimeter [纸面]光泽计
glass batiste 透明细薄织物
glass bead 玻璃小珠
glass ceramic 玻璃陶瓷
glass cloth 玻璃纤维织物;涂玻璃粉织物(擦光用);玻璃器皿揩布
glass curtains 窗纱
glass effect 玻璃效应
glass electrode 玻璃电极
glass fibre〈GF〉玻璃纤维
glass fibre fabric 玻璃纤维织物
glass fibre felt 玻璃纤维毡(作绝热用)
glass fibre reinforced plastic 玻璃纤维强化塑料,玻璃钢
glass filament yarn 玻璃纤维长丝
glass filter cloth 玻璃纤维滤布
glassine 玻璃纸
glass marble 玻璃球
glass optical fibre 玻璃光学纤维
glass plate test 玻璃片试验(测试还原染料染色时的还原状态)
glass reinforced plastics〈GRP〉玻璃强化塑料
glass silicide coating 玻璃硅化物涂层
glass textile 玻璃纤维纺织品
glass transition point 玻璃化转变点
glass transition temperature〈GTT, Tg〉玻璃化[转变]温度
glass tube heat exchanger 玻璃管热交换器(一种气—气热交换器,置于烘房排风管道中利用余热来预热新鲜空气)
glass wadding 玻璃纤维填料,玻璃纤维衬料
glass water gauge 玻璃水位计
glassy finish 玻璃光泽整理
Glauber's salt 元明粉,芒硝
glauconite 海绿石,海绿沙(天然软水剂)
glaucous 淡灰蓝色的,淡灰绿色的
glaze 色泽,光泽;加光,轧光,上光;极光
glazed calico 摩擦轧光细布
glazed chintz 摩擦轧光艳丽花布
glazed chintz finish 摩擦轧光整理
glazed cotton 上光棉线,加光棉线
glazed finish 高光泽[轧光或摩擦轧光]整理
glazed silk 上光丝,加光丝
glazed thread 上光线,加光线
glazed yarn 上光纱,加光纱
glazer 抛光轮;轧光机
glazing 上光,轧光
glazing agent 上光剂
glazing calender 轧光机,摩擦轧光机
glazing & finishing auxiliary 上光整理助剂
glazing & flatting calender 上光机,轧水机
glazing machine 上光机,轧光机
GLC(gas liquid chromatography) 气液色谱[分析]法
Glicotin A 格里科廷 A(非离子性硫二甘醇衍生物,配制印花浆时作染料的增溶剂,商名,意大利制)
glide rods 滑杆(挂布用,如用于挂环蒸化机)
gliding property 光滑性,平滑性;(织物表面的)滑移性

glitch 失灵，小故障
glitter 闪光[效应]；发光[效应]；(呢绒久磨后出现的)极光
glitter effect 闪光(或发光)效应
glitter free effect 消辉效果
glittering fabric 闪光织物
glittering finish 闪烁整理
glittering printing 闪烁印花
glitter pigments 闪光涂料
glitters 闪烁片
glitter threads 闪光纱线
Global C 格罗布 C(反光涂料浆，用于涂层布，商名，日本林化学)
Global coating 玻璃微珠反光涂层
Global Organic Textile Standard〈GOTS〉 全球有机纺织品标准
Global P 格罗布 P(反光涂料浆，用于印花布，商名，日本林化学)
Global Positioning System〈GPS〉 全球定位系统(美国)
global printing 玻璃微珠反光印花
globe digester 球形蒸煮釜
globe thermometer [干湿]球形温度计
globe valve 球阀
globoidal worm gear 曲面蜗轮，球面蜗轮
globular proteins 球状蛋白质
globule 小球；液滴
globulins 球蛋白
Glolal 格洛拉尔织物反光整理技术
gloss [表面]光泽
gloss decatizing machine 轧光蒸呢机
gloss finish 平光整理(使织物表面平滑而产生光泽)
glossiness 光泽性；光泽度
glossing 抛光[工艺]《丝》
gloss-iron 烫光

gloss meter 光泽计
gloss number 光泽指标
Glossop prints 格鲁索印花布(英国制)
gloss value 光泽值
glossy 有光泽的
glossy printing 闪烁印花
glossy printing ink 有光印花浆
glossy wool 有光羊毛(氯化羊毛)，丝光羊毛
glove presser 手套压烫机
glove press machine 手套压烫机
glow discharge 辉光放电(等离子体术语)
glow discharge plasma〈GDP〉 辉光放电等离子体
glow-discharge treatment 辉光放电处理
glowing 无焰燃烧
glowing resistance 耐烧灼性，耐灼热性
glowing time 无焰阴燃时间，余辉时间
glow plug 电热塞
glow-proof 耐烧灼的
glow-proofing agent 耐烧灼剂
glow starter 起辉器，点灯管
glow tube 辉光管
glucan 葡聚糖
gluconic acid 葡萄糖酸
glucopyranose 吡喃葡萄糖
glucosamine 氨基葡[萄]糖，葡糖胺
glucose 葡萄糖
glucose oxidase 葡[萄]糖氧化酶
glucose residue 葡[萄]糖残基
glucosidase 葡萄糖苷酶
glucoside 葡糖苷；糖苷
glucosidic bond 苷键，配糖键
glucosidic linkage 糖苷键[合]，苷键[合]
glucuronic acid 葡糖醛酸
glucuronic acid residue 葡[萄]糖醛酸残基

glue [动物]胶;胶水;胶合,黏合
glue application device 上胶装置
glued carpets 胶合地毯(黏合法地毯)
glued composite fabric 胶黏复合织物
glue-down 黏合
glued seam 胶合缝接
glue film 浆膜,胶膜
glueing 胶合;粘贴
glueing agent [台板]黏着剂
glueing carriage 上胶行车(或走车,筛网印花贴布用)
glueing device 上胶装置
glueing system 上胶系统
glueing system for printing 印花上胶系统
glueing unit 上胶单元,上胶装置
glue-line temperature [热熔胶]胶层温度,粉末成胶时的温度
glue pump 胶泵
glue reservoir 胶储槽
glue roller 上胶辊
glue size 浆液;浆料
glue squeegee 胶刮刀
gluey 胶粘的;胶合的
glutamic acid 谷氨酸,2-氨基戊二酸
glutaminase 谷氨酰胺酶
glutaraldehyde 戊二醛
glutaria aldehyde 戊二醛(丝胶固定剂)
glyceraldehyde 甘油醛
glyceride 甘油酯
glycerin 丙三醇,甘醇,甘油
glycerin chlorohydrin 氯甘油
glycerine 甘醇
glycerin triacetate 甘油三乙酸酯,三醋精
glycerol 甘油
glycerol diglycidyl ether 甘油二缩[水]甘油醚

glycerol monooleate 甘油单油酸酯(防锈油、整理剂)
glycerol monostearate ⟨GMS⟩ 甘油单硬脂酸酯(树脂润滑剂)
glycerol trismercaptoacetate 三巯基乙酸甘油三酯(羊毛纤维防缩整理剂)
glyceryl ester surfactant 甘油酯表面活性剂
glyceryl monoacetate 甘油醋酸酯,醋精
glyceryl monooleate 甘油单油酸酯(防锈油、整理剂)
glyceryl phthalate 邻苯二甲酸甘油酯,甘油酞酸酯
glycidol 缩水甘油(用作天然油稳定剂,匀染剂)
glycidyl acrylate 丙烯酸缩水甘油酯(辐射整理用剂)
glycidyl compound 甘油缩水化合物
glycidyl methacrylate ⟨GMA⟩ 甲基丙烯酸缩水甘油酯
glycidyl trimethyl ammonium chloride ⟨Glytac⟩ 缩水甘油三甲基氯化铵(纤维素纤维胺化剂,商名,法国制)
glycine 甘氨酸,氨基乙酸
glycine fibre 氨基乙酸纤维(由大豆蛋白制得的再生蛋白纤维)
glycocholic acid 甘氨胆酸(五倍子浸出液)
glycocoll 氨基乙酸
glycogen 糖原
glycol 乙二醇
glycoladehyde 羟乙醛
glycol cellulose 羟乙基纤维素,纤维素乙二醇醚(耐水洗的浆料)
glycol derivatives 乙二醇衍生物
glycol dimethyl ether ⟨GDME⟩ 乙二醇二

甲基醚
glycol ester surfactant 乙二醇酯表面活性剂
glycol ether 乙二醇醚,一缩二乙二醇
glycol ether of cellulose 羟乙基纤维素,纤维素乙二醇醚(耐水洗的浆料)
glycolipid 糖脂
glycollic acid 羟基醋酸,乙醇酸
glycolysis 糖酵解
glycosidase 糖苷酶
glycoside 苷,配糖物
glycoside link 苷键
glycosylase 糖基化酶
glycylalanine 甘氨酰丙氨酸
Glyezin A 古来辛 A(硫代双乙醇,印花助溶剂,商名,德国巴斯夫)
Glyezin CD 古来辛 CD(醚的衍生物,纤维素纤维膨化剂,可用于分散染料涤棉织物印花,商名,德国巴斯夫)
glyoxal 乙二醛
glyoxal crosslinking 乙二醛交联剂
glyoxal finishing agent 乙二醛类整理剂
glyptal resin 丙三醇邻苯二甲酐树脂,丙苯树脂,甘酞树脂
Glytac(glycidyl trimethyl ammonium chloride) 缩水甘油三甲基氯化铵(纤维素纤维胺化剂,商名,法国制)
gm(=gm., gram, gramme) 克
GM(galactomannan, galacto-mannoglycan) 半乳甘露聚糖(多糖类糊料主成分)
GMA(glycidyl methacrylate) 甲基丙烯酸缩水甘油酯
Gminder fabrics 格曼达织物(各种耐磨织物,德国格曼达)
g-mol(gramme-molecule) 克分子
GMS(glycerol monostearate) 甘油单硬脂酸酯(树脂润滑剂)
GMW(gram molecular weight) 克分子量
goat hair 山羊毛
Gobelin blue 暗青绿色,深蟹灰[色]
Godamo 哥达摩(涤/黏仿麻色织布)
godown 仓库
gofer(=goffer) 拷花,轧花;起皱;皱褶
goffered cloth 波纹织物
goffering 褶裥处理
goggles 防护眼镜,护目镜
golbute 印花布,染色布(巴基斯坦制,做男衬衫或女衣用)
gold 金色;浅橄榄棕[色]
gold and silver damask 金银线织锦缎
gold and silver lace 金银线花边
gold bronze powder 金[铜]粉
gold brown 金褐色
gold compound 含金化合物
golden apricot 土黄[色]
golden fleece 浅米黄[色]
golden ocher 土黄[色]
golden yellow 金黄[色]
gold foil manufacturing 金箔制造
gold foil printing 金箔印花
gold foil printing machine 金箔印花机
gold foil specification 金箔规格
gold number 金值(保护胶体作用值)
gold powder printing 金粉印花
gold print 金粉印花
gold purple reaction 金紫色反应(用氯化亚锡溶液测试氧化纤维素)
Goldthwait test 格德斯维特(人名)测试法(利用两种直接染料的差异染色测试棉花成熟度)
goldtone effect 金枪效应(毛织品)
GOLEM【德】用电子数据处理系统存储

和检索文档和信息的方法
golfers 从套衫和针织物中回收的羊毛
golf green 中绿[色],纯绿[色]
goniometer 角度计测角计
goniophotometer 测角光度计
goodness of fit value 适度值《测色》
good rinsibility 易洗净性能（指活性染料浮色易于洗去而不沾污白地）
goods 成品;[成品]织物
goods liquor ratio 坯料染浴比;织物染浴比
good solvent 优良溶剂
good sudsing 易起泡沫的
goof proof 防滑跌（地毯用语）
goolbuti 东南亚印花粗布
goose down 鹅绒
goose grey 鹅灰色
GOR(gas oil ratio) 气油比
Gore-Tex 戈尔特克斯（用聚四氟乙烯微孔膜复合的防水透气织物,商名,美国制）
Gore-Tex fabric 戈尔特克斯[透气防水]布（用聚四氟乙烯微孔膜复合而成的织物,美国专利）
Gorrie type 边浆式（指染袜机式样）
GOST 【俄】俄罗斯国家质量标准
GOTS (Global Organic Textile Standard) 全球有机纺织品标准
gouache print 水彩颜料印花,水彩画印花
goureux 毛条印花
Gov.(government) 政府
governing motion [卷绕]调节装置
government〈Gov.〉政府
government ordinance 政府法令
governor 调节器,调速器;控制器;稳定器
GP (German Patent) 德国专利

GP(green products) 绿色产品
GPC(gas partition chromatography) 气相分配色谱[法]
GPC(gel permeation chromatography) 凝胶渗透色谱法
GPC(general purpose computer) 通用计算机
g. p. d.(grams per denier) 每旦纤维质量（用克表示）;克/旦,克每旦
gpl(gram per litre) 克/升
GPS(Global Positioning System) 全球定位系统（美国）
gpt(grams per tex) 每特斯克数
GR (genetic regression) 遗传退化
grab take-up 卷布装置
gracopolymerization 接枝共聚
gradation [颜色]层次;云纹,晕色;分等,分级
gradation of tone 色阶,色级
grade 等级;程度,阶段
grade assignment 定级
grade diplomate 资格认证,资格登记
graded temperature process 分段升温染色法
graded temperature raising 分段升温
grade labelling 质量标签
grader 分等工;分类器,分级机
grade standard 等级标准,品级标准
gradient 梯度;（温度,压力等的）变化率
gradient elution 溶剂梯度洗提法
gradient tube of density 密度梯度管
grading 分等,分级;（裁剪样板的）放大缩小
grading sieve 分级筛
graduated cylinder 量筒
graduated patterns 色彩渐变花纹

graduated plate 刻度板

graduated rheostat 分级变阻器,分级电阻箱

graduated scale 分度尺

graduated shade [雕刻]云纹

graduated-shrinkage fibres 逐步收缩性纤维(在聚酯分子中加入适当的共单体,使纤维在制作中收缩。有膨松起绒感,较好的染色亲和力和不易起球性)

graduates 量杯,量筒

graft 接枝;移植

graft block copolymer 接枝嵌段共聚物

graft copolymer 接枝共聚物

graft copolymerization 接枝共聚[作用]

graft copolymerization fibre 接枝共聚纤维

graft degree 文化度

grafted cellulose 接枝纤维素

grafted fibre 接枝纤维

grafting 接枝,文化;(针织衣片的)缝接

grafting reaction 接枝反应

graft polymer 接枝聚合物

graft polymerization 接枝聚合[作用]

graft polymerization of silk 丝纤维接技共聚

graft polymerization on surface 表面接枝聚合

graft printing 接枝印花

Graham's salt 格莱汉姆盐(可溶性偏磷酸钠,作软水剂用)

grain 格令(英制重量单位,7000格令等于1磅);纹理;颗粒,粒度

grain size 粒度

grain soap 粒皂

gram ⟨gm, gm.⟩ 克

gram atom 克原子

gram-calorie ⟨g-cal⟩ 克卡路里

gram equivalent 克当量

gramme ⟨gm, gm.⟩ 克

gram molecular weight ⟨GMW⟩ 克分子量

gram molecule 克分子

Gram-negative bacteria 革兰氏阴性细菌

gram per litre ⟨gpl⟩ 克/升

Gram-positive bacteria 革兰氏阳性细菌

grams/litre 克每升,克/升

grams per denier ⟨g/d, G/D, g. p. d.⟩ 克每旦,克/旦;每旦纤维质量(用克表示)

grams per tex ⟨gpt⟩ 每特克斯克数

Gram staining 革兰氏染色法(细菌染色法)

grandrelle fabric 双层防雨布(英国名称)

grandrelle yarns 混色花线(用异色单纱并捻)

granite 花岗石呢

granny print 古色小花印花布

granted patent 批准的专利

granular dyes 颗粒染料

granulate 粒料,切片

granulation 成料状;造粒

granule 粒子,颗粒

granule dyeing 原液着色

grape 葡萄紫[色](深暗红紫色)

graph 图解,[曲线]图;图表

graphic 图解的;印刷的;雕刻的

graphic agency 图片社

graphical analysis 图解分析

graphic arts 图片刻印艺术,书画刻印艺术

graphic interface 图形接口,图像接口

graphic meter 自动记录仪[器]

graphic print 广告衫印花

graphic processor 图形处理机《计算机》

graphic programming 图形程序编制,图形

程序设计《计算机》
graphic recorder ［计算机］图形记录器
graphic tufting machine 花式簇绒机
graphite 石墨；石墨灰色（暗蓝灰色，暗灰蓝色）
graphite-ceramic fibre reinforced phenolic resin 石墨—陶瓷纤维增强酚醛树脂
graphite fibre 石墨纤维
graphite gliding strips 石墨衬条
graphite grease 石墨膏；石墨［润滑］脂（含有2%～10%的无定形石墨，宜用于潮湿环境中的轴承）
graphite panel 石墨衬条
graphite prints 石墨印花布
graphite ring 石墨环
graphite stains 石墨沾污
graphitized fibre 石墨化纤维
graphoscope 电子计算机显示器
Graphsize 格拉夫赛兹（整理剂，不削弱纤维制品多孔性抗水性能的氨基甲酸乙酯分散体树脂，商名，美国制）
grass bleaching 草地暴晒漂白法《麻》；日光漂白
grass fibres 植物粗纤维
grass green 草绿［色］
grassing 草地暴晒漂白法《麻》；日光漂白
grass stains 植物沾污
grate 炉栅，炉篦
graticule 标线，量板《物》
grating 光栅；栅；格子
graver ［手工］雕刻刀
gravimetric 重量［分析］的
gravimetric analysis 重量分析
gravimetric dispensing system 重量调配系统
gravimetric method 重量法

gravimetry 重量［测定］法
graving depth 雕刻深度
gravity bottle 比重瓶
gravity circulation 重力循环，自然循环
gravity drainage 重力排水
gravity feed lubrication device 重力进油润滑装置
gravity flow 重力液流
gravity steam trap 重力式疏水器
gravity suit 宇宙服
gravity supply tank 重力进料槽
gravure applicator 刻纹辊给液机（低给液设备）
gravure coating 刻花［凹点］涂层
gravure copy screen 凹版网屏
gravure engraving 照相凹版雕刻
gravure printed paper ［热转移印花用］照相凹版印花纸
gravure printing 凹版印花；照相凹版印刷
gravure process 照相凹版制版法；照相凹版印刷法
gravure proof-press 凹版样张印刷机
gravure roller 凹凸轧花辊
gravure roller coating 凹版滚筒涂层
gravure rotary press 照相凹版轮转印刷机
gravy-proofed cloth 漆布，油布
gray allocation 坯布分批
gray inventory 坯布在库
grayish 带灰色的
grayish green 浅石绿［色］
grayish lavender 中莲灰［色］
gray lilak 珍珠白［色］，芡实白［色］
gray morn 银灰［色］
gray sand 米色
gray scale 灰度标尺；灰色分级卡
gray sour 漂布吃酸（弱酸处理以中和碱

性)
gray stock 坯布在库
gray yarn 原色纱
grease 滑脂,润滑脂
grease content 油脂含量
grease dyeing 生坯染色,含油脂染色(毛哔叽或棉经毛织物未经精练的染色)
grease fulling 含油脂缩绒,生坯缩呢
grease gun 油脂枪
grease lubricator 牛油杯;油脂枪
grease milling 含油脂缩绒[工艺],生坯缩呢[工艺]
grease nipple 滑脂嘴
grease [wool] price 原毛价格
grease proof 防油脂的,抗油脂的
greaser 加油工
grease recovery [羊毛]油脂回收
grease smudge 油渍
grease spot 油渍
grease stain 油渍,油污
grease trap 集油器《环保》
grease wool 原毛,含脂毛
greasy 带油污的
greasy blowing 生坯半蒸呢
greasy fabric 呢坯
greasy handle 油脂状手感
greasy luster 油脂光泽
greasy wool 含脂原毛
Great Britain〈G. B.〉英国;大不列颠
Greek alphabet 希腊字母表
Green Barrier 绿色壁垒,绿色屏障
Green cotton 绿色棉花(无污染棉花)
green discharge 绿色拔染
greendyes 绿色染料(天然蒽醌化合物,由真菌产生的次级代谢物)
green eater 吃绿染料(混色时会引起绿色染料催化褪色的黄色还原染料)
greener [色光]偏绿[色]
green fibre 绿色纤维,无污染纤维
green-glazing 绿色上光
green-ground prints 绿地印花布(一种爪哇蜡防花布)
green house 暖房,温室
greenhouse effect 温室效应
greening [使]成绿色
greenish 略呈绿色的
green off [使]成绿色
green process 绿色工艺,无污染加工工艺
green products〈GP〉绿色产品
green reflectance 绿色反射率
green resin 生树脂,未固化树脂
green sand 海绿石,海绿沙(天然软水剂)
Green Seal 绿色认证(美国)
green soap 钾皂(用植物油与KOH制成的肥皂)
green surfactant 绿色表面活性剂
green textiles 绿色纺织品(不含有害物质纺织品)
green tide 绿色浪潮
green unripe cotton 绿色未成熟棉
green verditer 铜蓝[颜料]
greige 灰褐色
greige cloth 坯布,本色布
greige goods 坯布
greige silk 生丝;生坯绸
grenadine red 石榴汁红[色](浅红橙色)
grey 灰色;坯布
grey balance 灰平衡(印刷分色用术语)
grey body 灰体(对任何辐射波长的吸收率保持不变并恒小于1)
grey cloth 坯布,白坯布
grey cloth inspection machine 坯布验布机

grey component replacement〈GCR〉 灰成分替代(印刷分色用术语)
grey dyeing 生坯染色
grey examination 坯布检验
grey fabric 坯布,白坯布
grey finish 坯布整理
grey goods 本色布,坯布
greying 发灰色的
greying inhibitors 沾污抑制剂
greying of textiles in drycleaning 织物干洗中的沾污
greying of textiles in washing 织物洗涤中的沾污
grey inspecting 坯布检验
greyish 带灰色的
grey mercerization 坯布丝光
grey radiator 灰体辐射体
grey receiver 灰体吸收体
grey room 坯布间
grey scale 灰色分级卡;灰色标度
grey scale for assessing change in colour 变色评级灰色样卡
grey scale for assessing staining of colour 沾色评级灰色样卡
grey scale for staining 灰色沾色样卡
grey scale of colour change 色泽变化灰色分级卡
grey series 灰色系列
grey setting 坯布定形
grey shirtings 本色细平布
grey sour 初酸洗(煮练后第一次酸洗)
grey stock 坯布库
grey store room 原坯储藏库
grey tube cloth 圆筒坯布
grey tufted carpets 原坯布簇绒地毯
grey washing 坯布水洗

grey weight 坯布重量
grey wool [套毛中的]浅花毛,灰色毛
grey yarn 原纱;原丝;本色[未漂白]纱
grid 炉栅;栅极;网格
grid dryer 帘子烘燥机
gridelin 深紫红色
grid sheet 格子表,格子纸
grid stirrer 框式搅拌器
grin 间隔
grind 研磨;研碎,粉碎
grinder 研磨机;碎木机;磨针布机
grinding burrs 磨刮刀毛口
grinding lathe 磨皮辊机
grinding machine 磨针布机;研磨机;粉碎机;磨花筒机;磨毛机;磨刀机
grinding machine for shearing knives 磨剪呢(毛)刀机
grinding mill 研磨机
grinding roll[er] 磨辊
grinding stone for printing rollers 印花滚筒磨石
grinning of ground 露地
grinning of white ground 露白(印疵)
grin-through [印浆]透过
grip 钳;布铗;夹,夹具
gripper Axminster carpet 夹片式阿克明斯特地毯
grip performance[of the fibre] [纤维的]抱合性能,黏附性能
gripping chain 布铗链
gripping clamp in screen printing 筛网印花用夹持器
grip sack 手提包,旅行包
grit 砂粒,粒度(砂纸)
grit covered paper roll 砂皮辊
grit size(= grit number) [砂纸]粒度号

数,粗细
grommet 环管;金属孔眼;垫圈;橡胶密封圈
grooved breaking machine 沟辊揉布机
grooved cylinder 有槽滚筒,槽纹辊
grooved drum washing machine 带沟槽圆筒洗涤机(针织品加工)
grooved expander roll 扩布辊
grooved nylon roll 有槽尼龙辊
grooved pattern 波纹花纹,凹凸花纹
grooved plate 沟槽板,凹型板
grooved pulley 槽轮,三角皮带轮
grooved roller 沟槽罗拉;有槽滚筒;有槽罗拉
grooved roller printing 凹刻滚筒印花
grooved timing rubber roller 凹凸绷布辊
grooving 切槽
grosgrain 罗缎(一种横棱纹织物);茜丽绸
gross calorific value [煤]高位发热量,总热值
gross efficiency 总效率
gross grain 湿格令(含水重量)
gross heat-conductivity 总热传导
gross mass 毛重,带皮重量
gross weight (gr. wt., GW) 毛重,带皮重量
gross yards 总码数
ground 地色(指印花布的染色部分)
ground bleaching 白地漂白(指印花布)
ground colour 地色;地色染料
ground dyeing 染地色(指印花布)
grounder 打底剂,浸渍剂
ground fabric 底布,地布
grounding 打底;接地《电》
grounding agent 打底剂,浸渍剂
grounding bath [色酚]打底浴
grounding in 打底(印花)
grounding machine 打底机
grounding with iron 铁盐打底
groundlines 地纹线(雕刻)
groundnut fibre 花生纤维
ground shade 地色
ground shade dyes 地色染料
ground state 基态
ground thread 地线,底线;芯线
ground tint 地色,地色调;[花布]浅地色
ground water 地下水
ground work 基础
group drive 集体传动
group number 基数(计算亲水亲油平衡值)
growth regulators 生长调节剂
GRP(glass reinforced plastics) 玻璃强化塑料
GR reagent(guaranteed reagent) 特级试剂
grub screw 平头螺丝,无头螺丝
grumment 衬垫,垫圈
gr. wt.(gross weight) 毛重,带皮重量
GSA(General Services Administration) 总务管理局(美国)
GSB test 棉花成熟度测试(用两种红色或绿色直接染料染色进行测定)
GTT(= Tg, glass transition temperature) 玻璃化[转变]温度
guaiacol 愈疮木酚,邻甲氧基苯酚
guaiac resin 愈创树脂,愈创木胶
guaiacum 愈创树脂,愈创木胶
Guanaco 驼毛(南美产)
guanamine-formaldehyde resin 胍胺—甲醛树脂(荧光颜料用树脂)
guanidine 胍

guanine 鸟粪素（珠光色素成分）
guanyl urea 脒基脲,胍基甲酰胺
guaranteed reagent〈GR reagent〉特级试剂
guard 金银花边束带；防护罩；防护器
guarded hot plated method 遮蔽热板法
Guardian Angel 阻燃衣料（商名）
guarding device 安全防护装置
guard mask 防护罩
guard rail 防护栏杆
guar gum 瓜耳豆胶（制印花浆或上浆用）
guar gum thickening 瓜耳豆胶糊（制印花糊料或上浆用）
gudgeon 耳轴,轴柱；螺栓；托架
gudgeon pin 轴头销；活塞销；十字头销
guest towel 高档细薄小手帕；揩指尖小巾,小方巾
guibert 漂白重亚麻布（法国制）
guidance system 导布装置,导布系统
guide board 挡板,导板
guide bracket 导座
guide cloth 导布,头子布
guide distance 导程
guide frame 导布架
guide line 标志线；（造纸毛毯的）标准线；准则
guide mark 标记,记号
guide piece 导布,头子布
guide pin ［转移印花机］循环导向凸头；链节隔距销子；导向杆,导纱棒
guide plate 导板
guide pulley 导轮
guider 导辊,导盘；导纱辊；导布辊；导布器
guide rod 导杆
guide roller 导辊；导纱辊
guide roller bearing 导布辊轴承

guide roller with under liquor seal 带液下封口的导布辊
guide stops ［手工印花］筛网固定装置（对花用）
guide tray 导盘
guide tube 导纱管；引导管
guideway 导向槽,导向套筒
guide wheel 导轮
guiding device 引导装置
guiding groove 导槽,键槽
guiding nut 导螺母,活令
guluronic acid 古罗糖醛酸（海藻酸钠糊料主结构）
gum acacia 阿拉伯树胶（印花糊料）
gum Arabic thickening 阿拉伯树胶糊
gum content 胶含量
gum dragon 龙［须］胶,黄蓍［树］胶（用作印花糊料）
gum elastic 弹性树胶
gum gatto 刺槐豆胶
gum karaya 刺梧桐树胶（可用作印花糊料）
gum-mastic 乳香树胶
gumming 黏接
gumming force 胶着力
gumming machine 涂胶机；胶合机
gum percha 杜仲胶（天然胶乳,类似橡胶）
gum precipitate 皂液沉淀
gum resin 树胶［树］脂
gum roller 清水浆滚筒,白浆滚筒
gums 树胶（植物胶类）
gum senegal 类似龙胶的印花糊料
gum tape 松紧带
gum tragacanth 龙［须］胶,黄蓍胶（印花糊料）

gum turpentine 松节油
gum yarn 宽紧线,橡筋线
guniting 喷涂；喷浆；喷镀
gun oil burner 枪式喷油燃烧器（采用机械雾化）
Gurlitzer steamer 一种星形蒸化机（从侧面进星形架）
gusset 角撑板
gusset plate 角撑板
gusset stay 角撑条

gusset trough 三角形槽（卧式轧辊间形成的一种低容量轧槽）
gut of insect 昆虫内脏（蛀虫）
gutter 沟,槽
GW(gross weight) 毛重,带皮重量
gymnastic uniform 体操运动服
gypsum 石膏,硫酸钙
gyro-horizon 回转水平仪,陀螺地平仪
gyroscope 陀螺仪；旋转仪
gyroscopic moment 回转力矩

H

H_B（=H_b, H. Br., Brinell hardness） 布氏硬度

haarlem checks 亚麻彩格窗帘布（荷兰名称）

habberley carpet 低级染色毛纱地毯（英国名称）

habilatory art 服装艺术

habilimentation 服装艺术；穿着打扮

habotai（=habutae, habutaye） 纺绸,电力纺（日本名称）

hachure 晕线,影线；刻线；痕迹

hachure graving with pentagraph 缩小机拉斜线（花筒雕刻）

H-acid H-酸,1-氨基-8-萘酚-3,6-二磺酸

Hacoba bulking system 哈科巴连续膨化变形法（商名,德国制）

haddat 棉印花方巾

haematein 氧化苏木精（黑色植物染料）

haematin 羟基血红素

haematoxylin 苏木精

haemostatic fibre 止血纤维

haircord carpet 发毛起圈地毯

hair dyes 毛发染料

hair felt 毛毡

hair hygrometer 毛发湿度计

hairiness 羽毛感,毛羽

hairiness index 毛羽指数

hair-in-filling ［布面］纱线起毛（织疵）

hairless 无毛［发］的

hair like 似毛的

hair linings 动物纤维材质衬里

hair processing machine 毛绒处理机

hairy 多毛的,毛状的（织物表面毛糙多毛）

hairy acrylic fibre 毛型聚丙烯腈系纤维

hairy appearance 外观发毛（纱线或织物表面毛头突出）

hakistery 黑白地色印花（伊朗衣料印花）

halation 晕光作用,晕影

half-bath-finishing 半浴整理（即面轧法整理）

half bleach 半漂白；半漂白亚麻布

half boiling 半练,半脱胶

half chintz 半擦光印花布（地色为白色或乳白色）

half coupling 半联轴器

half-crystallization time 半结晶期

half decay time 半衰期

half degummed silk 半脱胶丝

half degumming 半脱胶

half discharge effect 半拔染效应

half discharge printing 半拔印花

half-drop design 半格组织图案

half-drying 半干

half-dyeing time 半染时间

half-emulsion thickener 半乳化糊

half-finished management 半成品管理

half-life 半周期,半衰期

half made 半完工的定制服装；半制成品

half mercerizing 半丝光处理

half-migrating time 半泳移时间

half nut 对开螺母
half period 半周期,半衰期
half-power 半功率
half resist effect 半防染效应
half salts of dicarboxylic acids 二羧酸半盐（共反应型催化剂）
half set 半定形
half-shadow photometer 半阴光度计
half tone 半色调,中间[色]调;浓淡点图（在网屏上用网点来决定色调的深浅）
halftone autotype screens [用感光法制成]半色调筛网
half-tone block 网目凸版
half-tone colour 中间色
halftone design 照相网点设计（由网点大小与密度决定色调层次）
halftone dot 网点
half tone effect 半色调效应
halftone negative 网目阴图片;半色调底片
halftone positive 网目阳图片;半色调片
halftone process 半色调工艺
halftone resists 半色调防染
halftone screen 网屏;网点版
half-value decay time of the charge〈HVDT〉电荷半衰期
half wash off 半净洗（炭化后洗除残液, pH 在 2.0～2.5）
half-wave potential 半波电位
half-wave rectifier 半波整流器
half width 半幅
half-wool 棉毛交织物
half wool dyeing 棉毛混纺织物的染色（用两种不同性质染料）
half-worsted [羊毛]半精梳
halide lamp 卤化物灯

Hall effect 霍尔效应《电》
hallucination 幻觉
halo.（halogen） 卤素
halo.（haloid） 卤[族]的
haloalkylphosphate 卤代烷基磷酸酯
haloalkylphosphine 卤代烷基膦
halocarbon 卤化碳
halochromism 加酸显色,卤色化[作用]
haloes 晕疵,渗开圆点《印》
halo fibre 光晕纤维
halogen〈halo.〉 卤素
halogen addition 卤素加成
halogenated acrylic ester 卤代丙烯酸酯
halogenated alkyl phosphate ester 卤代烷基磷酸酯
halogenated aromatic compound 卤代芳族化合物
halogenated ethylenic compound 卤代乙烯化合物
halogenated paraffin 卤代烷[属]烃
halogenation 卤化作用
halogen compound 卤素化合物
halogen family 卤族
halogen-free photochemical bleaching 无氯光化学漂白
halogen lamp 卤素灯
halohydrin 卤代醇
halohydrocarbons 卤代烃（溶剂）
haloid〈halo.〉 卤[族]的
haloing 晕渗现象
halo method 抑菌环宽度法（抗菌试验法）
Halon 海伦（高反射白色物质）
halophosphoric acid 卤化磷酸
halophosphorus compound 卤化磷化合物
halopolymer 含卤聚合物
halo printing 晕纹印花

halo test method 抑菌圈试验法
halt 停留,停止;停机
hammer clip 槌式布铗(拉幅机链条竖行时握持布边用)
hammer fulling mills 槌式起绒机
hammer washer(= hammer washing machine) 槌式洗涤机
hand [织物]手感;手工的;手控
hand bag 手提包,旅行袋
hand blocked fabric 手工模版印花织物
hand block printing 手工模版印花
hand brake 手制动器(一般指紧急制动器或停车制动器)
hand builder 改进手感的整理剂
hand calculator 手摇计算机
hand coater 手工涂浆装置
handcraft method 手工方法
hand drawing 手工描绘;手工印花
hand-drawn 抽花刺绣品,抽绣品
hand dyeing 手工染色
hand engraving 手工雕刻,手工修理
hand evaluation 手感评定
Hand Evaluation and Standardization Comitee(H. E. S. C.) 手感评定与标准化委员会
hand feeling 手感
hand-felling 手工缝边
hand flocking device 手工植绒装置
hand handle 手感;手柄
hand held calculator 携带式计算器,手提计算器
hand hemming 手工缝边
handhole 手孔
handing 手感
handkerchief 手帕
handkerchief print 手绢印花花样,百家衣花样(补缀般组合的印花样)
hand knife cloth cutting machine 手工割布机
hand knitted hosiery 手工针织品
hand knitting 手工针织,手工编织
hand knitting wool(= hand knitting yarn) 手编绒线,手编毛线
hand knob 手动按钮
handle [织物]手感;手柄
handle determination 手感测定
handle lever 手柄杆,开关杆,调节柄
handle modifiers 手感改良剂
handle-o-meter 织物手感测定器
handle tester 风格[试验]仪
handling 手感;搬运;装卸
handling equipment 搬运设备
handling property 使用性能,操纵性能
hand mercerizing machine 手工丝光机
hand meter 手感测定仪
hand microtome 哈氏切片机,手工切片机
hand modification 手感改良
hand modifier [树脂整理]柔软剂;毛感改良剂
hand-operated oil pump 手揿油泵
hand painting 手绘,手工印花
hand printing 手工印花
hand properties 手感
hand pump 手泵
handrail 扶手,栏杆(印染机走台扶手、栏杆)
hand raising 手工起毛
hand raising frame 手工起毛机
handrub 手搓
hand sample 小样
hand screen-printed fabric 手工筛网印花织物

hand screen printing 手工筛网印花
hand sewing 手工缝纫
hand shake 信号交换(通过专用线而不是数据线进行的信号交换)《计算机》
hand shrink 手工收缩
hand table printing 手工台板印花
hand tester 手感测试仪
hand touch 手感
hand towel 手巾
hand value 织物手感值
hand wash 手洗
handwheel 手轮;操纵盘
handwork 手工操作;精致工艺
hand wrinkle meter 织物折皱手感仪
hang degumming 挂练,吊练
hang dry 悬干,挂干
hang dyeing 吊染,挂染
hanging dryer 悬挂式烘燥机
hanging hank dyeing machine 悬挂式绞纱染色机
hanging nozzle drier 悬挂式喷气烘燥机
hanging property 悬挂性
hanging rooms 悬化室《染整》
hanging rotary dyeing machine 悬挂式旋转染色机
hanging rotary dyer 悬挂式旋转染色机
hangings 帘布,挂布
hanging silk dyeing machine 悬挂式染绸机
hang scouring 吊练,竿练
hang scouring bowl 挂练槽
hang tag [商品]使用,保养说明标签
hang thread 纱头吊痕(染疵)
hang-up 中止,暂停;意外停机
Hangzhou silk gauze 杭罗,杭州纱罗
hank 绞;亨克,亨司(英制长度单位,棉纱线为840码❶,毛纱线为560码)

hank brushing machine 手工刷毛机
hank bulking technique 绞丝膨化技术
hank can 绞纱桶
hank cylinder 绞纱辊,绞纱滚筒
hank drier 绞纱烘燥机
hank drying machine 绞纱烘燥机
hank dyed 绞[装]染[色]的
hank dyeing 绞纱染色
hank filling machine 绞纱上浆机
hank finishing 绞纱整理
hank handling 绞纱搬运,绞纱处理
hank lustring machine 绞纱上光机
hank mercerizing machine 绞纱丝光机
hank pole 绞纱棒,绞丝棒
hank printing 绞纱印花
hank printing machine 绞纱印花机
hank rod 绞丝棒;绞纱棒
hank scouring 绞纱精练
hank scouring machine 绞纱煮练机,绞纱洗涤机
hank sizing 绞纱上浆
hank sizing machine 绞纱上浆机
hank spraying washing machine 绞丝淋洗机
hank spreading device 绷绞装置
hank squeezer and dryer 绞纱轧水烘燥机
hank stand 绞纱架
hank washer(=hank washing machine) 绞丝洗涤机;绞纱洗涤机
hank washing train 洗[绞]线机组
hank yarn bleaching machine 绞纱漂白机
hank yarn washing machine 绞纱洗涤机

---

❶ 1码=3英尺=91.44厘米

Hansa yellow 汉沙黄(黄色颜料)
HAP(hydrolytic animal protease) 水解动物蛋白酶[通称]
hapten 半抗原《生物》
harbour blue 蓝绿[色]
hard base 硬碱(电负性高,难极化,不易失去电子)
hard block 刚性嵌段(合成纤维分子结构)
hard bowl 硬滚筒,硬轧辊(指金属轧辊)
hard-burned 强烈烧毛的[织物]
hard cure 剧烈焙烘,硬焙烘
hard detergent 硬洗涤剂(难以生物降解的洗涤剂)
hardener 硬化剂,固化剂
hardening 硬化;固化,凝固;焙烘;淬火
hardening agent 硬化剂,固化剂
hard feel 硬挺感,粗硬感
hard filled bowl 硬质层压轧辊
hard [face] finish [精纺呢绒]光面整理
hard finishing 厚浆整理,硬挺整理
hard handle 粗硬手感
hardimeter 水质硬度测试器
hardness 硬度
hardness-scale 硬度标度
hardness test 硬度试验
hardness tester 硬度试验计
hardometer 硬度计
hard segment 硬段(高分子结构)
hard soap 硬皂
hard-to-handle fabrics 难以处理的织物(娇嫩的织物)
hardware 硬件(计算机的电子、机械、电、磁等元件或部件);金属器具,金属附件
hardware check 硬件检验

hardware monitor 硬件监视器;硬件监督程序
hard waste reclaiming line 硬质废料回用生产线
hard water 硬水
hard waxing [纱线]上硬蜡整理
hard-wearing 耐磨的
hard winding 卷绕过紧
harmful metals 有害金属
harmful pollutants 有害污染物
harmful substance tested textiles 纺织品有害物测试
harmonic drive 谐波传动
harmonic oscillation 谐振
harmonic speed changer 谐波减速装置
harmonious colour 和谐色
harmonization 协调
harmony of analogy of hue 色相类比调和
harmony of analogy of tone 色调类比调和
harmony of gradation of hue 色相连续调和
harmony of succession of hue 色相连续调和
Harrisset process 哈里塞特羊毛[中性氯化]防缩法(商名,美国)
Harris tweed 海力斯粗花呢
harsh [手感]粗糙
harsh hand 粗糙手感
harshness [手感]粗糙度
harsh yarn 硬丝,粗糙丝(黏胶丝疵点);糙纱线
Hart moisture meter 哈特湿度计
hartshorn black 鹿角黑[色]
harvest gold 芥末黄[色],橄黄[色]
hasp 搭扣
hat 帽

Hatay stabilization process 哈泰定形法
hatch 天窗,格子门;开口,升降口
hatching lines 阴线(雕刻印花滚筒时,雕刻成与滚筒呈一定角度的平行线,以保证获得印浆)
hatching screen 阴线筛网(又称影线、网格线)
hat-forming machine 帽坯机
hat proofing 帽子硬挺整理,帽檐上浆
hat steamer 蒸呢帽锅
Hawaiian print 夏威夷印花花样(有大而鲜艳的热带植物)
hawking 液下浸轧染色(毛织物染靛蓝的旧方法)
hawking machine [毛织物]平幅靛蓝染色机(织物在液下循环浸轧)
hazard 公害;危险[性]
hazardous decomposition products 危险的分解产物;有害的分解产物
Hazardous Materials Indentification System〈HMIS〉危险品鉴定方法
hazardous substance 危险物质
hazardous substance values 有害物质数值,危险物质数值
haze 混浊;烟雾
haze blue 烟雾蓝[色]
hazel 榛子棕[色](浅黄褐色)
Hazen-colourimeter 哈森比色计
haziness printing 朦胧印花
hazy effect(＝hazy view effect) 模糊效果,朦胧效果《印》
HBCD(hexabromocyclododecane) 六溴环十二烷(涤纶阻燃剂)
H bonding 氢键合
HBP(hyper-branched polymer) 超支化聚合物

HB yarn(high bulk yarn) 高膨体纱
HD(＝H. D., heavy duty) 厚型;牢固型;高效型;重型;重级;重载
HD(＝H. D., high density) 高密度
HDPE(high density polyethylene) 高密度聚乙烯
HDT(heat distortion temperature) 抗变形温度
HE(＝H. E., high efficiency) 高效率
HE(high elasticity) 高弹性
head amplifier 前置放大器;摄像机放大器
head box 压力盒;流料箱;高位调浆箱
head colours 着色染料
head end 车头
head-end drive 车头传动
header 头部;主管道;顶盖,端板
header box 高位箱;高位槽
head framing 车头机架
headless setscrew 无头止动螺丝
headliner 车顶衬里
head plate 封头,端板
headscarf(＝head shawl, head square) 头巾;围巾
headstock 车头;主轴箱
headstock side frame 车头墙板
head-tail shading 头尾色差
head tank 高位槽
head water 水源
Health and Safety Regulations〈HSR〉卫生与安全管理条例(英国)
health care textile(＝health protection textile) 保健纺织品
health-promoting effect 健康促进作用
healthy fashion 健身时装
heaping machine 自动堆布机

hearing aids 助听器
heart cam(=heart-shaped cam) 心形凸轮
hearth rug 壁炉地毯
heat absorbent 热吸收剂
heat absorbing dyes 吸热染料（吸红外染料）
heat-absorbing reaction 吸热反应
heat absorption 吸热,热吸收
heat account 热平衡计算
heat accumulator 蓄热器
heat-activatable 遇热起作用的（反应）
heat ageing 热老化
heat and power generation 热电联产
heat balance 热平衡
heat balance diagram 热平衡图
heat bonded fabric 热黏合织物
heat-bonded wadding 热熔絮棉
heat bonding 热黏合
heat booster 加热器
heat build-up 热积聚
heat calculation 热力计算,热平衡计算,热工计算
heat capacity 热容[量]
heat carrier 热载体,载热体,载热剂
heat carrier boiler 热载体锅炉
heat-carrying agent 热载体,载热体,载热剂
heat change 热交换
heat channel 热桥;热沟,热槽;热通道
heat colour 热色;暖色
heat conduction 热传导
heat conductivity 导热性;导热率
heat consumption 耗热量
heat content 焓;热函
heat contraction 热收缩
heat conveyance system 传热系统

heat cooling effect 蒸发冷却效应
heat cure 热硫化;热固化
heat deterioration 热恶化,热退化
heat diffusivity 导热性;热导率
heat-dispersing surface 散热面
heat dissipation 热消散,热散逸
heat dissolution 溶解热
heat distortion 热畸变,热扭变
heat distortion temperature〈HDT〉 抗变形温度
heat durability 耐热性
heat economizer 热交换器,省热器
heated iron shoe 靴形热压板
heated lips 加热唇封
heat effect 热效应
heat efficiency 热效率
heat effluent 温排水《环保》
heat electric couple 热电偶
heat endurance 耐热性
heat energy 热能
heat engineering 热力工程
heater 加热器,发热器;热源
heater coils 加热盘香管;盘管加热器
heater plate 加热板
heat escaping 热逸散
heat exchange 热交换
heat exchange function 热交换功能
heat exchanger 热交换器
heat fading 热变色
heat fast 耐热的
heat fastness 耐热性,耐热牢度
heat filter 滤热片
heat fixation 热定形
heat flow diagram 热流图
heat flow meter 热流量计
heat flow transducer 热流量传感器

heat flux 热通量
heat flux transducer 热通量传感器
heat fusion press 热熔转印机,热熔压烫机
heat generator 热发生器
heather dyeing 染后混色
heather effect 混色效应
heather mix carpet 混色地毯
heather mixture 混色毛纱
heather violet 石南紫色
heat inactivation of enzyme 酶的热钝化
heating 加热
heating by high frequency 高频加热
heating chamber 加热室,燃烧室
heating coil 加热蛇管;蛇管加热器,盘管加热器
heating-cooling cycle 加热—冷却循环
heating element 加热元件,发热体
heating equipment 加热装置
heating fabric 供暖纺织品(如电热褥)
heating generator 热发生器
heating installation 暖气设备
heating jacket 加热夹套
heating medium 载热体,加热介质,热媒体
heating pipe 热气管,暖气管
heating press 热压机
heating radiator 热辐射器
heating rate [发]热率
heating sludge digestion tank 加热污泥消化池《环保》
heating surface 加热面
heating test 加热试验
heating value equivalent to primary energy 等价热值
heating ventilation and air condition 〈HVAC〉采暖通风和空调系统
heat input 输入热量
heat-insulated 隔热的,绝热的
heat insulated board 绝热板
heat insulating ability 保暖性
heat insulating felt 隔热毛毡
heat insulating jacket 绝热夹套
heat insulating wadding 绝热填料,绝热衬垫
heat insulation 热绝缘
heat insulation board 隔热板
heat insulation material 绝热材料,保温材料
heat lamination 火焰胶合,焰熔层压
heat liberation 放热
heat loss 热消耗
heat memory feature 潜热性,储热性
heat modification 热变性
heat moisture comfortable clothing 热湿舒适性服装
heat moisture comfortable clothing evaluating method 热湿舒适性服装评价方法
heat-moisture property of garment 服装的热湿性能
heat of activation 活化热
heat of adsorption 吸附热
heat of association 缔合热
heat of combustion 燃烧热
heat of crystallization 结晶热
heat of decomposition 分解热
heat of desorption 脱热,减热
heat of diffusion 扩散热
heat of dilution 稀释热
heat of dissociation 解离热
heat of dissolution 溶解热
heat of dyeing 染色热函,染色热

heat of explosion 爆炸热
heat of friction 摩擦热
heat of fusion 熔化热
heat of hydration 水合热
heat of ionization 电离热
heat of linkage 键合热
heat of melting 熔融热
heat of mixing 混合热
heat of oxidation 氧化热
heat of polymerization 聚合热
heat of precipitation 沉淀热
heat of radiation 辐射热
heat of reaction 反应热
heat of reduction 还原热
heat of solution 溶解热
heat of sublimation 升华热
heat of swelling 溶胀热
heat of transition 转化热
heat of vapourization 汽化热
heat of wetting 润湿热
heat output 热功率,发热量,输出热量
heat pickup 吸热
heat pipe 热导管,热管
heat pipe exchanger 热管式热交换器
heat plate 加热板
heat [transfer] printing 热转移印花
heat production 热生产
heat proof 防热的
heat proof agent 耐热剂,防热剂
heat protective glove 防热手套
heat pump 热泵
heat quantity 热量
heat radiation 热辐射
heat reactive 热反应性的
heat reactive polyurethane 热反应性聚氨酯

heat reclamation(＝heat recovery) 余热回收,热回收
heat recovery boiler 余热锅炉,废热锅炉
heat recuperation 热量回收
heat reflective finish 热反射整理
heat relaxation 热松弛
heat release kinetics 热释放动力学
heat reserve 蓄热性
heat reserve and retention fibre 蓄热保温纤维
heat reserve and retention function 蓄热保温功能
heat resistance 耐热性,耐热度
heat resistant fabric 耐热织物
heat resistant fibre 耐热性纤维
heat-resistant finish 耐热整理
heat resistant polymer 耐热聚合物
heat resistant screen 耐热性圆网
heat retaining power 保温性,保温能力
heat retention 保暖性
heat retention fabrics 保温织物,保暖织物
heat run 热试车(指通入蒸汽、热水等进行工艺试车)
heat saving 节热
heat seal 热封;熔焊,熔接
heat-sealable garment interlining 服装保暖衬
heat seal adhesive 热封黏着剂
heat sensitive discolour fabrics 热敏变色织物
heat sensitive dyes 热敏染料
heat sensitive paper strips 热敏试纸,测温纸
heat sensitive printing 热敏印花,热变色印花
heat sensitive sealing 热敏密封

heat sensitive sensor 热敏传感器
heat sensitivity 热敏[感]性
heat sensitizer 热敏剂
heat sensitizing agent 热敏[感]剂
heat sensor 热传感器,热感元件
heat set 热定形
heat-set clothing (＝heat-set garments) 热定形服装
heat set pleat 热定形褶裥
heat set stretch yarn 热定形弹性丝
heat settability 热定形性,热变定性
heat setter 热定形机
heat setting 热定形
heat setting machine 热定形机
heat shield 热屏;热保护层;隔热罩
heat shock 热振荡,热冲击
heat shrinkable 可热收缩的
heat-shrinkable fibre 热收缩纤维
heat shrinkage 热收缩率;热收缩
heat shrink ratio 热收缩比
heat sink 吸热层(吸收热量的任何气体、固体或液体区域)
heat source 热源
heat stability 热稳定性,耐热性
heat stabilizer 热稳定剂
heat storage 蓄热
heat-storage and thermo-regulated textile 蓄热调温纺织品
heat stress 热应力
heat stretching 热拉伸
heat stretching machine 热拉伸机
heat transfer 传热,导热
heat transfer coefficient 传热系数
heat transfer fluid 热传送液体(热载体)
heat transfer loop 传热回路
heat transfer medium 传热介质,导热介质

heat transfer oil 导热油
heat transfer paper 热转移纸
heat transfer printing 〈HTP〉热转移印花
heat transfer printing machine 热转移印花机
heat transfer rate 传热率
heat transmissibility 传热性
heat transmission 传热,热传递
heat transmission testing apparatus [织物]传热性试验器
heat transmitter 传热器,热传递器
heat transport 热传递,传热(穿着性能)
heat trap 吸热器
heat treatment 热处理
heat-up rate 升温速度
heat up time 加热时间
heat utilization ordinance 热能利用条例
heat utilization rate of enterprise 企业热能利用率
heat value 热值,卡值
heat welding 热熔接
heat weld non-woven fabric 热[熔]黏[合]非织造布
heat wheel 热轮
heat wheel exchanger 热轮式热交换器
heat-yellowing 加热泛黄
heat-yellowing test 加热泛黄试验
heat zone dwell system 热堆系统
heavily consolidated cloth 重缩绒织物
heavily felted cloth 重缩呢织物
heavily filled 上重浆的
heavily milled 重缩绒的,重缩呢的
heaviness 沉重感
heavy 深色的
heavy air 压缩空气
heavy chemicals 重化学品(大量生产的酸

碱等无机化学品)
heavy cloth 厚织物
heavy coating 高效涂层(防水整理)
heavy colour 浓色,深色
heavy cotton drill 厚卡其,厚棉斜纹布
heavy coverage [printing] 大面积覆盖印花
heavy duty〈HD,H.D.〉厚型;牢固型;高效型;重型;重级;重载
heavy duty bearing 重型轴承
heavy duty material 厚料
heavy duty padder 重型轧车(压力4~10吨)
heavy duty printing blanket 高负荷印花橡胶毯
heavy duty rotary screen 高负荷圆网,高强度圆网
heavy duty service 高负荷运行
heavy duty soap 强效肥皂
heavy duty synthetic detergent 强效合成洗涤剂
heavy duty water proof finish 高效防水整理
heavy fabric 厚重织物
heavy filling 粗纬(织疵)
heavy finishing 重浆整理,加重整理
heavy goods 厚重织物(指工业用布和涂层织物等)
heavy hydrogen 重氢
heavy impregnant 强效浸渍剂
heavy impregnation 强效浸渍
heavy laundering 强效洗涤
heavy liquid 浓液
heavy metal 重金属
heavy metal content 重金属含量(生态学—毒理学评估参数)

heavy metal pollutant 重金属污染物
heavy metal residues 重金属残余(生态学—毒理学评估参数)
heavy metal soap 重金属皂(除钾、钠外的金属皂)
heavy metal toxicity 重金属毒性
heavy oil 重油
heavy padder 重型浸轧机
heavy padding 重轧
heavy pad mangle 重型轧车
heavy print 饱满深色印花
heavy repair 大修
heavy shade 深色
heavy singeing 强烧毛
heavy sizing 重浆
heavy solution 浓溶液
heavy squeezing 重挤轧
heavy stock solution 浓母液
heavy water 重水
heavy wax 重质蜡
heavy weight 厚重
Heberlein 黑贝莱因整理(细薄织物耐久硬挺整理,商名,美国制)
HEC(hydroxyethyl cellulose)羟乙基纤维素
Hecospan 卷曲合纤丝及聚氨基甲酸乙酯弹性丝的包芯加捻丝(商名)
Hecowa 海科华耐洗整理(针对仿麻织物,商名,瑞士制)
Hecowa transfer process 海科华转移印花法(用于涤棉混纺织物)
hedge green 灰绿[色]
HEDTA(N-hydroxyethyl ethylene diamine triacetic acid) N-羟乙基乙二胺三乙酸(有机螯合剂)
height 高度

heighten [颜色等的]加深
height of the liquor 液面高度,液位
helianthin[e] 半日花素,酸性黄,甲基橙
helical 螺旋形的,螺旋线的
helical conformation 螺旋构象
helical crimp 螺旋[形]卷曲
helical fan 螺旋式风机
helical gear 斜齿轮,螺旋齿轮
helical scroll 螺旋形扩幅器
helical spur gear 斜齿轮
helical wheel 斜齿轮;螺旋轮
helical worm gear [螺旋]蜗轮
helicoidal roller 螺旋扩幅辊
helicoidal structure 螺旋[形]结构
helicoid screw 丝杠,蜗杆
heliocentric-type reducer 行星齿轮减速器
Heliofast printing machine 赫利奥法斯特印花机(一种多背辊式圆筒印花机)
Helio-Klischograph gravure 照相制版;电子雕刻
heliotrope 天芥菜紫(红青莲色)
helix type stirrer 螺旋型搅拌器
Helizarin Binder ET 海立柴林黏合剂ET(丙烯酸类黏合剂,牢度优良,手感柔软,商名,德国巴斯夫)
Helizarin Binder FWT 海立柴林黏合剂FWT(手感柔软,耐干洗,适用于涂料染色,商名,德国巴斯夫)
Helizarin Binder TS 海立柴林黏合剂TS(手感柔软,耐干洗,商名,德国巴斯夫)
Helizarin Binder TW 海立柴林黏合剂TW(丙烯酸类自交链黏合剂,商名,德国巴斯夫)
Helizarin Binder UD new 海立柴林黏合剂UD新(运行性能好,可用于防拔染印花,商名,德国巴斯夫)
Helizarin EVO P-100 海立柴林印墨EVO P-100(喷墨印花用颜料型印墨,商名,德国巴斯夫)
Helizarin pigment colours 海立柴林涂料(印染用涂料,商名,德国巴斯夫)
helmets 头盔
Helmholtz free energy 赫尔姆霍兹自由能
help desk 求助台《机》
hematein 氧化苏木精(黑色植物染料)
hematin 羟基血红素
hematine 苏木浸膏,苏木萃,苏木因
hematoxylin 苏木精
hemi- (词头)表示"半"
hemiacetal 半缩醛
hemiacetal group 半缩醛基
hemiacetal linkage 半缩醛键[合]
hemicellulase 半纤维素酶
hemicellulose 半纤维素
hemicolloid 半胶体
hemicrystalline 半结晶的,半晶状的;半结晶性
hemihydrate 半水合物
hemimatt fibre 半消光纤维
hemispherical head 球形封头
hemodialysis 血液透析
hemoglobin 血红素,血红蛋白
hemostatic effect 止血效应
hemostatic textiles 止血纺织品
hemp 大麻
He-Ne laser 氦—氖激光器
henry 亨利(电感单位)
Henry's partition law 亨利分配定律
heptane 庚烷
herbicide 除莠剂,除草剂
Hercosett (HERC) 赫科塞特(聚酰胺表

氯醇树脂,商名)
Hercosett 57 赫科塞特 57(阳离子型反应性聚酰胺表氯醇的水溶液,羊毛及其混纺织物的防皱、防缩、抗静电整理剂,商名,美国制)
Hercosett process 赫科塞特羊毛防缩法
Hercosett treatment 赫科塞特防缩处理
hermetic seal 气密,密封
heron 深紫灰[色]
herringbone 海力蒙,人字呢
herringbone effect 人字花型
herringbone gear 人字齿轮
herringbone type evapourator V 形管蒸发器
hertz〈Hz〉赫[兹](频率单位,周/秒)
H. E. S. C.(Hand Evaluation and Standardization Committee) 手感评定与标准化委员会
hetero- (词头)表示"异""杂"
heteroaggregation system 不匀凝聚法
hetero bireactive dyes 异双活性基染料
hetero-chromatic stimulus 异色刺激
heterocyclic 杂环的
heterocyclic compound 杂环化合物
heterocylic dyes 杂环(结构)染料
hetero fibre 异质[复合]纤维,共轭纤维
hetero filament 异质丝,共轭丝
heterofil bicomponent fibre 异质双组分纤维
heterofilm 异相薄膜,非均相膜;复合薄膜
heterogeneity 不均匀性,多相性;异种,杂质
heterogeneous 不均匀的;多相的;异类的
heterogeneous colours 多色性染料(染料中含有多种颜色)

heterogeneous oxidation 非均相氧化
heterogeneous phase 不均一相
heterogeneous polymerization 多相聚合[作用],非均相聚合[作用]
heterogeneous ring compound 杂环化合物
heterogeneous-slurry polymerization 非均相悬浮聚合[作用]
heterogeneous X-ray 多色 X 射线
heterolysis 异种溶解;外力溶解
heteromorphic 多晶[型]的
heteromorphism 多晶[型]现象
heteronuclear 杂环的
heteropolysaccharide 杂聚糖化合物
heuristic method 探索法
Hevea latex 天然胶乳
Hevea rubber〈H rubber〉天然橡胶
hex[a]- (构词成分)表示"六""己"
hexabiose 己二糖(属二糖类)
hexabromocyclododecane〈HBCD〉六溴环十二烷(涤纶阻燃剂)
hexachloroendo methylene tetrahydrophthalic acid 氯菌酸(阻燃剂,增塑剂,防霉剂)
hexadecyl trichlorosilane 十六[烷]基三氯硅烷(有机硅中间体)
hexadecyl trimethyl ammonium bromide〈HTAB〉十六烷基三甲基溴化胺(表面活性剂,杀菌剂)
hexads 六色配色(四个色相再加黑、白两个色相)
hexafluoroethane 六氟乙烷(冷冻剂)
hexafluoroisopropanol 六氟异丙醇(聚酯纤维溶剂)
hexagonal nut 六角螺母
hexagonal oil screw 六角油眼螺钉
hexagonal roller 六角辊
hexagon nipple 六角外螺丝

hexahydrobenzoic acid 六氢化苯[甲]酸,环己基甲酸
hexalin 环己醇,六氢苯酚
hexametaphosphate 六偏磷酸钠
hexamethoxymethyl melamine 六甲氧甲基三聚氰胺
hexamethylene adipamide 己二酰己二胺（锦纶66的结构单元）
hexamethylene diamine〈HMD,HMDA〉己二胺（锦纶66原料）
hexamethylene diammonium adipate 己二酸己二胺盐,锦纶66盐
hexamethylene diisocyanate 二异腈酸己二酯
hexamethylene glycol 1,6-己二醇
hexamethylene tetramine 六亚甲基四胺
hexamethylol melamine〈HMM〉六羟甲基三聚氰胺
hexamethyl phosphoric triamide 磷酰六甲基三胺（紫外线吸收剂）
2-hexanone 2-己酮;甲基丁基甲酮
hexavalent chromium 六价铬
hex nut 六角螺母
hexosan 己聚糖
hexuronic acid 己糖醛酸
HF(＝hf,h.f.,high frequency) 高周波,高频[率]
HF(hydrogen fluoride) 氟化氢
HFTA(hot flammability test apparatus) 热燃性试验仪
hibulking 高膨体[变形]
hidden dress 隐形服装
hidden heat 潜热
hide 兽皮,皮革
hiding power [织物]遮盖[能]力
Hifac process 希法防缩防皱整理法（脲醛树脂140℃蒸汽焙烘法）
high amylose starches 高直链淀粉
high black-panel temperature 高黑板温度（耐晒机测温）
high blending ratio 高混纺比
high bulk fibre 高膨体纤维
high bulking 高膨化[变形工艺]
high bulk yarn〈HB yarn〉高膨体纱
high capacity 高容量
high capacity squeezer 高效轧车
high carbon steel 高碳钢
high class treatment 高级处理《环保》
high consistency viscometer 高稠度黏度计
high contrast 高反差
high contrast dye 高反差度染料
high count cloth 精细织物,高支纱织物
high counts 高支,细支
high crimp high wet modulus rayon 高卷曲高湿模量黏胶纤维
high denier fibre 粗纤度纤维,粗旦纤维,高旦纤维
high density〈HD,H.D.〉高密度
high density bleacher 高浓度漂白池
high density downproof fabrics 高密防羽绒织物
high density foam 高密度泡沫
high density polyethylene〈HDPE〉高密度聚乙烯
high duty 高效;重负载
high duty detergent 高效洗涤剂
high duty steel 特殊钢
high efficiency〈HE,H.E.〉高效率
high efficiency extraction mangle 高效轧车
high efficiency mangling 高效轧水
high efficiency open width washing machine 高效平洗机

high efficiency water mangle 高效轧水轧车
high efficient column 高效柱《试》
high elastic deformation 高弹形变
high elasticity〈HE〉高弹性
high elasticity yarn 高弹弹力丝
high energy irradiation 高能照射
high energy photo-electron spectroscopy 高能光电子光谱学
high energy radiation 高能辐射
high energy type dye 高能型[分散]染料
higher alcohol 高级醇
higher preorientation 高预取向
high evapourative drying unit 高效蒸发烘燥机
high explosive 烈性炸药
high expression roll 高效压榨滚筒
high extraction roll 高效轧辊
high fashion 最新款式
high fidelity 高逼真度,高保真度
high finish 高级整理
high fire 强火,高火
high fixation dyes 强固着染料（活性染料的一种）
high foot 高踵,长踵
high frequency〈HF，hf，h.f.〉高周波,高频[率]
high frequency analysis 高周波分析,高频分析
high frequency analyzer 高周波浓度计,高频分析器,高频分析程序
high frequency bonding press 高频黏合压烫机
high frequency converter 高频转换器
high frequency discharge 高频放电（一种等离子体技术）

high frequency drying 高频[率]烘燥
high frequency drying machine 高频烘燥机
high frequency dye fixation 染料高频烘燥固着
high frequency heating 高周波加热,高频加热
high frequency preheater 高频预热器
high frequency sealing 高频熔接
high frequency seal machine 高频熔接机
high gloss finish 高光泽整理
high grade energy 高品位能
high grade finish 高级整理
high humidity 高湿度
high hydroscopic fibre 高吸水纤维
high hydroscopic polymer 高吸水聚合物
high hydroscopic towel 高吸水毛巾
high impact rayon 高冲击强度人造丝
high intensity discharge lamp 高强度放电灯
high key 亮色调图像调节键
high LAD potential（LAD = laundry air dry）高的洗涤晾干优势（去污整理剂的优良性能）
high level control valve 高水位控制阀
high level waste〈HLW〉高放射性废料
high light 高度光泽
highlight 亮点;最重要的部分
high liquor ratio 大浴比
high loft fabric 高蓬松度织物
high-low water alarm 高低水位报警
high luster finish 高光泽整理
highly combustible fabric 易燃织物
highly-conductive synthetic fibre 高导电性合成纤维
highly cross-linked etherified starch 高交链醚化淀粉

highly oriented fibre 高[度]定向纤维,高[度]取向纤维
highly sulfated oil 高度硫酸化油
highly swollen network structure 高膨化网状结构
highly valued material 高附加值材料
highly water soluble dextrin phosphate 高水溶性糊精磷酸盐
high-melting fibre 高熔点纤维
high-modification ratio 高变性比
high modulus fibre 高模量纤维
high modulus weave 高模量织物
high moisture content fibre 高含湿纤维;高吸湿纤维
high moisture content measuring and controlling unit 高含水率控制装置(监控轧车轧液率或预烘后织物含水率等)
high molecular compound 高分子化合物
high molecular surfactant 高分子表面活性剂
high performance 高效的,高速的,高性能的
high performance antistatic fibre 高性能抗静电纤维
high performance fibre 高性能纤维
high performance liquid chromatography〈HPLC〉 高效液相色谱分析[法]
high performance protective fabrics 高性能防护织物
high performance squeezing system 高性能挤压系统[轧车]
high performance textile 高性能纺织品
high performance thin layer chromatography〈HPTLC〉 高效薄层色谱分析法
high pH finisher 高 pH 油剂
high pile 长毛绒

high pile fabric 长毛绒织物,长绒头织物
high polymer 高聚合物
high polymer coagulant 高聚合凝聚剂
high power〈HP〉 大功率
high powered halogen light source 高功率卤素光源
high pressure〈HP〉 高压
high pressure atomizers 高压雾化器
high pressure column 高压柱《试》
high pressure drier 高压烘燥器(用于筒子纱设备)
high pressure dyeing 高压染色法
high pressure dyeing machine 高压染色机
high pressure kier 高压煮锅,高压釜,高压精练锅
high pressure liquid chromatography〈HPLC〉 高压液相色层分析仪
high pressure mangle 高压轧液机,重型轧车
high pressure mercury lamp 高压汞灯
high pressure nip 高压轧面;高压轧点
high pressure polyethylene 高压聚乙烯(相当于低密度聚乙烯)
high pressure sodium lamp 高压钠灯
high pressure spray 高压喷嘴
high pressure spray washing machine 高压喷射洗涤机
high pressure squeeze〈HPS〉 高压挤压
high pressure squeezing 高压挤压
high pressure steamer 高压汽蒸机
high pressure steaming〈HPS〉 高压汽蒸[法]
high pressure water scrubber 高压洁净器
high pressure winch 高压绞盘染色机
high quality fabrics 优质织物
high quality finishing 高级整理

high resolution 高分辨本领;高分辨率(印花、印刷术语)

high resolution diffraction 高分辨衍射

high resolution electron microscopy 高分辨电子显微术,高分辨电子显微镜检验法

high resolution mass spectrometer 高分辨质谱仪

high resolution nuclear magnetic resonance spectroscopy 高分辨核磁共振光谱学

high shear 高剪切

high shear measuring system 高剪切测定法(印花糊料测定)

high shrinkage fibre〈HSF〉 高缩率纤维,高收缩纤维

high solid coating 高固含量涂层

high solids 高固含量[的]

high solvent oil in water system 高溶剂含量油/水相乳化糊(溶剂含量75%~80%)

high speed cage 高速尘笼

high speed centrifuge 高速离心机

high speed chain-type mercerizing machine 高速布铗丝光机

high speed coagulative precipitation unit 速效凝结沉淀装置

high speed flat screen washer 高速平[筛]网洗涤机

high speed, high temperature and pressure package dyeing machine 筒子纱高温高压快速染色机

high speed hot flue stenter 高速热拉机

high speed impact 高速撞击

high speed membrane osmometer 高速膜渗透计

high speed open width washing machine 高速平洗机

high speed printing machine 高速印花机

high speed sand filtration 高速砂滤法

high speed setting machine 高速定形机

high speed steel 高速钢

high speed stenter 高速拉幅机

high speed trickling filter 高速淌[滴]流过滤

high stack plaiting machine 大堆装折布机

high strength composite〈HSC〉 高强度复合材料

high strength fibre 高强度纤维

high strength flame resistant fibre 高强耐火纤维

high stretched 高倍拉伸的,高度伸张的

high stretch fabric 高弹织物

high stretch yarn 高弹丝[纱]

high style fabric 高级整理织物(美国名称)

high sudsing detergent 高泡沫洗涤剂

high-tech 高新技术

high technic textiles 高技术纺织品

high tech zone 高新技术区

high temperature〈HT〉 高温

high temperature ageing 高温老化

high temperature and pressure beck 高温高压绳状染色机

high temperature and pressure degumming 高温高压精练

high temperature and pressure dyeing 高温高压染色

high temperature and pressure dyeing machine 高温高压染色机

high temperature baker 高温焙烘机

high temperature bleaching process 高温漂白工艺

high temperature carbonization 高温炭化

high temperature chemical vapour deposition 〈HT-CVD〉 高温化学蒸汽沉淀法

high temperature contact process 高温接触式黏合工艺（非织造布）

high temperature curer 高温焙烘机

high temperature dry dye fixation 轧染高温干固法，轧染焙烘法

high temperature dyeing 高温染色

high temperature dyeing autoclaves 高温高压釜式染色机

high temperature dyeing machine 高温染色机

high temperature fibre 耐高温纤维

high temperature high pressure open-width continuous steaming equipment 高温高压平幅连续汽蒸设备

high temperature high pressure scouring and bleaching range 高温高压练漂机

high temperature jig 高温卷染机

high temperature loop steamer ［常压］高温悬挂式长环蒸化机

high temperature lubricating oil 耐高温润滑油

high temperature maintenance 高温下保温

high temperature package dyeing apparatus 高温筒子染色机

high temperature paddle machine 高温桨叶染色机（成衣染色用）

high-temperature polymer 〈HTP〉 高温聚合物

high-temperature polymerization 〈HTP〉 高温聚合

high temperature protective clothing 高温保护服，防高温服

high temperature resistance fibre 耐高温纤维

high temperature setting 高温定形

high temperature steamer 高温蒸箱

high temperature steaming 高温汽蒸

high temperature steaming process 〈HTS process〉 高温汽蒸法

high temperature thermo setting 高温热定形

high temperature winch beck 高温高压绳状染色机

high tenacity 〈HT〉 高强度，高强力

high tenacity fibre 高强力纤维

high tenacity rayon 高强力人造丝，强韧人造丝

high tenacity viscose fibre 高强力黏胶纤维

high tenacity yarn 高强力［合成纤维］纱线

high tension 高压《电》；高张力

high thread count 紧密织物

high torque yarns 高捻度纱

high turbulence nozzle 高涡流喷嘴

high twist 高捻度

high vacuum 高度真空

high vacuum metallizing 高真空镀金属处理

high viscosity 高黏度

high viscosity index 〈HVI〉 高黏度指数

high-visibility clothing 高能见度服装；信号服

high visibility material 〈HVM〉 高亮度材料

high voltage 〈HV, hv〉 高压《电》

high warp tapestry 立经挂毯

high water level pump 高水位排水泵

high water temperature concept 〈HWT concept〉 高水温理念

high water temperature washing box 高水温洗涤槽
high wet modulus 高湿模量
high wet modulus fibre 高湿模量纤维
high wet modulus staple fibre 高湿模量短纤维
high wet pick-up 高带液率
Highzitt 印花用防染剂(商名,日本松井色素)
hiking dress 步行服,郊游服
Hill's diffusion equation 希尔扩散公式
hi-lo corduroy 间隔条灯芯绒
Hiltone pigment colours 希尔托恩涂料(印花用的水/油相涂料,商名,美国制)
hindered settling 受阻沉降,干扰沉降
Hindley worm gear 曲面蜗轮,球面蜗轮
hindrance 阻碍作用
hinge 铰链
hinged copolymer 铰链型共聚物
Hi-pile fabric 长毛绒织物
histidine 组氨酸
histochemical staining 组织化学着色
histochemistry 组织化学
histogram 直方图;矩形图;频率分布器
histolysis 组织溶解作用
Hi-Torc raising machine 海托克高转矩起毛机(一种起毛设备,采用液压控制,起毛均匀,美国制)
HLB(=H.L.B.,hydrophilic-lipophilic balance,hydrophile-lipophile balance) 亲水亲油平衡[值]
HLTD(home laundry and tumble dry) 家庭洗涤和转笼烘干
HLTD'S(home laundry/tumble dry cycles) 家庭洗衣机/转笼烘干周期
HLW(high level waste) 高放射性废料

HMD(=HMDA,hexamethylene diamine) 己二胺(锦纶66原料)
HMIS(Hazardous Materials Indentification System) 危险品鉴定方法
HMM(hexamethylol melamine) 六羟甲基三聚氰胺
HMPS(hydrogen methyl polysiloxane) 甲基含氢聚硅氧烷
HN-50 HN-50 芳族聚酰胺纤维(日本帝人)
hoar 灰白[色]
hob 蜗口杆,螺旋杆;齿轮滚刀
Hoffmann's presser 霍夫曼[蒸汽]烫衣机
hoist 起重机,吊车;绞车;升降机
hoisting equipment 吊装设备,起重设备
hoisting medium 吊装工具,起重工具《机》
hoisting work 吊装工作,起重工作《机》
hold-down device 压布装置
hold-down screw 止动螺钉
holder 把,柄;夹;支架;托,座;容器;间隔圈
holdfast 支架;夹,钳
holdfast coupling 夹紧联轴节
holding bar 支持棒
holding device 夹持装置
holding range 同步范围《电》
holding time 保留时间
hold out [织物]拉幅,扩幅
holdout 得色量,上染率(美国名称)
hold time 持续时间,停留时间,占用时间
hole 孔,洞,眼;破洞(织疵)
hole corrosion 孔眼腐蚀(聚酯纤维的碱减量整理)

hole damage on bleaching 漂白破洞
Holland finish 荷兰整理(布匹先经油及填料处理,再经轧光的暗光整理)
hollow cored fibre 空心纤维
hollow cut velveteen 凸条灯芯绒(各行绒头有高低)
hollow fabric 管状织物,中空织物
hollow fibre 空心丝,空心纤维
hollow fibre seperating membrane 中空纤维分离膜(水处理技术)
hollow polyester fibre 中空聚酯纤维
hollow section guide roller 空心导布辊(由中空的圆柱形金属管制成)
hollow shaft 空心轴
hollow squeegee 圆弧形刮浆器(平网双刮刀)
holocellulose 全纤维素
hologram 全息图
holographic foil 全息摄影用箔片
holographic grating 全息光栅
holography 全息摄影术
Holterhoff process [moiré] 霍耳德霍夫波纹整理法
home furnishing 家具织物,家具布,装饰织物
home launder 家庭洗涤
home laundering 家庭洗涤
home laundry and tumble dry〈HLTD〉家庭洗涤和转笼烘干
home laundry/tumble dry cycles〈HLTD'S〉家庭洗衣机/转笼烘干周期
home machine washing 家用洗衣机洗涤,家庭机洗
homespun 钢花呢,火姆司本;手工纺织呢
home textiles [家用]装饰织物
homo- (构词成分)表示"高""均""固"

homoaggregation system 均质凝聚法
homochromatic 同色的
homochromo-isomer 同色异构体
homochromo-isomerism 同色异构[现象]
homocyclic 碳环的;同素环的
homofil 单组分纤维
homogeneity 均匀性,均一性
homogeneous combustion 均匀燃烧
homogeneous dyes 单一染料
homogeneous foam 均匀泡沫
homogeneous material 均质物质
homogeneous oxidation 均相氧化
homogeneous padder 均匀轧车
homogeneous phase 均相
homogeneous pigment 匀质颜料,匀质涂料
homogeneous polymerization 均相聚合[作用]
homogenization 均化[作用]
homogenizer 高速搅拌器,均化器
homogenizing 均化[作用]
homologous series 同族系列
homologue 同系物,相似物,同族体
homomixer 高速搅拌机
homomorphism effect 异质同晶效应
homopolar bond 无极键
homopolymer 均聚物
homopolymerization 均相聚合[作用]
hone 油石;细磨石
honey [yellow] 密黄[色]
honeycomb backing 蜂窝状衬底
honeycomb fabric 蜂窝状织物
honeycomb laminates 蜂窝状层压
honeycomb screen 蜂窝状筛网
hoods for dyeing machines 染色机罩盖
hook disc 钩盘
hooked fibre 折钩纤维

hooker 码布机
hooking machine 色纱紧绞机
Hook's coupling 胡克接头,万向接头
hoop iron 铁箍
hooter 汽笛
hopper 箱;给料斗;加料[漏]斗
hopper agitator 料斗搅拌器
Höppler viscosimeter 霍普勒落球黏度计
hopsack 板司呢,席纹呢;方平织物
hopsack tweed 粗纺板司花呢
horizontal 横的,水平的,横向的
horizontal boiler 卧式锅炉
horizontal cloth run 横导辊穿布[方式]
horizontal coordination 横向协调
horizontal cylinder dryer 卧式圆筒烘燥机
horizontal drying machine 卧式烘燥机
horizontal emerizing machine 卧式磨毛机
horizontal end-ring glueing device 卧式[圆网]闷头胶黏设备(圆网印花机的附属设备)
horizontal exchanger 卧式热交换器
horizontal flame test 平置式阻燃试验
horizontal flash ager 卧式快速蒸化机
horizontal frame dryer 卧式拉幅烘燥机
horizontal frequency 行频
horizontal jet dryer machine 卧式喷射烘干机
horizontal kier 卧式煮练锅
horizontal lattice 水平帘子
horizontal migration 纬向泳移
horizontal nip 水平轧点
horizontal organization 专业[生产]机构,专业工厂(指专纺、专织、专印染等)
horizontal padder 平轧车,卧式轧车
horizontal paddle machine 卧式桨叶染色机(成衣染色用)

horizontal piece dyeing unit 卧式匹染机
horizontal plane〈HP〉水平面
horizontal plane folder 出布小平台(筒状针织物整理设备的一种平幅折叠出布装置)
horizontal repeat 横向循环(图案重复)
horizontal return pin chain 卧式往返针链
horizontal return tubular boiler 卧式回火管锅炉
horizontal roller bed 卧式辊床[结构]
horizontal rotary screen printing machine 卧式圆网印花机
horizontal screen steamer 卧式履带汽蒸机(用于丝织物松弛加工)
horizontal spindle dyeing machine 卧式锭轴染色机(染纱线用)
horizontal squeezer 平轧车,卧式轧车
horizontal star 卧式星形架(平幅洗涤用)
horizontal star-dyeing machine 卧式星形架染色机
horizontal steamer 卧式汽蒸机
horizontal stripe 横条
horizontal sueding machine 卧式磨毛机,卧式仿麂皮起绒机
horizontal two bowl pad mangle 双辊平轧车,卧式双辊轧车
horizontal type pneumatic padder 卧式气压轧车
horizontal type tube jets 卧式管状喷射染色机
horizontal warp dyeing machine 卧式经丝染色机
horizontal washing machine 横导辊水洗机
horn 报鸣器(喇叭)
horned nut 冠形螺母,花螺母
hornification 角化

horny handle 手感坚硬
horse-hair interlining 马尾衬
horse power 马力(功率单位,1 马力约等于 0.7355kW)
horse shoe washer 开口垫圈,马蹄形垫圈
horticultural textiles 园艺用纺织品
horticulture 园艺学
hose 袜子,袜类；软管,蛇[形]管,水龙带,挠性导管
hose connection (＝hose coupling) 软管接头
hose duck 水龙带帆布,软管帆布
hose dye boarder 袜子染色定形联合机
hose-examining machine 验袜机
hose pressure tester 袜子压力试验仪
hose setting machine 袜子定形机
hosiery 袜子；针织袜类
hosiery abrasion machine 袜子耐磨损试验器
hosiery board 袜子定形板
hosiery boarding machine 袜子定形机
hosiery calender 针织品轧光机
hosiery classification 袜子分级，袜子分类
hosiery dyeing 袜子染色
hosiery dyeing machine 袜子染色机
hosiery form [烫]袜板
hosiery paddle unit [袜类]桨叶式染色机；桨叶式整理机
hosiery shapes 袜子定形板
hospital gauze 医用纱布
hospital textiles 医用纺织品
Hostalux 霍斯塔卢克斯系列(荧光增白剂,具有不同色光,适应多种工艺,商名,瑞士科莱恩)
host computer [宿]主机
host fabric 基质织物,基底织物(用于转移印花)
hot air balloon 热气球
hot air chamber 热风干燥室,烘房
hot air circulating system 热风循环系统
hot air drum dryer 热风圆网烘燥机
hot air dryer 热风干燥机
hot air duct 热风风管
hot air fan 热风风机
hot air heating 热风加热
hot air jet dryer 热风喷射烘燥机
hot air jig tenter 热风摆动布铗拉幅机
hot air pin stenter 热风针板拉幅机
hot air setting 热空气定形
hot air shrinkage 热风收缩,干热收缩
hot air stenter 热风烘燥拉幅机
hot air suction drying machine 热风抽吸式烘燥机
hot-bonding tape 热黏合带
hot brittleness 热脆性
hot calendering 热轧光
hot calendering bonding 热轧黏合
hot colours 热色,暖色
hot cylinder 热滚筒
hot cylinder setting 热滚筒定形
hot dissolving method 热熔法
hot drawing 热拉伸
hot drop 热排放；热排液
hot dwell process 热堆法《染》
hot dyeing dye 高温染色染料
hot dyeing exhaust process 高热竭染法
hot dyer 高温染色工
hot flame 高温火焰
hot flammability test apparatus〈HFTA〉 热燃性试验仪
hot flue (＝hot flue dryer) 热风烘燥机
hot flue predryer 热风预烘机

hot flues 热风烘干机
hot flue stenter 热风烘燥拉幅机,热拉机
hot flue tenter 热风拉幅机
hot head press 热压头压烫机(用于耐久性压烫织物)
hot head pressing 热压头压烫
hot impregnation 热浸渍
hot impregnation process 热浸工艺
hot iron 高温熨烫
hot junction temperature 热接点温度(热电偶)
hot melt 热熔
hot melt adhesive 热熔性黏着剂
hot melt coating 热熔涂层
hot melt granulate 热熔粒料(制领衬布原料)
hot melt gravure coating 热熔刻花辊涂层
hot melt gravure printing 热熔[粉点]凹版印花(用于刮热熔领衬布)
hot melt gravure printing machine 热熔凹版印花机(用于纺织品涂层制热熔领衬布)
hot melt interlinings 热熔衬
hot melt laminating 热熔黏合叠层
hot melt powders 热熔粉料
hot melt screen coating 热熔圆网涂层
hot melt [sizing] system 热熔浆纱机
hot melt yarn 热熔性纱线
hot mercerizing 热[烧碱]丝光
hot oil [dyeing] process 热油轧染法(还原染料连续染色法,利用高温矿物油代蒸汽蒸化)
hot-pad-dry 〈HPD〉 热轧烘干
hot plasma 热等离子体
hot plate 热板,电炉热板
hot plate gassing frame 热板烧毛机

hot plate press 热板压呢机
hot press 热压机(利用热金属板,使布面发亮)
hot pressed finish 热压整理
hot printing 仿蜡防印花
hot process 热处理
hot roll 热辊;热轧
hot roll setting 热滚筒定形
hot set 热固化;热定形
hot setting adhesive 热固型黏着剂
hot spot 热点
hot spraying 热风喷射烘燥
hot stamp ink 热印型油墨
hot stenter setting 热拉幅定形
hot stretching 热拉伸
hot tenter 热风拉幅机
hot test 毛织物[衬垫]热收缩试验
hot transfer 热转移
hot transfer operation 热转移操作
hot type reactive dye 热固型活性染料
hot water resistance 耐热水性
hot water washing 热水洗涤
hot wet stretching 湿热拉伸
hot wetting 热润湿
hot wire probe 热线测头
hour 〈hr.〉 小时
hours 〈hrs.〉 小时数
housecoat 妇女家庭服;便服;工作罩裙
house dustmite protection 防家庭螨虫
house fabric 家用织物
household detergent 家用洗涤剂
household dyes 家用染料
household laundering 家庭洗涤
household textiles 家用纺织品
house moths 房屋蛀虫
house shoes 拖鞋

house wrap 房屋包覆层
housing 套,壳体,罩,屏蔽罩;机架,机罩
Housing and Urban Development (HUD) 城市和住房发展部(美国)
HP(high power) 大功率
HP(high pressure) 高压
HP(horizontal plane) 水平面
HPD(hot-pad-dry) 热轧烘干
HPD dyeing process 热轧烘干染色法
HPLC(high performance liquid chromatography) 高效液相色谱分析[法]
HPLC(high pressure liquid chromatography) 高压液体色层分析仪
HPS(high pressure squeeze) 高压挤压
HPS(high-pressure steaming) 高压汽蒸[法]
HPTLC(high performance thin layer chromatography) 高效薄层色谱分析法
hr.(hour) 小时
hrs.(hours) 小时数
H rubber(Hevea rubber) 天然橡胶
HS(=Hs, Shore scleroscope hardness) 肖氏硬度
HSC(high strength composite) 高强度复合材料
H-screen H 型圆网(用以印制半色调效应的特制圆网)
HSF(high shrinkage fibre) 高收缩纤维,高缩率纤维
H-shift 氢位移
HSR(Health and Safety Regulations) 卫生与安全管理条例(英国)
HT(high temperature) 高温
HT(high tenacity) 高强度,高强力
HTAB(hexadecyl trimethyl ammonium bromide) 十六烷基三甲基溴化铵(表面活性剂,杀菌剂)
HTC(humidity test control) 调湿试验
HT-CVD(high temperature chemical vapour deposition) 高温化学蒸汽沉积法
HTP(heat transfer printing) 热转移印花
HTP(high-temperature polymer) 高温聚合物
HTP(high-temperature polymerization) 高温聚合
HTS process(high temperature steaming process) 高温汽蒸法
HTTP(Hyper Text Transfer Protocol) 超文本传送协议
Huachun habotia 华春纺(涤纶丝与涤黏混纺纱交织织物)
Huada damask 花大缎
hub [轮]毂;衬套;插座;集线器
hub-like ring 套环
HUD (Housing and Urban Development) 城市和住房发展部(美国)
hue 色彩,色光;色相,色调
hue angle 色相角
hue circle 色相环
hue contrast 色相对比
hue dependency 色相依赖性
hue difference 色相差异
hue difference angle 色差角
hue of colour 色相
hue,value,chroma〈HVC〉 色相、明度、彩度(孟塞尔色彩系统)
human engineering 人体工程学
human error 人为误差
human tissue burn tolerance 人体组织耐燃度
humectant 润湿剂,保湿剂
humectation 润湿,增湿

humic acid 腐殖质,腐殖酸
humid air 湿空气
humidification 给湿,增湿[作用]
humidifier 给湿器,增湿器
humidifying and heating system 采暖给湿系统
humidiometer 湿度计
humidistat 恒湿器,保湿箱,湿度调节器
humidity 湿度
humidity control 湿度调节
humidity measuring instrument 湿度测试仪
humidity meter 湿度计
humidity moniters 湿度监控器
humidity penetration 透湿
humidity ratio 湿度比[率]
humidity resistance 耐湿性
humidity sensitive printing 湿敏变色印花
humidity sensor 湿度传感器
humidity test control〈HTC〉调湿试验
humidizer 增湿剂
humidor 蒸汽饱和室
humidostat 湿度调节仪
humus 腐殖质,腐殖酸
Hunter colour 亨特色度(亨特反射仪读数)
Hunter colour difference formula 亨特色差公式
Hunter coordinate system 亨特配色系统
Hunter Lab colour difference equation 亨特 Lab 色差公式
Hunter Lab system 亨特 Lab 测色系统
Hunter α-β system 亨特 α—β 测色系统
hurricane drier 快速热风烘燥机《毛》
husbandry 饲养
hush cloth 吸音绒布(双面起绒的厚棉布)
hush pipe 消声套管(衬有吸声材料的导管,气体通过它时吸收声能)
husks 棉籽壳
Hussong type dyeing machine 胡桑式染色机(悬挂式绞纱染色机,染色时染液通过绞纱倒顺循环)
HV(=hv, high voltage) 高压《电》
HVAC(heating ventilation and air condition) 采暖通风和空调系统
HVC(hue, value, chroma) 色相、明度、彩度(孟塞尔色彩系统)
HVDT(half-value decay time of the charge) 电荷半衰期
HVI(high viscosity index) 高黏度指数
HVM(high visibility material) 高亮度材料
HVP(hydrolytic vegetable protease) 水解植物蛋白酶[通称]
HWT concept(high water temperature concept) 高水温理念
hyacinth blue 风信子蓝[色](深紫蓝色)
hyacinth red 风信子红[色](暗红橙色)
hyacinth violet 紫罗兰
Hyamines 季铵盐杀菌剂
hybrid 混杂的
hybrid composites 复合材料
hybrid computer [模拟数字]混合型计算机
hybrid enzymes 杂交酵素,杂交酶;混合酶
hybridized yak fluff 杂交牦牛绒
hybrid jet ink 混合型喷墨墨水
hybrid vibration [红外光谱的]混合振动
hydr- (构词成分)表示"水""氢化的""氢的"

hydrant 消防栓
hydratation 水化[作用],水合[作用]
hydrate 水合物
hydrated cellulose 水合纤维素
hydrated micelle 水化胶束
hydrated salt 水合盐
hydrate water 结合水
hydration 水化[作用],水合[作用]
hydration rate 水合率
hydraulically operated 液压操纵[的]
hydraulically operated controller 液动调节器
hydraulic automatic jigger 液压式自动染缸
hydraulic baling press 液压式打包机
hydraulic barrier [汽水]折流挡板
hydraulic batching machine 液压式大卷装卷布机
hydraulic brake 水压制动器
hydraulic bursting strength 液压法顶破强力
hydraulic calender 液压轧光机
hydraulic center-shaft winder 液压中轴驱动卷布机
hydraulic control 液压控制;液力控制
hydraulic damping device 液压缓冲装置
hydraulic drive 液压传动
hydraulic gradient 水压梯度
hydraulic hose 液压软管,水龙带
hydraulic jack 液压缸,油缸(液压执行元件)
hydraulic jig 液压传动卷染机
hydraulic lifter 油泵推布车
hydraulic mangle 液压浸轧机
hydraulic method 水力方法
hydraulic motor 液压发动机,水力电动机

hydraulic power scouring 水力洗毛
hydraulic press 液压机;水压压呢机
hydraulic pressing machine 液压式打包机
hydraulic pressure 水压力,液压力
hydraulic pump 水压泵,液力泵
hydraulic seal 水封口,液封
hydraulic system 液压系统
hydraulic test 水压试验
hydraulic winder 液压式卷布机
hydrazide 酰肼
hydrazine 肼,联氨
hydrazine hydrate 水合肼
hydrazine hydrogen chloride 氯氢化肼,盐酸肼
hydrazine sulphonate 磺酸肼
hydrazinolysis 肼解作用
hydrazo- 亚肼基,次联氨基
hydrazone 腙
hydride 氢化物
hydrinde azo dyes 茚满偶氮染料,苯并茂偶氮染料
hydrion 氢离子,质子
Hydritex D 海德里泰克斯 D(氢硼化钠,商名,美国制)
hydro- (构词成分)表示"水""氢化的""氢的"
hydro batching-on arrangement 液压式卷布装置
hydroboration 硼氢化作用
hydrocage 甩水机,脱水机
hydrocarbon 碳化氢,碳氢化合物
hydrocarbon polymer 碳氢聚合物,烃类聚合物
hydrocarbon resistance test 抗烃类试验法,防油(抗油)试验法
hydrocarbon sulfonate 磺化烃

hydrocellulose 水解纤维素
hydrochloric acid 盐酸
hydrocolloid 水解胶体
hydrocracking 加氢裂解，氢化裂解
hydrocyanic acid 氢氰酸，氰化氢
hydrodealkylation 加氢脱烃作用
hydrodepolymerization 加氢解聚[作用]
hydrodynamic[al] 流体动力[学]的
hydrodynamical interaction 流体力学的相互作用
hydrodynamic behaviour 流体动力学性状
hydrodynamic circulation machine 水力循环染色机(成衣染色用)
hydrodynamic lubrication 液体动力润滑，流体润滑
hydrodynamic power transmission 液力传动
hydrodynamic pressure 流体压力
hydrodynamic processes in printing 印花中的流体动力学过程
hydrodynamics 流体动力学
hydrodynamic water resistance test 水力式防水试验
hydroejector 水力喷射器
hydro-entanglement 射流缠结(水刺法)
hydroethyl triazone 羟乙基三嗪酮(耐氯漂的棉织物树脂整理剂)
hydro-extracting 脱水
hydro-extracting cage 甩水机，脱水机
hydroextractor 脱水器，离心机
hydrofixation 湿固着
hydrofluoric acid 氢氟酸
Hydrofuga 哈特罗富加脱水机(一种高效脱水设备，商名，德国)
hydrofuge structure 防水结构
hydrofuge surface 防水表面

hydrogel 水凝胶
hydrogel fibre 水凝胶纤维
hydrogen 氢
hydrogen acceptor(＝hydrogen accepter) 氢受体
hydrogenase 氢化酶
hydrogenation 氢化[作用]，加氢[作用]
hydrogen bomb 氢弹
hydrogen bond 氢键
hydrogen-bond breaking agent 氢键断裂剂
hydrogen bonding 氢键作用，氢键键合
hydrogen bonding energy 氢键合能
hydrogen bridge 氢桥
hydrogen carbonate 重碳酸盐，碳酸氢盐
hydrogen chloride 氯化氢
hydrogen cyanide 氢氰酸，氰化氢
hydrogen discharge lamp 氢灯
hydrogen-donating accelerator 供氢促进剂
hydrogen donor 氢供体
hydrogen electrode 氢电极
hydrogen [ion] exponent 氢离子指数
hydrogen flame detector 氢焰检测器
hydrogen flame ionization detector 氢焰电离检测器
hydrogen fluoride〈HF〉 氟化氢
hydrogen halide 卤化氢
hydrogen ion 氢离子
hydrogen ion activity 氢离子活化度
hydrogen ion concentration 氢离子浓度
hydrogen ion indicator 氢离子[浓度]指示剂，酸碱指示剂，pH指示剂
hydrogen lamp 氢灯
hydrogen methyl polysiloxane〈HMPS〉 甲基含氢聚硅氧烷
hydrogen peroxidase 过氧化氢酶
hydrogen peroxide 过氧化氢，双氧水

hydrogen peroxide bleaching 过氧化氢漂白,双氧水漂白

hydrogen peroxide bleaching reaction chamber 双氧水漂白汽蒸箱

hydrogen peroxide bleaching regulators 过氧化氢漂白调节剂

hydrogen peroxide decomposition 过氧化氢分解

hydrogen peroxide dry-in process 过氧化氢烘干法(用于羊毛漂白)

hydrogen sulfate 重硫酸盐,硫酸氢盐

hydrogen sulfite 重亚硫酸盐

hydrogen sulphide 硫化氢

hydro-jet 水力喷射;水力喷射管

hydrolase 水解酶

hydrolysate 水解产物

hydrolysed dyes 水解染料

hydrolysed reactive dye 水解活性染料

hydrolysed starches 水解淀粉

hydrolysis 水解[作用]

hydrolysis-addition reaction 水解—加成反应

hydrolysis constant 水解常数

hydrolyst 水解催化剂

hydrolytic animal protease〈HAP〉 水解动物蛋白酶[通称]

hydrolytic cleavage 水解分裂

hydrolytic degradation 水解降解

hydrolytic degree [活性染料]水解率,水解度

hydrolytic depolymerization 水解解聚

hydrolytic enzyme 水解酶

hydrolytic polycondensation 水解缩聚[作用]

hydrolytic polymerization 水解聚合

hydrolytic ratio [活性染料]水解率

hydrolytic stability 水解稳定性

hydrolytic vegetable protease〈HVP〉 水解植物蛋白酶[通称]

hydrolyze 水解

hydromechanics 流体力学

hydrometer [液体]比重计,比重表

hydrometric method 比重计[测定]法

hydromotor 液压电动机,油电动机;射水[水压]发动机

hydronapper 湿式起毛机

Hydron blue 海昌蓝(一种含硫的还原染料)

Hydron blue vat 海昌蓝还原体

hydronium ion 水合氢离子,水合质子

Hydron Stabilosol 海昌稳定素染料(特选的硫化及硫化还原染料组合产品,超细粉性质,适用于悬浮体连续轧染,商名)

hydroperoxidation 过氧化氢[作用]

hydroperoxide bleaching 双氧水漂白

hydrophile 亲水物;亲水胶体

hydrophile-lipophile balance〈HLB,H.L.B.〉 亲水亲油平衡[值]

hydrophilic 亲水的

hydrophilic chain 亲水链

hydrophilic coating 亲水涂层

hydrophilic fibre 亲水性纤维

hydrophilic film 亲水性薄膜

hydrophilic finish 亲水性整理

hydrophilic group 亲水基团

hydrophilic-lipophilic balance〈HLB,H.L.B.〉 亲水亲油平衡[值]

hydrophilic property 亲水性

hydrophility 亲水性

hydrophilization process 亲水化法(用表面活性剂溶液的分离工艺)

hydrophobe 疏水物;疏水胶体
hydrophobic 疏水的
hydrophobic association 疏水缔合作用
hydrophobic associative polyacrylic salts 疏水缔合型聚丙烯酸盐
hydrophobic bond 疏水键
hydrophobic chain 疏水链
hydrophobic dyes 疏水性染料
hydrophobic fibre 疏水性纤维
hydrophobic finishing 疏水整理
hydrophobic group 疏水基团
hydrophobic hollow structure 疏水空腔结构
hydrophobic interaction 疏水作用
hydrophobicity 疏水性
hydrophobic property 疏水性
hydrophobic region 疏水区
Hydrophobol 海德罗福博尔拒水整理剂（商名,瑞士制）
hydroplanning effect 吸水平衡效果
hydroplastic 水塑性
hydro-pneumatic control 液压气动控制
hydropneumatic pressure 气水压力
hydro-pneumatic pressure system 液压气动装置
hydropress 水压机
hydropulsation 水脉冲
hydroquinone 氢醌,对苯二酚（照相显影剂）
hydros 保险粉
hydroscopic agent 吸湿剂
hydroscopic fibre 吸湿纤维
hydroscopic hysteresis 吸湿滞后性
hydroscopicity 吸水性
hydroscopic property 吸湿性
hydrosetting 湿[热]定形

hydrosetting process 湿定形工艺
Hydrosist ClO_2 海德罗西斯特 ClO_2（亚氯酸钠漂白时,二氧化氯气体对机械腐蚀的抑制剂与稳定剂,商名,美国制）
hydroslot squeegee 液压条缝刮刀
hydroslot squeegee system 液压条缝式刮刀系统（地毯印花用）
hydroslot system 液压条缝式给浆刮浆装置（地毯网印机用）
Hydrosoft 海特罗佐夫特柔软剂（商名,美国制）
hydrosol 水溶胶
Hydrosol 海德罗素（可溶性硫化染料,商名,德国制）
hydrosoluble 水溶的
hydrostatic head correction 静压头校正
hydrostatic head test of water repellency 静水压防水性测试法
hydrostatic loading 静液压
hydrostatic power transmission 液压传动
hydrostatic pressure [流体]静压,[静]水压[力]
hydrostatic pressure test 流体静压试验,水压试验（测拒水性）
hydrostatics 流体静力学
hydrostatic sludge removal 流体静压污泥排除法《环保》
hydrostatic stress 静水应力
hydrostatic test 静水压测定
hydrostropy 助溶
hydrosuction wringing machine 吸水轧干机
hydrosulfite bleaching 亚硫酸氢盐漂白
hydrosulfite discharge 亚硫酸氢盐拔染
hydrosulfite-glucose vat 保险粉—葡萄糖还原缸

hydrosulfite vatting 保险粉还原[工艺]
hydrosulfuric acid 氢硫酸,硫化氢
hydrosulphite 保险粉
hydrosulphite ager 还原蒸箱
hydrosulphite formaldehyde 连二亚硫酸钠甲醛,甲醛合低亚硫酸钠(雕白粉)
hydrosulphite indigo vat 靛蓝保险粉还原染液;保险粉靛蓝[还原]缸
hydrothermal 水热的,热液的
hydrothermal delustring 热液消光
hydrothermal treatment 热液处理
hydrothermal wetting 热液润湿
hydrothermal yellowing [羊毛]湿热泛黄
hydrotimeter 水硬度计
hydrotrop(＝hydrotrope, hydrotroph) 水溶增溶剂
hydrotropic 水溶的
hydrotropic agent(＝hydrotropic solubilizer) 水溶增溶剂
hydrotropy agent 助溶剂
hydroxamic acid 氧肟酸,异羟肟酸
hydroxide coagulation method 氢氧化合物凝集法《环保》
hydroxide compound 氢氧化合物
hydroxyacetic acid 羟基乙酸
hydroxyacetone 羟基丙酮
o-hydroxy benzene carbonic acid 邻羟基苯甲酸
hydroxyethylated starches 羟乙基化淀粉
hydroxyethylating amylose 羟乙基化直链淀粉
hydroxyethylation 羟乙基化作用
hydroxyethyl cellulose〈HEC〉 羟乙基纤维素
hydroxyethyl derivatives 羟乙基衍生物
N-hydroxyethyl ethylene diamine triacetic acid〈HEDTA〉 N-羟乙基乙二胺三乙酸(有机螯合剂)
N-(2-hydroxyethyl) ethyleneimine N-(2-羟乙基)乙烯亚胺
hydroxyethyl locust bean gum 羟乙基皂荚胶糊
hydroxyethyl starch thickening 羟乙基淀粉糊
hydroxy glutamic acid 羟基谷氨酸
hydroxy group 羟基
hydroxylamine 羟胺,胲,羟基胺
hydroxylamine sulfate 硫酸胺,硫酸羟胺
hydroxylation 羟基化[作用]
hydroxyl group 羟基
hydroxyl ion 羟离子
hydroxyl value 羟基值
hydroxymethane sulfonic acid 羟基甲磺酸(树脂交联的酸性催化剂)
hydroxymethyl 羟甲基
hydroxymethyl compound 羟甲基化合物
hydroxy naphthoic acid 羟基萘甲酸
hydroxypropylated starches 羟丙基化淀粉
hydroxypropyl derivatives 羟丙基衍生物
hydroxy substituted carboxylic acid 羟基代羧酸(含有羟基的有机酸)
hyg.(＝hygr., hygroscopic) 吸湿的
hygiene protection 卫生保护
hygienic-absorbent products 吸湿性卫生用品
hygienic finishing 卫生整理
hygienic finishing agent 卫生整理剂
hygienic laundering care 卫生洗烫护理
hygienic protective fabrics 卫生防护功能织物
hygienic textile 卫生纺织品,保健纺织品
hygr.(＝hyg., hygroscopic) 吸湿的

Hygra fibre 哈格拉高吸水性纤维(皮芯型复合纤维,商名,日本制)
hygral change 湿度变化
hygral expansion 湿膨胀
hygral expansion index 湿膨胀指数
hygral shock 湿冲击
hygral shrinkage 润湿收缩
hygral stress 湿应力
hygristor 湿敏电阻
hygro- 构词成分,表示"湿""湿气"
hygroautometer 自记湿度计
hygrograph 湿度记录表,湿度计
hygrometer 湿度计
hygrometric state 湿态
hygrometry 测湿法,湿度测定法
hygroscopic〈hyg., hygr.〉 吸湿的
hygroscopic agent 吸湿剂
hygroscopic capacity 吸湿量
hygroscopic clothing 吸湿性衣服
hygroscopic coeffieient 吸湿系数
hygroscopic degree 吸湿度
hygroscopic expansion 吸湿膨胀
hygroscopic finish 吸湿整理
hygroscopicity 吸湿性
hygroscopic property 吸湿性能
hygroscopic softener 吸湿柔软剂
hygroscopic synthetic fibre 吸湿性合成纤维
hygroscopic treating agent 吸湿性处理剂
hygrosensor 湿度探测器
hygrostat 恒湿器,保湿箱,湿度调节器
hygrothermograph 湿温仪,温湿自记器
Hypalon 海帕龙合成橡胶(化学成分为氯磺化聚乙烯,宜作高温高压设备的辊封材料,商名,美国杜邦)
hypchlorous acid bleaching 次氯酸漂白

hyper- (构词成分)表示"超出""过于""极度""过"
hyper allergenic 超过敏的
hyperboloidal gear 双伞齿轮,双曲面齿轮
hyper-branched polyester 超支化聚酯
hyper-branched polymer〈HBP〉 超支化聚合物
hyperchrome 浓色团
hyperchromic agent 增深剂
hyperchromic effect 吸收增强效应,增色效应
hyperconjugation 超共轭效应
hyperelastic 超弹性
hyperfiltration 超[过]滤
hyperfine 超精细
hyperfine structure 超微细结构
hypergeometric distribution 超几何分布
hypermolecular 超分子的
hyperoxide bleaching 过氧化物漂白
hyper-seal 超密封
hypersonic flow 超声流
hypersonic sound 特超声
hypersorption 超吸附[法]
hyper text 超文本《计算机》
Hyper Text Transfer Protocol〈HTTP〉 超文本传送协议
hypo 海波(硫代硫酸钠)
hypo- (构词成分)表示"次""低"
hypobromite 次溴酸盐
hypochlorite 次氯酸盐
hypochlorite bleaching 次氯酸盐漂白
hypochlorous acid 次氯酸
hypochrome 淡色团(会使颜色变淡的基因)
hypochromic effect 淡色效应
hypoid gear 偏轴伞齿轮

hypothesis 假说,假设;前提
hypsochromatic effect 浅色效应
hypsochrome 浅色团
hypsochromic effect 色光变浅效应,向紫效应
hypsochromic shift 向紫移;向短波长移
hypsoflore 浅荧光团
hyrogen electrode 氢电极
hysteresis 滞后现象
hythergraph 温[度与]湿[度关系]图
Hz(hertz) 赫[兹](频率单位,周/秒)

I

IACD(International Association of Clothing Designer) 国际服装设计师协会

IAQ (International Association for Quality) 国际质量协会

IARC(International Agency for Research on Cancer) 国际癌症研究会

IB(input bus) 转入总线《计算机》

I-bar ［窄型］工字钢

I-beam 工字梁,工字钢

IBS(interconnection bus signal) 互联总线信号《计算机》

IBS(internet bus) 联系总线《计算机》

ICBM(intercontinental ballistic missile) 洲际弹道导弹

ICC (ignition control compound) 控制引火化合物,控制点燃化合物

iceberry green 浅青灰［色］,湖灰［色］

ice blue 冰蓝［色］(淡绿光蓝)

ice boom 拦冰栅

ice colour bottoming agent 冰［染］染料打底剂

ice colour coupling agent 冰［染］染料偶合剂

ice colour developer 冰［染］染料显色剂

ice colour dip dyeing 冰［染］染料浸染

ice colour direct printing 冰［染］染料直接印花

ice colour discharge 冰［染］染料拔染

ice colour dyeing 冰［染］染料染色

ice colour impregnating agent 冰［染］染料浸渍剂

ice colour resist dyeing 冰［染］染料防染

ice colours 冰［染］染料

ice dyeing 冰染

Iceland lichen 冰岛地衣(含胶植物,可作浆料)

Iceland moss 冰岛苔藓(含胶植物,可作糊料用)

Iceland spar 冰洲［晶］石(双折射透明方解石)

ice wash 冰洗(牛仔裤处理)

ICFF(Information Council on Fabric Flammability) 织物耐燃性情报理事会

ICI(＝I. C. I., Imperial Chemical Industries Limited) 帝国化学工业公司(俗称卜内门公司)

ICI(＝I. C. I., International Commission on Illumination) 国际照明委员会

ICP (inductively coupled plasma) 电感耦合等离子体

ICP-MS method (inductive coupled plasma mass spectrometry method) 电感耦合等离子体质谱法

ICRM(International Carpet and Rug Market) 国际地毯展览会

ICS(Instrumental Colour System) 仪器测配色系统(国际羊毛咨询处采用)

ICTRD (International Center for Textile Research and Development) 国际纺织研究和发展中心

ideal blackbody 绝对黑体

idealized drawing 示意图

ideally perfect crystal 理想完整晶体
ideal solution 理想溶液
ideal temperature 理想温度(弗洛里温度或 θ 温度)
ideal viscous fluid 理想黏性流体
ideal white 理想白
identical period 等同周期
identification 鉴定,鉴别
identification belt 识别标志带
identification code 识别代码《计算机》
identification method of dye 染料鉴定法
identification method of dye on fibre 纤维上染料鉴定法
identification of dyestuffs 染料鉴定法
identification of textile fibre 纺织纤维鉴别
identification test 鉴别测试
identification thread 识别线,标志纱(织入布边的色经)
identifier 识别符,标志符
identify code 识别码
identity 同一性,等同
identity period 等同周期
idle 空转的;空载的;空载
idle gear 惰轮,过桥齿轮
idle motion 空程运动
idle period 停运时间
idle positon 空转位置;无效部位
idle pulley(＝idler,idler cylinder) 空转轮,跨轮,惰轮,调节皮带轮
idle roller 惰辊;空转导辊
idle run 空转,跨转
idle speed 空载转速
idle time 空载时间,停机时间
idle unit 闲置设备
idle wheel 空转轮,惰轮

idling condition 空载状态
idling gear 空转轮,跨轮,惰轮;调节皮带轮
idling time 停运时间,空转时间
IDRC(International Dry-cleaning Research Committee) 国际干洗研究委员会
IEC(International Electrotechnical Commission) 国际电工委员会
IEP(isoelectric point) 等电点
IES(Institute of Environment Service) 环保研究所
IFAI(Industrial Fabrics Association International) 国际工业用织物协会
IFATCC(International Federation of Association of Textile Chemists and Colourists) 国际纺织化学家及染色家协会联合会
IFP(＝I. F. P. ,interfacial polymerization) 界面聚合
Igepal Co 436 胰加泊尔 Co 436(非离子表面活性剂,耐硬水及金属离子,商名,德国制)
Igepon T 胰加漂 T(油酸酰胺乙烷磺酸钠阴离子型表面活性剂,商名,德国拜耳)
ignitability 可燃性,着火性
igniter 点火器
ignition 点火;着火,点燃
ignition control compound〈ICC〉控制引火化合物,控制点燃化合物
ignition electrode 点火电极
ignition plug 点火火花塞
ignition point〈ip,i. p.〉着火点,燃点
ignition resistance 抗引燃性,抗点燃性
ignition temperature 着火温度
ignition temperature of dust cloud 尘雾的燃着温度

ignition test 着火点试验
ignition time 点燃时间(指最短时间)
ignitor 点火器
ignitron 点火管,引燃管
IIC(International Institute for Cotton) 国际棉业协会
ikat 纱线扎染布
IK process [还原染料]丙种染色法(染色温度为室温)
ILA (International Laundry Association) 国际洗涤协会
Ile (isoleucine) 异白氨酸(蛋白质纤维结构成分之一)
ill. (illumination) 照明;照度,亮度,照明度
ill. (illustrated) 举例说明的;图解说明
ill. (illustration) 插图;注解;实例;图解
ill-defined 不明显,不清晰;不明确
illuminance 照度
illuminant 光源;发光装置;着色染料(指拔染或防染着色的染料)
illuminant A [国际照明委员会]标准光源 A(色温 2856K)
illuminant B [国际照明委员会]标准光源 B(相当于直接日光,色温 4870K)
illuminant C [国际照明委员会]标准光源 C(相当于平均日光,色温 6770K)
illuminant D65 [国际照明委员会]标准光源 D65(相当于平均日光,色温 6500K)
illuminant metamerism 照明条件等色《测色》
illuminated area 照明区(反射法测定织物折皱用语)
illuminated discharge 着色拔染[印花]
illuminated discharge style printing 着色拔染印花
illuminated inspection machine 照明验布机
illuminated mixture [毛织物]色线增艳
illuminated resist 色防[印花]
illuminated resist style printing 色防印花
illuminating colours 着色染料(拔染或防染着色的染料)
illuminating dyes [防拔染]着色染料
illuminating source 光源
illumination ⟨ill.⟩ 照明;照度,亮度,照明度
illumination photometer 照度计
illuminator 发光器,照明装置;反光镜
illuminometer 照度计
illumi yarn 光亮丝,晶光丝(单丝较粗,较坚韧,有近似珍珠的光泽)
illusion pattern 错视图案
illustrated ⟨ill.⟩ 举例说明的;图解说明
illustration ⟨ill.⟩ 插图;注解,实例;图解
Imacol C liq. 依玛可 C 液(非离子型,低泡,对碱稳定,防止折皱,适用于纤维素纤维及其混纺织物的染色,商名,瑞士科莱恩)
Imacol S liq. 依玛可 S 液(弱阳离子型,低泡,防止折皱,适用于合成纤维染色,商名,瑞士科莱恩)
image [图]像,影像
Image-3000 依迈奇 3000(印花图案处理机牌号,商名,荷兰斯托克)
image analysis 图像分析
image analysis in fabric inspection 纺织品检验图像分析
image data 图像数据
image edit 图像编辑,图像剪切
image element 像元,像素(表示图形图像

的最小单位,印刷术语)
image intensifier 图像增强器,影像增亮器
image processing 图像处理《计算机》
imagescope [图]像镜
image transfer 图像转移
image transmission 图像传输《计算机》
imaginary line 想象线(缝纫)
imbalance 失衡
imbedment 埋置[法];包埋,埋封
imbibe 吸液膨润
imbibition 吸液;吸液膨润
imbrication 鳞片;瓦状叠覆
imbue 浸透;染透
imcomplete combustion 不充分燃烧
imidazole 咪唑,1,3-二氮杂茂
2-imidazolidinone(＝2-imidazolidone) 2-咪唑啉酮
imidazoline 咪唑啉(无醛固色剂的化学结构)
imidazoline carboxylate 咪唑啉羧酸盐
imidazoline surfactant 咪唑啉表面活性剂
imide [酰]亚胺
imidization [酰]亚胺化反应
imido group 亚氨基
imine 亚胺
imino acid 亚氨基酸
imino alkyl starches 亚氨基烷基淀粉
imino group 亚氨基
imitating wax resist printing 仿蜡防印花
imitation 模拟,模仿
imitation antique carpet 仿古地毯
imitation batik dyeing effect 仿蜡染效应
imitation batik prints 仿蜡防花布
imitation beetle calender 仿捶布轧光,仿叠层轧光

imitation beetle finish 仿捶光整理
imitation embroidered fabric 仿绣织物
imitation fabric 仿生织物
imitation flower 人造花
imitation fur [fabric] 仿毛皮[织物]
imitation fur finish 仿毛皮整理
imitation leather 仿革织物,人造革
imitation leather processing 仿皮革加工
imitation leather technology 仿皮革技术
imitation linen finish 仿麻整理,充麻整理
imitation linen finish with gelatine solution 明胶溶液仿麻整理
imitation organdie 仿奥甘迪织物,仿蝉翼纱(合成纤维薄织物经合成树脂整理而成)
imitation part 仿制配件
imitation peach fabric 仿桃皮绒织物
imitation rabbit hair 充兔毛皮织物
imitation rabbit skin 仿兔皮长毛绒
imitation silk 充丝,仿丝
imitation silk finish 仿丝整理
imitation silk processing 仿丝加工
imitation silk technology 仿丝技术
imitation silver yarn 充银线,仿银丝
imitation suede 仿麂皮
imitation wax print〈imi-wax〉 仿蜡防印花
imitation wool processing 仿毛加工
imitation wool technology 仿毛技术
imitative finish 模拟整理
imi-wax(imitation wax print) 仿蜡防印花
immature cotton 未[成]熟棉,不成熟棉
immature fibre 未成熟纤维
Immedial dyes 益米狄耳染料(硫化染料,商名,德国制)
immediate defoamation 快速脱泡

immediate elastic deformation 即时弹性变形

immediate elastic recovery 瞬间弹性回复

immediate oxygen demand〈IOD〉 直接需氧量

immediate picture 直接图像

immediate recovery 即时回复

immersant 浸渍剂

immersed roller 浸没辊,液下导辊

immersion 浸渍,浸入

immersion accumulator 液下储布器

immersional wetting 浸润[作用]

immersion bleach 浸没式漂白

immersion bleach process 浸漂工艺

immersion coating process 浸渍涂层法

immersion colourimeter 浸没式比色计

immersion heater 浸没式加热器

immersion jig 液下卷染机

immersion length 浸渍长度

immersion lens 浸没透镜

immersion method 浸渍法,液浸法

immersion passage 浸渍道数

immersion refractometer 浸液折射计

immersion rod method 棒浸没法（测试助剂润湿性方法）

immersion roller 浸没辊

immersion squeezing 液下挤压

immersion test 浸液试验

immersion trough 浸渍槽

immersion wettability test 可浸润性试验

immiscibility 不溶混性,不混合性

immiscible liquid 不溶混液体

immiscible solvent 不溶混溶剂

immobile phase 固定相

immobilization 固定[作用]

immobilized anti-microbial agents 固着型抗菌剂

immobilized bed Bio-Reactor 固定床生物反应箱,固定床生物反应池（用于降低废水中 COD 和色度）

immune 免疫的;被免除的,不受影响的

immune cotton[yarn] 防染处理纱,变性防染棉纱

immune defense system 免疫防护系统

immunisation 防染处理

immunised cotton 变性防染棉（一种经过化学处理能防染直接染料的原棉）

immunization 防染处理

immunizing agent 变性防染剂

immunochemistry 免疫化学

impact 冲击;冲力;脉冲

impact mill 冲击式研磨机,冲压式粉碎机

impact penetration test 冲击渗透试验（测拒水性）

impact protection fabric 抗冲击防护织物

impact protection textile 抗冲击防护纺织品

impact reistance 抗冲击性,冲击抵抗

impact resilience 冲击回弹性

impact spray test 冲击喷射式渗透性试验

impact strength 冲击强度

impalpable 感触不到的;摸不着的

impedance 阻抗（以 Z 表示）

impeded drying 受控[恒温]烘燥

impeller 叶轮;推动器;压缩器

impeller blade 叶轮片

impeller tumble abrasion testing machine 叶轮翻滚耐磨测试机

impenetrability 不可渗透性

imperfect adhesion 贴布不良,贴布浮起（台板印疵）

imperfect colour step 花纹分段色变(印花花纹)
imperfect combustion 不完全燃烧
imperfect dyeing 染色不良
imperfect elastic behaviour 不完全[的]弹性性状
imperfect embossing 轧花不良
imperfection 疵点,缺点
imperfect selvage 坏边(织疵)
imperfect set 不良定形
imperfect tension 张力不适当
imperial blue 景泰蓝
Imperial Chemical Industries Limited (ICI, I. C. I.) 帝国化学工业公司(俗称卜内门公司)
imperial coating 防水精纺细呢
Imperial green 巴黎绿,翡翠绿(醋酸铜偏亚砷酸铜复盐)
imperial sizing 英制尺寸
impermeability 不透[气或水]性,不渗透性;防水性
impermeable 不渗透的
impermeable water barrier 不透水隔离层
Imperon pigment colours 印漂牢涂料(印染用涂料,商名,德国制)
impervious 透不过的,不可渗透的
imperviousness 不渗透性
impervious water proofing 不透气性防水
impinge(=impingement) 冲撞,冲击
impinging flow 染液循环流动
implement 工具;仪器
Imprafix SI 因普兰菲克斯 SI(有机硅化合物,涂层工艺交链剂,提高涂层织物的水洗,干洗牢度,商名,德国拜耳)
Impranil AM 因普兰尼尔 AM(弱阴离子型,聚丙烯酸酯分散液,织物涂层剂,商名,德国拜耳)
Impranil CA 因普兰尼尔 CA(有机硅涂层剂,商名,德国拜耳)
Impranil CHW 因普兰尼尔 CHW(涂层整理剂,双组分的聚氨酯,商名,德国拜耳)
Impranil DLN 因普兰尼尔 DLN(阴离子脂肪族酯型,聚氨基甲酸乙酯分散体50%的溶液,羊毛防缩剂,商名,德国拜耳)
impregnant 浸渍剂
impregnate 浸渍,渗透,浸透
impregnated fabric 浸渍织物,浸轧织物
impregnated nonwoven fabric 浸渍法非织造布
impregnated yarn 浸渍纱线
impregnating agent 浸渍剂
impregnating bath 浸渍浴
impregnating compartment 浸渍槽,浸渍区
impregnating drying machine 浸轧烘燥机
impregnating mangle 浸染机,浸轧机,压液机
impregnation 浸渍,浸轧,浸染
impregnation centrifuge 浸渍式离心分离机
impregnation channel 浸渍沟槽
impregnation process 浸渍工艺;浸渍黏合法(非织造织物成布工艺)
impregnator 浸轧机
Impregnole 伊姆普拉诺耳整理(一浴法防水整理,商名)
impress 印出,印上
impressing machine 花筒打样机(供雕刻完毕的花筒进行试样用)
impression 印痕;印象;模槽;压印,压痕

impressioning 压印
impressioning machine 压印机,印花[平板]打样机
impression mark 印痕[疵]
imprime 印花的(法国名称)
imprime marble 大理石纹花样
imprinting 盖印,压印
improper finish 修整不良,修剪疵
improper printing 偏印,歪印
improper tension 张力不当
improvement 改善,改进,改良;调质处理
improvement of contours 改变轮廓(电脑图案处理,电脑分色用术语)
improver 改进剂;添加剂
impulse 脉冲;冲击;冲量
impulse counter 脉冲计数器
impulse drying 脉冲干燥
impulse generator 脉冲发生器
impulse signal 脉冲信号
impulse transmitter 脉冲发送器
impulse type 冲击式;推进式;脉冲式
impulse voltage 脉冲电压
impurities 杂质,不纯净物
imputrescibility 不腐性,防腐性
IMS (industrial methylated spirit) 工业用甲醇变性酒精
in. (inch) 英寸
inaccurate 不精确的
inactivation 钝化[作用]
inactivator 钝化剂
inactive filler 惰性填料
inboard bearing 内置轴承
in bulk 大批,大量;散装;批发
inca gold 老黄[色]
incandescent lamp 白炽灯,白热灯
incandescent light 白炽光,白热光

incandescent lighting 白炽灯照明
incarnadine 肉色的,淡红色的
incentive 诱发的,诱因
inch ⟨in.⟩ 英寸
inches ⟨ins.⟩ 英寸(inch 的复数)
inching 寸动,微动,蠕动;微调整
inching button 微动按钮
inching device 微动装置
inching motion 寸动,微动,蠕动
inching speed 极慢的速度,微速
incidence 入射
incident beam 入射[光]线
incident intensity 入射[光]强度
incident light 入射光
incident ray 入射线
incineration 煅烧;烧烬;焚化《环保》
incineration for waste disposal 废物焚化处理
incineration test 烧烬试验,灰化试验
incinerator 煅烧炉,焚化炉(用于垃圾、污物处理)《环保》
incinerator unit 焚烧装置
inclination 倾斜;斜坡;倾向
inclination to soiling 沾污倾向
inclined flammability test 倾斜式燃烧性试验(阻燃试验用)
inclined nip 倾斜轧点
inclined-nip padder 斜轧点轧车
inclined segment 斜面弧形板
inclined type three bowl pneumatic padder 倾斜式三辊气压轧车
inclined viscometer 斜管黏度计
inclined washer 倾斜式水洗机
inclining padder 斜轧车
inclusion of air 含有空气
incohesive 无凝聚力的

incombustibility 不燃性
incombustible fabric 不燃性织物
incoming beam 入射光束
incoming order 进入指令
incompatibility 不相容性,不配伍性
incomplete degumming 精练不良,脱胶不良
incomplete scouring 精练不良
incompressibility 不可压缩性
incompressible fluid 不可压缩流体
incompressible void volume 不可压缩空间
incondensable 不冷凝的,不凝缩的,不能浓缩的
incontinence textile 失禁用纺织品
incorporation 编入;掺合,混合;公司
incorrodible 不腐的,抗腐蚀的
increasing ratio 增速比
increment 增量;增加
incrustation 结壳,结痂;矿渣;水垢,水锈
incubation 培育;保温处理
incubation method 气相保温培育法(树脂整理织物的游离甲醛测定法)
incubator 恒温箱,保温箱
incubator test [污水的]稳定性试验
ind.(index) 索引,目录;指数,系数;标志;符号;下标
ind.(=indus.,industrial) 工业的
ind.(industry) 工业
Indalca gum 印达卡胶(醚化甘露半乳聚糖,印花糊料,商名,意大利制)
Indalca PA 印达卡 PA(天然增稠剂,醚化甘露半乳聚糖,用于丝绸、锦纶、腈纶及羊毛印花,得色率高,清晰,光亮度高,手感柔软,商名,意大利宁柏迪)
indamine and indophenol colouring matters 吲达胺和靛酚染料

indan 茚满,2,3-二氢化茚
Indanthrene blue 阴丹士林蓝(染料)
indanthrene dip dyeing 士林染料浸染
indanthrene direct printing 士林染料直接印花
indanthrene discharge printing 士林染料拔染印花
indanthrene dyeing 士林染料染色
Indanthrene dyes 阴丹士林染料(蒽醌及靛族还原染料,商名,德国巴斯夫)
Indanthrene label 阴丹士林标记
indanthrene resist dyeing 士林染料防染
Indanthrene steamer 阴丹士林蒸箱(用于还原染料印花的导辊式蒸箱)
Indanthren yellow paper 阴丹士林黄试纸(用于测定染浴内保险粉是否充分)
indanthrone 靛蒽醌(商名为 Indanthrene blue)
indefinite⟨indef.⟩ [色泽]模糊的,不清楚的;无定限的,不确定的
indelible ink 不灭墨水
indelible marking ink 不灭墨印
indemnifying measure 补偿措施
indene 茚
indentation 刻痕,压痕,凹痕;成穴作用;呈锯齿形
indentation hardness 压痕硬度
indentation index 压痕指数
independent 无关,独立
independent component 独立组分
independent drive 单独传动
independent inverter 无源逆变器,自励逆变器
index⟨ind.⟩ 指数,系数;索引;标志;符号;下标;目录
index chuck 分度卡盘

index cock 指示开关；刻度旋塞
index compound 母体化合物；索引化合物
index disc 指示盘，刻度盘
index drum 控制滚筒
index error 分度误差；指示误差，指标差
index finger 指针
index gear 分度盘，分度齿轮
index hole 索引孔
indexing ［缩小机］拨格；索引
indexing cam 刻度凸轮
indexing mechanism 刻度机构，索引机构
indexing ring 刻度环
index mark 刻度，分度；指标；商标，品号
index number 指数
index of correlation 相关指数，相关指标
index of refraction 折射率
index point 指标（例如每染色或整理1千克织物所消耗的时间、用水、能量等）
index test 指标性试验
index wheel 分度轮，刻度盘
index word 下标字；索引字
IndiAge 一种纤维素酶（商名，杰能科）
Indian blue 靛蓝［色］（深暗蓝色）
Indian hog gum 印第安胶（一种半水溶性糊料，可作为龙胶代用品）
Indian madras 印第安色织物
Indian red 印第安红［色］（红棕色）
Indian Textile Journal〈Ind. Tex. J.〉《印度纺织杂志》
Indian tragacanth 印度黄芪胶，刺梧桐树胶
Indian yellow 印第安黄［色］（老黄色）
India print 带有印度图案特色的印花布，印度式花纹印花布
indicated power 指示功率
indicating finger 指针，指示爪

indication lag 指示滞后
indicator 指示器；指示剂；指针；指盘；计数器
indicator balance 指针天平
indicator current 指示电流
indicator disc 指示盘
indicator electrode 指示电极
indicator lamp 指示灯
indicatrix 特征曲线；指标图形
indicia 标记，象征
indienne 印花棉布（19世纪印度花布统称）
indifference 惰性；中性
indifferent 惰性的；不反应的
indifferent solvent 惰性溶剂
indigestible 难理解的；难消化的
indigo ［blue］ 靛蓝［色］，深紫蓝色；靛蓝植物染料；靛蓝地色花布
indigo auxiliary 靛蓝［还原］助剂
indigo blue prints 靛蓝花布
indigo bottom （＝indigo bottoming） 靛蓝打底
indigo brush printing 靛蓝刷印
indigo carmine 酸性靛蓝；靛胭脂蓝（绿光蓝色）；可溶性靛蓝
indigo dyeing 靛蓝染色
indigo dyeing machine 靛蓝染色机
indigo dyeing /sizing machine 靛蓝染色/上浆机
indigo dyer 靛蓝染色技师
indigo fermentation vat 靛蓝发酵缸，靛蓝发酵还原染液
indigoid ［vat］ colours 靛族还原染料
indigoid dyeing 靛属染料染色
indigoid dyes 靛族染料
indigo paper 靛蓝纸

indigo plant 靛蓝植物
indigo printing 靛蓝印花
indigo prints 靛蓝印花织物
indigo recovery and reuse 靛蓝染料的回收和再利用
indigo reduction vat 靛蓝还原缸
indigosol developer 印地科素染料显色剂
indigosol dip dyeing 印地科素染料浸染
indigosol direct printing 印地科素染料直接印花
indigosol discharge printing 印地科素染料拔染印花
indigosol dyeing 印地科素染料染色
Indigosol dyes 印地科素染料(可溶性还原染料,商名,瑞士及德国制)
indigosol resist printing 印地科素染料防印
indigo styles in Africa prints 靛蓝风格的非洲印花布
indigotin[e] 纯靛蓝
indigo vat 靛缸;靛蓝还原浴
indigo white 靛白(靛蓝还原后的隐色体)
indirect calorimetry 间接热量测定法
indirect circulation hot oil system 间接热油循环系统
indirect contact desuperheater 表面式减温器,间接接触式减温器
indirect fired tenter frame 间接燃烧加热拉幅机
indirect jacket heating 夹套间接加热
indirect radiation singeing 间接辐射式烧毛
indirect singeing 间接烧毛(辐射烧毛)
indirect sizing 间接上浆
indirect steam 间接蒸汽,二次蒸汽
indirect take-up 间接卷取

indirect type damping device 反射式给湿器
indiscerptible 不能分解的
individual design 单独花型
individual drive 单独传动
individual operation 单独操作
indocarbon 应得元(即硫化还元黑)
indocarbon dyestuff 应得元染料(硫化还原染料,商名,德国制)
indole 吲哚,氮[杂]茚
indoor furnishings 室内装饰织物
indoor shoes 室内鞋
indoor wear 室内服装,家用衣着
Indosol CR liq. 坚牢素 CR 液(阳离子性,坚牢素染料活性固色剂,商名,瑞士科莱恩)
Indosol E-50 坚牢素 E-50(阳离子性有机氮衍生物,坚牢素染料用固色剂,商名,瑞士科莱恩)
Indosol SF dyes 坚牢素 SF 染料(通过反应固色的直接染料,湿牢度佳,特别适合涤/棉或毛/黏等一浴染色,商名,瑞士科莱恩)
indoxyl 吲羟,3-吲哚酚
Ind. Tex. J. (Indian Textile Journal)《印度纺织杂志》
induced activity 感应放射性;诱导活动性
induced colour 被诱导色,被感应色
induced dipol-force 诱导偶极力
induced draft 引风,强制通风
induced draft cooling tower 强制通风式冷却塔
induced force 诱导力
inducing catalyst 诱导催化剂
inducing colour 诱导色,感应色
inducing flow 引流

induction 感应[现象];归纳[法]
induction charging 感应充电
induction coil 感应线圈
induction effect 诱导效应;感应效应;邻位效应
induction heating 电感应加热
induction motor 感应电动机
induction switch 感应开关
induction tachometer generator 感应测速发电机
induction voltage regulator 感应调压器
inductive colour 诱导色
inductive coupled plasma mass spectrometry method〈ICP-MS method〉电感耦合等离子体质谱法
inductive effect 诱导效应;感应效应;邻位效应
inductively coupled plasma〈ICP〉电感耦合等离子体
inductor 诱导物;感应器
inductosyn 感应式传感器
inductothermy 感应电热器
induline(=indulin) 引杜林染料,对氮蒽蓝
induration 硬化
industrial〈ind., indus.〉工业的
industrial alcohol 工业酒精
industrial automation 工业自动化
industrial business 工业企业
industrial carbon 工业碳
industrial chemistry 工业化学
industrial clothing 工业用服装
industrial diamonds 工业用钻石
industrial engineering 工业管理学,企业管理学,工业工程学
industrial fabric 工业[用]织物

Industrial Fabrics Association International〈IFAI〉国际工业用织物协会
industrial felt 工业用呢
industrial gum 工业胶
industrial hygiene 工业卫生
industrial instrument 工业测量仪表
industrialization 工业化
industrial laundering 工业洗涤
industrial lubricant 工业润滑油
industrial methylated spirit〈IMS〉工业用甲醇变性酒精
industrial noise pollution 工业噪声污染
industrial organization 工业组织
industrial pollution 工业污染
industrial process control〈IPC〉工业过程控制
industrial process control computer〈IPCC〉工业过程控制计算机
Industrial Safety and Health Law 工业安全卫生法(日本法)
industrial sewage 工业污水
industrial sewing machine 工业用缝纫机
industrial soap 工业皂
industrial standard 工业标准
industrial textiles 工业用纺织品
industrial waste 工业废弃物《环保》
industrial waster water 工业废水
industrial water 工业用水
industrial wiping cloth 工业用揩布
industry〈ind.〉工业
industry waste 工业废物;工业废水
inequality of width 不等幅
inert 惰性的;不活泼的;无光学活动性的
inert additive 惰性添加剂
inertance 惰性

inert gas 惰性气体
inertia 惯性,惰性;惯量
inertia force 惯性力,惰性力
inertial deposition 惯性沉降
inertial effect 惯性效应
inertial system 惯性装置
inertness 惰性
inert plasticizer 惰性增塑剂
inextensible 不可延伸的
inf.(infinite) 无限的;不定的
inf.(infinity) 无穷大
inf.(information) 信息,资料,情报
infant's wear 幼儿服
infection 感染,污染
infeed device 进布装置
in-feeding 进料,加料
infeed opening 进布口
infeed roller 进布辊
inferior 低级的,劣质的
Inferol MSA 因费罗尔 MSA(以烷基硫酸盐为基础的阴离子型无甲酚丝光润湿剂,易生物降解,商名,德国波美)
infiltration 渗滤;渗入物;地下水
infinite〈inf.〉 无限的;不定的
infinite bath 无限浴比,恒浴比(指纤维浸入液的浴比无限大)
infinite dyebath 无限染浴
infinite dyebath concentration 无限染浴浓度
infinitely variable adjustment 无级调整
infinite reflux 无限回流,全回流
infinite swelling 无限溶胀
infinite variable speed mechanism 无级变速装置,无级变速器
infinitive variable gear box 无级变速齿轮箱

infinity〈inf.〉 无穷大
inflammability 易燃性,可燃性,燃烧性,点燃性
inflammable 易燃的
inflammable fabric 易燃织物
inflammable gas 可燃性气体
inflatable 可膨胀的,可充气的
inflatable buildings 充气建筑物
inflatable fabric 可充气织物
inflatable life jacket 充气救生衣
inflatable matrix 充气模子(圆网雕刻)
inflatable pressure body 膨胀体(印花刮刀上用)
inflatable sac 可膨胀的囊,可充气的囊
inflatable structure 充气结构
inflated garment 充气服
inflating agent(＝inflating medium) 发泡剂;充气剂
inflation 充气,膨胀
inflection 偏移;弯曲,拐折;拐点
inflection point 拐点,弯曲点,转折点
inflection temperature 拐点温度
inflexibility 无伸缩性,无调节性
inflexible 不可弯曲的,刚性的
inflexion 偏移;弯曲,拐折;拐点
inflow 进气,流入
influence 影响
influent 进水,流入水
influx 流入[量]
INFO 信息,信息显示
informal dress(＝informal suit) 日常服,便服
informal wear 日常服装,非正式服装
informatics 情报学
information〈inf.〉 信息,资料,情报
information bit 信息位

Information Council on Fabric Flammability 〈ICFF〉织物耐燃性情报理事会
information label 产品说明标签
information of super highway 信息高速公路
information polymer 信息聚合物
information retrieval 信息检索,情报检索,资料检索
information systems 信息系统
information technology〈IT〉信息技术
information technology industry 信息技术工业
information technology products 信息技术产品
infra-audible sound 次声;超低声
infrared〈IR,IR.〉红外线;红外线的
infrared absorber 红外[线]吸收体
infrared absorbing dyeings 红外吸收染色
infrared absorbing dyes 红外吸收染料
infrared absorption 红外[线]吸收
infrared absorption anisotropy 红外[线]吸收[的]各向异性
infrared absorption band 红外吸收[光谱]带
infrared absorption spectroscopy 红外[线]吸收光谱学
infrared absorption spectrum 红外[线]吸收光谱
infrared camouflage dyes 红外伪装染料
infrared camouflage system 红外伪装系统
infrared camouflage technology 红外伪装技术
infrared chromatograph 红外光谱仪
infrared device 红外装置
infrared dichroism 红外二[向]色性
infrared directional radiation heater 红外定向辐射加热器
infrared discrimination of camouflage 红外鉴别伪装
infrared drying 红外[线]烘燥,红外[线]干燥
infrared drying machine 红外线烘燥机
infrared electronic feeler 红外电子探测器
infrared emissivity 红外线放射率
infrared emitter 红外辐射源
infrared fixation of reactive dyeing or printing 活性染料红外[线]固着印染法
infrared functional group 红外[吸收]官能基团
infrared gas burner 煤气红外线燃烧器（常用作织物的预烘装置）
infrared generator 红外发生器
infrared heater 红外加热器
infrared heater bank 红外加热组件
infrared heating 红外[线]加热
infrared heating dryer 红外加热烘燥机
infrared lamp 红外线灯
infrared laser 红外激光器
infrared measurement 红外[光谱]测定
infrared microscopy 红外显微术
infrared microspectrometer 红外[线]显微分光仪
infrared moisture measurement 红外线测湿[法]
infrared moisture tester 红外线测湿仪
infrared photo-electric stopmotion 红外光电制动
infrared photography 红外线照明
infrared pickoff 红外传感器
infrared predryer 红外预烘机
infrared radiation 红外[线]辐射
infrared radiation proof 防红外辐射

infrared radiation protection textile 防红外辐射纺织品

infrared radiation pyrometer 红外辐射高温计

infrared radiation resistance 抗红外辐射

infrared radiator 红外辐射器

infrared ray 红外线

infrared ray detecter (IRD) 红外线检测器

infrared ray gas burner 煤气红外燃烧器

infrared ray tenter 红外[线]拉幅机

infrared reflecting dyeings 防红外染色;红外反射染色

infrared reflection coating 防红外涂层

infrared scanning 红外扫描

infrared sensor 红外传感器

infrared source 红外[辐射]源

infrared spectrogram 红外[线]光谱图

infrared spectrometer〈IRSP〉红外线分光计,红外线分光仪

infrared spectrophotometer 红外分光光度计

infrared spectroscope 红外[线]分光镜

infrared spectroscopy 红外[线]光谱学

infrared spectrum 红外[线]光谱

infrared spectrum analysis 红外光谱分析

infrared technique 红外技术

infrared thermometer 红外[线]温度计

infrared transmitting fibre 红外透射纤维

infrastructure 下部结构,底层结构,基础结构

infusibility 不熔性,难熔性

infusible 不熔的,难熔的

infusion 浸入,浸渍,浸渍液,浸剂

infusorial earth 硅藻土

in gear 啮合

ingrain 原纱染色;原料染色的产品

ingrain colouring matters 原地显色染料（指直接在被染物上成色的染料）

ingrain dyeing 原纱染色(织物上产生混色效应的一种方法)

ingrain dyes 原地显色染料

ingrain knitting yarn 色纱;染色针织纱

ingrain yarn 混色纱,多色混合纱

ingredient 拼分,拼料,配料;[混合物的]组成部分

ingress 进入;入口处;通道

inhalation 吸入法,吸入

inhaler 吸气器;滤气器

inherent 原有的,内在的

inherent adhesion 固有黏着

inherent colour 固有颜色,本身颜色

inherently flame resistant 固有抗燃的

inherently flame retardant polyester 内在阻燃性聚酯[纤维]

inherent moisture 固有湿度,固有水分

inherent quality 内在质量

inherent regulation 自动调节

inherent viscosity 比浓对数黏度,特性黏度

inhibiting degradation 抑制降解

inhibition 抑制,禁止;阻化

inhibition phenomena 阻抑现象

inhibitor 阻聚剂,抑制剂,阻化剂;防锈剂

inhibitor of oxidation 氧化抑制剂

inhomogeneity 不匀性,多相性

in-house 机构内部的,近距离的

in-house quality control 车间质量控制

in-house use 自用

initial absorption 初期吸收,初始吸收

initial acidity 初始酸度

initial bath 头缸,初染浴

initial come-up time [设备]从开工到正常操作的时间
initial creep 初始蠕变
initial elasticity 初始弹性
initial elastic modulus 起始弹性模量
initial load 初期负荷,初始负荷
initial modulus 初始模量,初始模数
initial product 起始产品
initial program loading 初始程序调入
initial rate of dyeing 初染速率
initial rinse 初次冲洗
initial slope 起始斜率
initial state 初态,起始状态
initial stress 初应力
initial tension 初张力
initial velocity 初速度
initiating power 引发能力
initiation 引发[作用];起爆[作用]
initiative reaction 起步反应
initiator 引始剂,引发剂,起爆药;起始器
injection 注入,射入;注射,喷射
injection condenser 喷水凝汽器,喷射冷凝器
injection dyeing machine 喷射式染色机
injection moulding 注射模塑法
injection orifice 喷嘴
injection port 进样口
injection printing 喷射印花,喷墨印花
injection pump 注液泵
injection rinsing machine 喷射冲洗机
injectomat 自动喷粉机(糊料制备)
injector kier 注射式煮布锅
ink blue 蟹青[色]
ink characteristics 印墨特征
ink dyestuff 墨水色料
inked 修补着色,墨水涂染

inking 打墨印;修补着色
inking agent 修补剂;剥色剂(修复染疵,回染用)
ink jet dyeing 色料喷射染色
ink jet printing 喷墨印花
ink jet screen engraving 喷墨制网
ink mask exposing 喷墨掩蔽曝光
ink printed handkerchief 石印手帕
ink receptor 印墨接受器
inlay printing 凹纹着色轧花
inlet 入口,进口
inlet assembly 进布装置
inlet frame 进布架
inlet material 被套絮料
inlet/outlet combination 进/出布组合
inlet side 入口侧,入口端
inlet valve 进给阀
in line 在线的;在管道的;串联的;轴向的
in-line measurement 在线监测
in-line meter 管线上的仪表
inner casing 内罩
inner chelate 内络合物
inner complex salt 分子内错盐
inner decorating materials for automobile 汽车内饰面料
inner diameter 内径
inner indicator 内指示剂
inner phase [乳化体的]内相
inner quality 内在质量
inner race 内座圈,内环
inner salt 内盐
inner screw 阴螺旋,内螺旋
inner tube 内管
inner viscosity 结构黏度
innocuous 无[毒]害的

innocuousness 无害性；无毒性
innovation 革新，改革；新技术；合理化建议
inoculation 防水垢工艺
inodorous 无气味的
inorganic〈inorg.〉 无机的
inorganic acid 无机酸
inorganic analysis 无机分析
inorganic coagulant 无机凝聚剂
inorganic colours 无机颜料
inorganic compound 无机化合物
inorganic filler 无机填充剂
inorganic nano antibacterial agent 无机纳米抗菌剂
inorganic pigment 无机颜料
in parallel 并联
in phase 同相
in plane permeability 导水率
in-plant system 近距离[控制]系统
in-plant transport 厂内运输
in-plant transport system 厂内运输系统
IN process [还原染料]甲种染色法（染色温度60℃）
input 输入；输入功率或电压；输入端；输入信号
input bus〈IB〉 转入总线《计算机》
input channel 输入通道
input device 输入设备，输入装置
input drive 输入传动
input-output〈I/O〉 输入/输出
input quantity 投入量
input shaft 输入轴
ins.(inches) 英寸（inch 的复数）
ins.(insulated) 绝缘的
ins.(insulation) 绝缘
ins.(insulator) 绝缘子，绝缘体

in-scour recovery system 内洗涤式回收装置
insect growth regulator 昆虫生长抑制剂
insecticidal activity 杀虫活性
insecticidal finishing agent 防虫整理剂，杀虫整理剂
insecticide 杀虫剂
insecticide finish 抗虫整理
insect larva 昆虫幼虫
insect mark 虫渍
insect pest deterrents 制止害虫生长剂，防虫剂
insect proof 防蛀的，防虫的
insectproof finish 防蛀整理，防虫整理
insect protect net 防虫网
insect repellent 防蛀剂；防虫剂
insect repellent finish 防虫整理
insect resistance 抗蛀性
insect resistance finish 防虫整理
insensible perspiration 微量排汗
insensitive to heat 耐热的
inseparability 不可分离性
in series 串联，串接
inserter 插件
insertion group 嵌入基团
insertion hygrometer 插入式湿度计
insertion loss 介入损失
in-service inspection 运行中检查
in-service use 实际使用
inside caliper 内卡钳
inside engaged gear 内齿轮
inside-garment microclimate measuring system 内衣微气候测量系统
inside micrometer 内径测微计，内径千分表
inside oil retaining cup 内护圈油杯

inside width 净宽
in-situ developed dyes 原地显色染料
in-situ formation 原地形成
in-situ measurement 原位测量,现场测量
in-situ reactive dyeing 原地反应染色(活性染料新机理)
in-situ test 现场试验,原位试验
insol.(insoluble) 不溶的
insolation 日光暴晒
insole 鞋垫
insolubility 不溶性
insolubilization 不溶解
insolubilizing process 不溶性处理
insoluble〈insol.〉 不溶的
insoluble azo dyes 不溶性偶氮染料
insoluble azo pigment 不溶性偶氮颜料
insoluble soil 不溶性沾污物
insoluble starch 不溶性淀粉
INSP(inspect,inspection) 检验,检查;验布观察
Insp.(inspector) 检查员
IN special A dyes 热染特别A型还原染料(还原温度71.11~76.6℃,染色温度48.89~60.0℃)

IN special B dyes 热染特别B型还原染料(还原温度48.89~60.0℃,染色温度71.11~76.67℃)
inspect〈INSP〉 检验,检查
inspectability 可检查性
inspecting and measuring machine 验布量布机
inspecting board 检验板
inspecting machine 验布机
inspecting measuring and rolling machine 验布量布卷布机
inspecting standard 检验标准

inspection〈INSP〉 检查,检验;验布观察
inspection bureau 检验局
inspection by attributes [品质]项目检验
inspection by variables 对样差异检验《染》《整》
inspection certificate 检验证书
inspection frame 验布机,验布架
inspection hole 检查孔
inspection item 检查项目,检验项目
inspection light 检验灯;验布灯
inspection mirror 验布镜
inspection port 观察孔
inspection probe 检查针;探针
inspection stand 检验台(如检验圆网或平网雕刻质量的一种专用设备)
inspection table lighting 检验台灯光
inspector〈Insp.〉 检查员
inspiration 进气
inspissated 浓缩;凝结
inst.(instant) 即时的
inst.(institute) 学会;协会;学院
inst.(institution) 学会;院;制度
instability 不稳定性
Instacolor liquid dye dispensing system 伊斯顿染料调配系统(商名,美国制)
installation 安装,装置,装配;设备;整套装置;机群(由多台同种设备组成)
installation and commissioning 安装和调试
installation dimension 安装尺寸
installed 安装
instant〈inst.〉 即时的
instantaneous acting relay 瞬时继电器
instantaneous deformation 瞬时形变
instantaneous elasticity 瞬时弹性
instantaneous modulus 瞬时模量

instantaneous recovery 瞬时复原；瞬间回收
instantaneous set 瞬时变定；瞬时凝固
instantaneous stopping 瞬时停顿
instantaneous value 瞬时值，即时值
instant chromatography 瞬息色谱法
instant recovery 瞬间回复
institute〈inst.〉学会；协会；学院
Institute of Environment Service〈IES〉环境研究所
Institute of Textile Technology〈ITT〉纺织工艺研究所(美国)
institution〈inst.〉学会；院；制度
institutional laundering 规定洗涤法
Instron apparatus 英斯特朗电子强力机
Instron capillary rheometer 英斯特朗毛细管流变仪
Instron tensile tester 英斯特朗电子强力测试仪(可测试纤维材料、塑料、橡胶、皮革、纸张等的载荷延伸及耐撕、耐压等性能)
instruction 说明，说明书；指示，指令
instruction manual 使用说明[书]
instrument 仪器，仪表；器具
instrument air 仪表[用]气
instrumental analysis 仪器分析
Instrumental Colour System〈ICS〉仪器测配色系统(国际羊毛咨询处采用)
instrumental constant 仪器常数
instrumental error 仪器误差
instrumental mannequin 仪器式人体模型(耐燃试验用)
instrumental match (= instrumental [colour] matching) 仪器配色
instrumental match prediction 仪器配色测试法

instrumentation 测试设备；检测仪表；仪表化
instrumentation and controls 控制仪器
instrument board 仪表面板
instrument cluster 仪表组(仪表板上所装的各种组合仪表)
instrument fan 仪表风扇
instrument panel 仪表盘，仪表面板，仪控制板，仪器操纵盘
instrument range 仪表量程
insulant 绝缘材料
insulated〈ins.〉绝缘的
insulating ability 绝缘能力；保暖性能
insulating cap 绝缘帽《电》
insulating disc 绝缘片
insulating group 绝缘基
insulating jacket 保温夹套
insulating material 绝缘材料，绝缘物
insulating panel 绝热板
insulating side panel [烘房]绝热门板
insulating sleeve 绝缘套《电》
insulating tape 绝缘包布，绝缘带
insulation〈ins.〉绝缘
insulation against cold, heat and noise 隔冷、隔热和隔声
insulation layer 绝热层
insulation panelling [蒸箱或烘房的]绝热墙板
insulation potential 绝缘能力
insulation resistance 绝缘电阻
insulation wall 绝热墙
insulator〈ins.〉绝缘子，绝缘体
intact 完整的
intaglio 凹纹雕刻
intaglio pattern 凹纹雕刻模版
intaglio printing 凹纹雕刻印花(如滚筒印

花）

intake 输入,吸入,引入
intake silencer 进气减音器
intake valve 进气阀
in tandem 一前一后,串联的
integral detector 积分检测器
integral diffusion coefficient 积分扩散系数
integral drain cooler 内置式疏水冷却器
integral economizer 内置式省煤器
integral enthalpy 总体焓,总体热函
integral heat of adsorption 积分吸附热
integral part 整装部分
integral type 整体式
integrant 组成部分;组成的
integraph 积分仪
integrated 一体化的,集成的
integrated biological-chemical process 生物—化学联合法
integrated circuit 集成电路
integrated company 联营公司
integrated console 联控台
integrated data processing 集中数据处理
integrated dyeing 综合染色法;一体化染色法
integrated environmental control 综合环境控制
integrated finishing 多功能整理
integrated finishing and shrinking range 织物综合整理联合机（将树脂整理与机械防缩整理合并为一个过程）
integrated frequency 综合频率
integrated mill 联合［工］厂（如包括纺、织、印、染等）,全能［工］厂,综合［工］厂
integrated paste preparation system〈IPS〉［印浆］自动配装系统,印花糊料制备综合系统
integrated printing factory 联合印花厂,综合印花厂
integrated process 联合工艺,综合工艺
integrated production 综合性生产,全能性生产,一体化生产
integrated quality control 综合质量控制
integrated straightening unit 综合式整纬单元
integrated system 综合性加工系统,集合性加工系统
integrated waste water treatment 污水综合处理
integrating circuit 积分电路《电》
integrating instrument 积分仪
integrating sphere 积分球,累计球
integrating sphere photometer 积分球光度计
integration 一体化
integrator 积分仪,积分装置
integrator circuit 积分电路,积分电路
integrity 完整性,完全性;完善;完整保持性
intelligence quotient〈IQ〉智商
intelligent clothing 智能化服装
intelligent comfort finish 智能舒适性整理
intelligent composite material 智能复合材料
intelligent drives 智能化驱动
intelligent fabric guidance 智能化导布
intelligent materials 智能材料
intelligent sensor 智能型传感器
intelligent textile 智能纺织品
intensification 加深作用,放大作用,增强作用

intensified deep 增深
intensified test 强化的[洗涤]试验
intensifier 加深剂;加厚剂;增强器
intensity 强度
intensity level 强度级
intensity of illumination 照明强度
intensity of irradiance 照射强度
intensive developing 充分发色,充分显色
intensive steaming 充分汽蒸
inter- (构词成分)表示"在……中""在……间""在……内""相互"
inter. (intermediate) 中间的;中间体;中间色;中间产品
interaction 相互作用,交互作用;干扰,干涉
interaction of dyes and finishes 染料与整理剂的相互作用
interaction parameter 相互作用参数
interactive add 人工添加,人工补充
interactive discussion 相互讨论,交流
interactive mode 交互方式
interactive software 交互软件
interactive textiles 互感纺织品;功能性纺织品
intercalation 夹层;插入,夹杂;隔行扫描
interception 拦截作用(过滤机理)
interceptor 油分离装置;截流器,窃听器
interchain force 键间结合力
inter-chain reaction 交链反应
interchange 互换,交替
interchangeability 互换性,交换性
interchangeable parts 可互换零件
interchange impregnation 交换浸渍,交换浸轧
interchanger 交换器,换热器
interchange reaction 互换作用,互换反应

inter colour 国际流行色
interconnection bus signal 〈IBS〉互联总线信号《计算机》
interconnection network 互联网络
intercontinental ballistic missile 〈ICBM〉洲际弹道导弹
interconversion 互换,相互转换;变换
intercooler 中间冷却器
intercorrelation 组间相关
inter-cross-links 分子间交联键
inter-diffusion 相互扩散
interdiffusion coefficient 相互扩散系数
interesterification 酯交换[作用]
interface 界面;接触面;接口;接口设备
interface temperature 界面温度
interface transfer coefficient 界面传递系数
interfacial 界面的
interfacial activity 界面活性
interfacial adhesion 界面黏合,界面黏附
interfacial agent 界面活性剂
interfacial area 界面面积
interfacial bond 界面黏合,界面黏附
interfacial concentration 界面浓度
interfacial degradation 界面破坏
interfacial electrical phenomenon 界面电气现象
interfacial electric double layer 界面二重电荷层
interfacial energy 界面能
interfacial esterification 界面酯化
interfacial failure 界面破坏
interfacial film 界面薄膜
interfacial free energy 界面自由能
interfacial graft polymerization 界面接枝聚合

interfacial layer 界面层
interfacial migration 界面泳移
interfacial migration theory 界面移染理论
interfacial phenomenon 界面现象
interfacial photoreaction 界面光反应
interfacial phototreatment 界面光处理
interfacial polarization 界面极化
interfacial polycondensation 界面缩聚
interfacial polymerization 〈IFP,I.F.P.〉界面聚合
interfacial polymerization process 界面聚合工艺(用于羊毛防毡缩整理)
interfacial potential 界面势能
interfacial pressure 界面压力
interfacial reaction 界面反应
interfacial shear stress 界面剪切应力
interfacial strength 界面强力
interfacial stress 界面[摩擦]应力
interfacial surface area 界面表面积
interfacial synthesis 界面合成
interfacial tensimeter 界面张力计
interfacial tension 界面张力
interfacial tension ring test 界面张力环试验
interfacial viscosity 界面黏性
interfacing line 分界线
interfacing materials 粘衬材料
interfelting 毡化,毡合;缩绒,缩呢
interference 干扰,干涉,相互影响
interference colour 干扰色,干涉色
interference colour chart 干涉色图谱
interference filter 干扰滤光片
interference fit 干涉配合,静配合;压配合
interference microscope 干涉显微镜
interference microscopy 干涉显微术
interference spectrum 干涉光谱

interferometer 干涉仪
interferometric method 干涉测量法
interferometric microscopy 干涉测量显微术
inter-fibrillar amorphous region 原纤间无定形区
inter-fibrillar coupling 原纤间联结
inter-fibrillar H-bonding 原纤间氢键合
inter-fibrillar swelling 原纤间溶胀[作用]
inter-fibrillar void 原纤间空隙
interflaw 瑕疵间[的]
interflow 合流,混流;互通
interfusion 渗透;混合,融合
intergrated diagnosing system 集成诊断系统
interim 间歇;中间
interim storage [机器]暂时存放
inter-instrument agreement 仪器间测量一致性
interior 内部;内部的;室内装饰用料
interior craft 室内装饰品,室内工艺品
interior lace 室内装饰花边
interior textiles 室内装饰纺织品
interior trim 内部装饰
interlaboratory study 实验室间的研究
interlaboratory testing 实验室之间的试验
interlace 交织,经纬交错
interlaced fabric [纤维交缠的]非织造布
interlacing 交叉存取《计算机》;交替;交织;隔行扫描
interlacing point 交织点,组织点
interlacing yarn 交络丝
interlayer 中间层;层间的
interlayer agent 层间化学剂
interlayment [地毯]隔垫
interlining 衬头,中间衬料

interlining fabric 衬里布,垫布
interlining material 衬料
interlock 联锁,闭锁;联锁装置
interlock circuit 联锁电路
interlock fabric 双罗纹针织物,棉毛布
interlocking 咬合;连结,联锁
interlocking device 联锁装置
interlocking motor 自动同步机
interlocking scales theory 鳞片毡合理论《毛》
interlock pants 棉毛裤
interlock relay 联锁继电器
interlock rib 双罗纹组织;棉毛布
interlock singlet 棉毛衫
interlock switch 联锁开关
interlock T-shirt 棉毛短袖圆领衫
interloop 咬合,连结,连锁
intermediary substance 中间产品
intermediate〈inter.〉 中间色;中间的;中间体;中间产品
intermediate colour 中间色
intermediate cooling 中间冷却
intermediate dryer 中间烘燥机(平网印花机的附属设备)
intermediate drying 中间烘燥
intermediate filtrate tank 中间过滤槽
intermediate finish 中间整理
intermediate frequency 中频
intermediate gear 中间齿轮,过桥齿轮
intermediate hue 中间色
intermediate image 中间[图]像
intermediate infrared 中红外
intermediate inspection 中间检查
intermediate lining material 中间衬料
intermediate of dyestuff 染料中间体
intermediate printing [新花样]中间插入印花
intermediate product 中间产品;半制品
intermediate pulley 中间皮带轮
intermediate reduction clear 中间还原清洗
intermediate relay 中间继电器
intermediate roller 中间辊,腰辊
intermediates 中间产品
intermediate shaft 中间轴,过桥轴
intermediate squeezer 中间轧车
intermediate squeezing 中间轧液
intermediate storage 中间存储器;中间存布器,容布箱
intermediate superheater 中间过热器,再热器
intermediate supporting bowl 中支轧辊(将辊体与辊轴的两端闷头深入辊体内相当距离,使辊体支点内移,以减少辊体的挠度)
intermediate wheel 过桥齿轮
intermediate zone 中间区(拉幅机喂入区和加热区间施加张力的区域)
intermedium 中间体,媒介物
intermeshing [齿轮]相互啮合
intermicellar reaction [微]胞间反应;胶束间反应
intermicellar swelling 微胞间溶胀;胶束间溶胀
intermicellary space 微胞间空隙;胶束间空隙
intermiscibility [相互][溶]混性
intermitted defect 周期性病疵;间隔染疵,周期性染色不匀
intermittent crabbing machine 间歇煮呢机
intermittent current 间歇电流
intermittent drive 间歇式传动装置
intermittent dyeing machine 间歇式染色机

intermittent filter 间隔过滤《环保》
intermittent gearing 间歇传动装置
intermittent motion (= intermittent movement) 间歇运动
intermittent operation 间歇运行
intermittent pattern 间歇性图案
intermittent pressing 间歇压呢
intermittent printing 间隔印花法
intermittent printing machine 间歇[式]印花机,圆网间歇接版印花机
intermittent process 间歇性工艺过程
intermittent rope washing machine 间歇式绳状水洗机
intermittent rotary screen printing 间隔式圆网印花
intermittent rotary screen printing machine 间歇[式]圆网印花机
intermittent spot effect 间隔点纹效应
intermittent stress relaxation 间歇应力松弛
intermittent stripe 断续条纹
intermittent take-up motion 间歇[式]卷布装置
intermixing gas burner 预混式气体喷燃器
intermixture 混合物
intermolecular 分子间的
inter-molecular attraction 分子间吸引力
intermolecular bonding energy 分子间键[合]能
intermolecular cohesion 分子间内聚力
intermolecular condensation 分子间缩合[作用]
intermolecular cross-linking 分子间交联
intermolecular interaction 分子间相互作用
intermolecular rearrangement 分子间重排[作用],分子重排
intermolecular repulsion 分子间斥力
intermolecular transfer 分子间转移
intermolecular transposition 分子间重排[作用],分子间转位[作用]
internal antistat 内用抗静电剂
internal arithmetic 内部运算
internal bond strain 内部粘接应力
internal combustion 内燃[烧]
internal cross-linking agent 内交联剂(涂料黏合剂)
internal cross-linking system 内交联系统
internal diameter 内径
internal dyeing 纺着染色;原液着色《化纤》
internal electrolysis 内部电解
internal energy 内能
internal expansion brake 内涨式制动器
internal fitting 内部配件,内部装置
internal fluorescent lighting 内设[验布]荧光灯
internal friction 内摩擦
internal gauge 内径规
internal gear 内接齿轮
internal heating 内部加热
internal heat source 内部热源
internal indicator 内部指示剂
internal interference 内干涉
internal lighting 内设照明灯
internal microstructure 内部微细结构
internal plasticization 内增塑[作用]
internal pressure 内压力
internal reflectance (= internal reflection) 内反射
internal reflexion spectroscopy 内反射光谱法

internal resistance 内摩擦
internal ribbed tube 内肋片管,内螺纹管
internal standard 内部标准,内控标准
internal standard line 内部标准线
internal standard method 内部标准方法
internal standard substance 内部标准物质
internal steam water baffle 锅内蒸汽挡板
internal strain 内应变
internal stress 内应力
internal stretcher [圆筒形针织坯布的]内拉幅装置
internal surface 内表面
internal viscosity 内[部]黏性
internal volume 内部容积
international〈intn.〉国际的,世界的
International Agency for Research on Cancer〈IARC〉国际癌症研究会
International Association for Quality〈IAQ〉国际质量协会
International Association of Clothing Designer〈IACD〉国际服装设计师协会
International Carpet and Rug Market〈ICRM〉 国际地毯展览会
International Center for Textile Research and Development〈ICTRD〉国际纺织研究和发展中心
International Commission on Illumination〈ICI,I. C. I.〉国际照明委员会
International Cubex Machine 国际寇贝克斯仪器(用以测试毛织物的毡缩度)
International Dry-cleaning Research Committee〈IDRC〉国际干洗研究委员会
International Electrotechnical Commission〈IEC〉国际电工委员会
International Fashion Colour 国际流行色
International Federation of Association of Textile Chemists and Colourists〈IFATCC〉国际纺织化学家及染色家协会联合会
International Federation of Cotton and Allied Textile Industries 国际棉花与棉纺织工业联合会
international gray scale 国际灰色样卡
International Institute for Cotton〈IIC〉 国际棉业协会
International Laundry Association〈ILA〉国际洗涤协会
International Organization for Standardization〈IOS〉国际标准化组织
International Sericultural Commission 国际蚕丝业委员会
International Silk Association 国际蚕丝协会
International Standardization Organization〈ISO〉国际标准化组织
International Standards Organization 国际标准协会
International Textile Bulletin〈ITB〉《国际纺织通报》(瑞士季刊)
International Textile Instilute〈ITI〉 国际纺织学会(英国)
International Textile Machinary Exhibition〈ITME〉国际纺织机械展览会
International Textile Manufactures Association〈ITMA〉国际纺织业联合会
International Textile Manufactures Federation〈ITMF〉国际纺织业联合会
International Trade Organization〈ITO,I. T. O.〉国际贸易组织
International Wool Secretariat〈IWS〉国际羊毛局
International Wool Secretariat Technical

Center〈IWSTC〉 国际羊毛局技术中心
International Wool Textile Organization〈IWTO,I. W. T. O.〉 国际毛纺织协会
internet 网络,互联网
internet bus〈IBS〉 联系总线《计算机》
internet explorer 网页浏览器《计算机》
interoperability 互用性
interpenetrating polymer network〈IPN〉 互穿聚合物网络（黏合剂乳液聚合的一种原理）
interpenetration 互相穿透,互相渗入
interphase 中间相；界面
interphase interaction 相间[的]相互作用
interphase tension 界面张力
inter-ply shifting 布层间的移动
interposition 中间状态
interpretation 说明,解释；翻译,注释
interreaction 相互反应,相互作用
interreflection 相互反射
interreflection ratio 相互反射率
interrupt 中断,断开；阻止,妨碍；间歇
interrupted steam supply 进汽管堵塞
interruption 断路《电》；中断；断续
intersecting jet 交叉喷嘴（蒸汽与液体混合喷出）
Inter-Society Colour Council,USA〈ISCC〉 美国色彩协会
interstage desuperheater 级间减温器
interstage gland 级间密封,级间汽封
interstice 细隙,间隙,裂缝
interstitial 空隙的,间质的
interstitial micro-dyebath [织物]纱线间空隙染液微浴比（湿转移印花的染液浴比在1∶1以下）
interstitial volume 间隙体积
intertexture 交织；交织物

intertwine 相互缠结
interval scale 间隔尺度
Inthion dyes 英锡洪染料（缩聚硫化染料,商名,德国制）
intimate apparel 贴身内衣
intn.（international） 国际的,世界的
intolerable contamination 超过容许值的污染
intolerance 不耐性
intra- （构词成分）表示"在内""内部"
intra-chain potential energy 链分子内位能
Intracron dyes 英彩克隆染料（活性染料的一种,含α-溴代丙烯酰胺基团,商名,英国制）
intra-cross-links 分子内交联键
intractable 难处理的,难加工的,难操作的
intrafibre 纤维内[的]
intramicellar reaction 微胞内反应；胶束内反应
intramicellar swelling 微胞内溶胀；胶束内溶胀
intra-molecular condensation 分子内缩合[作用],成环[作用]
intramolecular cyclization 分子内环化作用
intramolecular reaction 分子内反应
intra-molecular rearrangement 分子内重排[作用]
intramolecular stiffness 分子内硬挺性
intramolecular transfer 分子内转移
intramolecular transposition 分子内转位[作用],分子内重排[作用]
intricate design 复杂的花样[设计]
intrinsic 内在的,固有的,本质的,特性的
intrinsic anisotropy [纤维的]内在各向异

性
intrinsic deposite dyeing 内沉积染色
intrinsic dynamic viscosity 特性动态黏度
intrinsic humidity 内在湿度
intrinsic ignition time 特性点燃时间
intrinsic mechanical properties 内在机械性能
intrinsic non-flammability 内在的不燃性，本质的不燃性
intrinsic numeric 本质数值，特性数值
intrinsic property 内在特性
intrinsic sensing 本能感受
intrinsic strength 内在强度
intrinsic viscosity 特性黏度
intro- （构词成分）表示"在内""进入""向内"
introfaction 加速浸饱[作用]，浸透化
introfier 加速浸饱剂
intrusion 侵入，干涉
intumescence 膨胀[现象]；泡沸[现象]
intumescent 膨胀[现象]的，发泡的
intumescent coating 发泡涂层
intumescent effect printing 发泡印花
intumescent flame retardant 发泡性阻燃剂
intumescent system 膨胀系统(燃烧机理)
intussusception 吸收，摄取
in-use condition 使用条件
Invadine JFC 艳维典 JFC(聚乙二醇醚衍生物，渗透剂，商名，瑞士制)
Invalon OP 艳维隆 OP(邻苯基苯酚，弱阴离子型，聚酯及其混纺织物的染色载体，商名，瑞士制)
invariant property 不变性质
Invatex CRA 艳维泰克斯 CRA(膦酸衍生物，阳离子型螯合剂，适用于冷轧堆漂白工艺，商名，瑞士制)

inventory 库存量；在库品，库存
inventory control 在库管理，库存管理
inventory turnover ratio 库存周转率
inverse emulsion 逆相乳液
inverse emulsion system 逆向乳液型(一种制合成增稠剂的方法)
inverse flow 回流
inverse gas chromatography 反气相色谱法
inversion 转化
inversion temperature 转化温度
inversion zone 逆转区，逆转层
inverted bucket steam trap 倒吊桶式阻汽排水阀；倒吊桶式疏水器
inverted repulsion motor 反常的推斥式电动机
inverter 变频器，交流器，反相器，倒相器；换流器；逆变器
invert soap 转化皂
invert sugar 转化糖
investment 投资
investment cost 投资费用
investment factor 投资系数(产量增加时，投资的增加倍数)
invigilator 监视器
inviscid 不黏[滞]的，无黏性的
invisible loss 无形的损失，风耗
invisible thread 暗线
in vitro 在玻璃器内，在试管内
in vitro test 试管试验
in vivo 在体内；自然条件下的[实验]
in vivo test 活体试验，活体检验
invoice 发票；装货清单
involute gear 渐开线齿轮
inwall [锅炉]内部砖衬
I/O(input-output) 输入/输出
IOD(immediate oxygen demand) 直接需氧

量

iodeosin 四碘荧光素（化学分析指示剂）

iodide 碘化物

iodine adsorption 碘素吸收

iodine compound 碘化物

iodine lamp 碘灯

iodine number 碘值

iodine solubilization method 碘加溶法

iodine sorption method 碘吸附法，碘吸着法

iodine sorption value 碘吸附值，碘吸着值

iodine stain 碘染色（纤维用检定）

iodine value 碘值

iodine-wolfram lamp 碘钨灯

iodometric 碘量滴定的

iodometry 碘量滴定法

iodonium compound 碘鎓化合物

iodophor 碘载体（碘的有机络合物，作消毒用）

ion 离子

ion abundance 离子丰度

ionamine colour 离胺染料

ionamines 离胺染料

ion beam 离子束

ion beam sputtering target 离子束溅射目标

ion cattering rod 离子风棒（用于静电消除）

ion etching 离子蚀刻[法]，离子浸蚀[法]

ion exchange 离子交换

ion exchange chromatography 离子交换色层法

ion exchange fibre 离子交换纤维

ion exchange filter 离子交换过滤器

ion exchange membrane 离子交换膜

ion exchanger 离子交换剂

ion exchange reaction 离子交换反应

ion exchange resin 离子交换树脂

ion exchange treatment 离子交换处理法

ion exchange water 离子交换水

ion exclusion 离子排斥

ion-free water 无离子水

ionic adsorption 离子[性]吸附[作用]

ionic atmosphere 离子氛

ionic attraction 离子吸引

ionic bond 离子键；离子键合

ionic conduction 离子传导，离子导电

ionic dissociation energy 电离能；离解能

ionic dye 离子染料

ionic expansion 离子膨胀

ionic initiator 离子型引发剂

ionic linkage 离子键

ionic liquid dyeing 离子液体染色（一种无水染色工艺）

ionic micelle 离子胶束

ionic mobility 离子淌度，离子迁移率

ionics (= ionic surface active agent, ionic surfactant) 离子型表面活性剂

ionization 电离[作用]，离子化[作用]

ionization constant 电离常数

ionization detector 离子化检测仪

ionization pen 离子风棒（用于静电消除）

ionization potential 电离势

ionization rod 电离棒（用于检测燃烧器火焰的熄灭）

ionization wires 电离线（连接电离棒的电线）

ionizator 电离器

ionizer 电离剂

ionizing radiation 致电离辐射，电离性辐射；离子辐射

ionizing solvent 离子化溶剂

ion linkage 离子键
ionogen 电解质；离子化低聚物；可离子化的基团
ionogenicity 离子性
ionogenic linkage 离子化键
ion pair 离子对
ion selective electrode 离子选择电极
ion tolerence 耐离子性
IOS(International Organization for Standardization) 国际标准化组织
ip(＝i. p. , ignition point) 着火点，燃点
IPA(isopropyl alcohol) 异丙醇
IPA repellency 耐异丙醇性（防油性的一种指标）
IPC(industrial process control) 工业过程控制
IPCC(industrial process control computer) 工业过程控制计算机
IPDI(isophorone diisocyanate) 异佛尔酮二异氰酸酯
IPN(interpenetrating polymer network) 互穿聚合物网络（黏合剂乳液聚合的一种原理）
IPS(integrated paste preparation system) ［印浆］自动配装系统，印花糊料制备综合系统
IQ(intelligence quotient) 智商
IR(＝IR. , infrared) 红外线；红外线的
iraser 红外辐射；红外激光；红外激光器
IRD(infrared ray detecter) 红外线检测器
Irgalan dyes 依加仑染料（1∶2金属络合染料，用于羊毛、丝等纤维的染色，商名，瑞士制）
Irgaphor TBT 依加带TBT（喷墨印花用涂料印墨，用于各种纤维，商名）
Irgapodol AS 依加波杜尔AS（阴离子型匀染剂，适用于锦纶、羊毛、涤纶的连续染色及印花，商名，瑞士制）
Irgasol CO new 依加索CO新（阴离子性型纤维素纤维染色多功能助剂，有渗透、分散及保护胶体作用，商名，瑞士制）
Irgasol HTW 依加索HTW（羊毛保护剂，咪唑啉衍生物，用于毛涤混纺高温染色，商名，瑞士制）
iridescent 虹彩，闪色，闪光
iridescent effect 闪光效应
iridescent fabric 闪光织物
iridescent lustre 虹彩光泽；荧光光泽
iris 彩虹色；可变光阑；虹膜
iris action 膜片作用；可变光阑作用
irisated 虹彩的，彩虹色的
irisated print 彩虹［印］花布
iris blue 鸢尾蓝［色］（淡紫色）；蓝色酸性染料（也作着色剂用）
irised prints 彩虹［印］花布
Irish 爱丽纱，爱尔兰纱
Irish beetle finish 爱尔兰捶布整理
Irish finish 爱尔兰整理（棉织物的充亚麻整理）
Irish moss 爱尔兰苔（作糊料用）
iris orchid 紫藤色
iris print 彩虹［印］花布
iron 烙铁，熨斗；熨烫，熨平
ironability 耐熨烫性
iron alloy 铁合金
iron alum 铁［铵］矾，铁明矾
iron binding power 铁结合力
iron binding power test 铁结合力试验
iron blue 铁蓝［色］（带蓝灰色）；深蓝［色］
iron bottoming 铁媒染剂，铁盐打底剂（指

乙酸亚铁溶液)
iron buff 铁黄[色](淡棕黄色);铁黄[矿物]染料(使水合氧化铁沉淀于棉纤维内);黄色颜料
iron chelate of indigo 靛蓝的铁螯合物
iron compound 铁化合物
iron-constantan thermocouple 铁—康铜热电偶
iron core 铁心
iron damp 湿烫
iron dry 熨[斗烫]干;干熨
ironer 轧液机,轧布机;烫衣工
iron fabric 硬挺整理的加光织物,增强摩擦轧光棉布
iron-free 免烫
iron-free finish 免烫整理
iron grey 铁灰[色](带绿光深灰色)
iron hoop band 打包铁皮
ironing 熨平,熨烫
ironing and preboarding machine [袜子]熨烫定形联合机
ironing board 烫衣板
ironing board cover 烫衣板盖布
ironing cylinder 熨烫滚筒
ironing fastness 耐熨烫牢度
ironing machine 熨烫机
ironing resistance 耐熨烫性
ironing test 耐熨烫牢度试验
iron liquor 粗制乙酸亚铁溶液(作媒染剂用)
iron mordant 铁媒染剂
iron out 熨平(衣服)
iron oxide 氧化铁
ironproofing 耐熨烫的
iron setting 熨烫定形
iron stain 锈迹,锈斑

iron tester 熨烫[收缩]试验仪
iron tie 铁扣(打包用)
irradiance 辐照度;照射度
irradiate 照射;辐照
irradiation 光渗;用紫外线照射;辐射
γ irradiation γ辐射;γ照射
irradiation effect 辐射效应
irradiation technology 辐射技术
IR-radiometer 红外辐射计
irregular 不匀
irregular bleach 漂白不匀
irregular creping 起绉不匀
irregular cutting pile 剪绒不正
irregular distribution of twist 捻度不匀
irregular dyeing 染色不匀(疵点)
irregular folding 不规则折叠
irregularity 不匀率
irregularity control 不匀率控制
irregularity index 不规则性指数,不匀率指数
irregularity of singeing 烧毛不匀(疵点)
irregularity test 条干均匀度试验,不匀率试验
irregular pattern 稀密不匀,稀密档
irregular selvedge 边不整齐
irregular stitch 针迹不良
irregular twist 松紧捻
irregular width 门幅不齐
irreversible 单面的(指织物);不可逆的
irreversible coagulation 不可逆凝固
irreversible deformation 不可复形变
irreversible process 不可逆过程
irreversible reaction 不可逆反应
irreversible swelling 不可逆溶胀
irritant 刺激的
irritate 刺激

irritating smog 刺激性烟雾
irritation test 刺激性试验
IRSP(infrared spectrometer) 红外线分光计,红外线分光仪
isatin 靛红,吲哚满二酮
ISCC(Inter-Society Colour Council, USA) 美国色彩协会
Isenokatagami 伊势印花纸版(日本式友禅印花用)
isinglass 鱼胶;云母;白云母薄片
island coating 局部涂层法
iso- (构词成分)表示"同""等""均匀""同分异构""异构"
ISO(International Standardization Organization) 国际标准化组织
isobar 等压线
isobutanol 异丁醇
isobutene(=isobutylene) 异丁烯
isobutyraldehyde(=isobutyl aldehyde, isobutyric aldehyde) 异丁醛(异丙醇的原料,抗氧化剂的中间体)
isochromatic 等色的;等色线
isochromatic stimuli 等色刺激
isochrome series 等色系列
iso-component 异构化组分
isocyanate 异氰酸盐;异氰酸酯
isocyanate adhesive 异氰酸酯黏合剂
isocyanate polymer 异氰酸盐聚合物
isocyanine 异花青
isodiazotate 反重氮酸盐;稳定重氮酸盐;反偶氮酸盐;异重氮酸盐
iso-elastic [名同]等弹性的
iso-elastic temperature 等弹性温度
isoelectic polymer 等规聚合物
isoelectric dyeing method [羊毛]等电点染色法

isoelectric heating 等电位加热
isoelectric point〈IEP〉 等电点
isoelectric scouring 等电点洗净
isoelectric wash 等电点洗涤
isoelectric wool scouring 等电点洗毛
isogonism 等角
isogradient 等梯度
isogram 等值线
isoionic point 等离子点(两性电解质)
isokinetic temperature 等动力学温度
Isolan K dye 依索伦 K 染料(无磺酸基型 1:2 金属络合染料,商名,德国制)
Isolan S dye 依索伦 S 染料(单磺酸型 1:2 金属络合染料,商名,德国制)
isolated colour 孤立色
isolated double bond 孤立双键
isolated pattern 大花纹
isolation 分离,隔离;绝缘
isolation function 隔离功能(如微胶囊)
isolation garment 绝热服
isolator 隔离开关;断路阀;绝缘体;单面波导管
isoleucine〈Ile〉 异白氨酸(蛋白质纤维结构成分之一)
isomer [同分]异构体,[同]异构物;同质异能素
isomerase 异构酶
isomeric [同分]异构的
isomeric colours 同质同谱色
isomeric match 无条件配色(分光反射率曲线一致的配色)
isomeride 异构体,异构物
isomerisation 异构化[作用]
isomerism 同分异构[现象]
d-isomerism 旋光异构[现象](光学异构现象,d指右旋性)

l-isomerism 旋光异构[现象]（光学异构现象,*l* 指左旋性）
isomerization 异构化[作用]
isomerization polymerization 异构化聚合[作用]
isomerizing agent 异构化剂
isomorphism [类质]同晶型[现象]；同晶型性,同型性
isonitrile 异腈；胩
isooctylphenyl polyoxyethanol 异辛基苯基聚氧化乙醇
isoosmotic pressure 等渗压
isoosmotic solution 等渗[压]溶液
isopentane 异戊烷
isopentene 异戊烯
isopeptide bond 异肽键
isophorone diisocyanate〈IPDI〉异佛尔酮二异氰酸酯
isophthalate 间苯二甲酸酯
isophthalic acid 间苯二甲酸,异苯二甲酸
isophthalic ester 间苯二甲酸酯,异酞酯
isopiestic distillation 等压蒸馏
isoprene 异戊间二烯
isopropanol 异丙醇（溶剂,灭菌剂,防腐剂）
isopropyl alcohol〈IPA〉异丙醇
isopropyl ether 异丙醚（染料溶剂,除油渍剂）
isopropyl myristate 肉豆蔻酸异丙酯；十四[烷]酸异丙酯（非离子柔软剂,润滑剂）
isoreactive dyeing 等速吸尽染色,等反应性染色
isoreactivity 等反应性
isosbestic point 等吸收点；等浓度点；等吸光点

isosmotic 等渗[压]的
isosteric heat of adsorption 等量吸附热
isotactic 等规,全同立构
isotemperature line（＝isotherm）等温线,恒温线
isothermal 等温的；等温线的
isothermal adsorption 等温吸附[作用]
isothermal change 等温变化
isothermal diffusion 等温扩散
isothermal dyeing method 等温染色法
isothermal process 等温法
isothermic dyeing 等温染色
isothermic dyeing method 等温染色法
isothiocyanate 异硫氰酸盐
isotints 等白系列（奥斯瓦尔德表色系中,白色含量相等的色系列）
isotones 等黑系列（奥斯瓦尔德表色系中,黑色含量相等的色系列）
isotonic solution 等渗溶液
isotope 同位素
isotope dilution analysis 同位素稀释分析
isotope effect 同位素效应
isotope exchange 同位素交换
isotope measurement 同位素测定法
isotope radiation 同位素放射
isotope static elimination 同位素静电消除器
isotope tagging detection 同位素示踪探测
isotope tracer technique 同位素示踪技术
isotron 同位素分析器
isotropic 各向同质性；各向同性的
isotropic fibre 各向同性纤维
isotropic material 各向同性物质
isotropic solution 各向同性溶液
isotropy 各向同性[现象]
isovalent colour circle 等值色环《测色》

isovalent colours 等价色(奥斯瓦尔德表色系中,具有相等白色和黑色量而色相不同的各种色)

isovel 等速线

isoviolanthrone 异紫蒽酮(阴丹士林紫R的化学名称)

IT(information technology) 信息技术

itaconate 衣康酸盐(或酯),亚甲基丁二酸盐(或酯)

itaconic acid 衣康酸,2-亚甲基丁二酸,亚甲基丁二酸

ITB(International Textile Bulletin)《国际纺织通报》(瑞士季刊)

itching feeling [苎麻]刺痒感

item 项目,条目,条款

iterative calculation 渐近计算法

ITF【法】法国纺织研究院

ITI(International Textile Institute) 国际纺织学会(英国)

ITMA(＝I. T. M. A.)【德】国际纺织机械展览会

ITMA(International Textile Manufactures Association) 国际纺织业联合会

ITME(International Textile Machinary Exhibition) 国际纺织机械展览会

ITMF(International Textile Manufactures Federation) 国际纺织业联合会

ITO(＝I. T. O., International Trade Organization) 国际贸易组织

ITT(Institute of Textile Technology) 纺织工艺研究所(美国)

ivory [yellow] 象牙黄[色]

ivory white 乳白色

ivy [green] 常春藤绿[色](暗橄榄绿色)

IW process [还原染料]乙种染色法(染色温度为45～50℃)

IWS(International Wool Secretariat) 国际羊毛咨询处,国际羊毛局

IWS fastness specifications 国际羊毛局色牢度标准

IWS superwash process 国际羊毛局超级洗涤工艺

IWSTC(International Wool Secretariat Technical Center) 国际羊毛局技术中心

IWTO(＝I. W. T. O., International Wool Textile Organization) 国际毛纺织协会

IWV【德】国际羊毛协会

Ixtle fibre 伊斯特尔硬质纤维

J

J-acid J酸(2-氨基-5-萘酚-7-磺酸的别名)
jacinth[e] 红锆石橙(橙色)
jack 千斤顶,起重器;插口,插座,塞孔
jack cord 塞绳
jacket cooling 夹套式冷却
jacketed 加夹套的,加套管的,外套覆盖的
jacketed kettle 夹套煮浆锅
jacketed multichamber vessel 夹套多室容器
jacketed reactor 夹套反应器
jacketed wall 双层壁,水汀夹板(内通蒸汽保温并防止产生水滴)
jacketing 外皮;种籽皮;套
jacking [roll] calender 卷布式轧光机
jacking finish 卷布式轧光整理
jacking screw 螺旋千斤顶
jack screw 千斤顶螺杆,起重螺杆
Jackson kier 杰克逊平幅交卷煮练锅
jack switch 插接开关,插头开关
jacquard blanket 提花毯
jacquard fabric 提花织物
jacquard pattern 提花式样
jacquard towel 提花毛巾
jade gray 灰粉绿[色]
jade green 翠绿[色],绿玉色
jade lime 果绿[色]
jado uniforms 柔道装
JAFCA(Japan Fashion Colour Association) 日本流行色协会

jam nut 锁紧螺母,保险螺母
Japan Chemical Fibres Association〈JCFA〉 日本化学纤维协会
Japan Chemical Industry Ecology Toxicology & Information Center〈JETOC〉 日本化学物质安全情报中心
Japanese acid clay 酸性白土
Japanese Industrial Standard〈JIS〉 日本工业标准
Japanese lacquer 日本漆
Japanese Patent〈JP〉 日本专利
Japanese Standards Association〈JSA〉 日本标准协会
Japan External Trade Organization〈JETRO〉 日本贸易振兴会
Japan Fashion Colour Association〈JAFCA〉 日本流行色协会
Japan tallow 日本蜡(用作柔软、防水整理剂)
Japan Textile News〈JTN〉《日本纺织新闻》(月刊)
Japan wax 日本蜡(用作柔软整理剂)
Japan Wool-product Inspection Institute Foundation〈JWIF〉 日本毛制品检查协会
japonette 印花棉绉布
jasmine 茉莉黄[色](淡黄色)
jaspe 混色效果(毛针织)
jasper 墨绿[色]
jasper red 碧玉红[色]
jaspe style printing 仿毛条印花(利用碱

皂化在化纤上的印花效果）
Javanese Batik prints 爪哇蜡防印花布（一种以绿地为主的深色花布）
java print 爪哇蜡防印花
javelle water 次氯酸盐消毒液
javellization 次氯酸钠消毒净水法
Javel water 次氯酸盐消毒液
jaw 爪；虎钳；夹紧装置；销键
jaw clutch 颚形离合器，爪型离合器
jaw coupling 爪形联接器
jaw spanner 爪形扳手
J-box J形箱，伞柄箱
J-box cold reaction chamber J形箱式冷反应室
J-box hot reaction chamber J形箱式热反应室
J-box scouring and bleaching range J形箱练漂机
J-box steamer J形汽蒸箱
JCFA (Japan Chemical Fibres Association) 日本化学纤维协会
jeaning 牛仔裤化
Jeanmaire process 贾美尔［还原染料］印花法（印浆中使用硫酸亚铁与氯化亚锡作还原剂，不用蒸汽蒸化的一种还原染料印花法）
jean pants 牛仔裤；紧身裤；粗斜纹棉布裤
jeans 牛仔裤；紧身裤；粗斜纹棉布裤
jel 冻胶，凝胶
jellification 胶凝作用；冻结
jelly 胶质物
jelly like 胶质状
jenappe 烧毛
jenappe yarn 烧毛纱线
jerry 简易的

jersey 乔赛（针织物统称），平针织物；紧身运动套衫
jersey cloth 匹头针织布（统称）；全毛或毛交织斜纹织物；有伸缩性的绒毛布；仿针织布
jersey crepe 仿平针组织绉绸
jersey dress 针织服装
jersey finish on wool 羊毛缩呢整理
jersey flannel 针织法兰绒
jersey velour 平针丝绒
jet aeration process 射流曝气法
jet ager 喷射汽蒸机
jet air circulation driers 喷射式空气循环烘燥机
jet air exhauster 喷射式排风机
jet beck 喷射式绳状染槽
jet black 烟黑［色］；乌黑［色］
jet blower 喷气鼓风机
jet bulking 喷射膨化加工
jet circulation dyeing machine 喷射循环染色机（成衣染色）
jet condenser 喷射冷凝器
jet cooker 喷射式煮浆锅
jet cylinder 喷热风式烘筒（用热风吹去织物表面蒸汽）
jet dryer 喷射式烘干机
jet dyeing 喷射染色
jet dyeing machine (＝jet dyer) 喷射式染色机
jet ejector 喷射器
jet engraving 喷射［雕刻］制网，喷蜡［雕刻］制网
jet engraving machine 喷射［雕刻］制网机，喷蜡［雕刻］制网机
jet flow 射流
jet hot air dryer 喷射式热风烘燥机

jet hot air drying 热风喷射式干燥
jet impingement 射流冲击
jet ink printing 喷墨印花
jet irradiation 喷射辐照
jet multi-layer dryer 多层热风烘呢机
jet nozzle 喷嘴
JETOC（Japan Chemical Industry Ecology Toxicology & Information Center） 日本化学物质安全情报中心
jet orifice 喷孔，喷嘴
jet overflow dyeing machine 喷射溢流[绳状]染色机
jet patterning machine 喷浆印花机（不用印花滚筒，不用筛网，而用喷头，将印花色浆喷至织物）
jet printer 喷墨印花机
jet printing 喷墨印花，喷射印花
jet printing machine 喷墨印花机，喷射印花机
jet printing system 喷射印花装置（最早应用于地毯印花，现应用于一般织物）
jet propulsion 喷气推进
jet pump 射流泵
jet quenching 喷射淬火，喷射冷却
jet rapid washer 喷射式快速水洗设备
JETRO（Japan External Trade Organization） 日本贸易振兴会
jet rope washing machine 喷射式绳状水洗设备
Jetscour DA conc. 精练剂 DA 浓（具有低泡、渗透、洗涤功能，商名，英国 PPT）
jet scouring 喷射精练
Jetscour LF conc. 精练剂 LF 浓（适用于棉及混纺品的煮漂，喷射染色机中使用，商名，英国 PPT）
jet screen 喷射制网，喷蜡制网

jet screen machine 喷射制网机
Jet Screen System 喷蜡制网体系（商名，日本东伸工业开发）
jet spray printing 喷液印花
jet squeezer 喷气轧车
jet stenter 喷射式拉幅机
Jettex A 捷特克司 A（酸性染料印墨，压电式喷墨印花机用，商名）
Jettex R 捷特克司 R（活性染料印墨，压电式喷墨印花机用，商名）
jet type hot flow dryer 喷射式热风烘干机
jet-type hot flue 喷射式热风
jet washer 喷射水洗机
jet washing 喷射洗涤
jewelry blue 宝蓝[色]
jeypore print 手工模板印花方巾（印度制）
jig 卷染机；夹具
jig dyeing 卷染[染色]
jig dyeing machine 卷染机，染缸
Jiget 一种间歇式平幅染色机（由卷染机和喷射染色机结合而成，商名，瑞士制）
jigger 卷染机；夹具
jigger raising 卷装起绒
jigging frame（= jigging stenter）摆动式拉幅机（能调节纬斜和使织物柔软）

jig machine 卷染机
jig motion device 往复装置；横动装置；摆动机构
jig padding 卷染机浸轧
jig scouring 卷染机洗涤
jig stenter 摆动式拉幅机（能调节线斜和使织物柔软）
jig winch 绳状染色机
jin-xiang satin 锦香缎（色织提花仿丝织

物)

JIS(Japanese Industrial Standard) 日本工业标准

J-method J形箱处理法

job 作业;职务

job analysis 职务分析

jobbing work 代加工工作

job description 职务规定,职务说明

job-dyeing 衣件零料染色

job finisher 代加工染整厂

job goods 整批廉价物

job lot 定购商品的批号

job rate 生产定额

jobs 零头布

job sequence 加工顺序

job step 工作步骤,加工步骤;作业段

job training time 培训时间

jockey 导轮

jockey pulley 张力盘;导轮;张紧轮

jockey pump 薄膜泵

jockey roller 导辊;张力辊

jockey sprocket wheel 过桥链轮

Jodd's USC system 乔特 USC 测色系统

jog 寸行,缓慢进行

jog backward 寸行向后

jog buttons 寸行钮

jog forward 寸行向前

jog function 寸行功能

jogging 寸行,微动

jogging suits 休闲服

jog key 齿合键

Johnson joint 约翰逊进汽头(美国生产的无填料、不加油的烘筒进汽装置)

joining shading 接版深浅[疵点]

joining stencil mark 接[头压]版印(手工网版印花疵)

join piece 联匹

joint 接头,接缝,接点

joint coupling 活接联轴节,万向联轴节

joint pin 连接销,接合针

joint products 联合产品

joint stencil mark 接版印,印花搭脱,框板印(平网印花疵)

joint ventures 合资经营,联营;联营企业

jonquil [yellow] 长寿花黄[色](带红光黄色)

Joule's equivalent 热功当量

journal〈Jour.,Journ.〉 会刊,期刊,杂志;日报;轴颈,辊颈

journal bearing 轴颈轴承,径向轴承

journal brass 轴颈筒套

Journal of the Society of Dyers and Colourists〈J. Soc. Dyers Col.〉《染色家协会会志》(英国月刊)

Journal of the Textile Institute〈J. T. I, J. Text. Inst.〉《纺织学会会志》(英国月刊)

Journal of the Textile Machinery Society of Japan 〈J. Textile Machy. Soc. Japan〉《日本纺织机械学会会志》(月刊)

journal rest 轴颈支承

joy stick 操纵杆,控制杆

JP(Japanese Patent) 日本专利

JSA(Japanese Standards Association) 日本标准协会

J-scray 小型 J 形堆布箱

J-shaped chute J形箱

J. Soc. Dyers Col.(Journal of the Society of Dyers and Colourists)《染色家协会会志》(英国月刊)

J. Textile Machy. Soc. Japan(Journal of the Textile Machinery Society of Japan)《日

本纺织机械学会会志》(月刊)
J. T. I. (=J. Text. Inst., Journal of the Textile Institute)《纺织学会会志》(英国月刊)
JTN(Japan Textile News)《日本纺织新闻》(月刊)
J-tube J形箱
judgement 评价,鉴定
juglone 胡桃酮(5-羟萘对醌)
jumbled ground 混色地
jumbo batch winder 大卷装卷布机
Jumbo jigger 特大卷染机(重型,多功能,商名,德国寇斯脱)
jumbo package 特大卷装
jumbo sized roll [织物]大卷装
jumper 背心装,工作服
jumper printing machine 间歇[式]印花机,圆网间歇接版印花机
jumper tube 快接管
jumper underpress 下热式熨烫机
jumper wire 跨接线
jumping device 跳跃装置
jumping sheet [失火时救援用的]接跳布
junction 接头,接点;连接

junction box 接线盒《电》
junction point 联结点
jungle green 丛林绿[色](深墨绿色)
jury pump 备用泵
just appreciable fading 始感褪色
just-in-time 即时化[生产]
just-in-time concept 即时化生产理念
just noticeable difference 细微差异,色差阈限(恰可察觉的差异)
just perceptible difference limen 最小识别阈(可感觉的最小神经刺激)
just size 正确尺寸
jute 黄麻
jute bowl 黄麻轧辊(将黄麻纤维薄片套入辊芯中,由压机加压,再经车磨而成,可作轧水辊筒)
jute packing 油麻丝;麻丝(嵌水管、煤气管等用)
juxtapose 并列循环图案
juxtaposition 并列,并置;邻近,接近
JWIF(Japan Wool-product Inspection Institute Foundation) 日本毛制品检查协会
JZS 【南】南斯拉夫标准协会

K

Kaai finish 卡艾整理(棉毛交织物无张力丝光整理,丝光后,棉纱收缩而形成绉纹)

K-acid K-酸,1-氨基-8-萘酚-4,6-二磺酸

Kagotsuko printing 卡格苏克防染印花(商名,日本)

kalimeter 碳酸定量器(碳酸盐中二氧化碳定量测定装置)

Kaloz process 卡洛兹工艺(用石灰和臭氧净化污水)

Kandar 坎达树脂处理法(挺爽性处理,耐洗,商名)

Kanekalon 可耐可龙(丙烯腈与氯乙烯的共聚物短纤维,商名,日本制)

Kannegieser flat-bed press 卡尼盖塞平板热压机(转移印花用)

kaolin 高岭土,白陶土,瓷土

kapillar-analysis 毛细分析[法]

kapok fibre 木棉纤维(天然保暖纤维)

Kapron 卡普纶(锦纶6纤维,商名,独联体制)

karaya gum 刺梧桐树胶,印花糊料

Karl Fischer method 卡尔·费歇尔法(快速测定含水量的方法)

Karl Fischer reagent 卡尔·费歇尔试剂

Karl Fischer titration method 卡尔·费歇尔微量水分滴定法

kasawari 印花平布(印度制)

Kashmir 开司米,山羊绒;开司米织物,山羊绒织物;精纺毛纱针织物

Kasuri dyeing 卡苏芮防染(商名,日本)

katabolism 分解代谢

Katanol mordant dyeing 卡他诺(合成媒染剂)媒染染色

Katanol ON 卡他诺ON(合成媒染剂,用于碱性染料染色)

kataphoresis 电泳

Kata thermometer 冷却[率]温度计,卡他温度计

katharometer 热导计

kauri 贝壳松脂(一种天然树脂,能完全溶解在丁醇中,用于检查烃类产品成分)

kauri-butanol value 贝壳松脂—丁醇(溶解)值

Kaurit M-90 考里特M-90(羟甲基三聚氰胺衍生物,纤维素织物的防皱整理剂,具有永久轧光效果,合纤织物的硬挺剂,商名,德国巴斯夫)

Kawabata Evaluation System 川端织物风格评定仪(用于测定织物手感等风格的测试仪,日本制)

Kayacelon E dyes 卡雅赛隆E染料(专用分散染料,用于涤棉中性一浴法染色工艺,商名,日本化药)

Kayacelon React CN dyes 卡雅赛隆反应CN染料(烟酸均三嗪活性基团的活性染料,商名,日本化药)

Kayacelon react colours 卡雅赛隆反应性染料(烟酸均三嗪型活性基团的活性染料,商名,日本化药)

Kayacion dyes 卡雅西昂染料(活性染料,

商名,日本化药)
Kayacryl dyes 卡雅克里尔染料(阳离子染料,商名,日本化药)
Kayacryl ED dyes 卡雅克里尔 ED 染料(分散性阳离子染料,商名,日本化药)
Kayacyl dyes 卡雅西尔染料(酸性染料,商名,日本化药)
Kayalon dyes 卡雅隆染料(分散染料,商名,日本化药)
Kayalon Microester 一种超细涤纶纤维专用分散染料(商名,日本化药)
kaya oil 椰子油
Kayaphor 卡亚福尔(荧光增白剂,商名,日本化药)
Kazakhstan carpets 哈萨克斯坦地毯
KC.(=Kc.,kilocycle) 千周
kcal.(kilocalorie) 千卡,大卡
keel mark 匹印,码印
keeper 定位螺钉;锁紧螺母;夹头,夹子
keep-off time 让开时间(机器开动前的间隙时间)
keep sample 存档样本,留底样本
Keesom force 取向力(偶极间引力)
Keffiyeh(=Kefieh) 凯菲手织印染布(全棉或丝绒棉纬,染色或印花,用作阿拉伯头巾)
Kelly green 凯利绿[色](深黄绿色)
kelp 海草;海草灰[色]
Kelpie process 克尔派整理(使羊毛或成品色泽鲜艳,并改善手感、弹性及防污性)
kelp pigment 海带色素(天然色素)
Kelvin 开尔文,开(开氏绝对温度)
Kelvin scale [开氏]绝对温标
Kelvin solid 开尔文固体(黏弹固体)
kemp 死毛,饻毛

Kemp's singeing machine 凯普烧毛机
keratin 角蛋白,角朊
keratinase 角蛋白酶
keratin cotton 角蛋白棉
keratin denaturing 角蛋白变性
keratin fibre 角蛋白纤维
keratin gene-transferred cotton 角蛋白转基因棉
keratinization 角蛋白化[作用]
keratinolytic reagent 角蛋白分解剂
keratoses 角质物
keratrans process 羊毛转移印花法
Kermel fibre 聚酰亚胺纤维(耐高温合成纤维,商名)
kermes 虫胭脂,虫红(最古的昆虫染料)
kermes scarlet 胭脂红
kernel 核,仁(果实籽内)
kerosene 煤油,火油
KESF system KESF 织物风格仪
kesis 条子布,色布(色泽鲜艳,印度、巴基斯坦制)
Kestner evapourator 凯斯特纳[式]蒸发器
ket- (构词成分)表示"酮"
ketene acetylation 乙烯酮法乙酰化[工艺]
keto- (构词成分)表示"酮"
keto-aldehyde resins 酮醛树脂
γ-ketobutanol 3-酮基-1-丁醇,单羟甲基丙酮
ketohexose 己酮糖
ketone 酮
ketone formaldehyde resin 酮类甲醛树脂
ketone resins 酮类树脂,酮树脂
ketose 酮糖
kettle 槽,釜,锅
kettle rub mark [织物的]染缸擦痕(疵点)

Kevlar 凯夫拉(聚对苯二甲酰对苯二胺纤维,也称作纤维B,商名,美国制)
Kevlar IT 凯夫拉IT(浸渍聚四氟乙烯乳液的凯夫拉纤维,商名,美国制)
key 键;信息标号;关键码
key bed 键
keyboard 键盘;电键板
keyboard entry 键盘输入
key colour 基本色
key componont 关键件
key factor 关键因素
key groove 键槽
key-in 键盘输入
Keymo finish 凯莫整理(用硫酸溶液处理的毛织物防缩整理,商名,英国制)
key number 索引号
key operation 关键作业
keyway 键槽
KF-dyeing process 小浴比染色工艺
kg(=kg., kilogram) 千克,公斤
kgf(kilogram force) 千克力,公斤力
khahua 印度匹染粗厚布
khaki 卡其黄[色],土黄[色]
khaki drills 卡其布
khaki grey 卡其灰色
Khangas 肯加围巾布(东非国家印制的图案)
kheetee 东南亚轧光布
khum 染色粗平布(土耳其名称)
kickback 回弹,退还;逆转,倒转
kick back pinion 换向小齿轮
kicker mill 锤式缩绒机;(制帽用)锤式缩呢机
kick-off 放液,排液
Kidderminster carpets 基德明斯特地毯(亦称苏格兰地毯)

kid-finished cambric 细薄棉布(用高级埃及棉纱织造,经柔软处理,作为少女夏季服装衬里)
kier 煮布锅,精练锅
kier bleach [煮布锅煮练后]漂白
kier blowing machine 釜式干法蒸呢机,密闭式蒸呢机
kier boiling 煮布锅[加压]精练
kier boiling aids 煮练助剂
kier boiling assistant 煮练助剂
kier-boiling fastness 煮练牢度
kier-boiling of coloured woven 色织物煮练
kierboil-peroxide bleach 煮练—氧漂
kier decatizing machine 罐蒸机
kiering 煮布锅煮练
kier jig 交卷式煮布锅
kier lining 煮锅衬里(煮锅内涂层,防止出现煮练斑)
Kierlon jet 奇利龙喷射(用于涤纶织物前处理,有去除油污、漂白功能,商名,德国巴斯夫)
Kierlon OLB conc. 奇利龙OLB浓(特殊精练净洗剂,非离子/阴离子,对去除油脂、蜡等杂质非常有效,商名,德国巴斯夫)
Kierlon OL conc. 奇利龙OL浓(精练净洗剂,主要用于棉及其混纺织物的精练,商名,德国巴斯夫)
kier mark 煮练斑渍
kier piler [煮布锅]象鼻甩布器
kier scouring 煮布锅[加压]精练
kier stain 煮练斑渍
kieselguhr 硅藻土
killer 去除剂;断路器
kilo- (构词成分)表示"千"
kilocalorie⟨kcal⟩ 千卡,大卡

kilocycle〈KC.，Kc.〉 千周
kilogram〈kg，kg.〉 千克，公斤
kilogram force〈kgf〉 千克力，公斤力
kilometre〈km，km.〉 千米，公里
kilo-newton〈kN〉 千牛[顿]
kilopascal〈kPa〉 千帕[斯卡](压力单位)
kilopoise〈kP〉 千泊(黏度单位)
kilotex〈ktex〉 千特[克斯](线密度单位)
kilovolt〈kV〉 千伏[特]
kilovolt-ampere〈KVA，kva〉 千伏安
kilowatt〈KW，K.W.，kw.〉 千瓦(特)
kilowatt-hour〈KWH，K.W.H.，kwh，kw-h，kw-hr〉 千瓦小时
kilted cloth 褶裥织物
kilting machine 打裥机
kimono silk 和服绸
kin[e]-(构词成分)表示"运动"
kinanthropometry 动态人体测量法
kinase 致活酶;激酶
kindling point 着火点,燃点
kindling temperature 着火温度
kinematic 运动学
kinematic fluidity 运动流度
kinematic viscosity 运动黏度
kinetic current 反应电流,动力电流
kinetic energy 动能
kinetic friction 动摩擦
kinetic friction coefficient 动摩擦系数
kinetics of dyeing 染色动力学
kinetic viscosity 动力黏度
kinetic Young's modulus 动态杨氏模量
kingfisher 翠鸟色
king-pin bearing 止推销轴承;止推框轴承
king's blue 品蓝[色],钴蓝[色](青光蓝色);蓝色矿物颜料

king's gold 雌黄(三硫化二砷)
king's green 巴黎绿[色](黄光绿色);砂绿(乙酰亚砷酸铜)
king's yellow 雌黄(三硫化二砷)
king tiro satin 金雕缎
kink stress 扭结应力
kinky 扭结的,卷曲的
kint goods pretreatment 针织品前处理
Kipp's apparatus 基普气体发生器
Kipp's gas generator 基普气体发生器
kiss coating 单面给胶涂层法
kiss dyeing process 单面给液染色工艺
kiss printing 凸纹滚筒印花;单面给液印花
kiss printing machine 凸纹滚筒印花机
kiss roll（=kiss roller） 单面给液辊;单面上胶辊
kiss roll padder 面轧车(用于单面给液)
kiss roll padding 单面给液法;单面上胶法
kiss roll pad dyeing 面轧辊轧染(用于织物单面染色或双面异色染色)
kissuto 棉印花布(东非名称)
kit 成套工具;用具箱
kitchen cloth 厨房用布
kitchen towel 揩巾
Kiton red G test 基通红G试验(测试羊毛性能)
Kiton Red test 基通红试验法(用于测试羊毛氯化的均匀性)
Kjeldahl analysis 凯达尔分析法(测定含氮量的方法)
KK process 短时轧卷[堆放]法
Klarit method 克拉里特软水法(酸性黏土与锰化合物制成的软水剂,商名,日本制)

Klauder-Weldon type [hank dyeing] machine 克惠氏旋转式绞纱染色机
klaxon 电喇叭
km(＝km.，kilometre) 千米，公里
kN(kilo-newton) 千牛[顿]
knap surface 绒面
Knecht process 克内希特纱线丝光鉴别法(用苯并红紫 4B 染色鉴别)
knee bend 直角弯头
knee-cap 护膝(运动员用)
knee joint 弯接头
knee over hose 舞袜
kneepad 护膝(保健用品)
knee-piece 斜撑
knife 刀；刀片
knife belt coater 刮刀带型涂布器
knife block 刀框，刀架子
knife breaker 刮刀式揉布机(用于丝绸或人造丝织物)
knife carrier 刀架，刀座
knife coater 刮刀涂布机，刮刀涂层机
knife coating 刮刀涂布
knife cut stencil 刀刻型纸，刀刻型版(花版)
knife edge 刀口，刀刃
knife edge mark 刀口印(码折疵)
knife frame 刀架
knife guard 刀[片]护罩，刀[片]护板，护刀罩；刀[片]自停装置
knife on-blanket coating 毯上刀涂层
knife over air 悬浮刮刀
knife over air coating 空气刮刀式涂层
knife over blanket coater 刮刀胶毯式涂布器
knife-over-roll coater 滚筒刮刀涂层机
knife-over-roll coating 刮刀滚筒式涂层

knife-over-roller method 滚筒上刮刀加工法(泡沫给液法)
knife roll coater 刮刀滚筒涂布器
knife switch 闸刀开关
knifing [地毯]割绒
knit 针织；编织；针织物
knit-de-knit [method] [印花]编织物拆编法(编织物经印花后拆散，以获得纱线的间隔染色效果)
knit-de-knit printing 假编法印花，编织拆散法印花
knit-de-knit space dyeing 假编法间隔染色，编织拆散法间隔染色
knit-de-knit space dyeing system 假编法间隔染色体系
knit fabric finishing 针织物整理
knit fabric shrinkage gauge 针织物收缩试验仪
knitgoods 针织品
knit goods resin-finishing 针织物树脂整理
knit guider 针织物导布器
knit picker 针织品缩水率测试仪器(美国制)
knitpicker test instrument 针织物收缩测试仪
knitted carpet 针织地毯
knitted composite 复合针织物
knitted curtains 针织窗布
knitted fabric 针织物，针织坯布
knitted gloves 针织手套
knitted goods 针织品
knitted neck tie 针织领带
knitteds 针织物
knitted stretch 弹力纱针织物
Knittex EPC 利特色 EPC(非离子性黏稠

液体,环氧树脂系产物,工业用毯、合纤绢网耐磨整理剂,对绢网具有防披、防变形作用,商名,瑞士制)

Knittex FF 利特色 FF(无甲醛树脂,防皱免烫整理剂,商名,瑞士制)

Knittex FRG conc. 利特色 FRG 浓(低甲醛反应性树脂,防皱免烫整理剂,商名,瑞士制)

knit tie 针织领带

knitting 针织;针织品

knitting pile carpet 针织圈线地毯

knitting wool balling machine 绒线成团机(团装绒线,卷绕设备)

knitting wool skein reeling machine(=knitting wool skein rolling machine) 绒线成绞机

knitting yarn 绒线

knit tubing 圆筒形针织物

knit vest 针织背心;针织内衣

knitwear 针织品

knitwear press 针织品熨烫机

knitwear setter 针织服装定形机

knit-weaving fabric(=knit-woven fabric) 针织机织织物

knob 节,瘤节;旋钮;按钮

knocking-off motion 碰撞式自停装置

knock-off 撞停;[碰撞]自停装置

knock-off finger 撞停触指

knock-off mechanism [碰撞]自停机构,关车机构

knock-off motion 碰撞式自停装置

knock over (染液)倒空

knock-wave 冲击波,撞击波

Knoop microhardness test 努普微硬度试验

knop 节;[割断的]绒头

knop cloth 毛圈织物

knoppy 有结节的

knop tweed 结子粗呢

knot breaking strength 接结强度,打结断裂强度

knot detector 布结探测器(安装在缩绒机等设备上,如遇织物打结,即会自动停车)

knot dyeing 扎染

knot effect 结子效应,绒结效应

knot knitting yarn 结子绒线

knotless 无结的

knot strength 打结强力,结节强力

knotted carpet 结子地毯,栽绒地毯

knotted pile carpet 打结绒头地毯

knotted point 打结点

know-how 技术诀窍;专门技能;实际知识

know-how textiles 专用织物;技能性纺织品,高技术含量纺织品

knowledge economy 知识经济

knowledge intensive industry 知识密集型产业

known number 已知数

known sample 已知样品

knurled nut 滚花螺母

knurled rod 滚花磁棒(圆网印花机上一种磁棒刮刀)

knurling 滚花

Kodel 科代尔(聚酯纤维,商名,美国制)

kokjak 蒟蒻粉(魔芋粉)

Kombi 科姆比试验法(测试染料扩散的方法)

kombi steamer 组合式汽蒸机

Koratran finishing 科拉特隆耐久压烫整理的延迟焙烘工艺

Koratran process 科拉特隆耐久压烫整理法（用二羟甲基亚乙基脲和催化剂硝酸锌的一种延迟焙烘整理法，美国科累特公司开发）

Kordofan gum 科尔多凡胶（即阿拉伯树脂）

kP(kilopoise) 千泊（黏度单位）

kPa(kilopascal) 千帕［斯卡］（压力单位）

KPR-180 plasma applicator equipment KPR-180型［光辉放电］等离子体处理机（商名，意大利与俄罗斯合作研究产品）

Kraft point 克拉夫特点（说明离子性表面活性剂性能）

Krostzewitz steamer 克劳斯威兹蒸化机（双回绕式蒸化机，印花布反面接触导辊）

Kroy chlorination process 克罗伊氯化法（羊毛氯化防缩整理法）

Kroy Deepim 克罗伊迪平工艺（赫科塞特羊毛制品连续氯化防缩工艺）

Kroy machine 克罗伊机（羊毛制品连续氯化防缩机）

Kroy process 克罗伊［毛织物］氯化防缩整理法（加拿大商名）

K/S (absorption coefficient/scattering coefficient) 吸收系数/散射系数（用以表示颜色深度的值）

K/S value K/S值，表面色深值

KTDF (Knitted Textile Dyers' Federation) 英国针织染色业联合会

ktex (kilotex) 千特［克斯］（线密度单位）

KTK process 浸轧热溶冷显色法

Kubelka-Munk equation 库贝尔卡—蒙克方程

Kubelka-Munk function 库贝尔卡—蒙克函数

Kubelka-Munk theory 库贝尔卡—蒙克理论

Kuralon 仓敷纶（聚乙烯醇短纤维，商名，日本制）

Kuraray 仓敷聚酯（聚酯纤维，商名，日本制）

kurkuma 姜黄素（天然染料）

kurtosis ［峰态］峭度，尖峰值

Kusters carpet dyeing range 库斯特地毯连续染色联合机（利用刮刀刀片将染液淌上地毯表面进行染色）

Kusters colour 库斯特着色器（地毯染色用）

Kusters mangle 库斯特轧车（一种均匀染色轧车，德国寇斯脱）

Kusters TAK process 库斯特 TAK 法（多色液流染色用于地毯加工）

kuteera gum 刺槐树胶

kV(kilovolt) 千伏［特］

KVA(＝kva,kilovolt-ampere) 千伏安

K value K值（黏度值）

KW(＝K.W.,kw.,kilowatt) 千瓦［特］

KWH(＝K.W.H.,kwh,kw-h,kw-hr,kilowatt-hour) 千瓦小时

kymograph 波动曲线记录器

Kynol 阻燃防化纤维（商名，日本制）

L

lab 试验室,实验室
lab assistant 实验助手
lab coat 试验室工作服
labdanum [resin] 南丹树脂,劳丹胶
lab dyer 试验室染色机
label 标签
label adhesive 标签胶黏剂
label applying machine 缝商标机
label attaching machine 钉商标机
label cloth 标签[用]布
label cutting machine 商标切割机
labelled size 标码
labelling 示踪,标记
labelling machine 贴商标机,上标签机
labelling scheme 商品标志编排法
labelling unit 上标签单元
label marking machine 打商标机,贴标签机
label printing 标签印花(用狭带印花作标签用)
label requirement 标签要求
label ribbon 商标带
label sewing machine 缝标签机
labile 不稳定的,不安定的,易变化的
labile bond 不稳定键
labile group 不稳定基团
labile hydrogen 不稳定氢
lability 不稳定性,易变性
laboratory 实验室,试验室
laboratory appliance 实验室仪器
laboratory automation system 实验室自动化系统
laboratory calender 实验室用轧光机
laboratory coating unit 实验室涂层装置
laboratory dispensers 实验室自动配液系统
laboratory drying cylinder 试验室用烘筒,小烘筒
laboratory dyeing machine 试验室染色机
laboratory jig 试验室用卷染机,小染缸
laboratory machine 试验室用机台
laboratory mangle(=laboratory padder) 试验室用轧车,小轧车
laboratory printing machine 试验室用印花机,小印花机
laboratory scale test 实验室规模试验
laboratory steam chest 试验室用蒸箱,小蒸箱
laboratory stirrer 试验室用搅拌器
laboratory test 实验室试验
laboratory test sample 实验室试验样品
laboratory-to-bulk reproducibility 小样大样的重演性
laboratory trial 试验室试验
labour intensity 劳动强度
labour intensive enterprise 劳动集约型企业,劳动密集型企业
labour intensive industry 劳动密集型产业
labour intensive machinery 劳动密集型机械
labour intensive process 劳动密集型加工
labour productivity 劳动生产率

labour saving 节省劳力,减轻劳力
lab size machine 试验室用机械
lab swatch 试验样本
lab test 实验室试验
labyrinth bearing 迷宫式轴承,曲径式密封轴承
labyrinth grease seal 迷宫式润滑脂封闭器,曲径式润滑脂封闭器
labyrinth ring 迷宫环
labyrinth seal 迷宫式密封,曲径式密封
labyrinth valve 迷宫阀
lac 高强力细绳;虫胶,虫漆,虫脂
Laccaprint 拉克普林(现成的罩印印花浆,商名,意大利宁柏迪)
laccase 虫漆酶(用于靛蓝染料的脱色)
lac dyes 虫胶天然染料(类似胭脂虫,呈黄棕色)
lace 花边,网眼织物;饰带;编带;鞋带
Lacet 拉塞特防缩整理(商名,美国制)
lachrymatory effect 催泪作用,催泪效果
lack 不足,缺乏
lacking levelness 染色不匀
lack lustre 无光泽;无光泽的
lack-paste in ground 印花露地,印地漏浆(印疵)
lacquer 漆;真漆;中国漆;清喷漆;漆器
lacquer cloth 树脂涂层织物
lacquer coated fabric 上漆织物,涂膜织物
lacquer finish 表面涂膜整理,上漆整理
lacquer layer [镍网上的]涂漆层
lacquer nickel screens 涂[感光]胶镍网
lacquer printing 漆料印花
lacquer prints 漆印花布(用漆料印花)
lacquer screen 涂漆筛网,上胶筛网(圆网雕刻)
lacquer suction machine [印花筛网]吸漆机

lacquer thinner 漆稀释溶剂
lacquer varnish 喷涂清漆,光漆
lact-(＝lacti-,lacto-)(构词成分)表示"乳""乳酸""乳糖"
lactalbumin 乳清蛋白
lactam 内酰胺(氨基酸脱去一分子水后所得的环状酰胺)
lactic acid 乳酸
lactone 内酯
lactone dyes 内酯染料
lactone process 内酯法(以环己烷为原料通过己内酯制取己内酰胺的工艺名称,美国研制)
lactonization 内酯化[作用]
lacto phenol cotton blue 乳酚棉蓝(用于显微镜检测羊毛损伤)
lactose 乳糖
Lactron 聚乳酸纤维(商名,日本制)
lacy 花边状的,网状的;有花边的
LAD(laundry air dry) 洗涤晾干
ladanum 南丹树脂,劳丹树脂
ladder-chain 梯形链条
laddering [毛圈]纵向脱散
ladder proof finish 防脱散整理
ladder stitch 梯缝;梯纹刺绣;脱散线圈
lading 装载,载重
ladle 勺;铁桶
laevo- (构词成分)表示"左的""左旋的"
laevo-configuration 左旋构型
laevulose 左旋糖
lag 滞后,延迟;惯性,惰性
lager 延迟器
lagged pipe 绝热管道,护热管道
lagging [汽锅等的]外套,保温套;滞后,延迟

lagging behind 滞后[运动]
lagging bow 中凹弓纬(织疵,从进布方向看去,纬纱呈凹形)
lagging casing 绝热外壳
lagging material 隔热材料;防护材料
laharia 印度扎染布
laid-in fabric 衬垫针织品
laid-in selvedge 折边
lake 色淀,沉淀色料
lake[red] 胭脂红[色]
lake blue 湖泊蓝[色](浅暗蓝色)
lake colour 色淀染料
lake green 湖绿[色],摩洛哥绿[色]
laky 胭脂红色的
Lambda process 拉姆达[毛织物]氯化防缩整理法(商名,加拿大)
Lambert-Beer's Law 朗伯—比尔定律
Lambert's Law 朗伯定律
Lambert's law of absorption 朗伯吸收定律
Lambicol A 宁必高 A(阴离子型和非离子型,聚丙烯酸分散体,涂料印花用合成增稠剂,商名,意大利宁ाद्रि)
Lambicol DK/T 宁必高 DK/T(发泡印花浆,有立体效果,商名,意大利宁迪迪)
lambsdown 驼绒,棉背毛绒面针织物
Lamé 拉梅(金属薄膜丝,商名,美国制);金银锦缎(金银丝交织)
lamellar arrangement 层状排列
lamellar fibril 片状原纤
lamellar structure 片晶结构,片状单晶结构
Lamepon A 雷米邦 A(油酰氯与蛋白质水解产物的缩合物,属阴离子型表面活性剂,商名,德国制)
Lameprint A9 拉玛普林 A9(甘露聚糖半乳糖醚化物,适用于丝绸印花,具有优良的轮廓清晰度、鲜艳度、渗透性,商名,德国科宁)
Lameprint T6 拉玛普林 T6(混合糊料,适用于涤纶直接印花,成糊率及得色率高,渗透好,商名,德国科宁)
laminar flow 层流,片流
laminariae 昆布属海藻(制糊料和海藻酸盐纤维用)
laminate calender 叠层轧光机
laminated board 层压板
laminated cloth 多层黏合布,胶合布
laminated coating 叠合涂层
laminated fabric 叠层织物,胶合织物
laminated film 层压薄膜
laminated finishing 层压整理,层压后处理
laminated jersey 层压针织品
laminated knit 层压针织品
laminated material 层压制件,胶合制件
laminated plastics 层压塑料
laminated press felt 层压毡毯
laminated product 层压制品
laminated synthetic leather 层压合成革
laminated water-proof fabric 叠层防水织物
laminating 涂层;层压;贴合
laminating line with calender technology 用压辊技术的层压生产线
laminating line with ultrasonic technology 用超声波技术的层压生产线
laminating machine 层压机
laminating of dry bonding 干法层压黏合
laminating system 层压系统
laminating technology 层压技术
lamination 叠层,层压;织物胶合工艺

lamination by kiss roll application 面轧法层压

lamination by knife coating 刮浆层压

lamination machine 层压机

laminator 层压机,胶合机

lamp black 灯烟,油烟,烟炱;灯烟色

lamp probe 灯探头

lampwick 管状织物;灯芯

lamé yarn 金银线

Lanacron S dyes 兰纳克隆 S 染料(单磺酸化 1∶2 金属络合染料,一种中性染料,着色强度高,耐日晒和水洗色牢度高,适合于羊毛连续染色及毛条印花,商名,瑞士制)

Lanalbin B Pdr. 兰纳宾 B 粉(阴离子型,增艳剂,用于中浅色染浴,商名,瑞士科莱恩)

Lanalux 兰柔适剂(阳离子型,含活性基团改性有机硅超微乳液,可增加毛织物光泽、柔软、弹性,用于各种工艺,商名,英国 PPT)

Lanaperm VPA liq. 兰纳平 VPA 液(弱阳离子型,特殊硅弹性体,用于羊毛防毡缩整理,商名,瑞士科莱恩)

Lanaperm VPO liq. 兰纳平 VPO 液(不含氯,用于羊毛氧化前处理的防毡缩整理,商名,瑞士科莱恩)

Lanasan dyes 兰纳山染料(酸性染料,经挑选的染毛优良品种,配伍性和坚牢度优良,商名,瑞士科莱恩)

Lanasan LT 兰纳山 LT(非离子型,羊毛低温染色助剂,商名,瑞士科莱恩)

Lanaset dyes 兰纳晒脱染料(改良型中性染料及活性染料,用于羊毛染色,色谱齐全,牢度高,商名,瑞士制)

Lanaset resin 兰纳晒脱树脂(羊毛防缩树脂)

Lana shield 地毯抗污整理技术(商名,加拿大)

Lanasol dyes 兰纳素染料(含有活性溴代丙烯酰胺基团,与羊毛纤维形成共价键,具有极高的湿牢度,适合国际羊毛局的机可洗羊毛染色要求,商名,瑞士制)

Lanasyn dyes 兰纳芯染料(1∶2 金属络合染料,用以染毛和聚酰胺纤维,日晒牢度优良,中性或弱酸性染色,商名,瑞士科莱恩)

Lancashire beetling machine 兰开夏打布机

landfil disposal 地埋处理[环保]

Langmuir adsorption isotherm 朗缪尔吸附等温线(表示一定温度下,固体单位表面积吸附气体量与气相压力的关系)

Langmuir's adsorption 朗缪尔吸附

laniferous 有柔毛的,有羊毛的;羊毛似的

Lankrolan SHR 3 兰克罗兰 SHR 3(官能团为过硫酸烷基酯,羊毛防毡缩剂,商名,美国制)

lanolin(=lanoline,lanolinum) 羊毛脂

lanometer 羊毛测定仪

lantern gear 针齿轮

lanthanum nitrate 硝酸镧

lanthionine 羊毛硫氨酸

LAO(=L.A.O.,low add-on) 低加重率;低添加量

lap calendering machine 絮层压平机,毡层压光机

lapis 雀绿[色],宝石绿[色];靛蓝印花[工艺]

lapis lazulilblue 天青石蓝[色]

lap joint 搭接合

lap leucine aminopeptidase 白氨酸氨肽酶
lap over 花样交搭,花样重叠《印》
lapped folding [织物]卷板折叠
lapped joint 搭接合
lappet fabric 浮纹织物
lapping 印花[用]毛衬布;[经轴染色]芯轴两端包布;搭接
lapping and rolling machine 布卷取机
lapping cloth 蒸呢用[棉]衬布
lapping machine 卷板机(织物包装机);研磨机
lapping with damping 布卷给湿
lapping yarn 包覆纱
lap roller 卷布辊
lap strap 绑腰安全带
lap top 笔记本电脑
lap up 卷取
lap welding 搭焊
Laresa finish 拉里萨整理(醋酯纤维织物的特种整理,可产生不同光泽似提花图案的效应,商名)
large batch 大布卷
large cloth batcher 大卷装卷布机
large contrast 大色差
large coverage design 覆盖面大的[印花]图案
large diameter drum 大直径滚筒
large diameter reel 大直径框架
large field 大视野
large field colour matching 大视场颜色匹配
large matrix 大容量矩阵存储器《计算机》
large package 大卷装
large repeat pattern 大循环花纹
large repeat rotary screen 大花回印花圆网
large roll 大卷装

large scale integration 大规模集成化;大规模集成电路
large scale pattern 大花型
large vent〈lv,LV〉大眼;大口径
large yardage production 大批量生产
lark 泥灰[色]
larva [蛀虫的]幼虫,幼体
larvicide 杀幼虫剂
LAS(light activated switch) 光敏开关
LAS(linear alkylbenzene sulfonate) 直链烷基苯磺酸盐
lasecon 激光转换器
laser 激光,激光器
laser beam diffraction 激光束衍射
laser beam recorder 激光束扫描
laser blasting 激光喷射(牛仔裤一种整理方法)
laser bounce 激光反射
laser cloth inspector 激光验布器
laser coating 激光镀膜
laser cutting 激光裁剪;激光切割
laser cutting system 激光切割系统
laser diffractometer 激光衍射仪
laser diode 激光束,激光二级管
laser disc 激光视盘,光盘
laser Doppler velocimeter(＝laser Doppler velocity meter) 多普勒式激光速度计
laser driller 激光钻孔机
laser dyes 激光染料
laser electrophotography 激光电子照相术
laser engraver 激光雕刻机
laser engraving 激光雕刻,激光蚀刻
laser engraving of rotary screen 圆网激光雕刻
laser etching 激光雕刻,激光蚀刻
laser fabric inspection system 激光验布系

统
laser fibre 激光光导纤维
laser film output machine 激光成像机
laser-induced thermal etching 激光蚀刻
laser induction chemical vapour deposition method 激光诱导化学气相沉淀法(制纳米微粒的方法)
lasering 用激光焊接
laser light 激光
laser marker 激光打标机
laser material 激光材料
laser microscope 激光显微镜
laser monitor for pollution 激光测污仪
laser optical scanning head 激光光学扫描探头
laser perforation 激光镂孔(一种用激光加工的印花方法)
laser photo-type setter 激光照相排版机
laser plotter 激光绘迹器
laser positioning unit 激光定位系统
laser printer 激光打印机
laser processing 激光加工
laser pyrolysis 激光热解
laser pyrolysis gas chromatography 激光热解气相色谱法
laser radiation 激光辐射
laser-Raman spectrometry 激光—拉曼光谱法
laser ray 激光
laser register 激光对花(圆网印花机)
laser registering 激光对花
laser registering rotary screen printing machine 用激光对花的圆网印花机
laserscan display 激光扫描显示
laserscan fabric inspection system 激光扫描验布装置

laser scanning 激光扫描
laser scanning analyser 激光扫描分析器
laser slicer 激光划片机
laser thermo-bonding 激光热黏合
laser tripler monochrometer (= laser tripler monochromator) 激光三联单色光度计
laser velocimeter 激光测速计
laser welder 激光焊接机
lash 齿隙
lasing 激光作用;产生激光的
lasting effect 耐久效果
lasting rupture 永久断裂
last-off [最后]退卷《染》《整》
Lastrilea 丙烯腈—二烯类共聚纤维(商名)
latch 插锁,止动锁;犁子,卡齿
latchet 鞋带
latent acid 潜酸
latent acid catalyst 潜在酸性催化
latent catalyst 潜酸催化剂
latent crimp 潜在卷曲;潜卷缩性(浸湿时发生卷缩现象)
latent crimpability 潜在卷曲性
latent defect 潜在疵点
latent heat 潜热
latent heat of vapourization 汽化潜热
latent high shrinkage 潜在高收缩性
latent image 潜象
latent solvent 潜溶剂,惰性溶剂
latent strain 潜应变,内应变
latent stress 潜应力,内应力
lateral adjustment 横向[对花]调节《印》
lateral chain 侧链
lateral contrast 侧向对比(视觉)
lateral deformation 侧向变形
lateral diffusion 横向扩散

lateral group 侧基
lateral movement 横移,横向运动
lateral order 侧序,横向有序,横移次序
lateral order distribution 侧序分布
lateral-oval paddle dyeing machine 横向椭圆形桨叶染色机(成衣染色机)
lateral refractive index 侧向折射率
lateral register [链分子排列的]横向定位,侧向定位;横向对花《印》
lateral repeat adjustment 横向[花回]调节《印》
laterals 丁字接管
lateral section 横截面断面
lateral swelling 侧向溶胀
latest fashion 时装
latex 胶乳,橡浆
latex applicator 胶乳施加装置
latex backed carpet 胶乳涂底地毯
latex backing 胶乳底背
latex-bonded fabric 胶乳黏合非织造织物
latex bonded fibre 胶乳黏合纤维
latex dipping process 胶乳浸渍法
latex foam 泡沫胶乳
latex-free tufting carpets 无乳胶簇绒地毯
latexing 胶乳整理(地毯底布涂胶)
latex joint 胶乳黏结
latexometer 胶乳比重计
latex thickener 胶乳增稠剂
latex yarn 橡筋线,伸缩线,松紧线
lath 板条
lather 泡沫;起泡沫
lather booster 增泡剂
lather collapse 泡沫破裂
lather quickness 起泡速度
lather value in the presence of dirt 污垢[存在时的]起泡值

lathery 起泡沫的;泡沫造成的
lat. ht. (latent heat) 潜热
lattice charge 帘子喂给量,输送帘子负荷
lattice conveyer 平板履带(平幅织物的堆置和输送装置,常用于织物前处理汽蒸设备)
lattice-conveyer reaction chamber 平板履带汽蒸箱(一种平幅连续前处理设备,可供织物煮练和漂白用)
lattice dryer 帘式烘燥机
lattice groups 点阵组(喷墨印花术语)
lattice roller 帘子滚筒(输送带上的滚筒),对中滚筒(用于针织品扩幅)
lattice structure 晶格结构
lattice work 格子,格子花样
Latyl dyes 拉提尔染料(染聚酯纤维的分散染料,商名,美国制)
laughing gas 笑气,一氧化氮气体
launder 洗涤;浆洗;洗烫
launderability 耐洗性能
launderette 自动洗衣店,快速洗衣店
laundering 洗涤;洗烫;浆洗
laundering performance 洗涤性能
launderometer 耐洗色牢度试验仪
launderproof 耐洗性能
laundry 洗衣;洗衣房,洗衣店
laundry air dry〈LAD〉洗涤晾干
laundry-air-dry potential 洗涤晾干性能[优势]
laundry bag 洗衣袋
laundry blue [漂洗时用的]绀青,洗涤蓝
laundry cycle 洗涤周期
laundry degradation [织物强度]洗涤递降
laundry duck 洗衣滚筒包覆用帆布
laundry mark 洗衣店标签
laundry nets(＝laundry netting) 洗衣用网

袋;染色用网袋

laundryproof 耐洗性能(指不褪色、不缩水性能)

laundry quality control 洗涤质量控制

laundry resistance 耐洗[烫]性

laundry shrinkage 洗涤收缩率

laundry soap 洗涤用皂

laundry softeners 洗涤柔软剂

laundry washing 机洗

laundry wheel 洗衣滚筒

laurate mixed ester 月桂酸混合酯

Laurel cloth 劳雷耳丙纶土建布(商名)

laurel green 月桂树绿[色](黄绿色)

laurel pink 月桂粉红[色](淡红色)

lauric acid 月桂酸,十二烷酸

lauryl alcohol 月桂醇,十二醇

lauryl dimethylamine oxide 十二[烷]基二甲胺氧化物(非离子净洗剂)

lauryl mercaptan 月桂基硫醇

lauryl pentachlorophenate 五氯苯酚十二烷盐(耐久性杀霉菌剂)

lauryl pentachlorophenol 月桂基五氯苯酚(防霉剂)

lauryl pyridium chloride 月桂基氯化吡啶鎓(阳离子表面活性剂)

lava lava 重浆印花布(英国制);花腰布;短裙

lavender 熏衣草紫[色](淡紫色)

Laventin LNB 莱伐汀 LNB(润湿、净洗剂,非离子型,用于低温精练,商名,德国巴斯夫)

lawn 上等细布

lawn finish 上等细布整理(使织物具有爽挺手感的上浆整理)

law of diffusion 扩散定律

law of intermediary colours 中间色律

lawsone 指甲花醌色素(天然色素)

layered composite 叠层复合材料

layered silicates 层状硅酸盐

layer separation 层离,脱层

lay off 停止;卸荷;下料

lay-on-air dryer 气垫式烘干机

layout 布置图;线路图;草图;[机器的]排列

lay-out of pattern 构图(布局)

lay-up 敷层;树脂浸渍增强材料;接头,绞合

lay up unit 折布装置(用于平折的筒状针织物,以供包装)

L-box L形蒸箱(一种平幅连续前处理设备,供织物煮、练和漂白汽蒸用,堆置并输送织物的装置通常为平板履带式辊床)

L-box reaction chamber with a lattic conveyer 单板履带L形汽蒸箱

L-box steamer L形汽蒸箱

lbs.(librae) 磅(复数)

L/C(=l/c,letter of credit) 信用证

LC$_{50}$(lethal concentration 50%) 半致死浓度《环保》

L-calender L形轧光机

LCD(liquid crystal display) 液晶显示器《计算机》

LD(lethal dose) 致死[剂]量

LD$_{50}$(lethal dose 50%) 半致死剂量《环保》

LDF(level dyeing factor) 匀染指数

LDPE(low density polyethylene) 低密度聚乙烯

LDPE fusible interlining 低密度聚乙烯热熔衬

leachate 淋滤物

leaching 淋沥(织物受水脱色或脱浆);沥滤

leaching antimicrobial agent 浸出型抗菌剂(抗菌剂通过浸出而毒死微生物)

leaching rate 滤出率,沥滤率,浸提率

leach resistant antimicrobial fabrics 耐淋沥的抗微生物织物

lead 先导,引导;导线,引线《电》;铅;铅锤

lead chromate 铬酸铅

lead chromate treated cotton 铬酸铅处理棉布(增强耐晒性)

lead compound 铅化合物

lead end 前端,导前

leader(＝leader cloth) 导布;导布带;引头布

leader mark 接头[布]印(染疵)

leader tapes 导布;导布带;引头布

lead grey 铅灰色

leading bow 中凸弓纬(织疵,从进布方向看去,纬纱呈凸形)

leading cloth 导布

leading-in wire 引入电线

lead naphthenate 环烷酸铅

lead oxychromate 金鸡纳红,铬氧酸铅

lead poisoning 铅中毒

lead roll 导辊;导布辊

lead screw [车床]丝杠,导螺丝

lead time 研制周期(产品设计至实际投产时间,订货至交货时间)

lead white 铅白

leaf green 叶绿[色](黄绿色);绿色颜料

leaflet structure 叶状结构(丝纤维结构中的尖端)

leaf spring 板簧

leakage 漏,泄漏,漏损

leakage flow 漏流

leakage reactance 漏磁电抗,漏抗

leakage test 泄漏试验,密封性试验

leak-free 密闭,密封

leak-off 泄漏,渗漏

leak-off pipe 溢流管

Leamington Axminster 利明顿阿克明斯特小地毯

lean handle [涂层]干燥手感

lean production 效益差的生产

least significant difference 最小显著差异

least significant digit 最低[有效]位,最低位有效数字

leather 皮革,熟皮

leather brown 皮革棕[色](棕色)

leather cloth 漆布;人造革;棉毛麦尔登

leather coupling 皮带联轴节,皮质联器

leather dyeing 皮革染色

leather dyes 皮革染料

leather fabric 漆布;人造革;棉毛麦尔登

leather fat-liquoring 皮革上油

leather finishing 皮革整理

leatheroid 人造革

leather printing 皮革印花

leather tanning 皮革鞣制

leather washer 皮垫圈

lecithin 卵磷脂,大豆卵磷脂,磷脂酰胆碱(制合成增稠剂用的乳化剂)

Lectrolite fabrics 累克屈罗赖特织物(商名,织物能导电,无毒,有防菌、防焰、抗过敏、抗微生物等作用)

LED(light emitting diode) 发光二极管

ledge 突出物

ledger blade [剪毛机的]固定刀片

lees 沉淀物;凝结物;残余物

left elevation 左视图

left-handed machine 左手机(国际标准规定,染整机器的左右手,以操作者面对进布方向而定,与我国现行规定的正好相反)

left hand thread 左螺纹;倒牙(逆时针方向旋进)

left-over material 废弃物

left/right hand machine 左/右手车

left side view 左视图

legal weight 法定重量

legend 边印(表达商名、织物图案或染整加工特征等);图例,图代符号

leghorn 意大利麦秆编制的草帽;草帽黄[色](淡橘黄色)

legibility 清晰度

legislation on environmental protection 环境保护法规

legislative control 立法控制,法规控制

Leipzig yellow 铬黄,贡黄,莱比锡黄

leisure wear 家常服装,假日便装,休闲服

lemon [yellow] 柠檬黄[色]

lemon chrome 柠檬铬黄(黄色颜料)

length 长度;范围

length counter 测长器

length diameter ratio 长径比(长度与直径之比)

length felting shrinkage 缩绒纵向收缩

length indicator 测长器

length measurement 长度测量

length meter 测长器

length of bath 浴比

length of charred zone 炭化区长度(易燃性测定)

length of cut 匹长

length of dyeing 染浴大小,染布长度

length of liquor 浴比

length of piece 匹头长度

length of repeat [花纹]循环尺寸

length recorder 长度记录仪

length stopping 定长自停

length×width×height〈L×W×H〉 长×宽×高

lengthwise 纵向,纵长的

lengthwise cutting 纵向切割

lengthwise fold [织物]纵向折叠

lengthwise streaking 纵向条痕(疵点)

lengthwise stripe pattern 纵条花纹

lengthwise tear 纵向撕破

leno 纱罗,纱罗织物

leno and gauze 纱罗

leno cloth 纱罗织物

lens 透镜

lens wipes 擦镜布

lentil-headed screw 扁平螺钉

Lenzing modal 兰精莫代尔纤维(商名,奥地利制)

Leomin 莱奥明(不同化学组成的柔软剂,用于改善各种织物的手感,商名,瑞士科莱恩)

Leonard drive 伦纳德变速传动装置

Leonil 莱奥尼(各种纤维的渗透、洗涤等耐酸耐碱精练及染色助剂,商名,瑞士科莱恩)

Leophen M 利奥芬M(中性磷酸酯,用于精练工艺的高效润湿,渗透剂对酸碱稳定性好,商名,德国巴斯夫)

Leophen U 利奥芬U(润湿剂,复配物,用于精练工艺,耐酸碱稳定性好,尚有洗涤、分散效果,商名,德国巴斯夫)

Lepidoptera 鳞翅目昆虫(蛀虫名)

leptons 轻子,轻粒子(比质子更小的粒子)

less batch changes/less machine stoppage 少换卷/少停车(大卷装的优点)
lesso 浅色印花布
less ordered region 低序区,有序度较差区[域]
lethal concentration 致死浓度
lethal concentration 50%〈LC_{50}〉 半致死浓度《环保》
lethal dose〈LD〉 致死[剂]量
lethal dose 50%〈LD_{50}〉 半致死剂量《环保》
let off 放掉(水,汽等)
let-off pipe 出水管,排水管,排汽管
letter of credit〈L/C,l/c〉 信用证
letter press printing 凸版印刷术
letting down 改长
letting out 改宽
lettuce green 莴苣绿[色](黄绿色)
leu(=leucine) 白氨酸,亮氨酸,异己氨酸
leuc[o]- (构词成分)表示"白色""无色""极淡色"
leuco [染料]隐色体
leuco acid [还原染料]隐色酸,酸性隐色体
leuco acid dyeing process 隐色酸染色法
leucobase 无色母体
leuco colour 染料隐色体
leuco-compound 隐色体
leuco derivative 无色衍生物
leuco dye 染料隐色体
leuco ester [还原染料]隐色酯(指可溶性还原染料)
leucoindigo 靛蓝隐色体,靛白
leuco-pad steam process [还原染料]隐色体轧染汽蒸法

Leucophor 雷可福(荧光增白剂,商名,瑞士科莱恩)
leuco potential〈l. p.〉 [还原染料]隐色体电势
leucorosaniline 隐色品红
leuco salt 隐色[体]盐
leucoscope 色光光度计;光学高温计;感光计
leuco stability 隐色体稳定性
leuco sulphur dye 隐色硫化染料(含有隐色体的硫化染料)
Leucotrope O 柳科托罗普 O,助拔剂 O,咬白剂 O
Leucotrope W 柳科托罗普 W,助拔剂 W(氯化对甲苯磺酸与二甲基苯胺间磺酸所成的钙盐缩合物,拔染促进剂,商名,德国巴斯夫)
leuco vat dyes 隐色还原染料
leuco vat ester dyes 隐色酯化还原染料
leuco vat padding process 还原隐色体浸轧法
Levafix CA 高固着率和低盐型活性染料(商名,德国制)
Levafix dyes 丽华实(活性染料的一种,内含磺酰胺活性基团,商名,德国制)
Levafix EA dyes 丽华实 EA 染料(活性染料,含二氟一氯嘧啶基团,商名,德国制)
Levafix E dyes 丽华实 E 染料(活性染料,含二氯喹噁啉基团,商名,德国制)
Levafix PN dyes 丽华实 PN 染料(活性染料,含氟氯甲基嘧啶活性基团,具有较好的印浆稳定性,适用于纤维素纤维的印花,商名,德国制)
Levagal HTC 丽菲格 HTC(匀染剂,载体,高温工艺极佳的匀染性,也适用于

修色,商名,德国拜耳)

Levalan N dyes 莱法兰 N 染料(2∶1金属络合染料,具有两个磺酸基,商名,德国拜耳)

Levasol DG 利伐素 DG(以二甘醇为基础的非离子性溶剂,商名,德国拜耳)

Levasol PO 利伐素 PO(酞菁染料助剂,非离子型溶剂,为氨基、酰胺基、乙二醇类的混合物,商名,德国拜耳)

Levasol TR 利伐素 TR(以三甘醇为基础的酞菁染料溶剂,商名,德国拜耳)

Levegal FTS 丽菲格 FTS(阴离子型的有机复合物,用于聚酰胺纤维的匀染剂,商名,德国拜耳)

Levegal HTN 丽菲格 HTN(匀染剂,分散剂,非离子型,适用于分散染料染聚酰胺纤维,有极高的缓染作用,商名,德国拜耳)

Levegal PEW,PR 丽菲格 PEW,PR(匀染剂,载体,不含有害氯化芳香族等物质,易生物降解,不影响日晒牢度,羊毛防染性好,商名,德国拜耳)

level 水平面;水准;水准仪;均匀的;校平

level absorption 均匀吸收

level alarm 液位报警

level control box 液面控制[浆]槽

level controller 液面调节器

level detector 电平检测器;液面检测器

level dye index 匀染指数

level dyeing 匀染,均匀染色

level dyeing factor〈LDF〉匀染指数

level dyeing property 匀染性

leveler 匀染剂;均匀剂;校平器;水准仪

level gauge 液面指示器;水准仪

level glass 液位玻管

level indicator 液面指示器,水平指示器

leveling(=**levelling**) 匀染的;校平;水准测量

leveller 匀染剂;均匀剂;校平器;水准仪

levelling acid dyes 匀染性酸性染料;匀染型酸性染料

levelling agent 匀染剂,均化剂

levelling dryer 平幅烘燥机

levelling dye 匀染染料

levelling dyes for rayon 黏胶丝匀染染料

levelling effect 匀染效应;匀整效应

levelling grade 匀染度

levelling phase in dyeing [染色过程中]匀染阶段

levelling power 匀染力

levelling property 匀染性

levelling screw 校平螺钉,校准螺钉

levelling test 匀染性试验

levelling type acid dye 匀染性酸性染料;匀染型酸性染料

level list 匀整布边(均匀而光滑的布边)

levelness 匀染性;水平度

levelness index 匀染指数

levelness of printing 印花均匀度

levelness tester 匀染度仪

level out 匀染;使相等;使平整

level pile thickness 平齐绒头厚度

level pressure mangle 均匀轧车

level probe 液面探测器

level regulator 液位调节器,液面调节器

level sensor 液面传感器

level shade 均匀色泽

level touch 匀整的触感(如地毯剪毛)

level up 拉平,平衡,成水平

lever 杆,杠杆,柄

lever and weight pressing 杠杆加压

lever effect 杠杆效应

lever type selvedge suction device 杠杆式吸边器
lever valve 杠杆阀
lever weighting 杠杆加压
leviathan 大型洗毛机
levigate 磨光
levis 牛仔裤
Levi shade 利惠色（深蓝牛仔裤色泽）
levo- （构词成分）表示"左的""左旋的"
Levogen DR,WRD 丽富坚 DR,WRD（不含甲醛固色剂，可提高直接、活性染料染色湿牢度，商名，德国拜耳）
Levogen FL 丽富坚 FL（不含甲醛固色剂，适用于活性染料的染色、印花，可改善活性染料耐酸分解能力，商名，德国拜耳）
levoisomer 左旋异构体
levorotatory 左旋的（有机化合物结构）
levulinic acid 乙酰丙酸
levulose 左旋糖，果糖
Lewis acid 路易斯酸（能接受电子[对]的化合物）
Lewis base 路易斯碱（能给出电子[对]的化合物）
ley [洗涤]碱液；废皂碱水
ley boil 纯碱煮练
liability fraying 磨损倾向
Liaoning Pongee 辽宁柞丝绸
liberation 释放，放出
liberty prints 手工模板印花布；自由式印花花样
libra 磅；镑
librae ⟨lbs.⟩ 磅（复数）
licence 许可；许可证
licence affiliation [专利]技术合作
licenced products 专利产品

license 许可；许可证
license affiliation [专利]技术合作
licensed products 专利产品
lichen 青苔色
lichen starch 地衣淀粉，地衣聚糖
licker roller 粘浆辊
lick padding 面轧
lick roller 舐液辊，给液辊
lidar 激光雷达
Liebermann-Storch test 利伯曼—斯托奇试验（检定织物上树脂方法）
life belt 安全带，保险带；救生圈
life buoy 救生衣，救生带，救生圈
life cycle 生命期，流行期
life duration 耐用期，使用寿命
life expectancy 预期寿命
life lubrication 永久润滑
life saving vest 救生背心
life sciences chemical 生命科学化学
life span [机器]使用寿命
life test 使用寿命试验
life time 使用期限，使用寿命
lifetime grease 长效润滑脂
lifted winch 提布辊
lifter 升降机构，升降机
lifter bevel wheel 升降伞齿轮
lifter cam 升降凸轮
lifter lever 升降杠杆
lifter motion cam shaft 升降凸轮轴
lifter wheel 升降轮
lifting adjuster 升降调节器
lifting arm 升降臂
lifting arm of plaiter 落布架手臂
lifting device 起重装置（搬运装置）
lifting device for doctors 印花刮刀升降装置

lifting equipment 升降设备
lifting eye 吊眼,起吊环
lifting gear 起重装置;升降装置,升降齿轮
lifting lever 升降杆
lifting motion 升降装置;升降运动
lifting point position 吊装位置
lifting rod 升降杆,提升杆,起重杆
lifting screw 升降螺杆;千斤顶螺杆
lifting shaft 升降轴
lifting speed 升降速度
lifting truck 堆布车;起重车
lift pump 提升泵
lift reel 提布轮(绳状间歇式染色机中提起并传动绳状织物的导轮)
lifts 花筒吊车(滚筒印花机车头装卸花筒轴等部件);跳刀(滚筒印花疵,由花筒刮刀跳动而造成);[网印]框架垫片
lift station [污水]升液站
lift truck 铲车;堆桩车
ligand 配位体;配合基;向心配合体
ligand field theory 配位场理论,配合基场理论
ligasoid 液气悬胶(液相悬浮于气相中而成)
light 光,光泽;轻的;浅色;冲淡[程]度
light absorption 光吸收
light actived switch〈LAS〉 光敏开关
light adaptation 明视适应性(色度学用语)
light ageing 光老化
light and shade contrast 明暗反衬,明暗对照
light and shade from lines 线状明暗效果(雕刻用语)
light and weather fastness tester 日光兼气候牢度试验仪
light barrier 光栅
light blocking effect 遮光效应
light booth 光源箱
light clear blue 浅湖绿[色]
light clear series 明澄色系列《测色》
light collecting capability 集光能力
light colour 浅色
light colour effect 浅色效应
light damage 光损伤(纤维降解、变黄)
light degradable 光[可]降解的
light degradation resistance 耐光降解性
light duty 轻型的,轻效应的
light duty detergent 轻效型合成洗涤剂
light duty soap 轻效型肥皂,纯肥皂
light edges 边浅
light effect finish 光效应整理(主要指轧波纹整理)
light emitting diode〈LED〉 发光二极管
lighten [色彩]调淡;发亮,变亮
lighten end product 轻质产品;轻质成品
lighter 点火器,引燃器
light exposure apparatus 曝光设备
light fabric 轻薄织物
light fading and tendering 光褪色和脆损
light fading test 光褪色试验
light fast 耐晒的
light fast direct dyes 耐光直接染料
light fastness 耐光性,耐光色牢度
light fastness rating 耐光色牢度等级
light fastness standard 耐光色牢度标准
light fastness tester 耐日晒色牢度试验机
light filter 滤光片
light finishing 轻浆整理
light flux 光通量
light fuel oil 轻油

light gray 浅灰色
light grayish green 灰豆绿[色]
light grazing 光线擦过
light gun 光笔
light-induced ageing 光致老化
lighting 照明,采光
lighting hood 照明头罩
lighting point 照明点
lighting switchboard 照明配电屏
lighting torch 点火炬,点火棒,点火喷燃器
light inspection board 验光板
light interference 光干涉
light ironing 微烫
light lime green 浅灰绿[色]
lightly printed pattern 浅[色]印花图案
light metal 轻金属
lightmeter 光度计,照度计
light microscopy 光[学]显微[镜]术
lightness 亮度;明度(非自身发光体反射亮度);轻;浅(指色泽)
lightness axis 明度轴《测》
lightness index 明度指数
lightning arrester 避雷器
lightning rod 避雷针
light nip 轻轧点
light oil 轻油;轻石油馏分
light path 光程
light pen 光笔
light petroleum 石油醚
light pink 浅粉[色]
light pipe 光管;导光管
light pipe assemblies 光管组合
light print 浅色印花
light proof 耐光的
light protection 防晒性

light protective agent 防晒剂
light quantity gauge 光量测定仪
light reflector 反光镜,反光碗
light reflector coated with dielectric diaphragm 电介质膜反光镜(用于放样机)
light resistance 耐光性
light retroreflective coating finishing 回复性反射涂层整理,回归反射涂层整理
light rolling 轧光,压光
light run 空转
light scattering 光散射
light scattering coefficient 光散射系数(抗沾污性能测试)
light scattering photometer 散射光度计
light scattering profile [布面]光散射分布
light sensitive agent 光敏剂
light sensitive coating 光敏涂层
light sensitive colour printing 光敏印花
light sensitive discolour fabrics 光敏变色织物
light sensitive dyes 光敏染料
light sensitive gel 感光胶
light sensitive paper 光敏试纸
light sensitivity 敏感度
light sensor 感光器
light shade 浅色
light sizing 轻浆
light source 光源
light source A A光源(钨丝灯光源,色温 2856K)
light source C C光源(平均昼光,色温 6774K)
light source colour 光源色
light source CWF CWF光源(冷白荧光灯)

light source for process camera 照相制版光源

light source TL84 TL84 光源(节电荧光灯)

light splitter 分光器

light squeezing 轻轧

light stability agent 光稳定剂

light stabilizer 光稳定剂

light stabilizing treatment 光稳定处理

light starching 轻浆(上浆用语)

light stimulus 光刺激

light switch 照明开关

light tight 不透光的,防光的

light transmission 光传输;光透射

light transmission measuring instrument 光透射测试仪

light transmittance 透光率,光透射比;光透射系数

light transmittance pattern 光透射图像

light transmitter 光发射器

light trap 光阱

light two bowl calender 轻型双辊轧光机

light up plug 点火塞,火花塞

light value 光值

light wavelengths 光波长

light weight 轻量

light weight coating 轻质涂层

light weight fabric 轻质织物,薄型织物

light wringing [染纱]轻度绞干,抹绞(纱绞加工)

lignan 木聚糖,木酚素

lignification 木质化[作用]

lignified tissue 木质化组织

lignin 木质素,木质,木素

ligninase 木质素酶

lignin number 木质素值

lignin resin 木质素树脂

lignin sulfonate 木质素磺酸盐

ligno cellulose 木质纤维素

lignosulfonic acid 木质素磺酸

ligroine 石油英,石油醚

lilac 丁香紫[色](淡紫色);淡雪青[色]

lilac grey 淡紫灰[色]

liliquoid 乳状胶体

Liloc-overflow 利乐克溢流染色机(高温染色机,商名)

lim.(limit) 极限,限度;范围;限制

lim.(limited) 有限[责任]的,有限制的;极限;时限

lime 石灰

lime-baryta process 碳一氧化钡工艺(软水处理)

lime-base grease 钙基润滑脂

lime blossom shade 椴绿[色](浅黄绿色)

lime boil 石灰水煮练

lime green 酸橙绿[色](暗黄绿色)

lime hydrate 熟石灰

lime process 石灰法

lime soaking 石灰水浸泡

lime soap 石灰皂,钙皂

lime soap dispersing power [of surfactant][表面活性剂的]钙皂分散力

lime soap stains 钙皂污渍

lime-soda water softening process 石灰—纯碱软水工艺

lime sour 去钙皂的酸洗(用淡酸分解钙皂)

lime stain 石灰斑,钙斑

lime stone 石灰石

lime vat [靛蓝]石灰还原浴

liming machine 石灰浸渍机

limit⟨lim.⟩ 极限,限度;范围;限制

limitation 限度,限定;界限,极限,局限性
limit controller 限制控制器
limited〈lim.,Ltd.〉 有限[责任]的,有限制的;极限;时限
limited bulking 适当膨化
limited swelling 有限泡胀,有限溶胀
limited use 有限次数使用
limited wash fastness 有限洗涤牢度
limiter 限幅器,限制器
limiting amplifier 限幅放大器
limiting breaking velocity 极限破损速度
limiting concentration 极限浓度
limiting current 极限电流
limiting diffusion coefficient 极限扩散系数
limiting friction 极限摩擦
limiting oxygen index〈LOI〉 限氧指数(测定纤维可燃性值)
limiting value 极限值
limiting viscosity number 特性黏度
limit of dertermination 定量限度
limit of detection 检出限度
limit of elasticity 弹性极限
limit of fatigue 疲劳极限
limit of reliability 可信限度
limit oxygen index 极限耗氧指数(易燃性测试)
limit size 极限尺寸,极限量
limit strength 极限强力
limit switch 限位开关,行程开关
limp handle value 手感柔软值
limpness 柔软度
limp state 柔软态;无弹[性状]态
limy 胶黏的;含石灰的
lin.(linear) 直线[的],线型[的];一次[的]

linden green 椴绿[色](浅黄绿色)
line 线,直线;管路;画线
linear〈lin.〉 直线[的],线型[的];一次[的]
linear alkyl benzene sulfonate〈LAS〉 直链烷基苯磺酸盐
linear amplifier 线性放大器
linear analysis 线性分析
linear burning rate 线性燃烧速度
linear chain molecule 直链分子
linear compounds 直链化合物
linear condensation 线型缩合
linear control system 线性控制系统
linear correlation 线性相关
linear density 线密度
linear equation 一次方程,线性方程
linear extruder 线性挤压机(用于泡沫染整,液体从缝隙中流出,保持规定的形状、厚薄及数量)
linear high polymer 线性高聚物
linear induction motor 直线感应电动机
linear integrated circuit 线性集成电路
linearity 直线性;线性;直线度
linear law 线性定律
linear light source 线型光源
linear macromolecule 线型大分子
linear oligomer 线型低聚物
linear optimization 线性优化
linear polymer 线型聚合物
linear pressure 线压力(指单位长度上的压力)
linear printing method 一次完成印花法(用合成黏合剂作印花糊料,印后通过热溶或高温汽蒸,不需水洗就可完成印花工艺)
linear program 线性规划《数》

linear reactant 线型反应物,链状反应物
linear relationship 线性关系
linear speed 线速度
linear surface velocity 表面线速度
linear theory of viscoelasticity 黏弹线性理论
linear variable〈lv,LV〉线性变异
linear velocity 线速度
linear viscoelasticity 线性黏弹性
linear yard [统幅]每码
line balancing 设备平衡,生产平衡
line blender 管道混合器
line bonding 条状黏合,行式黏合
line check 小检修
line commutated inverter 有源逆变器《电》
line cutting stand 画线机《雕刻》
line dry [悬挂]晾干
line engraving 线条雕刻
line gratings 织物密度测量板
linen 亚麻布;亚麻纱线;仿亚麻制品
linen batiste 漂白亚麻平布
linen bleaching cycle process 亚麻织物漂白循环工艺
linen bowl 麻纤维轧辊(用下脚麻纤维作原料,经压机压制而成)
linen canvas 亚麻帆布
linen cloth 亚麻织物,亚麻布
linen crepe 亚麻绉布(利用织物组织,织成绉纹)
linen finish 仿亚麻整理
linen hands 亚麻的手感
linen industry 亚麻[纺织]工业
linen-like 仿[亚]麻的;仿麻型
linen-like finish 仿麻整理
linen like finishing agent 仿麻整理剂
linen-like polyester crepe 涤纶仿麻绉

linen look 亚麻感
linen oyster 半漂或浅色刺绣亚麻布
linen pretreatment 亚麻预处理
linen resin finishing 亚麻织物树脂整理
linen yarn 漂白亚麻纱;亚麻纱
line of reference 基准线,参考线
line on line 合线(借位)《雕刻》
line production 流水作业法
liner [纺织品]衬里
line-screen 网屏,网线版《雕刻》;线纹筛网
line shaft 总轴,天轴
lingerie laminates 内衣层压物
lining [衣服的]衬里,里子,材料
lining cloth 衬里布
lining fabric 衬里织物
lining felt 衬毡,隔热衬毡
lining plate 衬板
link 连接;连杆;环节;链节
linkage 连接,接线;联动[装置];键,键合;连杆;连锁;联动装置
linkage housing 联动箱
link drive 连杆传动
link guide 链节导向器
link lever 连杆
link lock 连锁杆
link motion 连杆运动;连杆装置
linneas 印花棉布(非洲市场用语)
linnet 仿麻织物
linolenic acid 亚麻酸;9,12,15-十八碳三烯酸
linseed oil 亚麻子油,亚麻仁油
lint 棉绒
lint ball 起球(疵点)
lint blade(＝lint doctor) 小刀,铜[刮]刀(用以刮除花筒上的棉纱、棉絮、棉屑

及带浆杂质)
linters and slubs 棉短绒和纱头(残留在织物上的杂质)
lint extraction device 吸尘装置,飞花收集装置
lint filter screen 棉绒滤网
Lintrak process 林曲莱克工艺(毛裤耐久褶裥整理法)
lint screen 棉绒滤网
lipase 脂[肪]酶
lipid 类脂[化合]物
lipid bilayers 油脂体双层
lipide 类脂物
lipidosome 类脂质体
lipin 类脂质,类脂[化合]物
Lipolase 列波酶(一种脂肪酶,商名)
lipolysis 脂[类分]解[作用]
lipophile 亲脂体
lipophilic 亲脂的,亲油的
lipophilic group 亲油基(疏水基)
lipophilicity 亲油性;亲油程度
lipophilic property 亲油性,亲脂性
lipophobic 疏油的,疏脂的
lipophobic group 疏油基(亲水基)
lipoprotein lipases 脂性蛋白脂肪酶
lip roll 舐液辊
lip seal 唇封(由柔软的充气袋和外包耐磨的衬垫组成,用作高压反应汽蒸罐的封口)
lip sealing method 唇封法
lip stick 唇红[色](玫红色);口红
liq.(=**lq.**,**liquid**) 液体;液体的
liquate 熔融;熔解
liquation 熔融;熔解
liquefaction 液化[作用]
liquefaction point 液化点

liquefied gas 液化气
liquefied natural gas 液化天然气
liquefied petroleum gas〈**LPG**〉 液化石油气
liquefied refinery gas 液化石油精练气
liquefier 液化器;液化剂;稀释剂
liquefying activity 液化力
liquescency [可]液化性
liquid〈**liq.**,**lq.**〉 液体;液体的
liquid ammonia 液氨
liquid ammonia dyeing 液氨染色
liquid ammonia finish 液氨整理(纤维素纤维织物经液氨处理后,可达到改善手感、防皱、耐磨、降低缩水率等效果,如美国的桑福瑟特整理)
liquid ammonia mercerization 液氨丝光
liquid ammonia treating system 液氨处理装置
liquid ammonia treatment 液氨处理
liquid ammonia treatment range 液氨处理机
liquid application system 给液系统
liquid bath 液浴,液槽
liquid chlorine 液氯
liquid chromatogram 液相色谱[图]
liquid chromatography 液相色谱分析法
liquid CO_2 application on dyeing process 液态二氧化碳在染色工艺中应用(无水染色技术)
liquid crystal 液晶
liquid crystal display〈**LCD**〉 液晶显示器《计算机》
liquid detergent 液体净洗剂
liquid disperse dyes 液体分散染料
liquid dyes 液态染料
liquid-filled thermometer 液体温度计

liquid film seal 液膜密封
liquid flow zone 液流区,流动区
liquid heating medium 液态加热介质
liquid infiltration 液体渗入,液体浸渍
liquid in glass thermometer 液体玻璃温度计
liquid junction potential 液体电位差,液体汇合势差
liquid level gauge 液位表
liquid level indicator 液面指示器
liquid level relay 液位继电器(容器中液体高低变化到预定位置时动作的继电器)
liquid level transmitter 液位遥测仪
liquid limit 液态极限;塑性上限
liquid-liquid chromatography〈LLC〉液—液色谱分析法
liquid medium 液体介质
liquid nitrogen 液氮
liquidometer 液面计
liquid pack 水封,液封
liquid paraffin 流动石蜡,液态石蜡
liquid phase 液相
liquid reactive dyes 液体活性染料
liquid removal system 轧液装置;除液装置
liquid repellency 拒液性,抗液性
liquid seal 液封;水封;油封;水封器
liquid-solid chromatography〈LSC〉液—固色谱分析法
liquid varnish 湿蜡
liquid volume viscosity 液体体积黏度
liquification 液化
liquified gas 液化煤气,液化(石油)气
liquified natural gas〈LNG〉液化天然气
liquifier 液化器,液化剂;稀释剂

liquor 液,溶液,煮液
liquor circulation 液体环流
liquor circulation rate 染液循环速度
liquor circulatory system 溶液循环装置
liquor collection vessel 液体收集器
liquor concentration measuring and regulating system 溶液浓度测量和调节系统
liquor distributor tube 配液管
liquor film 水膜
liquor flow dyeing machine 液流染色机
liquor formulation 染液拼料,浴液配方
liquor/goods ratio 浴比
liquor level 液面高度,液位
liquor level indicator 液面指示器
liquor level regulator 液面调节器
liquor phase heater 液相加热器
liquor ratio〈LR〉浴比;液比
liquor ratio dependency 浴比依存性
liquor retention 轧液率
liquor retention agent 液体保留剂
liquor reversing control 溶液倒顺控制
liquor stream 液流
liquor throughput 浴液通过能力
liquor trough 液槽,轧槽
liquor turbulence 洗液湍流
liquor turn over 液体转换
Lissapol YC 洗涤剂 YC(改性磷酸酯与聚磷酸酯的复合,阴离子型,用于直接染料、活性染料染色织物的后处理,商名,英国卜内门)
list 布边;边缘;饰带;表,表格;目录
listerine 防腐溶液
listing 布边;边深浅(染疵)
lit.(litre,liter)升
liter〈lit.〉【美】=litre
literature information 文献资料

litharge 一氧化铅
lithium alkyl 烷基锂
lithium alumina silicate 硅酸锂铝(合金)
lithium aluminium hydride 氢化铝锂
lithium aryl 芳基锂
lithium bromide 溴化锂
lithium chloride 氯化锂
lithium chloride hygrometer 氯化锂湿度计
lithium compound 锂化合物
lithium fluoride 氟化锂
lithium grease 锂基润滑脂(用高脂肪酸锂作为基础的、对热稳定而又防水的润滑脂)
lithium hydroxide 氢氧化锂
lithium thiocyanate 硫氰酸锂(强力氢键断裂剂)
lithochromic print 平板彩色印刷;彩色石印
lithographic felt 石印用毡
lithographic printing 平版印花,平版印刷
lithographic printing machine 平版印刷机(印刷转印纸)
lithography 平版印刷术;石印术
litho offset plate 胶印平版
lithopone 锌钡白(由硫化锌和硫酸钡混合制成的白色颜料)
litmus 石蕊(有机化合物,用作酸碱测定指示剂)
litmus paper 石蕊试纸(用于酸碱测定)
litre〈lit.〉升
little iron 微烫,稍烫(指织物具有免烫的一般性能,商业术语)
live line garment 带电操作服
liveliness 回弹性;滑爽;鲜明感性
liveliness of fabric 织物回弹性,织物滑爽性

live load 活载荷
livening treatment 丝鸣整理;增艳处理
liver [brown] 肝棕色
live steam 直接蒸汽,活蒸汽
live steam pressure 新汽压力
live steam reheater 新汽加热器,汽—汽热交换器,汽—汽再热器
livid 青灰色的,铅色的
living material 生活资料
living sewage 生活废水
lixivation 去碱作用
lixiviant 浸滤剂
lixiviate 浸滤
lixiviation 浸析,溶滤,浸滤作用
lizard 深秋香;军草绿[色];橄榄绿[色]
LL(loudness level) 音量标准,音量水平,响度级
LLC(liquid-liquid chromatography) 液-液色谱分析法
LMP(low molecular polymer) 低分子聚合物
LNG(liquified natural gas) 液化天然气
load 载荷,负荷;装入;装配
load bearing capacity 承载能力
load cell 载荷传感器;测力传感器
load coefficient 负荷系数
load cycles 负荷周期,[循环试验中的]循环次数
load deflection 载荷挠度
load deformation curve 载荷变形曲线
loaded 增重
load elongation curve 载荷伸长曲线,负荷伸长曲线
load elongation diagram 载荷伸长曲线图,负荷伸长曲线图
loader 输入器;装料器;输入程序《计算

机》
load extension curve 载荷延伸曲线,负荷延伸曲线
load factor 用电负荷率
loading 布匹加重(指布匹在整理时加入淀粉、陶土、糊精、氯化镁等);装料,装载,加料,填料;负载,负荷
loading agent 增重剂
loading capacity 负载容量
loading control 负载控制
loading duration 加载时间
loading lever 加压杆
loading port 进布口
loading routine 输入程序
loading wheel 加压手轮
load length 负荷长度,断裂长度(以千米为单位)
load off 卸[负]荷,卸载
load test 载荷试验,负荷试验
load-up stop motion 过载自停装置
lobbing [花筒脱芯]打滑;[花筒受压]偏心变形
lobed plate 凸轮板
local analysis 局部分析
local calendering 局部轧光
local chintzing 局部擦光[工艺]
local defect 局部性疵点
local deluster 局部消光
local effect 局部效应
local effect finish 局部效应整理
local embossing 局部压花[工艺]
local exhaust ventilation 局部排气装置
local flocking 局部植绒
local illumination 局部照明
localization 定位[作用]
localized defect 局部性疵点

local lustrous finishing 局部光泽整理
local overheating 局部过热
local pollution 局部污染
local raising 局部起毛
local sanding 局部磨毛
local stretching 局部拉伸
locater 定位器;测位器
locating bearing 止推轴承,定位轴承
locating hole 定位孔
locating key 定位键
locating lever 定位杆
locating pin 定位销
locating screw 定位螺钉
location 定位;配置;位置
location diagram 位置图(一种电气图,说明联合机、单元机和部件安装的位置)
location mark 定位标记,定位符号
location reagent 定位试剂(斑点或谱带显色剂)
locator 定位器;测位器
lock 封,封口;锁,关闭;闭锁装置
lock gear 制动装置;定位机构;闭锁机构
locking cap 固定帽,止动螺钉
locking device 锁紧装置;保险装置,安全装置
locking lever 联锁杆
locking lip 锁闭盖,[轴承衬圈的]制动唇
locking ring 锁环
locking screw 锁紧螺钉
locknut 锁紧螺母
lockpin 锁销,插销
lock screw 锁紧螺丝
lock washer 锁紧垫圈
lock yarn bouclé 结子线,毛圈线
locus 轨迹,轧线
locust bean gum 刺槐豆胶(印花糊料)

locust bean gum thickening 刺槐豆胶糊
lodgement 沉淀[物];沉积处
loft 蓬松;弹性(指毛织物);蓬松度
loft drying [室内]风干,[室内]晾干
lofted effect 蓬松效应
lofted yarn 膨体纱
Loft fibre 一种环保柔软蓬松纤维(用于织制毛巾)
loftiness 松软丰满手感,蓬松感
lofty 有弹性的;蓬松的
lofty hand(=lofty handle) 蓬松手感,弹性感
log [运行]记录
logarithmic paper 对数坐标纸
logarithmic viscosity number 对数黏[度]数,比浓对数黏度
log card 记录卡
logger 记录器
logo 边印
log off 注销《计算机》
log sheet 记录表
logwood 苏木
logwood black 苏木黑
logwood dyeing 苏木黑染色
logwood extract 苏木浸膏,苏木萃(植物性媒染染料)
LOI(limiting oxygen index) 限氧指数(测定纤维可燃性值)
LOI value 限氧指数值
lokao 绿胶(从中国鼠李属植物中提取的绿色染料,又名中国绿)
London dispersion force 伦敦分散力
London shrinking 伦敦预缩法(织物经充分润湿后,在不加张力下自然干燥,以达到预缩效果)
London shrunk 伦敦预缩法

London shrunk machine 伦敦预缩机
London smoke 暗灰色
lone electron pair 孤对电子
long 长流的(指浆糊的连续流动性)
long afterglow 长余辉
long bath 大浴比染浴
long-chain dye process [棉经纱靛蓝]连续染色工艺
long chain molecule 长链分子
long cuts 稀释(指印花浆染料浓度的冲淡)
long duration test 耐久性试验
long dwell steamer 长储蒸箱;大容量蒸箱
long ends 拖纱(印疵,指织物表面有浮纱,阻碍印花)
long flow 长流型,高切变型(印花糊料)
long flow thickeners 高[切变]糊料(网印中糊料黏度受到刮浆的剪切应力而呈现变化,变化小的称为高切变糊料)
long hair finish [毡帽坯]长毛修剪整理
longitudinal adjustment 纵向[对花]调节《印》
longitudinal air flow 纵向气流
longitudinal cutting device 纵向切割器
longitudinal dilatation 纵向膨胀
longitudinal elasticity 纵向弹性
longitudinal fluting 纵向凹槽
longitudinal fracture 纵向断裂
longitudinal relaxation 纵向松弛
longitudinal repeat adjustment 纵向花回调节《印》
longitudinal seal 纵向密封条
longitudinal section 纵断面,纵切面,纵截面
longitudinal shearing machine 纵向剪毛机
longitudinal shrinking 纵向收缩

longitudinal slitting 切条
longitudinal slot 纵向狭缝
longitudinal strain 纵向应变
longitudinal stress 纵向应力
longitudinal stripes 纵向条纹
longitudinal vibration 纵向振动
long-life grease 长效润滑脂
long liquor 大浴比
long liquor ratio 大浴比
long loop drying machine 长环烘燥机
long loop steamer 长环蒸化机,悬挂式蒸化机
long-oil 长油的
long-oil varnish 长油性清漆
long period 长周期
long pile carpet 长绒地毯
long piled fustian 长毛纬绒织物
long pile fabric 长绒织物
long pile shag blowing and drying machine 长毛绒蒸烘机
long pile shag carding machine 长毛绒梳理机
long range 长距离的,大幅度的
long range elasticity 高弹性,大幅度弹性
long range interaction 远程相互作用
long rheology 长流变性(印花糊料性能)
long runs 大批量
long run screen 耐用筛网(指圆网)
long run test 使用寿命试验,连续负荷试验,长期运转试验
long slot burner 长缝式火口,长缝式燃烧器
long spacing 长间距,长周期
long standing 久置
long stitch course pattern 长线迹花纹
long stroke compensator 升降辊松紧架

long term repeatability 长期重复性
long term stability 长期稳定性
long term storage 长期储存
long term test 长期试验
long term thermal stability 长期热稳定性
long-time cycle 长工作周期
look 外观
look-over(=looker-over) 验布工,检验工
look-through 透过性,透视度
loom finished [落机]坯布
loom finished linen 原纱漂白亚麻布
loom goods 坯布
Loom printer 印经纱装置(商名,法国制)
loom sizing 织机上浆
loom state [fabric] [落机]坯布
loom-state printing 坯布印花
loop 线圈;毛圈;绒圈;圈,环;循环
loop ager 悬环式蒸化机
loop cloth 毛圈织物,毛巾布
loop construction 线圈结构
loop conveyor 环形输送带,环形运输装置
loop cutting machine 剪绒机,割绒机
loop drier 悬环式烘布机
loop dyeing 悬挂式染色,环状染色
loop dye machine for denims 牛仔布纱线经轴环状染色机(意大利制)
loop dye process 纱线经轴环状染色法
looped fabric 毛圈织物,毛巾织物
looped pile 毛圈
looped plush 毛圈式长毛绒
loop effect 环圈效应;卷曲效应
loop fabric 毛圈织物,毛巾织物
loop folding machine 圈码机
loop formation system [蒸化机]成环系统
loop-forming device 成环装置(长环蒸化机的一个部件)

loop knot 扭结,环结
loop method 环形法
loop pile carpet 毛圈式地毯,起圈地毯
loop pile fabric 毛圈绒头织物
loop scouring 挂练,吊练
loop stability 成环稳定性
loop steamer 长环蒸化机,悬挂式蒸化机
loop strength 钩接强力
loop towel 毛巾,面巾
loop transfer process 环圈移液法(用长环形输液带包绕轧辊,在运行中从浸液槽带液上来和加工织物同时受轧,起到移液作用)
loop type 悬挂式;悬环式;环式
loop yarn 环圈线,起圈纱线
loose cheese 松式筒子,松式筒子纱
loose colour 浮色;易褪色的染料
loose dye 未固着染料
loose edge 松边(织疵)
loose fibre dyeing 散纤维染色
loose fibre dyeing machine 散纤维染色机
loose fill type insulant 松散绝热材料
loose fit 松配合
loose installation [地毯]覆盖铺设
loose material 散纤维(泛指粗梳或精梳前的纺织纤维)
looseness 松散性
loose pulley 游滑轮,惰轮,活动皮带轮
loose rewinding 毛球倒松机
loose running fit 松转配合
loose steam style 浮蒸印花法(蒸后不洗,色鲜易褪)
loose stock dyeing 散纤维染色
loose stock dyeing machine 散纤维染色机
loose structure 松散结构
loose winding 松卷,松绕

loose wool dyeing machine 散毛染色机
loose woven goods 稀松织物
loss in strength 强力损失
loss modulus 损耗模量
loss of colour 褪色
loss of drying 干燥减量,干燥失重
loss of weight 重量损失,[加工时]失重
loss on scouring 精练失重
lot 批;批量;组,套
lot card 批号卡
lot inspection 全批检验
lotion 洗涤剂;洗涤
lot number 批数,批号
lot product 批量产品
lot production 成批生产
lot sample 批样(一批中抽出的试样)
lot size 批量大小
lot tolerance fraction defective 容许不合格批
lot to lot colour variation 批与批色差
lot to lot shading problem 批间色差问题
lotus effect 荷叶效应(指荷叶的拒水自洁效应)
loud [色彩]刺目的,耀眼的;刺耳的
loudness 响度
loudness level〈LL〉音量标准,音量水平,响度级
lousiness [织物的]起毛
louvre damper 百叶窗挡板
louvre drier 百叶帘式烘燥机
lovat 带灰光色(如绿色);洛伐脱混色法(优质毛织物用,混合后色泽不互相抵消)
Lovibond tintometer 洛维邦调色计
low add-on〈LAO, L. A. O.〉低加重率;低添加量

low add-on applicator 低给液量给液装置（用于织物的整理和染色,可节约染化料,减少烘燥所需的能量）

low add-on process 低给液工艺

low add-on with curved blade applicator 弧形刮刀低给液装置

low add-on with gravure-roll padding 刻纹辊低给液工艺

low add-on with kiss-roll padding 单面辊低给液工艺

low add-on with quetch-suck process QS 低给液工艺（转移轧吸法低给液工艺）

low add-on with spray method 喷雾法低给液工艺

low add-on with transfer padding 转移轧吸低给液工艺

low bed process camera 下座式照相机（照相雕刻用）

low carbon life (= low carbon living) 低碳生活

low carbon steel 低碳钢

low carbon textile wet processing 低碳纺织湿处理,低碳印染

low chrome dyeing method 低铬染色法

low coverage design 覆盖面小的［印花］图案

low crock latex 低脱色胶乳物（加在涂料色浆中,以提高涂料印花的摩擦脱色牢度）

low density polyethylene 〈LDPE〉 低密度聚乙烯

low duty 轻型的

low-dyeing polyamide 低染色性能聚酰胺,差异染色聚酰胺

low elastic yarn 低弹丝

low elongation 低伸度

low end 低级织物

low end mill 低级纺织品工厂,低档产品工厂

low energy surface 低能量表面

low energy type dye 低能量型［分散］染料

lower bushing 下轴瓦

lowering of freezing point 冰点下降（测相对分子质量用）

lowering of roll 落刀；辊子降落

lower limit 下限

lower variation of tolerance 下偏差

low flammability 低可燃性,不易燃性,难燃性

low flash product 低燃点产品

low foamer 低起泡剂

low foaming detergent 低泡沫洗涤剂

low foaming surfactant 低泡沫表面活性剂

low formaldehyde resin 低甲醛树脂

low formaldehyde resin finishing 低甲醛树脂整理

low frequency discharge 低频放电（一种等离子体技术）

low frequency heating 低频加热

low grade 低级的

low grade energy 低品位能

low ink warning 缺墨警告（喷墨印花）

low level open washer 低水位平洗槽,低水位平洗机

low level washing cistern 低水位水洗槽

low liquor ratio 小浴比

low liquor ratio dyeing machine 小浴比染色机

low molecular polymer 〈LMP〉 低分子聚合物

low neatness 劣等净度,劣等洁净

low pile 低绒头、低绒圈
low polymer 低聚物
low polymerized precondensate 低聚合度预缩合物
low power stretch 低[功能]弹性
low pressure gas 低压煤气
low pressure gas infrared heater 低压煤气红外加热器
low pressure kier 低压煮布锅,开口蒸布锅
low pressure plasma 低压等离子体
low pressure polyethylene 低压聚乙烯(相当于高密度聚乙烯)
low refractive index 低折射率
low scouring 低温精练,轻度精练
low shear measuring system 低剪切测定法(印花糊料)
low shrinkage fibre 低收缩纤维
low soiling finish 防污整理,抗污整理
low solids 低含固量的
low solids thickeners 低固含量增稠剂,低固含量糊料
low solvent oil-in-water system 低溶剂含量油水相乳化糊(溶剂含量10%～15%)
low stretch fabric 低弹织物
low stretch yarn 低弹丝
low sudser 低起泡剂
low sulphur heavy oil 低硫重油
low temperature bleaching 低温漂白
low temperature characteristic 低温特性
low temperature cross-linking 低温交联
low temperature cure finish 低温烘焙整理
low temperature curing 低温焙烘
low temperature dyeing 低温染色
low temperature flexibility 低温柔性
low temperature medium 低温介质,低温溶剂;低温手段
low temperature moldability 低温模塑性能
low temperature oxidation 低温氧化法
low temperature oxygen plasma 低温氧等离子体
low temperature performance 低温性能
low temperature plasma〈LTP〉 低温等离子体
low temperature polymerization 低温聚合
low temperature preparation 低温前处理
low temperature resistance 耐寒性,耐低温性
low temperature scouring 低温精练
low temperature silk dyeing 低温丝绸染色
low temperature solidifying binding agent 低温固化黏合剂
low temperature solution polycondensation 低温溶液缩聚[作用]
low temperature tempering 低温回火
low temperature wool dyeing 低温羊毛染色
low tension 低张力
low tension batch 低张力卷轴
low tension dryer 低张力烘干机
low twist 低捻度
low velocity〈lv,LV〉 低速
low visibility 低能见度
low volume〈lv,LV〉 小体积,低容积
low volume dispensing system 低容量配料系统
low volume/high velocity〈LV/HV〉 小体积高速度的
low wet pickup 低轧液率
Loynes 洛尼斯[浅色印花]薄呢
lozenge 菱形,菱形纹

lozenge effect 菱形效应

l. p. (leuco potential) [还原染料]隐色体电势

LPG(liquefied petroleum gas) 液化石油气

lq. (=liq.,liquid) 液体;液体的

LR(liquor ratio) 浴比;液比

L rating 耐光色牢度等级

LSC(liquid-solid chromatography) 液—固色谱分析法

L-square 直角板,角尺

Ltd. (=lim.,limited) 有限[责任]的,有限制的;极限;时限

LTP(low temperature plasma) 低温等离子体

lu. (lumen) 流量(光通量单位)

lube 润滑;润滑油

lubricant 润滑剂

lubricant consumption 润滑剂消耗

lubricating 润滑,减摩

lubricating agent 润滑剂

lubricating agent of fibre 纤维润滑剂,纤维油剂

lubricating cup 油杯

lubricating effect 润滑效果

lubricating felt 润滑用毡

lubricating felt cup 油毡润滑杯

lubricating grease 润滑油脂

lubricating gun 润滑油枪

lubricating hole 注油孔

lubricating oil 润滑油

lubricating wick 油芯

lubrication 润滑[作用];润滑法

lubrication chart 润滑图表

lubrication free chain 自润滑链

lubrication free slide rails 自润导轨

lubrication interval 润滑周期,加油周期

lubrication period 加油周期

lubricator 润滑器,润滑装置;润滑剂;加油工

lubricity 平滑性,润滑能力,润滑性质

lucite 2-甲基丙烯酸(合成荧光树脂)

Ludigol 卢迪戈尔(硝基苯磺酸钠,弱氧化剂,在印花汽蒸时防止过度还原作用,商名,德国制)

Ludigol treatment 卢迪戈尔处理

Lufibrol 路快宝系列(除杂剂,纤维保护剂,用于纤维素纤维及其混纺织物的碱煮练工艺,商名,德国巴斯夫)

lug 吊耳,耳状物

luisine 英国优质丝夹布(用埃及棉制)

luke warm water 温水

Lumacron E-RD dyes 路马克朗 E-RD 染料(快速染色型涤纶用分散染料,也适用于超细旦纤维,商名,韩国 LG 化学)

Lumacron S-HW dyes 路马克朗 S-HW 染料(涤纶用分散染料,高湿牢度,商名,韩国 LG 化学)

lumen〈lu.〉 流量(光通量单位)

lumen staple fibre 空心人造短纤维

Lumicron MFB 路马克朗 MFB(高超细旦涤纶纤维用分散染料,商名,德国 M. Dohmen 公司)

Lumimagic 一种高效抗菌防臭新材料(日本东丽公司)

luminaire 光源,发光体

luminance 发光率;亮度,照度,发光度,光亮度

luminance factor 亮度因数

luminance purity 亮度纯度

luminescence 发光

luminescence analysis 荧光分析

luminescence emission 光发射
luminescence spectrum 发光光谱,荧光光谱
luminescent 发光的,荧光的
luminescent dyes 荧光染料
luminescent effect 发光效应
lumineux ribbon 有光丝带
luminometer 光度计,照度计
luminophore 发光团,发光体
luminosity 亮度,发光度;可见度
luminosity factor 视感度
luminous density 光密度
luminous efficiency 发光效率
luminous flux 光流量,光通量
luminous flux density 照度,光通量密度
luminous intensity 发光强度
luminous paint 发光涂料
luminous pigment 夜光涂料,发光颜料
luminous pigment printing 荧光涂料印花
luminous printing 荧光印花
luminous reflectance 光反射率,光反射比
luminous sensitivity [感]光灵敏度
luminous transmittance 光透射率
lumps 团,块;坯布,双幅布(织后剪开)
luobo flax 罗布麻
Luprimol 路皮莫系列(涂料印花手感改进剂,商名,德国巴斯夫)
Luprintol 路平托儿系列(涂料印花用乳化剂,商名,德国巴斯夫)
luster【美】=lustre
lusterer 光亮织物
lustering 上[柔]光整理(美国用语)
lustering agent 上光剂,光亮剂
lustering calender 上光轧光机
lusterness 光泽度
lustre 光泽,亮光;发光

lustre cloth 有光呢(棉经,有光毛纬)
lustre decatizing 光泽蒸呢
lustre finish 上[柔]光整理
lustre finished rug 上光整理地毯
lustre lacking uniformity 光泽不匀
lustreless 无光泽
lustre lining 亮光里子呢,羽纱
lustre meter 光泽[测定]计
lustre paper [毛织物整理用的]压光纸板
lustre plate 上光板(滚筒式压呢机用上光金属板),压光托板
lustring 上[柔]光整理;光亮绸;加光丝带
lustring machine 上[柔]光机
lustrous colours 鲜艳色
lustrous fibre 有光纤维
lustrous finish 有光整理,光泽整理
lustrous furniture fabric 有光家具布
lustrous rayon 有光人造丝
lustrous thread 有光线
Lusynton SE 路新通 SE(有机萃取剂,用于特殊退浆工艺,商名,德国巴斯夫)
luteous 深橘黄色的
Lutexal HEF 路得素 HEF(高分子合成增稠剂,涂料印花用,商名,德国巴斯夫)
Luther condition 卢瑟条件(在特定光源下,使测色仪三个受光器的波长综合感度与相应的 CIE 光谱三刺激值达到一致的条件)
lux〈lx.〉勒[克司](照度单位)
lux gauge 照度计,勒克司计
luxometer 勒克司表,照度表,流明计
luxurious hand 丰满手感
luxury 豪华,华贵
lv(=LV,large vent) 大眼;大口径

lv（=LV,linear variable） 线性变异
lv（=LV,low velocity） 低速
lv（=LV,low volume） 小体积,低容积
LV/HV（low volume/high velocity） 小体积高速度的
L×W×H（length×width×height） 长×宽×高
lx.（lux） 勒[克司]（照度单位）
lyase 裂解酶
lycopene [colour] 蕃茄色素（天然植物染料,食用色素）
Lycra 莱克拉（聚氨基甲酸酯弹性纤维纱,橡筋线代用品,即氨纶,商名,美国杜邦）
lye [洗涤]碱液;废皂碱水
lye boil 碱液煮练
lye boiling 碱液煮练
lye cleaning system 碱液清洁装置
lye leaching 浸碱
lye limit concentration 碱液极限浓度
lye recovery unit 碱液回收装置
lye recuperator 碱回收蒸箱（用于丝光机）
lye recycling 烧碱循环利用
lye stain 碱液污渍,练浴污渍
Lyocell fibre 天丝棉[以木浆粕为原料,用氧化甲基吗啉（NMMO）为溶剂制成的新型再生纤维,无环境污染,称"绿色"纤维,商名,英国考陶尔]
lyocell fibre finishing 天丝棉整理
Lyocol 赖可均系列（具有溶解染料及分散作用的助剂,商名,瑞士科莱恩）
lyoenzyme 细胞外酶

Lyofix MLF new 利奥弗斯 MLF 新（烷基改性三聚氰胺甲醛衍生物,非离子型,纤维素纤维及合成纤维织物用树脂整理剂,也可作硬挺整理剂,商名,瑞士制）
lyogel 水乳胶体,水凝胶
Lyogen 赖可匀系列（烃基聚乙醚,羊毛、涤纶、锦纶的分散匀染剂,商名,瑞士科莱恩）
lyolipolarity 双亲性
lyolysis 液解[作用]
lyophilic 亲液[性]的
lyophilic colloid 亲液溶胶,亲液胶体
lyophilic property 亲液性
lyophilization 冷冻干燥,升华干燥
lyophobic 疏液[性]的
lyophobic colloid 憎液溶胶,疏液胶体
lyophobicity 疏液性
lyophobic property 疏液性
lyosol 水溶胶
lyosorption 吸取溶剂[作用],溶剂吸附
lyotrope 感胶离子;易溶物
lyotropic 亲溶剂的
lyotropic polymer 溶致性聚合物
lyotropic solution 溶致性[液晶]溶液
lypase 脂肪酶
lysine〈LYS〉 赖氨酸,2,6-二氨基己酸
lysinoalanine 赖氨酸丙氨酸
lysis 溶解
lysozyme 溶菌酶
lytic effect （细胞）溶解效应

M

m(meta-) (构词成分)表示"间[位]" "介""偏"

M(＝m,mass) 质量

M(＝m,min,min.,minute) 分[钟];微小的

M(＝m,m.,metre,meter) 米;表,仪,计

MA.(＝ma.,microampere) 微安[培]

MA.(＝ma.,milliampere) 毫安[培] (10^{-3}安培)

MA(methacrylate) 甲基丙烯酸酯

MA(methylacrylate) 丙烯酸甲酯

MA(methyl alcohol) 甲醇

MAA(methacrylic acid) 甲基丙烯酸,异丁烯酸

Mäander 迈思德短横直导辊水洗机(德国制)

MAC(marginal aqua control) 临界水质控制《环保》

MAC(modacrylic) 改性聚丙烯腈[纤维]

MacAdam's ellipse 麦克亚当椭圆(色品图上用以评价视觉对色差敏感性的一种经验椭圆)

MacAdam U-V system 麦克亚当 U-V 测色系统

macaroni fibre 中空纤维

macaroni yarn 空心丝

Macbeth SpectraLight 麦克贝斯标准光源仪

Macegard CDH 马塞加特 CDH(自交链型丙烯酸酯共聚物,聚酯双面织物的防勾丝整理剂,商名,美国制)

maceration 浸渍[作用]

Mach 马赫(速度单位,当速度等于音速时为 1 马赫)

machine〈M/C,Mc,mc.,m/c〉 机器,机械;机组;设备

machine alignment 机器校准

machine allocation 机台配置,看台数

machine arrangement 机器流程

machine assembly 机器装配

machine assignment 看台数

machine base 底座

machine bed 底板

machine bleaching 机械漂白

machine block printing 机器木版印花

machine bolt 螺栓

machine building 机器制造

machine building works 机器制造厂

machine calender 凹凸轧花机,拷花轧压机,浮雕轧压机

machine clipping 机械修剪

machine coating 机械涂层

machine construction 机器结构

machine contact time 装置与织物接触时间

machine design 机械设计

machine dimensions 机器尺寸

machine direction〈MD〉 机器运行方向

machined washer 机器加工垫圈

machine dyeing 机械染色

machine element 机械零件

machine-engraved roll 网纹雕刻辊(低给

液轧辊用);雕刻微孔轧辊(一种高效轧辊,轧辊表面刻有许多微孔,微孔形状有倒三角锥形、四边形等,雕刻面与轧辊成 45°角)
machine erection 机器安装
machine error 机器错误,机器误差
machine exploitation 设备利用率
machine finishing 机械整理
machine foundation 底脚,机座
machine frame 机架
machine hour [机器的]台时工作量
machine-hour costs 台时成本
machine knitting 机器针织,机器编织
machine made 机器制成的
machine-made wrapped hose 机制夹布软管
machine maker 机器制造者,机器制造厂商
machine manufacturer 机器制造者,机器制造厂商
machine operator 机器操作员
machine performance 机械[运转]状态
machine printing 机器印花
machine productivity index 〈MPI〉 机器生产能力指数
machine room 印花车间(俗称)
machinery 机械,机器
machinery cleaner 机器清洁剂,机器清洁助剂
machinery noise 机器噪声
machinery raising 机械起毛
machinery steel 〈M.S.〉 结构钢
machine sewing of grey 坯布机器缝头
machine sewing thread 缝纫机用线
machine shop 机械工场
machine shrink 机械预缩

machine sizing 机械上浆
machine stitch seam 机器线缝
machine stop(=machine stop bar) 停车色档(印染疵)
machine stoppage 停机
machine testing 机械测试
machine utilization 机械效率,机械利用率
machine variant 机器变体
machine vice 机用老虎钳
machine wash 机械洗涤
machine washability 耐机洗性;可机洗性
machine washable finished wool 机洗整理羊毛制品,经机可洗整理的羊毛制品
machine washable silk 机洗丝绸
machine washable wool 可机洗羊毛制品;机可洗羊毛制品
machine washing severity 机器洗涤考验
machine word 计算机字;机器字
machining 机械加工
machinist 机工;机械师;机器操作工人
Machnozzle 马赫喷嘴(超音速喷气脱水装置)
Machnozzle washing machine 马赫喷嘴水洗机(一种高效水洗机,由两组横导辊组成,每组出布口处装有一个马赫喷嘴脱水装置,代替轧水辊,荷兰制)
macintosh 防水胶布;橡皮布;胶布雨衣
macintosh blanket 防水橡皮衬布,防水胶布
macintosh cloth 橡胶防雨布
mack(=mackintosh) 轻薄防水织物;防水胶布;雨衣
mackintosh blanket(=mackintosh rubber blanket)[印花机]防水橡皮衬布,防水胶布

macr-(=macro-)（构词成分）表示"长""大""宏""粗视"
macro-analysis 常量分析
macro-Brownian motion 宏观布朗运动
macrocapillary 大毛细管
macroconfiguration 宏观结构
macroconformation 大构象
macro-creep 宏观蠕变
macro-dispersoid 粗粒分散胶体
macro emulsification 粗滴乳化[作用]
macrofibril 大原纤（纤维状的微细组织，其直径在200纳米左右）
macro-fibrillation 大原纤化，初级原纤化
macro-flow 宏观流动
macromole 大分子，高分子
macromolecular alignment 大分子排列
macromolecular compound 大分子化合物，高分子化合物
macromolecular dispersion 大分子分散
macromolecular solution 大分子溶液
macromolecule 大分子，高分子
macroporous structure 宏观多孔结构
macroradical 大游离基
macroscopic deformation 宏观形变
macroscopic examination 粗量检定
macrostructure 宏观结构，粗视结构
macular 有斑点的，不清洁的
maculation 污点，斑点
MAD（multiple applicator device） 多路并联施料设备（在毡毯上施加染液和胶料）
madder 茜草[植物染料]
madder bleach 茜草印花法的漂白工艺（织物在茜草媒染染料印花前，经过加强的漂练处理）
madder red 茜红

madder style 茜草染料印花法（先印媒染剂，固着后再用茜素染色）
madder style printing 茜草染料印花法
made-in-order pallet 定制的货盘，集装盘（物料搬运器具，仓库等处供铲车装卸及搬运货物用）
made-to-measure polymers 定向聚合物
made-to-order 定制的；定制服装
magazine 机器装置的储存库
magazine boiler 自动加煤小锅炉
magazing screen printing machine 老式平网印花机
magenta 品红；碱性品红
magenta filter 品红滤片（圆网雕刻制半色调用）
magenta filter adjustment 品红滤色片调节
magenta raster 品红光栅网板，品红滤色片（用于圆网印花半色调制网）
magic-eye controlled guider 光电导边器
magma [稀]糊；稠液；乳浆剂
magnaflake bronze powder 含镁青铜粉
Magnasoft HSSD 麦格纳柔软剂HSSD（氨基及聚醚双重改性有机硅，水分散体，具有抗静电、抗起球、低泛黄等功能，良好的亲水效果，商名，美国威科）
Magnasoft UE 麦格纳柔软剂UE系列（氨基改性有机硅，透明微乳液，商名，美国威科）
Magnasoft XE 麦格纳柔软剂XE系列（氨基改性有机硅，透明微乳液，商名，美国威科）
magnesia 氧化镁，镁氧
magnesite 菱镁矿
magnesium carbonate 碳酸镁
magnesium chloride 氯化镁

magnesium compound 镁化合物
magnesium hardness〈MgH〉镁硬度
magnesium hydrogen phosphate 磷酸氢镁
magnesium lauryl sulfate 月桂基硫酸镁，十二烷基硫酸镁（阴离子净洗剂，发泡剂）
magnesium nitrate 硝酸镁
magnesium oxide 氧化镁
magnesium silicate 硅酸镁
magnesium sulphate 硫酸镁
magnet 磁铁，磁体
magnet- （构词成分）表示"磁力""磁性""磁电"
magnet bar 磁棒（用于圆网印花或平网印花的磁性刮印机构）
magnet beam 磁性轴（用于控制印花导带）
magnet coil 磁性线圈
Magnet-Combi-squeegee 磁性混合刮浆装置，刮刀与磁辊的组合（圆网印花用，商名，奥地利齐玛）
magnet core 磁铁芯
magnetic 磁[性]的
magnetic bearing 磁性轴承；磁方位
magnetic brake 磁性闸
magnetic card 磁卡片《计算机》
magnetic clutch 电磁离合器
magnetic contactor 磁性接触器，电磁开关
magnetic core 磁芯
magnetic data carrier 磁性数据承载器
magnetic disc 磁盘
magnetic disc storage 磁盘存储器
magnetic drum 磁鼓《计算机》
magnetic dyeing 磁性染色
magnetic field generator 磁场发生器
magnetic flowmeter 磁性流量表

magnetic flux 磁通（磁力的量度）
magnetic gearing 磁性传动
magnetic lens 磁[性]透镜
magnetic memory 磁存储器
magnetic microscope 磁[性]显微镜
magnetic moment 磁矩
magnetic permeability 导磁系数，磁导率
magnetic powder clutch 磁粉离合器
magnetic printing technique 磁性印花技术
magnetic rating 磁性率
magnetic resistance 磁阻
magnetic roller squeegee 电磁刮浆辊（圆网和平网印花机用）
magnetic separation 磁性分离
magnetic shielding 磁屏蔽
magnetic squeegee 磁性刮刀
magnetic stirrer 磁性搅拌器
magnetic storage(＝magnetic store) 磁存储器
magnetic susceptibility 磁化率，磁化系数
magnetic valve 磁阀
magnetic weighting 磁性加压
magnetite 磁铁矿
magnetization 磁化；磁化强度；（电动机）激励
magnetizing current 磁化电流
magneto 永磁电动机，磁电动机，磁石发电机
magneto- （构词成分）表示"磁力""磁性""磁电"
magneto chemistry 磁化学
magnet roller 磁性加压辊，磁性罗拉
magnet-roll-system 磁力辊装置《印》
magnetron 磁控管
magnet-squeegee-system 磁性刮浆系统
magnet tension device 磁性张力装置

magni- (构词成分)表示"大"
magnification 放大,放大倍数
magnifier 放大镜
magnifying glass 放大镜
magnifying power 放大率
magnitude 大小,尺寸;数量;程度
magnitude of detonation [粉尘]起爆值
magnoknife 磁性刮刀
magnoroll 磁性辊棒(用于印花刮浆)
magnoswitch 磁性开关
mahagony acid [石油]磺酸
mahagony gum 漆树胶
mahagony soap [油溶性]石油磺酸皂
mahagony sulfonate [石油]磺酸盐
Mahlo 马洛检测仪(商名)
Mahlo textometer 马洛测湿仪(织物湿度测定和控制仪器,采用电导法测湿,并反馈控制调速伺服电动机)
mahogany brown 柳桉木棕[色]
Maifoss Process Maifoss 工艺(一浴染二色羊毛)
maillot [体操运动员穿的]紧身衣;[衣裤连在一起的]游泳衣(法国名称)
mail-order 邮购
mail-order catalogue 邮购目录
main burner 主燃烧器
main chain 主链,主链条;主链节
main component 主成分
main controller 主控制器
main distillate fraction 主馏分
main drive 总传动,主传动
main fuse 主熔断器,主保险丝
main header 主联箱,主管道
main motor 主令电动机
main-process stream (工艺)主流程线;总流程

main relay 主继电器
mains 电源;馈电线,电力线
main shaft 主轴
main steam pipe 主蒸汽管
main supports 主架,横梁
main switch board 总开关板
maintenance〈maint.〉维修,保养,保全;维持
maintenance and repair 维修(日常维护和大小检修的统称)
maintenance facilities 维修设备,维修工具
maintenance-free operation 不需维护的运行
maintenance manual 保养指南,保养手册
maintenance method 保全工作法
maintenance worker 保全工,保养工
main test 主要检验
main transformer 主变压器
maize [yellow] 玉米黄[色]
maize fibre 玉蜀黍蛋白纤维,玉米蛋白纤维
maize flour 玉米粉,苞米粉
maize starch 玉米淀粉
major brand 主要牌号,名牌
major overhaul 大检修
major repair 大修理
make alkaline 碱化,碱处理
make pliable 半脱[丝]胶,半练丝;[蚕丝]半练工艺
make-up 包装;补足;装配;修理
make-up air 补充空气
make-up float 浮子式给水调节器
make-up heat 补充加热
make-up machine 卷取机
make-up of chemical loss 化学药品补充[量]

make-up rolled on card board 卷板成包
make-up rolled on tube 卷筒成包
make-up room 包装间
make-up tank 调配槽
make-up water 补充水,补给水
make-up yard lapped 折叠成包
make weight 补足重量,增重;平衡力
making 制造
making a range of samples 系列取样
making-up 织物包装;包装标志;生头
making-up machinery 织物包装机械
MAK list 【德】工作场所有害物最高允许浓度名单(德国公布禁用染料等的名单)
malabar 东南亚印花手帕料(色泽鲜艳并有反衬色);马拉巴印花布(英国和印度向东非洲出口印花布的总称)
Malabar tallow 马拉巴油脂
malachite green 孔雀石绿[色](鲜艳黄绿色);[碱性]孔雀绿[染料]
maladjustment 调节不良,失调
maleamic acid 马来酰胺酸
male and female calendering 凹凸辊轧花工艺
male and female flange 凹凸面法兰
male die 阳模
male female embossing machine 轧花机;凹凸辊轧花机
male fittings 外螺纹连接件
male flange 凸法兰
maleic acid 马来酸,顺丁烯二酸
maleic anhydride 马来酐,顺式丁烯二(酸)酐,马来酸酐(无甲醛防皱整理剂)
maleimide 马来酰亚胺,顺丁烯二酰亚胺
N-maleoyl degradation chitosan N-马来酰化降解壳聚糖(用于防皱整理)
male pattern 凸纹图案
male screw 阳螺旋
male T 外螺纹三通管接头
male thread 阳螺纹,外螺纹
malformed fish 畸形鱼[环保]
malfunction 出错,故障,事故,错误动作,失灵
malfunction of lint doctor 小刀铲色
malic acid 苹果酸,羟基丁二酸
MA liquor applicator 最低给液量给液装置(指单面给液装置,备有给液自控系统,可达到极低的给液量)
mallow [purple] 锦葵红(浅紫红色)
malodorous substance 恶臭物质《环保》
malonic acid 丙二酸
malt 麦芽
malt amylase 麦芽糖化酶;麦芽淀粉酶
maltase 麦芽糖酶
malt diastase 淀粉麦芽糖化酶(麦芽退浆剂)
malt extract 麦芽精,麦芽膏
malting 用麦芽制剂退浆
maltose 麦芽糖
mammalian toxicity 哺乳动物毒性
MAN (methacrylonitrile) 甲基丙烯腈,异丁烯腈
management 经营;管理,管理法;管理部门
management information package (MIP) 管理信息包《计算机》
management information system (MIS) 管理信息处理系统
management principles 经营方针
management strategy 经营策略
managing staff 管理人员

Manchester brown 曼彻斯特棕,碱性棕
Manchester yellow 曼彻斯特黄,马休黄
mandarin orange 柑橘色
mandarin red 朱红[色],橙红[色]
mandatory 强制性的
mander washing machine 曲流式水洗机（一种非直、非横导辊式高效平幅水洗设备,德国制）
mandrel 印花轧辊[铁]芯子,心轴
mandrel-cambered bowl 中高轧辊,凸芯轧辊
mandrel press 花筒装拆机
manfr.（＝manuf., mfr., mfre., manufacture）制造；制品；制造厂
manfr.（＝manuf., manufacturing）制造的,生产的
manganese bronze（＝manganese brown）锰棕色；锰棕（矿物染料）
manganese brown discharge 锰棕拔染
manganese compound 锰化合物
manganese dioxide 二氧化锰
manganese dyeing 锰金属染色
manganese permutite 锰泡沸石（软水剂）
manganese zeolite 锰沸石
mangle 压呢机,轧液机；轧光机；[滚筒]轧车
mangle crease 轧车折痕,轧皱印
mangle drier 轧水干燥机
mangle expression 轧液率
mangle finish 轧光整理
mangle lustring 卷轴轧光（织物卷成轴,在压力下回转,布层经受强烈挤压,产生似麻的光泽）
mangle trough 轧槽
mangling 轧液,轧水
manhole 人孔,检修孔

manhole cover 人孔盖
man-hour 工时
manifest image 显像
manifold yarn 多股并合线
maniford 歧管,集合管,多支管；复式接头
manioc starch 木薯淀粉
manipulated variable 被控变量
manipulation 处理,操作；操纵,控制
man-machine communication 人—机交流
man-made fibre〈MMF〉化学纤维,人造纤维
man-made leather 人造皮革,合成皮革
man-made protein fibre 人造蛋白质纤维
manna flour 稻子豆粉,刺槐豆粉
mannan 多缩甘露糖,甘露聚糖（多糖类糊料主要成分）
mannanase（＝mannase）甘露聚糖酶
manning 人员配备
mannite（＝mannitol）甘露糖醇
mannogalactan ether 甘露聚糖半乳糖醚（一种天然印花糊料的化学成分）
mannogalactan gum 甘露半乳糖胶
mannose 甘露糖
mannuronic acid 甘露糖醛酸
manoeuverable 操纵灵敏的；机动的,可调动的
Manofast 曼诺法斯特（还原剂,二氧化硫脲,商名,英国哈特门）
manograph 压力自记器,流[体]压[力、强]记录器,压力计
manometer [流体]压力计,压力表
manometric method 测压法
mansard 热风干燥室,烘房
mansard steamer 回绕式蒸化机（印花织物背面接触导辊,可免搭浆）

mantle 外壳,套,罩
manual 手动的,手工的,人工的
manual adjusting valve 手动阀
manual adjustment 手调
manual control〈M/C〉人工控制,手控
manual digitizer 手控数字转换器
manual mode 人工模式
manual operation 人工操作
manual press 手工印花机(印T恤衫用)
manual push button control 手揿控制装置
manual regulation 手动调整,手工调整
manual repairing 手工修理,手工雕刻
manual repeat adjustment 手工花回调节（手工对花）
manual screen printing 手工网印
manual starter switch 手控起动开关
manual valve positioner 手动阀门定位器
manufacture〈manuf.,manfr.,mfr.,mfre.〉制造;制造品;制造厂
manufactured fibre 再生纤维
manufactured graphite 人造石墨
manufacture flexibility 生产灵活性
manufacturer〈Mfr.〉制造商;厂主;生产者
manufacturing〈manfr.,manuf.〉制造的,生产的
manufacturing assembly drawing 制造装配图
manufacturing cost 制造成本
manufacturing flow chart 制造过程表,工艺流程图
manufacturing process 制造过程
Manutex F 麦努特克斯F(低黏度型海藻酸钠,印制精细线条用,商名,英国凯尔可)
Manutex RS 麦努特克斯RS(高黏度低固含量的海藻酸钠,增稠剂、防止泳移剂及印花糊料,商名,英国凯尔可)
manway 人孔,人通道
map 映象;变换;图
map cloth 地图布
maple 槭木棕[色](淡棕色)
maple sugar 槭糖棕[色](黄棕色)
MA process（minimum application process）低给液工艺
marbled ground 混色地
marble melt 玻璃球熔体
marble wash 云纹石洗
marbling printing 斑点印花
margin 限度,容许极限,边缘
marginal aqua control〈MAC〉临界水质控制《环保》
marginal check 边缘检查
marginal test 边缘试验
margin phenomena 边缘现象
margin to seam 缝头
Marglass 玻璃纤维(商名,英国制)
marigold yellow 金盏花黄[色](橙色)
marine[blue] 藏青[色]
marine insurance policy〈MIP〉海运保险单
marine product 航海用品
mark〈mk.〉标记,标志,记号,符号;特征;型号;污渍;斑疵;着色沾污
marked line 标线
marker 标记;打印工;打标记机;划衣片样板;排料,排料图《裁》
marker buoy 指示浮标
marker making 描样(裁剪前将衣样描在织物上)
market analysis 市场分析
market and sales fore casting 市场和销售

预测

market bleach 普通漂白[布](不作印染用的一般性漂白或漂白布)
market channel 市场渠道
market demand 市场需求
market development 市场开发
market economy 市场经济
market information 市场信息,营销信息
marketing 营销,买卖
market promotion 市场宣传推销
market research 市场调研
marking 打印;标志;刷唛头;描样(裁剪前将衣样描在织物上)
marking back 反面搭色(印疵)
marking colour 打印染料
marking fabric 商标标记织物
marking ink 标记墨水
marking machine 打印机;刷唛头机
marking off 搭色(印疵)
marking off on the rollers 滚筒搭色
marking out 划线,标记
marking system inspection 记分制检验法
Mark No. 唛头号数
mark-off 标志,搭色(印疵)
mark-off ghosting 搭色虚印(转移印花疵)
marks 唛头
maroon 紫酱[色]
marquise finish 棉贡缎加光整理,缎纹整理
marresistance 表面抗损性
marron marrow 暗棕灰[色]
marseilles soap 马赛皂,丝光皂
marshaller 司仪,号令员,调度员
marsh gas 沼气,甲烷
mars red 樱红[色]

Marthieson style continuous scouring & bleaching range 马瑟森型连续练漂机
Martindale abrasion machine 马丁代尔耐磨试验仪
Martindale tester 马丁代尔支数试验仪
Martius yellow 马休黄(2,4-二硝基萘酚钠,可作为防蛀剂)
mask 屏蔽,掩蔽;防护面具
masked reactive group 掩蔽的反应基团
masking 掩蔽,遮蔽;伪装;蒙版修正[色调](指摄像中加蒙版以调整色调)
masking agent 屏蔽剂
masking of screens 筛网的覆盖
masking retouching 蒙版修整(照相雕刻用)
mass ⟨M,m⟩ 质量
mass absorption coefficient 质量吸收系数
Massachusetts Institute of Technology ⟨MIT⟩ 麻省理工学院(美国)
mass action 质量作用
mass action law 质量作用定律
mass analyzer 质谱分析器
mass balance 物质收支平衡
mass burning rate 物质燃烧速率
mass by electron charge 电荷质量
mass chromatography 质量色谱法
mass colouration (= mass coloured dyeing) 纺前着色,本体着色
mass conservation 质量守恒,质量不变
mass conservation law 质量守恒定律
mass defect 质量亏损,质量损失
mass dyeing by injecting 注射法本体染色
mass flow 质量流量,重量流量
mass fraction 重量分率
mass number [原子]质量数
mass pigmentation 纺前着色,本体着色

mass production 大量生产,批量生产
mass production trial 成批生产试验
mass retention 剩余质量
mass specific heat 质量比热
mass spectra〈MS〉 质谱
mass spectrogram 质谱图
mass spectrography 质谱法
mass spectrometer〈MS〉 质谱仪;光谱仪
mass spectrometer tube 质谱仪管
mass spectrometric analysis 质谱分析
mass spectrometry 质谱测定法
mass spectroscope analysis 质谱分析
mass spectroscopy 质谱法;质谱分析法
mass tone 主色,浓色
mass transfer 质量传递,质量转移
mass transfer rate 质量转移速度,质量转移率
mass transfer theory 质量转移理论
master [电]主单元
masterbatch 高浓度染色配料
master control 主控制
master controller 主令控制器(用来频繁地换接多回路控制线路的主令电器)
master dividing gear 标准分度齿轮
master gear 主齿轮;标准齿轮
master meter 标准仪表
master motor 主令电动机(带有测速发电机,决定机器的运行速度,其他电动机均跟从主令电动机实现同步运行)
master pattern 原始图案
master plan 总计划;总平面图
master programme clock 主程序控制装置
master switch 总开关
master tool 基础工具,标准工具
master valve 总阀,主阀
master viscometer [毛细管型]主黏度计,标准黏度计
mastic 涂料;胶黏剂;淡黄色;乳香棕[色](橄榄棕色)
mat 席;垫;消光,无光;闷光织物
match 配色;配色打样;[地毯]拼接配合
matched pairs 配对染料(两种不同类型染料混合,供混纺纤维染同一色泽)
matching 拼色,配色,仿色,微调,配合
matching mark 啮合标记,配合标记(配件安装时的记号)
matching parts 配合件
match test for flammability 火柴法可燃性试验
match-to-print concept 符合印花的概念
match to shade 配色
material 材料;物质;织物
material balance 物料平衡
material consumption 物质消耗
material distribution flaps 原料分配板
material flow 原料周转
material handling 物料搬运(指厂内各工序之间的搬运);物资搬运管理
maternity dress 孕妇服
Matex-colour roller system 智能型染色用二辊均匀轧车(具连续可变锥度的轧辊,控制色差较好)
Matexil 马特克锡尔系列(印染助剂系列,商名,英国卜内门)
Matexil DA-AC 马特克锡尔 DA-AC(甲二萘磺酸二钠盐,分散染料分散剂,商名,英国卜内门)
Matexil FN-PC 马特克锡尔 FN-PC(氰胺或双氰胺,含磷酸基团活性染料的固着催化剂,商名,英国制)
Matexil PN-AD 马特克锡尔 PN-AD(涤纶印花用碱拔助剂,多元醇类化合物,

商名,英国卜内门)
Matexil PN-DG 马特克锡尔 PN-DG(涤纶印花用碱拔助剂,芳香醇聚氧乙烯醚类表面活性剂,商名,英国卜内门)
mat finish 消光整理
math.(mathematical) 数学的
math.(mathematics) 数学
mathematical analysis 数学解析
Mather Platt ager 马瑟泼拉特布匹连续蒸化机(商名,英国制)
Mather & Platt type steamer 马瑟泼拉特型蒸化机
mating surface 接触面,接合面,啮合面
matrix [复合材料的]基体
matrix-fibril bicomponent fibre 基质—原纤型双组分纤维(截面呈天星式的双组分复合纤维)
matrix memory 矩阵存储器
matrix particles 填料颗粒(微胶囊中颗粒)
Matsumin Fine Colour 麦芝明微胶囊染料(适于斑点染色,商名,日本松井色素)
Matsumin MR-96 麦芝明 MR-96(涂料印花黏合剂,皮膜不泛黄,手感柔软,更适用于荧光涂料印花,商名,日本松井色素)
matt 无光[泽]的,消光的
matt calender 消光轧光机(使用木制轧辊,进行织物消光整理)
matted material 毡制品;草垫制品
matt fabric 无光织物
matt finishing 消光整理,无光整理
matt gold 黯金色的
matting 席子;织物毡并;消光
matting agent 消光剂
matting power 缩绒性,缩呢性,毡合性

matt printing 消光印花(指白涂料印花)
matt prints 无光印花布
mattress 褥,垫子
mattress flammability standard 褥垫可燃性标准
matt white effect [白涂料]消光白色效应
matt yarn 无光纱
maturity 成熟度
mauve(=mauveine) 紫红色;[碱性]木槿紫染料,苯胺紫染料
mauve wine 紫酱[色],酒红[色]
max.(maximum) 最大值;极大值;最大的
Maxilon dyes 麦西隆染料(染聚丙烯纤维的阳离子染料,商名,瑞士制)
Maxilon M dyes 麦西隆 M 染料(迁移性阳离子染料,适用于各类聚丙烯腈纤维的染色,商名,瑞士制)
Maxilon Pearl 珍珠状阳离子染料(商名,瑞士制)
maximal colour 最全色
maximum〈max.〉 最大值;极大值;最大的
maximum allowable concentration 最大允许浓度
maximum allowable pressure drop 最大容许压力降
maximum and minimum thermometer 最高最低温度计
maximum demand 最高需用量《电》
maximum emission concentration 最高排放浓度
maximum explosion pressure 最大爆炸压力
maximum explosive concentration 最高爆炸浓度

maximum load 最高负荷《电》
maximum pattern area 最大花型范围
maximum permitted concentration 最大容许浓度
maximum pick-up 最大带液量
maximum potential strength 最大潜在强度
maximum value 最大值
Max Spelio process 马克斯斯比洛法(意大利湿转移印花法)
Maxwell colour triangle 麦克斯韦颜色三角
Maxwell disk 麦克斯韦混色盘
mazarine blue 深紫蓝[色]
mb. (millibar) 毫巴(10^{-3}巴,压强单位)
MBAS (methylene blue activated substance) 亚甲蓝活化物质(污水处理测试用)
MBR (microbe bio-reactor) 微生物反应箱,微生物反应池《环保》
M/C (manual control) 人工控制,手控
M/C (=Mc, mc., m/c, machine) 机器,机械;机组;设备
MCAA (=MCA, monochloroacetic acid) 氯醋酸,一氯醋酸
MCB (monochlorobenzene) 一氯苯,氯苯
MCP (microcapsulated) 微胶囊[型]
MCP Colour AP conc. 微胶囊染料 AP 浓型(适于聚酰胺织物的斑点染色,商名,日本松井色素)
MCP HP dyes 微胶囊分散染料(商名,日本材化学)
MCPP (multicolour printing process) 多[套]色印花工艺
MCT dyes (monochlorotriazine dyes) 一氯三嗪型染料(一种活性染料)
MD (machine direction) 机器运行方向

MDI (4, 4'-diphenylmethane diisocyanate) 4,4'-二苯基甲烷二异氰酸酯
mdl. (model) 模[型];型号
Me. (methyl) 甲基
MEA (monoethanolamine) 一乙醇胺
meadow green 草绿[色]
mealy 有小浓斑的;小浓斑疵(印染疵)
mealy appearance 斑点状《印》
mean 平均值;平均的,中间的
meandering 蛇行
meandering flow 曲流
mean free path 平均自由行程
mean heat flux 平均热流,平均热负荷
mean length 平均长度
mean prickle estimate〈MPE〉平均刺痛估计
mean radiant temperature 平均放射温度
means 方法,方式,手段;工具,设备
mean skin temperature 平均皮肤温度
mean-square deviation 均方差
mean temperature〈T_m〉平均温度
mean value 平均值;平均数
MEAS (monoethanolamine sesquisulfide) 一乙醇胺倍半硫化物(羊毛定形剂)
M. E. A. S. (monoethanolamine sulphite) 乙醇胺亚硫酸盐(毛纤物的耐久防皱剂)
measurand 被测对象,被测量,待测量
measure 测量;尺寸,大小;度量制,测量法
measured range 测量范围,测量幅度
measured value 测定值
measure frame 拉幅框(圆筒形针织物加工用)
measurement 测定
measurement of friction 摩擦力测定

measuring 测定
measuring amplitude 测量范围,测量幅度
measuring and examining machine 测长验布机
measuring and folding machine 码布折布机
measuring and lapping machine 码布叠布机
measuring and rolling machine 码布卷筒机
measuring and stop motion 测长自停装置,定长自停装置
measuring cylinder 量筒
measuring device 测长装置(测布的长度)
measuring disk 测量指示盘
measuring double rolling machine 量布折布卷布机,对折卷布机
measuring error 测量误差
measuring feed roller 测长进布辊
measuring flask 量瓶
measuring instrument 测定器,计量器具,计量仪器,计量仪表
measuring machine 码布机
measuring method 测定法
measuring motion 测长装置;测长运动
measuring pipet 吸管,吸量管
measuring pump 计量泵
measuring range 测试范围,量程,测量幅度
measuring roller 计量辊,测长辊
measuring rolling and doubling machine 码布卷筒卷板机
measuring sensors 测量传感器
measuring tape 卷尺
measuring technology 测量技术
measuring transducer 测量传感器
mec.（＝mech.,mechanical） 力学的;机械的
mec.（＝mech.,mechanics） 力学
Mecerol 丝光油(丝光渗透剂,阴离子型,烷基硫酸盐,不含甲酚,低泡,商名,瑞士科莱恩)
mechan- (构词成分)表示"机械"
mechanic 技工,技师;机械的
mechanical〈mec.,mech.〉 力学的;机械的
mechanical adhesion 机械黏附
mechanical aeration 机械曝气,机械通风
mechanical agitation 机械搅动
mechanical automatic cloth guider 机械式自动吸边器
mechanical brake 机械制动器
mechanical centring system 机械定中心系统
mechanical clack valve 机械瓣阀
mechanical compensating balance 机械补偿天平
mechanical damage 机械损伤
mechanical deformation 机械变形
mechanical device 机械装置
mechanical draft 机械通风
mechanical drive 机械驱动装置
mechanical drum scanner 机械转筒扫描装置
mechanical equilibrium 力学平衡
mechanical equivalent of heat 热功当量
mechanical fabric 工业[用]织物
mechanical finishing 机械整理
mechanical flocking 机械植绒
mechanical frothing device 机械发泡装置
mechanical lubrication 机械润滑
mechanical means of water extraction 机械脱水法

mechanical piler 机械堆布器
mechanical property 机械性能
mechanical seal 机械密封
mechanical shock 机械冲击
mechanical shrink proofing process 机械防缩工艺
mechanical stop motion 机械自停装置
mechanical stretch finish [织物]弹力整理
mechanical surface property 表面机械性能
mechanical test 仪器检验
mechanical type exhauster 机械式排气风机
mechanics〈mec.,mech.〉力学
mechanism 机理；历程；机构，装置
mechanism of photofading 光褪色机理
mechanochemical degradation 机械化学降解
mechanochemical method 机械化学法
mechanochemical reaction 机械化学反应
mechatronics 机电一体化
med.(medium) 介质；中间的；中级的
median lethal concentration 50 半致死浓度《环保》
median lethal dose 50 半致死剂量《环保》
median tolerance limit〈TLm〉[急毒性]中位容许限度《环保》；半数生存界限浓度
medical absorbent cotton 医用脱脂棉
medical and hygienic textile 医药卫生用纺织品
medical chemistry 医药化学
medical fibre 医用纤维
medical finish 医疗整理
medical gauze 医用纱布
medical protective wear 医用防护服
medical suture 医用缝合线

medical textile 医疗用布，医疗用纺织品
medicated cotton 脱脂棉，药棉
medicated finishing 疗效整理（在织物上添加对人体疾患具有治疗作用的药物的加工技术）
medium〈med.〉介质；中间的，中级的
medium boiling solvent 中沸点溶剂(100~150℃)
medium carbon steel 中碳钢
medium colour 中色
medium energy type dyes 中能型[分散]染料
medium finish 中度上浆整理
medium gray 中灰[色]
medium infrared 中红外线
medium length fibre 中长纤维
medium oil 中油（从煤焦油蒸馏而得）
medium pattern printing 中型花纹印花
medium runs 中批量
medium solvent oil-in-water system 中溶剂含量油水相乳化糊法（溶剂含量40%~50%）
medium speed〈ms〉中速；平均速度
medium staple fibre 中长纤维
medium steel〈MS〉中碳钢，中硬钢
medium tenacity 中等强度
medium thickeners 中切变糊料
medium tolerance limit〈MTL〉半数生存限制浓度
medium weight fabric 中厚织物
medium weight fancy suiting 中厚花呢
medulla 毛髓，髓[质]
medullameter 毛髓测定器（采用光电设备）
medullary cell 髓细胞
medullated 有髓的

medullated fibre 髓质纤维
medullation 髓[质]化
meg.(megohm) 兆欧[姆]
meg[a]-(构词成分)表示"大""强""兆""百万"
mega 兆(10^6)
megacycles〈megc.〉兆周
Megafix B dyes 美佳菲克斯 B 型活性染料(双活性基团,商名,上海万得化工)
Megafix BPS dyes 美佳菲克斯 BPS 型活性染料(印花用活性染料,商名,上海万得化工)
megc.(megacycles) 兆周
megger 兆欧表,摇表(测量 $10^5 \Omega$ 以上高电阻的仪表)
megohm〈meg.〉兆欧[姆]
Meigleam G-500 迈格里玛 G-500(夜光印花浆,商名,日本明成)
MEK(methylethyl ketone) 丁酮,甲基乙基酮
Melafix Process Melafix 工艺(毛的氯化处理,商名,瑞士)
melamine 三聚氰[酰]胺,蜜胺(俗名)
melamine formaldehyde acid colloid 三聚氰胺甲醛酸性胶体
melamine formaldehyde precondensates 蜜胺甲醛预缩合物(纤维素纤维防皱防缩整理剂)
melamine formaldehyde resin 蜜胺甲醛树脂;三聚氰胺甲醛树脂
melamine resin 三聚氰胺树脂,蜜胺树脂(织物防缩防皱用剂)
melamine-toluene sulfonamide formaldehyde resin 三聚氰胺—甲苯磺酰胺甲醛树脂(荧光颜料用树脂)
melange cloth 混色织物

melange effect 混色效应;混合效应
melange printing 毛条印花
melange yarn 混色纱
Melatex H-25 米拉坦克斯 H-25(非离子型,自交联丙烯酸类聚合物,防起毛起球助剂,商名,日本东海制药)
meldable fibre 热熔纤维
Melliand Textilberichte〈Melliand Textilber.,MTB.〉《梅利安德纺织学杂志》(德国月刊)
mellow finishing 柔软整理
mellow hand 柔软手感
mellowing 柔软处理;揉布
mellow yellow 芽黄[色]
melt 熔化,熔融;熔化物
melt-additive brightener 熔体增白剂
melt adhesives 热熔型黏合剂
melt balls [烧毛时的]熔球
melt blowing 熔喷法
melt blown line 熔喷法生产线
melt coating 热熔涂层
melt crystallization 熔融结晶化
melt drip 熔滴(阻燃试验)
melt dyeing [再生纤维]熔融着色
melted size 熔浆
melt embossing process 熔膜轧纹成形法
melten sizing 热熔上浆
melt extrusion 熔体挤出
melt-flow index 熔融指数
melt fracture 熔融成分
melt index〈MI〉熔体指数;熔融指数(热熔胶)
melt indexer 测熔融指数的仪器
melting 熔融
melting curve 熔化曲线
melting dilatation 熔胀[作用]

melting hole behaviour 熔孔性
melting point〈mp,M.P.,m.p.〉熔点
melting temperature〈T_m〉熔融温度,熔点
melt laminating process 热熔层压法
melton 麦尔登呢
melton finish 麦尔登整理(重缩绒,不露底纹)
melt release system 熔［融］释［放］体系(转移印花用语)
melt release system transfer printing 熔融法转移印花
melt resistance 抗熔性
melt roller process 热熔辊法
melt sizing 热熔上浆
melt transfer printing 熔融转移印花
melt viscometer 熔体黏度计
melt viscosity 熔体黏度
Mem.(member)［机器］构件,部件;成员,会员;组成部分
mem.(=memo,memorandum) 备忘录;摘要
member〈Mem.〉［机器］构件,部件;成员,会员;组成部分
membrane 膜,隔膜,膜片
membrane cleaning［过滤］膜清洗(水处理技术)
membrane configuration 隔膜结构
membrane displacement［过滤］膜的更换
membrane electrode 膜电极
membrane equilibrium 膜［渗］平衡
membrane extract 薄膜萃取
membrane filter method 膜过滤法
membrane filtration 膜过滤(水处理技术)
membrane flux［过滤］膜的流量
membrane flux maintenance 维持［过滤］膜流量

membrane flux recovery［过滤］膜流量的恢复
membrane penetration 薄膜穿透
membrane permeability 膜渗透性,膜透气性
membrane potential 膜电位差
membrane pump 隔膜泵
membrane separation technology 膜分离技术
membrane squeezer 薄膜加压轧车
membrane structure 膜结构
membranometer 薄膜式压力计
memo(=mem.,memorandum) 备忘录;摘要
memoir 论文集,研究报告
memorandum〈mem.,memo〉备忘录,摘要
memory 记忆装置;存储,记忆;焙烘织物的回复记忆性(织物具有恢复其焙烘时形态的功能);存储器
memory board 存储板
memory capacity 存储容量
memory colour 记忆色(视觉)
memory crease 耐久褶裥
memory dump 存储器信息转储;存储器清除打印
memory effect 记忆效应
memory store 存储库《计算机》
mender 织补工,保全工长;退修染整疵点织物
mending 缝补［用］纱线;［有疵织物］回修复染;织补,缝补;修理
mending defect 修补疵
mending department 修布间
mending fault(=mending mark) 修补疵,织补疵,织补痕

mending stitch 缝补线迹

mending zone ［膨化机的］热后成形区

meniscus ［液柱的］弯月面,弯液面;凹凸透镜

menthol 薄荷醇

mephitis 恶臭,臭气

mercaptan 硫醇

mercaptobenzothiazole 巯基苯并噻唑（俗称快热粉,防霉剂）

mercaptoethanol 巯基乙醇（染料的溶剂及水溶性还原剂）

mercerisation 丝光作用,丝光工艺

mercerising 丝光［工艺］

mercerization 丝光作用,丝光工艺

mercerization style 丝光织物

mercerize 丝光处理

mercerized and printed bed sheet 丝光印花床单

mercerized cotton 丝光棉布;丝光棉纱

mercerized cotton thread 丝光棉线

mercerized finish 丝光整理

mercerized sewing thread 丝光缝纫线

mercerized thread 丝光线

mercerized towel 丝光毛巾

mercerized wool 丝光羊毛

mercerized yarn 丝光纱线

mercerizer 丝光机

mercerizing 丝光［工艺］

mercerizing machine 丝光机

mercerizing machine for tubular fabrics 筒状针织布丝光机

mercerizing range 丝光联合机

mercerizing resistance 耐丝光性

mercerizing strength 丝光浓度

mercevic 烘干法丝光

Mercevic process 默瑟维克干丝光法（适用于针织物,先浸轧碱液,然后在气垫式拉幅机中烘燥）

merchandising 经商,推销

mercurial manometer 水银压力计

mercuric acetate 醋酸汞

mercuric chloride 氯化汞

mercuric oleate 油酸汞

mercurimetry (=mercurometry) 汞滴定法

mercurous chloride 氯化亚汞

mercury-arc lamp 汞弧灯

mercury column 水银柱,汞柱

mercury compound 汞化合物

mercury-filled thermometer 水银温度计

mercury gauge 水银压力计

mercury-in-glass thermometer 玻璃管水银温度计

mercury-in-steel thermometer 钢壳水银温度计

mercury intrusion method 汞注入法（测定织物渗透性）

mercury lamp 水银灯

mercury manometer 水银压力计

mercury penetration method 汞透入法,压汞法

mercury switch 水银开关,水银继电器

merino 美利奴绵羊;美利奴羊毛;美利奴毛织物;棉毛混纺针织物;长弹毛织物;菲律宾窄幅细棉布

merino wool 美利奴羊毛

merit 优点,特征;指标,标准

merit rating 质量评级,性能评价;考绩

Merkyl PM-TL 默吉尔 PM-TL（乳酸苯汞三乙醇铵,耐久性抗菌剂和防霉剂,商名,美国制）

mesh 齿轮啮合;网眼,筛孔,网状物;目（筛孔密度单位）

mesh analysis 筛选分析
mesh bag 网袋(服装、袜子等小件物品染色时用)
mesh cloth 网眼布
mesh count 网目密度
mesh decorative cloth 网眼装饰布
mesh fabric 网眼织物
mesh fineness 网目细度
mesh line roll 网纹辊(表面刻有网纹的镀铬钢辊,常用作给液辊)
mesh negative 网目负片(胶片)
mesh number 网目数,筛目,筛号
mesh screen 筛子;筛网
mesh size 网目,目数
mesh structure 网眼组织
mesh tester 网目测计器
Mesitol HWS liq. 迈西托尔 HWS 液(阻染剂,用于直接染料对羊毛和纤维素纤维混纺织物的染色,可防止直接染料对羊毛的沾色,商名,德国拜耳)
meso- (构词成分)表示"内消旋""中间""中位"
mesomeric resonance 中介共振
mesomerism 中介现象;稳定异构
meson 介子,重电子
message 信息
message exchange 信息交换装置(通信线和计算机之间的缓冲设备)
Messaline finish 梅萨林整理(摩擦轧光整理,使织物有光泽和柔软的手感)
messuring tape 卷尺,皮尺
mesylation 甲磺酰化[作用]
mesyloxy 甲磺酰氧基(接在纤维素纤维上,能起阻燃作用)
meta- ⟨m⟩ (构词成分)表示"间[位]""介""偏"

meta-acid 偏酸
metabisulphite 偏亚硫酸氢盐
metabolic fate 代谢行径,代谢结局
metabolic process [新陈]代谢过程;变化过程
metabolism 新陈代谢
metabolite 代谢物
metachromasy 广义的变色现象
metachromatism 广义的变色现象,变色反应
metachrome dyeing 同浴媒染染色
metachrome dyes 同浴铬媒染料
metachrome mordant 铬媒染剂
metachrome process 同浴铬媒染色法
metaformaldehyde 变甲醛,三噁烷,三聚甲醛
metal 金属
metal alcoholates 金属醇化物
metal arc welding 金属电弧焊
metal bellows 金属波纹管
metal binder 金银粉印花黏合剂
metal bowl 金属[轧]辊
metal chelating dyes 金属螯合染料
metal chelating finish 金属螯合整理(金属离子与纤维螯合后,可提高阻燃性)
metal colouration 金属着色
metal colouration of polyolefin textiles 聚烯烃纺织品的金属着色
metal colouration on fibre [纺织纤维]用金属着色(尤指不易染色的合成纤维)
metal-complex 金属络合体
metal complex acid dyes 金属络合酸性染料
metal complex dyes 金属络合染料
metal-complexing agent 金属络合剂
metal-contained fabrics 金属纤维面料

metal detector 金属探测器
metalepsis 取代[作用]
metal fibre〈MTF〉金属纤维
metal flocking fibre 金属植绒纤维
metal foil transfer printing 金银箔转移印花
metal-free phthalocyanine 不含金属酞菁
metal halogen lamp 金属卤素灯(常用作照相雕刻设备的光源)
metal indicators 金属指示剂
metal infrared heater 金属红外加热器
metal iodine lamp 金属碘灯
metallic bond 金属键
metallic cloth 金属线织物
metallic coating 金属涂层
metallic contaminant 金属沾污物
metallic fabric 金属线织物
metallic gauze 金属纱罗(印花用)
metallic grey 金属灰[色](浅红灰色)
metallic luster 金属光泽
metallic pigment 金属粉末颜料
metallic pollutant 金属污染物
metallic printing 金属[粉]印花;金银粉印花
metallic prints 金属[粉]印花布
metallic salt 金属盐
metallic sensitivity 金属敏感性
metallic soap 金属皂(一般指铁、铜、镁、钙皂)
metallic soap water proofing 金属皂防水整理
metallic stamping foil 金属压印箔片,烫金箔
metallic teazle raising machine 金属刺果起绒机
metallic weighting [丝织物的]增重整理

metalliferous dyes 含金属染料,金属络合染料
metallisable dyes 可金属络合的染料
metallized dyeing 金属络合染料染色
metallized dyes 金属络合染料
metallized fabric 金属涂层织物
metallized phthalocyanine 含金属酞菁
metallized thread 金属线
metallography 金相学
metalloid 准金属;非金属
metallurgy 冶金学
metal matrix composite〈MMC〉金属基复合材料
metal organic compound 有机金属化合物
metal-oxide-semiconductor 金属—氧化物—半导体
metal powder printing(= metal printing) 金银粉印花
metal salt after treatment 金属盐后处理
metal salt water proofing 金属盐防水整理
metal screen 金属筛网
metal-sequestering agents 金属螯合剂
metal signalling system 金属标志系统(布边上打金属标志)
metal spangles dyes 金属闪烁染料
metal spray 金属喷镀(用高压空气将烧熔的金属丝喷成雾状,喷镀到金属表面上)
metal stencil plate 金属型板
metal toners 染料与金属盐制备的颜料
metal work 金加工
metamer 位变异构体;条件等色
metameric 位变异构的
metameric colours 条件等色;异谱同色
metameric colour stimuli 条件等色色刺激(不同光谱功率分布的色刺激,在同样

观察条件下产生同样的颜色)
metameric difference 条件等色差,异谱同色差
metameric [colour] match 条件[等色]配色(指两个染样在某一光源下显示配色效果相等,而在另一光源下则不相等)
metamerism 异谱同色[现象];条件等色[现象]
metamerism index 〈MI〉条件等色指数;同色异谱指数
metamorphosis 变态,变质,蜕变;变形[作用]
metaphosphate 偏磷酸盐(软水剂的主要成分)
meta-phosphoric acid 偏磷酸
meta-position 间位《化》
metascope 红外线指示器
metasilicate 硅酸盐
metastability 亚稳性,亚稳度
metastable 亚稳定性的
metastable ion 亚稳离子
metastable state 亚稳状态
metathesis 复分解[作用],置换[作用];易位[作用]
metathetical reaction 复分解反应,置换反应
Metax 一种电磁波屏蔽原料(利用电解涂层方法制造金属纺织纤维原料,商名)
metaxylene 间二甲苯
meter 〈M,m,m,〉【美】= metre
-meter (构词成分)表示"计""器"
metered addition technology 计量加液技术
metering 计量,测量
metering device 计量装置
metering disperser 计量分配器(调配染液或印糊装置)
metering pump 计量泵
metering rod for machine coating 机器涂层用计量杆
metering roll 长度计量辊
metering systems for dyes & chemicals 染料和化学品计量系统
metering time 加料时间
metering valve 计量阀
metering volumetric pump 容积计量泵
methacrylamide 甲基丙烯酰胺
methacrylate 〈MA〉甲基丙烯酸酯
methacrylate resin 甲基丙烯酸树脂
methacrylic acid 〈MAA〉甲基丙烯酸,异丁烯酸
methacrylic ester 异丁烯酸酯,甲基丙烯酸酯
methacrylonitrile 〈MAN〉甲基丙烯腈,异丁烯腈
methanation 甲烷化
methane sulfonic acid 甲磺酸
methanesulfonyl chloride 甲磺酰氯化物
methanolysis 甲醇分解[作用]
methaphosphate 磷酸甲基酯
methenamine 六亚甲基四胺,乌洛托品
methine dyes 甲川型染料
methionine 蛋氨酸,甲硫基丁氨酸
methnol 甲醇
method for determination 分析方法
method for determining coating adhesion 涂层黏附强力测试法
method for determining low temperature performance of coating fabric 涂层织物低温性能测试法
method for measuring 测定方法

method of comparative judgement 对比判断法

methoxyl group 甲氧基

methoxymethylated polyamide 甲氧基甲基化聚酰胺

methoxymethyl compound 甲氧基甲基化合物

methoxymethyl melamine 甲氧基甲基蜜胺,甲醚化羟甲基三聚氰胺

methyl〈Me.〉甲基

methyl acetylene 甲基乙炔

methylacrylate〈MA〉丙烯酸甲酯

methylal 甲醛缩二甲醇;甲缩醛(俗称)

methyl alcohol〈MA〉甲醇

methylamine 甲胺

methylated dimethylol uron 甲基化二羟甲基乌龙(棉布防皱整理剂)

methylated methylol melamine 甲基化甲基蜜胺,甲醚化羟甲基蜜胺

methylated methylol urea 甲基化羟甲基脲

methylation 甲基化作用

3-methyl-2-benzothiazoline hydrazone 3-甲基-2-苯并噻唑腙(测定空气中游离甲醛用的比色剂)

3-methyl-2-benzothiazolinone hydrazone 3-甲基-2-苯并噻唑啉酮腙(测试游离甲醛用,纳希试剂的代用品)

methyl cellulose(=methyl cellulose ether) 甲基纤维素(制自碱纤维素和甲基氯,作毛纱或混纺纱浆用或涂料印花浆用)

methyl cellulose thickening 甲基纤维素糊

methyl chloride 甲基氯,氯代甲烷

methyl cyclo hexanone 甲基环己酮

methylcyclopentanone oxime 甲基环戊酮肟

methyl dyes 甲基染料

methylenation 亚甲基化[作用]

methylene 亚甲基

methylene blue 亚甲基蓝(碱性蓝色染料,定量分析指示剂)

methylene blue absorption 亚甲蓝吸收

methylene blue absorption number 亚甲蓝吸收值

methylene blue activated substance〈MBAS〉亚甲蓝活化物质(污水处理测试用)

methylene blue method 亚甲蓝法(测阴离子表面活性剂含量)

methylene blue test 亚甲蓝试验(检查毛纤维损伤程度及鉴别纤维用)

methylene chloride 二氯甲烷(发泡剂和染色助剂)

methylene dyes 亚甲基染料

methylene ether link 亚甲基醚键

methylene succinic acid 亚甲基丁二酸,衣康酸

methylene urea 亚甲脲

methylethyl ketone〈MEK〉丁酮,甲基乙基酮

methyl group 甲基

methyl hydrogen polysiloxane 甲基氢基聚硅氧烷

methyl isothiocyanate 异硫氰酸甲酯

methyl laurate 十二酸甲酯(净洗剂,乳化剂,润湿剂的中间体)

methylmagnesium bromide 甲基溴化镁(格利雅试剂)

methyl methacrylate〈MMA〉甲基丙烯酸甲酯

N-methylmorpholin-N-oxide〈NMMNO〉N-甲基吗啉-N-氧化物

methylnaphthalene 甲基萘(染色载体)

methylol 羟甲基
methylol acetamide 羟甲基乙酰胺（用于纤维素纤维防皱整理）
methylol acetylene diurea 羟甲基乙炔双脲
methylol acrylamide 羟甲基丙烯酰胺（用于纤维素纤维接枝纤维的硬挺整理）
N-methylol acrylamide〈NMA〉 联氮羟甲基丙烯酰胺（羊毛混纺制品辐射接枝改性剂）
methylol amide 羟甲基酰胺
methylolated dye 羟甲基化染料
methylolation 羟甲基化
methylol carbamate 羟甲基氨基甲酸酯
methylol compound 羟甲基化合物
methylol dyes 羟甲基型染料
methylol formamide 羟甲基甲酰胺
methylol group 羟甲基
methylol imidazolidone 羟甲基咪唑并烷酮，羟甲基乙烯脲
methylol melamine 羟甲基蜜胺，羟甲基三聚氰胺
methylol methacrylamide 羟甲基甲基丙烯酰胺
methylol methyl triazone 羟甲基甲基三嗪酮（用于纤维素纤维防皱整理）
methylol triazines 羟甲基三嗪（用作棉织物整理剂）
methylol ureas 羟甲基脲（用作纤维素接枝剂）
methyl orange 甲基橙（酸碱测定用指示剂）
methyl orange test 甲基橙测试
methyl red 甲基红（酸性红色染料，指示剂）
methyl salicylate 水杨酸甲酯（聚酯纤维的导染剂）

methyl starch thickening 甲基淀粉糊
methyltauride surfactant 甲基牛磺酸类表面活性剂
methyl triazone 甲基叠氮
methyl vinylpyridine grafted cotton 甲基乙烯基吡啶接枝的棉纤维
methyl violet 甲基紫（碱性紫色染料）
METL (minimum effective treatment level) 最小有效处理量
metol 对甲氨基酚，米吐尔（照相显影剂，商名）
metre〈M, m, m.〉 米；表，仪，计
metre clock 计长器
metre constant 仪表常数
metre counter 米制测长器，米尺计长表；公制测长器
metre dial hand 测量盘[指]针
metre-in system 进口节流系统
metre-out system 出口节流系统
metre ruler 米尺
metres per hour〈m. p. h.〉 米/小时
metres per minute〈m/min, m. p. m., mpm〉 米/分[钟]
metres per second〈m/s〉 米/秒[钟]
metric 米制的，公制的
metrical garn count〈N_m〉 公制支数[制]
metrical instrument 公制计量仪器
metrication 公制化
metric chromaticity 米制色度
metric counts 公制支数
metric lightness 米制明度，米制亮度
Metric No. 公制纱线支数
metric numbering system〈N_m〉 公制支数[制]
metric quantities 米制量度
metric sizing 公制尺寸

metric system 米制,公制(国际上通用的单位制,长度、质量和时间的公制单位分别为厘米、克、秒或米、公斤、秒)

metric ton〈MT〉 公吨

Meypro gum 美宝胶系列(印花糊料,瓜尔豆胶的醚化物,适用于丝绸、合成纤维织物的印花,商名,瑞士制)

mezzanine 调色间,染料调配室(美国俗称,常在一楼与二楼之间)

Mezzera continuous scouring machine 美塞拉连续煮练设备(商名,意大利制)

MF(microfiltration) 微米级过滤(水处理技术)

MFDA(multifunctional dispersing agent) 多功能分散剂

MFN status(most favoured nation status) 最惠国待遇

M formulation(microgranular formulation) [染料]形成微粒

Mfr.(Manufacturer) 制造商;厂主;生产者

mfr.(＝mfre,manuf.,manfr.,manufacture) 制造;制品;制造厂

mg(＝mg.,mgr,milligram) 毫克

MgH(magnesium hardness) 镁硬度

mgr(＝mg,mg.,milligram) 毫克

M-G set 电动发电机组;整流机(将交流变成直流)

mho 姆[欧](电导单位)

MHW(Ministry of Health and Welfare) 厚生省(日本)

MI(melt index) 熔体指数;熔融指数(热熔胶)

MI(metamerism index) 条件等色指数;同色异谱指数

MI(migration index) 泳移指数;移染指数

mica 云母

mica ice fibre 云母冰凉纤维

micellae 胶束,微胞,晶子

micellar 胶束的,微胞的

micellar aggregate 胶束凝集体,微胞凝集体

micellar catalysis 胶束[体系]催化

micellar emulsion 胶束乳液(指微滴乳液)

micellar film 微胞膜,胶束膜

micellar force 微胞间[的]力

micellar hydration 胶束水化作用

micellar orientation 胶束定向[作用]

micellar phase 胶束相

micellar solubility 胶束溶解度,微胞溶解性

micellar structure 胶束结构,微胞结构

micellar theory 胶束理论,微胞理论

micell colloid 微胞胶体

micelle 胶束,微胞,晶子

micell formation 微胞形成,胶束形成

Michael addition reaction 迈克尔加成反应(乙烯砜型活性染料对纤维素纤维的染色反应)

micr- (构词成分)表示"小""微""微量""百万分之一""扩大""放大"

Micrex process 米克里克斯法(针织物的化学机械联合整理工艺)

micro- (构词成分)表示"小""微""微量""百万分之一""扩大""放大"

micro-acid car bonizing 微酸炭化

microadjuster 微调,微量调整器

microadjustment 微量调整,微调

microampere〈MA.,ma.〉 微安[培]

microanalysis 微量分析

micro and semimicro viscometer 微量和半微量黏度计

microbalance 微量天平,微量秤
microbar 微巴(压强单位)
microbe 微生物,细菌
microbe bio-reactor〈**MBR**〉 微生物反应箱,微生物反应池《环保》
Microbe Shield 抗菌处理检测技术(商名,美国)
microbial activity 微生物活性
microbial attack 细菌侵袭,微生物侵袭
microbial decolourization 细菌脱色;微生物脱色《环保》
microbial degradation 微生物降解
microbial dyestuff 微生物染料
microbial film 微生物膜
microbial isolated protein 微生物分离蛋白质
microbial methylation 微生物甲基化作用
microbial population 微生物种群
microbial resistance 抗微生物性
microbial transglutaminase〈**MTG**〉 微生物谷氨酰胺转胺酶(用于羊毛抗皱整理)
microbicide 杀微生物剂
microbiological degradation 微生物降解
microbiological resistance 抗微生物性
microbiological resistance test 抗微生物试验
microbiological synthesis 微生物合成
microbistatic 抗微生物的
micro-Brownian diffusion 微观布朗扩散
micro-Brownian movement 微[观]布朗运动
microbulking 微膨化
microcapillarity 微毛细管作用
microcapsulated〈**MCP**〉 微胶囊[型]
microcapsulated disperse dyes 微胶囊分散染料
microcapsulated phase change material 微胶囊相变材料
microcapsulate technology 微胶囊技术
microcapsule 微胶囊
microcapsule dyes 微胶囊染料
microcapsule flame retardant technology 微胶囊阻燃技术
microcapsule functional finishing agent 微胶囊功能整理剂
microcapsule perfumed finishing 微胶囊香味整理
microcapsule perfumed finishing agent 微胶囊香味整理剂
microcapsule perfume printing 微胶囊香味印花
microcapsule photochromic dye printing 微胶囊光致变色染料印花
microcapsule photochromic pigment 微胶囊光致变色颜料
microcapsule pigment 微胶囊颜料
microcapsule printing 微胶囊印花
microcapsule technique 微胶囊技术
microcapsule technique for antimicrobial and insecticide finish 微胶囊技术应用于抗菌和杀菌整理
microcapsule technique for oil repellent finish 微胶囊技术应用于拒油整理
microcellular structure 微孔结构,微泡[沫]结构
microchromatography 微量色谱法
microclimate 微气候,小气候(试验用语)
micro computer 微型计算机
microconfiguration 微观结构
microconjugate 微观共轭[型]
microcontrol 精密调节器,精密控制器

microcosm 微观世界
microcrack 微[细]龟裂,微[细]裂纹,微裂
micro crystalline 微晶
microcrystalline cellulose 微晶纤维素
microcrystalline structure 微晶结构,[纤维]微晶结晶
micro crystalline wax 微晶蜡
microcrystallite 微晶
microdensitometry [纤维分子结构的]微密度测定[法]
microdial 精密刻度盘
microdispersed 微粒分散的
microdispersoid 微[粒]分散胶体
microdyeoscope 染色显微观察仪(用以连续观察在染色过程中染料与纤维的相互作用)
micro-ecology 微生态学
microemulsion 微乳液
microemulsion binder 微乳液黏合剂
microemulsion copolymerization 微乳液共聚
microencapsulated 微囊密封的,用微胶囊包起来的
microencapsulated flame retardancy 微胶囊阻燃整理
microencapsulation 微囊化;微胶囊化
microfibre 微纤维(直径为 0.3～1.0 分特的化学纤维)
micro fibre dyeing 超细纤维染色
microfibre procedures 超细纤维加工方法
microfibre processing 超细纤维加工步骤
microfibril 微纤维,微纤丝,微原纤[维](纤维状的微细组织,其直径为6～8微米,长度10～50微米)
microfibrillar structure 微纤结构

microfibrillation 微原纤化
microfilm 微胶片
microfilter 微型过滤器
microfiltration (MF) 微米级过滤(水处理技术)
microfine denier fibre 超细纤维
microfoam dyeing 微泡沫染色
microgap switch 微动开关
microgel 微粒凝胶
microgel particle 凝胶微粒[子];凝胶微质点
micro glass bead 玻璃微珠
microgranular formulation (M formulation) [染料]形成微粒
microgranulated organic pigments 微颗粒化有机颜料
micrography 显微照相术,显微绘图术
micro-heterogeneity 微观不均匀性
microheterogeneous 异质微量[体]
microhole 细孔,微孔
microhomogeneity 微观[结构]均匀性
microlamp 显微镜用灯
microlatex resin 微型乳树脂
microlength stretch 微伸[长];微拉幅
microline counter 织物密度测定器
microlitic structure 微晶结构
microlustre method 微光泽测定法(测试织物光泽)
micro-marketing 极小批量销售
micromatic setting 微调装置
micro membrane 微膜
micrometer 测微计,千分尺,分厘卡
micrometer calipers 千分卡尺
micrometer microscope 测微显微镜
micrometer screw 微调螺丝
micrometer thickness gauge [织物]厚度测

定仪
micromicro 皮[可],微微(10^{-12})
micromicron 皮[可]米,微微米(10^{-12}米)
microminiaturization 微小型化
Micro Modal Air 新型超细再生纤维(最柔软的纤维素纤维,奥地利兰精集团)
micromonitor system 微监控装置
micromorphology 微观形态学;微观形态
micron 微米(10^{-6}米)
microniser(= **micronizer**) 微磨机,粉碎机
microorganism 微生物
microorganism identification 微生物鉴定
microorganism resistance 抗菌性
micro peach fabric 微细桃皮绒织物
micropollution 微量污染
micropore 微孔隙(直径1~100微米范围的孔隙)
micropore structure [纤维截面的]微孔结构
microporosity 微孔性;微孔度
microporous calcium silicate heat insulator 微孔硅酸钙隔热材料
microporous coated fabric 透气性上胶[防雨]织物,微孔胶[防雨]织物
microporous coating 微孔涂层
microporous fibre 微孔纤维
micro porous membranes 微孔膜
microporous polymer 微孔聚合物
microporous PTFE film 微孔聚四氟乙烯薄膜
microporous roller 微孔轧辊
micro-prism 微棱晶片
microprocessor-controlled colour dispensing system 微机控制染料调配系统
micro processor systems 微处理器系统

microscope 显微镜
microscope slide 显微镜用载玻片
microscope testing of fibre 纤维显微镜检验
microscopical determination 显微镜检验
microscopic analysis 显微镜分析
microscopic cross-section 微观截面
microscopic examination 显微镜检验
microscopic observation 显微镜观察
microscopic pick counter 织物分析显微镜,测织物经纬密度显微镜
microscopic study 显微镜研究
microscopy section 显微切片
micro-second 微秒
micro-separation 微裂离
microslide [显微镜的]载玻片
microslit 微裂化
microsolubility 微溶性(在显微镜下观测,可用于鉴别纤维)
microspectrophotometer 显微分光光度计
microsphere 空心微粒,[充以溶剂或气体的]微粒
microstretch 微伸[长];微拉幅(通过机械的方法使织物在幅度方向上分成若干个微小的区域并对其施加均匀的拉力,以达到将织物均匀扩幅的效果)
microstretch expander 微伸扩幅装置
microstretching 微伸
microstretch roll 微拉幅辊(用于微拉幅的一种齿形扩幅辊,调节两只微拉幅辊的啮合深度,可以控制扩幅程度)
micro-stretch unit〈**M-S unit**〉 微拉幅装置
microstructure 微观结构,显微组织
micro-suspension dyeing 微悬浮体染色(一种毛低温染色法)

microswelling 微膨胀
microswitch 精密开关,微动开关
microtest 精密试验,[高倍]显微试验
microtome [显微术用]切片机,切片刀
microtoming 显微切片技术
microtorsion balance 微量扭力天平
microviscosimeter 微型黏度计
microvoid 微空隙
microvolt 微伏[特](10^{-6}伏特)
microwave 微波
microwave absorption cell 微波吸收管
microwave alkali-peeling finishing 微波碱减量整理
microwave device 微波器
microwave drying 微波烘燥
microwave drying machine 微波烘燥机
microwave dyeing 微波染色
microwave dyeing machine 微波染色机
microwave heating 微波加热
microwave heat setting 微波热定形
microwave humidity meter 微波测湿仪
microwave incineration 微波灰化法
microwave instrument 微波仪(可用于测量湿度)
microwave low temperature plasma 微波低温等离子体
microwave method 微波方法
microwave moisture content measuring 微波含湿测定
microwave moisture detection 微波测湿
microwave plasma 微波等离子体
microwave radiation 微波辐射
microwave radiation protective coverall 防微波辐射工作服
microwave radiation shield technology 微波辐射屏蔽技术
microwave radiation shield textile 微波辐射屏蔽纺织品
microwave radiometer 微波辐射计
microwave response 微波反应
microwave source 微波[能]源
microwave spectroscopy 微波波谱学
middle ball point 中球针尖
middle shade 中间色
middle soap 凝块皂,中性皂
middle tone 中间色调
midget 小型的
midget motor 微型电动机
midnight blue 午夜蓝[色](深暗蓝色)
midnight shutdown 停机过夜
midsole 鞋底夹层
mid-value 中值
mid-wale corduroy 中条灯心绒
midway deflection 中间垂度(指两个皮轮中间测定的皮带垂度,以确定皮带是否过紧或过松)
midwife's gown 助产士服
Mignone NS 米格农 NS(防泳移剂,用于分散染料及还原染料热熔连续染色工艺,商名,日本染工)
mignonette 灰绿色
migrating dye 泳移性染料;移染性染料
migration [染料]泳移;移染;迁移,移动
migration current 泳动电流
migration effect 移染效果
migration ghosting 漂移虚印,迁移虚印(转移印花疵点)
migration index〈MI〉泳移指数;移染指数
migration inhibitor 防泳移剂
migration of dye 染料泳移,渗色
migration of the double bond 双键移位

migration [transfer] printing 泳移转移印花，湿态转移印花

migration test [染料]泳移试验；移染性试验

Mikethrene dyes 米盖士林染料(还原染料，商名，日本三井)

Miketon dyes 米盖通染料(分散染料，商名，日本三井)

mil. (=mL, ml., millilitre) 毫升

mild alkaline scour 温和碱煮

mildcure 温和焙烘

mild curing process 温和焙烘法

mildew 霉；生霉，发霉；灰(丝绸疵点)

mildew inhibitor 防霉剂

mildew preventive 防霉剂

mildew proof finish(= mildew proofing) 防霉整理

mildew proofing agent 防霉剂

mildew resistance 防霉性

mildew resistance finish 防霉整理

mildew resistant 防霉的，抗霉的

mildew retarding agent 抑霉剂

mild finishing 柔软整理；轻度整理

mild scouring 轻洗；轻度煮练

mild steel angle bar 三角铁

mild steel ⟨ms⟩ 软钢，低碳钢

mild steel plate 钢板

military and defense textiles 军事国防用纺织品

military protective clothing 军队防护服

military standard ⟨MIL-Std⟩ 美国军用标准

military uniform 军服

milk casein fibre 乳酪[蛋白]纤维

milkness of fibre 纤维乳白度

milk protein fibre 牛奶蛋白纤维

milk white 乳白色

mill 子型钢芯(也称阳模)，子模；工厂；[刻纹滚模]钢芯；缩绒，缩呢

Millclean IBA 缩绒清洗剂 IBA(阴离子型，适用于棉及毛制品的湿整工艺，去浮色，防沾色，商名，英国 PPT)

milled cloth 缩绒织物，缩呢织物

milled finishing 缩绒整理

mille-fleur print 千朵花印花花样，小花印花花样

mill ends 零头布；零绸；厂零

mill engraving 钢芯雕刻

mill engraving machine 花筒轧纹机，钢芯雕刻机

mille point 小花斜纹呢(英国制)

mill-finished fabric 色织布

milli- (构词成分)表示"毫""千分之一"

milliampere ⟨MA., ma.⟩ 毫安[培](10^{-3} 安培)

millibar ⟨mb.⟩ 毫巴(10^{-3} 巴，压强单位)

millicron 毫微米，纳米(10^{-9} 米)

milligram (= milligramme) ⟨mg, mg., mgr⟩ 毫克

millilitre ⟨mL, ml., mil.⟩ 毫升

millimetre 毫米

millimetres of mercury ⟨mmHg⟩ 毫米汞柱(压强单位，等于 133.33 帕斯卡)

milli-micron 毫微米，纳米(10^{-9} 米)

millinewton 毫牛[顿](10^{-3} 牛顿)

milling 缩绒，缩呢；毡合；[染料]研磨

milling agent 缩绒剂，缩呢剂；毡合剂

milling and scouring machine 洗缩联合机

milling calender 缩绒轧光机

milling contraction 缩绒收缩

milling [acid] dyes 耐缩绒[酸性]染料

milling faults 缩绒疵点，缩呢疵点

milling/felting theory 缩绒理论
milling flocks 缩呢毛屑
milling in the length 纵向缩绒,纵向缩呢
milling machine 缩绒机,缩呢机;刻线机
milling power 缩绒性,缩呢性;毡合性
milling property 缩绒性
milling resist effects 防缩效果
milling scrimps 缩呢折痕
milling shrinkage 缩绒收缩
milling soap 缩绒用皂
millisecond 毫秒(10^{-3}秒)
millitex 毫特[克斯](10^{-3}特)
Millitron machine 米里特朗印花机(计算机控制的喷射印花机,商名,美国制)
Millitron microjet colour injection system 米里特朗彩色微型喷液印花法(地毯及长毛绒织物不接触印花法)
Millitron microjet printing machine 米里特朗微型喷液印花机(用于地毯印花)
mill notch 铣槽
Millon's reaction 米隆反应(用米隆试剂测试,丝、羊毛纤维因含蛋白质而呈红色反应)
Millon's reagent 米隆试剂(用以鉴别天然丝和羊毛)
Millon's test 米隆测试法(测试整理剂成分)
mill ruling machine 刻线机,摇线机(一种雕刻设备,可在上蜡花筒表面压刻斜纹、网纹及雪花细点)
mill run [织物的]等外品,次货
mill wrinkles 折痕(疵点)
Milori blue 米洛丽蓝(蓝色颜料)
Milsize 米尔赛兹(氧化淀粉,商名)
Milstat C-29 米尔斯塔特 C-29(季胺盐类化合物,高效抗静电剂,商名,英国制)

MIL-Std(military standard) 美国军用标准
mimosa 含羞草黄[色](黄色)
min.(minimum) 最小值,极小值;最小的
min(=min., M, m, minute) 分[钟];微小的
min.(minor) 较少的,少数的;不重要的
mineral acid 矿物酸
mineral colour dip dyeing 矿物染料浸染
mineral colour discharge 矿物染料拔染
mineral colour printing 矿物染料印花
mineral cotton 矿棉,石棉
mineral dyeing 矿物染料染色
mineral dyes 矿物染料
mineral fibre 矿物纤维
mineral hardness [水的]含矿物质硬度
mineralization 矿化作用
mineral khaki 矿物卡其,矿物草黄(染料)
mineral oil 矿物油
mineral pigments 矿物颜料
mineral spirit 矿油精
mineral turpentine 矿物松节油
mineral wax 地蜡,矿物蜡
mineral weighting 矿物增重(丝绸)
ming green 明绿[色](艳绿色)
mini- (构词成分)表示"小"
miniature dots 小点子(涂层)
miniature motor 微型电动机
miniature pattern 微型花样
Minicare 易维护织物,洗可穿织物(商名)
minicomputer 小型计算机
mini-emulsion polymerization 微型乳液聚合法
minification 缩小[率]

Mini-fluid 低给液设备(商名,德国高乐)
mini-foam 微泡沫体
minimal application of liquor 低给液
minimal down time 最短停产时间
minimal value 最小值,极小值
Minimerc machine 紧凑型小丝光机(德国制)
minimum〈min.,min.〉 最小值,极小值;最小的
minimum add-on 最低给液量
minimum application process〈MA process〉低给液工艺
minimum applicator 最低给液量装置(给液率10%～30%)
minimum care 易维护织物,洗可穿织物
minimum effective treatment level〈METL〉最小有效处理量
minimum energy state 最低能量状态
minimum film-forming temperature 最低皮膜形成温度
minimum ignition energy 最小燃着热
minimum inhibition concentration 最小抑制[细菌]浓度
minimum iron finish 免烫整理
minimum speed 最低速度
minimum standard 最低标准
minimum strength 最小强度
minimum tension 最低张力
minimum value 最小值
mining and manufacturing industry 矿产品工业
mini-stenter 小型布铗拉幅机
Ministry of Health and Welfare〈MHW〉厚生省(日本)
Ministry of International Trade Industry〈MITI〉通商产业省(日本)

Ministry of Labour〈ML〉劳动省(日本)
minium 朱红色
mink finish 貂皮整理
minor〈min.〉 较少的,少数的;不重要的
minor maintenance 小修
minor overhaul 小平车,小检修
mint green 薄荷绿[色]
minus alkali ion 负碱性离子,即水合羟基离子(空气中存在的负离子,对人体健康有益)
minus ion fibre 负离子纤维(负碱性离子)
minus ion finishing 负离子整理(负碱性离子)
minus ion generator 负离子发生器
minute〈M,m,min,min.〉 分[钟];微小的
minute adjustment 精调
MIP(management information package) 管理信息包《计算机》
MIP(marine insurance policy) 海运保险单
mi-parti [织物的]层叠效应
mire 淤渣;淤泥
Mirodye RE 一种弱氧化剂(阴离子型,印花和染色时用于抗还原,商名,意大利米洛)
miroir 光亮表面(指整理效果)
Mirokal 54H 米洛卡尔54H(分散剂,螯合剂,阴离子型,有机酸类络合成分,对硬水和金属离子有效,有溶解钙皂能力,可防止机器上硅酸盐沉淀,商名,意大利米洛)
Mirolube PAC 米洛罗勃PAC(无泡润滑剂,阴离子型和非离子型聚酯共聚物水溶液,用于减少纤维间及与机器间的摩擦,商名,意大利米洛)

Mirooxy 432 new 米洛克西 432 新(氧化剂,催化混合物,用于还原染料、硫化染料的氧化,可代替双氧水,商名,意大利米洛)

Miropon RES 米洛旁 RES(防沾污净洗剂,阳离子型,改性脂肪族胺,适用于各种染料染色后的洗涤,可防止未固着染料的沾污,商名,意大利米洛)

Miroprint ORO 32 米洛普林特 ORO 32 ("金粉"效果印花浆,树脂混合物,具有金属闪光效果,取代金属粉的环保产品,适用于各种织物印花,商名,意大利米洛)

Mirosyl VE 米洛西尔 VE(有机硅微乳液,弱阳离子型,氨基改性有机硅,可使纤维膨松柔软,改善缝纫性、抗皱性及回弹性,商名,意大利米洛)

Mirowet A52 米洛回特 A52(低泡耐碱润湿剂,阴离子型,用于还原染料和活性染料的皂洗,也适用于纤维素纤维及其混纺织物的练漂加工,商名,意大利米洛)

mirror 镜,反光镜,反衬镜

mirror finish 光亮整理

mirror image 镜像,反射像(指左右相反的影像)

mirror image molecule 镜像分子

mirror inspecting machine 反光镜式检验机

Mirror OS 米洛尔 OS(氧化稳定剂,阴离子型,不含硅酸盐,适用于连续和非连续以及冷轧堆练漂工艺,商名,意大利米洛)

Mirror Print 双面印花圆网印花机(商名)

mirror symmetry 镜面对称

mis- (构词成分)表示"坏""错""误"

MIS(management information system) 管理信息处理系统

misalign 对花不准

misaligned feeding 送料不准,送布歪斜

misalignment 定线不准,未校准

misc.(miscellaneous) 各种的;杂项

misc.(miscible) 可混溶的

miscellaneous〈misc.〉 各种的;杂项

miscellaneous fibre 野杂纤维

miscibility [溶]混性,可混性,掺混性

miscible〈misc.〉 可混溶的

misclip 脱边,荷叶边,扇形边(拉幅时脱针夹疵点)

miscut 误切

misfeed device 意外停车装置(因进布不良能自动停车)

misfeeding 加错印浆(印疵);颜色不符规定,错色(印疵)

misfit 不适合,配错;[零件的]不配合;[衣服的]不配身;[套版对花的]不准

misfit of printing pattern(＝misfitting) 对花不准(印疵)

misfitting of the stencil frame 模版套歪(手工印疵);花版套歪(筛网印花疵)

misfurnish 脱浆(机印疵)

mishandling in cleaning 清洁的错误处置

mismatch 失配,不匹配;对花不准(印疵)

misplacing of stencil 脱版(手工印花疵)

misrating 错评[级]

misregister 对花不准

miss 失误,失败

miss cut fibre 漏切纤维

miss point 漏点

mis-stitching 错缝

mist[grey] 雾灰色(淡红灰色)

mist cushion 湿垫
mis-timing 时间失调
mistletoe 豆绿[色]
mist spray conditioning unit 喷雾给湿设备
mist spray damping machine 喷雾给湿机
MIT(Massachusetts Institute of Technology) 麻省理工学院(美国)
mite-proof fabrics 防螨织物
mite-proof finishing agent 防螨整理剂
miter gear 等径伞齿轮
miter joint [地毯]斜面对接
MITI(Ministry of International Trade Industry) 通商产业省(日本)
miticide 杀螨剂
Mitin FF 灭丁 FF(阴离子型,羊毛防虫蛀剂,耐日晒牢度和耐湿摩擦牢度高,商名,瑞士制)
Mitter rotary screen printing machine 米特[地毯]圆网印花机(使用电铸镍网)
mix 混合的
mixed 混合的
mixed acid 混合酸
mixed catalysts 混合催化剂
mixed cellulase 混合纤维素酶
mixed colour effect 混色效应
mixed colours 混合色
mixed combined brushing machine 混合型刷绒机
mixed crystal 混合晶
mixed dyeing 拼色染色
mixed dyes 混合染料(用于混纺织物染色)
mixed fabric 交织物;混纺织物
mixed fibre 混合纤维
mixed filament yarn 混纤丝
mixed-flow pump 混流泵

mixed indigo vat 混合靛蓝还原缸
mixed liquor suspended solid(MLSS) 浮游物质浓度,混合液中悬浮固体量
mixed loading 混合增重法
mixed satin 软缎(交织缎类丝织物)
mixer 混合机
mixing 混合,混和
mixing cistern 调浆桶
mixing colour 拼色
mixing compartment 混合仓
mixing ratio 混合比
mixing room 混料间;混棉间
mixing vessel 调浆桶
mixture〈mixt.〉 混合物,混合剂;混纺纱;混纺织物,交织织物;混色纱;混色织物
mixture dyes 混合染料
mixture indicator 混合指示剂
mixture ratio 混用率(包括混纺率、交织率)
mk.(mark) 标记,标志;记号,符号;特征;型号;污渍;斑疵;着色沾污
ML(Ministry of Labour) 劳动省(日本)
mL(=ml., mil., millilitre) 毫升
MLSS(mixed liquor suspended solid) 浮游物质浓度,混合液中悬浮固体量
MLSS loading 浮游物质浓度负荷《环保》
MMA(methyl methacrylate) 甲基丙烯酸甲酯
MMC(metal matrix composite) 金属基复合材料
MMF(man-made fibre) 化学纤维,人造纤维
mmHg(millimetres of mercury) 毫米汞柱(压强单位,等于133.3帕斯卡)
m/min(metres per minute) 米/分[钟]

MMT(monomethyl terephthalate) 对苯二甲酸单甲酯
mo.(month) 月
mo.(monthly) 按月的
mobile batching-up chamber 可移动的封闭卷布箱
mobile device 移动装置
mobile phase 流动相
mobile reaction chamber 移动式单卷汽蒸反应箱
mobile source 移动污染源《环保》
mobile textiles 汽车用纺织品
Mobile Thermo 自动保湿型服装(一种智能型服装,商名,日本制)
mobility 流动性;湎度;迁移率
mocha 深咖啡［色］(深褐色)
mock burnt-out printing (= mock burn printing) 假烂花印花,仿烂花印花(涂料罩印的一种方法)
mock discharge printing 仿拔染印花(涂料罩印)
mock dyed fabric 假染织物(不加染料,仅用于实验对照),未染色对照布《试》
mock dyeing ［纱线］假染定形(指纱线在卷装染色筒中不用染料和药品而仅受水、温、压处理而获定形效果)
mock gold printing 仿金粉印花
mock printing 仿印花
mock-up 样品,模型(与实物一样大小)
modacrylic〈MAC〉改性聚丙烯腈［纤维］
modacrylic fibre 改性聚丙烯腈纤维
Modal fibre 莫代尔纤维(高强和高湿模量纤维素纤维)
modality 形态,样式,方式
model〈mdl.〉模［型］;型号
modelling 模型化

model machine 样机
model test 模型试验
modem 调制—解调器《计算机》
mode of deformation 变形种类;变形格式
moderate 中等
moderate speed〈ms〉中速;平均速度
moderator 缓和剂;减速剂;调节器
mode shades 流行色泽,时髦色泽
modifiable factor 可变因素
modification 变性,改性,修改
modification of motifs 花纹图案修改(如几何图案晶形用 UV 偏振光技术进行转化、添加)
modification process 改性工艺
modified 改性的,变性的;改进的
modified acryl fibre (= modified acrylic fibre) 改性聚丙烯腈［纤维］
modified acrylics 改性聚丙烯腈系［纤维］
modified biochemical oxygen demand test 改进的生化需氧量测试
modified β-CD(modified β-cyclodextrin) 变性 β-环糊精
modified cellulose fibre 变性纤维素纤维
modified condensation resin 改性缩合树脂
modified cross-section 异形断面
modified cross-section fibre 异形截面纤维
modified β-cyclodextrin〈modified β-CD〉变性 β-环糊精
modified dimethylol diglyoxal monour-eine 变性二羟甲基二乙二醛单脲(洗可穿整理反应剂)
modified enzyme 改性酶,修饰酶
modified fibre 改性纤维
modified nano titanium oxide 改性纳米二氧化钛
modified polyamide fibre 改性聚酰胺纤维

modified polyester fibre 改性聚酯纤维
modified resin 改性树脂
modified soda 变性苏打（纯碱与小苏打的混合物）
modified starch 变性淀粉
modified system 改型装置，改进的装置
modified viscose staple 改性黏胶短纤维
modified wool 改性羊毛
modified yarn 加工丝
modifier 变色剂，改性剂，调节剂
modify 改性，变性；修改，改变
modular building 模块建筑物
modular construction 标准组件结构，部件[单元]结构，积木式结构
modular design 积木化设计
modular desizing washing machine 退浆洗涤联合机
modular system 标准系统；定型设备
modulation 调幅；调整；调制
modulator 调制器
module 模块，程序块；组件《计算机》
modulus 模数
modulus of rupture〈MR〉断裂模数
modutrol 调节控制[器]
mohair 马海毛，安哥拉山羊毛
mohair plush 马海毛长毛绒
Mohr bleaching process 摩尔练漂法（一种用次氯酸盐和双氧水的联合棉漂练法）
moiré 波纹，云纹；云纹绸；松板印，云斑（加工疵）
moiré antique 波纹轧光条影丝织物
moiré calender 波纹轧光机
moiré cloths 波纹布，云纹布
moiré dyeing 波纹染色
moire effect 云纹效果，波纹效果

moiré effect fashion 波纹效果时装
moiré finish 波纹轧光整理
moiré-free pattern 无云纹的花型
moiré-free screen 无波纹筛网
moiré lisse 无波纹轧光织物
moiré pattern 波纹花样；波纹图像；[电子显微照片上的]干涉图纹
moist crosslinking 湿交联，潮态交联
moist cure 湿焙烘
moist cure catalyst 湿焙烘催化剂，湿交联催化剂
moist cure resin 湿焙烘树脂，湿交联树脂
moist dweel process 湿堆置工艺
moistening 给湿
moistening chamber 给湿箱；给湿室
moistening roller 给湿辊
moist heat setting 湿热定形
moist heat transfer 湿热转移
moist lubricator 喷雾润滑器（给油方式）
moist[ure]-o-graph 含湿曲线
moisture 水分
moisture absorbency 吸湿性
moisture absorbing equilibrium 吸湿平衡
moisture absorbing synthetic fibre 吸湿性合成纤维
moisture absorption 水蒸气吸收，吸湿
moisture apparatus 测湿计，湿度计
moisture barrier 湿汽隔离层
moisture-carrying capacity 携带水分能力
moisture catcher 水分分离器
moisture content 含水量，含湿量，含水率
moisture content control 湿度控制
moisture content tester 测湿仪
moisture-cured 湿态固化
moisture determination balance 吸湿测定秤，水分测定仪

moisture equilibrium 湿度平衡,吸湿平衡
moisture-excluding efficiency 去湿效率
moisture-free 干燥的,不含水分的
moisture-free weight 烘干重量,干重
moisture-heat permeability [织物的]湿热透气性
moisture instrument 测湿仪
moisture-laden air 含湿空气
moisture measuring and monitoring system 水分测定监控系统
moisture meter 湿度计,测湿器
moisture monitor 回潮监控器
moisture percentage 含水率
moisture permeability 渗潮性,透湿性,透湿度
moisture permeability index 水分渗透指数,透湿指数
moisture permeable coating 透湿涂层
moisture permeable finish 透湿整理
moisture permeable waterproof 透湿防水性
moisture permeable waterproof fabric 透湿防水织物
moisture permeable waterproof finish 防水透湿整理
moisture pick-up 吸湿
moisture profile 水分分布图
moisture proof 防潮的
moisture recorder〈MR〉湿度记录器
moisture regain〈MR〉回潮率
moisture register 湿度记录器
moisture releasing equilibrium 放湿平衡
moisture resistant 防潮
moisture sensibility 湿度敏感性
moisture sensitive discolour fabrics 湿敏变色织物

moisture sensor 湿度传感器
moisture sorption 水蒸气吸着
moisture sorption equilibrium 水蒸气吸着平衡
moisture/steam barrier lining 防湿/蒸汽隔层衬料
moisture-strength curve 含水率—强力曲线
moisture teller 水分[快速]测定仪
moisture tester 测湿器
moisture testing 水分测定
moisture testing oven [测湿]烘箱
moisture transmission 透湿,水分渗透
moisture uptake 吸液量
moisture vapour permeability 通气透湿度
moisture vapour permeable waterproof effect 透湿防水效果
moisture vapour transmission〈MVT〉透湿[汽]性,湿汽透过[作用]
moisture vapour transmission rate〈MVTR〉透湿率,湿汽转移率
moisture wicking property 水分芯吸性能
moisturising capacity 润湿能力
moist warm 湿热的
moka 深咖啡[色](深褐色)
mol.（molecular）分子的
mol.（molecule）分子
molality 质量摩尔数
molal solution 重模溶液
molar 摩尔的
molar absorbance (= molar absorption) 摩尔吸光度
molar concentration 物质的量浓度
molar dye strength 染料力份
molar fraction 摩尔分率
molar solution 容模溶液

molar specific heat 摩尔比热
molar volume 摩尔体积
molar weight 摩尔质量
molasses 糖蜜
mold 霉,霉菌;型,模;造型,铸模;模塑,模压,压制;样板
moldability 塑模性,成型性
moldable fabric 可[模塑]变形织物,造型织物
molded fabric 模压织物(一种模压成形的非织造织物);模塑[合纤]织物(经热压定形的合纤织物)
molded hose 模制软管
molded laminated plastics 模型层压塑料
molded rubber back [地毯]模压橡胶底板
molding 发霉;造型,制模,压模,模塑,模制,压制;压制件
mold lubricant 离型剂,模子润滑剂
mold releasing agent 离型剂,脱模剂
mole 摩尔(表示物质的量)
mole brown 地鼠棕[色]
molecular〈mol.〉分子的
molecular arrangement 分子排列
molecular association 分子缔合[现象]
molecular asymmetry 分子不对称[性]
molecular attraction 分子吸引
molecular bond 分子间键合
molecular chain 分子链
molecular chain movement 分子链运动
molecular cohesion 分子内聚[作用]
molecular configuration 分子构型
molecular conjugation 分子共轭,分子结合
molecular diffusion 分子扩散
molecular dispersivity 分子分散性
molecular distillation 分子蒸馏

molecular entanglement 分子缠结
molecular flexibility 分子柔曲性
molecular formula 分子式
molecular interaction 分子间相互作用
molecular link 分子间结合
molecular orbit theory 分子轨道理论
molecular orientation 分子定向,分子取向,分子定位
molecular orientation of fibre 纤维的分子定向,纤维的分子取向,纤维的分子定位
molecular polarization 分子极化作用
molecular preorientation 分子预取向
molecular rearrangement 分子重排作用
molecular sieve 分子筛
molecular size 分子大小
molecular size distribution 分子大小分布
molecular spectrum 分子光谱
molecular structure 分子结构
molecular symmetry 分子对称性
molecular vibration 分子振动
molecular volume 分子容量
molecular weight〈mol. wt〉分子质量
molecular weight determination 分子质量测定
molecular weight distribution 分子质量分布
molecule〈mol.〉分子
mole fraction 摩尔分率
mole percentage 摩尔百分率
molik 摩立克(仿毛色织布)
mollifier 软化剂;软化器
molten metal 熔态金属
molten metal dyeing process 熔态金属染色法
molten-metal fixation method 熔态金属固

着法
molten salt 熔态盐
molten solvent 熔化溶剂
mol. wt(molecular weight) 分子质量
molybdate orange 钼橙(无机颜料)
molybdate red 钼红(无机颜料)
molybdenised lubricants 钼化润滑剂(耐高压高温的润滑油)
molybdenum blue 钼蓝(颜料)
molybdenum compound 钼化合物
molybdenum disulphide lubricants 二硫化钼润滑剂
Molykote 摩力克润滑脂(含二硫化钼)
momentary-contact actuator 短暂接触激发器(自动测缝头用部件)
momentary over-loading 瞬时过载
moment compensation roller 力矩补偿辊筒
moment of rotation 转[动]矩
mon- (构词成分)表示"单""一"
monascourarin 红曲色素(一种食用色素)
monascourarin colour 红曲色素(天然微生物色素)
Monforisator 蒙福茨热辊定形机(一种接触式热定形设备,其热辊由内筒和夹套组成,内筒通入煤气直接燃烧,夹套用真空泵抽到 130～150 毫米汞柱,内存有1/3水,沸腾蒸发为饱和蒸汽,热辊表面温差小于 2℃,商名,德国蒙福茨公司)
Monforts reactor 蒙福茨快速蒸化机(用于两相法印花,商名,德国蒙福茨公司)
monitor 监视器;监控器;监听器;监测器
monitored control system 监控系统
monitoring 监控;监视;监听;监测

monitoring device 监测装置
monitoring station 监测站
monitoring unit 检测装置
monitor panel 监控面板
monitor screen 监控屏
monitor system 监控系统,监测系统;操作系统(计算机等)
monkey spanner 活络扳手,万能螺旋扳手
monkey wash 马骝洗,喷马骝(牛仔裤整理)
monkey wrench 活络扳手,万能螺旋扳手
mono- (构词成分)表示"单"—"
monoalkyl carbamate formaldehyde 单烷基氨基甲酸酯甲醛化合物
monoazo dye 单偶氮染料
monobasic acid 单盐基酸
monochloroacetic acid〈MCAA,MCA〉 氯醋酸,一氯醋酸
monochlorobenzene〈MCB〉 一氯苯,氯苯
monochlorohydrin 一氯乙醇
monochlorotriazine dyes〈MCT dyes〉 一氯三嗪型染料(一种活性染料)
monochroic 单色的
monochromat 全色盲
monochromatic 单色的
monochromatic analyzer 单色分析器
monochromatic colour 单色
monochromatic harmony 单色协调(由深浅不同的色调配合)
monochromatic illumination 单色光照明
monochromatic light 单色光
monochromatic light beam 单色光束
monochromatic light stimulus 单色光刺激《测色》
monochromatic radiation 单色辐射
monochromatic triangle 单色三角形《测

色》

monochromatism 全色盲

monochromator 单色器,单色仪,单色光镜

monochrome 单色[的];单色画;单色照片

monochrome process [媒染染料]同浴铬媒染色法

monochrometer 单色器,单色仪,单色光镜

monoethanolamine〈MEA〉 一乙醇胺

monoethanolamine bisulfite 一乙醇胺次亚硫酸盐

monoethanolamine carbonate 一乙醇胺碳酸盐

monoethanolamine sesquisulfide〈MEAS〉 一乙醇胺倍半硫化物(羊毛定形剂)

monoethanolamine sulphite〈M.E.A.S.〉 乙醇胺亚硫酸盐(毛织物的耐久防皱剂)

monofil(=monofilament) 单[根长]丝,单纤[维]丝

monofunctional fixation 单功能团固着(固色剂结构)

monofunctional molecule 单官能分子

Monogen 毛能净(洗涤剂,高级醇硫酸酯钠盐,用于羊毛、羊绒的净洗,商名,天津达一奇)

monogenetic dye 单色染料(用任何染法只能染出一种颜色)

monoglyceride sulfate 单甘油酯硫酸盐

monograph 专题论文;记录;图

monohydric alcohol 一元醇

monolayer 单层;单分子层

monolayer[molecules]adsorption 单分子层吸附,单分子吸着

monolithic film 单组分薄膜

monomer content 单体含量

monomethyl terephthalate〈MMT〉 对苯二甲酸单甲酯

monomolecular layer 单分子层

monomolecular membrane 单分子膜

monomolecular solution 单分子溶液

monophosphonate 磷酸盐

Monopol brilliant oil 玛瑙珀油;太古油(蓖麻油的低度磺化物,渗透剂,匀染剂,柔软剂,商名,德国拜耳)

Monopol oil 马诺波油

Monopol soap 马诺波皂

mono pump 莫诺泵(采矿等用)

monorail 单轨;吊轨

monosaccharide 单糖类

monosodium glutamate 谷氨酸一钠

monotone 单色调,黑白调

monotone design 单色调图案

Monsanto crease recovery tester 孟山都织物折皱回复仪

Monsanto method 孟山都试验方法

Monsanto testing method of crease recovery 孟山都折皱回复角测试法

monthly〈mo.〉 按月的

month〈mo.〉 月

Montirama 250 蒙迪腊马 250 型多用途联合机(包括拉幅、烘燥、热定形、蒸呢、转移)

montmorillonite 蒙脱土

mood 风格;情调(指织物的花样图案等)

moon beam 月灰[色]

moon light 浅米灰[色]

mop towel 揩巾

mordant agent 媒染剂

mordant bath 媒染浴

mordant dip dyeing 媒染染料浸染

mordant dyeing 媒染染色

mordant dyes 媒染染料
mordanting 上媒染剂［工艺］
mordanting assistant 媒染助剂
mordant lake 媒染染料色淀
mordant printing 媒染印花
mordant resist dyeing 媒染染料防染
mordant rouge 红媒染剂
morin 桑色素（一种桑科黄色植物染料）
morning crepe 晨绉（热轧整理消光丝制成的绉织物）
morning mist 淡湖蓝［色］,清水蓝［色］
morphine 吗啡（麻醉剂）
morpholine 吗啉（1,4-氧氮杂环己烷）
morpholine soap 吗啉皂
morphological change 形态［学的］变化
morphological property 形态学性能
morphological structure 形态［学］结构
morphology 形态学；表面形状
morsaic print 镶嵌式花型印花
mosquito resistance 防蚊性
mosquito resistant finishing 驱蚊整理,防蚊整理
moss [green] 苔藓绿［色］,秋香［色］（黄绿色）
moss effect 苔绒效果（一种轻度起绒效果）
moss effect finishing 苔绒整理
moss finish (= mossing finish) 苔绒式整理（呢绒经轻度起毛后底纹仍可看出的一种整理）
moss touch 苔绒触感（一种刷绒效果）
most favoured nation status 〈MFN status〉 最惠国待遇
mot.(motor) 电动机,马达
mote 棉籽屑,尘屑
moth 蛀虫,蠹虫

moth-eaten 虫蛀的
mother colours 基本色
mother liquor 母液
mothicide 杀蛀虫剂
moth-infested goods 虫蛀品,虫蛀织物
moth proofing 防蛀；防蛀处理
moth proofing agent 防蛀剂
moth proofing fastness 防蛀牢度
moth proofing finish 防蛀整理
mothproofness test 防蛀性试验
moth proof treatment 防蛀处理
moth repellency 抗蛀性
moth repellent 防蛀剂
moth resistance 防蛀虫性
moth resistance finish 防蛀整理
moth test cloth 蛀虫测试布
motif 花纹图案,基本花纹图案,主题花纹
motif lace 花卉花边
motif pattern 花纹图案,基本花纹图案
motif printing 主题花纹图案印花
motionless mixer 静态混合器
motion study 动作分析
motive 花纹图案,基本花纹图案,主题花纹
motive power 原动力
motometer 转速计,转数表
motor〈mot.〉电动机,马达
motor actuated valve 电动阀
motor commutator 电动机整流器
motor control relay 电动机控制继电器
motor driven pump 电动泵
motor driven throttle flaps 电动节流阀瓣（用以调整风机的进风率）
motorgenerator [set] 电动发电机［组］
motor guard rails 电动机防护栏杆

motorized detwister 电动退捻装置(将绳状织物退捻,以便开幅)
motorized register setting 电动对花(用于滚筒印花机)
motorized winch 电动绞车
motor mechanical characteristics 电动机机械特性
motoroperated valve 电动阀
motor overload protection 电动机过载保护装置
motor reductor 电动机减速器
motor winch 电动绞车
mottled carpet 斑纹地毯
mottled colour 斑点色
mottled effect 色点效应,斑点效应
mottled yarn 斑点纱线,异色合股花式线
mottling 斑点染色,色点染色
mould 霉,霉菌;型,模;造型,铸模;模塑,模压,压制;样板
moulded parts 模压部件
mould fungus 霉菌
mouldiness 霉状;霉性
moulding 发霉;造型,制模,压制;模塑,模制,压制;压制件
moulding material 模塑材料;成形材料
mould inhibitor 抑霉剂
mould proofing 防霉
mould resistance 抗霉菌性
mould shrinkage 模压收缩;成形收缩
mouldspores 霉菌孢子
mould stain 霉斑
mounting 安装;框架;装置;调整;封固
mounting plate 安装板,装配平台
mounting position 安装位置
mounting tolerance 安装公差
mount-type threebowl pneumatic padder 联轧式三辊气压轧车
mourning 丧服
mourning colour 丧服色
mouth 口,孔;[缩绒机的]给布口;输入端,进入管
movable crank 活动曲柄
movable enclosed batching chamber 可移动的封闭卷布箱
movable member 可动部件
movable reaction chamber 单卷汽蒸反应箱,可移动的布卷汽蒸车
movable table dryer 移动式台板烘燥机
movable type trough 移动式轧槽
movement accumulator 移动式存布器,移动式储布器
Movin DC 莫文 DC(酚聚氧化甲烯的衍生物,消除传染病干洗助剂,商名,德国拜耳)
moving bed 流化床
moving boundary 界面移动
moving perforated belt 移动的多孔输送带
Movyl 莫维尔(聚氯乙烯纤维,商名,意大利制)
mp(=M.P.,m.p.,melting point) 熔点
MPC(multimedia personal computer) 多媒体个人计算机《计算机》
MPE(mean prickle estimate) 平均刺痛估计
m.p.h.(metres per hour) 米/小时
MPI(machine productivity index) 机器生产能力指数
mpm(=m.p.m.,m/min,metres per minute) 米/分[钟]
MR(modulus of rupture) 断裂模数
MR(moisture recorder) 湿度记录器
MR(moisture regain) 回潮率

M. S.(machinery steel) 结构钢
MS(mass spectra) 质谱
MS(mass spectrometer) 质谱仪;光谱仪
ms(medium speed, moderate speed) 中速;平均速度
MS(medium steel) 中碳钢,中硬钢
m/s(metres per second) 米/秒[钟]
M-S (micro-stretch) 微拉伸扩幅装置
ms(mild steel) 软钢,低碳钢
M-S unit(micro-stretch unit) 微拉幅装置
MT(metric ton) 公吨
MTB.(Melliand Textilberichte) 《梅利安德纺织学杂志》(德国月刊)
MTF(metal fibre) 金属纤维
MTG(microbial transglutaminase) 微生物谷氨酰胺转胺酶(用于羊毛抗皱整理)
MTL(medium tolerance limit) 半数生存限制浓度
mucic acid 黏酸,半乳糖二酸
mucilages 植物黏液(用于整理);胶水;黏质
mucous membarmes 黏膜
mud cracking 霜白疵,起花花(穿着后局部脱色发白);霜白现象(厚织物印染疵);[涂层]喷白霜;大龟裂
muddy colour 土色,浊色
mud pipe 排泥管
muff 袖窿;袖窿卷装;联轴节槽;轴套;衬套,保温套
muff dyeing 袖窿绞纱染色(适用于弹力纱或膨化纱的绞纱染色)
muff expander 袖窿绞纱扩张器
muff joint [管端的]套筒接合
muffler 消声器
muff rapid dyeing machine 袖窿纱快速染色机(用于弹力纱或膨化纱的绞纱染色)
muff winder 袖窿绞纱卷绕
muisance 公害
mulatto 黄褐色的
mulberry fibre 桑皮纤维
muller 研磨器,粉碎器;研杵;搅棒
mulmul 麦尔纱,薄纱
multi- (构词成分)表示"多"
multiapron dryer 多帘式烘燥机
multiaxial warp knit〈MWK〉多轴向经编
multiaxial warp knitting fabrics 多轴向经编织物
multibath process 多浴工艺,多浴法,多浴操作
multibox 多槽平洗机,平幅皂洗机
multibox open-width washing machine 多格平洗机
multibranched polymer 多支链聚合物
multicellular structure 多孔结构
multichannel 多通道(扫描装置)
multichannel oscillator 多道示波器
multi coaramide〈F3-fibre〉芳纶 III,多元共聚芳纶(商名,中国)
multicolour 彩色,多色
multicolour cloth 多色织物
multicolour-designer [单圆网]多色印花设备
multicolour dyeing 多色[效应]染色
multicolour dyeing machine 多色染色机
multicolour dye stream 多色染液流
multicoloured damask 彩色花缎
multicolour effect 多色效应
multicolour laboratory dyeing machine 实验室多色染色机
multicolour pattern 彩色图案
multicolour printing 多色印花

multicolour printing machine 多[套]色印花机

multicolour printing process 〈MCPP〉 多[套]色印花工艺

multicolour stream dyeing 多色液流染色

multicolour warp 多色经纱

multicolour weaving imitation 仿色织

multicolour weaving mill 色织厂

multicomponent 多组分的

multicomponent fibre 多组分[复合]纤维

multicomponent formulation 多成分处方

multicomponent polymer fibre 多组分共聚物纤维,共聚物纤维

Multicromat 玛尔蒂克罗玛特多色地毯印花机(利用电子计算机控制液流型印花设备,德国制)

multicyclone 多段旋风分离器

multi-deflection continuous ink jet 多偏转连续喷射[喷墨印花]

multi-effect evapouration 多效蒸发[作用]

multi-end open-width bleaching range 多头平幅连续漂白联合机

multienzyme complex 多酶复合体

multifibre 复型纤维(皮芯型纤维的组合体)

multifibre adjacent fabric 多种纤维贴衬织物(沾色试验用)

multifibre fabric 多纤维交织布(测试用)

multifibre test cloth 试验用多种纤维织物

multifil 复丝

multifilament 复丝;多纤维丝

multifil strip 复丝条(用于印花机刮刀)

multifil tape 复丝胶带(用于印花机刮刀)

Multifine 25 马蒂芬 25(螯合剂,具有分散性,可去除水中钙、镁、铁等金属离子,提高产品等级,商名)

multiflux theory 多[光]通道理论

multiformity 多种形式

multifuel burner 多燃料复合喷燃器

multifunctional additive 多功能添加剂

multifunctional auxiliary 多功能助剂

multifunctional catalyst 多功能催化剂

multifunctional clothing 多功能服装

multifunctional coating 多功能涂层

multifunctional compound 多功能化合物

multifunctional dispersing agent 〈MFDA〉 多功能分散剂

multifunctional fabric 多功能织物

multifunctional finishing 多功能整理

multifunctional processes 多功能加工

multifunction calender 多功能轧光机

multihand 综合性手感

multi-jet spray 多喷口喷雾[嘴]

Multi-krome 多色效应[毛织品](美国商品,指脱脂和不脱脂羊毛的混纺织物在同一染浴中染色后所得的双色效应)

multi-lap continuous dyeing 多圈连续染色(在松弛状态下进行,宜于染细薄织物)

multi-lap continuous dyeing machine 多圈连续染色机(在松弛状态下进行染色,适用于细薄织物)

multilayer 多层

multilayer adsorption 多[分子]层吸附

multilayer copolymer 多层共聚物

multilayer drying and heat setting stenter 多层烘燥热定形拉幅机

multilayer fabric 多层织物

multilayer fibre 多组分纤维;多层纤维

multilayer molecules adsorption 多分子层

吸附
multilayer stenter 多层拉幅机
multilevel dryer 多层烘干机
multiloop superheater 盘管过热器
multimedia personal computer〈MPC〉多媒体个人计算机《计算机》
multimeter 万用计,万用表
multimolecular 多分子的
multimolecular layer 多分子层
multimolecular layer adsorption 多分子层吸收
Multinal C-320 马蒂诺 C-320(精练漂白助剂,碱、硅酸盐、非离子活性剂的复配物,用于针织物的精练漂白,增加白度,商名)
Multinal F-26 马蒂诺 F-26(精练漂白渗透剂,非离子表面活性剂,低泡,有乳化抗沉淀作用,商名)
multipassage dryer 多程式烘房
multipass airlay dryer 多路气流式烘燥机
multipass pin stenter 多层针板拉幅机
multipass stenter 多次往复式拉幅机
multipattern 复合花样
multiphase heat setter 多相热定形机
multiphase treatment 多相处理
multiple 多层的;多路的;多级的;复式的
multiple applicator device 多路并联施料设备(在毡毯上施加染液和胶料)
multiple apron dryer 多帘式烘干机
multiple-bath process 多浴法
multiple bond 重键
multiple calendering 多道轧光
multiple camera 多重成像,多角摄像机
multiple cloth 多层组织,多层织物
multiple colour printing machine 多[套]色印花机

multiple conveyer dryer 多级传送带烘干机
multiple-disc clutch 多层圆盘离合器
multiple dispersion 多分散性
multiple-effect evapourator 多效蒸发器
multiple-effect multiple-stage 多效多级
multiple exposure 多次曝光
multiple fabric 多层织物
multiple flame burner 多焰燃烧器
multiple gas burners 复式煤气燃烧器
multiple hit 多点射击
multiple images 复杂图像,多样图像
multiple layer fabric 多层织物
multiple membrane 多层膜
multiple microflash 多点高强度瞬时光源
multiple structure 多层织物组织
multiplex channel 多路通道
multiplexer 多路调制器
multiplex gummer 多能浆边机(包括浆边、开幅、汽蒸)《针》
multiplexor 多路调制器
multiplication system 放大系统
multiplicity 多重性;复合;大量
multiplicity of strata 多层(植绒用语)
multiplied tone 多层次色调
multiplier phototube 光电倍增管
multipollutant 多种污染物
multipolymer 共聚物
multiport valve 多路阀,多向阀
multiprint 多色印花
multipurpose calender 多能轧光机(通用轧光机,采用不同的穿布方式,可进行平轧光、软轧光、摩擦轧光、叠层轧光等多种整理)
multipurpose decatizing machine 多用途蒸呢机

multipurpose device 通用装置
multipurpose finish 多用途整理
multipurpose meter 多用途测试表
multipurpose raising machine 多功能起毛机,通用起毛机
multipurpose shading paste 多用途调色浆
multiroll calender 多辊轧光机(指五辊或五辊以上的轧光机)
multiroller machine 多辊缩呢机（美国名称）
multisection stenter 多节拉幅机
multishift 轮班制的
multispace dyeing 多间隔染色(在纱线上采用多种段长进行的间隔染色)
multispace dyer 多间隔染色机
multistage 多级的
multistage biological oxidation column 多级生物氧化塔《环保》
multistage bleaching 多级漂白法
multistage flash evapourator 多级急骤蒸发器,多级闪蒸蒸发器
multistage reaction chamber 多层翻板式汽蒸反应箱
multistage sampling 多级抽样法
multistage stenter 多级拉幅机
multistage stretching 多级拉伸
multistage system of continuous scouring and bleaching range 多层翻板式连续练漂联合机
multistrand dryer 多股烘干机《针》
multi-tier conveyer dryer 多层输送带式烘燥机
multi-tier reaction chamber 多层翻板式汽蒸反应箱
multi-tier stenter（＝multi-tier tenter）多层拉幅机

multitube piece dyeing machine 多管匹染机
multitubular heater 多管加热器
multitubular reactor 多管式反应器
multi-unit 多单元装置
multi-unit electric heater 多组电热器
multi-wash test 多次洗涤试验(测试洗涤剂的抗再沉积能力)
mummy 木乃伊棕(黄褐色);褐色天然颜料
municipal sewage（＝municipal sewerage）城市下水,城市污水《环保》
municipal water system 城市水道系统
Munsell chroma 孟塞尔彩度
Munsell Colour of Book 孟塞尔色卡
Munsell colour tree 孟塞尔色树
Munsell hue 孟塞尔色调
Munsell notation 孟塞尔标志
Munsell notation system 孟塞尔表色系统
Munsell renovation system 修正孟塞尔表色制
Munsell [colour] solid 孟塞尔色立体
Munsell [colour] system 孟塞尔[表色]制（以亮度、色相、彩度为颜色的基本标志）
Munsell value 孟塞尔亮度值(以纯白为10,纯黑为0)
Munsell value function 孟塞尔亮度值函数《测色》
Munsell value scale 孟塞尔亮度标尺
muriatic acid 粗盐酸(盐酸的俗称)
murrey 桑果紫红[色]（深暗紫红色）
musahri 蚊帐布(印度名称)
mushroom apparel flammibility tester 蘑菇式服装易燃性试验仪
mushroom ironing press 蘑菇式压烫机

mushroom-shaped probe 蘑菇状探针
muslin 平纹细布,薄纱织物;稀薄毛织物
muslin finish 薄纱整理,细布整理
muss [织物]捏皱
mussiness 轻微折皱
mussing [织物]捏皱
mussing resistance 防捏皱性
muss-resistant finish 防捏皱整理
mustang 熟褐色
mustard [yellow] 芥末黄[色](浅暗黄色)
mustard brown 芥末棕[色](棕色)
mustard gas 芥子气
mustard gas protective finish 芥子气防护整理
mustard gold 稻草黄[色]
mustiness 霉状;霉性
MUT [消费者和环境保护纺织品协会颁发的]环境友好纺织品标签
mutagen 变异源《环保》
mutagenic activity 诱变活性
mutagenic agents 诱变剂
mutagenicity [生物的]诱变性
mutagenic substance 诱发物
mutamer 旋光异构物;变构物
mutation 变更;突变
muted gray 柔和的灰色

muted stripe 晕色条纹
mutual correlation 互相关《数》
mutual diffusion 相互扩散
mutual induction 互感《电》
mutual interaction 相互作用
mutual solubility 互溶性
mutual solvent 互溶溶剂
mutual viscosity 互黏性
MVT(moisture vapour transmission) 透湿[汽]性,湿汽透过[作用]
MVTR(moisture vapour transmission rate) 透湿率,湿汽转移率
MWK(multiaxial warp knit) 多轴向经编
mycelium 菌丝体
mycin 霉菌素
Mycocks expander 迈科克斯扩幅器(由三根弯辊组成的扩幅装置)
myelin 髓磷脂
myrabolams (= myrabolans, myrobolams, myrobolans) 樱桃李(产于中国和东南亚国家,其果实用于制单宁酸,可用来染黑色棉布和丝绸)
myricyl alcohol 蜂花醇
myristyl alcohol 十四烷醇,肉豆蔻醇
myrtle [green] 爱神木绿[色](墨绿色)
mysore 迈索尔染色棉布(平纹)

N

N_d (denier number) 旦数
N_e (English garn count) 英制纱线支数
N_f (French garn count) 法制纱线支数
N_m (metric numbering system, metrical garn count) 公制支数[制]
N (national form) 美国[标准螺纹]牙形
N (Newton) 牛顿(力的单位)
N (=n, normal) 正常的,标准的;法线;当量浓度的
N (normality) 当量浓度;规定浓度
N (normalized) 标准的;标准化的
N (north) 北
n (note) 摘录;注解;记号;草稿
N (=No., no., number) 数,号;支数;编号;数量;第……号
N (=nt, number of teeth) 齿数
N. A. (=n/a, non available) 不详;无可用数字
NA (not applicable) 不适用的
nacarat 鲜艳橘红色
Nacconal 纳科纳尔(烷基芳基磺酸盐阴离子型净洗剂,商名)
NaCMC (sodium carboxymethyl cellulose) 羧甲基纤维素钠盐
nacre 珠光,闪光效应
nacreous finish 珠光整理
nacre pigment printing 珠光涂料印花
nacre printing 珠光印花
nacre velvet 珠光丝绒(绒头与地纹异色)
Nafka gum 纳夫加胶(特殊印花用低黏度糊料)

NAFTA (North Atlantic Free Trade Area) 北大西洋自由贸易区
nail 钉;钉住
Nailamide dyes 纳伊拉米德染料(适用于聚酰胺纤维的酸性染料,商名,意大利制)
nail bearing 滚针轴承
nain 现代波斯地毯(有手织风格)
nainsook finish 奈恩苏克整理(不上浆的柔软整理,两面略有光泽)
naked wool 全毛薄呢
Namasan WSN 南马桑 WSN(甲磺酸,阴离子型毛织物表面定形剂,商名,德国制)
name cloth 标记带
name label 标签名
name of article 品名
name of commodity 商品名称,商名
name plate 名牌,商标
name selvedge 字边(在织物边上标明商标、厂名或其他字样)
name-weaving machine 织字边机
NAMSB (National Association of Men's Sportwear Buyers) 全美男式运动服装购买者协会
nankeen twill 斜纹色布(染黄褐色,作袋布用)
nankeen yellow (=nankin yellow, Nanking yellow) 南京黄(浅灰黄色)
nano- (构词成分)表示"纳[诺]""毫

微(10^{-9})"
nano binder 纳米黏合剂
nano biofibre 纳米生物纤维
nano biosensor 纳米生物传感器
nano carbontube 纳米碳管
nano-ceramic powder 纳米级陶瓷粉
nano composite material 纳米复合材料(用氧化锌、陶瓷粉等纳米级材料混入纤维原料制成,具有特殊功能)
nano device 纳米器件
nano ecodyes 纳米生态染料
nano-effect 纳米效应
nanofarad 纳法[拉],毫微法[拉](电容单位,10^{-9}法拉)
nano-fibre 纳米纤维
nano film 纳米薄膜
nano filter technology 纳滤技术
nanofiltration〈NF〉 纳米级微粒过滤器
nanofiltration membrane technology 纳米过滤膜技术(可用于烧碱回收)
nano finishing 纳米整理
nano functional fibre 纳米功能纤维
nano functional films 纳米功能薄膜
nanolidar technology 纳米激光雷达技术
nano metal film 纳米金属薄膜
nano-metaloxide-based electro-optical films 纳米金属氧化物光电功能薄膜
nanometer〈nm〉 纳米,毫微米(10^{-9}米)
nanometre【美】= nanometer
nanometre material 纳米材料
nanometric grade 纳米级,毫微米级
nano paint 纳米漆
nanoparticle 纳米粒子
nanoparticle-measuring technology 纳米粒子测量技术
nanoplankton 微型浮游生物

nanopore 纳米孔
nano powder material 纳米粉体材料
nano-radiation-resistant material 纳米抗辐射材料
nano-Sb_2O_3 纳米级氧化锑(用作聚酯缩聚催化剂或阻燃剂)
nano-science 纳米科学
nanosecond〈NS,ns〉 纳秒,毫微秒(时间单位,10^{-9}秒)
nano self-cleaning material 纳米自清洁材料(有防油、防水、防污效果)
nano-size 纳米浆料(纳米材料用在纺织浆料中)
nanosol 纳米溶胶
nano-structure 纳米结构
nano structure assembling system 纳米组装体系
nano structure system 纳米结构体系
nano technology 纳米技术
nano-TiO_2 纳米级氧化钛[粉体](用于抗菌、抗紫外线等纺织品)
nano-ZnO 纳米级氧化锌[粉体](用于抗菌、抗紫外线等纺织品)
Nansa SL30,SS30 南萨 SL30 和 SS30(耐酸性的十二[烷]苯磺酸钠润湿剂,商名,英国制)
nap 绒毛;起绒;拉绒;拉毛
nap bar 橇档,毛档(绸缎织疵)
nap cloth 起绒织物,绒面织物
nap fabric 起绒织物,起毛织物,绒面织物
nap finish 搓绒整理(毛织物经起毛、剪毛,再搓擦成珠状、波状或卷曲状的整理)
naphtha 石脑油;粗汽油;粗挥发油
naphthalate polyester filament 聚萘二甲酸

酯长丝
naphthalene 萘
naphthalene derivative 萘衍生物
naphthalene formaldehyde sulfonates 萘甲醛磺酸盐
1,4,5,8-naphthalenete tracarboxylic acid 1,4,5,8-萘四甲酸
Naphthanilide stabiliser CGR 纳夫塔尼利德稳定剂 CGR(不溶性偶氮染料染色时,用此稳定剂后,可使色酚、色基或色盐同浴染色,商名,瑞士制)
naphtha soap 石脑油皂
naphthenate 环烷酸盐,环烷酸或酯
naphthene 环烷
naphthenic acid 环烷酸,环酸;环己烷甲酸
naphthenic acid soap 环烷酸皂
naphthenic oil 环烷油
naphthol 萘酚
Naphthol 纳夫妥(不溶性偶氮染料偶合组分,商名,德国制)
Naphthol AS 纳夫妥 AS(纳夫妥打底染料,商名,德国制)
naphtholated articles (= naphtholated goods) 纳夫妥印花织物;纳夫妥染色织物
naphtholate printing 色酚印花
naphtholating (= naphtholation) 色酚打底,纳夫妥打底
naphthol colour 纳夫妥染料
naphthol dyeing range 纳夫妥染色联合机
naphthol dyes 纳夫妥染料,萘酚染料
naphthol nitrite printing process 纳夫妥亚硝酸钠印花法
naphthol preparing 色酚打底,纳夫妥打底
naphthol printing 纳夫妥印花

naphthoquinone 萘醌
Naphtol 纳夫妥(不溶性偶氮染料偶合组分,商名,德国制)
Naphtol AS 纳夫妥 AS(纳夫妥打底染料,商名,德国制)
Naphtol AS thermosol development process 〈NTD process〉 色酚 AS 热溶显色法
napkin 手帕;头巾;尿布(英国名称);餐巾
napkin cloth 餐巾布
napless [呢绒上]没有绒毛的;磨破了的
napless finish 光洁整理
Naples yellow 那不勒斯黄
nap[finished] overcoating 珠皮呢
nap pattern 毛结花纹
nappe 台布;抹布;餐巾;水舌《环保》
napped cotton fabric 棉绒布
napped fabric 起绒织物,拉绒织物,拉毛织物
napped-finish goods 起绒织物,拉绒织物,拉毛织物
napped jersey 起绒针织物,起绒乔赛
napper 起绒机,拉绒机,拉毛机;搓呢机;拉绒工
nappiness [棉纱的]毛羽,发毛程度
napping 起绒,拉绒,拉毛;搓呢
napping clothing 起毛机针布
napping cotton 起绒性能好[的]棉花
napping finish 起绒整理
napping in the reverse direction 反向起绒
napping machine 起绒机,拉绒机,拉毛机;搓呢机
napping mill 起绒机,拉绒机,拉毛机
napping printing 起绒印花
napping property 起绒性能,拉绒性能
napping resist 抗拉绒[印花](色浆中有

形成坚牢薄膜的物质)
nappy 起绒的,拉绒的,起毛的;尿布
naps 珠皮呢
nap warp 经毛(绸缎织疵)
nap weft 纬毛(绸缎织疵)
narrow cloth 狭幅毛织物
narrow fabrics (= narrow goods) 带状织物;狭幅织物(一般指幅宽27英寸以下的织物)
narrow sheeting 狭幅床单布(40英寸以下)
narrow tee 小口径三通
narrow wares 带;狭幅织物
narrow width 布幅不足(织疵)
NASA(National Aeronautics and Space Administration) 国家航空和航天局(美国)
nascent oxygen 初生[态]氧,新生氧
nascent state 初生态,新生态
Nash pump 纳氏泵(一种水环式真空泵,功率较大,常用于织物的真空脱水)
Nash reagent 纳希试剂(用N-羟甲基交联剂整理织物时,测定游离甲醛的比色剂)
nasturtium red 旱金莲红(暗红橙色)
nasturtium yellow 旱金莲黄(橙黄色)
National Aeronautics and Space Administration 〈NASA〉 国家航空和航天局(美国)
National Association of Men's Sportwear Buyers 〈NAMSB〉全美男式运动服装购买者协会
National Bureau of Standards 〈NBS〉 国家标准局(美国)
National Cancer Institute 〈NCI〉国家癌症研究所(美国)

national costume 民族服装,国定服装
National Electrical Safety Code 〈NESC〉国家电气安全规范(美国)
National Engineering Research Centre 〈NERC〉国家工程研究中心(上海)
National Environmental Policy Act 〈NEPA〉国家环境政策法案
National Environmental Protection Act 〈NEPA〉国家环境保护条例
National Environmental Protection Association 〈NEPA〉国家环境保护协会
National Environmental Research Council 〈NERC〉国家环境研究委员会(英国)
National Fire Protection Association 〈NFPA〉国家防火协会(美国)
national form 〈N〉美国[标准螺纹]牙形
National Institute for Occupational Safety and Health 〈NIOSH〉国家职业安全卫生研究所(美国)
National Institute of Drycleaners 〈NID〉国家干洗商协会(美国)
National Institute of Health 〈NIH〉国家健康研究所(美国)
National Joint Committee 〈NJC〉全国联合委员会(英国)
National Joint Committee for the Carpet Industry 国家地毯工业联合委员会(英国)
National Knitted Outwear Association 〈NKOA〉国家针织外衣厂商协会(美国)
National Physical Laboratory 〈NPL〉国家物理实验室(英国)
National Special Thread 〈NST〉国家特种螺纹(美国)
National Sporting Goods Association 〈NSGA〉国家运动制品协会(美国)

national standard 国家标准
National Technical Information Service〈NTIS〉国家技术情报服务处(美国)
National Translations Centre〈NTC〉国家翻译中心(美国)
native 国产的,土产的,本地的;天生的;朴素的
native cellulose 天然纤维素纤维
NATO(North Atlantic Treaty Organization)北大西洋公约组织
natrium lamp 钠光灯
natte 色织席纹绸(经和纬色泽显著不同)
nattier [blue] 淡蓝色
natty 整洁的;漂亮的;灵巧的
natural 天然的,自然的;原色毛纱
natural abundance 自然丰度
natural ageing 自然老化
natural asphalt 天然沥青
natural bleaching 天然漂白
natural blended [fabric] 天然纤维[为主的]混纺织物
natural caoutchouc 天然生橡胶
natural cellulose 天然纤维素
natural circulation 自然循环
natural colour 自然色
natural coloured cotton 天然彩色棉
natural colouring matter 天然色素
natural colour system 自然颜色系统,自然表色系统
natural conditioning 自然温湿度调节,自然空气调节
natural condition test 自然条件下试验
natural contamination 自然污染
natural convection〈NC〉自然对流
natural convection dryer 自然对流干燥机

natural crimp 天然卷曲,自然皱缩
natural day light 天然日光
natural draft 自然通风
natural draught〈ND〉自然通风
natural draw ratio 自然拉伸比,自然延伸率
natural durability test 自然耐久性试验
natural dyes 天然染料
natural fibre 天然纤维
natural fibre colour 天然纤维色
natural firmness 固有坚牢性
natural functional agent 天然功能剂
natural gas 天然气
natural gas exposure 自然曝气法
natural gray yarn 天然本色毛纱
natural gum 天然胶
naturalia 原色薄呢
natural indigo 天然靛青
natural light 天然光
natural lighting 自然采光,日光照明
natural look 自然外观
natural lustre finishing machine [毡辊]烫光机
naturally coloured cotton 天然彩色棉
naturally coloured cotton fabrics 天然彩色棉织物
naturally pigmented wool 天然有色羊毛
natural mineral pigment 天然矿物颜料
natural moiré 天然波纹;天然木纹
natural moistening 自然给湿
natural plant dyeing 天然植物[染料]染色
natural polymer 天然高分子
natural regain 自然回潮率
natural resin 天然树脂
natural rubber〈NR〉天然橡胶

natural selection 自然选择,自然淘汰
natural selvedge 织物布边(自然边)
natural silk 天然丝,蚕丝,真丝
natural tinted fabric 天然本色织物
natural twist 天然转曲
natural ventilation 自然通风
natural weight 湿重(试验用语)
nature 性质,特性;种类
Nature Colour System 〈NCS〉 自然颜色系统(瑞典)
Nature Colour System brilliant 〈NCS brilliant〉 自然颜色系统的高光泽版
nature fabric 本色织物
nature-identical colour 天然染料等同体(与天然染料结构相同的合成色素)
nature inspection 性能检查
naught 零
Navajo print 印地安式印花花纹
Navimeter 纱线卷装硬度测定计
navy 藏蓝[色]
navy blue 海军蓝[色]
navy cloth 海军制服呢
navy serge 海军哔叽
n. b. (nominal bore) 公称内径
NB (＝N. B., n. b., nota bene) 注意,留心
NBC (nuclear biological and chemical) 核生物和化学
NBR (nitrilebutadiene rubber) 丁腈橡胶
NBS (National Bureau of Standards) 国家标准局(美国)
NBS unit of colour difference 国家标准局色差单位(美国)
NC (natural convection) 自然对流
NC (noise criteria) 噪声标准
Nc (numerical control) 数[字]控[制],数字计算机控制
NCI (National Cancer Institute) 国家癌症研究所(美国)
NCS (Nature Colour System) 自然颜色系统(瑞典)
NCS brilliant (Nature Colour System brilliant) 自然颜色系统的高光泽版
ND (natural draught) 自然通风
NEAC (Nippon Electric Automatic Computer) 电气公司生产的电子计算机(日本)
neadend 布头(英国名称)
near-infrared absorbable dyes 近红外吸收染料
near infrared dyes 近红外染料(作为光存储材料,如用于光盘信息储存)
near infrared radiation 近红外线辐射
near infrared ray 〈N. I. R.〉 近红外线
near-net-shape 近似网状形态
near ultraviolet ray 近紫外线[的]
neatening 修整,修补
neat fashion 雅致的时装
neat's foot oil 牛脚油(用于制革、防水和羊毛加油等)
neat-to-wear 防皱整理织物
nebulosity 云雾状态
neck 轴颈,辊颈
neck bush 内衬套,轴颈套
neck cloth 领巾,颈巾
neck collar 轴颈环
neckdown 颈[状收]缩,细颈[现象]
necked yarn 竹节纱
neckerchief 围巾,领巾,妇女颈部服饰
neck handkerchief 围巾;三角头巾
neck journal bearing 轴颈轴承
necklet 皮围巾;小项圈

neck piece 领饰;皮围巾;领圈
necktie 领带
neck wear 颈部服饰（如领子、围巾、领带等）
needle 针,织针
needle bar 针板
needle bearing 滚针轴承
needle clamp 针铗
needle covered roller 针状覆盖辊
needle craft 缝纫;刺绣
needle crystal 针状结晶
needle defect 针疵
needle density 针密度
needled fabric 针刺非织造织物,针刺织物
needled weftless felt 针刺无纬毛毡
needled wool sweater 针织羊毛衫
needled woven felt cloth 针刺毡合织物
needle eye 针孔
needleizing 易缝纫处理（减少织物对缝针的阻力）
needleloom carpeting 针刺机制地毯
needle penetration 针穿刺
needle-punched carpet 针刺地毯
needle punched nonwoven fabric 针刺非织造布
needle trap ［洗涤机］针形汽水阀
needle valve 针阀
NEFA(non-esterified fatty acid) 非酯化脂肪酸
negative〈neg.〉 底片,负片;负的;消极的;阴性的
negative absorption 负吸收
negative air ion 负氧离子
negative air pressure 负气压
negative assistant 消极式助剂

negative catalyst 负催化剂,缓化剂
negative charged particle 负电荷粒子
negative feed back 负反馈
negative film 负片,底片
negative group 阴根,负[性]基
negative ion 负离子
negative ion fibre 负离子[功能]纤维
negative ion textile 负离子纺织品
negative making 制负片（照相用语）
negative meniscus ［液面的］凹形面,负液面
negative pressure 负压
negative print 花纹留白
negative result 阴性结果
negative staining technique 负片着色法（电子显微技术）
negative stretch 负拉伸
negative synergism 反协同效应
negative take-up motion 消极式卷取运动
negative temperature coefficient 负温度系数
negative thermal fabrics 消极式保温织物
negative thixotropy 负触变性;负摇溶现象
négligé（=negligee） 妇女长睡衣;便服
negligible change 无变化,最高级（指色牢度评级）
negrepellise（=negrepilisse） 黑色缩绒织物
neige 白色效应
neigeuse 斑点纹粗呢
Nekal BX 拉开粉 BX（二丁基萘磺酸钠,阴离子型表面活性剂,商名,德国制)
Nekanil C 纳克尼尔 C（脂肪醇与环氧乙烷缩合物,非离子型净洗剂,具有良好的润湿性,商名,德国制）

Nekanil LN 纳克尼尔 LN(烷基苯酚聚乙二醇醚,非离子型净洗剂,商名,德国制)

Neokal 尼奥卡尔(阴离子渗透剂,商名,德国制)

Neolan P dyes 宜和仑 P 染料(改良型 1∶1 金属络合染料,适用于羊毛、丝绸、锦纶,商名,瑞士制)

Neolan salt P 宜和仑盐 P(聚乙氧基脂肪醇类,1∶1 金属络合染料染色的匀染剂,商名,瑞士制)

neon indicator〈NI〉 氖灯指示器

Neopat binder 尼欧帕黏合剂(涂料染色用黏合剂,运行稳定性好,商名,意大利宁柏迪)

Neopat pigment paste 尼欧帕涂料色浆(染色用涂料色浆,颗粒细,商名,意大利宁柏迪)

Neopolar dyes 尼奥普拉染料(新型弱酸性染料,商名,瑞士制)

Neopralac pigment colours 纳奥普拉克涂料(印花用涂料,商名,法国制)

neoprene 氯丁橡胶,氯丁二烯橡胶

neoprene coated fabric 氯丁橡胶涂层织物

neoprene latex 氯丁橡胶乳液

Neoprint binder 尼欧普林黏合剂(阴离子型,丁二烯共聚体或聚氨酯,手感极柔软,弹性、牢度好,商名,意大利宁柏迪)

Neoprint white 尼欧普林白(白色罩印用涂料浆,商名,意大利宁柏迪)

Neo pynamin 胺菊酯(无毒防蛀剂,商名,日本住友)

Neorate NA 纽雷特 NA(阴离子,涤纶织物碱减量用渗透剂,商名,日本日华)

Neovadine AN 尼奥凡定 AN(非离子型,胺的乙氧基产物,羊毛染色的匀染剂,剥色剂,商名,瑞士制)

nep 棉结,白星;毛粒,毛结;麻粒,麻结

NEPA(National Environmental Policy Act) 国家环境政策法案

NEPA(National Environmental Protection Act) 国家环境保护条例

NEPA(National Environmental Protection Association) 国家环境保护协会

nep and moits 棉结杂质

NEPA regulations 国家环境保护协会法规

nephelometer 烟雾计,能见度测定仪,比浊计

nephelometry 散射测浊法,能见度测定法

neplune green 浅粉绿[色]

nep potential 起毛潜力,拉毛性能

NERC(National Engineering Research Centre) 国家工程研究中心(上海)

NERC(National Environmental Research Council) 国家环境研究委员会(英国)

Nernst equation 能斯特方程

Nernst's adsorption 能斯特吸附

Nernst's partition law 能斯特分配定律

Nervanaid DP 纳尔范奈特 DP(二亚乙基三胺五醋酸五钠,铁铜络合剂,用于稳定过氧化物漂白浴,商名,英国制)

nerve 回缩性,[弹性]复原性

NESC(National Electrical Safety Code) 国家电气安全规范(美国)

nesslerization 等浓比色法

Nessler's reagent 奈斯勒试剂

Nessler tube 奈斯勒比色管,奈氏比色管

net 网;网状物;净的

Net.(network) 网眼织物;网络;网格

net calorific value [煤的]低位发热量

net conveyer 网状导带
net dryer 网带式烘燥机
net effect 网眼效应;净效应,总效应
net highpolymer 网状高分子
net knit fabric 针织网眼织物
net mass 净重
net pattern 格子花纹
net polymer 网状聚合物
net shaped structure 网状结构
netted 网状的
nettle cloth 苎麻织物;漆布(一种涂漆或上釉的粗厚棉布)
net weight〈nt. wt.,n. wt.〉净重
net weight test 净量检验
network〈Net.〉网眼织物;网络;网格
network bond 网状结合
network structure 网络结构,网状结构
net yards[扣除疵布的]布匹净长
neu.(=neut.,neutral)零点,零位;零线;中性的,中和的
neucleophilic addition reaction 亲核加成反应(指乙烯砜型活性染料和纤维素纤维的反应机理)
neucleophilic displacement reaction 亲核取代反应(指卤代杂环类活性染料和纤维素纤维的反应机理)
neucleophilic reagent 亲质子试剂
neutral〈neu.,neut.〉零点,零位;零线;中性的,中和的
neutral activating method 中性活化法(应用于亚氯酸钠漂白)
neutral ageing 中性蒸化
neutral axis 无彩色轴《测色》
neutralbath dyeing 中性浴染色
neutral chromate solution 中性红矾液
neutral colour 中和色,不鲜明的颜色

neutral detergent 中性洗剂
neutral dyeing 中性染色
neutral dyeing acid dye 中性染色酸性染料
neutral dyeing metal complex dye 中性染色金属络合染料
neutral dyeing wool dye 中性染色羊毛染料
neutral dyes 中性染料(1∶2型金属络合染料)
neutral filter 中性滤光器《测色》
neutral-fixing reactive dyes 中性固色活性染料
neutral grey 中性灰
neutralise 中和
neutralization 中和作用
neutralization agent 中和剂
neutralization tank 中和槽
neutralize 中和
neutralizer(=neutralizing agent) 中和剂
neutral medium dyeing 中性染料
neutral packing 中性包装
neutral point 中性点,中和点;白色点,非彩色点(色度学用语)
neutral position 零点位置,不作用位置
neutral resins 中性树脂,碱不溶树脂
neutral scouring 中性洗毛
neutral setting 中性调定
neutral soap 中性皂
neutral steam developing dyes 中性汽蒸显色染料
neutral steam printing[method] 中性汽蒸印花法
neutral step wedge 灰色分级比色楔样
neutraltone 中和色调(指灰度)
neutrase 中性蛋白酶

Neutrogen 中性素(一种不溶性偶氮染料,色酚和色基的稳定型组合,可在中性介质中发色)
neutron 中子
Never-press 免烫整理(商名)
new blue 新蓝,二红光碱性暗蓝
New Cotton 免烫新棉布(商名)
New Dyfoam W 新迪芳 W 系列(发泡印花浆,有立体效果,商名,日本大日精化)
New Dyfoam W-400 新迪芳 W-400(发泡印花浆,商名,日本大日精化)
new energy technology 新能源技术
new generation synthetic fibre 新一代合成纤维,新合纤
new material technology 新材料技术
newness retention 保持永新
Newplex HC-75H 纽帕莱克斯 HC-75H(现成的闪烁印花浆,商名,日本林化学)
new polyfunctional ceramic coating 新多功能陶瓷涂层
New Serial Titles〈NST〉《新刊题录》(美国国会图书馆出版物)
New Source Performance Standard〈NSPS〉新原始特性标准(污水处理的依据参数)
new style〈n. s.〉新式,新型;时髦式样
Newton〈N〉牛顿(力的单位)
Newtonian behaviour 牛顿黏度性能
Newtonian flow 牛顿[型]流动
Newtonian flow property 牛顿流动性
Newtonian fluid 牛顿流体
Newtonian liquid 牛顿液体
Newtonian shear viscosity 牛顿剪切黏度
Newtonian viscosity 牛顿黏度
Newton system 牛顿制(黏度单位)

new wave 新浪潮
new wool 新羊毛(未曾使用过的)
New Zealand Wool Board〈NZWB〉新西兰羊毛管理局
next-to-skin wear 贴身衣着
NF(nanofiltration) 纳米级微粒过滤器
N fading(nitrogen oxide gas fading) 二氧化氮褪色试验;氧化氮气体褪色
NFPA (National Fire Protection Association) 国家防火协会(美国)
NGR (nuclear gammaray resonance adsorption) 核 γ 射线共振吸收(即穆斯鲍尔效应)
NHE(normal hydrogen electrode) 标准氢电极
NHP(=nhp,n. h. p., nominal horse power) 额定马力,标称马力
NI(neon indicator) 氖灯指示器
niced 裹胸布;围巾
nichrome wire 电热丝(镍铬合金)
nick 裂口;切口;槽口
nicked yarn 竹节纱
nickel 镍灰(红光浅灰色)
nickel catalyzator 镍催化剂
nickel chloride 氯化镍(电镀用化学品)
nickel compound 镍化合物
nickel gauze 镍网
nickel green 镍绿(浅暗绿色)
nickeline 尼克林合金(镍、铜、锌合金);红镍矿
nickel modified polypropylene fibre 镍改性聚丙烯纤维
nickel mordant 镍媒染剂
nickel screen 镍网
nickel steel〈ns〉镍钢
nickel sulfate 硫酸镍(电镀用化学品)

nicotine stain 尼古丁污斑
NID（National Institute of Drycleaners）国家干洗商协会（美国）
nigger （黑色）皂脚
niggerhead 疙瘩花纹、花圈花纹或卷曲花纹的织物；织物表面起球（疵点）
niggerhead curl 混纺珠皮呢
niggerhead granite weave 花岗石花纹
night attire 睡衣
nightblindness 夜盲
nightdress 睡衣（妇孺用）
night full dress 燕尾服、晚礼服
night gown 妇女长睡衣
night shift 夜班
night shirt 男长睡袍
night suit 睡衣（短衣、长裤成套）
night wear 晚间家常服；睡衣
nighty 睡衣（俗称）
nigre （黑色）皂脚
nigrosine 酸性黑色染料
NIH（National Institute of Health）国家健康研究所（美国）
Nikka Gum MA-39 尼卡胶 MA-39（醚化淀粉，印花糊料，适用于碱性防拔染印花工艺，商名，日本日华制）
Nikkanon AB 尼卡农 AB（卫生整理剂，商名，日本制）
Nile blue 尼罗河蓝（浅绿蓝色）
Nile green 尼罗河绿（淡绿色）
ninhydrin ［水合］茚满三酮
ninon 尼龙绸，薄绸（丝或化纤制，平纹织物）
niobium carbonitride fibre 碳氮化铌纤维
NIOSH（National Institute for Occupational Safety and Health）国家职业安全卫生研究所（美国）

nioxam 镍铵溶液
nip 轧点，轧面
nip absorption technique 轧吸技术
nip and crown paper 钳口印纸法（碳示踪法，检查罗拉圆整度的方法）
nip coating 轧液涂层（双面涂层）
Nipco roller 尼普戈滚筒（瑞士生产的一种均匀轧辊，外套塑料弹性套筒，筒体内有活塞装置，向套筒的内壁施加液压，补偿挠度，滚筒可有两个工作面，互成90°，可构成两个轧点）
nip creases 小折边（由于边松经过轧点时造成）
nip dyeing 面轧染色
nippadding 面轧，单面轧液
nipped-up edge 封边
nipping point 握持点，夹持点
nipple 螺纹接套，短管接，连接管；油嘴
Nippon Electric Automatic Computer 〈NEAC〉电气公司生产的电子计算机（日本）
nip pressure 轧点压力，钳口压力
nip resistance time 轧点耐压时间
nip roll（＝nip roller）轧辊
nip roller seal 辊封（高温高压设备的封口）
nip width 轧点宽度
N. I. R.（near infrared ray）近红外线
niris rugs 尼里斯地毯（波斯羊毛地毯）
nit 杂物（纱或布上的）；尼特（亮度单位）
NIT（nonionic tenside）非离子表面活性剂
nitrate 硝酸盐
nitrate mordant 硝酸盐媒染［剂］
nitrate white discharge 硝酸盐拔白
nitration 硝化［作用］
nitric acid 硝酸

nitridation 氮化[作用];渗氮
nitrification 氮的硝化作用
nitrile 腈
nitrile alloy fibre 腈合金型纤维(用混抽等特殊方法制取的腈纶)
nitrile binder 腈类黏合剂
nitrilebutadiene rubber〈NBR〉丁腈橡胶
nitrile butadiene rubber roll 丁腈橡胶辊
nitrile group 氰基,腈基
nitrile latex 腈类胶乳
nitrile rubber〈NR〉丁腈橡胶
nitrile synthetic fibre 腈系合成纤维
nitrite 亚硝酸盐
nitrite printing process 亚硝酸钠法印花(可溶性还原染料印花法的一种)
nitrite process 亚硝酸钠染色法(可溶性还原染料染色法的一种)
nitroaniline 硝基苯胺
nitrobenzene 硝基苯
nitro cellulose 硝酸纤维素,硝化纤维素
nitrocellulose rayon（＝nitrocellulose silk）硝酸纤维素人造丝,纤维素硝酸酯人造丝
nitro-cotton 硝化棉(用棉纤维制的硝酸纤维素)
nitro dyes 硝基染料
nitrogen cycle 氮循环
nitrogen fading 氮气褪色
nitrogen gas 氮气
nitrogenization 氮化[作用]
nitrogenous matter 含氮物质
nitrogen oxide gas fading〈N fading〉二氧化氮褪色试验;氧化氮气体褪色
nitrogen oxides 氮氧化物
nitrogen phosphorus detector〈NPD〉氮磷检测器(纺织品生态检测)

nitroglycerine 硝化甘油(炸药)
nitrometer 氮测定仪
nitromuriatic acid 硝基盐酸,王水(俗名)
nitrophenolic dyes 硝基酚染料
nitrorayon 硝酸[纤维素]人造丝
nitrosamine 亚硝胺
nitrosation 亚硝化[作用]
nitrosoamide 亚硝基酰胺
nitroso dyes 亚硝基染料
nitroso group 亚硝基
nitrosyl 亚硝酰[基]
nitrosyl chloride 亚硝酰氯
nitrosylsulfuric acid 亚硝基硫酸
nitrous acid 亚硝酸
nitroxyl 硝酰基
NJC（National Joint Committee）全国联合委员会(英国)
NKAO（National knitted Outwear Association）国家针织外衣厂商协会(美国)
nm（nanometer, nanometre）纳米,毫微米（10^{-9}米）
NMA（N-methylol acrylamide）联氮羟甲基丙烯酰胺(羊毛混纺制品辐射接枝改性剂)
NMDR（nuclear magnetic double resonance）[核磁]双共振
NMMNO（N-methylmorpholine-N-oxide）N-甲基吗啉-N-氧化物
NMR（＝n. m. r., nuclear magnetic resonance）核磁共振
No.（＝no., N, number）数,号;支数;编号;数量;第……号
no bleach 勿用漂白剂;无漂白
no contrast 无色差
no crush finish 无折皱整理
nodding action 摆动

node 结,节,瘤;节点,结点;节瘤花线织物(英国制)
node-fibril structure 结点—原纤结构
node index 节点索引《计算机》
no draw press 无牵引压榨
nodular graphite cast iron 球墨铸铁
nogs 大麻
noil cloth 绵绸,䌷丝织物
noil poplin 绵绸(䌷丝制)
noil stripes 䌷丝条子布,精纺短毛纱条子呢
noily wool 含短纤维低级羊毛
no-iron cotton 免烫棉布
no-iron finish 免烫整理
no-ironing 免烫(具有良好的洗可穿性能,商业术语)
no-ironing suits 免烫服
no-iron shirts 免烫衬衣
Noir reduit 苏木黑水
noise 噪声
noise-abatement equipment 消声装置
noise barrier 噪声屏障
noise criteria〈NC〉噪声标准
noise elimination 消声
noise-free performance 低噪声作业
noise immunity 抗干扰度
noise level 噪声水平
noise limitor 杂音抑制器
noise measuring apparatus 噪声测定器
noise meter 噪声计;声级计(声学用语);噪声测试器;噪声级表(无线电用语)
noise nuisance 噪声;噪声公害
noise pollution 噪声污染
noise pollution level〈NPL〉噪声污染级
noise rating number 噪声[分]级数,NR值
noise silencer 消声器
noise standard 噪声标准
noise suppression control〈NSC〉噪声抑制控制
no-load 空载[荷]
Nom.（**nomenclature**）命名法;术语;命名原则
Nom.（= **nom.**,**nominal**）公称的,标称的;额定的;名义的
no machine washing 不可机洗
Nomad carpet 诺曼德地毯(波斯羊毛地毯)
nomenclature〈Nom.〉命名法;术语;命名原则
Nomex 诺梅克斯(一种芳香族聚酰胺耐高温材料,有时用来包覆滚筒)
Nomex endless blanket 诺梅克斯环状毯(能耐高温)
nominal〈Nom.,nom.〉公称的,标称的;额定的;名义的
nominal bore〈n.b.〉公称内径
nominal count 公称支数,名义支数
nominal gage length [强力试验机两钳口间的]标称隔距长度
nominal horse power〈NHP,nhp,n.h.p.〉额定马力,标称马力
nominal load 额定负荷,标称负荷
nominal output 公称产量
nominal size 公称尺寸;公称细度
nominal weight 标称重量
nominal [machine] width 机器公称宽度(用印染机械中的导布辊幅度作为机器宽度的一种表示方法)
non- (词头)表示"非""无""不[是]""不重要的""无价值的"
nonaggregated dye 不聚集染料
nonapparel fabric 非服饰用织物

nonaqueous dyeing technology 无水染色技术

nonaqueous polymerization 非水相聚合

nonaqueous processing 无水加工工艺（主要指溶剂加工，包括漂白、染色、整理等）

nonaqueous solution 非水溶液

nonaqueous solvent 非水溶剂

nonaqueous solvent dyeing 非水溶剂染色

nonaqueous titrimetry 非水滴定［分析］法

non available 〈N. A. , n/a〉 不详；无可用数字

nonbarry dyeing 无条花染色

non-Binghaon fluid 非宾汉流体

nonbiodegradable material 不能生物降解的物质

noncarrier dyeing 非载体染色（聚酯快速染色）

noncatalytic polymerization 无触媒聚合

noncellulosic fibre 非纤维素纤维，合成纤维

noncellulosic ultra filtration membrane 非纤维素超滤膜

nonchlorine retentive finish 耐氯整理

noncircular filament 非圆截面丝

noncolours 无色的颜色（黑灰白的表现法）

noncombustible 不可燃的，不燃的

noncombustible fabric 不燃性织物

non-conducting 不导电

nonconductive fibre 绝缘纤维

noncongealing paste 不冻膏，不冻糊

non-conjugated 非共轭的

non-conjugated bond 非共轭键

noncontacting edge spreader 非接触式剥边器

noncontacting measurement 非接触性测定

noncontacting pickup 无触点传感器

noncontacting temperature measurement 非接触式测温

noncorrodibility 不腐蚀性

noncorrodible steel 不锈钢

noncorrosive fibre 不腐蚀纤维

noncorrosive material 耐腐蚀材料

noncreasable 耐［揉］皱的

noncrease rayons 抗皱人造丝绸

noncreasing fabric 不皱织物

non-crimp 无卷曲，无皱缩

noncrushable 耐［揉］皱的

noncrushable linen 耐［揉］皱亚麻布（平纹组织，用强捻纬纱织成，或经树脂整理，以增强弹性）

noncrystalline 非晶性的

noncrystalline region 非晶区

noncylindrical filament 非圆柱形长丝

nondeflecting roll 均匀轧辊

nondeformed fibre 不变形纤维

non-depositing plasma 非沉积性等离子体

nondestructive method 非破坏性测试法

nondestructive test 非破坏性试验，无损试验

nondetergent fatty matter 非洗涤剂性脂肪物

non-directional sound source 无指向性声源《环保》

nondischargeability 不可拔染性

nondischargeable dye 不可拔染的染料

nondiscolouring 不变色，不脱色

nondriven bowl 被动辊

nondrying oil 不干性油

nonductile 无延性的

nondusting 不扬尘

nondyeing fibre 不染性纤维
nonelastic 非弹性的
nonelastic fabric 无弹力织物
nonelectrifiable 不起电的
non-equilibrium plasma 非平衡等离子体
non-equilibrium thermodynamics 非平衡热力学
non-esterified fatty acid (NEFA) 非酯化脂肪酸
nonfading 不褪色
nonfelting 不毡合性；不毡合的
nonfelting property 不毡合性，不缩绒性
nonferrous metal 有色金属
nonfibre content 含杂量，非纤维含量
non-fibrillating fibre 非原纤化纤维
non-fire-retardant fabrics 无阻燃剂织物（纤维本身有阻燃作用）
nonflame fibre 不燃性纤维
nonflame properties 不燃性，防火性
nonflammability 不燃性
nonflammable 不可燃的，不易燃的
nonfoaming scouring 无泡沫精练
nonfoaming test 不起泡沫试验
non-fogging coating 不起雾的涂层，防雾涂层
nonformaldehyde DP finish 无醛耐久压烫整理
nonformaldehyde fixing agent 无甲醛固色剂
nonformaldehyde reactant 无甲醛试剂
nonformaldehyde resin finishing 无甲醛树脂整理
nonfouling 不污的
nonfrosting 不局部脱色
nonfusible 不可熔的
nonglare shielded fluorescent lamp ［有］防眩［罩的］荧光灯
nonglitter effect 无极光效应
nonglossy 无光
nongrafted 未接枝的
nonhomogeneous 不均匀的
nonhygroscopic 不吸湿的
nonideal solution 非理想溶液
nonignitibility 不可燃性，不着火性
nonimage area 非图纹部分，［图案中的］空白部分
noninflammability 不燃性
nonionic compound 非离子［性］化合物
nonionic detergent 非离子型洗涤剂
nonionic disperse dyes 非离子型分散染料
nonionic dyes 非离子型染料
nonionic emulsifier 非离子［型］乳化剂
nonionics 非离子［型］
nonionic softener 非离子型柔软剂
nonionic surfactant (= nonionic surface active agent) 非离子型表面活性剂
nonionic tenside (NIT) 非离子表面活性剂
noniron finish 免烫整理
nonironing 免烫
nonirritating 无刺激性的（如对皮肤）
nonisothermal condition 非等温条件
nonisotropic body 非各向同性体，各向异性体
nonius 游标，游尺
nonkeratinous fibre 非角蛋白质纤维
nonknit fabric 非针织物
nonlaminar flow 非层流
non-leaching antimicrobial agent 非浸出型抗菌剂（抗菌剂与纤维有较牢固的结合，不会浸出）
nonleveling dye 不均染染料

nonlinear adsorption isotherm 非线性吸附等温线
nonlinear correlation 非线性相关
nonlinear elastic relations 非线性弹性关系
nonlinear viscoelasticity 非线性黏弹性
nonluminous 无光的,不发光的
nonmandatary(＝nonmandatory) 非强制性的
non-mechanized 非机械化的
nonmercerized cotton fabric 本光布
nonmetallic bowl 非金属轧辊
nonmetameric 无条件等色的(指一对染色体在任何光源下,显示出相同的光谱反射曲线和相同色坐标值)
nonmetameric match 无条件等色配色(指在任何一种光源下,一对染色体比色相等)
nonmicellar solution 非胶束溶液
nonmiscible 不混溶的
nonneedle knitted fabric 无针针织布
non-Newtonian flow 非牛顿流动
non-Newtonian fluid 非牛顿流体
non-Newtonian liquid 非牛顿液体
non-nitrogen resin 不含氮树脂,无氮树脂
nonoperating expense 营业外支出
nonoriented fibre 未取向纤维
nonpareilles 仿驼毛呢(用纯毛纱、毛与麻或毛与山羊毛混纺纱制成,法国制)
non-pathogenic mould 非病原体霉菌
nonperfect register 对花不准
nonpermanent finish 非耐久性整理
nonpersistent 不持久;不稳定
nonpile floor covering 无绒头地毯
non-pilling 不起球
non-planar structure 非平面结构
nonpolar adsorption 非极性吸附

nonpolar bond 非极性键
nonpolar compound 非极性化合物
nonpolar fibre 非极性纤维
nonpolar force 非极性力
nonpolar polymer 非极性聚合物
nonpolar solvent 非极性溶剂
nonpollution technology 无污染工艺
nonporous film coating 无孔薄膜涂层
nonporous material 无孔结构材料
nonporous PU film 无孔聚氨酯薄膜
nonporous rubber roll 无孔橡胶辊
nonpressure 无压,常压低压(与高压相对而言)
nonpressure steamer 无压力汽蒸箱,常压汽蒸箱
nonracking 空转,空撑
non-radiating 无辐射的
nonreactive 不反应的
nonrecoverable deflection 不可回复挠度,不可回复变形,不可逆变形
nonrecoverable elongation 不可回复伸长
nonregenerable energy source 非再生能源
non-regulatory use 不规范使用
nonrepeating design 不重复花纹
non-restrictive 不受限制,没有绷紧感
nonreturn valve 止回阀
nonreversibility 不可逆性
nonreversible crimp 不可逆卷曲
nonreversible reaction 不可逆反应
nonrewetting 非再湿性
nonrigid plastics 非刚性塑料
nonround filament 非圆形长丝
nonrun 不脱散,不抽丝
nonrun finish 防脱散整理,防抽丝整理
nonrun hosiery 防脱散袜子;防脱散针织物

non-self-crosslinking binder 非自交联型黏合剂
nonselfluminous stimulus 非自发光刺激
nonsewn selvedge 漏缝[布]边(缝纫疵)
nonshaatnez 洁净织物,洁净衣服(指无亚麻和羊毛混入的织物或衣服)
nonshatterable glass 不碎玻璃,安全玻璃
nonshift finish 防滑移整理
nonshrink 防缩,抗缩
nonshrinkable wool 防缩羊毛,氯化不缩羊毛
nonshrinkage 防缩性
nonshrink treatment 防缩处理
nonskinned fibre 无皮层纤维,全芯纤维
nonslip finish 防滑整理
nonsneezing 不呛鼻
nonsoap detergent 非皂洗涤剂
nonsoil retentive softener 抗污柔软剂
nonsolvent water 不起溶解作用的水(对聚合物起溶胀作用的水)
nonspiral 非螺旋形的
nonstage transmission 无级变速器
nonstaining property 防污[染]性;不着色性
nonstationary 非静止状态,不稳定的
nonsteady state 非稳定态
nonsteady state diffusion 非稳态扩散
nonstick coating 防黏涂层(四氟乙烯涂层)
nonstop 连续不停的(指机台不断运转)
nonstop bale press 连续打包机
nonstop infeed device 连续进布装置
nonstop winder 连续卷布机
nonstretch 非伸缩
nonstretch bulked yarn 非伸缩性[膨化]变形纱

nonstretchy 非伸缩性
nonstretch yarn 非弹力丝,非伸缩性丝
nonswelling (=nonswollen) 不溶胀的
nonsymmetrical design 不对称花纹
nonsystematic error 偶然误差,非系统误差
nontainting 未腐败的,未污染的
nontension open-width washing machine 松式平幅水洗机
nontension washing machine 松式水洗机,无张力水洗机
nontextile 非纤维制品
non-thermal plasma 非热等离子体
nonthermoplastic filament yarn 非热塑性长丝
nonthermoplastic textured yarn 非热塑性变形丝
nonthixotropic 非触变性的,非摇溶性的
nontorque yarn 非捻回弹力纱,无捻回纱
nontouch dryer 无接触烘燥机
nontoxic cotton 无毒棉
nontoxicity 无毒性
nontransparent 非透明的
nontype 不标准的
non-uniform engraving 不均匀雕刻
nonuniformity 不均匀性,非匀质
nonviable 不能生存的,不可行的
nonvolatile matter〈NVM〉不挥发性物质
nonvolatile memory 非易失性存储器《计算机》
nonwater process 无水加工
nonwettable 不可润湿的
nonwoven 非织造的,无纺织的
nonwoven durables 耐久性非织造织物
nonwoven fabric 非织造织物,无纺布
nonwoven finishing range 非织造布整理生

产线
nonwoven floor covering 非织造地毯
nonwoven geotextiles 非织造土工布
nonwovens 非织造布
nonwoven scrim 非织造纱布（用作纤维网,有较好的强力和形稳性）
nonyellowing 不泛黄的
nonyellowing softeners 不泛黄柔软剂
nonyl alcohol 壬醇（印花用消泡剂）
nonylbenzene sulfonates 壬基苯磺酸盐
nonylnaphthalene sulfonates 壬基萘磺酸盐
nonylphenol 壬基苯酚（净洗剂、分散剂、润湿剂的原料）
nonylphenol ethoxylate〈NPE〉羟乙基壬基酚（非离子表面活性剂）
nonyl polyethylene oxide〈NPEO〉壬基酚聚氧乙烯醚
No-op instruction 无操作指令《计算机》
nopster 修呢工
Norane 诺兰恩防水处理（商名）
Nordhausen acid 发烟硫酸（别名）
Nordic Environmental Label 北欧环境标签（从纤维原料到纺织最终产品的生态标记）
norm 标准,规格,定额
normal〈n,N〉正常的,标准的；法线；当量浓度的
normal ageing 正规蒸化法
normal chain 正链
normal colour vision 正常色觉
normal combustion 完全燃烧
normal component 法向分量；正常组分
normal conditions 标准状况,常规条件（0℃,760毫米汞柱压力）
normal consistency 标准稠度
normal correlation 正态相关

normal cycle 标准循环时间
normal direction flow 正常流向（指从上到下,从左到右）
normal distribution 正态分布,常态分布
normal force 法向力,正交力
normal hydrogen electrode〈NHE〉标准氢电极
normality〈N〉当量浓度；规定浓度
normalization 标准化,统一化,正规化,规格化
normalized〈N〉标准的；标准化的
normal mixture ［棉毛混纺针织物的]常规混合（黑白两色相混而得灰色）
normal mode frequency 正常［振荡］频率,简正频率
normal nylon 正规锦纶（指锦纶66）
normal operating conditions 正常操作的条件
normal pressure〈n.p.〉常压,正常压力；正压力
normal pressure and temperature〈N.P.T,npt,n.p.t.〉标准压力与温度；常压常温
normal production programme 正常生产程序
normal running 正常运转；正常操作
normal solution 当量溶液,规度溶液
normal temperature〈NT,N.T.〉常温；标准温度
normal temperature and pressure〈NTP,ntp〉常温常压
normal test position 正常试验位置
normal trichromation 正常色的视觉（三原色性）
normal twist 低捻,弱捻（一般指人造丝）
norminal width 名义幅度（拉幅机针或夹

间的最大距离）
norm parts 标准件,规范件
north〈N〉 北
North Atlantic Free Trade Area〈NAFTA〉 北大西洋自由贸易区
North Atlantic Treaty Organization〈NATO〉 北大西洋公约组织
north light roof 北向采光屋顶
north skylights 北向天然光,昼光
nose 头部,鼻凸头
no-seam 无缝（指圆筒形针织物）
nose bar 凸杆,撑杆
nose key 鼻形键,凸形键,钩头楔
noserag 手帕（俗称）
nose roll assembly 鼻辊式防缩装置（筒状针织物机械防缩联合机用）
no significant counts〈NSC〉 无显著放射性
no spin 停喷水并脱水（洗衣机用语）
nostalgia 重新流行（在流行趋势中）
nostalgic fashion 怀旧款式
no-swell finish 抗760胀整理
nota bene〈NB、N. B.、n. b.〉 注意,留心
not applicable〈NA〉 不适用的
notation 符号,标记;代号
notch 缺口,凹口选择器标记
notched belt 齿形皮带,同步传动带
note〈n〉 摘录;注解;记号;草稿
not traceable 不含痕迹（化学品）
Novalon dyes 露华浓染料（染聚酰胺纤维的分散染料,商名,瑞士制）
Novanyl dyes 诺瓦尼尔染料（染聚酰胺纤维的酸性或直接染料,商名,英国制）
Nova screen 诺佛镍[圆]网（高目数,高开孔率,商名,荷兰斯托克制）
novel effect 花式效应

novel textile design 新颖的织物设计
novelty 花式组织;花式纱;新颖;新产品
novelty pattern 新颖花样
novelty suitings 花式衣料（指各种组织花纹,特别是锦缎或提花组织）
novolac [线型]酚醛清漆
novolac resin 诺沃拉克树脂（酚醛树脂名称）
novolak [线型]酚醛清漆
novolak resin 诺沃拉克树脂（酚醛树脂名称）
Novolan 诺和兰（生物酶,用于羊毛的蛋白酶,商名,丹麦诺维信）
Novoloid 诺沃洛伊德纤维（内含酚醛树酯）
Novozym 诺和唑（生物酶,用于丝绸砂洗的蛋白酶,商名,丹麦诺维信）
noxious 有害的;有毒的
noxious gas 有毒气体,有害气体
noxious substance 有害物质
nozzle〈noz.〉 喷嘴,喷口;管接头
nozzle applicator 喷嘴给液器（用于泡沫整理的一种给液装置）
nozzle box 喷风管,风盒
nozzle casing 喷嘴罩
nozzle check 喷嘴检查
nozzle drier（＝nozzle dryer） 热风喷嘴烘燥机
nozzle dyeing machine 喷射染色机
nozzle head 喷嘴
nozzle manifold 多支喷嘴
nozzle paddle dyeing apparatus 喷射式桨翼染色机
nozzle stub 短管
nozzle suction drier 抽吸式干燥机
nozzle vane 排气叶片

n. p. (normal pressure) 常压,正常压力；正压力

NPD(nitrogen phosphorus detector) 氮磷检测器(纺织品生态检测)

NPE(nonylphenol ethoxylate) 羟乙基壬基酚(非离子表面活性剂)

NPEO(nonylpolyethylene oxide) 壬基酚聚氧乙烯醚

NPL(National Physical Laboratory) 国家物理实验室(英国)

NPL(noise pollution level) 噪声污染级

N. P. T. (=npt, n. p. t. , normal pressure and temperature) 标准压力与温度；常压常温

NR(natural rubber) 天然橡胶

NR(nitrile rubber) 丁腈橡胶

n. s. (new style) 新式,新型；时髦式样

ns(nickel steel) 镍钢

NS (=ns, nanosecond) 纳秒,毫微秒(时间单位,10^{-9}秒)

NSC(noise suppression control) 噪声抑制控制

NSC(no significant counts) 无显著放射性

NSGA (National Sporting Goods Association) 国家运动制品协会(美国)

NSPS (New Source Performance Standard) 新原始特性标准(污水处理的依据参数)

NST(National Special Thread) 国家特种螺纹(美国)

NST(New Serial Titles)《新刊题录》(美国国会图书馆出版物)

NT (=N. T. , normal temperature) 常温；标准温度

nt (=N, number of teeth) 齿数

NTC (National Translations Centre) 国家翻译中心(美国)

NTD process (Naphtol AS thermosol development process) 色酚 AS 热溶显色法

NTIS(National Technical Information Service) 国家技术情报服务处(美国)

NTP (=ntp, normal temperature and pressure) 常温常压

nt. wt. (=n. wt. , net weight) 净重

N-type of fibre 抗缩纤维

nuage [织物的]云层效应,斑点花纹,暗影效应,迷雾效应

nuance 色彩微差

nubby yarn 竹节纱(疵点)；结子花线

nubs 粗节(纱疵)

nuclear biological and chemical 〈NBC〉核生物和化学

nuclear biological and chemical weapon 核生物和化学战争武器

nuclear energy 核能

nuclear explosions 核爆炸

nuclear flash 核爆炸光辐射

nuclear gammaray resonance adsorption 〈NGR〉核γ射线共振吸收(即穆斯鲍尔效应)

nuclear interaction 核的相互作用

nuclear magnetic double resonance 〈NMDR〉[核磁]双共振

nuclear magnetic relaxation 核磁弛豫；核磁松弛

nuclear magnetic resonance 〈NMR, n. m. r. 〉核磁共振

nuclear magnetic resonance absorption 核磁共振吸收

nuclear magnetic resonance spectroscopic analysis 核磁共振谱分析

nuclear magnetic resonance spectroscopy 核

磁共振谱法
nuclear magnetic resonance spectrum 核磁共振谱
nuclear magneton 核磁子
nuclear power 核能
nuclear power for peaceful utility 核能的和平利用
nuclear power plant 核能工厂
nuclear radiation 核辐射
nuclear spin 核自旋
nuclear static elimination 核静电消除器
nuclear waste 核废料
nuclear weapon 核武器
nuclei （nucleus 的复数）核；晶核
nucleophilic addition reaction 亲核加成反应（活性染料反应机理）
nucleophilic group 亲核性基团
nucleophilicity 亲核性
nucleophilic promotor 亲核性促进剂，亲核性助催化剂（如用于氯乙烯的低温聚合）
nucleophilic reaction 亲核反应；给电子反应
nucleophilic reactivity 亲核反应性
nucleophilic substitution 亲核取代
nucleophilic substitution reaction 亲核取代反应
nucleus 核；晶核
nude 象牙黄[色]，蛋壳黄[色]
nudes 透明袜，肉色针织物
nugget 金棕[色]，柑橘橙[色]
nuisance 公害
nuisanceless closed loop technique 无公害闭环式工艺
nuisanceless technology 无公害工艺
nujol mull 石蜡糊

null 零；零的；无效的
null adjustment 零位调节，零调
nullhypothesis 原假设
nullify 取消，废弃，使无效
null indicator 零位指示器
null point 零点
null string 空行《计算机》
number 〈N, No., no.〉数，号；支数；编号；数量；第……号
numbered duck 编号帆布
numbering 支数测定；编号
numbering system [纤维和纱线的]细度计量制（分定长制和定重制两类）
number of monofilaments 单丝根数
number of passage 道数；(循环试验中的)循环次数
number of teeth 〈N, nt〉齿数
number of turns (= number of twists) 捻数，捻度
number range 数值范围
number shirt 胸前印有数目字的圆领衫
numeral 数字；数字的
numeration 计数法，读数法
Numeri 纽默里特性（毛针织物摩擦特性）
numeric 数字；数字的，数值的
numerical analysis 数值分析
numerical colour difference 数值色差
numerical control 〈Nc〉数[字]控[制]，数字计算机控制
numerical controlled machine tool 数字程序控制机床，数控机床
numerical data 数据
numerical data processing 数据处理
numerical rating system 数字分级法
numerical value 数值

numeroscope 数字记录器,示数器
nun tuck 裙子横褶
nurse cloth 护士布(用做护士制服或工作服)
nursing wear 护士服
nut 螺母,螺帽
nut bolt 带螺帽的螺栓
nut cap 盖帽,封紧帽,死螺帽
nutgall 五棓子,五倍子,没食子
nutria 海狸;海狸毛皮(南美洲产)
Nuva 诺沃系列(拒水拒油剂,弱阳离子型,有机氟化合物,有耐久拒水、拒油效果,易去污,商名,瑞士科莱恩)
NVM (nonvolatile matter) 不挥发性物质
n. wt. (=nt. wt. , net weight) 净重
Nyanthrene dyes 尼恩士林染料(还原染料,商名,美国制)
Nyesta 耐斯太填塞箱变形锦纶丝,耐斯太填塞箱变形锦纶织物(商名)
Nyliton dyes 尼利汤染料(染聚酰胺纤维的酸性染料,商名,美国制)
Nylocet dyes 尼洛塞特染料(分散染料,商名,瑞士制)
Nylofixan PM liq. 尼龙固PM液(阴离子型固色剂,能改进锦纶染色和印花的湿牢度,商名,瑞士科莱恩)
Nylomine A dyes 尼龙明A染料(酸性匀染性染料,用于锦纶染色,商名,德国巴斯夫)
Nylomine C dyes 尼龙明C染料(酸性耐缩绒染料,用于锦纶染色,商名,德国巴斯夫)
Nylomine D dyes 尼龙明D染料(金属络合染料,用于锦纶染色,具有高的湿牢度和耐晒牢度,商名,德国巴斯夫)
Nylomine dyes 尼龙明染料(染聚酰胺纤维的染料,有A、B、C、D、P型,商名,德国巴斯夫)
nylon 锦纶,尼龙(聚酰胺纤维)
nylon 6 锦纶6,尼龙6
nylon 13 锦纶13,尼龙13(聚十三内酰胺纤维)
nylon 66 锦纶66,尼龙66
nylon acetone test 锦纶的丙酮测试[法]
nylon bearing 尼龙轴承
nylon bush 尼龙衬套(用尼龙制成,无需加油,无噪声并有耐久性)
nylon coated fabric 锦纶涂覆织物
nylon dye 锦纶用染料
nylon hose 锦纶袜,尼龙袜
nylonic acrylic fibre 锦纶[化]丙烯腈系纤维
nylonized polyester fibre 锦纶化[的]聚酯纤维
nylonizing 锦纶处理(一种黏合锦纶织物的方法)
nylon modified phenolic resin fibre 锦纶改性的酚醛树脂纤维
nylon mousse 锦纶弹力丝(低特)
nylon MXD-6 锦纶MXD-6(以间苯二甲胺与己二酸所制的芳族聚酰胺纤维)
nylon palace 尼龙纺,锦纶派力斯
nylon puckered fabric 锦纶绉纹织物
nylons 锦纶长袜,尼龙长袜
nylon seersucker taffeta 锦纶绉条纹薄塔夫绸;锦纶塔夫泡泡纱
nylon shioze 锦纶纺,尼丝纺
nylon-spandex bicomponent fibre 锦纶-聚氨基甲酸乙酯双组分纤维(弹性纤维)
nylon stretch tights 弹力锦纶紧身连袜裤
nylon tricot 锦纶经编织物
nylon tulle [经编]锦纶薄纱;锦纶绢网

Nylon type 91 锦纶 91 型（添加荧光增白剂的锦纶 66 纤维，商名，美国制）

Nylon type 472 锦纶 472 型（环脂族聚酰胺纤维，现称奎阿那纤维，商名，美国制）

nylon velvet 锦纶丝绒（绒头系锦纶，有防水性能）

Nyloquinone dyes 尼洛基农染料（分散染料，商名，法国制）

Nylosan dyes 尼龙山染料（染聚酰胺纤维的阴离子染料，有 E,F,M,N,S 等型，大多为弱酸性染料，商名，瑞士科莱恩）

Nylosan E dyes 尼龙山 E 染料（酸性染料，阴离子型，匀染性好，特别适用于锦纶染色，商名，瑞士科莱恩）

Nylosan F dyes 尼龙山 F 染料（酸性染料，阴离子型，特别适用于锦纶染色，湿牢度优良，商名，瑞士科莱恩）

Nylosan N dyes 尼龙山 N 染料（酸性染料，阴离子型，特别适用于锦纶染色，拼色性好，商名，瑞士科莱恩）

nysilk 超声变形处理法（对锦纶针织品处理，变更其分子结构而成无光型）

Nysil process 耐赛尔工艺（提高锦纶吸湿性，使具有类似真丝性能的处理）

nytril fibre 奈特里尔纤维（聚偏氰乙烯纤维）

NZWB（New Zealand Wool Board） 新西兰羊毛管理局

O

o- （词头）表示"正""邻[位]""原"
O(ohm) 欧姆
O(oil) 油;石油
O(order) 指令;调配;级,数量级,位;次序;有序
o.(ortho) 正;原;邻[位]
O(output) 输出;产量,生产率
oad (＝o.a.d.,overall dimension) [机台] 总尺寸;外形尺寸
oak leaves 柞叶(植物染料)
oak silk 柞蚕丝
oakwood 栎木棕(棕色)
oasis [沙漠]绿洲绿(浅暗黄绿色);浅秋香色
OBA（oxy-benzoic acid） 羟基苯[甲]酸,羟基安息香酸
Obermaier dyeing machine 奥氏[循环染液]染色机
Obermaier jet beck 奥伯梅厄喷射染色机（小浴比染色设备,德国制）
Obis printing 奥比斯印花[法]（使用单辊筒获得多色效果的特殊印花方法）
object carrier 载玻片
object code 结果代码,目标代码（汇编程序或编译程序所产生的代码）
object colour 物体色
object glass 物镜
object illumination 目标照明,局部照明
objective 目标,对象;物镜
objective aperture 物镜孔径
objective astigmatic corrector 物镜像散校正器
objective characteristic 客观特征,仪器检验的特征
objective evaluation 客观评定,仪器评定
objective lens 物镜
objective measurement 客观测定,仪器测定
objective method 客观试验方法,仪器试验方法
objective ranking 客观分级,仪器评定
objective test 客观试验,仪器试验
object language 目标语言,结果语言《计算机》
object micrometer 物镜测微计,物镜测微尺
object program 目标程序,结果程序
oblate ellipsoid 扁[圆]椭球体
oblique ironing 熨烫歪斜,熨烫不正(疵点)
oblique pillowblock bearing 斜支座止推轴承
oblique weft 斜纬
obliquity 倾斜,斜度
observation 观测,观察;遵守,实行;检查
observational error 观察误差
observation data〈OD〉观测数据
observer〈obsr.〉观测员,观察员;评论员
observer metamerism 观察者条件等色性《测色》
obsolete equipment 陈旧设备
obsr.（observer） 观测员,观察员;评论员

obstacle indicator 故障指示器
obstruction 障碍,阻塞,干扰;障碍物
OC (organic carbon) 有机碳
OC (oxygen consumption) 耗氧量,氧消耗量
ocb (=o. c. b. ,oil circuit breaker) 油断路器,油开关
OCC (open circuit characteristic) 开路特性,空载特性[曲线]
occlusion 吸着,吸留;堵塞;包藏
occupational disease 职业病
occupational hazard 职业危险,职业公害
Occupational Safety and Health Act 〈OSHA〉 职业安全与保健条例(美国)
Occupational Safety and Health Administration 〈OSHA〉 职业安全与保健管理局(美国)
ocean disposal 海洋废弃物
ocean green 海洋绿[色](淡黄绿色)
ocher 赭石
ochraceous (=ochre yellow) 赭色的,赭石黄(浅暗橘黄色)
OCR(optical character recognition) 光学字符识别(用计算机录入文档的功能)
oct. (octagon) 八角形,八边形
octadecyl ethylene urea 十八烷基亚乙基脲(耐久性柔软剂,防水剂)
octagon 〈oct.〉 八角形,八边形
octal 八;八进制的
octalobal cross section 八叶形横截面
octamethyl cyclic tetrasiloxane 八甲基环四硅氧烷(简称D4,有机硅原料)
octane [正]辛烷
octane number 辛烷值
octanol 辛醇
octet 八位[二进制数的]位组;八重线

octyl benzene sulfonate 辛基苯磺酸
octyl phenol ethoxylate 〈OPE〉 辛基酚聚氧乙烯醚
octyl sulfate 辛基硫酸盐
ocular 目镜;视觉的
ocular micrometer 目镜测微计
OD (observation data) 观测数据
OD (optical density) 光密度
OD (out draw) 外拉伸
OD (outside diameter) 外[直]径
odd electron 奇[数]电子;多余的电子
odd-frequency motor 畸频电动机(在非标准频率下工作)
oddments 残次品;零批(羊毛)
odd-pitch screw 非标准螺纹螺钉
odd scraps 碎布
odd-shaped cross section 异形截面
ODM(original design manufacturer) 原始设计制造商
odometer 速度计;里程计;测距器
odor 【美】=odour
odorant 恶臭物质,有气味物质;有气味的
odoriferous 有气味的
odorimetry 气味测定法,嗅觉测量法
odour [香或臭]气味
odour adsorbing finish 吸臭整理
odour-causing bacteria 产生臭气细菌
odour concentration 气味浓度;臭气浓度
odour control 臭气控制
odour intensity index 臭气强度指数
odour proof 防臭
odour remover 消臭剂
odour resistant 防气味的;防臭的
odour test 硫醇测定,气味测定
odour treatment by fire 燃烧脱臭《环保》

odour unit〈OU〉 臭气单位；气味单位
Oe.（oersted）奥斯特（磁场强度单位）
OECD（Organization for Economic Cooperation and Development）经济协作与开发组织
O. E. E. C.（Organization for European Economic Cooperation）欧洲经济合作组织
Oeko-Tex 国际纺织生态领域研究及检验协会
Oeko-Tex certification 国际上［纺织品］生态标准证书
Oeko-Tex label 生态纺织品标签
Oeko-Tex Standard 100 国际上认可的纺织品生态标准
OEM（original equipment manufacturer）原始设备制造商
OEM（own equipment manufacture）自主设备制造
Oersted〈Oe.〉奥斯特（磁场强度单位）
OE yarn（open end yarn）气流纺纱线
O-fading 臭氧褪色
OFDA（optical fibre diameter analyser）纤维直径光学分析仪
off 停车，关着，断路
offal 垃圾
off-black 黑色不黑，黑色偏离
off centre 跑偏
off-clip 脱边（拉幅机脱铗疵）
off colour 色差（尤指染色织物的左、中、右和头梢色差）
off-colour wool 色污毛，尿污毛，变色毛
off-contact 脱开接触（机器停运装置）
off end 车尾
off-gas 废气；抽气
off-gauge 非标准的，不合规格的
off-grade 级外，等外，副次品

off-grain 纬斜（织疵）
office of pesticide programs〈OPP〉农药项目办公室
official acceptance test 正式验收试验
Official Gazette of the U. S. Patent Office〈O. G.〉《美国专利局公报》
official moisture regain 公定回潮率
official moisture regain of union products 混纺品公定回潮率
official regain 公定回潮率
off-key 不协调
off-limit 禁止进入
off-line memory 脱机存储器《计算机》
off-line test 离线［间接］试验，机台外试验
off-pattern 花形不符（印疵）
off-patterned cloth 漏花织物《针》
off-period 停运时间
offposition 断路位置，关闭位置
offpressing 最终烫平，最终压烫
offscum 废渣
offset 抵销，补偿；偏移，偏置；橡胶版印刷；调整偏差
offset lithography 平版胶印术
offset press 胶版印刷机
offset printing 胶印，胶版印刷
offset printing ink 胶版印刷油墨
off shade 色差（尤指染色织物的左、中、右和头梢色差）
off-shade yarn 黄白纱（疵品）
off size 尺寸不对
off-specification 不合规格
off-specification goods 不合规格产品
off-square 经纬缩率差（经线缩率与纬线缩率间的差异）
off-square fabric 不方正织物（经密与纬

密不同);不平衡织物(经纬缩率不一)
off-standard 不符合标准;等外级
off-tone 色光不一致(染疵)
off water 废水
off-white 黄白色,灰白色
Ofna-pon AS 奥夫纳邦 AS(脂肪酸及蛋白质缩合物,纳夫妥 AS,奥夫纳兰染料及海灵登染料染色的保护胶体及分散剂,商名,德国制)
O. G.(Official Gazette of the U. S. Patent Office)《美国专利局公报》
ogee washer S 形垫圈
Ohm〈O,Oh,Oh,,oh〉欧姆(电阻单位)
OI(=O. I.,oxygen index)[需]氧指数
oil〈O〉油;石油
oil absorbency 吸油性
oil and fat content 含油脂率
oil and IPA(isopropyl alcohol) repellency 拒油和拒异丙醇性
oil and soil 油污
oil and soil repellent test 拒油污试验
oil arving [丝或黏胶长丝织物]上油光泽整理
oil atomizer 油雾化器
oil-based ink 油基印墨
oil-based wet coagulation 油基湿法凝固涂层
oil bath 油浴
oil belt duck 重型防水帆布
oil boom 拦油栅
oil borne soil 带油污质
oil-borne stains 油媒[介]污渍
oil bowl 油杯
oil brightening 上油增艳处理《丝》
oil buffer 油缓冲器
oil burner 油燃烧器

oil circuit breaker〈ocb,o. c. b.〉油断路器,油开关
oil circulating heating 油循环加热
oil circulation heating system 油循环加热系统
oil circulation lubrication 循环润滑
oil cleaner 滤油器
oilcloth 油布,漆布,防水油布
oil colours 油溶性染料
oil content 含油率
oil control ring 刮油环,油环
oil cup 油杯,油罐
oil disk 油盘
oil drain plug 放油塞
oil dripper 滴油器
oil drip tray 滴油盘
oiled pick 油纬(织疵)
oiled silk 油绸(经亚麻籽油处理的薄丝织物,有防水性能)
oiled threads 油污丝(生丝疵)
oiled yarn 润滑纱线
oil emulsion 乳化液,乳化油,乳胶,油乳液
oil equivalent 标准油
oiler 加油器
oil expansion tank 油膨胀槽
oil feed component 给油元件,给油部件
oil fence 油网《环保》
oil filter 滤油器
oil finish 上油处理
oil-fired boiler 燃油锅炉
oil fog 油雾
oil grease spotting fastness 油污牢度《丝》
oil groove 油槽
oil hole 油眼
oil hydrosol 油水溶胶,油水乳液

oil immersion 油浸
oiliness 油性
oiliness compound 润滑油增效剂;油性化合物
oiling 加油,上油;油化;润滑
oiling agent 上油剂
oiling motion 加油装置
oiling roller 给油辊
oil in water emulsion〈O/W〉水包油乳化液,油水相乳化液
oil-in-water system 水包油系统(涂料印花体系)
oil keeper 油承
oil leak 漏油
oilless bearing 无油轴承
oil level 油位
oil looking fabric 油光布
oil loving 油溶的,亲油性的
oil misting 油雾[化]
oil mist lubrication 喷雾式加油
oil mist spray 油雾喷射器
oil mordant 油媒染剂
oil of bitter almond 苦杏仁油(苯甲醛)
oil of winter green 冬青油(水杨酸甲酯)
oil oxidation stability 油的氧化稳定性
oil pad 油垫
oil pick up 上油率
oil pipe 油管
oil pollution 油污染
Oil Pollution Act〈OPA〉防油类污染条例
oil pot 油壶
oil pressure 油压
oil proofing 防油[整理]
oil proofing synergist 拒油整理增效剂
oil repellency 抗油性
oil repellency rating 防油等级

oil repellency test 防油试验
oil repellent 抗油[脂]剂(一般为氟的化合物);抗油的
oil repellent finishing 拒油整理
oil reservoir 油箱,储油器
oil-resin 含油树脂
oil-resist 防油
oil resistance 抗油性,耐油性
oil-resist finish 防油整理
oil retaining felt washer 护油毡垫圈
oil seal 油封
oil seepage 油渗出
oil separator 油分离装置
oil-silk 油绸(经上油处理的轻薄电力纺,做雨衣等用)
oilskin 防水油布,油布雨衣
oil slabs 油飞花织入(织疵)
oil slick boom 拦集浮油栅
oil slinge 抛油环
oil-soluble dyes 油溶性染料
oil spill 漏油,浮油
oil spots 油渍,油污
oil spotted wale 油针(针织疵)
oil-spotted yarn 油污纱
oil sprayer 喷油器
oil stain 油沾污
oil stain cleaning agent 油沾污清洁剂
oilstone 油石
oil strainer 滤油器,滤油网
oil streak 油污条痕
oil sump 油槽
oil tight 不透油的
oil-treated fabric 涂[亚麻籽]油织物
oil trough 油槽
oil tube bracket 油管托架
oil viscosity rating 机油黏度号数

oil waterproofing [织物]上油防水法
oilway 油槽,油路
oily condensate 冷凝油滴
oily hand 油滑手感
oily matter 油垢物
oily mixture 油性混合物
oily soil 油污
oily wastewater 含油废水
ointment 软膏
OIS (order information system) 定单信息系统;指令信息系统《计算机》
old fustic 染色桑膏(天然金黄色植物染料,美国用于染毛)
oldham coupling 十字滑块联轴器
old moss green 老苔绿[色](橄榄色)
old rose 老玫红[色](浅暗红色)
old wash [牛仔衣]怀旧洗
oleate 油酸盐
olefine [链[烯[烃],烯属烃
olefine fibre 烯烃类纤维(含85%以上乙烯、丙烯或其他烯类的合成纤维)
olefinic hydrocarbon 烯烃
olefinsulphonate 烯(烃)磺酸盐(乳化剂)
oleic acid 油酸,十八烯酸
oleine oil 和毛油(以油酸为基质)
oleometer 油比重计
oleophilic group 亲油基
oleophobic 疏油的
oleophobic effect 疏油作用,憎油作用
oleophobic finisher 疏油整理剂
Oleophobol C 奥利氟宝 C(有机氟系聚合物,阳离子型,赋予织物高的防水、防油及防污功能,主要用于纤维素纤维,商名,瑞士制)
Oleophobol SL 奥利氟宝 SL(有机氟系聚合物,阳离子型,赋予织物高的防水、防油及防污功能,主要用于人造纤维,商名,瑞士制)
oleoresinous 含油树脂的
oleyl alcohol 油醇
olfactometry 气味测定法,嗅觉测量法
oligo- (构词成分)表示"寡""少""低"
oligoacrylonitrile 丙烯腈低聚物
oligoamide 酰胺低聚物
oligoamino acid 氨基酸低聚物
oligoesters 酯类低聚物
oligomer 低聚物,低分子量聚合物
oligomeric condensation 低聚物缩合[作用]
oligomeric cyclic phosphonate 环膦酸酯低聚物(阻燃剂)
oligomer removers [聚酯]低聚物去除剂
Oligomex N 一种高温缸清洗剂(季胺盐化合物,阳离子型,用于清除高温染色缸内的低聚物、油渍或残余染料等,商名,德国波美)
oligosaccharide 低聚糖
oligoterephthalic acid glycol ester 对苯二甲酸乙二醇酯低聚物
oligourethane 氨基甲酸酯低聚物
Olinor KW 66 奥列诺 KW 66 浆纱平滑剂(丙烯酸酯共聚物的混合物,阴离子型,用于经纱上蜡,可消除纱线表面绒毛,减少摩擦,商名,德国科宁)
olive [green] 橄榄[绿]色
olive drab 草黄[色],草绿[色],灰橄榄色
olive oil 橄榄油
olive oil potash soap 橄榄油钾皂,橄榄油软皂
ombré 由深到浅的色条纹布;由深到浅的色调
ombré check 深浅色格纹,彩虹方格花纹

ombre dyeing 深浅色染色,虹彩染色(在被染物的横向产生由深到浅的特殊效应的染色方法)
ombré effect 云纹效应(雕刻用语,指色泽由深到浅的效应)
ombré fabric 深浅条纹织物
ombré moire renaissance 深浅色波纹织物
ombré printing 虹彩印花[工艺](具有彩虹效应或由浅逐渐趋深的条纹效应)
ombré raye 虹彩与地色交错条纹
ombré silk 月华色丝(用绞丝染色法染成深浅不一的色调)
ombré stripe 彩虹条纹(深浅晕色条纹花样,有色织的,也有印花的)
omission of examination 漏验
omni- (构词成分)表示"全部""总""一切""遍及"
omnibus design 多用途花样设计
O. M. S. (= o. m. s. , output per manshift) 每人每班产量;单人生产率
on 开车;开着,导通
on-and-off controller 开关控制器
once through yield 单程得率
ondé 异色棉毛交织呢;波纹整理织物
ondometer 频率测量仪;波长测量仪
on-duty 值班,当班
one-address instruction 一地址指令《计算机》
one-bath chroming method 铬媒一浴法
one bath dyeing 一浴法染色
one bath dyeing and finishing 染整一浴法
one bath fixing agent 一浴法固色剂
one bath method 一浴法
one bath method of dyeing 染色一浴法
one-bath process 一浴法,单浴法
one line production 单线生产,单线流水作业
one-point method 单点法(测定黏度的方法之一)
one process 单程式,一次加工法
one repeat 一完整的花型循环
one's complement [二进制]反码,一的补码《计算机》
one-shot 一揽子
one-shot bleaching agent 一揽子漂白助剂(多种漂白助剂的综合)
one-shot lubrication system 中央润滑系统,集中加油系统
one-sided 单面的
one-sided drying machine 单面烘燥机
one-sided frame 单面机
one-sided spiral transport system 单面螺旋状运行织物
one-sided terry 单面毛圈织物
one side printing 单面印花
one side printing machine 单面印花机
one step dyeing 一步法染色
one step dyeing and finishing 染色—整理一步法
one-up-one-down weave 平纹组织
one-way circulation 单向循环
one-way clutch 单向离合器
one-way stretch 单向拉伸
one-way valve 单向阀
onium compounds 鎓类化合物,[负性元素]最高正价化合物
onium dyes 鎓类染料(含有离子化的鎓类基团的水溶性阳离子染料)
onium surfactant 鎓类表面活性剂
on-line 联机;在线
on-line analyzer 在线分析仪《自动》,[生产]线上分析仪

on-line colourimetric metering system 在线测色系统
on-line console 联机控制台
on-line control 在线控制,直接控制
on-line detection 在线检测
on-line instrumentation 在线仪表检控
on-line maintenance 不停产检修
on-line monitoring 在线监测
on-line operation 在线操作,联机操作
on-line test 在线[直接]试验,联机试验;工艺过程中测试
on-line titrators 在线滴定仪
only-coat fibre 全皮纤维
only-core fibre 全芯纤维
on-off control 开关式控制,起停控制,继电器式控制
on-off lever 开关连杆
on-off servo 继电伺服系统,继电随动系统
on-off switch 双位式开关
on-off system 双位调节系统
on-off valve 双位阀,截止阀
on-position 工作位置
on shade 近似颜色
on stream 在运转中;在操作中
on-the-dot registration 准确的对花
on-tone 色光一致
on weight of bath〈OWB〉 按浴重[计算]
on weight of fabric〈owf〉 按织物重量[计算]
on weight of goods〈OWG〉 按物重[计算]
on weight of resin〈OWR〉 [催化剂]按树脂重[计算]
on weight of solution〈ows〉 按液体重量[计算],按溶液重量[计算]
OOO(＝O.O.O.,o.o.o.,out-of-order) 故障,不正常
ooze leather 仿麂皮织物(花式毛呢或经编棉织物经过仿麂皮整理而成)
OP(＝op,o.p.,over proof) 超过额定的
OPA(Oil Pollution Act) 防油类污染条例
opacifier 不透明剂,遮光剂
opacifying agent 遮光剂
opacifying effect 不透明效果,暗的效果,遮光效果
opacity 不透明性;浑浊度;暗度
opalescence 乳[白]光;乳[白]色
opal finished georgette crepe 仿烂花乔其,仿烂花绡
opal finishing 仿烂花整理;仿烂花印花;消光整理;消光印花
opaline green 蛋白石绿(浅灰绿色)
opal printing 消光白色印花(用可溶性钨酸盐或钼酸盐浆料印花,通过钙、镁、钡盐溶液处理,生成不溶性白色)
opal quartz tube 乳白石英管(红外辐射加热器的元件材料)
opaque 不透明的,无光泽的
opaque body 不透明体
opaque camera [雕刻]放样机,放影机
opaque colour 覆盖色;不透明色
opaque fibre 不透明纤维
opaque fluid 不透明流体
opaqueness 不透明性,不透明度
OPE(octyl phenol ethoxylate) 辛基酚聚氧乙烯醚
open-air weathering test 耐室外气候牢度试验
open area [圆网]开孔面积
open bearing 开式轴承,对开轴承
open belt 开式皮带(驱动皮带轮轴和被动皮带轮轴平行,同一方向旋转的皮

带装置）

open blowing machine 开式蒸呢机

open bobbin 松式筒子[纱]（用于筒子染色）

open boil 开盖煮练，不加压煮练

open circuit 断路《电》；开式流程；开式回路

open circuit characteristic〈OCC〉 开路特性，空载特性[曲线]

open coil 有孔盘管

open decatizing machine 开式蒸呢机

open depot 露天仓库

open design 现存花样

open duster 笼式除尘机

open-ended 可扩充的，无终止的

open end yarn〈OE yarn〉 气流纺纱线

opener 开布机

opener roller 开松辊

open fabric 组织稀松织物

open faller gill spreader 平幅针板扩幅机

open flame 明火

open fold [平幅]折叠

opening and carding machine for imitation fur fabric 人造毛皮开松梳整机

opening and slitting machine 剖幅开幅机

opening area ratio of screen gauze 开孔率（单位面积圆网中孔目总面积所占的百分率）

opening, mangling and drying machine 开幅轧水烘燥机

opening rail [金属]扩幅板

opening roller 开幅辊；开松辊

opening scroll roller 螺旋扩幅辊

open kier 开口式煮布锅，低压煮布锅，常压煮练锅

open kier boiling 开盖煮练，不加压煮练

open-knit 网眼针织物

open line 明线《电》

open loop 开式回路

open-loop control 开环控制，无反馈控制

open loop driving system 开环调速系统

open-meshed fabric 网眼织物

open nonwoven fabric 疏松非织造布

open-ocean barrier 海洋隔离人工堤

open pan 常压煮布锅；常压容器

open place 稀弄，松档（织疵）

open porosity 表面多孔性

open reed 筘痕（织疵）

open return bend U形管

open routine 开型程序，直接插入程序《计算机》

open scourer 平幅煮练机

open-set mark 稀弄，松档（织疵）

open setting 稀经稀纬

open slack washing 开幅松式洗涤

open slack washing machine 平幅松式水洗机

open-slit type nozzle 狭缝式喷嘴

open soaper [多槽]平洗机，平幅皂洗机

open steam 直接蒸汽

open structure 松散结构

open texture [织物]稀松结构，稀松组织

open type 开式

open type decatizing machine 开式蒸呢机

open type frame 开启式机架（指轧车或轧光机的墙板上部呈L形，轧辊装卸方便）

open vat 还原液[浸染]槽

open vessel 开口容器

open washer 开口垫圈，开缝垫圈

open washing 开幅洗涤，平洗

open washing range 平幅显色皂洗联合机

open weave 稀薄组织,稀松组织,纱罗组织,多孔组织
open weave knits 网眼针织物
open width 平幅,开幅
open-width bleaching 平幅漂白
open-width bleaching range 平幅漂白联合机
open-width boil-out 平幅煮练
open-width continuous degumming 平幅连续精练
open-width developing and soaping range 平幅显色皂洗联合机
open-width draw drum washing machine 平幅吸鼓水洗机
open-width dyeing 平幅染色
open-width dyeing range 平幅染色联合机
open-width [fulling] finisher 平幅缩绒整理机
open-width fulling machine 平幅缩绒机
open-width hydroextractor 平幅脱水机
open-width kier 平幅煮布锅
open-width microwave dyeing machine for silk 丝绸用平幅微波染色机
open-width poly-stream washing machine 平幅多流水洗机
open width scouring and bleaching 平幅练漂
open-width scouring machine 开幅煮练机
open-width suction dryer 平幅真空吸水干燥机
open-width treatment 平幅处理
open-width washer 开幅洗涤机,平洗机
open width washing 平洗
open-width washing and relaxing range 平幅洗涤松弛联合机
open width washing machine 平幅水洗机

open winding 平幅卷绕
openwork knitting stitch 网眼组织,纱罗组织《针》
open work pattern 透孔织物花纹
oper.(＝opn,operation) 操作;运算;运转;经营
oper.(＝opr.,operator) 操作者;运算符
opera flannel 浅色全毛法兰绒
operating button 操作按钮
operating characteristic 运转特性
operating costs 业务开支,生产费用,运转费用,周转费用
operating cycle 操作周期,运转周期
operating instruction 操作说明
operating mechanism 运转机构;操作机理
operating personnel 操作人员,挡车工
operating rate 工作速度;运算速度
operating schedule 作业计划
operating sequence 操作程序
operating skills 操作技术
operating speed 布速;车速
operating system〈OS,os〉操作系统
operating temperature range〈OTR〉操作温度范围
operation〈oper.,opn〉操作;运算;运转;经营
operational amplifier 运算放大器
operational directive 操作指令
operational relay 操作继电器
operational suitability testing〈OST〉运转适应性试验
operation analysis 运行分析
operation code 操作码
operation control 操作控制,运转控制
operation register 操作码寄存器《计算机》
operation test 操作试验,运算试验

operative temperature 操作温度
operator〈oper.,opr.〉 操作者;运算符
operator console 操作员控制台
operator support system〈OSS〉 操作者支持系统
opinion〈opn.〉 意见,主张;评价
opium 鸦片,罂粟
opn.(=oper.,operation) 操作;运算;运转;经营
opn.(opinion) 意见,主张;评价
OPP(office of pesticide programs) 农药项目办公室
opp.(opposed) 对立的,反对的
opp.(opposite) 相反的,相对的
opp.(opposition) 反对,对立
opposed〈opp.〉 对立的,反对的
opposite〈opp.〉 相反的,相对的
opposite colour 反对色《测色》
opposition〈opp.〉 反对,对立
opr.(=oper.,operator) 操作者;运算符
opt.(optical) 光学的;旋光的
opt.(optics) 光学
opt.(optimum) 最佳,最优
optic 光[学]的;视觉的;旋光的;镜片
optical〈opt.〉 光学的;旋光的
optical activity 旋光性,旋光度
optical analysis 光学分析
optical anisotropy 光学各向异性
optical bleach 荧光增白
optical brightener 荧光增白剂
optical character 光学特性
optical character recognition〈OCR〉 光学字符识别(用计算机录入文档的功能)
optical comparator 光学比较仪,光学比色计

optical densitometer 光密度计
optical density〈OD〉 光密度
optical detection 光学检测
optical dichroism 光二色性
optical dispersion 光色散
optical electronic fabric positioning device 光电[织物]定位系统
optical-electronic weft-feeler 光电探纬装置
optical-electronic weft straightener 光电整纬装置
optical feeler 光学探测器
optical fibre 光导纤维;光学纤维,光纤
optical fibre diameter analyser〈OFDA〉 纤维直径光学分析仪
optical filter 滤光器
optical geometry 光学几何结构
optical geometry 0/45 0°照明/45°探测
optical geometry 45/0 45°照明/0°探测
optical geometry 0/d 0°照明/漫反射探测
optical geometry d/0 漫反射照明/0°探测
optical glass 光学玻璃
optical grating 光栅
optical illusion 视错觉
optical isomer(=optical isomeride) 旋光异构体,旋光异构物
optical isomerism 光学异构,旋光异构[现象]
optical method 光学方法
optical microscope 光学显微镜
optical pyrometer 光测高温计
optical rotation 旋光性,旋光度
optical scanning 光学扫描
optical-scan response 光学扫描特性曲线
optical selvedge feeler 光电探边器
optical sensing device for spotting thick

seams 光电接头探测装置(在轧光机等设备上,用于探测织物接头,及时将信号传至自控装置,使轧辊稍许松压,以免轧坏)

optical sensing lead 光电探头

optical sensor 光敏元件

optical smoke density test 光学烟雾密度试验

Optical Society of America 〈OSA〉 美国光学学会

optical solution 光测溶液

optical solvent 光测溶剂

optical spectrum analyzer 光谱分析仪

optical storage materials 光存储材料

optical thickness 光学厚度

optical wedge 光楔

optical whitening 荧光增白[工艺],增白工艺

optical whitening agent 荧光增白剂

optics 〈opt.〉 光学

Optidye™ 一种用于多种纤维及染料浸染工艺优化设计的计算机软件(商名,德国制)

Optifix F liq. 爱德素 F 液(爱德素染料专用固色剂,阳离子型,商名,瑞士科莱恩)

optima 最优,最佳

optimal value 最优值

Optimax super impregnator (= Optimax system) 高液量浸轧系统(商名,德国)

optimeter 光度计;光电比色计,光学比较仪

OPTIM fibre 一种羊毛纤维(用特殊拉伸技术,改变羊毛纤维细度和性能,澳大利亚制)

optiminimeter 光学测微计

optimization 最优化,最佳化;优选[法]

optimization technique [最]优化技术

optimized formula 最佳处方

optimized process 最佳工艺过程

optimizing control 最佳控制

optimum 〈opt.〉 最佳,最优

optimum concentration 最适浓度

optimum condition 最佳条件

optimum dyeing 最佳染色

optimum estimate 最优估计

optimum feed location 最佳进料位置

optimum humidity 最佳湿度

optimum performance 最佳性能

optimum programming 最佳程序设计

optimum seeking method 优选法

optimum successive over-relaxation 最优逐次超松弛

optimum temperature 最佳温度

optimum water consumption 最佳水耗

optimum working frequency 〈OWF〉 最佳工作频率

optional attachment 任选附件

optional test 选项试验

Optisal dyes 爱德素染料(色泽鲜艳的直接染料,阴离子型,用爱德素 F 固色后牢度极佳,特别适用于涤棉一浴法染色,商名,瑞士科莱恩)

opto-electronic monitoring systems 光电监控系统

opto-electronic scanning system 光电扫描系统

opto-electronic sensing device 光电探测装置

OR. (= or., orn., orange) 橙色,橘色;黄色颜料;橙色的

Orafix PF 奥腊菲克斯 PF(丙烯酸树脂

与羟甲基化合物的乳化液,用作涂料和植绒印花的黏合剂,商名,瑞士制)

oral toxicity 经口毒性

orange〈**OR,or.,orn.**〉 橙色,橘色;黄色颜料;橙色的

Oraprint 欧拿普林(金银粉印花用黏合剂,光亮效果好,不会变黑,牢度优良,手感柔软,商名,意大利宁柏迪)

Orbis method of printing 奥比斯[单辊筒多色]印花法(辊筒周围敷以各色涂料,干后在车床上车平,然后将湿布印上,可得色彩缤纷的图案)

orbit 轨道

orbital 轨道的

orbital velocity 轨道速度

orbiter 轨道飞行器

orcein 苔红素,地衣红(红棕色植物染料)

orchid 兰花紫(浅红紫色)

orchid pink 兰花粉红

orchil 苔色素(由地衣制取的红紫色植物染料)

orcin(=**orcinol**) 苔黑酚;地衣二酚;5-甲基间苯二酚

order〈**O**〉 指令,调配;级,数量级,位;次序;有序

order code 指令码

order information system〈**OIS**〉 定单信息系统;指令信息系统《计算机》

order of addition 加料顺序

order of magnitude 数量级

order of reaction 反应顺序

ordinary bleach 常规漂白

ordinary bright cone 普通有光丝[宝塔]筒子,普通有光[宝塔]筒子丝

ordinary maintenance 日常维护

ordinary temperature 常温

ordinary twill 正规斜纹

ordinary wool 粗纺用羊毛

ordinate 纵坐标

ore 矿,矿石

ore dressing 选矿

Oremasin pigment colours 奥雷马辛涂料(印染用涂料,商名,瑞士制)

org.(**organic**) 有机的

org.(**organization**) 组织,机构,构造

organ cell 器官细胞

organdy finish 奥甘迪(蝉翼纱)整理(用浆料、树脂或化学品处理,使棉、丝或合纤的细薄平纹织物产生暂时的或耐久的挺爽效应)

organic〈**org.**〉 有机的

organic acid 有机酸

organic acid-releasing agents 有机释酸剂

organic base soap 有机碱肥皂

organic carbon〈**OC**〉 有机碳

organic colouring matter 有机色素

organic compound 有机化合物

organic halogen compound 有机卤化合物

organic hydroperoxide 有机过氧化氢,过氧化氢有机[化合]物

organic impurity 有机杂质

organic load(=**organic loading**) 有机[物]负荷

organic luminescent materials 有机荧光物质

organic matrix 有机[聚合物的]基体,有机[聚合物的]基质

organic nitrogen 有机氮

organic peroxide 有机过氧化物

organic phosphite 有机亚磷酸酯

organic pigments 有机颜料

organic pollutant analysis 有机污染物分析
organic promoter 有机助催化剂;有机促进剂;有机助聚剂
organic solvent fastness 有机溶剂牢度
organic stabilizer 有机稳定剂(用于双氧水漂白)
organic synthetic dye 有机合成染料
organic synthetic pigment 有机合成颜料
organization〈org.〉组织,机构,构造
Organization for Economic Cooperation and Development〈OECD〉经济协作与开发组织
Organization for European Economic Cooperation〈O.E.E.C.〉欧洲经济合作组织
organochlorine 有机氯
organofunctional group 有机官能[基]团
organ of vision 视觉器官
organogel 有机凝胶
organohalogen content 有机卤素含量(生态学、毒理学评估参数)
organohalosilane 有机卤化硅烷(有机硅防水剂)
organoleptic evaluation 感官检验,感官评定
organoleptic test (=organoleptics) 感官试验
organolite 有机碱交换料,离子交换树脂(软水处理用)
organomercury compound 有机汞化合物
organometal 金属有机化合物
organometallic catalyst 有机金属催化剂
organometallics 金属有机化合物
organophilic 亲有机性的,亲有机物质的
organophobic 疏有机性的,疏有机物质的
organophosphate 有机磷酸酯

organophosphite 有机亚磷酸酯
organophosphorus compound 有机磷化合物
organopolysiloxane 有机聚硅氧烷
organosilicon 有机硅[化合物]
organosilicon compound 有机硅化合物
organosilicon polymer 有机硅聚合物
organosiloxane 有机硅氧烷
organosilyl cellulose 有机硅与纤维素纤维的结合体
organosol 有机溶胶(用有机溶剂作为连续相的胶态溶液)
organotincarboxylate 有机锡羧酸盐
organotin compound 有机锡化合物
organza 透明硬纱(丝或人丝制)
organzari 硬挺整理的丝织物(类似印度妇女莎丽服织物)
oriental blue 东方蓝(暗蓝色);蓝色颜料(普鲁士蓝与群青的混合物)
oriental carpet 东方块毯
oriental crepe 东方绉,重双绉
oriental rugs 东方块毯
orientation 取向,定向;校正方向;定向度
orientation birefringence 取向双折射,定向双折射
orientation effect 取向效应,定向效应
orientation force 取向力;偶极间引力
orientation uniformity 取向均匀性
oriented adsorption 取向吸附
oriented crystallization 取向结晶化
oriented polyester film 取向聚酯薄膜
oriented reinforcement 取向强度
oriented shear 定向剪切
orifice 孔,孔板;出口,孔口
orifice flowmeter 孔板流量计
orifice plate 细孔板(流量测定装置用)

orifice steam trap 节流孔疏水器
orifice viscometer 细孔式黏度计
orig.(original) 原文[的];原来的;起源的
origin 原点;原始地址
original〈orig.〉 原文[的];原来的;起源的
original colour 原色
original design manufacturer〈ODM〉 原始设计制造商
original equipment manufacturer〈OEM〉 原始设备制造商
original film 黑白稿,黑白片
original pattern 花样原稿
original piece 原布
original sample 原样
original solution 供应液,原液
O-ring "O"形密封圈,"O"形环
orn.(=OR.,or.,orange) 橙色,橘色;黄色颜料;橙色的
ornamentation 装饰,装饰品
orpiment 雌黄(黄色或橙色颜料)
Orr's white 奥尔白(即锌钡白)
orseille 苔色素(由地衣制取的红紫色植物染料)
ortho- (构词成分)表示"正""原""邻"
ortho-acid 正酸,原酸
orthochromatic 正色的(摄影用语)
orthochromatic film 正色胶片
orthocortex of wool 羊毛的正皮质层,正外皮层
ortho effect 邻位效应
orthogonal 正交的,直角的,直交的
orthogonal coordinate 直角坐标《数》
orthogonal coordinate system 正交坐标系
orthogonal design 正交设计
orthogonal test 正交试验
orthogonal trial 正交试验
ortho mixture 不同纤维交织物
orthophosphate 正磷酸盐
ortho-phosphoric acid 正磷酸
ortho-position 邻位
Ortolan dyes 奥尔托兰染料(染羊毛的1∶2金属络合染料,商名,德国制)
Ortol dyes 奥尔托染料(酸性染料,商名,德国制)
OS (=os,operating system) 操作系统
OS (=os,oversize) 尺寸过大;特大型;加大尺寸(指一种大于标准的修理尺寸)
OSA(Optical Society of America) 美国光学学会
osage orange 桑橙(植物染料)
OSA Uniform Colour Scales System 美国光学学会颜色系统
osc.(oscillation) 摆动,振荡,振动
osc.(oscillator) 摆动器,振荡器
osc.(oscilloscope) 示波器
oscillate 振荡,振动,摆动
oscillating airjet roller 摆动式空气喷射辊(地毯多流染色机用)
oscillating arm 摆动臂,拐臂
oscillating brush 振动刷
oscillating centring head 振荡式定中心头子
oscillating-cylinder viscometer 振动圆柱黏度计
oscillating detector 振荡检测器,检波器
oscillating disk method 摆动盘法(测界面黏度)
oscillating-disk viscometer 摆动盘式黏度计
oscillating doctor 摆动式刮刀

oscillating jet technique 振荡射流法(测动表面张力)
oscillating lever 摆动杆
oscillating motion 摆动装置;摆动运动
oscillating roller 振动滚筒
oscillating traverse motion 往复摆动运动
oscillating washer 摆动式洗涤机
oscillation ⟨osc.⟩ 摆动,振荡,振动
oscillation frequency 振荡频率
oscillator ⟨osc.⟩ 摆动器,振荡器
oscillatory load 摆动负荷(测疲劳用)
oscillatory viscoelastometer 振动黏弹计
oscillogram 波形图,示波图
oscillograph 示波器
oscillometer 示波计
oscilloscope ⟨osc.⟩ 示波器
oscilloscope screen 示波器荧光屏
OSHA(Occupational Safety and Health Act) 职业安全及保健条例(美国)
OSHA(Occupational Safety and Health Administration) 职业安全与保健管理局(美国)
osmiophilic layer 亲锇层(彩棉成分)
osmium tetraoxide treatment 四氧化锇处理(提高耐高温纤维光稳定性的方法)
osmometer 渗透压力计
osmometry 渗透压测定法
osmosis 渗透[作用]
osmotic balance 渗透平衡;渗透天平
osmotic bleaching 渗透漂白
osmotic method 渗透方法(测定平均聚合度)
osmotic pressure 渗[透]压[力]
OSS(operator support system) 操作者支持系统
OST(operational suitability testing) 运转适应性试验
Ostwald purity 奥斯瓦德纯度
Ostwald ripening 奥斯瓦德成熟度
Ostwald's colourmetre 奥氏比色计
Ostwald's universal-photometre 奥氏万能光度计(测定物体光泽的仪器,用以检验纤维染色牢度)
Ostwald's viscosimetre 奥氏[毛细管]黏度计
Ostwald [colour] system 奥斯瓦德[表色]制
OTR(operating temperature range) 操作温度范围
OTR(overload time relay) 过载限时继电器
ottoman 粗横棱纹织物
ottoman cord (=ottoman rib) 粗直棱纹织物
OU(odour unit) 气味单位,臭气单位
O.U.(Oxford University) 牛津大学
ounce ⟨oz,OZ.,·oz.⟩ 盎司(质量单位,约等于28.3克)
ounces ⟨ozs,ozs.⟩ 盎司的复数
outage 停止,中断;停运,断电;排出口
outboard bearing 外接轴承(在电机与风机之外)
outcome 结果;输出
outdoor furnishings 户外用布,野外用布(如各种篷布、帆布等)
outdoor weathering 室外气候老化,室外风化,室外风蚀
out draw ⟨OD⟩ 外拉伸
out end 机尾,机末
outer casing 外罩壳
outer coat fibre 刚毛,发毛

outer defect 外观疵点
outer garment 外衣
outer phase ［乳化体的］外相
outer wear 外衣
outfit 服装，全套衣服；装置；备用工具；配备
outflow resistance 流出阻力，排放阻力
outgassing 渗气（真空镀金属时挥发的蒸气）；除气作用
outgoing beam 出射光束
outgoing quality limit 出厂品质限制
outing cloth 户外运动服织物
outing flannel 软绒布
outing shirt 旅游衫
Outlast fibre 一种调温纤维（腈纶型）
outlet 出口，排出口；出线；电源插座
outlet position 出布端位置；出线位置《电》
outlet roller 输出辊，送出辊
outlet section 出布处，［拉幅机］出布铁链处
outline 轮廓，线条；提纲；略图
outline of pattern 花纹轮廓
outlook express ［微软公司出品的］电子邮件客户端《计算机》
out-of-balance 不平衡，失衡
out of colour sample 不合色样，与色样不符
out of control 失控
out of fashion 不流行
out-of-order 〈OOO，O.O.O.，o.o.o.〉故障，不正常
out of register 对花不准，对版不准；不对齐
out-of-repair 失修
out-of-roundness 不圆度，椭圆度

out of season 不合时令
out-of-service time 停台时间
out of shape ［服装］走样
out of style 不时兴
out of work 损坏（指机器）
output〈O〉输出；产量，生产率
output instruction 输出指令
output per manshift〈O.M.S.，o.m.s.〉每人每班产量；单人生产率
output shaft 输出轴
output signal 输出信号
output voltage〈OV〉输出电压
output work queue 输出排队《计算机》
output writer 输出程序
outshot 等外品（指原料）
outside caliper 外卡钳
outside diameter〈OD〉外［直］径
outside indicator 外指示剂
out workes 外包工
ouvre 小花格子织物
OV（output voltage）输出电压
ov（＝o.v.，OV，overvoltage）过［电］压
oval flowmeter 椭圆流量计
ovality 椭圆度
oval paddle machine 椭圆槽桨翼漂染机（用于针织品、袜子的精练、漂白和染色）
oven 烘箱；炉
oven ageing 烘箱老化
oven curing 高温焙烘
oven dry tensile strength 绝干拉伸强力
ovendry weight 烘干重量
oven exhausts 烘房排气
oven method 烘箱测含水方法
oventest 炉热试验，耐热试验
over-（构词成分）表示"在……上面""在

……上空""越过""超过""过分""外加""额外"

overaeration 过量曝气
overageing 过度老化
overall 外衣,罩衫;工作服;防护服;工装裤;总的,军衣
overall coefficient of heat transfer 总传热系数
overall crystallinity 总结晶度
overall dimension 〈oad, o. a. d.〉[机台]总尺寸;外形尺寸
overall flocking 满地植绒
overall rate of consumption 总消费率
overall reduction 总还原作用
overall roller 满地滚筒
overall scheme 总示意图;总方案,总规划
overall size 总尺寸;外廓大小
overall view 全面图,全貌图
overall wear characteristic 总磨损特性
overall yield 总得率
overbleach 过漂,漂白过度
overcapacity 超负荷
overcasting 包边缝纫(使毛边不致松散)
overchroming 铬[媒]处理过度(可造成色变、发脆等疵病)
overchurning [碱纤维素]过酯化,过黄酸化
overcoat 外套,大衣
overcoating 大衣料,外套料
over coating finish 表面涂层整理
overcompression 过度压缩(指机械预缩过程中造成的橘皮状起皱疵病)
overcondensed resin 过度缩聚树脂
overconditioning 过度给湿
overcure 过度焙烘;过[度]熟化;过硬化;过[度]硫化

over-current 过电流
over dense 超密(指经纱过密)
overdevelopment 显影过度
overdosage 过量
overdrying 烘燥过度,过干
overdry weight 绝对干重(烘干至不变重量)
over dyeing 套染,罩染(如直接印花后套染地色);过度染色
overdyeing stability 过度染色的稳定性(对色织物染料的要求)
overedger 包缝机
overedge stitch 包缝线迹
overedging machine 包缝机
over-elongation 过度伸长
over-end winding 轴向卷绕
over-end withdrawal 轴向退绕
over-esterification 过酯化
over-etched 腐蚀过度的
over exposure 过分曝晒,曝光过度
overfeed 超喂[装置]
over feed assembly 超喂送布装置
over feed delivery 超喂输送;超喂供应
over feed device 超喂装置
over feeder 超喂装置
over feed expanding machine 超喂扩幅机
overfeed fabric [经编]超喂织物
over feeding 超喂
over feed of colour paste 色浆超喂
overfeed pinning roller 超喂上针辊
overfeed pin stenter 超喂针板拉幅机
over feed pulley 超喂皮带盘;超喂滑轮
over feed rate 超喂率
overfeed roller 超喂辊
overfeed tenter 超喂拉幅机
overfire air 二次风,上部引入的燃烧空气

overflow 溢流,溢出,上溢
overflow beck 溢流染槽
overflow dyeing 溢流染色
overflow dyeing machine 溢流染色机
overflow gate 溢水隔板,溢水口
overflow indicator 溢出指示器
overflow jet dyeing machine 溢流喷射染色机
overflow outlet 满溢出口
overflow outlet gutter 溢流排液沟道
overflow position 溢出位
overflow rinsing 溢流冲洗
overflow shower and suction unit [丝光机的]冲吸碱液装置
overflow tap 溢流栓
overflow trough 溢流槽,排水槽
overflow valve 溢流阀(采用浮子装置,限制液体溢流)
overflow weir 溢流堰(丝光机冲碱装置)
overfoaming 溢泡[现象]
over garment 罩袍,大衣
over-gassed 烧毛过度
over-gassed yarn 焦纱(烧毛过度而灼焦的纱)
over hair [盖在表面的]硬毛
overhand (= overhanding) 平式缝接,锁缝
overhanging edge 松边
over-hardening 过度硬化
overhaul 大修;彻底检查
overhead driving 架空传动
overhead float dryer 架空式悬浮烘燥机(烘房架空,减少占地)
overhead paddle machine 顶桨式染色机,架式桨翼染色机
overhead process camera 上梁式摄影机(照相雕刻用)
overhead rail 吊轨
overhead system 架空系统
overhead tackle 上挂的滑车
overhead take-off mechanism 架空式落布装置,机顶落布装置
overhead tenter frame 架空式拉幅机(烘房架空,减少占地)
overhead track 架空导轨
overhead travelling crane 桥式起重机,桥式吊车
overheated steam in continuous dyeing 过热蒸汽连续染色
overheating 加热过度
overhung bearing 悬吊轴承
overlaid seam 搭缝
overlap 跳版(平网印花用语),重叠;印花布交搭
overlapping 重叠,交叠;搭接
overlapping seam 搭接缝
overlay 覆盖;[键盘用的]覆盖板;涂层;印刷垫版
over line 反分线(缩小雕刻制锌版时,将两色邻接部位的轮廓线稍稍重叠的一种处理法)
overload 过载,超负荷
overload time relay〈OTR〉过载限时继电器
overlooker 工长,管理员,检查员
over mature fibre 过成熟纤维
over-mercerization 过度丝光
over oxidation 过度氧化
over pressure 超压力,剩余压力
over pressure thin layer chromatography 超压薄层色谱
overprint [地色上]罩印;[印花上]套印;

[织纹上]盖印
over-printed resist 罩印防染
over printing 罩印,套印,叠印
over-printing displacement technique 盖印置换技术(地毯印花)
over proof〈OP,op,o.p.〉超过额定的
overpull 套衫
over-raising 起绒过度(起绒疵病)
over reduction 过度还原
over-reduction sensitive dyes 过还原敏感的染料
overrun 超限运行;越程;超越
over sack 短大衣
overscouring 精练过度
over-seaming 包缝
over-sensitive dyes 过灵敏染料
over-setting 过度定形(因定形温度过高或时间过长而造成的疵点)
overshoot(=overshot) 跳花(织疵);浮纬花纹;溢出
oversize〈OS,os〉尺寸过大;特大型;加大尺寸(指一种大于标准的修理尺寸)
overspill 溢出物
overstrain 张力过大,超应变
over stretch 过度伸长,拉伸过度
overtuft 超量簇绒
over view 综述,总的看法
overvoltage〈OV,ov,o.v.〉过[电]压
over voltage protection 过电压保护
overvoltage relay 过电压继电器
overweight 超重
over-wraping 搭接;外包装
O/W(oil in water emulsion) 水包油乳化液,油水相乳化液
OWB(on weight of bath) 按浴重[计算]
owf(on weight of fabric) 按织物重量[计算]
OWF(optimum working frequency) 最佳工作频率
OWG(on weight of goods) 按物重[计算]
own brand 自主品牌
own equipment manufacture(OEM) 自主设备制造
OWR(on weight of resin) [催化剂]按树脂重[计算]
ows(on weight of solution) 按液体重量[计算],按溶液重量[计算]
OX.(oxidizer) 氧化剂
oxalic acid 草酸
oxamide 草酰胺,乙二酰二胺(稳定剂)
Oxamine colours 奥萨明染料(直接染料,商名,德国制)
oxanthrone(=oxanthrol) 蒽酚酮
oxazine dyes 噁嗪[型]染料
oxazoline surfactant 噁唑啉表面活性剂
oxblood [red] 牛血红(棕红色)
oxford 牛津[衬衫]布;深灰混纺针织纱;灰色合股线毛呢
oxford chambray 牛津切姆布雷织物(平纹,色经白纬)
oxford cloth 牛津布
Oxford University〈O.U.〉牛津大学
oxidant 氧化剂
oxidase 氧化酶
oxidation base 氧化[发色]基
oxidation bleaching 氧化漂白
oxidation channel 氧化沟《环保》
oxidation cleavage 氧化分裂,氧化裂解
oxidation colourising 氧化显色
oxidation colours 氧化染料(如苯胺黑)
oxidation discharge [printing] 氧化拔染[印花]

oxidation ditch 氧化沟《环保》
oxidation dyeing method 氧化染色法
oxidation dyes 氧化染料
oxidation number 氧化值
oxidation pond 氧化塘《环保》
oxidation pond process 氧化塘法，自然曝气法《环保》
oxidation pretreatment 氧化预处理
oxidation printing 氧化印花
oxidation-reduction catalyst 氧化还原催化剂
oxidation-reduction indicator 氧化还原指示剂
oxidation-reduction potential 氧化还原电位
oxidation-reduction reaction 氧化还原反应
oxidation-reduction titration 氧化还原滴定
oxidation-resin process 氧化树脂[羊毛防缩]工艺(泛称)
oxidation rinsing 氧化冲洗
oxidation stability 氧化稳定性
oxidation strip 氧化剥色
oxidation susceptibility 氧化敏感性
oxidation treatment 氧化处理
oxidative coupling 氧化偶合
oxidative decolourization 氧化脱色《环保》
oxidative degradation 氧化降解[作用]
oxidative desizing 氧化退浆
oxidative desizing agent 氧化退浆剂
oxidative fission 氧化分裂
oxidative polymerization 氧化聚合[作用]
oxidative self-purification 氧化自净[作用]
oxidative thermostability 热氧化稳定性
oxide 氧化物

oxides of sulphur 硫氧化物
oxidimetric analysis 氧化滴定分析
oxidimetry 氧化滴定[法]；氧化测定[法]
oxidized cellulose 氧化纤维素
oxidized colour 氧化[显色的]染料
oxidized oil stain 氧化油污斑
oxidized starch 氧化淀粉
oxidized wool 氧化羊毛
oxidizer 〈OX.〉氧化剂
oxidizing ageing 氧化蒸化
oxidizing agent 氧化剂
oxidizing anion exchanger 氧化阴离子交换器；氧化阴离子交换剂
oxidizing compartment [染色]氧化槽
oxidizing environment 氧化环境
oxidizing flame 氧化焰
oxidizing material 氧化物质
oxidizing process 氧化过程
oxidizing-reducing reagent 氧化—还原试剂
oxido-absorption column 氧化吸收塔
oxido-reductase 氧化还原酶
oxido-reduction 氧化还原[作用]
oximation 肟化[作用]
oxime 肟
oxirane dyes 环氧乙烷[型]染料(一种活性染料)
oxirane value 环氧乙烷值
oxo-acid 氧络酸，含氧酸
oxonium ion 水合氢离子，水合质子
oxo-process 羰基合成过程
oxo-synthesis 氧化合成
ox-red indicator 氧化还原指示剂
oxy-acetylene cutting 氧乙炔切割
oxy-acetylene welding 氧乙炔焊接
oxy-acid 含氧酸

oxy-benzoic acid〈OBA〉 羟基苯[甲]酸,羟基安息香酸

p-oxybenzoyl copolyester fibre 对氧基苯甲酰共聚酯纤维(以对羟基苯甲酸、对苯二甲酸二苯酯和对苯二酚等为基本原料研制的耐高温纤维)

oxycellulose 氧化纤维素(漂白疵)

oxycellulose ester 氧化纤维素酯

oxydant [双组分推进剂中的]含氧成分,氧化剂

oxydase 氧化酶

oxygen acid 含氧酸

oxygenated bleaching compound 氧化型漂白剂

oxygenation 充氧作用

oxygenation capacity 充氧能力

oxygenation coefficient 充氧系数

oxygen bleaching 氧漂白

oxygen bridge 氧桥

oxygen carrier 载氧体,导氧剂

oxygen consumption〈OC〉 耗氧量,氧消耗量

oxygen cycle 氧循环

oxygen deficit 缺氧值,亏氧值(溶氧饱和量与实际溶氧量之差值)

oxygen depletion 缺氧,亏氧

oxygen index〈OI,O.I.〉 [需]氧指数

oxygen index method 氧指数试验法,含氧指数法(阻燃性试验用)

oxygen sag curve [水流]溶氧下垂线《环保》

oxygen scavenger 氧清除剂,[锅炉供水]除氧添加剂

oxymeter 量氧计,血氧定量计

oxymonoazo dyes 羟基单偶氮染料

oxypropyl ether of starch 淀粉的氧化丙醚

oxy-soap 含氧皂

oyster grey 牡蛎灰[色](淡灰绿色)

oz(＝OZ.,oz.,ounce) 盎司(质量单位,约等于28.3克)

ozocerite(＝ozokerite) 地蜡

ozokerite yellow 地蜡黄

ozonation process 臭氧氧化法《环保》

ozone 臭氧

ozone aging 臭氧老化

ozone bleaching 臭氧漂白

ozone crack 臭氧裂解

ozone degradation 臭氧降解(作用)

ozone exposure 臭氧泄漏

ozone fading 臭氧褪色

ozone generator 臭氧发生器

ozone oxidation 臭氧氧化(用于污水处理)

ozone resistance 抗臭氧性

ozone treatment 臭氧处理

ozone value 臭氧值

ozonides 臭氧化物(臭氧与不饱和有机物质的不稳定化合物)

ozonization 臭氧化[作用]

ozonized oxygen 臭氧

ozonizer 臭氧发生器

ozonolysis 臭氧分解[作用]

ozonolytic aging 臭氧[分解]老化

ozonometer 臭氧计

ozs(＝ozs.,ounces) 盎司的复数

P

p. (para) 对[位]

P(=P., p, p., poise) 泊(黏度单位);平衡杆

P(=P., p, p., press., pressure) 压力;压强

PA(=P. A., partially acetylated) 部分乙酰化的

PA(=P. A., polyamide) 聚酰胺

Pa(Pascal) 帕[斯卡](压强、应力单位)

PA(pattern analysis) 图案分析,花样分析

pa(per annual) 每年

PAA(polyacrylic acid) 聚丙烯酸

PAA treatment(peracetic acid treatment) 过醋酸处理(棉织物处理后能提高染料吸收与牢度)

pace maker 起搏器[医];步速器

pachometer(=pachymeter) 测厚计

pack 包;组装;组件

packability 填充性

package〈pkg〉 卷装;包装;数据包《计算机》

package ability 可装入性

package build 卷装成形

package carrying belt 件货输送带

packaged boiler 组装锅炉

package density 卷装密度;包装密度

package-drying machine 筒子纱干燥机;卷装[纱]干燥机

packaged unit 小型装置,可移动装置

package dyeing 卷装[纱]染色;筒子纱染色;轴经染色

package dyeing machine 筒子纱染色机

package-forming machine 筒子整形机

package hardness [纱线的]卷装硬度

package holder 筒子座,筒子锭座,卷装座

package scouring 卷装[纱]精练,筒子纱精练

package size 卷装大小

packaging bag 包装袋

packaging equipment 包装设备

pack cloth 打包[粗麻]布

pack duck 打包帆布(黄麻制)

pack dyed yarn 筒子染色纱

pack dyeing 筒装染色

packed column 填充塔,填充柱

packed dryer 充填式干燥机

packer 打包工;打包机

packing 打包,包装,堆砌;存储《计算机》

packing bolt 填密螺栓

packing cloth 打包[粗麻]布

packing damage 包装损伤

packing density 堆砌密度,排列密度;存储密度,记录密度(指信息)《计算机》

packing design 包装设计,包装图案

packing dimension 包装尺寸

packing factor 充填因子

packing felt 毡垫,毡衬

packing inspection 包装检验

packing joint 密封连接

packing list 装箱单

packing machine 打包机,包装机

packing material 包装材料
packing press 打包机,包装机
packing ring 填塞环垫圈,填密环,胀圈
packing room 打包间
packing scale 包装秤
packing sheet 包装布
packing sleeve 衬垫套子
packing specification 包装规格
packing textiles 包装用纺织品
packing tower 填充塔
packing twine 打包绳;包装麻线
pack system [丝绸]成匹[挂练]系统
pack thread 打包绳;捆扎线
paco 羊驼,羊驼毛,阿尔帕卡织物;半毛破布;再生毛
pad 浸轧,轧染;衬垫
pad-bake dyeing [浸]轧—[焙烘]染[色]法(即热溶染色法)
pad-batch cold dyeing 轧卷冷堆染色法(包括浸轧活性染料与碱剂、打卷、搁置及皂洗)
pad-batch technique 浸轧—堆放工艺
pad bleaching 轧漂
padded back lining 黑背[印花]里子布(正面印花,背面印黑色的斜纹布)
padder 浸轧机,轧染机,打底机,轧车
pad-develop 浸轧显色
padding 浸轧,轧染,打底;衬里织物;衬垫;填充料;(对记录信息)加[字符]《计算机》
padding and hot flue drying range 热风打底机
padding bath 浸轧浴,轧染液,打底液
padding binder 轧染用黏合剂
padding cloth 衬布,垫布;衬垫织物
padding liquor 轧染液

padding mangle 浸轧机,轧染机;打底机,轧车
padding roller 轧辊
paddings 西装麻衬布(黄麻或大麻制)
padding solution 浸轧溶液
padding thread 加固线,衬垫线,填充线
paddle 桨;(桨状)搅拌器
paddle agitator 桨式搅拌器
paddle dyeing machine(=paddle dyer) 桨叶式染色机
paddle stirrer 桨式搅拌器
paddle wheel 桨轮;桨轮式染色机
paddock [coating] 帕多克防水细呢(英国制,一般为棕色)
pad-dry-bake〈PDB〉(=pad-dry-cure〈PDC〉) 浸轧—烘干—焙烘[法]
pad-dry-cure technique 轧—烘—焙工艺
pad-dry dyeing equipment 轧烘染色设备
pad-dry technique 浸轧—烘干工艺
pad dyeing 轧染
pad dyeing machine(=pad dyer) 轧染机
pad finishing 浸轧加工法
pad-ink 打印色,打印油墨
pad-jig dyeing 浸轧卷染法(指还原染料悬浮体轧染,用卷染机还原的小批量生产法)
pad lock 挂锁
pad-ox dyeing process [硫化染料]轧染氧化工艺(商名,瑞士科莱恩)
pad pan 浸轧槽
pad-roll 轧卷染色(用于圆筒形针织物)
pad-roll bleaching range 轧卷式漂白联合机
pad roller 轧辊
pad-roll [dyeing] method 轧卷[堆置]染色法;浸轧—卷蒸法

pad-roll system dyeing machine 轧卷染色设备

pad-roll system scouring and bleaching range 轧卷式单幅练漂联合机（织物间歇式汽蒸练漂设备）

pad stamping machine 打印机

pad-steam dyeing 轧蒸染色工艺

pad-steam process 浸轧汽蒸染色法,轧蒸法

pad-store technique 浸轧堆积回苏工艺（毛染整用）

pad-thermofix dyeing 浸轧焙固染色法

pad thermosol dyeing 浸轧热溶染色

pad thermosol process 浸压热熔法

pad trough 浸轧槽

page addressing 页面寻址《计算机》

page boundary 页界《计算机》

page composer 页面编排器《计算机》

pagoda [blue] 塔蓝（暗绿光蓝色）,蟹青[色]

PA handle 聚酰胺纤维织物的手感

pahom (=pahone) 披巾（泰国男女作上衣用,色彩鲜艳）

pahpoon 闪光棉布

paillette 珠片（缝在织物上作装饰用的闪光小圆片）

pailette de soie 珠片绸,亮晶绸（绸面上缀以珠片等亮晶物质）

pailette satin 闪光缎

paint 涂料；油漆；颜料；[涂]漆；刷[上]涂料

painted blue 靛蓝染色（印度土法）

painted design 手描花样；花布图案

painted fabric 手描花布

PA interlining 聚酰胺（黏合）衬

paint film 涂料薄膜

painting 涂蜡《雕刻》

paint thinner 涂料稀释剂

paired-bond dissociation 双键离解[作用]

paired comparison 成双对比

paired-comparison technique ［一对之间］对比优选法

paired comparison test 成对比较试验

paisley 佩斯利涡旋纹花呢（苏格兰制）

paj 洋纺（全丝平纹薄织物）

pajama checks 格子睡衣布

pajamas 睡衣；[印度人等穿的]宽大服装

pajama stripes 条子睡衣布

pajunette 妇女睡衣

pakamas 帕卡玛围巾布（英国制）

pal 帕耳（固体上振动强度的无量纲单位）

palace 派力斯织物

palace brocade 花纺,提花全丝派力斯

palace crepe 派力斯绉

palace plain 全丝派力斯

Palagal SF 派莱格 SF（分散染料高温染色的分散匀染剂,有移染作用,商名,德国巴斯夫）

Palanil Carrier A 派拉尼尔载体 A（芳香醚,商名,德国巴斯夫）

Palanil CC dyes 派拉尼尔 CC 染料（分散染料,偶氮及杂环结构,属环保型,不含有害的有机胺物质,适用于涤纶及混纺织物染色,商名,德国巴斯夫）

Palanil dyes 派拉尼尔染料（专染聚酯纤维的分散染料,商名,德国巴斯夫）

Palanil Luminous dyes 派拉尼尔荧光染料（分散染料,杂环结构,环保型,不含有害有机胺物质,色泽鲜艳,部分具荧光性,适用于生产鲜艳流行色及易识别的用途,商名,德国巴斯夫）

Palanthrene dyes 派伦士林染料(还原染料,商名,德国制)

Palatine Fast/Palatine dyes 派拉丁坚牢/派拉丁染料(1∶1金属络合染料,羊毛专用,优良的匀染性和毛尖色差覆盖能力,特别适用于匹染,商名,德国巴斯夫)

pale 苍白的,灰白的

pale pinkish grey 藕色

pale shade 浅色,淡色

pallas fur 帕拉斯毛皮绒(法国制)

pallet 转动式筛网印花的台板(用于针织衣片印花)

Palm Beach cloth (= palm beige) 胖哗叽(商名)

Palmer calender 帕尔麦轧光机

palmering 柔软加光整理

Palmer machine 帕尔麦[毡滚筒]烘干机

Palmer stretching machine 帕尔麦[圆形针链]拉幅机

Palmer tenter 帕尔麦帘式烘燥拉幅机

palmette design [地毯]云彩图案

palmetto 矮棕榈绿[色](暗黄绿色)

palmitic acid 棕榈酸,十六酸,软脂酸

palm [kernel] oil 棕榈[核]油

pam [hair] 羊绒,山羊绒,开司米(克什米尔用语)

PAM(polyacrylamide) 聚丙烯酰胺

PAM(pulse-amplitude modulation) 脉冲幅度调制,脉幅调制

pan 锅;盘;槽

PAN (polyacrylonitrile) 聚丙烯腈

panache 彩色花纹效应

panama 巴拿马薄呢

Panama canvas 巴拿马帆布

Panama suiting 巴拿马西服呢

pan bobbin stand 盘式筒子座

pancake coil 盘香管

panchromatic 全色的(摄影用语)

panchromatic film 全色胶片

panchromatic plate 全色干板

pancrea diastase 胰淀粉酶制剂(退浆剂)

pancreatic amylase 胰淀粉酶

pancreatin 胰酶

Panduran dyes 潘杜兰染料(媒染染料,商名,瑞士制)

panel 女服嵌料、嵌条或饰条;板;翼片;仪表盘;配电板;护墙板

panelboard 配电板;控制屏

panel design 嵌花花纹

paneling 阔纵条花纹

Panex 帕内克斯(碳纤维,商名,美国制)

panne 平绒(棉底绢丝绒头,整理时用高压辊筒将绒头压平,使有光泽)

panne effect 平绒效果

panne satin 高级厚重平缎

panne velvet 平绒

pannonia leather 潘诺尼亚漆布(粗棉或黄麻纱织,表面涂漆)

panorama function 全景化功能

pan scale 盘秤

pansy 三色堇紫(深紫色)

pantagraph 缩放仪(绘图用);花筒缩小雕刻机

pantie hose 连裤袜,连袜裤;紧身裤

panting 裤料

pantistocking 连袜裤,连袜袜

pantograph 缩放仪(绘图用);花筒缩小雕刻机

pantograph engraving 缩小机雕刻[花筒]法

pantograph machine 花筒缩小雕刻机

pantomorphism ［结晶］全对称［现象］，全对称性
pantone 国际标准色卡
pantopat 缩放花纹试样机
pants 裤子,衬裤,短裤
pantsuit 裤套装(上衣与裤子配套的妇女服装)
panty hose 连裤袜;连袜裤,紧身裤
panung 潘衣布(流行于泰、越、柬诸国作衣料的色格或色条棉布)
paon velvet 帕昂天鹅绒(制女帽或作装饰用)
papain 木瓜酶,木瓜蛋白酶
paper bowl 纸粕轧辊
paper cambric 狭幅细纺;狭幅细薄布
paper chromatographic method 纸上色层分析法
paper chromatography ⟨PC⟩ 纸上色层分析法,纸色谱法(常用于染料分析)
paper cloth 纸布;涂纸浆棉布或麻布
paper dress 纸制衣服
paper drying finish ［高级丝织物］卷纸烘燥整理
paperiness 纸样感(织物风格)
papering 电压纸板(毛织物整理用)
paper lamination technique 叠纸法(叠层黏合)
paper machine cloth 造纸用布类材料
paper makers' felt (= **paper making felt**) 造纸毛毯
paper muslin 轧光硬挺细布
paper pattern 型纸,衣服纸样
paper press 电压机,纸板压呢机
paper pressing 纸板压呢
paper programmed tape 程序控制纸带
paper roll 纸粕轧辊
paper sheet 纸片
paper stencil ［刻花］纸版;型纸
paper stencil printing 纸版印花
paper substrate 转印纸坯
paper taffeta 硬挺塔夫绸(整理后挺薄如纸)
papery-finish 仿纸光滑整理
papping 上浆
Pappreserve printing 充蜡防热印花
paprica 辣椒红［色］(橙红色)
par. (**paragraph**) 节,段;短评
par. (**parallel**) 平行;并行;并联;类似;相同的;平行的
para ⟨p.⟩ 对［位］
parabola 抛物线
parabolic 抛物线的
parabolic distribution chamber 抛物线形分配室(用于泡沫输送)
parachute 降落伞
parachute fabric 降落伞绸,降落伞织物
paracortex ［羊毛］仲皮质
paracortex of wool 羊毛的仲皮质区
paracrystalline region 类结晶区
para-dyes 偶合染料
paraffin 石蜡
paraffin duck 石蜡防水帆布
paraffin wax 石蜡
paraformaldehyde 多聚甲醛
Paraform T 派拉丰 T(含铝盐石蜡乳化液,防雨剂,商名,德国制)
Paragium AQ 派拉吉尤姆 AQ(阳离子性,锆型防水剂,商名,日本制)
Paragium RC 派拉吉尤姆 RC(十八烷基亚乙基脲,非离子型,防水剂,柔软剂,商名,日本制)
Paragium SP 派拉吉尤姆 SP(N-羟甲基

硬脂酰胺衍生物,非离子型,可增加织物滑爽手感,有防水性,商名,日本制)
paragraph〈par.〉 节,段;短评
Paraguard O-300 派拉加德 O-300(碳氟化合物,防水防油剂,弱阳离子性,商名,日本制)
para-isomer 对位异构体
para-linkage 对位键[合]
parallax 视差;倾斜线
parallel〈par.〉 平行;并行;并联;类似;相同的;平行的
parallel access 并行存取《计算机》
parallel-by-character 字符并行《计算机》
parallel circuit 并行电路,并联电路
parallel connection 并联[结法]
parallel feedback 并联反馈《电》
parallel flow 顺流,平行流[动]
parallel laid 平行铺叠
parallel port 并行接口
parallelprobe 平行试样;平行探索
parallel tests 平行试验,替换检定
paramagnetic nickel chelate 顺磁性镍螯合物(光稳定剂)
paramagnetic resonance absorption 顺磁共振吸收
paramagnetism 顺磁性
paramatta 毛葛;棉毛呢
parameter 参数,参变数,参量
parameterization 参数化
paramorph 同质异晶体
paramount 最高的,至关重要的,头等的
paranitraniline red 对硝基苯胺红(俗称毛巾红、巴拉红)
para-orientation 对位取代;对位取向
para-oriented ring 对位取代环;对位取向环

paraphase 倒相;分相放大器,倒相放大器
para-red 巴拉红,毛巾红;红色颜料
para-red developing range 巴拉红显色机,纳夫妥显色机
para-xylene〈PX,px〉 对二甲苯
para-xylylene acid〈PXA〉 对苯二甲酸
para-xylylene diamine〈PXD〉 对苯二甲胺
parchment finish 酸丝光整理
parchmentising 酸丝光,羊皮化处理(细特棉织物经冷浓硫酸液处理)
parements 装饰织锦
parentheses-free notation 无括号标序法,无括号标序记号《计算机》
parenthesis 插入语,圆括号
par example〈p. ex.〉 举例;例如
Parex system 帕勒克斯高速烧毛法
pari 生丝原重(指未脱胶前重量)
paring 刮屑,碎屑;削皮
paring-off 修边
Paris blue 巴黎蓝[色](深蓝色)
Paris green 巴黎绿[色](黄光绿色);砂绿(乙酰亚砷酸铜)
parisian silk 巴黎丝(铜氨纤维旧称)
parisienne 巴黎黑色薄花呢(法国产,美利奴毛制);小花纹丝织物(法国名称);丝毛交织物
Paris yellow 巴黎黄[色](黄色);铬黄
pariweighting 生丝回重增量(生丝精练时减除的重量,经上浆染色后回复到生丝原重)
parka 风雪大衣,派克大衣(带风帽的厚大衣)
parrot green 鹦鹉绿[色](黄绿色)
parsley green 欧芹绿[色](橄榄绿色)
part〈pt.〉 部分;要素;零件

part consistency 零部件质量均齐性
parti 多色
partial acetolysis 部分醋[酸水]解
partial boiling bath 半练浴《丝》
partial combustion 部分燃烧
partial correlation 偏相关
partial degumming 部分精练
partial desiccation 半干燥
partial error 局部误差
partial fashion 部分成形
partial journal bearing 半围轴承
partially acetylated〈PA,P. A.〉部分乙酰化的
partially acetylated cotton 部分乙酰化棉,PA棉
partially aromatic polyamide fibre 部分芳基化聚酰胺纤维
partially drawn yarn 部分拉伸丝,预拉伸丝,低拉伸丝
partially etherified cotton 局部醚化棉,低醚化棉
partially oriented yarn 部分取向丝,预取向丝,低取向丝
partially substituted cellulose acetate fibre 低取代[度]醋酯纤维
partial maintenance 中修
partial miscibility 部分互溶
partial pattern cancellation 局部花纹消除
partial pressure 分压力(混合气体中的一种组分如果单独存在于容器中会产生的压力)
partial pressure suit 部分高压服,高空代偿服
partial saponification 部分皂解
partial saponification number 部分皂化值
partial scouring 部分精练

partial shrink 部分收缩
partial thixotropy 局部触变性,局部摇溶[现象]
particle 粒子,粉粒,颗粒;质点;极小量
particle charge 粒子电荷
particle filtration 颗粒过滤
particle matter [悬浮在空气或其他气体中的]微粒物质
particle measuring technology 粒子测量技术
particle size 粒度,颗粒大小
particle size distribution 粒度分布
parti-coloured 杂色的;多样色的
particulate 粒子,微粒;碎粒状
particulate activation 微粒激活作用(控制泳移)
particulate reinforced materials 颗粒增强复合材料
partition 分段;划分
partition chromatography 分配色谱法
partition coefficient 分配系数
partition panel fabric 间隔[用]织物
partons 部分子（带电小粒子,比质子更小）
partridge cloth（＝partridge cord）斑点灯芯绒
parts〈pts〉份
parts by weight〈pbw〉重量份数
part sectioned view 局部剖面图
parts list 配件目录
parts per billion〈PPB,ppb〉几亿分之几
parts per hundred million〈pphm〉亿分之几
parts per million〈ppm,P. P. M.,p. p. m.〉百万分之几
parts per thousand〈ppt〉千分之几

part width 部分幅度
partwool blanket 混纺毛毯
party clothes 社交服,交际服
party-coloured 杂色的;多样色的
party line〈Pl〉合用线
PAS(paste application system) 印浆应用系统
Pascal〈Pa〉帕[斯卡](压强、应力单位)
Pascalian body 帕斯卡体,非黏流体,不黏[滞]流体
pashim(＝pashm,pashmina) 羊绒,山羊绒,开司米(克什米尔用语)
pashmina shawl 羊绒披巾
pashmina tweed 羊绒粗呢(克什米尔制)
pass 扫描;通过,传送
pass.(passage) 道(卷染机上的布卷自一端移转至另一端);路径,通道,流道,走廊
pass.(passive) 无源的;钝态的;被动的
passage〈pass.〉道(卷染机上的布卷自一端移转至另一端);路径,通道,流道,走廊
passagecounter 计道器(卷染机用的道数计算器)
passage head 道数
passage number(＝passage of fabric) 道数(卷染时布卷每从一端移转至另一端为一道)
passameter 外径指示规
pass-by 旁路,旁通
passenger side bag 乘客座侧安全气囊
passer 检验工
pass-fail 合格—不合格
pass-fail criterion 合格—不合格标准
pass-fail judgment 合格—不合格判定
passimeter 内径指示规

passivation 钝化[作用]
passive〈pass.〉无源的;钝态的;被动的
passive component 辅助成分
passive dressing 辅助修饰(起绒辊)
passive drive 辅助传动
pass word 密码《计算机》
pass word protection 密码保护《计算机》
paste [浆]糊;浆状;增稠剂;玻璃状物质
paste application system〈PAS〉印浆应用系统
paste bucket monitor 浆桶监控器《印》
paste coating 浆涂布法,浆料涂层
paste dosing system 色浆计量系统
paste dot coating 浆点涂层
paste dye 浆状染料
paste grain [皮手套]丰满纹理整理
paste hose 浆管(印花机配用的自动给浆软管)
pastel 浅淡的;菘蓝染料
pastel dyeing 浅淡[色]染色
pastel peach 浅桃色
pastel shade 浅色,浅淡优美的色彩
paste overflowing 溢浆
paste point coating 浆点涂层
paste pump system 印浆运送系统
paste recovery system 印浆回收系统
paste transport system〈PTS〉印浆输送系统
pastille 点子花纹
pasting 糊化;双层黏合工艺;轧光布布端黏合
pasting point of starch 淀粉的糊化温度
pat.(patent) 专利;专利权;专利件
pat.(＝patt.,pattern) 花纹组织,图案,花样,式样;样板;[服装]纸样
patalo 帕托利绸(有轧染的或手工木版

印边纹的花纹）

patassium bichromate 重铬酸钾

patch 补丁，补片

patching 修补

patch pattern 补丁图案

patch test 织物贴附性试验

patchy 有深色染斑的

patent〈pat.〉专利；专利权；专利件

patent-back carpet 背面涂胶地毯；特制底地毯

patent beaver 防水海狸呢

patent cloth 蜡布

patent cord 毛或棉经毛纬的长毛绒织物

patent flannel 防缩薄法兰绒

patent leather 漆皮

patent selvedge 中央布边

patent specification 专利说明书

patent velvet 灯芯绒（纬起绒，有凸条效应）

patera 插座；接线盒

paternoster dye store 念珠式染料仓库（转动式设施，按间隔取用各种染料）

paternoster storage system〈PSS〉念珠式储存系统

path 路径，轨迹，通路

patina green 铜锈绿（浅黄光绿色）

patio carpet 游廊地毯

pat-it-out 轻拍熄灭法（耐燃试验）

patola 派多拉绸（印度丝织物）

patole 帕托利绸（有扎染的或手工木版印边纹的花纹）

Patone system 潘通色卡系统

patrol 巡回

patrolling cycle 巡回周期

patrolling operation 巡回操作

pattern〈pat.，patt.〉花纹组织；图案，花样；式样；样板；［服装］纸样

pattern acquisition 花纹探测

pattern adjustment 图案搭配，图案调整

pattern analysis〈PA〉图案分析，花样分析

pattern area 花纹面积，花型范围

pattern arrangement 纹样排列

pattern bonding 花纹黏合法

pattern bowel 印花花纹辊

pattern card 印染样本

pattern coat 印花涂布，花型涂布

pattern cutting 花型剪裁

pattern depth 花型深度

pattern design 花纹设计

pattern development process 花纹显像过程

patterned carpet 花纹地毯

patterned doctor blade 花纹刮刀（用于地毯多流染色机）

pattern fit allowance 对花公差

pattern fitting 对花

pattern flocking 图案植绒

pattern information 花纹信息

patterning 形成花纹，组成图案

pattern lay-out 花纹排列；花纹设计

pattern limitation 花型范围

pattern line 样板划线

pattern making 花型制作

pattern manipulation 花纹变换，花纹控制

pattern misfit 对花不准

pattern overlapping 接头印，叠版印

pattern paper 花样，图案设计纸

pattern register device 印花对花装置

pattern repeat 花样循环，花型循环，图案循环

pattern scanner 花纹扫描器

pattern shade difference 花纹色泽深浅不

符

pattern stability test 模型稳定性试验(人造革模型热稳定性试验法)

pattern thread cropping machine 剪花机

pattern transposition 花纹位移

pattern walewise transposition 花纹纵移

pattern work 提花织物;色织织物

paulin 防雨帆布,篷帆布

Pauly reagent 保丽试剂

pause 中止,暂停;停机时间

pawl 爪,棘爪;掣子,撑头

payback time 偿还期,回收期(投资费用能得到偿还或回收所需要的时间)

pay-off 松卷装置;清算,支付

PBB(polybrominated biphenyl) 多溴联苯(禁用的阻燃剂)

PBDE(polybrominated diphenyl ether) 多溴化联苯基醚(禁用的溴系阻燃剂)

PB finish (permanent bright finish) 耐久性光泽整理

PBI(pile burning index) [长毛地毯的]绒面燃烧指数

PBI(polybenzimidazole) 聚苯并咪唑

PBO fibre(poly-p-phenylene benzobisthiazole fibre) 聚对苯亚甲苯并双噻唑纤维(新型高强度合成纤维)

PBT fibre (polybutylene terephthalate fibre) 聚对苯二甲酸丁二酯纤维(弹性涤纶),PBT 纤维

pbw (parts by weight) 重量份数

PC(paper chromatography) 纸上色层分析法,纸色谱法(常用于染料分析)

PC (=PC., pc, p. c., pct, p. ct., percent) 百分比

P/C(polyester/cotton) 涤棉混纺织物

PC (personal computer) 个人计算机

PC(photocathode) 光电阴极

PC (photo cell) 光电池;光电管

PC (photo conductor) 光导电体;光敏电阻

Pc(phthalocyanine) 酞菁染料

pc. (piece) 匹;接头;拼合;块,片,段

PC(polycarbonate) 聚碳酸酯

PC(pressure controller) 压力控制器

pc. (price) 定价,价格

PC(pulsating current) 脉冲电流

PCB (=p. c. b., printed circuit board) 印刷电路板

PCB (polychlorinated biphenyl) 多氯联苯(有毒禁用的防腐剂)

PC blend (polyester and cotton blend) 涤棉混纺织物

PCC (Pretema Colour Computer) 普雷特马配色计算机(瑞士制)

Pc-developing dyes 酞菁显色染料

PCI (program controlled interrupt) 程序控制中断

PCL(polycaprolactam) 聚己内酰胺

pcm (= p. c. m. , percentage of moisture) 含水[百分]率

PCM (phase change material) 相变材料

PCM(pulse code modulation) 脉冲编码调制

PCM (punch-card machine) 轧纹板机,踏花机;穿孔机

P controller P调节器,比例调节器

PCP(pentachlorophenol) 五氯苯酚(毒性大的防腐剂,国际上禁用)

PCP(primary control program) 主控制程序《计算机》

PCP(process control program) 过程控制程序《计算机》

Pc-reactive dyes 酞菁活性染料
pcs(=pcs., pieces) 块;件,片;匹
PCS(poly-carbamoyl sulfonates) 聚氨基甲酰磺酸盐(羊毛防缩剂)
Pc-surfur dyes 酞菁硫化染料
pct(=p. ct., PC., pc, p. c., percent) 百分比
PCT(plasma chemical treatment) 等离子体化学处理
PCT(pre-clarification tank) 预澄清槽
P. D.(=pd, pitch diameter) [螺纹的]中径,[齿轮的]节径
pd(polar distance) 极距
pd(potential difference) 位差,势差
pd(pulley drive) 滑轮传动
PDA(pump drive assembly) 泵传动装置
PDB(pad-dry-bake) 浸轧—烘干—焙烘[法]
PDC(pad-dry-cure) 浸轧—烘干—焙烘[法]
PDF(print document formulation) 打印文件模式《计算机》
PDM(product data management) 产品数据管理
PDM(pulsator [hank] dyeing machine) [绞纱]脉动染色机
P ds.(=p. ds., potential differences) 位差,势差
PDSCL(Poisonous and Deleterious Substances Control Law) 有毒和有害物质控制法(日本)
PE(permissible error) 允许误差
PE(photo-electric) 光电的
PE(polyethylene) 聚乙烯;聚乙烯[系]纤维
PE(potential energy) 势能,位能
PE(power equipment) 电源设备,动力设备
PE(pressure element) 压力元件
Pe(=pe, p. e., probable error) 概率误差
peach 浅桃色(浅红橙)
peach-face touch 桃皮面触感
peach red 桃红[色](黄光绯色)
peach skin 桃皮织物,桃皮绒
peach skin touch 桃皮织物手感
peach velvet 桃皮绒
peacock blue 孔雀蓝[色](暗绿光蓝)
peacock green 孔雀绿[色](黄光绿色)
pea green 青豆绿[色](黄绿色)
pea jacket 厚呢上装(海员服装)
peak 波峰;顶点;引入脉冲点
peak concentration 最高浓度
peak ratio 最大比例
peak temperature 峰值温度,最高温度
peak-to-valley ratio 峰谷比
peak value 最高值,最大值,峰值
peanut fibre 花生蛋白纤维
pearl 珍珠
pearlescent 珠光般的
pearlescent colour 珠光色
pearl essence 珠光粉;鱼鳞粉(俗称);珍珠粉
pearl fabric 双反面针织物
pearl grey 珍珠灰(淡灰灰色)
pearling-off 水珠滴流(防雨试验)
pearlization 珠光化;灰白化
pearl luster printing 珠光印花
pearl moss 角叉菜,鹿角菜(红藻类,用以制印花浆料)
pearl pigment 珠光颜料
pearl plant 紫草(植物染料)
pearl polymerization 成珠聚合,珠状聚合

pearl powder 珍珠粉
pearl printing 珠光[效应]印花
pease cloth 比斯呢(间隔型提花条格色织布)
peat fibre 泥炭纤维
peat mat 泥煤薄层(用于印染厂污水处理)
peau 起绒织物(用粗纺毛纱、再生纤维素纤维或蚕丝织制,常采用金刚砂起绒工艺加工)
peau d'ange 天使缎(商名,经无光平滑整理的醋纤绸)
peau de chamois 麂皮绉(小卵石绉纹柔软丝织物)
peau de crepe 光滑软绉(商名)
peau de cygne 天鹅皮绸
peau de gant 白丝花缎
peau de mouton 普德蒙顿仿羊皮女大衣呢
peau de poule 法国八枚哔叽(高级粗纺毛纱制,匹染)
peau de soie 双面横棱缎(法国制)
peau de souris 鼠皮绸(柔软丝织物)
peau de suede 格子花呢
PEB (poly-p-ethylene-oxybenzoate) 聚对苯甲酸乙氧[基醇]酯
pebble 绉纹,小卵石纹
pebble cheviot 绉面大衣呢
pebbled 满地绉纹,满地小卵石纹
pebble effect 泡泡效应
pebble mill 球磨机
PEC (= P. E. C., pec, photoelectric cell) 光电池;光电管;光电元件
pectase 果胶酶
pectate lyase 果胶酸裂解酶
pectates 果胶酸盐

pectic acid 果胶酸
pectic substance 果胶质,植物胶质
pectin 果胶质
pectinase 果胶酶
pectinesterase 果胶[甲]酯酶
pectin lyase 果胶裂解酶
pectisation 凝结,凝胶化[作用]
pectocellulose 果胶[质]纤维素
pectose 果胶糖,果蔬胶
pecul 担(中国重量单位,等于50千克)
pedal 踏板,脚盘
pedestal 轴架;支座,垫座
PEE fibre(polyester ether fibre) 聚醚酯纤维
peel 外壳;皮;剥离
peel adhesion [皮层]剥离黏着力
peel bond strength 剥离强度(剥离黏合织物所需的力)
peelimg test 剥离试验
peeling 剥离;去皮,剥皮
peeling angle 剥离角(印花布从印花导带脱开进入烘干机的角度)
peeling machine 剥离机
peeling reaction 剥离反应
peeling strength 剥离强度
peeling technique 剥离技术
peeling test 剥离试验
peel-off velocity 剥离速度
peephole 窥孔,视察窗,观测孔
PE fibre (polyester fibre) 聚酯[系]纤维
peg 木钉,栓钉;陪衬小花
PEG (polyethylene glycol) 聚乙二醇
peg coverage [每色花筒]印花面积
pegging [绒头织物]磨光工艺
peg printing 木栓滚筒印花(一种早期的凸纹印花法)

peg printing machine 凸纹滚筒印花机
peg rail 分布桩横条，分布栓，木钉导轨（用于绳状洗染机）
PEIW（population equivalent of industrial wastes）工业废水的人口当量
pekin 北京条纹（等宽的彩色直条纹）；北京宽条子绸（条子为等宽的缎纹或棱纹直条）
pekin stripes 北京条纹织物（等宽条纹）
pekin velour 棉丝交织绒头条子布（丝经棉纬）
pellet 丸，颗粒；小片，压片
pelletized dyes 丸粒[状]染料
Pellon 佩纶（天然和化学纤维不规则叠层后经化学及热处理黏合成的各种非织造织物的总称，商名，美国制）
pelote 毛线球（法国名称）
peluche 长毛绒（法国名称）
peluche argent 银线丝绒（法国制）
peluche double 双面长毛绒
peluche duvet 天鹅绒（丝经，棉纬，法国制）
peluche gaufrée 压花长毛绒，拷花长毛绒
peluche legère 轻薄长毛绒
peluche long poil 长毛丝绒（蚕丝经，绢丝纬，法国制）
peluche ombré 月华丝绒（同一颜色，色泽由深至浅）
peluche épinglée 毛圈长毛绒（法国名称）
pemosors 多层共聚物
pencil stripe 细点条子条纹；铅笔线条纹
pencil stripes 细点条子织物
pendant 吊锤；垂杆；下垂物
pendant drop method 悬滴法（测表面张力）
pendant group 侧基

pendant switch 拉线开关
pen drawing 钢笔素描
pendular arm 摇臂，摆臂，摇杆
pendulum 摆；张力辊
pendulum arm 摇臂，摆臂，摇杆
pendulum arm weighting 摇臂加压
pendulum balance 摆锤天平
pendulum compensator 摆式松紧架
pendulum impact system 摆锤式冲击方法（强力试验）
pendulum plaiter 摆动式落布架
pendulum roller 松紧辊，张力调节辊
pendulum spring 摆弹簧，悬挂弹簧
pendulum stock [缩绒机]摆锤
pendulum strength tester 摆锤式强力试验机
pendulum test 摆振试验
pendulum traverse motion 摆锤往复运动
pendulum type extracter 吊笼式脱水机
pendulum viscosimeter 摆锤式黏度计
pendulum weighting arm 加压摇臂
penetrability 贯穿性，穿透性；渗透性
penetrant 渗透剂，贯入料
penetrant test 渗透试验
penetrating agent 渗透剂
penetrating agent BX 渗透剂 BX
penetrating force [缝针对布的]穿透力
penetrating power 渗透能力
penetrating printing 渗透印花
penetrating property 渗透性
penetration 渗透；穿透率
penetration accelerator 渗透促进剂
penetration dyeing 透染
pen recorder 描笔式记录器
Pen Shan ssu silk 本山素绸（中国东北柞蚕丝织物）

pen steel washer 薄钢片垫圈
pent[a]- (构词成分)表示"五"
pentachlorophenol〈PCP〉 五氯苯酚(毒性大的防腐剂,国际上禁用)
pentad 五价物;五色相配色
pentagraph 缩放仪;比例绘图器
pentagraph engraving 缩小雕刻
pentagraph machine 雕刻花筒用缩放机
pentalin 五氯乙烷(油溶剂)
pentalobal cross section 五叶形[横]截面
Penta screen 潘塔网(一种特细圆网,可达125目、185目,甚至255目,可印制精细的花纹,荷兰制)
Penta special screen 潘塔特殊网(高目数特殊圆网)
Penton 彭通(聚醚纤维,商名,美国制)
pentosan 多缩戊糖,戊聚糖
pentose 戊糖
penumbra 半影
PEO (polyethylene oxide) 聚氧化乙烯,聚环氧乙烷
peony 牡丹红[色](暗红色)
peony brocade 牡丹绸
pepper and salt effect 芝麻呢;芝麻点纹
peppermint oil 薄荷油
pepper trash 大粒杂屑
peptidase 肽酶
peptide 肽[类],缩氨酸
peptide bond 肽键,缩氨酸键
peptide chain 多肽链
peptideses 溶胶化酶
peptisation (=peptization) 胶溶[作用];消化[作用]
peptizer (=peptizer) 胶溶剂
peptizing 胶溶[作用],胶束吸附[作用]
peptizing agent 胶溶剂

peptone 胨,蛋白胨
per- (构词成分)表示"通过""完全""十分""高""过""全"
peracetic acid 过醋酸,过乙酸
peracetic acid treatment〈PAA treatment〉过醋酸处理(棉织物处理后能提高染料吸收与牢度)

per annum〈per an,per. ann.〉 每年
Perapret D 比力白D(阴离子型丙烯酸聚合物的乳液,洗可穿整理中起防污作用,商名,德国巴斯夫)
Perble range 潘勃尔翻板式练漂联合机(日本制)
perborate 过硼酸盐
perborate bleaching 过硼酸[钠]漂白
perborate of soda 过硼酸钠
PERC (perchloroethylene) 全氯乙烯,四氯乙烯(除油渍剂、干洗剂)
percale 高级密织薄纱(匹染或印花)
percale [bed] sheeting 密织床单布
percaline 珀克林(丝光高级细薄纱;经上光柔软整理的棉衬里布)
percent〈PC,PC.,pc,p.c.,pct,p.ct.〉百分比
percentage balance 百分率刻度天平,比例秤
percentage by weight 重量百分率
percentage content 含量百分率
percentage cover 覆盖百分比
percentage moisture content 含水率
percentage of A-class goods 正品率
percentage of boil off 练减率《丝》
percentage of combustible ash and refuse 煤渣含碳率
percentage of crimp elasticity 卷曲弹性回

复率
percentage of deviation 偏差率
percentage of elongation 伸长率
percentage of exhaustion 上染百分率，吸色率，上色率
percentage of free water 附着水分率
percentage of moisture〈pcm，p. c. m.〉含水[百分]率
percentage of permissible deviation 容许偏差百分率
percentage of size 上浆率
percentage of trash content 含杂率
percentage of variation 不匀率
percentage of variation of strength 强力不匀率
percentage of water hold 含水率
percentage of wear and tear 折旧率
percentage open area〈POA〉开孔面积百分率
percentage point 1%差值(如5%与7%之差，可称为相差两个1%差值)
percentage size pickup 上浆率
percentage strain 应变率
percent defects 废品率
percent error 误差百分率
percent test 挑选试验
percent transmittance 透光百分率
percent weight in volume〈w/v〉单位容积中重量百分比
perceptibility 感知度
perch 验布架；验呢台
percher 检验工
perching 验布
perching machine 验布机
perching of grey good 坯布检查
perchlorate 过氯酸盐

perchlorinated polyvinyl chloride 聚过氯乙烯，氯化聚氯乙烯
perchlorizing 过氯化
perchloroethylene〈PERC〉全氯乙烯，四氯乙烯(除油渍剂、干洗剂)
percolation 渗滤；穿流[法]
percolator 渗滤器
Peregal O 平平加O(多羟乙基化脂肪醇，匀染剂，乳化剂，剥色剂，可生物降解，商名)
Peregal P 平平加P(低聚体缩聚物，非表面活性，还原染料匀染剂，商名，德国制)
Perepret 比力白系列(洗可穿免烫整理添加剂，提高织物缝纫性，减少自交联反应引起的纤维强力损伤，并有手感柔软作用，商名，德国巴斯夫)
perfect combustion 完全燃烧
perfect fluid 理想流体
perfectly oriented structure 完全取向[分子]结构
perfect picture imaging system 完美成像系统
perfect plasticity 理想塑性
perfect reflecting diffuser 完全漫反射面
perfect register 对花准确
perfect transmitting diffuser 完全透射，漫反射面
perfluoro alkyl compound 全氟烷基化合物
perfluoro carbon 全氟化碳
perfluoro-kerosene〈PFK〉全氟煤油
perforated 穿孔的，打眼的
perforated basket [离心脱水机的]多孔转笼
perforated beam 多孔经轴

perforated belt 多孔带
perforated belt steamer 多孔带式汽蒸箱
perforated cage 筛网圆筒
perforated cage filter 多孔笼式过滤器
perforated cage printer 筛网印花机
perforated conveyer belt steamer 多孔履带式汽蒸箱
perforated cylinder 多孔滚筒;蒸呢滚筒
perforated drum dryer 多孔转鼓烘燥机
perforated dye spool 多孔染色筒管
perforated fabric 网眼织物;仿纱罗织物
perforated former [筒子染色用]多孔筒管
perforated metal sheet 多孔金属板
perforated pipe 多孔管,筛眼管
perforated pipe holder 多孔管支架
perforated plate 多孔板
perforated roller 多孔卷轴;网眼辊
perforated screen 多孔筛网
perforated steaming cylinder 多孔汽蒸滚筒
perforated tube 多孔芯管(筒子纱染色用)
perform 预制坯件
performance 性能;表现;运行特性;运转效能
performance capability 工作能力
performance characteristics 表现特性,效能特性,工作特性
performance chart 性能图
performance curve 性能曲线
performance density 表观密度
performance evaluation 性能评价
performance figure 性能数字,质量指标
performance index〈PI〉性能指数
performance in wear 穿着性能

performance number 品值,品度值
performance property 性能特性
performance test 性能试验,应用试验,运转试验
perfumed fabrics 芳香织物
perfumed finishing 香味整理
perfumed printing 香味印花
per hour〈ph.〉每小时
per hundred parts resin〈p.h.r.〉每一百份树脂中的份数
perhydrol 浓双氧水(含30%过氧化氢)
Perigen 派力琴(广谱防虫蛀剂,拟除虫菊酯,毒性低,易生物降解,商名,英国制)
Periloc machine 毛纱缩绒机(商名,国际羊毛局开发)
perimeter 周长,圆周
perimeter wall 围护墙,周界墙
period 周期;时期
periodate oxidation 高碘酸盐氧化[作用]
periodate oxidized cellulose 高碘酸盐氧化纤维素
periodical 期刊,杂志
periodic cerification 周期性鉴定
periodic copolymer 嵌段共聚物
periodic irregularity 周期不匀率
periodicity 周期性
periodic proof test 定期的试用试验
periodic sample 周期性取样
periods per second〈pps,p/s〉周/秒[钟];赫[兹]
peripheral equipment 外围设备,外部设备
peripheral speed 圆周速度
periphery 周围,圆周;圆柱表面
peristaltic pump 蠕动泵(依靠蠕动作用产生液流的一种泵)

periwinkle 长春花蓝[色](浅紫光蓝色)

perkan 厚毛织物,厚呢

Perkin's violet 柏金氏青莲,佩金紫

Perlaprint G 佩拉普林 G(现成的珠光浆,有金属效果的涂料色浆,遮盖力高,可在地色上印花,商名,意大利宁柏迪)

perle 珍珠绒整理(将呢面上的绒毛整理成小球形);珠皮呢;珍珠呢

perlines 丝光细布

Perlitazol dyes 柏力塔措染料(染聚酰胺纤维的分散染料,商名,德国制)

Perliton dyes 柏力吞染料(染聚酰胺纤维的分散染料,商名,德国制)

Perlit SE 贝力 SE(含环氧树脂的硅油乳化液,作交链组分和催化剂,商名,德国拜耳)

Perlit SI-SW 贝力 SI-SW(不含金属盐的硅油乳化液,阳离子性,与贝力 SE 或贝力 VK 合用,作为合成纤维、纤维素纤维及其和羊毛混纺物的防水剂,商名,德国拜耳)

Perlit VK 贝力 VK(环氧树脂,在贝力 SI-SW 处理浴中作交链组分用,商名,德国拜耳)

Perlon fast dyes 贝纶坚牢染料(聚酰胺纤维用的分散、酸性两类染料,商名,德国制)

Perlon roll (= Perlon roller) 贝纶辊(一种高效弹性轧水辊筒,由贝纶纤维薄片加黏合剂层压而成,德国制)

Permachem treatment 珀马琴[防菌]处理(纺丝液中添加防菌药品,纺制耐洗的防菌纤维)

Permafresh EFR 派马弗来希 EFR(抗皱免烫树脂,改性咪唑啉酮,甲醛含量低于 0.1%,商名,美国旭日)

Permafresh LF-2 派马弗来希 LF-2(抗皱免烫树脂,反应性改性 2D 树脂,商名,美国旭日)

Permafresh TG 派马弗来希 TG(抗皱免烫树脂,水溶性,高浓度,甲醛含量极低,商名,美国旭日)

Permalose T 珀马罗斯 T(亲水性聚合物的水分散体,聚酯及其混纺织物的耐久去污整理剂,商名,英国卜内门)

Permalose TM 珀马罗斯 TM(聚酯聚醚多嵌段共聚物,亲水性水分散体,涤纶及其混纺织物耐久性去污整理剂、抗静电剂,商名,英国卜内门)

permanent adhesives 永久性[台板]黏着剂

permanent antistatic fibre 持久性抗静电纤维

permanent bright finish 〈PB finish〉 耐久性光泽整理

permanent creasing 永久褶痕

permanent deformation 永久形变

permanent elongation 永久伸长

permanent embossed finish 耐久性拷花整理,耐久性轧花整理

permanent finish 耐久整理

permanent fireproofing 耐久性阻燃工艺

permanent fireretardant rayon 永久性阻燃黏胶丝

permanent flame retardant 耐久性阻燃剂

permanent growth 第二潜伸,次级蠕变(纤维受负载变形后,负载虽经解除,但不能恢复原形)

permanent hardness 永久硬度

permanent hard water 永久硬水

permanent loss 永久损耗

permanent luster 永久性光泽

permanent lustered finish 耐久性上光整理

permanent pleating 耐久褶裥加工
permanent pleating machine 永久褶裥机，耐久褶裥机
permanent pleat set 永久[性]褶裥定形，耐久[性]褶裥定形
permanent press finish 耐久压烫整理
permanent press〈pp〉 耐久压烫，耐久定形
permanent set 永久定形
permanent setting of worsted fabric 精梳毛织品的永久定形
permanent sizing 耐洗上浆整理
permanent water-proofing agent 耐久性防水剂
permanganate bleaching 高锰酸盐漂白
permanganate number 高锰酸钾值
Permavel CT 一种透湿散热功能性助剂（商名，英国 PPT）
permeability 渗透性，渗透率；透气性；透水性
permeability index 透湿指数
permeability test 透湿试验
permeable 可渗透的；渗透性的
permeable fabrics 透气性织物
permeable to light 可透光的
permeable to vapour 可透蒸汽的
permeameter 渗漏仪《试》
permeate 渗透液，透过液
permeating degree 透气度；渗透程度
permeation 渗透
Permeometer 织物透气性测试仪
permethrin 苄氯菊酯（杀虫剂）
per minute〈pm〉 每分[钟]
permissible deviation 允许偏差
permissible error〈PE〉 允许误差
permissible motor 防爆电动机

permittivity 介电常数，电容率
permonosulphuric acid 过一硫酸，卡罗氏酸（羊毛防缩氧化剂）
permselectivity 渗透选择性
Permutite 泡沸石，人造沸石（含水的钙、钠、钡等硅铝酸盐的总称，软水剂）
permutoid 交换体
permutoid reaction 交换反应；交换体[沉淀]反应
permutoid swelling 交换溶胀
Pernambuco wood 伯南布哥木（巴西木的一种，含有红色植物染料）
pernyi silkworm 中国柞蚕
peroxidase 过氧化物酶
peroxidation 过氧化[反应]
peroxide 过氧化物
peroxide accumulation rate 过氧化物积聚率
peroxide bleached wool 过氧化物漂白羊毛
peroxide bleaching 过氧化物漂白
peroxide decomposer 过氧化物分解剂
peroxide killer 去除双氧水助剂
peroxide value〈POV〉 过氧化值
peroxidization 过氧化作用
peroxy- （构词成分）表示"过氧"
peroxy-acetic acid 过醋酸，过氧乙酸
peroxydiphosphate 过二磷酸盐
peroxyester 过氧化酯
perpendicular cut shape 直剪影（整理疵）
perpetual screw 蜗杆
per procurationem〈pp〉 由……所代表
Perrotine printing machine 佩罗廷木版印花机
per sample〈p.s.〉 每种试样
per second〈p.s.〉 每秒[钟]
persening 防水麻织物（黄麻或亚麻制）

Persian berries 波斯浆果(黄色植物染料)
Persian carpet 波斯地毯
Persian print 波斯印花布(花纹大,色鲜艳,英国制)
Persian red 天然红色染料;红色颜料
persians 波斯绸(轻质平纹里子绸,印有大花卉图形)
Persian shawl 波斯羊绒披巾(松树图案)
persimmon 柿棕[色](棕色)
persimmon red 柿红[色](红橘色)
persis 浆状苔色素
persistence 难分解性《环保》;持久性
persistence length 相关长度;持久长度;不变长度
Persoftal AFS 珀索夫塔 AFS(高分子脂肪酸酰胺,阳离子型柔软剂,可与阳离子染料染丙烯酸系纤维同浴伴用,商名,德国拜耳)
Persoftal FN 珀索夫塔 FN(脂肪酸酰胺和聚乙二醇醚的混合物,非离子型,纤维素纤维的柔软剂、抗静电剂,商名,德国拜耳)
Persoftal SWA 珀索夫塔 SWA(含有硅酮的阴离子型的柔软剂、润滑剂,商名,德国拜耳)
Persoftal WKF 珀索夫塔 WKF(脂肪酸衍生物的混合物,阳离子型抗静电剂、柔软剂、润滑剂,商名,德国拜耳)
personal computer〈PC〉个人计算机
personal error 人为误差
personal hygiene 个人卫生
personalization 个性化
personal protective equipment〈PPE〉个人保护器材
personal washing agent 皮肤洗涤剂
persorption [气孔]渗入吸附

perspective view 透视图
perspiration 汗渍
perspiration absorptive finish 吸汗整理
perspiration degradation 汗渍降解
perspiration fastness 汗渍牢度
perspiration fastness tester 耐汗渍[色]牢度试验仪
perspiration proof 防汗渍,耐汗渍
perspiration-resistant 耐汗渍的
perspiration-sunlight fastness 耐汗渍与日光综合牢度
perspiration test 耐汗渍牢度试验
perspirometer 耐汗渍牢度试验仪
persulfate 过硫酸盐
persulfate bleaching 过硫酸盐漂白[工艺]
persulfate desizing 过硫酸盐退浆
persulfate resist printing 过硫酸盐防染印花
persulfuric acid 过硫酸
perturbation 微扰;扰动;扰乱
per unit machine yield 每机台产,单机台产量
PES(=PEs, polyester) 聚酯
PES(plasma emission spectroscopy) 等离子体发射光谱[法]
PES film lamination 聚酯薄膜层压
pesticide 杀虫剂
pesticide protective clothing 防杀虫剂服装
pesticide residue 杀虫剂残余物
PET(=PETP, polyethylene terephthalate) 聚对苯二甲酸乙二[醇]酯
PET(photo emission tube) 光电发射管
PET(=PET, polyethylene terephthalate) 聚对苯二甲酸乙二[醇]酯
petcock 小旋塞
Petex 佩特克斯(以聚酯纤维为原料的纺

丝黏合法制成的非织造物,商名)
petits pois 小斑点
petranol 乙醇汽油(由乙醇与汽油混合而成)
Petra wet sueding machine 佩特拉湿磨毛机(商名,意大利白卡拉里)
Petri dish 佩氏培养皿
Petrocci method 抗菌试验法
petrochemicals 石油化学制品
petroleum aromatics 石油芳[香]族烃,石油芳[香]族化合物
petroleum cracking 石油裂解
petroleum, oil and lubricant〈pol〉石油、油类和润滑油
petroleum sulfonate 石油磺酸盐
petrol gas producing plant 汽油汽化器
petrolift 油泵,燃料泵
petticoat 衬裙,内裙
petunia 矮牵牛花紫(暗紫色)
pewter 锡镴灰(暗蓝光灰色)
p. ex.〈par exemple〉举例;例如
pF(=pf, pico-farad) 皮法[拉],微微法[拉](10^{-12}法拉)
pf(plastic-film capacitor) 塑料膜电容器
pf(power factor) 功率因数;力率
pf(pulse frequency) 脉冲频率
PFK(perfluoro-kerosene) 全氟煤油
PFK(programmed function keyboard) 程序操作键盘
PFM(pulse-frequency modulation) 脉冲频率调制
PFR rayon 阻燃黏胶纤维(商品)
PFZ process 氟化锆盐阻燃剂工艺
PGC(pyrolysis gas chromatography) 热解气相色谱[法]
PG method(phloroglucine method) 间苯三酚法(测织物上游离甲醛方法)
ph.(per hour) 每小时
ph.(phase) 相,相位,周相;阶段,方面
ph.(phot) 辐透(照度单位)
pH(potential of hydrogen) 酸碱度
pH adjustment pH 调整
phase〈ph.〉相,相位,周相;阶段,方面
phase adjuster 相位补偿器《电》
phase angle 相角,相位角
phase boundary crosslinking 相界交联
phase boundary limited cross linking process 相界有限交联法(羊毛防缩整理法)
phase boundary potential 接触电位差
phase-change fibre 相变纤维
phase change material〈PCM〉相变材料
phase-change material cooling 相变材料冷却
phase compensation 相位调整《电》
phase contrast 相反差
phase contrast microscope 相反差显微镜
phase detector 鉴相器
phase impulse 相位脉冲
phase inversion of emulsion type 乳液转相
phase inversion temperature〈PIT〉[乳状液的]相变温度;相转变温度
phase lag 相位滞后
phase lead 相位超前
phasemeter 相位计
phase microscope 相位显微镜
phase separation catalyst 相分离催化剂
phasesequence 相序
phase transfer catalyst 相位转移催化剂
phasing down 逐步减少,分阶段减少
phasing out 逐步中止,分阶段停止
PHBV fibre(poly hydroxybutyrate-co-hydroxyvalerate fibre) 羟基丁酸与羟基戊酸共聚

纤维
pH control 调节 pH
pH controlling agent pH 控制剂
pH correction pH 校正
phenol 酚；[苯]酚，石炭酸
phenolaldehyde resin 酚醛树脂
phenolfurfural resin 苯酚糠醛树脂
phenolic fibre 酚醛纤维
phenolic resin 酚醛树脂
phenolic substituents 酚的取代基（纤维中与染料起作用的基团）
phenol oxidase 酚氧化酶
phenolphthalein 酚酞（酸碱测定用指示剂）
phenol red 酚红，苯酚磺酞
phenomenon 现象，征兆
phenylalanine 苯基丙氨酸
phenylene diisocyanate 对苯二异氰酸酯
phenyl hydrazine 苯肼
phenyl isocyanate 异氰酸苯酯
phenylmercurials 苯基汞制剂（防霉腐剂）
phenyl mercuric acetate 苯汞醋酸盐（杀菌剂，用于织物消毒整理）
phenyl silane 苯基硅烷
pheron 酶蛋白
philanising 充毛棉布[硝酸]处理法
Philip screw 菲氏螺钉（带十字槽头的螺钉）
pH indicator pH 指示剂
phloroglucine method〈**PG mothod**〉 间苯三酚法（测织物上游离甲醛的方法）
phloxine 酸性红色染料
PHMB（**polyhexamethylene biguanide**） 抗菌整理剂[REPUTEX 20]，聚六甲双胍的盐酸盐
pH meter pH 测定仪；氢离子浓度计

Phobatex FTC new 新福博特克斯 FTC（以变性三聚氰胺甲醛为基础的缩聚物与石蜡的混合物，耐沸水洗涤的防水剂，商名，瑞士制）
Phobotone WS conc. 福勃通 WS 浓（聚硅酮烷，耐久性拒水整理剂，适用于各类织物，商名，瑞士制）
phoenix tail fabric 凤尾纱色织布
phon 方（响度级单位）
phonetic laminate 隔音板
phonometer 声位测量仪，音强度计
Phos. B（= **phos. bro.**, **phosphor bronze**） 磷青铜
phosgene 光气，碳酰氯
phosistor 光敏晶体管
phosph- （构词成分）表示"磷"
phosphate 磷酸盐
phosphate buffer 磷酸盐缓冲剂
phosphate surfactants 磷酸酯型表面活性剂（用作防静电剂、乳化剂）
phosphine 膦；磷化氢；碱性染革黄棕
phosphino polycarboxylic acid 膦基多元羧酸（免烫整理剂）
phospho- （构词成分）表示"磷"
phospholipid 磷脂体
phospholipid bilayers 磷脂双分子层（羊毛染色缓染剂）
phosphomolybdic acid 磷钼酸（与碱性染料相结合，可制成耐光的涂料）
phosphonium 鏻[基]
phosphonium compound 鏻化合物
phosphonium flame retardant 含鏻阻燃剂
phosphonium surfactants 鏻[盐]表面活性剂
phosphor 磷光体；黄磷；荧光物质
phosphor bronze〈**Phos. B, phos. bro.**〉 磷

青铜

phosphorescence 磷光[现象]
phosphorescence-emission 磷光发射
phosphoric acid 磷酸
phosphorimetry 磷光分析
phosphorizing 磷化,引入磷元素
phosphorous compound 含磷化合物
phosphorous-nitrogen synergism〈P/N synergism〉磷—协同效果(阻燃整理)
phosphorous pentaoxide 五氧化磷
phosphorus ionic liquid antipilling agent 磷离子液体抗起毛起球整理剂
phosphorus nitride dichloride 二氯化氮化磷(纤维用阻燃剂)
phosphorylase 磷酸化酶
phosphorylated cotton 磷酸化棉(棉纤维经尿素和磷酸处理而成)
phosphorylation 磷酸化[作用]
phosphotungstic acid 磷钨酸(与碱性染料结合,可制耐光涂料)
phot〈ph.〉辐透(照度单位)
phot.〈photograph〉照相,照片
Photine 芳丁(荧光增白剂,二苯乙烯三嗪衍生物,商名,英国制)
photo- (构词成分)表示"光""光电""照相[术]"
photobleaching 光漂白[作用]
photocatalysis of dyeing effluent treatment 用光催化分解染色废水处理《环保》
photocatalyst 光催化剂
photocatalytic decolouration 光催化脱色《环保》
photocatalytic degradation 光催化分解
photocatalytic deodorisation 光催化除臭
photocatalytic purification 光催化纯化
photocathode〈PC〉光电阴极

photo cell〈PC〉光电池;光电管
photocell control 光电管控制,光电池控制
photocell transducer 光电池换能器
photochemical activity 光化活性
photochemical addition 光化加成[作用]
photochemical cleavage 光化分裂
photochemical decomposition 光化分解
photochemical degradation 光化降解
photochemical development 光化显色[作用]
photochemical effect 光化学效应
photochemical induction 光化诱导[作用]
photochemical initiation 光化引发[作用]
photochemical inversion 光化转化[作用]
photochemical patterning 光制花样(采用光化学方法将图案通过分色传递到筛网上)
photochemical polymerization 光化聚合[作用]
photochemical printing 感光印花
photochemical reaction 光化反应
photochemical smog 光化学烟雾
photochemical tendering 光化学损脆
photochemical textile printing 感光[织物]印花
photochemistry 光化学
photo chromatic printing 光敏变色印花
photochromic 光致变色的
photochromic compound 光致变色化合物
photochromic dyes 光致变色染料
photochromic fibre 光致变色纤维
photochromism 光致变色[现象]
photocolourimeter 光比色计
photocolourimetry 光比色法
photoconductive cell 光导电池

photoconductive element 光敏元件
photoconductivity 光电导性
photoconductor 光[电]导体
photocopolymerization 光致共聚合[作用]
photocurable coating 光可焙固的涂层(环氧树脂类)
photocurable composition 光焙固成分
photocurable printing inks 光焙固印墨
photodecomposition 感光分解[作用]
photodegradable polymer 感光降解聚合物
photodegradation 感光降解[作用]
photodegradation of fibre 纤维的感光降解[作用]
photodensitometer 光密度计
photodepolymerization 光解聚[作用]
photodetection 光电探测
photodetector 光电检出器,光电探测器
photodetector diode array 阵列光电二极管
photodichroism 光二色性
photo diode 光电二极管
photo disintegration 光致蜕变
photoeffect 光电效应
photoelastic analysis 光弹性分析
photoelasticity 光弹性[学],光致弹性
photoelastic method 光[测]弹性法
photoelastic test 光弹性试验
photo-electric〈PE〉光电的
photoelectrical checking device 光电检查装置
photoelectrically scanned pattern drum 光电图案扫描圆筒
photoelectric bobbin feeler 光电探纬器
photoelectric cell 〈PEC,P.E.C.,pec〉光电池;光电管;光电元件
photoelectric centring system 光电定中心系统

photoelectric cloth inspection 光电验布
photoelectric colour difference meter 光电色差计
photoelectric colourimeter 光电比色计,光电色度计
photoelectric colourimetry 光电比色法
photoelectric compensator 光电松紧器(光电管检测织物松弛情况)
photoelectric edge feeler (= photoelectric edge sensor) 光电探边器
photoelectric engraving 光电雕刻
photoelectric follow-up 光电随动[装置];光电跟踪
photoelectric illuminometer 光电照度计
photoelectric meter 光电仪
photoelectric photometer 光电光度计
photoelectric pick counter 光电织物密度分析器(自动记录)
photoelectric reflectometer 光电反射计
photoelectric scanner 光电扫描器
photoelectric selvedge detector 光电探边装置
photoelectric sensing device 光电传感器
photoelectric servocontrol system 光电式伺服电动机控制系统
photoelectric spectrocolourimeter 光电分光测色计,光电分光色度计
photoelectric spectrophotometer 光电分光光度计
photoelectric switch 光电开关
photoelectric system 光电系统
photoelectric tentering of weft 光电整纬
photoelectric tube 光电管
photoelectric weft feeler 光电探纬器
photoelectric weft tentering instrument 光电整纬器

photoelectro-luminescence 光电致发光
photoelectro-magnetic effect 光电磁效应
photoelectronic energy spectrum 光电子能谱
photo emission tube〈PET〉 光电发射管
photoemulsion 照相乳剂
photoengraved cylinder printing machine 光雕滚筒印花机
photoengraving 照相雕刻
photoengraving method 照相雕刻法
photoetching 光刻[法],光蚀[技术],感光雕刻
photofading 光褪色
photo film 感光胶片
photo-functional dyes 光功能性染料
photogelatin 感光底片胶
photogenic [磷]光的;光产生的
photograph〈phot.〉 照相,照片
photographic darkroom 照相暗室
photographic dyeing technique 感光染色术
photographic engraving 照相雕刻
photographic film 感光胶片
photographic film with diazonium compound 重氮化合物感光胶
photographic hardening agent 感光坚膜剂
photographic intensifier 感光加厚剂
photographic plate〈pp〉 照相底片,胶片
photographic plotter 照相绘图机
photographic printing 照相印花
photographic reducer 感光减薄剂
photographic standards 样照
photographic stencil 照相制[网印]版
photographic visibility scale [烟雾]可见度相卡(耐燃性试验)
photogravure printing 照相凹版印刷
photo initiated polymerization 光引发的聚合作用
photoinitiation 光引发[作用]
photo-ionization detector〈PID〉 光致电离检测仪
photo isomerization 光致异构化
photolacquer 感光漆
photolithographic off-set process 照相平版胶印印刷
photoluminescence 光致发光,荧光
photo luminescent material 光致发光材料
photolysis 光[分]解[作用]
photolytic attack 光解破坏
photomechanical process 照相制版工艺
photometer 光度计
photometric 光度学的,光度计的,测[量]光[度]的
photometric range 光度范围
photometry 光度学
photomicrograph 显微照相;显微照片
photomicroscopic apparatus 显微照相设备
photomontage 照相拼版;集成照片[制作法]
photomultiplier [tube] 光电倍增管
photon 光子
photonic crystal 光子晶体
photooxidation 光氧化[作用]
photopatterning 光制花样,照相雕刻
photopic vision 亮视觉,白昼视觉
photopolymer 光[致]聚合物
photopolymerization 光[致]聚[合]作用
photo printing 照相印花,光照印花
photo printing machine 照相印花机
photo process 光学处理
photoprotection 耐光防护,光保护作用
photo proton 光质子
photo rapid dyestuff 感光显色染料

photo receptor 光接收器
photorectifier 光电检波器
photoresistive detector 光敏电阻探测器
photoresist material 光致抗蚀剂,光刻胶;光阻材料
photoresistor 光敏电阻器
photo scanning tunneling microscope〈PSTM〉 光子扫描隧道显微镜
photosensitive agent 光敏剂
photosensitive discoloured dyes 光敏变色染料
photosensitive discolour microcapsule 光敏变色微胶囊
photosensitive dyes 光敏染料
photosensitive emulsion 感光乳液
photosensitive lacquer 感光漆
photosensitive latex 感光胶
photosensitive paper 光敏纸,感光纸
photosensitive plastics 光敏塑料
photosensitive polymer layer 光敏聚合层
photosensitization 光敏作用
photosensitized initiation 光敏引发[作用]
photosensitizer 光敏剂
photosensor 光电传感器
Photoshop 图像处理软件《计算机》
photostability 耐光性,光稳定性
photostabilizer 耐光剂,光稳定剂
photoswitch 光电开关
photosynthesis 光合[作用]
phototendering 光致脆化(指纤维或织物)
phototropic dyes 光变性染料(在日光下变色,但在阴暗处放置后,又恢复原色)
phototropism 向光性,光变性
phototropy 光[致]色互变[现象]
phototube 光电管,光电池
phototube feeler 光电探纬器

phototube selvedge detector 光电探边装置
photo voltaic〈PV〉 光伏
photo voltaic cell 光伏电池
photo voltaic detector 光伏探测器
photo voltaic device 光伏器件
photo voltaic effect 光伏效应
photo voltaic industry 光伏工业
photo voltaic printing 光伏印花
photo voltaic system 光伏系统
photo voltaic technology 光伏技术
photovolt reflection meter 光电反射仪
photoyellowing 光致泛黄(指羊毛)
p. h. r.（per hundred parts resin） 每一百份树脂中的份数
pH regulator pH调节剂
pH sensitivity pH敏感性
Phtalogen dyes 酞酞罗近染料,酞酞素染料(原地显色染料的一种,含有酞菁结构,商名,德国制)
pH test paper pH试纸
phthalic acid 酞酸,邻苯二甲酸
phthalic anhydride 酞酐,邻苯二甲酸酐
phthalocyanine〈Pc〉 酞菁染料
phthalocyanine dyes 酞菁染料,苯二甲蓝染料,酞花青染料
phthalocyanine intermediate 酞菁素,酞菁染料中间体
phthalocyanine pigment 酞菁颜料
phthalocyanines 酞菁染料,苯二甲蓝染料,酞花青染料
Phthalogen K 酞酞罗近K(酞菁染料的铜络合剂,非离子型氨基酸的铜盐,商名,德国制)
PHV（primary hand values） 基本手感值
pH value 氢离子浓度指数
phys.（physical） 物理的

phys. (physics) 物理,物理学
phys. (physiologics) 生理学
physical 〈phys.〉物理的
physical adsorption 物理吸附
physical chemical properties 理化性能
physical dimension 外形尺寸
physical imperfection 物理疵点
physical index 物理指标
physical isomerism 物理[性]异构[现象];物理异性[现象]
physical measurement 物理测量
physical photometer 物理光度计,色彩计
physical protection 人体防护
physical testing 物理[性]试验
physicochemical modifications of fibre 纤维的物化改性
physicochemical treatment process 物化处理法《环保》
physics 〈phys.〉物理学,物理
physiological calorimetry 生理学热量测定法
physiological function 生理机能
physiologically active substance 生理活性物质
physiological three primary colours 生理三原色
physiologics 〈phys.〉生理学
physiology 生理学
physisorption 物理吸着
phytoprotein 植物蛋白质
PI(party line) 合用线
PI(performance index) 性能指数
P. I. (＝p. i., pounds per inch) 磅力/英寸(英制线压力单位)
PI(polyimide) 聚酰亚胺
PI(pressure indicator) 压力指示器

PIB(polyisobutylene) 聚异丁烯
PIC(pressure indicating controller) 压力指示控制器
pick bar 纬向条花(织疵)
pick count 织物经纬密度
pick counter 织物分析镜;织机产量表;织物密度镜
pick density 纬密
pick glass 织物分析镜
Pickle's taking-up motion 皮克尔式[七轮]卷布装置
pickling 酸浸;浸渍
pickoff 传感器,发送器;脱去;摘去
pick-o-meter 电子织物分析器(商名)
pickout mark 拆痕,拆毛,拆档(织疵)
pickouts 缺纬(织疵)
pickover 跨度,开间;跳花(织疵)
picks per inch 〈ppi, p. p. i.〉每英寸纬数
pickup 轧液率,吸液率;拾波器,拾音器;敏感元件;传感器
pickup electrode 传感电极(如用于测湿装置的检测头)
pick-up point 剥离点
pickup pump 真空泵,抽出泵
pick-up roller 给液辊;上胶辊
picnometer 比重瓶,比重管
pico-farad 〈pF, pf〉皮法[拉],微微法[拉](10^{-12}法拉)
PI controller PI调节器,比例—积分调节器(一种常用的工业调节器,不仅具有比例作用,而且具有积分作用,因此能指挥执行机构动作,直至最终消除输入偏差)
picot [刻花滚筒]对花点
picotage (起绒织物的)斑点效应(使绒毛头端变形,可引起光泽反射差异)

picrocarmine 苦[胭]脂红[染料]
pictograms 象形标记
picture 图画,图案;形象;图片
picture imaging system 成像系统
picture input device〈PID〉 图像输入设备
picture velvet 印经丝绒,画景丝绒
picul 担(中国重量单位,等于50千克)
PID(photo-ionization detector) 光致电离检测仪
PID(picture input device) 图像输入设备
PID(proportional integral differential) 比例微积分《计算机》
PID controller PID调节器,比例—积分—微分调节器(一种较完善的工业调节器,除比例作用和积分作用外,还有微分作用,调节器能超前动作,克服对象动态滞后,从而改善调节品质)
piece〈pc.〉 匹;接头;拼合;块,片,段
piece beam [织机]卷布辊
piece carte 匹头卷板
piece-dyed cloth 匹染色布
piece dyeing 匹染
piece dyeing machine 匹染机(织物的间歇式染色设备,如喷射染色机、溢流染色机等)
piece end 零头布;短码布,头子布
piece-end sewing machine 缝头机(将坯布连接起来,以利连续染整加工)
piece goods 匹头,布匹
piece goods dyeing 布匹染色,匹染
piece-length 匹长
piece number 匹号
piecer 接头工
piece rate 计件工资
piece roller 卷布辊
pieces〈pcs,pcs.〉 块;件,片;匹

piece scouring [布]匹[精]练
piece scouring machine 洗布机;洗呢机
piece to piece variation 匹间变化
piece-ups 接头数
piece work 件工,计件工作;单件生产
piecing(=piecing-up) 接头
pied 杂色花纹的,斑纹的
pier 柱,支柱
pierce 刺破;刺穿
pierced fabric 纱罗织物
piercing test 贯穿试验
piezo- (构词成分)表示"压力""压"
piezo-composite material 压电复合材料
piezoelectric crystal 压电晶体(显示出压电效应的晶体)
piezoelectric effect 压电效应
piezoelectric transducer 压电传感器(喷墨印花的喷印装置)
piezoink jet〈PIJ〉 压电式喷墨
piezometer 流压计;水压计;测压计
piezosensitive discolour fabrics 压敏变色织物
pigeon 鸽灰[色](紫光灰色)
pigeon blood 鸽血红[色](暗红色)
pigeon catch 鸽形掣子,燕形掣子
pigeon's neck 鸽颈蓝(浅蓝光灰色)
pig iron 坯铁,生铁
pigment 涂料;颜料
pigmentation 色素淀积;染料悬浮体染色(指用染料悬浮体浸轧);涂料制备
pigment binder 涂料胶黏剂,胶合剂
pigment colouration 涂料着色
pigment dispersion 颜料分散液
pigment dyeing 涂料染色,颜料染色
pigment dyes 涂料;色素染料

pigmented polyester ［纺丝前］着色聚酯
pigmented rayon 纺［前］着［色］人造丝，无光人造丝（纺丝液中加入有色颜料或起消光作用的白色涂料后纺成）
pigmented wool （天然）有色羊毛
pigmented yarn 纺［前］着［色］丝；无光化纤纱，无光丝
pigment exhaust dyeing 涂料吸尽染色，涂料浸染
pigment extender 涂料印花稀释剂
pigment fibre yarn 无光化纤纱；着色化纤纱
pigment injection machine 涂料喷射机
pigment melanin 黑素，黑色素（生物体中特种细胞的代谢产物，含有褐色或黑色色素）
pigment padding ［染料］悬浮体轧染；涂料轧染
pigment pad dyeing 悬浮体轧染［工艺］（织物经士林染料悬浮液轧染后，再在染缸或平洗机上进行还原处理）；涂料轧染［工艺］（涂料和黏合剂同时浸轧，再烘干固色）
pigment paper process 碳素纸印像法（用于照相凹版雕刻）
pigment precursor 色素母体
pigment printing 涂料［树脂］印花［工艺］
pigment printing concentrate 涂料印花浓缩浆（美国印花技术用语）
pigment rayon 无光人造丝；着色人造丝（纺丝液中加入起消光作用的白色涂料，如二氧化钛后纺成）
pigment resin colour〈PRC〉颜料树脂色
pigment resin emulsion colours 涂料乳胶色浆
pigment resin transfer process 涂料树脂转移［印花］法
pigments and additives 涂料及添加剂
pigment taffeta 涂料塔夫绸（用涂料着色的丝线织成，在绸面起无光效应）
pigment transfer 涂料转移［印花］
pigment volume concentration〈PVC〉颜料体积浓度

PIJ（piezoink jet）压电式喷墨
pile 绒头；绒毛
pile anchorage 绒毛的固定（绒毛的抗压）
pile and nap lifting machine 起绒机
pile bonding machine 绒毛黏合机，植绒机
pile burning index〈PBI〉［长毛地毯的］绒面燃烧指数
pile carpet 绒头地毯
pile cistern 堆布池
pile coating 植绒
pile crush 绒头抗压性能（地毯、丝绒等织物的绒头竖立效应）
pile cutting 割绒
pile cutting machine 割绒机
pile cutting motion 割绒装置
pile density 绒毛密度
pile dressing 绒头排直［整理］
pile fabric 绒头织物，割绒织物，起绒织物；针织长毛绒
pile face 绒面；毛面
pile floor covering 毛绒地毯
pile height 绒毛高度
pile knit fabric 毛圈针织物
pile lay 绒头倒伏方向
pile loop 毛圈，绒圈
pile material 绒毛材料
pile on pile 叠花丝绒
pile-on properties 染深性能
pile overcoating 立绒大衣呢

pile pressure 绒头压力
piler 堆布机；自动甩布装置；折叠机构
pile reversal 绒头反向
pile roller 起绒罗拉
pile root 绒头根
pile rug 绒头地毯
pile setting 起绒
pile shearing machine 剪绒机
pile thickness control 绒毛厚度控制
piling device(= piling machine) 堆布机；自动甩布装置；折叠机构
pill 球粒；纤维绒球（疵点）
pillar 支柱
Pill Box 箱式起球试验仪
pilled-in selvedge 紧边（织疵）
pilling [织物表面]起球（疵点）
pilling degree （织物表面）起[毛]球度
pilling propensity 起球倾向
pilling resistance 抗起球性
pilling tester 起毛起球试验仪
pilling testing method 起毛起球试验法
pillow 轴衬，轴枕；垫座，枕块
pillow block 轴台
pillow cases 枕套织物
pillow joint 球形接合
pillow tubing 圆筒形棉织物（管袋类织物、双层布）
pillprone textile 易起球纺织品
pill-resistant finish 防起球整理
pill wear off 毛球脱落
pilly 成团，起球
pilot 控制导线，辅助芯线；主控的；引导的；中间规模的，实验性的
pilot boiler 试验锅炉
pilot cloth 海员厚绒呢
pilot combustion chamber 引燃室

pilot controller 光导控制器，导频控制器
pilot ignition 引燃
pilot lamp〈PL〉指示灯，标灯，信号灯
pilot motor 伺服电动机
pilot-operated valve 导阀，控制阀
pilot plant 中试装置，实验工厂；示范工厂
pilot roller 导辊
pilot-scale 半工业规模，试验性规模
pilot test 小规模试验，典型试验
pilot valve 控制阀，引导阀（一种小型阀，用以控制大型阀的动作）
pilot wheel 操纵轮，导轮
pimelic ketone 环己酮
pin and clip stenter 针板布铗两用拉幅机
pin bar 针板（用于针板拉幅机）
pin bar carrier 针板托架
pinboard 插接板，接线板
pincers 钳，铁钳
pin chain stenter 针链拉幅机
pinch roll 摩擦[导布]辊
pin clip [拉幅机的]针铗
pinclip chain 针板布铗两用链
pinclip exchanger 针板布铗调换器
pin clip stenter 针[板布]铗[两用]拉幅机
pin connection diagram 出脚图，管脚图（表示电子管和集成电路等器件出脚与外部的连接关系）
pineapple cloth 菠萝纤维织物；手帕亚麻布；上浆全丝薄纱
pineapple leaf fibre 菠萝叶纤维（内含菠萝麻）
pine needle 松针绿[色]（暗绿色）
pine oil 松油
pine silk 人造丝纺绸（黏胶长丝和黏胶短纤维交织物）

pin frame 针板拉幅机
pin gear 针齿轮
pinhole 针洞,小洞,针眼(织疵)
pinholing 破洞(疵点)
pinion 小齿轮,副齿轮
pink ⟨pk⟩ 桃红色,妃色,粉红色
pinked edge 锯齿边,荷叶边(织疵)
pinking machine 锯齿边布样切裁机
pinking scissors (=**pinking shears**) 齿边布样剪刀
pink salt 锡盐(通常指二氯化锡或氯锡酸铵)
pin mark [布边]针孔疵(拉幅疵点)
pinning brush 上针转刷(针板拉幅机进布处的附属装置)
pinning device 上针装置(如上针转刷)
pinning monitors 上针监控器
pinny 针尖白点状(印疵)
pin plate (拉幅机的)针板
pinpoint 细点子
pin seaming 销钉式接口
Pinsprocket type weft straightener 针轮式整纬装置(商名,日本土谷)
pin stenter (=**pin-stenter frame**) 针板拉幅机
pin stripe 细条子,线条
pin stripe effect 细条效应
pin support 针板座
pint ⟨pt.⟩ 品脱(英美液量及容量单位,等于1/8加仑)
pin tenter 针板拉幅机
pin tentering 针板拉幅
pin truck 运输小车
pints ⟨pts⟩ 品脱
pin valve 针形阀
pin wheel hosiery boarder 针轮式袜子定形机
PIPD controller 比例—积分、比例—微分调节器,PI、PD 调节器(根据被调量与给定值的偏差大小、符号以及变化速度等,通过简单逻辑判断来控制 PI 与 PD 调节规律的切换)
pipe alignment 管道定线
pipe bends 管子弯头
pipe bomb 铁管炸弹,管状炸弹
pipe cap 管帽,管盖
pipe clip 管夹
pipe coil 盘管,蛇[形]管
pipe coupling 联管节,管接头
pipe dryer 管道式烘燥机
pipe fittings 管子配件
pipe hanger 管道支吊架,管架
pipe interior 管子内壁
pipe laying 铺设管线,配管
pipe line ⟨PL⟩ 管路,管道
pipeline heat transmission efficiency 热力管道输送效率
pipelining 管道铺设,管道安装
pipe network 管道网,管道系统
pipe plug 管堵,管塞
pipe sleeve 套管
pipe socket 管套
pipe stopper 管塞
pipe strap 管扣
pipet (=**pipette**) 吸量管,吸管
pipe thread 管[端]螺纹
pipe union 管子接头
pipework 管道系统
pipe wrench 管子扳手,管子钳
piping 管路,管系;导管
piping and instrument diagram 管路及仪表布置图

piping diagram(= piping drawing) 管系图,管路图
piping system 管系
piqué 凹凸织物
piqué crepe 棉绉;凹凸绉
piqué stripe 灯芯绒条纹,凸花条纹
Pirelli cylinder test 毕勒里圆筒试验(用于黏合力试验)
Pirle finish 珀尔整理(呢绒的防雨、防缩、防污整理,商名,英国制)
pirn barré 纬档(疵点);纬向条纹
pistache(= pistachio green) 阿月浑子果仁绿[色](淡黄绿色)
piston 活塞
piston boss 活塞销壳
piston crown 活塞顶,活塞头
piston cup 活塞皮碗
piston flow bioreactor 活塞流生物反应器(用于废水处理)
piston head 活塞顶,活塞头
piston pin 活塞销
piston pin boss 活塞销座
piston pump 活塞泵,往复泵
piston ring 活塞环
piston rod 活塞杆
pit 麻点;槽,坑;凹处
PIT(phase inversion temperature) [乳状液的]相变温度;相转变温度
pitch 节距,齿距;第一次打样处方
pitch circle 节圆
pitch diameter〈P. D. ,pd〉 [螺纹的]中径;[齿轮的]节径
pitching in [机印]对花
pitching pattern 对花定位点(印花)
pitch mark [机印]对花标记
pitch pattern 对花定位点(印花)

pitchpin 安全钉,安全针,保险销;[模版]对花小钉
pitch point 对花点
Pitot tube 皮托管(测量气流压力所使用的特制金属管)
pitting 麻点腐蚀;小孔;凹痕
PIV(= P. I. V. ,positive infinitely variable) 无级变速
PIV drive 无级变速传动
pivot 枢轴,支枢(支持其他机件旋转)
pivot bearing 枢[轴]承;回转轴承;摆动支座
pivoted arm 旋臂,旋轴
pivot joint 铰接
pivot pin 枢销
pK 电解质[的]离解度(电离常数的负对数)
pk(pink) 桃红色,妃色,粉红色
pkg(package) 卷装;包装;装箱
PL(party line) 合用线
PL(pilot lamp) 指示灯,标灯,信号灯
PL(pipe line) 管路,管道
PL(programming language) 程序设计语言《计算机》
placarder 防染(法国用语)
placement 键接;部位,方位;布局,安排
PLA fibre(polylactic acid fibre) 聚乳酸纤维(一种以玉米为原料的新型纤维,易生物降解,符合环保要求)
plaid flannel 格子法兰绒
plain 平的,平纹的;素的;简单的
plain all-over flocking 单色植绒
plain bearing 滑动轴承,普通轴承
plain blanket 素毯
plain carpet 素色地毯
plain cloth 平布,平纹织物

plain crepe 平绉
plain dyed 单色的,素色的
plain dyeing 单色染色,素色染色
plain edge 平边
plain embossed design 素色轧花
plain fancy suiting 素花呢
plain finish 不丝光整理,本色整理
plain finished yarn 本光纱线
plain-knitted fabric 平针织物
plain plush 素色平纹长毛绒,素色毛圈织物
plain roller 光面[印花]滚筒,光板[印花]辊
plains 重浆素色棉布;平布
plain satin 平缎,素缎;素累缎(中国手织)
plain seam 平缝
plain shade 单色,素色
plain shaft 光轴
plain towel 素色毛巾
plain tubular knit fabric 圆筒形平针织物
plain velvet 素丝绒,平丝绒
plain water 淡水
plait 褶边;褶裥
plait entry 平幅进布
plaiter 落布架;折布机,码布机
plait exit 平幅落布
plait form 折叠形式
plaiting 折布,码布
plaiting apparatus 落布装置,甩布架
plaiting arm 折布杆
plaiting device 平幅出布装置
plaiting machine 码布机,折布机,折叠机
planar 平面的
planarity of dye molecule 染料分子平面性
planar winding [平]面试缠绕

Planck constant 普朗克常数(量子理论基本定律中的一个常数)
Planckian radiator 普朗克放射体,完全放射体,黑体
plane 平面;投影;平面的
plane abrasion 平磨
plane coating 薄膜涂层
planetary reducer 行星减速器
planetary [gear] transmission 行星齿轮传动
planet carrier 行星齿轮架(行星齿轮系统中带轴的支架)
planet gear 行星齿轮
plangi 结扎防染染色
planimeter 测面仪,面积仪,求积仪,积分器
planking 毡合;缩绒
plankton 浮游生物
planography 平面印刷,平版印刷术
plant 工厂;车间;装置;植物
plant bulk 装置[外形]尺寸
plant cellulose 植物纤维素
plant dyes 植物染料
plant fibre 植物纤维
plant floor 现场
plant layout 工厂布置,车间布置
plant piping 工厂管道[系统]
plant trial 工厂试验,工厂条件下试用
plan view 平面图
plasma 原生质;等离子体;血浆
plasma arc acetylene process 等离子体弧[制]乙炔工艺
plasma arc welding 等离子弧焊
plasma atomic fluorescence spectrometer 等离子体原子荧光分光计
plasma body 等离子体

plasma chemical treatment〈PCT〉 等离子体化学处理
plasma-coated surface 等离子体涂层表面
plasma display 等离子体显示器
plasma emission spectroscopy〈PES〉 等离子体发射光谱[法]
plasma etching 等离子体刻蚀
plasma flame reactor 等离子体焰反应器
plasma generator 等离子体发生器
plasma grafting of polyester 聚酯纤维的等离子体接枝(处理后降低表面电阻率等性能)
plasma graft polymerization 等离子体接枝聚合
plasma guide 等离子体波导管
plasma-induced polymerization 等离子体诱导聚合
plasma jet 等离子喷射
plasma jet excitation 等离子喷射激发
plasma jet irradiation 等离子体喷射辐射
plasma light 等离子体光
plasma modification 等离子体改性
plasma modification of polyester fabric 聚酯纤维织物的等离子体改性
plasma polishing 等离子体抛光
plasma polymer 等离子聚合物
plasma polymerization 等离子聚合[作用]
plasma-protease combined treatment 等离子体—蛋白酶联合整理(羊毛防缩)
plasma science 等离子体科学
plasma spraying process 等离子体喷涂法(涂层工艺)
plasma stream reactor 等离子流反应器
plasma surface modification 等离子体表面改性处理
plasma surface treatment 等离子体表面处理
plasma technology 等离子体技术
plasma technology for cleaning 用于清洁的等离子体技术
plasma technology for coating 用于涂层的等离子体
plasma technology for pretreatment 用于前处理的等离子体
plasma torch 等离子电弧枪
plasma treated spun silk 等离子体处理绢丝(改进绢丝性能)
plasma treated wool 等离子体处理羊毛(改进羊毛性能)
plasma treatment 等离子体处理
plasma treatment technology 等离子体处理技术
plastein 类蛋白
plastelast 塑弹性物；弹性塑料
plaster 橡皮膏；膏药
plaster-of-paris 熟石膏
plastic 塑料；塑性的
plasticate 塑炼；增塑
plastication 塑炼[作用]
plastic bonding 塑性树脂黏合
plastic bulk storage tank 塑料储槽
plastic coated fibre 涂塑料纤维
plastic coating 塑料涂层；塑胶膜
plastic deformation 塑[性]变[形]
plastic elasticity 塑性弹性
plastic film 塑料薄膜
plastic-film capacitor〈pf〉 塑料膜电容器
plastic flow 塑性流动；黏[滞]流[动]
plastic flow behaviour 塑流性
plastic fluid 塑性流体
plastic fluidity 塑性流动性
plastic foam 泡沫塑料

plastic friction 塑性摩擦
plastic hammer 塑料锤
plasticised fabric 塑[性]化织物
plasticity 可塑性;黏性
plasticity-recovery number 塑性恢复值
plasticity retention index [可]塑性保持指数
plasticize 增塑
plasticizer (=plasticizing agent) 增塑剂
plastic-laminated 塑料胶合的,塑料层压的(指织物)
plastic limit 塑性极限
plastic mobility 塑性淌度;塑性流动性
plastic nylon 尼龙塑料
plastic optical fibre〈POF〉 塑料光导纤维
plastico-visco elasticity 塑性黏弹性
plastic range 塑性[温度]范围,黏程
plastics 塑料;塑料制品
plastic state 黏态,塑[性状]态
plastic viscosity 塑性黏度
plastic yieldpoint 塑[料]流点;塑性屈服点
plastilock 用合成橡胶改性的酚醛树脂黏合剂
plastimetry 塑性测定法,塑度测定法
plastisol 塑溶性印墨(美国),塑料溶胶
plastisolbased ink 塑料溶胶基油墨
plastoelastic body 塑弹性物体
plastomer 塑性体;塑料
plastometer 塑性计,塑度计(用以测定橡皮或塑料的硬度)
plastometry 塑性测定法,塑度测定法
plastosoluble dye 塑[料可]溶性染料
plate 板,片;(电容器与蓄电池的)电极板;照相底片,胶片;金属板
Plateau border 普拉特奥边界(泡沫整理术语,指泡沫间薄膜交界处形成的三角形液柱,其曲率半径为负值)
plate chromatography 板色谱[法](薄层色谱法)
plate coater [色谱]板涂布器
plate coupling 圆盘联轴带
plate cutting 刻锌板(花筒雕刻)
plate cutting stand 锌版雕刻台
plated 鸳鸯布(织物正反面的质料不同或颜色不同);拼包(不同等级品种合成一包);拼匹
plate fin cooler 翅片式冷却器
plate fin heat exchanger 翅片式热交换器
plate insulation 绝热板
platen 压板;压[印]盘(印刷用);台板
platen heat exchanger 屏式热交换器
platen press 台板压印机
platen superheater 屏式过热器
plate printing 平版印花
plate printing machine 平版印花机
plate singeing 铜板烧毛
plate singeing machine 铜板烧毛机,平板烧毛机
plate spring 板簧
plate tension device 板式张力器
platetype heat exchanger 板式热交换器
plate wheel 盘轮
platform [染整设备]平台
platform balance 台秤
platform bearer 走台搁座
platille 平纹细亚麻布
plating 镀,电镀
plating machine 折叠机;卷板机;折布机;码布机
platinizing 镀铂
platinum electrode 铂电极

platinum resistance temperature sensor 铂电阻感温计,铂电阻测温计
platinum resistance thermometer 铂电阻温度计
play 间隙,游隙；火焰闪动；摆动；启动（继电器衔铁）
PLC(programming language control) 逻辑控制；程序设计语言控制；可编程序控制器
please turn over 〈PTO,P.T.O.,p.t.o.〉见反面,见下页
pleat 打褶；褶裥
pleat abrasion 褶裥磨损
pleated fabric 褶裥织物
pleated skirt 百褶裙
pleat finish 打褶加工
pleat-hold after washing 洗后褶裥保持
pleating machine 褶裥机,打褶机
pleating seal 折叠式封口（用于热风拉幅机烘房的进布口和出布口）
pleat pressing machine 打褶机
pleat retention 褶裥保持性
pleionomer 同性低聚物；均低聚物
plenum 送气通风,压力通风系统；充实
plenum chamber 充实室
pleochroic 多色的,多向色的
pleochroism 多色[现象],多向色性
pleochromatic 多色的,多向的
pleochromatism 多色[现象],多向色性
pleomorphic 多晶的
pleomorphism [同质]多晶形[现象],同质异形[现象],多形性
plethora 过多,过剩
pleuche 金丝绒
plexiglass 有机玻璃
Plexon 普勒克松（外涂塑料薄膜的黏胶纤维或玻璃纤维,商名,英国和美国制）
Plexon process 普勒克松法（纤维的塑料浸渍及涂层工艺）
pliability 可挠性,柔韧性
pliable 易挠的,柔韧的
pliable and tough 柔韧
pliable hand 柔软手感
plied 合股的（指线）
pliers 钳,手钳
plissé 泡泡纱效应（在棉布上印浓碱以产生条状或点状绉缩）
plissé crepe ［条状］泡泡纱；细绉布
plissé fabric 褶皱织物
plissé printing 泡泡纱印花
plot 标绘,标记；绘图,画曲线,作图
plot nozzles 狭缝风口
plotter 普洛特快速花纹绘制器（英国制）；绘图机；图形显示器
ploughing 花筒严重刮伤
plucker roller 开松辊
pluck testing 抗拔试验（植绒用语）
plug 管塞,塞头；电插头
plugboard 接插板,转接板
plug cord 塞绳
plug gauge 塞规（检查孔眼的极限量规）
plugging 堵塞
plug-in connector 插入式接点
plug socket 插座,塞孔
plug valve 旋塞阀门
plum 梅红［色］（暗红青莲色）
plumb line 铅垂线；准绳
plumcolour 茄色,深紫色
plume 羽毛
plumetis 小花薄洋纱（棉织物）；薄花呢（毛织物）

plumety 透明印花亚麻纱;薄细布;点子薄细布

plummet 铅锤,线铊,垂线

plum purple 梅红[色](暗红青莲色)

plunger 柱塞,活塞

plunger pump 柱塞泵

plunging coupling 柱塞联轴器;滑动套万向接头

plural component coating 多组分涂层

Pluronic Polyols 普卢龙尼克多元醇类[聚合物](聚氧乙烯—聚氧丙烯嵌段共聚物,非离子型表面活性剂,有润湿、渗透、净洗、乳化、匀染等多种用途,商名,德国制)

plush 长毛绒;长毛绒织物

plush astrakhan 仿羔皮长毛绒

plush brushing machine 长毛绒刷清机

plush carpet 剪绒地毯

plush cockle back 长毛绒背弓(长毛绒疵病)

plushette 低级长毛绒织物

plush finish [地毯]去毛绒整理

plush loop 毛圈,长毛绒圈

plush over combing 长毛绒梳伤(疵病)

plush raising machine 长毛绒起绒机

plush range 长毛绒整理机组

plush roller 绒辊,清洁辊

plush touch 毛圈的触感

ply 纱线股数;织物层数;合股

plywood 胶合板

ply yarn 合股线

pm(per minute) 每分[钟]

pm(purpose made) 特制的

PMA(polymethacrylate) 聚甲基丙烯酸酯

PMA(polymethyl acrylate) 聚丙烯酸甲酯

PMAN (polymethacrylonitrile) 聚甲基丙烯腈

PMH(＝p.m.h., production per manhour) 每人每小时的产量,每工时产量

PMR(proton magnetic resonance) 质子磁谐振

pneu. (pneumatic) 气动的;气压的;空气的

pneu. (pneumatics) 气体力学,气动力学;气动装置

pneum- (构词成分)表示"空气""气体"

pneumatic〈pneu.〉 气动的;气压的;空气的

pneumatic actuator 气动执行机构

pneumatic bellows 气袋

pneumatic cellular fibre 含气空胞纤维(具有细胞状结构的聚酯纤维,相对密度仅为0.02,美国制)

pneumatic cloth feeder 气压导布装置,气动喂布装置

pneumatic compensator 气压[摆式]松紧架

pneumatic control 气动控制

pneumatic conveyor 气动输送装置

pneumatic cushion 气垫

pneumatic cut-off batcher (＝pneumatic cut-off winder) 气压自停卷布机

pneumatic cylinder 气缸

pneumatic cylinder mounting system 气动装花筒系统

pneumatic dancing roller 气动张力调节辊

pneumatic discharge apparatus 气动送料装置,风送装置

pneumatic drier 气流干燥机

pneumatic edge spreader 气动剥边器

pneumatic ejector 气压喷射器

pneumatic handling 气动输送

pneumatic loading device 气动加压装置
pneumatic measurement 气流测定［值］
pneumatic operator 气动执行机构
pneumatic press 气动加压
pneumatic pressure system 气压装置
pneumatics〈pneu.〉气体力学，气流力学；气动装置
pneumatic sensor 气流传感器
pneumatic squeezer 气动加压轧车
pneumatic stretching clamps 气动绷网夹
pneumatic stretching frame 气动绷网机（平网印花机的附属设备）
pneumatic valve 气动阀
pneumatic vibrating mechanism 气压振动机构（自动称料装置中用）
pneumatic weighting 气动加压
pneumatic width-control cock 气动调幅旋塞（用于烧毛机，根据不同的织物门幅调整火口的烧毛幅度）
pneumohydraulic control 气动—液动控制
pneutronic control 气动—电子控制
P/N synergism（phosphorous-nitrogen synergism）磷—氮协同效果（阻燃整理）
POA(percentage open area) 开孔面积百分率
pocket hardness metre 袖珍硬度计
pockets 袋；圆筒形织物
POF（plastic optical fibre）塑料光导纤维
pOH 氢氧根离子活度的负对数
P.O.H.（production per operative hour）每工时产量
point〈pt.〉点
point bond 点黏合
point coating 点涂布法，点子涂层
point contact detector 点接触探测器
pointer 指针；指示字；指示器；地址计数器《计算机》
point half tones effect 云纹效应
pointilistic effects 点子花纹效应，点画效应
pointille 小点子花纹
point light lamp 点光源灯（发光元件集中在一个极小的面积上的特殊灯泡，可用作连拍机等的光源）
point of grip 夹持点，握持点
point of tangency〈pt〉切点
point-paper pattern 意匠图案
point particle 质点
point sampling 点抽样法
point source 点光源《测色》
pois.（poisonous）有毒的
poise〈P，P.，p，p.〉泊（黏度单位）；平衡杆
Poiseuille flow 泊肃叶流动
Poiseuille's law 泊肃叶定律
poisoning 中毒
poisonous〈pois.〉有毒的
Poisonous and Deleterious Substances Control Law〈PDSCL〉有毒和有害物质控制法（日本）
poisonous substance 有毒物质
poisonous waste 有毒废物；有毒废水
poker bar 拨火棍，钎（清理风口棒）
pol(petroleum, oil and lubricant) 石油，油类和润滑油
pol.（polymeric）聚合的
pol.（polymerize）聚合
pol.（=polym.，polymer）聚合物，聚合体
POL（problem-oriented language）面向问题的语言《计算机》
polar absorption 极性吸收
polar activation 极性活化［作用］

polar binding force 极性键合力
polar bond 极性键
polar coordinates 极坐标《数》
polar cosolvents 极性共溶剂
polar distance〈pd〉极距
Polar dyes 普拉染料(弱酸性染料,商名,瑞士制)
polar forces 极性力(两个固体物质相黏附的力)
polargraph 极谱仪
polargraphic analysis 极谱分析
polar group 极性基
polarimetry 偏振测定法,旋光测定法
polarity 极性
polarity of chain [分子]链的极性
polarization 偏振光
polarization analyzer 检偏振[光]镜,检偏振片
polarization microscope 偏光显微镜
polarize 极化;偏振化
polarized fluorescence 偏振荧光;极化荧光
polarized infrared 偏振红外线
polarized light 偏振光
polarized microscope 偏振光显微镜
polarized shining [熨烫的]极光
polarizing photomicroscope 偏光摄像显微镜
polarograph 极谱记录器,极谱仪
polarographic analysis 极谱分析[法]
polarography 极谱[分析]法;极谱学
polar pigment 极性颜料
polar polymer 极性聚合物
polar reaction 极性反应
polar solvent 极性溶剂
polar weave 极向织造

polar winding 极向缠绕
polechangeable motor 换极电机
pole dryer 挂棒烘燥机
pole figure 极像图
pole tent 支柱结构帐篷
Polidene 33 含偏氯乙烯共聚物(商名,英国制)
Polisetile dyes 波利塞蒂尔染料(染不变性聚丙烯纤维的分散染料,商名,意大利制)
polish 加光,磨光,抛光;抛光剂
polished cottons 加光棉织物,光亮棉织物
polished finish 上光整理
polished part 磨光的部件
polisher 抛光机;打光剂,擦亮剂
polishing 抛光,上光[整理],打光
polishing machine 磨光机,抛光机
polishing machine for imitation fur fabric 人造毛皮烫光机,人造毛皮整理设备
polishing process 深度处理,高级处理《环保》
polishing stone 磨石
polka dot 色地白圆点花布图案
pollen-resistant fabric 防花粉织物
pollutant 污染物
pollutant emission 污物排放
polluted print mark 印记污染
polluted water 污水
pollution control process 污染控制方法
pollution distribution 污染分布
pollution exhaust criteria 排污标准
pollution-free technology 无污染工艺
pollution-free vent gas 无污染排放气
pollution level 污染程度
pollution load 污染负荷
pollution loading amount 污染负荷量《环

保》
pollution monitoring 污染监测
pollution source 污染源
pollution survey 污染调查
pollution treatment 污染处理
polo dots 印花大圆点
polo shirt 开领短袖衬衫,马球衬衫
Pol-Rotor [电热]烫光机(商名,德国制)
poly- (构词成分)表示"聚""多""复"
polyacetaldehyde 聚乙醛
polyacetals 聚缩醛类
polyacetylene 聚乙炔,乙炔聚合物
polyacid 缩多酸,多元酸
polyacrolein 聚丙烯醛
polyacrylacid ester 聚丙烯酸酯
polyacrylamide〈PAM〉聚丙烯酰胺
polyacrylate 聚丙烯酸酯
polyacrylic acid 聚丙烯酸
polyacrylic ester 聚丙烯酸酯
polyacrylic ester coating agent 聚丙烯酸酯涂层剂
polyacrylic resin 聚丙烯酸树脂
polyacrylonitrile〈PAN〉聚丙烯腈
polyacrylonitrile fibre 聚丙烯腈纤维
polyacrylonitrile-vinyl chloride fibre 腈氯纶(商名,中国)
polyaddition 加[成]聚[合][作用]
polyalcohol 多元醇
polyalkane 聚烷烃
polyalkylamine surfactant 聚烃胺表面活性剂
polyalkylation 多烷基化[作用]
polyalkylene sulfone 聚烯砜
polyaluminosiloxane 聚铝硅氧烷
polyamide 聚酰胺
polyamide calender bowl 聚酰胺轧光辊(由聚酰胺材料薄片压制而成,具有回弹性好、耐磨度高、可高速运转等特点)
polyamide-coated glass fibre 聚酰胺涂层的玻璃纤维
polyamide dye 聚酰胺纤维用染料
polyamide-ester 聚酰胺酯[类]
polyamide 6 fibre 锦纶 6(商名,中国)
polyamide 66 fibre 锦纶 66(商名,中国)
polyamide fibre 聚酰胺[系]纤维
polyamide-imide fibre 聚酰胺[酰]亚胺[系]纤维
polyamide-oxadiazole fibre 聚酰胺—噁二唑纤维(耐高温纤维,美国制)
polyamide treatment 聚酰胺树脂[界面聚合]处理(毛织物防缩工艺)
polyamidoester fibre 聚酰胺酯[系]纤维
poly-p-aminobenzhydrazine terephthaloyl fibre 聚对氨基苯甲酰肼对苯二甲酰纤维
polyaminotriazole fibre 聚氨基三唑纤维,聚氨基三氮杂茂纤维(由癸二酰肼制取)
polyammonium compound 多胺类化合物
polyamylose 聚直链淀粉
polyanhydrides 聚酐[类]
polyatomic acid 多元酸
polyatomic alcohol 多元醇
polyatomic phenol 多元酚
polyazo dyes 多偶氮染料
polyazo pigment 多偶氮颜料
polybasic 多碱价的;多元的;多代的
poly bath boiling-off 多浴精练
poly-p-benzamide fibre 聚对苯甲酰胺纤维(耐高温纤维,美国制)
polybenzimidazole〈PBI〉聚苯并咪唑

polybenzimidazole fibre 聚苯并咪唑纤维，PBI 纤维(耐高温纤维，美国制)
polybenzothiazole 聚苯并噻唑
polybenzothiazole amide 聚苯并噻唑酰胺
polybenzothiazole imide 聚苯并噻唑[酰]亚胺
polybenzoxazole 聚苯并噁唑
poly-2,2'-[*m*-phenylene]-5,5'-bibenzimidazole fibre 聚-2,2'-[间亚苯基]-5,5'-双苯并咪唑纤维
poly bis-benzimidazo-benzophenanthro-line fibre 聚双-苯并咪唑-苯并菲绕啉纤维(耐高温纤维，美国制)
polyblend 聚合物混合体
polyblend fibre 聚合物混[合]纺[丝]纤维，聚合物混合体纤维
polybrominated biphenyl〈PBB〉多溴联苯(禁用的阻燃剂)
polybrominated diphenyl ether〈PBDE〉多溴化联苯基醚(禁用的溴系阻燃剂)
polybutadiene 聚丁二烯
polybutene(＝polybutylene) 聚丁烯
poly-*n*-butyl acrylate 聚丙烯酸[正]丁酯
polybutylene terephthalate 聚对苯二甲酸丁二酯
polybutylene terephthalate fibre〈PBT fibre〉聚对苯二甲酸丁二酯纤维(弹性涤纶)，PBT 纤维
polycaprolactam〈PCL〉聚己内酰胺
poly-carbamoyl sulfonates〈PCS〉聚氨基甲酰磺酸盐(羊毛防缩剂)
polycarbonate〈PC〉聚碳酸酯
polycarbonate fibre 聚碳酸酯纤维
polycarboxylic acid 多元羧酸
polycationic compound 聚阳离子化合物
polycationic fixing agent 聚阳离子固色剂

Polychlal fibre 波莱克勒尔聚氯乙烯醇纤维(氯乙烯与乙烯醇的共聚物再与聚乙烯醇混合后通过乳浊液纺丝而得，日本制)
polychlorinated biphenyl〈PCB〉多氯联苯(有毒禁用的防腐剂)
polychloroprene rubber 氯丁橡胶
polychlorostyrene 聚氯苯乙烯
polychroism 多色[现象]
polychromatic 多色的
polychromatic colour 多色光
polychromatic dyeing 多色淋染印花，多色[流液]染色法
polychromatic dyeing machine 多色染色机，多色淋染印花机
polychromatic dyeing process 多色流染印法(商名，英国卜内门)
polychromatic fibre 热敏变色纤维(能随温度的变化而改变色泽的纤维)
polychromatic illumination 白色光照明《测色》
polychromatic process [地毯]多色[流染]印法
polychrome 多色的
polychrome printing 多色印花(印花滚筒由多个花纹色拼合并制成)
polychromy 彩饰法；多彩饰
polycondensate 缩聚物
polycondensation 缩聚[作用]
polycot fabric 涤棉混纺布
polycrosslinking 多步交联(防皱整理工艺)
polycrystal 多晶体
polycrystalline 多晶的，多晶质的
polycrystalline structure 多晶结构
polycrystalline zirconium dioxide fibre 多晶

二氧化锆纤维

polycyclamide 聚环酰胺[类]（含环烷基的聚酰胺总称）

polycyclic pigment 多环颜料

polycyclic ring 多核环

polycyclization 多环化[作用]

poly-2,6-dimethyl-p-phenyleneoxide fibre 聚-2,6-二甲基对苯醚纤维

polydisperse 多分散

polydispersibility 多分散性

polydispersity 多分散性，聚合度分布性

polydithiazole 聚双噻唑

polyelectrolyte 聚[合]电解质，高[分子]电解质

polyene 多烯[烃]

polyenic 多烯的

polyepoxide 聚环氧化合物

polyester 〈PES, PEs〉 聚酯

polyesteramide 聚酰胺酯

polyesteramide fibre 聚酰胺酯纤维

polyester and cotton blend 〈PC blend〉 涤棉混纺织物

polyester conveyer 涤纶导带（用于圆网印花机或平网印花）

polyester/cotton 〈P/C〉 涤棉混纺织物

polyester crepe 涤纶绉

polyester dye 聚酯纤维用染料

Polyesteren dyes 聚酯士林染料（用于染聚酯纤维的分散性还原染料，商名，德国制）

polyester ether 聚酯醚

polyester ether fibre 〈PEE fibre〉 聚醚酯纤维

polyester fabric weight reduction range 聚酯织物减量整理联合机

polyester fibre 〈PE fibre〉 聚酯[系]纤维

polyester film lumiyarn 聚酯[薄]膜金银皮

polyester gauze 涤纶薄纱

polyester habotai 涤丝纺

polyesterification 聚酯化[作用]

polyester J-box 聚酯塑料制的J形箱

polyester metallized fibre 涤纶金属化纤维

polyester photosensitive film 聚酯光敏胶片，聚酯感光胶片

polyesterpolyamide alloy fibre 聚酯—聚酰胺混合体纤维

polyester textured fabric 涤纶变形织物

polyether 聚醚

polyether ester fibre 聚醚酯纤维

polyethoxy alkylamine surfactant 聚乙氧基烷基胺表面活性剂（用作染色助剂）

polyethoxy ethers 聚乙氧基醚类

polyethoxy glyceride 聚乙氧基甘油酯

polyethoxylated castor oil 聚乙氧基蓖麻油

polyethylene 〈PE〉 聚乙烯；聚乙烯[系]纤维

polyethylene emulsion 聚乙烯乳化剂（树脂整理的柔软剂）

polyethylene fibre 聚乙烯纤维

polyethylene glycol 〈PEG〉 聚乙二醇

polyethylene glycol content 聚乙二醇含量

polyethylene glycol esters 聚乙二醇酯类（用作抗静电剂、增塑剂、润滑剂）

polyethylene oxalate 聚乙二酸乙二[醇]酯

polyethylene oxide 〈PEO〉 聚氧化乙烯，聚环氧乙烷

poly-p-ethylene oxybenzoate 〈PEB〉 聚对苯甲酸乙氧[基醇]酯

poly-p-ethylene oxybenzoate fibre 聚对苯甲酸乙氧[基醇]酯纤维（商名为A-Tell纤维，日本制）

polyethylene terephthalate〈PET,PETP〉 聚对苯二甲酸乙二[醇]酯

polyethylene terephthalate fibre 聚对苯二甲酸乙二酯纤维,PET 纤维

Polyfast dyes(＝Polyfast colours) 波力法斯特染料(涂料和分散染料的混合染料,用于涤棉混纺织物印花,商名,美国依蒙达)

polyfluoroethylene fibre 聚氟乙烯纤维

polyfunctional 多官能的；多功能的

polyfunctional carbamate 多官能性氨基甲酸酯

polyfunctional catalyst 多官能催化剂

polyfunctional compound 多官能性化合物

polyfunctional diaper 多功能尿布

polyfunctional fixation 多官能团固着

polyfunctional fixing agent 多功能性固着剂

polyfunctionality 多功能性

polyfunctional monomer 多官能团单体

polyfunctional reactive dyes 多官能团活性染料

polygalatic acid 多聚半乳糖酸

polygenetic dyes 多色性染料(指媒染染料,用不同媒染剂可得不同的色泽)

polyglycol 聚二醇[类]；聚乙二醇(任何二醇化合物经脱水缩合得到的聚合物)

polyglycolic acid 聚乙醇酸

polyglycollide fibre 聚乙交酯纤维,聚乙醇酸交酯纤维

polyhedron foam of substeady state 多面体泡沫

polyheterocyclic amide 聚杂环酰胺

polyhexamethylene adipamide 聚己二酰己二胺(聚酰胺 66)

polyhexamethylene biguanide〈PHMB〉 抗菌整理剂[PEPUTEX 20],聚六甲双胍的盐酸盐

polyhexamethylene terephthalamide fibre 聚对苯二甲酰己二胺纤维

polyhydric 多羟[基]的

polyhydric alcohol 多元醇

poly hydroxybutyrate-co-hydroxyvalerate fibre〈PHBV fibre〉 羟基丁酸与羟基戊酸共聚纤维

polyimidazopyrrolone fibre 聚咪唑并吡咯酮纤维(耐高温纤维,美国制)

polyimide 聚酰亚胺

polyimide fibre 聚酰亚胺纤维(耐高温纤维,美国制)

polyimide film 聚酰亚胺薄膜

polyisobutylene〈PIB〉 聚异丁烯

poly isophthaloyl metaphenylene diamide fibre 聚间苯二甲酰间苯二胺纤维(商名为 Nomex,美国制)

polylactic acid fibre〈PLA fibre〉 聚乳酸纤维(一种以玉米为原料的新型纤维,易生物降解,符合环保要求)

polylol 多元醇

polym.(＝pol. polymer) 聚合物,聚合体

Polymat labor dyeing machine 波力麦特实验室染色机(商名,瑞士制)

poly-m-benzamide fibre 聚间苯甲酰胺纤维(高强高模量耐高温纤维,美国制)

polymer〈pol.,polym.〉 聚合物,聚合体

polymer alloy(＝polymer blend) 聚合物混合体

polymer-blend fibre 聚合物混纺纤维(两种聚合物混合后纺得的纤维)

polymer builder 聚合体助剂(具有增效作用的添加物)

polymer coagulant 高分子凝集剂

polymer emulsion 聚合体乳液(织物柔软整理用)

Polymer G,GE 波力膜 G,GE(阳离子反应性树脂,用于氯化处理后的羊毛机可洗防毡缩处理,商名,英国 PPT)

polymeric〈pol.〉 聚合的

polymeric binder 高分子黏合剂

polymeric compound 聚合物

polymeric dyes 聚合染料

polymeric homologue 同系聚合物

polymeric jacket 聚合物(酯)外套(光导纤维的外层)

polymeric plasticizer 高分子型增塑剂,聚合物型增塑剂(比单体增塑剂分子链要长得多的增塑剂)

polymeric surfactant 高分子表面活性剂

polymeric thickener 聚合浆料,聚合增稠剂

polymeride 聚合物,聚合体

polymerisable dyes 可聚合染料(水洗、摩擦牢度较好)

polymerisate 聚合产物

polymerism 聚合[现象]

polymerizate 聚合物

polymerization 聚合[作用]

polymerization accelerator 聚合加速剂

polymerization activator 聚合活化剂

polymerization catalyst 聚合催化剂

polymerization-coupling reactant 聚合偶联剂

polymerization initiator 聚合引发剂

polymerization mechanism 聚合机理,聚合历程

polymerization rate 聚合速率

polymerization retarder 聚合抑止剂,阻聚剂

polymerize〈pol.〉 聚合

polymerized aluminum chloride 聚合氯化铝《环保》

polymerized hydrocarbon fibre 聚烯烃纤维

polymerized sulfur-aluminum chloride 聚合硫氯化铝《环保》

polymerizer 聚合剂;聚合机,高温焙烘机(织物经树脂处理后,经此机焙烘,可促进树脂聚合并固着于纤维上);[圆网]焙烘箱

polymerizing machine 树脂聚合焙烘机(热固着设备)

polymer-making autoclave 压热聚合釜,高压聚合釜

polymer modification 聚合物改性

Polymer TM 波力膜 TM(阳离子反应性树脂,用于氯化处理后,羊毛及混纺制品的机可洗防毡缩、抗起球处理,商名,英国 PPT)

polymer weighting 聚合物增重

polymetaxylylene adipamide fibre 聚己二酰间苯二甲胺纤维(锦纶 MXD-6)

polymethacrylate〈PMA〉 聚甲基丙烯酸酯

polymethacrylic acid 聚甲基丙烯酸(用作黏度调节剂、乳化剂等)

polymethacrylonitrile〈PMAN〉 聚甲基丙烯腈

polymethine dyes 聚甲炔染料

polymethyl acrylate〈PMA〉 聚丙烯酸甲酯

polymethylene 聚亚甲基;环烷烃(环式饱和烃的总称)

polymethyl galacturonate 果胶

polymethyl hydrogen siloxane 聚甲基氢硅氧烷

Polymode Process 波力·莫德方法(一种染色和纤维表面改性同时进行的方

法,有改进手感和仿真效果)

polymolecularity 多分子性,多分散性

polynosic(＝polynosic fibre, polynosic rayon) 波里诺西克纤维(高湿模量黏胶纤维),富强纤维,虎木棉

polynuclear 多核的,多环的

polyol 多元醇

polyolefin[e] fibre 聚烯烃纤维

polyorganosiloxanes 聚有机硅氧烷类

polyoxamide fibre 聚草酰胺纤维,聚乙二酰胺纤维(耐高温纤维)

polyoxy alkyl compound 聚氧烃基化合物

polyoxyethylene alcohol 聚氧化乙烯醇

Polypayre 波莱佩耶管道绝热法(新型的管道绝热法,制造厂预先在管道内部涂上聚氨基甲酸酯)

polypeptidase 多肽酶

polypeptide 多肽

polypeptide linkage 多酰键

polyphase system 多相制(由几个频率相同但电压的相位不同的交流电路所组成的电路系统)

polyphenol-aldehyde fibre 聚酚醛纤维(耐高温纤维,美国制)

poly-m-phenylene isophthalamide fibre 聚间苯二甲酰间苯二胺纤维(商名为Nomex,美国制)

polyphenylene oxide〈PPO, ppo〉聚苯氧化物

polyphenylene sulfide fibre〈PPS fibre〉聚苯硫醚纤维(优异的耐化学腐蚀性、热稳定性及电绝缘性)

poly-p-phenylene terephthalamide〈PPT〉聚对苯二甲酰对苯二胺

poly-p-phenylene terephthalamide fibre 聚对苯二甲酰对苯二胺纤维(商名为Kevlar,美国杜邦)

polyphenylene triazole fibre 聚亚苯基三唑纤维(耐高温纤维,美国制)

polyphenylether fibre 聚苯醚纤维

polyphosphate 多磷酸盐

polyphosphonate 聚膦酸酯(用于制取阻燃聚酯纤维的添加剂及作为螯合剂)

polyphosphoric acid 多磷酸

poly-p-phenylene benzobisthiazole fibre〈PBO fibre〉聚对苯亚甲基苯并双噻唑纤维(新型高强度合成纤维)

polypropylene〈PP, Pp〉聚丙烯

polypropylene dyeability 丙纶[纤维]可染性

polypropylene fibre 聚丙烯纤维,丙纶(商名,中国)

polypropylene fibre hosiery 丙纶袜

polypropylene glycol 聚丙烯乙二醇

polypropylene oxide 聚氧化丙烯(用作润滑剂、表面活性剂等)

polypyrrole 聚吡咯

polypyrrole direct coating 聚吡咯直接涂层

poly-α-pyrrolidone fibre 聚-α-吡咯烷酮纤维(即锦纶4)

polypyrron fibre 聚吡咯纤维(耐高温纤维,美国制)

polyreaction 聚合反应(加成聚合和缩聚的总称)

polyreactive group 多活性基团

polysaccharide 多糖类

polyset process [树脂整理]多次焙烘法(棉织物先浸轧树脂和弱催化剂,焙烘后,续浸轧强催化剂,烘干,最后焙烘以完成交链的整理工艺)

polysilicate 聚硅酸盐,聚硅酸酯

polysilicate ferric coagulant〈PSF coagulant〉聚

硅酸铁混凝剂（处理印染废水）

polysiloxane 聚硅氧烷（耐洗的防水剂，俗称硅酮）

polysiloxane micro emulsion 聚硅氧烷微乳液

polysoap 聚皂（高分子表面活性剂）

polystyrene〈PS〉聚苯乙烯

polystyrene-butadiene 聚苯乙烯—丁二烯聚合物

polystyrene fibre 聚苯乙烯纤维

polystyrene resin 聚苯乙烯树脂

polystyrol 聚苯乙烯

polysulfide 多硫化合物

polysulfide crosslinked structure 多硫化合物[的]交联结构

polysulfonamide〈PSA〉聚磺酰胺，芳砜纶

polysulfone 聚砜

polysulphonamide 聚磺酰胺，芳砜纶

polyterephthaloyl oxalamidrazone fibre 聚对苯二甲酰—草酰—双脒腙纤维（耐高温抗燃纤维，德国制）

polytetrafluoroethylene〈PTFE, p. t. f. e.〉聚四氟乙烯

polytetrafluoroethylene fibre 聚四氟乙烯纤维

polythene 聚乙烯

polythene pipe 聚乙烯管

polythene vessel 聚乙烯储槽

poly-1,3,4-thiadiazole fibre 聚-1,3,4-噻二唑纤维（耐高温纤维，美国制）

polytrifluorochloroethylene fibre 聚三氟氯乙烯纤维

polytrimethylene terephthalate〈PTT〉聚对苯二甲酸丙二醇酯

polytropy 多变性，多变现象

polyunsaturated compound 不饱和高分子化合物

polyurethane〈PU, PUR〉聚氨酯，聚氨基甲酸酯

polyurethane coating agent 聚氨酯涂层剂

polyurethane elastic fibre 氨纶（弹性聚氨酯纤维）

polyurethane fibre 聚氨基甲酸酯纤维，氨纶（商名，中国）

polyurethane fibre hosiery 氨纶丝袜

polyurethane finishes 聚氨酯整理剂

polyurethane finishing 聚氨酯整理

polyurethane foam 聚氨酯泡沫

polyurethane leather 聚氨基甲酸酯合成革

polyurethane roll 聚氨酯轧辊（一种高效轧水辊，采用聚氨酯橡胶包覆而成，具有耐磨、抗变形等优点）

polyurethane rubber roll 聚氨酯橡胶辊

polyvalency 多价

polyvalent alcohol 多元醇

poly-V-belt 多条三角皮带，凸面皮带

polyvinylacetal 聚乙烯醇缩乙醛（狭义）；聚乙烯醇缩醛类（广义）

polyvinyl acetate 聚乙烯醋酸酯

polyvinylacetate acrylate 聚乙烯醋酸丙烯酸酯

polyvinylacetate ethylene 聚乙烯醋酸酯乙烯

polyvinyl alcohol〈PVA, P. V. A.〉聚乙烯醇

polyvinyl alcohol fibre 聚乙烯醇纤维，维尼纶，维纶（商名，中国）

polyvinyl butyral〈PVB〉聚乙烯醇缩丁醛

polyvinyl chloride〈PVC, P. V. C., pvc〉聚氯乙烯

polyvinyl chloride acetate fibre 氯乙烯—醋酸乙烯共聚纤维

polyvinyl chloride acrylonitrile fibre 氯乙烯—丙烯腈[共聚]纤维
polyvinyl chloride coating agent 聚氯乙烯涂层剂
polyvinyl chloride fibre 聚氯乙烯纤维
polyvinyl derivative 聚乙烯衍生物
polyvinyl finish 聚乙烯整理
polyvinylidene chloride〈PVDC〉 聚偏氯乙烯(羊毛防缩整理剂)
polyvinylidene chloride fibre 聚偏氯乙烯纤维
polyvinylidene cyanide fibre 聚偏氰乙烯纤维
polyvinyl methyl ether〈PVME〉 聚乙烯基甲基醚
polyvinyl pyrrolidone〈PVP〉 聚乙烯基吡咯烷酮(保护胶体)
polyvinyl synthetic fibre 聚乙烯[基]类合成纤维
polyxyleneadipamide fibre 聚己二酰苯二甲胺纤维
Polyzine-N 波利扎因 N(退浆酶,商名)
pomegranate 石榴红[色](暗红色)
Pompeian red 庞贝红[色](浅棕红色);红色颜料
Pompeian yellow 庞贝黄[色](浅暗橙黄色)
ponceau 朱红[色];酸性红色染料
poncho 穗饰披巾(南美人所穿,形似毯子,中间开有领口);雨披(尤指橡皮制的)
poncho cloth 防雨厚毛毯,军用防雨披(宿营用)
pond 池,塘,水池
pongee imperial 厚重光亮丝府绸
pongee print 印花细棉布

pongee silk 柞丝绸,茧绸
Ponsol dyes 保晒霖染料(还原染料,商名,美国制)
Pontamine dyes 庞塔明染料(直接染料,商名,美国制)
Pontamine White 庞塔明白(荧光增白剂,商名,美国制)
pony cloth (＝pony skin) 仿马皮长毛绒(棉地,绒头为马海毛)
poodle 卷曲绒头织物
pool 池,水坑
pooled fabric 混纺织物
poor absorbency 吸水性差
poor combustion 不完全燃烧
poor feeling 手感不良
poor mark 印花不良
poor penetration 渗透不良
poor quality product 低档织物
poor selvedge 布边不良
poor smoothness 平滑性不良
poor solvent 不良溶剂
poor white 白度不白
popeline (＝poplin) 府绸;毛葛
poplin broche 花府绸;花毛葛
poplinette 纱府绸
poppet valve 圆盘阀
poppy [red] 罂粟红[色](朱红色)
pop tester [织物]破裂强度试验仪
population 总体,母体;组;群(用于统计)
population equivalent of industrial wastes〈PEIW〉 工业废水的人口当量
population mean 总体均值
population variance 总方差
populin 府绸
Poral fibre 一种超细中空低强涤纶(具有吸湿、排湿、柔软特性)

porcelain eye 瓷圈,瓷眼(供绳状织物练漂或洗涤加工时导布用)
porcelain guide 导布瓷圈
porcelain perforated plate 多孔陶瓷板(红外辐射器材料)
porcelain printing 仿涂料印花(用水溶性染料代替涂料印花)
pore 细孔,微孔;气孔
pore model of diffusion 孔道扩散模型(染料在纤维中扩散的一种模型)
pore radius 孔隙半径
porin 孔蛋白
poromer 透气性合成革
poromeric material 微孔性[人造革]材料,透气性[人造革]材料
poromerics 多孔人造革,透气性人造革
Porosimeter 织物透气性测试仪(商名)
porosimetry 孔隙度测定法,孔率测试法
porosity 多孔性,透气性;孔积率;孔隙度;松度
porosity test [织物]透气性试验
porosity testing machine [织物]透气性试验机
porous 多孔的,疏松的
porous antitoxic clothing 透气防毒服
porous ceramic plate 多孔陶瓷板
porous fabric 多孔织物
porous fibrous materials 多孔纤维材料
porous material 多孔性物质
porous metal 多孔金属(具有多孔性的金属粉末合金,常作轴套用)
porous polymeric material 透气性合成材料
porous water-proofing 透气性防水整理
port 口,孔;通路,门;汽门;水门
portable 〈pt.〉轻便的,可携带的,携带式

portable gas-powered blower 手持式气动风机
portable lamp 工作手灯,台灯
portable measuring 简易测定
portable mixing tank 便携式混合槽(用于混合和输送染化料的不锈钢槽,配有泵、电机及管路,置于一个灵活的小车上)
portable moisture meter 便携式测湿仪
portable pump 轻便泵
portable scray 轻便J形堆布器
portable tester 便携式试验仪
port glass 观察孔玻璃,观察窗
port-hole 视孔,口,门,汽门
portière 帷幔,门帘;厚装饰织物
portion-wise addition 分批添加
Portique method 波蒂克法(织物阻燃试验法)
port wine 葡萄牙酒红[色](深紫红色)
POS(=pos,pos.,posit,positive) 正片,阳图底片,正像;正的;阳性的;阳极
position drum 定位滚筒
positioner 定位器,夹具
positioning 定位,配置;位置控制
positioning controller 定位控制器
position signal 定位信号
position transducer 位置传感器
positive 〈POS,pos,pos.,posit〉 正片,阳图底片,正像;正的;阳性的;阳极
positive adsorption 正吸附
positive catalyst 正催化剂
positive charged group 阳荷性基团
positive circulation 强制循环
positive clutch 积极式离合器;齿状离合器
positive displacement pump 正排量泵

positive draft 人工[强制]通风
positive drive 积极传动
positive feedback 正反馈
positive film 正片
positive gearing 直接传动
positive group 阳根,正基;阳极组;阳性基
positive infinitely variable〈PIV,P.I.V.〉 无级变速
positive infinitely variable drive 无级变速传动
positive infinitely variable gear box 无级变速齿轮箱
positive pressure 正压
positive result 阳性结果
positive seal 正压密封
positive stop 限位挡块
positive temperature coefficient 正温度系数(当温度增加时物质的电阻、长度或一些其他特征量也随之增加)
positive thermal fabrics 积极式保温织物
POSS 多面体倍半硅氧烷纳米复合材料
post- (构词成分)表示"后""次"
post 柱,杆;岗位,台
post-activated 后[段]活化[的]
post-bleach 后漂白
post-boarder 后定形机
post-boarding 后定形(整理时的热定形工艺)
post-bonding 后黏合工艺
post-chlorization 后氯化[作用]
post-condensation 后缩合[作用]
post-cure 后焙烘,延迟焙烘;后硫化;后固化
post-curing process 延迟焙烘法,后焙烘法

post-dye 织后染色,后染色
post dyeing 坯毯染色
poster cloth 广告招贴[用]布(表面光滑,布背重浆整理)
post-extension swell 拉伸后膨胀
post formed laminates 二次成型层压
post-forming 热后成形,二次成形
post-ignition 后燃;燃后
post-impregnation 后浸渍
post-inflation finish 后膨松整理
post-irradiation effect 后辐照效应
post-mortem dump 停机后输出,算后转储,算后打印《计算机》
post-mortem routine 算后检查程序《计算机》
post-polymerization 后聚合[作用]
post printing 织后印花,处理后印花
post processor 后处理机
post-purge 后吹扫,后清扫
post scouring 后洗净;后精练
post-set 后定形(指织物在后整理时的热定形工艺)
post-shrinkage 后收缩
post-softening 后软化
post-treatment 后处理
post-washing vessel 后洗槽
post-weld heat treatment〈PWHT〉 焊后热处理
pot 多联匹(一般长 10000 码,亚麻布整理术语)
pot.(potential) 位,势;潜力;位能,势能;电位,电势,电压;潜在的
pot.(potentiometre) 电势器,电位计,电位器;分压器
potable water 饮用水
potash 碳酸钾

potash soap 钾皂；软皂
potassa 苛性钾，氢氧化钾
potassiotartrate of antimony 酒石酸氧锑钾，吐酒石
potassium antimony tartrate 酒石酸锑钾
potassium bitartrate 酒石酸氢钾
potassium carbonate 碳酸钾
potassium chloride 氯化钾
potassium cyanate 氰酸钾
potassium dichloroisocyanurate 二氯异氰尿酸钾（羊毛防缩整理剂）
potassium dichromate 重铬酸钾，红矾钾
potassium ferrocyanide 亚铁氰化钾
potassium fluoride 氟化钾
potassium fluozirconate 氟锆酸钾
potassium hydroxide 氢氧化钾
potassium iodide test paper 碘化钾［淀粉］试纸
potassium monopersulfate 过硫酸氢钾（漂白剂）
potassium permanganate 高锰酸钾
potassiumperoxy monosulfate 过一硫酸氢钾（羊毛防毡缩剂）
potassium persulphate （= potassium peroxydisulphate） 过［二］硫酸钾，高硫酸钾（氧化剂，漂白用）
potassium soap 钾皂
potassium sulphocyanide 硫氰酸钾，硫氰化钾
potato starch 马铃薯淀粉
potency 潜能
potential〈pot.〉位，势；潜力；位能，势能；电位，电势，电压；潜在的
potential acid 潜酸（能释放酸的物质），释酸剂
potential advantages 潜在优点

potential barrier 潜在屏障
potential buffer solution 潜在缓冲溶液
potential crimp 潜在卷曲
potential current transformer 变压变流器（变流器与变压器的组合）
potential difference〈pd〉位差，势差
potential drop 电位降，电压降
potential energy〈PE〉势能，位能
potential gradient 势梯度，位梯度
potential of hydrogen〈pH〉酸碱度
potential of hydrogen value pH［值］
potential pressure 静压
potential shrinkage 潜在收缩
potential temperature 潜在温度
potential transformer〈P.T.,p.t.〉变压器，测量用变压器
potentiometre〈pot.〉电势计，电位计，电位器；分压器
potentiometric analysis 电位滴定分析
potentiometric recorder 电势记录器
potentiometric titration 电势滴定［法］，电位滴定［法］
pot eye 导布瓷圈
potheater 罐式加热器，压热釜
pot life 有效时间；活化寿命；适用期
pot linen 多匹缝接的亚麻布
pottery duck 陶器工业用帆布（作过滤布用）
potting 煮呢；装入（罐内或器内）
pouce 法寸（法国古长度单位，约为2.7厘米）
pouce glass 织物分析镜（法国制）
poult［de soie］真丝横棱绸（质较厚重）；波纹绸，绉绸（法国名称）
pound 磅（英制重量单位，1磅等于453.6克）

poundal 磅达(英制力的单位)
pounds per gallon〈ppg〉磅力/加仑
pounds per inch〈P.I.,p.i,ppi〉磅力/英寸(英制线压力单位)
pounds per square foot〈psf〉磅力/平方英尺
pounds per square inch〈P.S.I.,psi.,p.s.i.,p.p.s.i.〉磅力/平方英寸(压强单位,约为 6.895 千帕)
pounds per square inch absolute〈psia,p.s.i.a.〉磅力/平方英寸(绝对值)
pounds per square inch gauge〈psig,p.s.i.g.〉磅力/平方英寸(表值)
pouring 浇注,浇铸;倒出,溢出
pour point 浇注点,流动点
POV(peroxide value) 过氧化值
powder adhesive 粉末黏合剂
powder clutch 粉末离合器(以磁性粉末为动力媒介体的摩擦式电磁离合器)
powder coating 粉末涂层
powder dot coating 粉点涂层
powder dyes 粉状染料
powdered bearing 粉末冶金轴承(含有石墨)
powdering 满幅点子花纹
powder metallurgy 粉末冶金(采用压力并加热将金属粉末制成零件)
powder method 粉末法
powder point coating 粉点涂层
powder silk 火药袋[用绢丝]绸
powder spraying coating 粉末喷送涂层,撒粉涂层
powder super fine dyes 超细粉染料
power〈pwr.〉功率;能量;动力;电源
power brake 机动制动器
power consumption 能耗,电力消耗

power conversion efficiency 功率转换效率
power drive 机械传动
power electronics 电力电子学
power equipment〈PE〉电源设备,动力设备
power fabric 弹性布,弹性织物
power factor〈pf〉功率因数;力率
power failure 电源故障
powerhouse 供电间,发电间
power line 动力线,动力网
powernet fabrics 弹性针织物
power-operated valve 机械驱动阀
power-operated widening motion 电动调幅装置
power perch 验呢机;验呢台
power plant 发电厂,动力厂
power pump 机械驱动泵,动力泵
power regulator 功率调节器,调功器
power selsyns 电轴(电气同步旋转系统,在这个系统中,几个机械上没有联系的工作机构,但有相同的转速或恒定的转速比例关系)
power socket 电源插座
power spectra 能谱
power station 动力站,供电站,发电厂
power stretch 强力弹力,高弹性伸缩(伸缩率在 50%~70%之间)
power stretch fabric 高弹织物
power switch 电源开关
power switchboard 电力配电屏
power-take-off〈pto〉功率输出端,动力输出端
power unit〈PU〉功率单位,动力单位;电源设备
pp(permanent press) 耐久压烫,耐久定形
pp(per procurationem) 由……所代表

pp(photographic plate) 照相底片,胶片
pp(power panel) 配电盘
PP(=Pp,polypropylene) 聚丙烯
PPase(pyrophosphatase) 焦磷酸酶
PPB(=ppb,parts per billion) 几亿分之几
PPE(personal protective equipment) 个人保护器材
ppg(pounds per gallon) 磅力/加仑
pphm(parts per hundred million) 亿分之几
ppi(pounds per inch) 磅力/英寸
ppi(=p.p.i.,picks per inch) 每英寸纬数
ppm(=P.P.M.,p.p.m.,parts per million) 百万分之几
PPO(=ppo,polyphenylene oxide) 聚苯氧化物
pps(=p/s,periods per second) 周/秒[钟];赫[兹]
pps(pulses per second) 每秒[钟]脉冲数
PPS fibre(polyphenylene sulfide fibre) 聚苯硫醚纤维(优异的耐化学腐蚀性、热稳定性及电绝缘性)
p.p.s.i.(=P.S.I.,psi.,p.s.i.,pounds per square inch) 磅力/平方英寸(压强单位,约为 6.895 千帕)
ppt(parts per thousand) 千分之几
ppt(precipitate) 沉淀物;沉淀
PPT (poly-*p*-phenylene terephthalamide) 聚对苯二甲酰对苯二胺
practicability 实用性,应用性
practical example 实例
practical notes 生产实践注意事项
practical property 实用性能
practical use test 实用试验

practice 实践,实习;操作规程
pram suit 婴儿装
prayer seam 平行缝接,普通缝接
PRC(pigment resin colour) 颜料树脂色
pre- (构词成分)表示"前""先""预先"
pre-acidifying dyeing method 预加酸染色法
preadduct 预加合物
pre-aeration 预曝气
preageing period 人工老化期
preamplifier 前置放大器《电》
preassemble 预装配
prebaking 前焙烘法
preboarder 预定形机
preboarding 预定形(指锦纶袜类的染前定形)
preboiling dyeing method 预沸染色法
precaution 预防措施
precautionary measures 预防办法;安全措施
prechrome dyes 媒染(预媒)染料
prechrome process 预媒染色法
precipitability 沉淀度;沉淀性;临界沉淀点
precipitant 沉淀剂,沉析剂;沉淀物,脱溶物
precipitate⟨ppt⟩ 沉淀物;沉淀
precipitating bath 沉淀浴;凝固浴
precipitation 沉淀[作用]
precipitation chromatography 沉淀色谱法,沉淀色层[分离]法
precipitation gelation 沉淀凝胶化[作用]
precipitation inhibitor 防沉淀剂
precipitation polymerization 沉淀聚合[作用]
precipitation separation 沉淀分离法

precipitation titration 沉淀滴定
precipitator 沉淀器;聚尘器
precise pattern 精确花样
precision 精确性;精密度;准确度
precision balance 精密天平
precision cut flock 定长切断短绒
precision cutting 定长切绒[屑](植绒用语)
precision instrument 精密仪器
precision regulating screw 微调螺丝
precision regulator 精密调节器,精密控制器
precision setting 精密调节,微调
pre-clarification tank〈PCT〉预澄清槽
precleaning 预净化,预清洗
pre-coat 预涂层
precoated filter 涂层过滤器;涂胶过滤器
precoating 预涂层
precoded card 编码卡片
precoiled spiral 预盘绕螺旋管
precombustion chamber 预燃室,预燃炉
pre-commissioning check 投入运转前检查
precondensate 预冷凝;预缩合;预冷凝物;预缩合物
precondensator 预冷凝器;预缩合器
precondition 预处理;预调湿处理;悬浮液处理
preconditioning 温湿度预调节;预调湿
precontraint process 预加应力过程
precrêping 预压花
precrêping calender 预压花机
pre-curing 预焙烘;预固化;预熟化;预硫化
precuring process 预焙烘法
precursor 母体
predefined process 预定的过程

pre-degumming 预练
predict 预测
pre-draft 预拉伸
pre-drier(=pre-dryer) 预烘机
pre-drying treatment (试样的)预干燥处理
predrying tunnel 预烘隧道
pre-drying unit 预烘燥装置
predye 预染色,织前染色
preemulsion 预制乳状液(未经匀化的粗乳状液)
pre-engineered 采用预制件的,用预制件建造的
prefab 预制的
prefabrication 预制
preferred orientation 择优取向,优先取向
pre-filtration 预过滤
prefixing machine (锦纶袜的)预定形机
preflush 预先冲洗
P regulator 比例调节器,P调节器
preheater 预热器
pre-heating chamber 预热室
pre-heating treatment 预热处理
prehumidifier 预给湿装置[器]
prehydrolysis 预水解
preignition 预燃;燃前
prelas 防水棉或亚麻织物(用油和蜡处理过的)
preliminary assessment of light fastness 初期耐光坚牢度评估
preliminary drying 预烘
preliminary experiment 预试
preliminary finish [布匹]预整理,前处理
preliminary heat setting 预热定形
preliminary operation 试运行
preliminary project 初步设计方案

preliminary screening 预先筛选
preliminary steeping 预浸《毛》
preliminary tension 初张力，预张力
preliminary test 预试
pre-manufacture 预加工
premature boiling 煮练不足，生块（丝绸练疵）
premature colour failure 染料过早失效
premature degumming 精练不足，脱胶不足（丝绸练疵）
premature oxidation 过早氧化
pre-mercerization 预丝光
pre-metallized acid dyes 金属络合酸性染料
premetallized dyes 金属络合染料
premium workwear concept 全优工作服理念
premix 预混合；[增强塑料]预混料
pre-orientation degree 预取向度
pre-oriented yarn 预取向丝（或纱），部分取向丝（或纱），低取向丝（或纱）
prep. (preparation) 前处理；制备；准备
prep. (preparatory) 准备的
prepadding 预浸轧
preparation〈prep.〉 前处理；制备；准备
preparation agent (= preparation medium) 前处理剂；油剂《化纤》
preparation tank （染料的）储备罐
preparatory〈prep.〉 准备的
preparatory process 准备工程，前处理工程
prepared extender 备用印花冲淡浆
prepared-for-print 待印花的半制品织物
prepared hank yarn 轧光绞纱（制花边用）
prepared material 经过前处理的织物
prepared reduction style 预还原法

pre-pigmentation dyeing 预悬浮体染色
prepolycondensate 预缩聚物
prepolymer 预聚物
pre-polymer gel 预聚物凝胶
pre-polymerization 前聚合，预聚合
prepreg 预浸渍体；预浸胶体（玻璃纤维增强的塑料半制品）
prepreg method [碳纤维网]预浸渍法，预浸胶法（复合材料成形法之一）
preprint 印前预准备
preprinted resist 先印防染法
pre-programmed chain 预定程序电路
pre-purge 预吹扫，预清扫
preradiation grafting 预辐射接枝
preraised 预起绒的
pre-reduction method [还原染料印花]预还原法
pre-relaxing treatment 预松弛处理
prerequisite 前提条件
prescour 预精练
pre-scouring agent 预精练剂
preselection 预选
presensitized 预敏化，预增感
presentation 表示；显示；图像
preservative 防腐剂
preset 预调定，预定形
preset capacitor 预调电容器，微调电容器
preset control 程序控制
preset parameter 固定参数
preset shrinkage 预定形收缩
preset time 提前时间《计算机》
presetting 预定形
presetting machine 预定形机
presharpen technique [活性染料]预加碱染色法（指染色初期，先加一部分碱剂，以缓和固色率的一种绳状染色技

术）

preshrink finish(=pre-shrinking) 预缩整理

preshrinking and presetting of yarn 纱线预缩定形

preshrinking corrugation 预缩折皱

preshrinking machine 预缩机

pre-shrinking rate 预缩率

preshrunk 已预缩的，预收缩的

press 压呢机；打包机；烫衣机；压辊脱水；烫平

press. (=P，P.，p，p.，pressure) 压力；压强

press board 电[热]压[呢]纸板；熨烫工作台；手工印花台板

press button 按钮

press cake 滤饼

press cloth 滤布

pressed air 压缩空气

pressed crease retention 熨烫褶裥持久性

pressed-in crease 熨烫折痕

pressed-in crease retention 熨烫折痕耐久性

pressed pile 瘪绒，倒绒（疵点）

pressed plush 拷花长毛绒，压花长毛绒

presser 压片，压针板，加压杆；承压滚筒；烫衣工；打包工

press finishing [针织品]热板压烫整理

pressfit 压配合，压力配合；压烫合身（服装）

press head 压头，压板；挡块

pressing 轧水，压榨；压呢；熨烫；压制[法]

pressing and setting machine 连续压烫机，压烫定形机《毛》

pressing bowl 轧辊，压辊

pressing dummy [成衣]蒸烫机

pressing factor 挤压因素，压缩因素

pressing & friction calendar 压烫摩擦轧光机（毛制品加工用）

pressing machine 压呢机；烫衣机

pressing shrinkage 压烫收缩

press iron 熨斗，烙铁

pressostat 稳压器，恒压器

press paper 电[热]压呢纸板

press plate 压板

press retention 压烫保持性

press roller 压辊；（预缩机的）压送辊

press sheet 垫布

press shrinkage 压缩收缩

press steaming 熨烫汽蒸

press switch 按钮开关

press table （小包机的）压板，底板

press tenter 压烫拉幅定形机（适用于圆筒形针织坯布）

pressure 〈P，P.，p，p.，press.〉 压力；压强

pressure atomization 加压喷雾

pressure beam 紧压辊（卷取调节装置部件）；张力[落下]辊

pressure beck 高压绳状染槽

pressure bladder 密封气囊

pressure bobbin dyeing 加压筒子染色

pressure boil(=pressure boiling) [棉布的]高压煮练

pressure bowl [印花机]承压滚筒

pressure-break （层压塑料内的）裂缝

pressure cake bleaching 丝饼加压漂白法

pressure casting method 压铸法

pressure container 压力容器

pressure controller 〈PC〉压力控制器

pressure control valve 压力控制阀

pressure cylinder [印花机]承压滚筒

pressure decatizing 加压蒸呢;全蒸;罐蒸
pressure distribution 压力分布
pressure drop 压[力下]降
pressure dyeing 高压染色
pressure dyeing machine 高压染色机
pressure element〈PE〉压力元件
pressure engraving machine 钢芯雕刻机;轧钢芯机,轧花[雕刻]机
pressure flow 压力流
pressure gauge 压力计
pressure gradient 压力梯度
pressure indicating controller〈PIC〉压力指示控制器
pressure jig 高压卷染机
pressure kettle 高压锅
pressure kier 高压煮布锅
pressureless pretreatment steamer 无压力前处理蒸箱
pressure lock sample device （高压染色机的）关压取样装置
pressure mark （布面发亮的）压痕,压斑;压印（整理疵,织物表面发亮）
pressure monitor 压力监控器
pressure nozzle 压力[式]喷嘴
pressure package dryer 卷装加压烘燥机
pressure pan 高压釜,高压罐
pressure pump 压力泵
pressure reducing valve 减压阀
pressure regulator valve 压力调节阀
pressure relay 压力继电器（流体压力变化达到预定值时动作的继电器）
pressure release valve 放压阀
pressure-relieving vessel 扩容器
pressure resistance 耐压性
pressure restraint layer 压力保持层
pressure retaining valve 保压阀

pressure roller 加压辊
pressure saturation 饱和压力
pressure-seal 气压封口
pressure-sealable cylinder 加压密闭圆筒
pressure-sensitive adhesive 压敏胶黏剂
pressure sensitive dyes 压敏染料
pressure sensitive tape 压敏胶带
pressure sleeve 压力套筒
pressure specific heat 定压比热
pressure steam ager 高压汽蒸箱
pressure steamer 高压汽蒸箱
pressure vacuum gauge 压力吸力计,真空压力计
pressure vessel 高压容器
pressure washing 压力冲洗
pressurization 增压
pressurized oil circulation 加压油循环
pressurized water 加压水（指压力容器中的水）
pressurizer 增压装置,稳压器
pre-steaming 预蒸化
Prestech roll 普雷斯泰克辊（一种微孔弹性轧辊,由纤维材料压制而成,轧水效果极好,美国制）
pre-steeping 预[浸渍]退浆;[纤维素的]预浸渍[碱化]
Prestogen 普莱斯托津系列（双氧水漂白助剂,商名,德国巴斯夫）
Prestogen EB 普莱斯托津 EB（不含表面活性剂的双氧水漂白有机稳定剂,适用于冷漂除杂工艺,商名,德国巴斯夫）
Prestogen W 普莱斯托津 W（有机酸和缓冲稳定剂混合物,羊毛在酸性双氧水浴中快速漂白的助剂,商名,德国制）
pre-store 预存

pre-straightener 预整纬器
prestrain 预应变
prestress 预应力
pre-stretching process 预拉伸过程
Pretema Colour Computer 〈PCC〉 普雷特马配色计算机(瑞士制)
Pretema Spectromat FS 普雷特马滤色片型测色仪(商名,瑞士制)
pretension 初张力,预张力,预拉伸,预加张力
pretreatment 预处理,前处理
pretreatment steamer 前处理汽蒸箱
prevention of static electricity 静电防止
preventive maintenance 预防性维修,预防性保养
prewasher 预洗器,预洗机
prewashing 预洗
prewetting device 预湿装置
pri. (=prim.,primary) 初级的;原始的
price 〈pc.〉 定价,价格
prickle 刺痛感
prill 颗粒
prim. (pri.,primary) 初级的;原始的
Primaloft 仿羽绒纤维(商名,美国制)
primary 〈pri.,prim.〉 初级的;原始的
primary air 一次风,一次空气
primary amine 伯胺
primary amine surfactant 伯胺表面活性剂
primary backing fabric [簇绒毯的]基底布
primary bond 主价键
primary cellulose acetate 初级醋酸纤维素,纤维素三醋酸酯
primary coat 底涂层;首次涂层
primary coil 初级线圈
primary colours 基色,原色(指红、黄、蓝三色)
primary control program 〈PCP〉 主控制程序《计算机》
primary creep 第一潜伸,初级蠕变(载荷解除后能缓慢地复原的变形)
primary degradation 初级降解[作用]
primary degumming 初练《丝》
primary energy 一次能源(亦称天然能源,来自自然界,不需加工或转换而直接加以利用的能源,如石油、天然气、煤、太阳能、水能、风能等)
primary fibrillation 初次原纤化,初级原纤化
primary hand values 〈PHV〉 基本手感值
primary hydration 初级水合作用
primary material 原料
primary pollutant 一次污染物
primary scouring 初练
primary sedimentation process 初级沉降法《环保》
primary standard substance 一级标准试剂
primary steam 一次蒸汽(直接从锅炉产生的蒸汽,又称直接蒸汽)
primary stimuli 原刺激(视神经的红、绿、蓝三种感光刺激)
primary storage 主存储器《计算机》
primary treatment 一级处理,初级处理
primary treatment of waste water 废水一级处理,废水简易处理,废水初级处理
primary valence 主[化合]价
Primasol 平马素系列(染色助剂,具有好的防泳移效果,提高渗透和匀染性,在浴中尚有防皱作用,商名,德国巴斯夫)
Primasol AMK 平马素 AMK(丙烯酰胺与丙烯酸的共聚物,用于分散染料染

聚酯纤维预烘时的防泳移剂,商名,德国巴斯夫

Primasol FP 平马素 FP(脂肪族磺酸化合物,阴离子型,渗透剂和防泳移剂,商名,德国制)

Primasol PAN 平马素 PAN(非离子型,渗透剂和固色催化剂,用于丙烯酸系纤维,商名,德国巴斯夫)

Primazin dyes 普里马青染料(活性染料的一种,含酰胺基活性基团,商名,德国制)

Primazin P dyes 普里马青 P 染料(活性染料的一种,含二氯哒酮基团,商名,德国制)

Primenit VS 泼力明脱 VS(天然与再生纤维织物的防水剂,树脂整理的柔软剂,商名,德国制)

primer 打底剂;底层黏结剂

prime urethane 初级尿烷

primrose yellow 樱草黄[色](黄色)

primuline yellow 樱草灵黄[色](黄色);黄色直接染料

prin.(principal) 主要的,重要的

prin.(principle) 原则,原理;规则

princesse cashmere 开司米单面棉绒(背面起绒,仿开司米,法国制)

principal⟨prin.⟩ 主要的,重要的

principal colour 主色

principal component 主成分

principal coordinates 主坐标

principal length 主体长度

principal valence 主[化合]价

principle⟨prin.⟩ 原则,原理;规则

principle of superposition 叠加原理

print 印花

3Dprint(three dimensional print) 三维立体印刷;三维立体打印;三维立体印花

printability 适印性,印花适用性

print base fabric 印花地布

print belt [自动网印机]无缝印花胶带

print block 印花头

print bonded fabric 印花黏合[非织造]织物

print bonding 印花黏合

print change system 印花不停机换网换色操作系统(减少停机,提高生产效率,特别对小批量有利,荷兰斯托克)

print coat 印花涂层

print-cold batch process 印花—冷轧堆法(羊毛印花用工艺)

print control character 打印控制符

print document formulation⟨PDF⟩ 打印文件模式《计算机》

print dress 印花服装

printed calico 印花[平]布

printed carpet 印花地毯

printed casement 印花窗帘布

printed circuit 印刷电路

printed circuit board⟨PCB, p. c. b.⟩ 印刷电路板

printed circuit diagram 印刷电路板图

printed cloth 印花布

printed cut pile carpet 印花割绒地毯

printed flannels 印花法兰绒

printed georgette velvet with flower 烂印乔其绒

printed ground 印花底色

printed knit fabric 印花针织物

printed lable tape 印刷商标带

printed moquette 印花绒头织物

printed stitch-bonded fabrics 印花缝编织物

printed top 印花毛条
printed wool sweater 印花羊毛衫
printed yarn 印花纱线
printer 印花工,印花工作者;打印机,印刷机;印刷器
printers 印花[用]平纹坯布(英国名称)
printer's hydro 雕白粉(美国名称)
Printex 普林坦克斯(天然增稠剂,有不同的成糊率,用于丝绸、锦纶、涤纶及羊毛,得色量高,清晰度、光亮度高,手感柔软,商名,意大利宁柏迪)
print fold 印花折皱
print gingham 印花格子布
print head 印花头
printing 印花[工艺]
3Dprinting (three dimensional printing) 三维立体印刷;三维立体打印;三维立体印花
printing adhesive 印花用黏着剂(印花台板贴布用)
printing and dyeing in a single process 印染一步法
printing apron 转移印花压毯(包覆在加热大滚筒上的毯子)
printing binder 印花黏合剂
printing blanket 印花橡胶衬布,印花导带
printing blanket aligning unit 印花导带整位装置
printing blanket transport system 印花导带输送装置
printing block 手工印花木模
printing blotches 印花[块面]不匀(疵点)
printing board 印花台板
printing carriage [网印]网版行车,平网印花走车
printing cell 印花单元

printing concentrates 印花浓浆(涂料印花现成印浆)
printing defect 印花疵点,印疵
printing doctor 印花刮刀
printing down machine 拷贝机,复印机
printing emulsifier 印花用乳化剂
printing faults 印花疵点,印疵
printing gum 印花用糊料
printing hydro 雕白粉(美国名称)
printing machine 印花机;印[麻]袋[唛头]机
printing methods 印花方式,印花方法
printing oil 印染油,土耳其红油,太古油
printing on demand 按需印制
printing on print 叠印
printing pads 印花垫块(经纱印花装置)
printing paste 印花[色]浆
printing paste modifier 印花糊改进剂;印花糊改性剂
printing position 印花位置
printing range 印花机组
printing recipe 印花处方
printing roller 印花滚筒
printing roller crane 花筒吊车(滚筒印花机机头处装卸花筒轴等部件)
printing roller engraving plant 花筒雕刻设备
printing roller grinding machine 磨花筒机
printing roller polishing machine 花筒抛光机
printing roller spindle 印花滚筒铁芯子
printing roller turning and finishing machine 花筒车磨机
printing roller turning machine 车花筒机
printing room 印花车间
printing screen 印花筛网

printing screen frame 印花网框
printing shell [印]花[滚]筒(不包括芯子)
printing spatula 印花刮浆板(手工印花用具);[网印]刮浆刀
printing squeegee 印花刮刀
printing station 印花位置
printing stencil 印花模板;印花筛网(手工印花)
printing styles 印花方式,印花方法
printing table 印花台板
printing trolley 筛网印花框机
printing unit 印花单元
printing viscosity index〈PVI〉 印花浆黏度指数
printing woollen blanket 印花机毛衬布
printing wrinkle 印花折皱
print motor (由印刷电路构成的)微电动机,印刷电动机
Printofix PD liq. 印花固 PD 液(弱阴离子型,用作金银粉及白色涂料印花的黏合剂,商名,瑞士科莱恩)
Printofix PF 印花固 PF(合成树脂聚合用的分散液,印花助剂,用以固着涂料,商名,瑞士制)
Printofix white colours 印花固白色涂料(即时用印花用涂料,商名,瑞士科莱恩)
Printogen E 印德匀 E(含有乳化剂的矿油,作印花油用,商名,瑞士制)
Printogen HDN liq. 印德匀 HDN 液(固色促进剂,非离子型,用于分散染料的过热蒸气或热风固色,商名,瑞士制)
print-on(＝print-on print) 直接印花
printout 晒印;复制;打印输出,印出
print pad 轧印[花筒](用于轧满地色浆或化学品)
print paste 印花浆
print paste rheology 印花浆流变性
print plate 印花模子,印花底板
print ratings 印花效果等级
prints 印花布
print stain 印花沾污
print wash system 在印花机上冲洗圆网的方法
printworks 印花厂
print yarn stripes 印经条子织物
priority 先,前;优先;优先级,优先权
prismatic colours 光谱色彩,棱镜色彩
prism spectrophotometer 棱镜型分光光度计
pristine surface 原始表面,原来表面
privileged instruction 特权指令《计算机》
privileged operation 特许操作
probability 概率;可能性
probability of error 误差概率
Proban 普鲁本(四羟甲基氯化磷,棉阻燃剂,商名,英国制);普鲁本阻燃整理法(英国)
Proban CC 普鲁本 CC(四羟甲基氯化磷—尿素初缩体,棉织物阻燃整理剂,商名,英国奥布赖)
Proban process 普鲁本阻燃整理法(用 THPC-脲预聚物溶液浸轧并经氨气处理的耐洗性阻燃整理工艺,商名,英国)
probation 检验,验证,鉴定
probe 探测;探测器;探头,探针;传感器
probing 探测,测试
problem definition(＝problem description) 题目说明

problem-oriented language ⟨POL⟩ 面向问题的语言《计算机》
proc.（proceedings） 研究报告集，记录汇编；会刊；学报；记录
procedure 程序；工序；步骤；工艺规程；方法；过程
procedure-oriented language 面向处理过程的语言《计算机》
proceedings ⟨proc.⟩ 研究报告集，记录汇编；会刊；学报；记录
process 方法；过程，流程，历程；工序；程序；作用；操作，加工，处理
processability 加工性能
process automation 工艺过程自动化
process calculation 工艺计算
process camera 制版照相机
process change 工艺变化（包括加重、失重、收缩和伸长等）
process colour 三原色套印色料
process computer 过程控制计算器
process conditions 工艺条件
process control ［生产］过程控制，［加工］程序控制
process control program ⟨PCP⟩ 过程控制程序《计算机》
process control unit 工艺控制装置（应用微型计算机对染整设备的工艺参数进行自动控制）
process design 工艺设计
processed starch 加工淀粉
process engraving 三原色雕刻，照相制版法
process flow diagram 工艺流程图
processing 处理；加工；操作，作业
processing aids 加工助剂

processing behaviour 加工性能
processing line 工艺流程线，流水线
processing loss 加工损耗
processing parameter 工艺参数
processing plan 工艺方案
processing property 工艺性能
processing system ［计算机的］处理系统
process lens 制版镜头
process line 工艺流程线，流水线
process monitoring 操作过程的监控，加工过程的监视
process optimization 过程最佳化，工艺最优化
processor 加工机械；处理机，处理器；处理程序
process plate 三原色套色版
process printing 三原色套印，彩色套印
process programmer 操作程序设计器
process route 工艺路线
process shrinkage 加工缩率（指织物在染整加工过程中产生的收缩率）
process silk （绢网印花用的）绢网
process steam 工艺用汽
Procilan dyes 普西兰染料（染羊毛的活性染料，含丙烯酰胺基或氯乙酰胺基团，商名，英国制）
Procilene PC dyes 普锡兰 PC 染料（由普施安 T 与迪司潘素 PC 染料复制而成，用于涤棉一步法热熔染色或直接印花，商名，英国卜内门）
Procinyl dyes 普施尼染料（活性分散染料，染聚酰胺纤维用，商名，英国制）
Procion 普施安（染料商名，英国卜内门）
Procion H dyes 普施安 H 染料（活性染料，含一氯三嗪基团，商名，英国制）
Procion HEXL dyes 普施安 HEXL 染料

（活性染料，含有两个一氯均三嗪活性基团，属环保型，不含有害有机胺物质，商名，德国巴斯夫）

Procion M dyes 普施安 M 染料（活性染料，含二氯三嗪基团，商名，英国制）

Procion-resin process 普施安树脂法（棉织物用活性染料和活性树脂混合浸轧并烘干、焙烘，同时获得耐洗的染色和防缩防皱效应）

Procion Supra dyes 普施安特级染料（活性染料，商名，英国制）

Procion T dyes 普施安 T 染料（活性染料，含膦酸基团，商名，英国制）

procurement scheme 采购方案

prod.（produce） 生产，制造

prod.（product） 产品，生成物；[乘]积

prode sampling 钩棒取样

produce⟨prod.⟩ 生产，制造

producer 发生器；发生炉；生产者；生产厂

producer-coloured 纺丝厂着色，纺前着色

producer colour polyester 原液染色聚酯，原液着色聚酯

producer gas 发生炉煤气

producer's dyed fibre 化纤厂生产的着色纤维，原液着色纤维

producer's textured yarn⟨PTY⟩ 化纤厂生产的变形丝，纺丝厂变形丝

product⟨prod.⟩ 产品，生成物；[乘]积

product data management⟨PDM⟩ 产品数据管理

product finishing roll 出布辊

production 生产；制造；产品

production bottleneck 生产薄弱环节，瓶颈现象

production capacity 生产能力

production constant 生产常数

production control 生产管理

production ecology 生产生态学

production efficiency 生产效率

production equipment 生产设备

production hours 生产小时数，生产时间

production monitoring 生产监控

production order sheet 生产指定书

production per capita 每人生产量

production [output] per machine hour 台时产量

production per manhour⟨PMH，p. m. h.⟩ 每人每小时的产量，每工时产量

production per operative hour⟨P.O.H.⟩ 每工时产量

production quantity 生产量

production scheme 生产流程图

production targets 生产指标

production time [有效]工作时间；（计算机的）运算时间

production volume 生产量

productivity 生产[能]力，生产量，生产性

Product NPC 普罗杜克特 NPC（烷基芳基聚乙二醇醚，热熔染色的浓染助剂，商名，德国制）

product performance 产品特性

proenzyme 酶原

professional clothing 工作服

profile 轮廓，外形；侧面图；（温度速度等的）分布图

profile chain link 模制链节

profiled fibre 异形[截面]纤维

profiled fibre cross section 异形纤维[横]截面

profiled filament 异形截面长丝

profile fibre 异形[截面]纤维
profile roll 咬合辊
profile steel structure 型钢结构
profoamer 泡沫促进剂
proform [纺织]预制品，[纺织]初步加工成品
prog.（**progress**）进展；进度
program 程序；大纲，方案；说明书
program abort 程序异常中止《计算机》
program card 程序穿孔卡片
program computer [卡片]程序计算机
program control 程序控制
program controlled interrupt〈PCI〉程序控制中断
program controller 程序控制器
programmable controller 可编程序控制
programmable memory 程序可控存储器
programme 程序；大纲，方案；说明书
programme chart 程序图
programmed check 程序检验
programmed function keyboard〈PFK〉程序操作键盘
programmed heating 程序加热
programmed switch 程序开关
programmed temperature gas chromatography 程序升温气相色谱法
programmer 程序设计器
program[m]ing 计划，规划；程序设计，程序编制
programming language〈PL〉程序设计语言《计算机》
programming language control〈PLC〉逻辑控制；程序设计语言控制；可编程控制器
program reader 程序读取器《计算机》
progress〈prog.〉进展；进度

progress control 进度管理
progression agent 增效剂；改进剂
progressive curing 自发焙固（延迟焙烘中的一种不良现象）
progressive metering 渐进式计量
progressive proofs 套色打样
progressive shrinkage 递进缩水性
progressive tentering roller 逐步扩幅辊，递进扩幅辊
progressive type transmission 级进式传动
prohibited dyestuffs 禁用染料
prohibited ingredients 禁用物质，禁用成分
prohibited use 禁用
project 投影；投射；计划，设计
projection 投影；投影图；投影法；凸出物
projection microscope 投影显微镜
projector 投影仪；放映机；幻灯
project site 工程现场
proliferation 增生，增殖；扩散；激增
promoter 促进剂，助催化剂；助聚剂
promulgation 散播，传播
proneness to snagging 易钩丝性《针》
proof 证明，证据；检验；样张；标准酒精度；耐久的，可防止的，能抵抗的
proofed breathable fabric 防水透气织物
proofing 防水处理；防护
proofing press [花筒]打样机
proofing sheetings 无疵坯布；标准坯布
proofing twills 防雨卡坯布
proof test 验证试验；校验
propadiene 丙二烯
propagation [链]增长作用；传播，蔓延
propanetriol 丙三醇
propeller 叶轮，推进器；螺旋桨
propeller fan 螺旋桨式风机，轴流式风机

propeller mixer 螺旋桨搅拌器
propeller pump 螺旋桨式泵
propeller stirrer 螺旋桨搅拌机
propensity 倾向
properties of mucilages 胶浆(增稠剂)性能(美国印花技术用语)
property 性质;性能,特性(专指材料、物质等的性质)
property sort [按]特性分类,[按]性能分类
propiolactone cotton 丙内酯处理棉(可增进棉的染色性)
propionic acid 丙酸
propionic anhydride 丙酸酐
prop lever 支杆,撑杆
proportional band 线性范围,成比例的范围
proportional control 线性控制,正比控制
proportional controller 比例调节器
proportional counter 比例计数器
proportional error 相对误差
proportional-integral controller 比例—积分调节器
proportional-integral-derivative controller 比例—积分—微分调节器
proportional integral differential〈PID〉比例微积分《计算机》
proportional-integral-proportional-derivative controller 比例—积分、比例—微分调节器,PI、PD调节器
proportional-plus-floating control 比例—无差调节,均衡调节
proportional relation 比例关系
proportional sampling 比例抽样法
proportional valve 比例调节阀
proportioner 比例调节器;定量配合器

proportioning 配合,定量,配量
proportioning box 配浆箱
proportioning damper 分配挡板
proportioning pump 配合泵,比例泵
proportion of upper and bottom tension 线紧率(缝纫)
proprietary brand 特殊品种
proprietary product 特定商品
propyl cellulose 丙基纤维素
propylene 丙烯
propylene carbonate 丙烯碳酸酯
propylene glycol 丙二醇
prorate value 按比例推算值,按比例分配值
prospectus 说明书;简介,提要
prosthesis 假体,修补术
prot.(protected) 有防护的
prot.(protein) 蛋白质
prot.(prototype) 样机;原型;标准;样板
protanopia 红绿色盲
protanopic chromaticity confusion 红绿色盲的错乱(对蓝—绿、蓝光红无视觉)
protease 蛋白酶
protectant 防护剂,保护剂
protected〈prot.〉有防护的
protecting cover 防护罩
protecting gown 防护衣
protection 防护,保护,防止
protection against cold weather 防寒性
protective ability 防护能力
protective agent 防护剂(保护丝、毛不受碱的损伤)
protective clothing 防护衣,防护工作服,护身服;宇宙航行服
protective coating 保护涂层,保护膜
protective colloid 保护胶体

protective cover 安全罩
protective device 防护装置
protective earwear 防噪声耳套
protective eyewear 护目镜
protective fabrics 防护织物
protective oxidizing agent 保护性氧化剂（一种温和氧化剂，如间硝基苯磺酸钠，应用于拔染印花，可防止产生地色浮雕等疵病）
protective paper ［转移印花用］衬纸
protective power ［织物的］保护能力
protective profile 安全管，保护装置
protective relay 保护继电器
protective ribbon 防护用带
protective textiles 防护用纺织品
protector 保护器，保护装置，保护设备；防毒具
protein〈prot.〉蛋白质
proteinaceous 蛋白质［状］的
proteinase 蛋白酶
protein disulphides isomerase 二硫键重排的蛋白酶
protein [base] fibre 蛋白质纤维
protein hydrolysate 蛋白质水解物
proteinmases 蛋白质水解酶
protein-splitting enzymes 蛋白水解酶，分解酶
proteolysis 蛋白水解［作用］
proteolytic enzyme 蛋白水解酶
proteolytic products 蛋白水解物
protochloride of tin 氯化亚锡
proto fibre 原［生］纤维
protofibril 原［生］原纤，原［生］微丝
proton acceptor 质子接受体
proton affinity 质子亲和力
protonation 质子化作用，加质子作用

proton donor 质子给予体
protonic pretreatment 质子化前处理
proton magnetic resonance〈PMR〉质子磁谐振
proton microscope 质子显微镜
proton nuclear magnetic resonance spectroscopy 质子核磁共振光谱学
proton transfer 质子传递［作用］
protopectin 原果胶
protopectinase 原果胶酶
protophilia 亲质子性
protophilic solvent 亲质子溶剂
protophobic solvent 疏质子溶剂
protoplasm ［细胞的］原生质
prototype〈prot.〉样机；原型；标准；样板
prototype design 原型设计；样机设计；标准设计
Prototype number 标准号数（专用于美国制的染料）
protozoa 原生动物
protracted test 疲劳试验
protruding 伸出的，突出的
protruding ends 伸出的纤维端；断纤维；毛丝
provision 预备，准备；设备；供给，供应
Provision Surgical Helmet System 预防外科面具系统
proximate analysis 近似分析
proximity suit 近程防护套装
proximity switch 近程开关
prune 梅干红［色］（暗红青莲色）
Prunell 普伦尼尔羊绒厚呢
Prussian blue 普鲁士蓝［色］（深蓝色）
Prussiate black 黄血盐精元
prussic acid 氢氰酸
pry 撬棒，杠杆

p/s(=pps, periods per second) 周/秒[钟];赫[兹]
p.s.(per sample) 每种试样
p.s.(per second) 每秒[钟]
PS(polystyrene) 聚苯乙烯
PSA(polysulfonamide) 聚磺酰胺,芳砜纶
PSA fibre 聚磺酰胺纤维,芳砜纤维(耐高温纤维)
pseudo- (构词成分)表示"伪""拟""假""准""充"
pseudocolours 伪彩色
pseudo core-spun yarn 仿包芯纱
pseudo-elasticity 伪弹性,假弹性
pseudo end point 似似终点(电气分析)
pseudo leather 仿皮革,人造革,合成革
pseudo-phase-change fibre 伪相变纤维
pseudoplastic flow 假塑性流动
pseudoplastic flow behaviour 假塑流动性
pseudo-plasticity 假塑性
pseudosolution 假溶液,胶体溶液
pseudostable state 假稳定状态
pseudo synergism 拟似协同作用
pseudotautomerism 假互变异构[现象]
pseudo-viscosity 假黏度;非牛顿黏度
pseudo-viscous behaviour 假黏度性能,非牛顿黏度特性
psf(pounds per square foot) 磅力/平方英尺
PSF coagulant(polysilicate ferric coagulant) 聚硅酸铁混凝剂(处理印染废水)
P.S.I.(=psi, p.s.i., p.p.s.i., pounds per square inch) 磅力/平方英寸(压强单位,约为6.895千帕)
psia(=p.s.i.a., pounds per square inch absolute) 磅力/平方英寸(绝对值)

psig(=p.s.i.g., pounds per square inch gauge) 磅力/平方英寸(表值)
PSS(paternoster storage system) 念珠式储存系统
PSTM(photo scanning tunneling microscope) 光子扫描隧道显微镜
psychological complementary 心理的补色《试》
psychology 心理学
psychrometer 干湿球湿度计,定湿计
psychrometic chart 温湿图
psychrometric chroma diagram 心理色品图
psychrometric ratio 湿度比
psychrometry 湿度测量法;测湿学
pt.(part) 部分;要素;零件
pt.(pint) 品脱(英美液量及容量单位,等于1/8加仑)
pt.(point) 点
pt(point of tangency) 切点
pt.(portable) 轻便的,可携带的,携带式
P.T.(=p.t., potential transformer) 变压器,测量用变压器
PTFE(=p.t.f.e., polytetrafluoroethylene) 聚四氟乙烯
PTFE film lamination 聚四氟乙烯薄膜层压
PTFE-linings 聚四氟乙烯衬料
PTFE non-oil bearing 聚四氟乙烯无油轴承
pto(power-take-off) 功率输出端,动力输出端
PTO(=P.T.O., p.t.o., please turn over) 见反面,见下页
pts.(parts) 份
PTS(paste transport system) 印浆输送系

统

pts. (pints) 品脱

PTT (polytrimethylene terephthalate) 聚对苯二甲酸丙二醇酯

PTT fibre 聚对苯二甲酸丙二醇酯纤维（新型聚酯纤维,可用常压沸染法）

PTY (producer's textured yarn) 化纤厂生产的变形丝,纺丝厂变形丝

PU (=PUR, polyurethane) 聚氨酯,聚氨基甲酸酯

PU (power unit) 功率单位,动力单位;电源设备

publication〈publ.〉 出版物;刊物

public hazard 公害

public hygiene 公共卫生

public nuisance 公害

puce 暗红色

pucker 皱纹;褶裥;起皱器《缝》

pucker crinkle 起皱

puckered 泡泡效应;起皱效应

puckered cloth 皱纹布

puckered fabric 绉纹织物

puckered selvedge 褶边（织疵,由于布边较松而成褶裥状）

puckerfree 无皱缩

puckering ［沿衣服缝线的］皱纹;皱褶;折叠;皱起

puckering severity index ［织物］折皱度指数

PU coagulation process 聚氨酯凝固法（湿涂层法,用于合成革制造）

PU coating agent 聚氨酯涂层剂

puddle 胶土

puebla cut 普布拉割绒法（一行左向割绒与一行右向割绒相间的方法）

puebla velveteen 普布拉棉绒

PU elastomer filament 聚氨酯弹性体长丝（弹力丝）

pufferpipe ［煮布锅中的］中心喷管

puffing ［织物的］蓬松装饰

puffing agent 发泡剂,膨松剂

PU film lamination 聚氨酯薄膜层压

puggaree (=puggree, pugree) ［印度人用的］薄头巾;［帽子后的］遮阳布

pulled down yarn 收缩纱（地毯用的特制纱,经过处理,可使织成的绒头收缩而成浮凸花纹）

pulled surface 皱裂［表］面

puller ［布］拉出器

pulley 皮带盘;滑轮;滑车

pulley block 滑车

pulley drive〈pd〉 滑轮传动

pull lever 拉杆,拖杆

pull-off roller 出布辊,拖布辊

pullon 套衫

pull-out fuse 插入式熔丝

pull over 套衫

pull switch 拉线开关

pull-up 层脱,层压板脱层;拉起

pulp 浆粕,纸浆;浆料

pulp digester ［制浆］煮浆器

pulping 打［成］浆

pulping by Celdecor-Pomilio process 塞尔代科尔—博米利奥法打浆

pulsating current〈PC〉 脉冲电流

pulsating dye bath ［纱染色］脉动染浴

pulsating tension 跳动张力;抖动张力

pulsation 脉动

pulsator [hank] dyeing machine〈PDM〉 ［绞纱］脉动染色机

pulsator hopper 脉动漏斗

pulse 脉冲
pulse air collector 脉冲空气除尘器
pulse-amplitude modulation〈PAM〉脉冲幅度调制,脉幅调制
pulse code modulation〈PCM〉脉冲编码调制
pulsed servo 脉冲伺服系统,脉冲随动系统
pulsed xenon 脉冲氙灯
pulse frequency〈pf〉脉冲频率
pulse-frequency modulation〈PFM〉脉冲频率调制
pulse generator 脉冲发生器
pulse height 脉冲高度
pulse jet 脉冲喷射
pulse motor 脉冲电动机
pulser 脉冲发生器
pulser fault 脉动发生器故障
pulses per second〈pps〉每秒[钟]脉冲数
pulse type steam trap 脉冲式疏水器
pulse width 脉冲宽度
pulse width modulated frequency changer 脉冲调宽式变频器
pulverization 粉碎[作用];研磨[作用]
pulverized soap stone 滑石粉
pulverizer 磨粉机;粉碎机;雾化器
Pumex 普麦克斯湿磨毛机(商名,意大利制)
pumice 浮石,轻石
pump 泵
pump block 泵座,泵套
pump cleaning device 泵清洁装置
pump delivery 泵输出量
pump drive assembly〈PDA〉泵传动装置
pump flushing system 泵冲洗系统
pump housing 泵壳

pumping head [泵的]扬程(俗称水头)
pump jacket 泵套
pumpkin 南瓜橙[色](橙色)
pump lift 泵扬程
pump shaft 泵轴
pump side 泵侧[圆网印花机输送印浆侧];[机器]受张力侧
punch 冲孔;冲头
punch card 穿孔卡,冲孔卡
punch card control panel 打孔卡片控制板
punch-card machine〈PCM〉轧纹板机,踏花机;穿孔机
punched card programme control 打孔卡片程序控制
punched tape 穿孔带
punching nozzle 冲孔喷风管
puncture resistance 抗刺破[强度]
pup stenter 短小拉幅装置
PUR(＝PU, polyurethane) 聚氨酯,聚氨基甲酸酯
purchase 采购
purdah 帘子,帏幔;印度高级面纱
pure dyed silk 不增重的染色丝绸
pure finish 清水整理(不用浆料药品)
pure new wool 纯新羊毛
pure purple 纯紫色
pure silk 纯丝绸(不用增重剂)
pure silk brocade 真丝花缎,真丝锦缎
pure silk goods 纯丝织物(染整中不加增重剂)
pure sizing 轻浆
pure starch finish 纯淀粉上浆整理
pure viscosity 纯黏性
purge 清洗,清除;换气
purge gas 清洗气;排气
purging [设备的]清洗

purification 提纯,净化;精练
purified cellulose 纯净纤维素
purified cotton 消毒棉;吸水棉
purifying finish 防臭整理
purine 嘌呤(珠光色素成分)
purity 纯度,纯净
Purkinje phenomenon 浦尔金耶现象(随光的亮度下降,视感觉的适应性由黄红色向蓝色转移的现象)
purl fabric 双反面针织物
purple 青莲
purple boundary 纯紫轨迹
purplish 略呈紫色的
purplish blue 藏蓝[色]
purply 略呈紫色的
purpose 目的;用途,效果
purpose made〈pm〉 特制的
purpurin[e] 红紫素,1,2,4-三羟基蒽醌(紫色媒染染料)
PUR roll 聚氨酯橡胶辊(一种高效轧水辊筒)
push button 按钮
push-button control 按钮控制
push button panel 按钮屏
push button switch 按钮开关
pusher 推杆
pusher arm 推动杆;推动臂
pusher block 推杆,撞杆
push fit 推入配合
pushing handle 推手柄
pushing prints 透印双面花布
pushmina 羊绒,山羊绒,开司米(克什米尔用语)
pushmina shawl 羊绒披巾
pussywillow 柔柳绸(一种软薄绸)
PU synthetic leather 聚氨酯合成革

putrefaction 腐烂[作用],腐败[化]
putrefactive bacteria 腐败细菌
putrefiable 易腐败的,会腐烂的
putrescibility 腐烂性,易腐性
putridity resistance 耐腐性
putty 油灰;填油灰;油灰色(淡灰褐色,淡褐灰色)
put up 卷装;(成品)包装;(成品)折叠
PV(photo voltaic) 光伏
PVA(=P.V.A.,polyvinyl alcohol) 聚乙烯醇
PVA size recovery system 聚乙烯醇浆料回收系统
PVA ultrafiltration recovery system 聚乙烯醇[浆料]超过滤回收系统
PVB(polyvinyl butyral) 聚乙烯醇缩丁醛
PVC(pigment volume concentration) 颜料体积浓度
PVC(=P.V.C.,pvc,polyvinyl chloride) 聚氯乙烯
PVC artificial leather 聚氯乙烯人造革
PVC coating 聚氯乙烯涂层
PVC coating agent 聚氯乙烯涂层剂
PVC conduit 聚氯乙烯硬管
PV cells 光伏电池
PVC flexible hose 聚氯乙烯软管
PVC-insulated copper wire 聚氯乙烯包覆铜芯线
PVC-sheathed cord 聚氯乙烯护套软线
PVDC(polyvinylidene chloride) 聚偏氯乙烯(羊毛防缩整理剂)
PVI(printing viscosity index) 印花浆黏度指数
PVME(polyvinyl methyl ether) 聚乙烯基甲基醚
PVP(polyvinyl pyrrolidone) 聚乙烯基吡

咯烷酮（保护胶体）
PV printing 光伏印花
PWHT(post-weld heat treatment) 焊后热处理
pwr.(power) 功率；能量；动力；电源
PX(＝px, para-xylene) 对二甲苯
PXA(para-xylylene acid) 对苯二甲酸
PXD (para-xylylene diamine) 对苯二甲胺
pycnometre 比重瓶，比重计
pyjama check 格子睡衣布
pyjamas （宽大的）睡衣裤；（印度和巴基斯坦伊斯兰教徒穿的）宽松裤
pyknometre 比重瓶，比重计
pyknometric method 比重瓶法
Pyracryl dyes 皮拉克吕染料（染腈纶纤维的阳离子染料，商名，瑞士制）
pyranthrone 皮蒽酮染料
pyrazole 吡唑
pyrazolone 吡唑啉酮
pyrazolone dye 吡唑啉酮染料
pyrethrum 除虫菊（防蛀用）
Pyrex AH 派勒克斯 AH（织物阻燃剂，商名，德国制）
pyrex glass 派莱克斯耐热玻璃（硼硅酸盐耐热硬质玻璃，美国制）
pyridine 吡啶，氮[杂]苯
pyridine derivative 吡啶衍生物
pyridinium compounds 吡啶季铵盐化合物（用作防水剂、杀菌剂）
pyrimidinone 嘧啶酮（六环树脂整理用剂）
pyro- （构词成分）"焦""火""热""高温"
pyro-acid 焦酸

pyrocatechol 焦儿茶酚，邻苯二酚（照相显影剂）
pyrocellulose 焦纤维素，高氮硝化纤维素
pyro-chemistry 高温化学
pyro-condensation 热缩作用
pyrodextrinisation 高温糊精化[作用]
pyroelectric coefficient 热电系数
pyroelectric detector 热电探测器，热释电探测器
pyroelectric effect 热电效应
pyrogen 发热源
pyrogenic decomposition 热解，高温分解[作用]
pyrogen removal 发热源的去除
pyrograph 烫画，烙画（在木或皮革上）
pyrology 热工学
pyrolysated 热分解的；热解的
pyrolysis 热解，高温分解
pyrolysis gas chromatography〈PGC〉热解气相色谱[法]
pyrolytic chromatography 热解色谱法；热解色层分析法
pyrolytic decomposition 高温分解
pyrometer 高温计
pyrophosphatase〈PPase〉焦磷酸酶
pyrophosphate 焦磷酸盐
pyrophosphoric acid 焦磷酸
pyrosulfite 焦亚硫酸盐
pyrotechnic 烟火[制造]技术
pyrrole 吡咯
pyrrolidone 吡咯烷酮
pyruvic acid 丙酮酸

Q

Q(quality factor) 品质因数,Q 值
Q(quality of heat) 热量
QC(quality control) 质量控制,质量管理;质量检查;密度控制
QED(quod erat demonstrandum) 证明完毕;已如所示
QI(quality index) 质量指标
QPL(qualified products list) 合格产品目录
Qr. (=qr.,quarter) 夸特(宽度单位,等于 1/4 码);四分之一;四等分;象限;方位角
QRS(quick response system) 快速反应系统
QRV(quick release valve) 快泄阀
QSP(quick stop) 快停
qua. (=qual.,qualitative) 定性的;品质的;合格的
qua. (quality) 品质,质量;性质
quad. (quadrangle) 四边形,四角形
quad. (quadrant) 象限;扇形体;扇形齿轮
quad. (quadruple) 四倍的;四路的
quadgauge 四角规矩
quadr- (构词成分)表示"四""第四""平方""二次"
quadrangle〈quad.〉 四边形,四角形
quadrant〈quad.〉 象限;扇形体;扇形齿轮
quadrant balance 扇形天平,细度秤
quadrant bracket 扇形托架
quadrant drive 扇形齿轮传动

quadrant gear 扇形齿轮
quadratic mean value 二次平均值
quadrature-lagging 后移;滞后 90°相位差
quadri- (构词成分)表示"四""第四""平方""二次"
Quadriga cloth 夸德里加高密薄织物(棉制,有时印花,经专门整理,色泽手感较好,商名,美国制)
quadrochromatic designs 四原色图案
quadrochromatic designs printing 四原色图案印花
quadruple〈quad.〉 四倍的;四路的
quadruple cloth 四层织物;四层组织
quadruple effect evaporator 四效蒸发器
qual. (=qua.,qualitative) 定性的;品质的;合格的
qualification test 质量鉴定试验,合格试验
qualified name 限定名
qualified products list〈QPL〉 合格产品目录
qualitative〈qua.,qual.〉 定性的;品质的;合格的
qualitative analysis 定性分析
qualitative electrophoretic technique 定性电泳技术
qualitative spot test 定性斑点试验
quality〈qua.〉 品质,质量;性质
quality audit 质量检查(纱线)
quality brand 名牌
quality class 品质级

quality classification 品质分级
quality coefficient 质量系数
quality control〈QC〉质量控制,质量管理;质量检查;密度控制
quality factor〈Q〉品质因数,Q值
quality improvement 品质改善
quality index〈QI〉质量指标
quality inspection 质量检验
Quality Label Association for Textile Finishing Industry 纺织品整理工业质量标签协会
quality level 质量水平
quality mark 品质标记,品质符号
quality number 品号;(羊毛)品质支数
quality of aspect 外观质量
quality of fit 配合等级
quality of heat〈Q〉热量
quality setting 密度调整《针》
quality standard 质量标准
quality symbol 品质标记,品质符号,品质标签
quan.(quantitative analysis) 定量分析
quant 量子;定量的
quant.(quantitative) 定量的;数量的
quanta 定量;总数;量子(quantum 的复数形式)
quantifiable 可量化的
quantify 用数量表示
quantitative〈quant.〉定量的;数量的
quantitative analysis〈quan.〉定量分析
quantitative analysis of end group 端基定量分析
quantity 数量;程度;大小
quantity of electric energy 电量
quantity of light 光量
quantize 数字转换;量子化

quantum 定量;总数;量子
quantum efficiency [荧光]量子效率(测定荧光染料的效率数据)
quantum mechanics fundamentals of colour 色的量子力学理论
quantum pump 定量泵
Quarpel finishing 夸佩尔[防水]整理(用吡啶化合物和氟化合物组成的耐久性防水整理,商名,美国)
Quarpel oil repellent finishing 夸佩尔拒油整理(商名,美国)
quart 夸脱(容量单位,等于1/4加仑)
quarter〈Qr.,qr.〉夸特(宽度单位,等于1/4码);四分之一;四等分;象限;方位角
quarter bleach 1/4漂白(指亚麻纱布的漂白程度)
quarter goods 夸特织物(指以夸特计算宽度的织物,每夸特为9英寸)
quartz 石英
quartz fabric 石英布
quartz fibre 石英纤维
quartz glass 石英玻璃
quasi- (构词成分)表示"类似""准""半""亚"
quasi-crosslink 准交联
quasi-fibrous 准纤维[状]的,似纤维[状]的
quasi-permanent deformation 似永久形变,准永久形变
quasi-viscous creep 准黏性蠕变
quaterized Pc dyes 季铵化酞菁染料
quaternary ammonium 季铵
quaternary ammonium anion exchanger 季铵盐阴离子交换剂

quaternary ammonium anion halide exchanger 季铵卤化盐离子交换剂
quaternary ammonium base cellulose 季铵纤维素
quaternary ammonium borohydride 硼氢化季铵
quaternary ammonium cellulose dichromate 季铵纤维素重铬酸盐
quaternary ammonium copolymer 季铵共聚物
quaternary ammonium hydroxide 氢氧化季铵
quaternary ammonium retardant 季铵缓凝剂
quaternary ammonium salt 季铵盐
quaternary ammonium surfactant 季铵类表面活性剂
quaternary ammonium trisiloxane 季铵化三硅氧烷
quaternary nitrogen 季铵
quaternization 成季碱反应
Quellin 奎林（一种马铃薯淀粉，具有冷水溶解性，上浆剂，商名，荷兰制）
quench 淬火；骤冷；断开；猝灭，熄灭
quench bath 骤冷浴
quench box 灭火槽（烧毛机部件）
quenching effect 骤冷效应《测色》
quercitron 栎皮粉；黑栎黄（黄色）；黄色植物染料
quetsch ［浆纱机的］上浆装置（包括导纱辊、浸没辊、压浆辊）；［浸轧机的］轧辊
queue 排队，队列
quick-access memory 快速存取存储器
quick check 快速试验
quick dryness 速干性
quick intervention 快速干预；快速介入

quick release 快速放泄
quick release valve〈QRV〉快泄阀
quick response 快速反应（市场经营）
quick-response control system 快速反应控制系统，高灵敏度控制系统
quick response system〈QRS〉快速反应系统
quick-stick test 快粘试验
quick stop〈QSP〉快停
quick-subliming dyestuff 快升华染料
quick view 快速浏览
Quickwash Plus 快洗法（快速测定水洗后织物收缩的方法，商名，美国）
Quickwash Plus technology 快洗测定技术（美国）
quick wringing 快绞
quiescent 静止的
quiescent zone 静止澄清区《环保》
quiet colour 素净色
quiet comfort 静止舒适性
quieter material 消音材料
quiet shade 素净色
quilled 褶叠的，褶裥的
quilter 绗缝机（用于缝制被褥、床垫等）
quilting 绗缝，衲缝；被褥料
quilting cloth 绗缝布
quilting goods 绗缝织物
quilting machine 绗缝机（用于缝制被褥、床垫等）
quinine treatment 奎宁处理（羊毛耐热变性处理）
quinochrome 醌色素
quinoline dyes 喹啉染料；氮萘染料
quinoline yellow ［酸性］喹啉黄
quinone 醌，苯醌
quinoneimine dyes 醌亚胺染料

quinoxaline colouring matters 喹噁啉染料
quintal 公担(等于100千克)
quod erat demonstrandum〈QED〉证明完毕;已如所示
quod vide〈q. v.〉参见
quot.(quotation) 引文;报价
quot.(quoted) 引用的
quota 定额,配额
quota system 配额制,定额分配制

quotation〈quot.〉引文;报价
quoted〈quot.〉引用的
quoted price 报价
quviut 麝牛绒(从驯化的麝牛下腹部取得)
q. v.(quod vide) 参见
Q value Q值(膨化值);品质因数;优值;动力效率的倒数

R

r. (＝rad, radiation absorption dose, roentgen absorbed dose) 拉德(吸收辐射剂量单位)
R.(Reaumur) 列氏[温]度, 雷默[温]度
R(receiver) 接收器；容器
R.(recorder) 记录表；记录器, 记录装置；记录员
R(recovery) 回收；回缩；恢复
R(＝r., rad, radical) 游离基, 自由基；原子团, 基, 根, 烷烃基
R(＝r., radius) 半径
R.(＝°R, Rankine) 兰氏[温度], 兰金[温度]；兰金温标
R(＝r., ratio) 比率, 比, 比例
R(＝r., res., resis., resistance) 阻力；电阻
R(＝r., röntgen, roentgen) 伦[琴](辐射剂量单位)
ra.(radioactive) 放射性的
RA(＝Ra, Rockwell hardness A-scale) 洛氏硬度A
rabbet 插孔, 插座；缺口, 切口
rabbit hair 兔毛
rabbit hair cloth 兔毛呢
race rotation 空转
race time 空转时间
race way 座圈；轨道
rack 齿杆, 齿条, 架；导轨, 滑轨
rack-and-pinion gear 齿条传动装置
rack drive 齿条传动
racked pattern 波纹花纹, 波纹图案

racking 转移, 位移, 动作变换
racking arm 摇臂, 撑杆
rack pinion 齿条小齿轮
rack post 支杆
rack wheel 齿条传动齿轮
Raco-yet system 雷柯叶特系统(喷液式退煮漂一步法工艺与设备, 商名, 德国克莱韦弗)
Ractogen KWC 雷克托津KWC(退浆助剂, 非离子型, 能退合成浆料和淀粉浆, 商名, 日本洛东)
rad (＝r., radiation absorption dose, roentgen absorbed dose) 拉德(吸收辐射剂量单位)
rad.(radial) 径向；半径的
rad.(radical) 基, 根
rad.(radio) 无线电
Rad.(radioactivity) 放射性；放射能力
rad.(radius) 半径
radar diagrams 雷达图
Radar method 雷达式摩擦系数测定法
radial〈rad.〉 径向；半径的
radial air stream 径向气流
radial arm 旋臂
radial bearing 径向轴承
radial-blade fan 径向叶片通风机
radial clearance 径向间隙
radial distribution 半径方向分布
radial dyeing apparatus 径向流动染色装置
radial flow machine 径向流动式[毛条]染

色机
radial-flow pump 辐流泵,径向流动泵
radial migration 径向位移
radial pleated filter 径向褶裥式过滤器
radial shrinkage 径向收缩
radial straining 径向应变
radial stress 径向应力
radial thrust bearing 径向止推轴承
radial tire 子午线轮胎
radian 弧度
radiance factor 辐射系数
radiant 辐射的
radiant boiler 辐射式锅炉
radiant burner 辐射式火口(烧毛机用)
radiant dryer 辐射式烘燥机
radiant emittance 辐射率
radiant energy 辐射能
radiant exposure 辐射暴晒
radiant heat 辐射热
radiant intensity 辐射强度
radiant panel test 辐射板试验
radiant power 辐射力
radiant singeing machine 辐射式烧毛机
radiant-target pyrometer 辐射高温计
radiating area 辐射面积
radiating capacity 辐射能力
radiating fin 散热片,辐射片
radiation 辐射热
γ-radiation γ-辐射;γ-照射
radiation absorption dose 〈r.,rad〉 拉德(吸收辐射剂量单位)
radiation catalysis 辐射催化作用
radiation coefficient 辐射系数
radiation crosslinking 辐射交联
radiation-curable system 辐射[能]固化法
radiation curing 辐射固化,辐射焙烘

radiation exchange 辐射交换
radiation finishing 辐射整理(利用高辐射能使高分子材料发生化学反应而改善材料性能的加工技术)
radiation fins 辐射片,散热片
radiation grafting 辐射接枝
radiation hazard 放射危害,辐射危害
radiation heater 辐射加热器
radiation heat load 辐射热负荷
radiation-induced grafting 辐射诱导接枝[反应]
radiation initiation 辐射引发[作用]
radiation intensity 辐射强度
radiation meters 辐射仪
radiation proof overall 防辐射工作服
radiation protection textile 防辐射纺织品
radiation protector 辐射防护器
radiation pyrometer 辐射高温计
radiation resistant finish 防辐射整理
radiation resist finishing agent 防辐射整理剂
radiation resist textile 防辐射纺织品
radiation sensitizer 辐射敏化剂
radiation shield 辐射[保护]屏蔽
radiation shield textile 辐射屏蔽纺织品
radiation singeing 辐射烧毛
radiation treatment 辐射处理
radiator 散热器;辐射体;辐射器;辐射源
radiator fins 散热片
radiator hose 散热器软管
radical 〈R,r.,rad.〉 游离基,自由基;原子团;基,根;烷烃基
radical-anion initiator 游离基—阴离子引发剂,自由基—阴离子引发剂
radical catalyst 游离基催化剂,自由基催化剂

radical coupling 游离基偶合
radical decomposition 游离基分解
radical electronegativity 游离基电负性,自由基电负性
radical inhibitor 游离基抑制剂
radical initiator 游离基引发剂
radical ion 游离基离子
radical polymerization 游离基聚合
radical scavenger 游离基清除剂
radical stripping 剥色(指完全剥色)
radical transfer 游离基转移
radio⟨rad.⟩ 无线电
radio- (构词成分)表示"放射""辐射""无线电""X射线"
radioactive⟨ra.⟩ 放射性的
radioactive compound 放射性化合物
radioactive contamination 放射性污染
radioactive density determination 放射性密度测定
radioactive emission 放射线辐射
radioactive indicator 放射性指示剂;辐射指示器
radioactive initiator 放射性引发剂
radioactive isotope 放射性同位素
radioactive static eliminator 放射性静电消除器
radioactive thickness gauge 放射性测厚计
radioactive-tracer 放射性示踪物
radioactive-tracer fibre 放射性示踪纤维
radioactive tracer technique 放射性示踪技术
radioactive waste 放射性废弃物
radioactivity⟨Rad.⟩ 放射性;放射能力
radioactivity determination 放射性测定
radioactivity measurement 放射性检测
radioactivity strength 放射性强度

radiochemical analysis 放射化学分析
radiochemical synthesis of dyes 染料的放射化学合成法(用紫外线或 X 射线或 ^{60}钴 γ-射线照射中间体而合成染料)
radio components⟨RC⟩ 无线电元件
radio frequency⟨RF, R. F., rf⟩ 射频,无线电频率;高频
Radio frequency dryer 射频烘燥机(用于筒子纱、毛条等的烘燥,商名,英国制)
radio frequency drying 射频烘燥;高频烘燥
radio frequency heating 射频加热;高频加热
radio frequency plasma 射频等离子体
radio-frequency presentation 射频显示
radio frequency welding 射频焊接
radio graphic X 射线照相的
radio graphy X 射线照相术,辐射照相术;射线检验学
radio high frequency dyeing 射频染色,高频染色;高频加热染色
radioisotope 放射性同位素
radiological finish 辐射整理(采用人工放射性同位素和电子加速器来改善高聚物的性能,从而获得防油、拒水、阻燃、抗静电及染化料固着等效果)
radiolysis 辐解作用
radiometer 辐射计
radiometric analysis 放射分析
radiometric chemistry 放射化学
radiometric control systems 辐射控制系统
radiometric detector 辐射探测器
radiometric technique 辐射技术
radiometric units 辐射测量单位
radiometry 辐射度量学;放射分析法

radiotracer method 放射指示剂法；[放射]示踪剂法
radius ⟨R，r.，rad.⟩ 半径
radius bar 半径杆
radius of curvature 曲率半径《数》
radiux chiffon 拉迪克斯雪纺（棉经人造丝纬平纹光亮织物）
radix point 小数点
radnor cloth 拉德纳提花装饰织物（用丝光棉纱制）
raffinate 残液；提余液
raffinate splitter column 残液裂解塔，残液分离塔
rag 碎布，碎呢，碎料；破衣
rag bolt 棘螺栓；地脚螺栓
ragged selvedge 破边（织疵）
rail 轨，轨道；横档；栏杆
rail for access door 移门轨道
railing post 栏柱
railroad stripe 黑白条纹
rails width adjustment [链]轨道宽度调整
railway repp 火车坐垫绒头织物，火车绒
rainbow colours 虹彩色
rainbow dyeing 虹彩染色
rainbow effect 虹彩效应
rainbow printing 虹彩印花
raincloth 防雨布
rain drop repellancy 防雨性能
rain drop test 雨淋试验
raining 淋降，滴漏
rain making machine 人工降雨装置（防雨性试验）
rain penetration test 雨水浸透性试验（测织物防水性）
rain proofing 防雨整理
rain proofness 防雨性

rainwear fabrics 雨衣布
raion 再生[纤维素]纤维，人造丝（意大利名称）
raised backs 单面绒布；起绒夹里，毛绒夹里
raised blanket 拉绒毛毯
raised checks 浮纹格子布
raised colours [后处理]显色固着的印花染料
raised crossover rib 十字凸花纹
raised effect 凸纹效应
raised fabric 起绒织物
raised face flange 凸面法兰
raised finish 起绒整理，拉绒整理
raised grey fabric 起毛织物坯布
raised knit 起绒针织布
raised pattern 浮雕花纹，凸纹
raised pile 立绒；丝绒
raised printing effect 立体印花效应
raised resist print 防起绒印花
raised stitch 起毛线圈，毛圈组织
raised stripe 形成凹凸的条纹；起绒凸条；起毛凸条
raised style printing 凸纹[方式]印花，凸版印花
raised surface structure 凸纹表面结构，起绒表面结构
raised velvet 凸花丝绒
raiser [经纬线]浮点
raising 起绒，拉绒，刮绒
raising against the hair 反[毛]向起绒
raising agent 起绒剂
raising apparatus 起绒装置
raising auxiliary 起绒助剂
raising brush 起绒刷辊
raising capacity 起绒能力

raising defect 刮绒伤,起毛疵
raising effect 起绒效果
raising elements 起绒元件
raising fillet 起绒钢丝针布
raising gig (=raising machine) 起绒机,拉绒机,刮绒机
raising machine version〈RMV〉起绒机形式
raising plains 平纹绒布
raising roller 起绒辊
raising satin 拉绒缎
raising streak 起绒条纹
raising teasel machine 刺果起毛机
raising with the hair 顺[毛]向起绒
RAM (random access memory) 随机存取存储器
RAM (rapid anhydrous method) 快速无水法(用液氨进行轧蒸工艺)
Raman spectrophotometry 拉曼分光光度测定[法]
Raman spectroscopy 拉曼光谱学；拉曼光谱测定法
Ramasit K 腊弥细 K(含胶和氧化铝的石蜡乳化液,非永久性防水剂,商名,德国制)
ramie 苎麻
ramie boiling kier 苎麻煮练锅
ramie cloth 苎麻织物
ramie degumming 苎麻脱胶
ramie linen 充亚麻[的]苎麻织物
ramie lining 苎麻衬布
ramie shirting 苎麻细平布,苎麻衬衫料
ramose 分支的；有支的；多支的
ramp 接线夹；滑轨,滑道；盘,平台；斜面
random 随机,任意,无规,偶然的；乱的
random access device 随机存取设备

random access memory〈RAM〉随机存取存储器
random arrangment 不规则排列
random check 随机抽查
random coating 不规则[点子]涂层
random copolymer 无规共聚物
random copolymerization 无规共聚
random crosslinking 无规交联
random cutting 不规则切绒[屑],任意切绒[屑](植绒用语)
random degradation 无规降解
random diffusion 随机扩散
random dyeing [纱线]多色间隔染色,多色局部染色
random error 随机误差
random fault 突发性疵点
random inspection 随机抽查
random manner 随机状态
random mesh screen 随机圆网(无规排列筛孔,新型圆网,荷兰斯托克)
random pattern 随机花纹
random pilling tester 乱翻式起球试验仪
random process 随机过程
random processing 随机处理
random sample 随机试样
random scatter 漫射
random-sheared 随机剪毛
random weaving 小样试织
range〈RG,rg.,RGE〉排列；分等；量程,范围,幅度；机组
range inspection 抽查
range of linearity 线性范围
range of oscillation 摆动范围
range of sensitivity 灵敏度范围
range of shades 色泽分布范围
rank 序列,秩；排列；等级；分类

Rankine〈R.,°R〉 兰氏[温度],兰金[温度];兰金温标
ranking 顺序;分等,分级
rank score 评级,评分
rapid-access 快速[存取]存储器
rapid ageing 快速老化;快速蒸化
rapid ager 快速蒸化机,还原蒸化机(用于还原染料印花后的快速蒸化)
rapid anhydrous method〈RAM〉 快速无水法(用液氨进行轧蒸工艺)
Rapidase 拉披德斯(分解淀粉与蛋白质的退浆剂,由淀粉酶与蛋白酶混合而成,商名,美国制)
Rapidazol dyes 拉披达唑染料(用亚硫酸钠稳定重氮盐与纳夫 AS 系混合而成,俗名快磺素,商名,德国制)
rapid balance 快速天平
rapid bleaching process 快速漂白工艺
rapid cross linking 快速交联
rapid cure 快速焙烘
rapid cure adhesive 速干黏着剂
rapid cycle digestion 快速蒸煮
rapid desizing 快速退浆
rapid dryer 快速烘燥机
rapid dyeing 快速染色
rapid dyeing disperse dyes 速染型分散染料
rapid dyeing method 快速染色法
Rapid dyes 快色素
rapid fashion response [对于时尚潮流的]快速反应(反馈)
Rapid fast dyes 拉披达染料(纳夫妥 AS 与稳定重氮盐的混合体,俗名快色素,商名,德国制)
rapid fill 快速加料
rapid fixation dyeing units 快速固色染色机

rapid flow method 快流[测定]法
rapid iron 快烫,稍烫(商业术语)
rapid mixers 快速混合器,快速搅拌器
Rapidogen Developer N 拉披达近显色剂 N(二乙氨基乙醇,可中性蒸化,使快氨素显色,商名,德国制)
Rapidogen dyes 拉披达近染料(纳夫妥 AS 与重氮氨基化合物混合而成,俗名快氨素,商名,德国制)
rapid press 快速熨烫;快速熨烫机
rapid process of resin finishing 快速树脂整理
rapid response 快速反应
rapid-responsible 快速响应性,快速反应性
rapid sand filter 快速沙过滤
rapid steamer 蒸化机,速蒸机
rapid steam generator 蒸汽快速发生器
rapid stirrer 快速混合器,快速搅拌器
rapid strike 快速初期染着
rapid surface area tester 表面积快速测试仪(用于石棉透气性试验)
rapid wash 快速清洗
rare earth cerium 稀土铈
rare earth chloride 氯化稀土
rare earth compounds dyeing 稀土染色
rare earth element 稀土元素
rare earth metal 稀土金属
rare earth metal complex 稀土金属络合物
rare earth oxide 稀土氧化物
rare earths〈RE,R.E.〉 稀土金属
rarefied air 净化空气
rarefy 抽真空;使纯净
raschel carpet 拉舍尔经编地毯
raschel double needle bar fabric 拉舍尔双

面经编织物
raspberry 木莓色(暗红)
raspberry red 木莓红(红色)
raster 网栅(筛网每厘米的线数),光栅;网版(照相制版用);扫描区
raster image processor〈RIP〉光栅图像处理软件(喷墨印花用)
raster scan 光栅扫描
ratch (=ratchet) 棘爪;棘轮机构
ratchet and pawl 棘轮及爪
ratchet bar 棘齿条
ratchet catch 棘轮掣子;擒纵器,卡子;停止挡
ratchet wheel 棘轮;锯齿轮
rate 率;速率,速度;等级,程度;定额,定值
rate action [仪表]速率作用
rate constant 速率常数
rate control 比例控制,速度控制
rated 定额的;计算的
rated capacity 额定能力,额定容量
rated conditions 额定参数
rated horse power 额定马力
rated life 额定寿命
rated speed 额定速度,额定转速
rated value 额定值
rateen 纯毛哔叽
rate feedback system 速度反馈系统
rate of absorption 吸收率
rate-of-burning test 燃烧速度试验
rate of change 变化率
rate of check-rating 等级符合率
rate of coagulation 凝固速度
rate of combined package 拼件率
rate of combustion 燃烧强度,燃烧速度
rate of crystallization 结晶化速度

rate of degumming 脱胶率
rate of diffusion 扩散速度
rate of dyeing 上色率
rate of dye-uptake 上染率,上色率
rate of effusion 流出速度
rate of evaporation 蒸发速度
rate of expression 轧液率
rate of fixation 固色率
rate of flow 流速,流量率
rate of heat diffusion 热扩散速度
rate of heat exchange 热交换率
rate of loading 载荷速度
rate of operation 运行率
rate of photofading 光褪色速度
rate of radiation 辐射率,辐射强度
rate of settling 沉降速度
rate of steaming shrinkage 汽蒸缩率
rate of strike 瞬染[上色]率
rate of sweating 出汗速度
rate of temperature rise 升温速度
rate of washing shrinkage 缩水率
rate of wear 磨损率
rate of weaving shrinkage 织缩率
ratiné 平纹结子呢;珠皮大衣呢(纬线绒头,经过整理,搓磨成珠球状)
rating 定额;额定值;评级
ratio〈R,r.〉比率,比,比例
ratio flow controller 流量比例控制器
ratiometer 比率计
ration 定量配合,限额,限定
rational analysis 理性分析
rational drying 合理的干燥
rationalization 合理化
rationalization proposal 合理化建议
rational utilization 合理利用
ratio of elongation 伸长比[率]

ratio of food to microorganism 食料与微生物重量之比,污泥负荷
ratio of liquor 浴比
ratio of reflected radiation 反射的辐射量比,辐射反射率
ratio of speed 速比
ratio scale 比例尺
ratio test 比率检验法
ratteening 毛绒成珠工艺,毛绒卷曲工艺
ratteening machine 毛绒搓磨[成珠]机
rave bolt 地基涨紧螺栓(机台固定用)
rave bolt plug 地基涨紧螺栓套(机台固定用)
ravelled hanks 绞纱凌乱
raw catalyst 新催化剂,未还原催化剂
raw cotton 原棉
raw data 原始数据
raw edge 毛边,坏边(织疵)
raw fabric 本色织物
raw flax 生亚麻,原亚麻
raw hair 原毛
rawkiness 条花,条痕(织疵)
rawl bolt 地基涨紧螺栓(机台固定用)
rawl bolt plug 地基涨紧螺栓套(机台固定用)
raw material 原料
raw material dyed yarn 散纤维染色纱
raw material dyeing 散纤维染色
raw raising 坯布拉绒;煮练坯布拉绒
raw ramie 生苎麻,原苎麻
raw sienna 浓黄土色(鲜黄棕色)
raw silk dipping 生丝浸泡
raw silk fabric 生坯丝绸
raw stock 原料,未加工纤维
raw stock bleaching machine 散纤维漂白机
raw stock dyeing 散纤维染色;原毛染色
raw stock dyeing machine 散纤维染色机,原毛染色机
raw thread milling 原坯缩绒,原坯缩呢
raw umber 棕土色(黄棕色);哈巴粉(棕色天然颜料)
raw-wool scouring 洗毛
raw-wool scouring machine 洗毛机
raw yarn 原色纱
ray 线条呢;光线;射线;辐照
γ-ray irradiation polymerization γ射线照射聚合[作用]
Rayolan TS 42,OVG 雷亚伦 TS 42,OVG(有机硅及蜡类混合物乳化液,另外添加柔软剂、乳化剂、抗静电剂,阳离子型,适用于缝纫线润滑及柔软加工,商名,德国波美)
rayon 再生[纤维素]纤维,人造丝
rayon brocade 细花绸,人造丝花绸,提花人造丝绸
rayon crepe 人造丝绉
rayon cut staple 黏胶切缎纤维,再生短纤维
rayon dye 人造丝用染料
rayon fabric 人造丝织物
rayon filament 人造丝
rayon filament yarn 黏胶长丝[纱],人造丝纱
rayon flannel 人造丝斜纹绒(黏胶丝与醋酯丝混纺后交染,轻度起绒)
rayon flock 黏胶短绒(用于植绒)
rayon georgette 人造丝乔其
rayon lining 人造丝羽纱,人造丝里子纺
rayon lining silk 人造丝斜羽绸,人造丝美丽绸,人造丝羽缎
rayon lining twill 美丽绸

rayonne 再生[纤维素]纤维,人造丝
rayon serge 人[造]棉哔叽
rayon shioze 人造丝无光纺
rayon staple[fibre] 黏胶短纤维,再生短纤维
Rayosan C,CF 雷奥山 C,CF(活性紫外线吸收剂,苯系衍生物,阴离子型,用于纤维素纤维和锦纶,耐日晒、耐水洗色牢度优良,商名,瑞士科莱恩)
Rayosan P liq. 雷奥山 P 液(紫外线吸收剂,苯系衍生物,阴离子型,用于涤纶织物,耐日晒、耐水洗色牢度优良,商名,瑞士科莱恩)
ray protection 射线防护罩
RB(=Rb,Rockwell hardness B-scale) 洛氏硬度 B
r. b. (roller bearing) 滚柱轴承;滚轴支座;[桁架]伸缩支座
R-box scouring and bleaching range R 箱式练漂机
RC(radio components) 无线电元件
RC(=r. c. , reinforced concrete) 钢筋混凝土
RC(=Rc,Rockwell hardness C-scale) 洛氏硬度 C
RC(=r. c. , remote control) 遥控
RC(resistance capacitance) [电]阻[电]容
RC(resistive-capacitive) [电]阻[电]容的
r. c. (rubber-covered) 包橡皮的
rclcc foundation [针布]地布(一层橡皮、一层棉布、一层麻布和两层棉布胶合而成)
RCM(reactive dye compatibility matrix) 活性染料配伍因子
R/D(=R & D, research and development) 研究与开发,研发

RD(regular dyeing) 正常染色
R&D center 研究与开发中心
re. (=ref. , reference) 参考,参照;基准;参考资料
RE(=R. E. , rare earths) 稀土金属
REACH laws and regulations(Registration, Evaluation, Authorization and Restriction of Chemicals) 化学品注册、评估、授权和限制法规(欧盟法规)
reactance 电抗
reactant 反应物
reactant cross-linker 反应性交联剂
reactant direct dye 反应性直接染料
reactant finish 活性树脂整理
reactant fixable dyes 反应性可固着染料
reactant polyure thane fixing agent 反应性聚氨酯固着剂
reactant resin 活性树脂
reactant-type resin finishing agent 反应性树脂整理剂
reaction chamber 反应箱
reaction intermediate 反应中间体
reaction kinetics 反应动力学
reaction mechanism 反应机理
reaction product 反应产物
reaction rate 反应速率
reaction velocity 反应速度
reaction vessel 反应容器;反应釜
reactivated carbon 再生活性炭
reactivation 复活[作用],再活化,重激活
reactive bath technique 反应浴(用于防缩处理)
Reactive Black liq. form 液体活性黑(印花用活性染料,可拔白,无副产品,对环保有利,商名,意大利宁柏迪)
reactive cationic dyes 活性阳离子染料

reactive cellulosic fibre 反应性纤维素纤维

reactive /disperse combination 活性—分散拼混染料

reactive disperse dyes 活性分散染料,反应性分散染料(印染聚酰胺和醋酯纤维织物用)

reactive dye compatibility matrix 〈RCM〉活性染料配伍因子

reactive dyeing 活性染料染色

reactive dye printing 活性染料印花

reactive dyes 活性染料,反应性染料

reactive dyes for wool 毛用活性染料

reactive dyes F type F 型活性染料

reactive dyes in discharge printing 活性染料用于拔染印花

reactive dyes in resist printing 活性染料用于防染印花

reactive dyes KD type KD 型活性染料

reactive dyes KE,KP type KE、KP 型活性染料

reactive dyes KM type KM 型活性染料

reactive dyes KN type KN 型活性染料

reactive dyes K type K 型活性染料

reactive dyes M type M 型活性染料

reactive dyes one-bath-one-process dyeing 活性染料一浴一步法染色

reactive dyes P type P 型活性染料

reactive dyes with fluorochloromethyl pyrimidine group 氟氯甲基嘧啶基型活性染料

reactive dyes with methyl sulphonyl chloromethyl pyrimidine group 一氯甲基甲砜嘧啶基型活性染料

reactive dyes with monofuorotriazine group 一氟均三嗪基型活性染料

reactive dyes with pyrimidine group 嘧啶基型活性染料

reactive dyes with triazine group 均三嗪基型活性染料

reactive dyes with trichloropyrimidine group 三氯嘧啶基型活性染料

reactive dyes with vinylsulphone group 乙烯砜基型活性染料

reactive dyes X type 活性 X 型染料

reactive fibre 反应性纤维

reactive group 反应基

reactive polymer 反应性聚合物

reactive prepolymer 反应性预聚合物

reactive sensitivity 反应灵敏度

reactive site 活性部位

reactive softening agent 反应性柔软剂

reactive-under-reactive resists 活性罩活性染料防染

reactive vat dyes 活性还原染料

reactivity 反应性;反应率

reactivity of monomer 单体反应性

reactivity ratio 反应率;竞聚率;活性率

Reactofil dyes 雷阿托菲尔染料(活性染料,含二氯三嗪基团,商名,瑞士制)

Reactolan dyes 雷阿托兰染料(毛用活性染料的一种,含二氟一氯嘧啶基团,商名,德国制)

Reactone dyes 雷阿通染料(活性染料的一种,商名,瑞士制)

reactor 反应器;电抗器

reactor drum steamer 圆筒蒸化机

reactor fixation 反应器固着

reader 读数装置;读出器,阅读机

read error 读数误差

re-adhesion 再黏附

reading 读数

reading microscope 读数显微镜

reading rate 读出速度
reading tachometer 读数转速表
readjust 再调整,重调;微调
readjustment 重新调整;微调
read only memory〈ROM〉只读存储器
read-out 读出
readymade 现成的(如服装)
ready-to-wear 现成的(指衣服)
ready-to-wear industry 成衣工业,服装工业
reaeration sludge 再生污泥
reagent 试剂;反应物;反应力
reagent bottle 试剂瓶
reagent identification 试剂鉴定
Realan dyes 丽雅伦染料(毛用含氟氯嘧啶和乙烯砜型活性基团的多磺酸基活性染料,不含金属,高的湿牢度,商名,德国制)
real cycle 实测循环
re-align 重新校准,重新排列
real silk 真丝,天然丝
real silk fabric 真丝织物
real specific heat 真实比热
real wax prints 真蜡防印花
reamer 绞刀;扩孔器
reaming 绞孔,扩孔
re-animalising 动物化加重处理(蚕丝在含有骨胶或乳酪素的磷酸钠浴液中处理,可增加强力或重量)
rear elevation 后视图
rear pantograph arrangement 后缩放尺(缩小机部件)
rearrangement [分子]重排[作用]
rear view 后视图
reasoning 推理,推论,论证
re-assemble 重新组合;重新装配

reassessment 再评估
Reaumur〈R.〉列氏[温]度,雷默[温]度
Reaumur thermometer 雷默温度计,列氏温度计
reaving 拆散[织物],拆布
rebatching steamer 双滚筒汽蒸机
rebleaching 复漂
reboiling of silk 丝的复练
rebound degree 回弹率
rebound elasticity(＝rebound resilience) 回弹性
rebound resiliometer 回弹性测定计
rebound test 回弹试验
Rec.(＝rec.,record)记录,记载
recalibration 重校准
recall 再调用《计算机》
receiver〈R〉接收器;容器
receiving inspection 接受检查
receiving tank 储槽
recemic 外消旋的
receptacle 插座;容器,接受器
reception 接收,接收法
recess 凹处,凹座;电机槽
recessed sleeve 凹槽套筒
recess shearing 凹式剪花[工艺](在丝绒或地毯上剪出凹形花纹)
recipe 处方,配方
recipe memory facility 处方记忆设施
reciprocal 相互的,互惠的;反商,倒数
reciprocal piler 往复堆布机
reciprocal relation 相反关系
reciprocal viscosity 黏度倒数,倒黏度
reciprocating comb devices 往复梳理装置(多色染色法分配色流装置)
reciprocating hoisery boarder 往复式袜子定形机

reciprocating motion 往复运动
reciprocating pump 往复泵,活塞泵
reciprocating yarn dyeing machine 往复式染纱机,悬挂式绞纱染纱机
reciprocity 互易性,可逆性,相关性
recirculating air filter 循环空气过滤器
recirculating tumble drier 转笼烘干机
recirculation 再循环
recirculation pump 循环泵
recirculator 再循环管,再循环器
reckoning 计算,估计,判断
reclaimed cotton 再生棉
reclaimed rubber 再生橡胶
reclaimed wool 再生羊毛
reclamation 回收,回用;再生
recognized method 公认的方法
recoil 重绕;反冲;弹回,退缩
recombination 再化合;复合
recommended Munsell renotation system 修正的孟塞尔表色系统
recondensation 再凝缩
reconditioning 修理,检修,整修,更新
reconstituted daylight 合成昼光《测色》
reconstitution 重新组成
recoppering 重新镀铜(印花滚筒铜层在用薄之后,重新镀铜使之加厚再用)
record〈Rec.,rec.〉记录,记载
record chart 记录纸
recorder〈R.〉记录表;记录器,记录装置;记录员
recording 记录,录音;存储
recording extensometer 自动记录伸长试验仪
recording hygrometer 记录式湿度计
recording instrument 记录装置
recording metre 记录式仪表,自记仪表

recording oscillograph 记录示波器
recording screen (电子显微镜的)记录荧光屏
recording spectrophotometer 自动记录分光光度计
recording thermometer 自动记录温度表
record of production〈ROP〉生产记录
recoverability 回复性能
recoverable deformation 可复形变
recoverable extension 可复伸长,弹性伸长
recoverable heat 可再生热,可回收热
recoverable strain 可复应变
recovered cotton 再生棉
recovered wool 再生毛
recovering of heat energy 热能回收
recovering tank 回收槽
recovery〈R〉回收;回缩;恢复
recovery boiler 废热锅炉
recovery characteristic 回复特性
recovery curve 弹性回复曲线
recovery from creasing 折皱回复度
recovery limit 回复极限
recovery of colour 色泽复原
recovery plant 回收装置
recovery pump 回收泵
recrystallization 再结晶[作用]
rect.(rectangular) 直角的;矩形的
rect.(rectifier) 整流器;检波器《电》;精馏器
rectangular〈rect.〉直角的;矩形的
rectangular distribution 矩形分布,均匀分布
rectangular duct 矩形风管
rectangular header 矩形联箱
rectangular waveguide 矩形波导
rectification 精馏;整流;矫频

rectification column 精馏塔
rectifier〈rect.〉 整流器;检波器《电》;精馏器
rectifying column 精馏塔;精馏柱
rectilinearity 直线性
recto〈ro.〉 书籍的右页(即单数页)
rectometer 挂布机,挂绸机(一种手工量布叠布工具);精馏计
recuperation 复原;回收
recuperative heat exchanger 热回收换热器
recuperator 去碱汽蒸箱(用于丝光机);同流换热器;间壁式换热器;能量回收器
recurability 复焙烘性
re-cure 再焙烘
recycleable 可循环利用的
recycleablility 回收性
recycled off-gas 循环尾气,循环废气
recycled wool 再生毛
recycle fibre 回收纤维
recycle gas water scrubber 循环气水洗塔
recycle plan 回收设备
recycle pump 循环泵
recycle stream 循环物料,循环[气、液]流
recycling 再循环《环保》
recycling of waste water 废水再利用
redampening 再给湿
redder [色光]偏红
reddish 带红色的,微红的
re-degumming 复练
redeposition 再沉积[作用]
redesigning 修改设计,重新设计
re-development 再开发
red,green,blue system〈RGB system〉 红、绿、蓝三原色
red indigo 红靛;紫花

redispersibility 再分散性
redistribution 再分布
redistribution baffle 再分布挡板
red liquor 红液(醋酸铝媒染液)
Redo 雷多(乙烯树脂涂胶织物,商名)
red oil 红油,油酸,十八烯酸
redox 氧化还原[作用]
redox activation 氧化还原活化作用
redox analyser 氧化还原分析器
redox catalyst system 氧化还原催化体系
redox indicator 氧化还原指示剂
redox initiate polymerization 氧化还原引发聚合[作用]
redox initiation 氧化还原引发[作用]
redox initiator 氧化还原引发剂
redox meter 氧化还原电位计
redox polymerization 氧化还原聚合[作用]
redox potential 氧化还原电势
redox reaction 氧化还原反应
redox resin 氧化还原树脂
redox system 氧化还原[引发]系统
redox titration 氧化还原滴定
red prussiate of potash 赤血盐,铁氰化钾
red purple 红紫[色]
red shift 红[向]移[动]
red shift effect 深色效果,红移效果
red stripiness 红筋(染疵)
red tide 赤潮(海洋污染造成的)
reduced air 减压风,减压气流
reduced bath 还原浴
reduced chrome mordant [还原]铬媒(毛染色时用的重铬酸盐)
reduced colour 还原过的染料
reduced concentration 对比浓度,降低的浓度

reduced density 对比密度
reduced modulus 对比模量，折合模量，换算模量
reduced pressure 还原压力
reduced printing colour 冲淡的印花浆
reduced sampling inspection 缩减抽样方法
reduced specific viscosity〈RSV〉 比浓黏度
reduced temperature 对比温度，换算温度
reduced transparency 对比透明度
reduced turbidity 比浓浊度
reduced viscosity 黏［度］数；比浓黏度（增比黏度对溶液浓度之比）
reducer 减压阀；减速器；节流器；还原剂；稀释剂；异径接头，渐缩管
reducibility 重演性，重现性
reducing ［花样］缩小
reducing agent 还原剂
reducing bleach 还原漂白
reducing bleacher 还原漂白剂
reducing dyes 还原染料
reducing elbow 异径弯头
reducing electrode 还原电极
reducing joint 异径接合；大小头
reducing nozzle 异径喷嘴
reducing pipe 渐缩管，异径管
reducing power 还原能力
reducing tee 异径三通
reducing temperature 还原温度
reducing valve 减压阀
reducing washing 还原清洗
reductant 还原剂
reductase 还原酶
reduction 还原；减速；减少，冲淡
reduction accelerator 还原加速剂，还原促进剂
reduction activation 还原活化作用
reduction activator 还原活化剂
reduction apparatus 减速装置
reduction bleaching 还原漂白
reduction carrier 导氢剂（利用蒽醌导氢，使地色拔除干净）
reduction clear ［印花］冲淡浆；稀释用［水油乳化］印浆
reduction clear-free dyeing 免还原清洗染色
reduction clearing 还原清洗（在涤纶染色后，用烧碱保险粉还原液清洗分散染料的浮色）
reduction clearing agent 还原清洗剂
reduction current 还原电流
reduction discharge printing 还原拔染印花法
reduction formula 换算公式
reduction gear 减速齿轮，减速装置
reduction gear box 减速箱，齿轮减速器
reduction padder 还原处理用轧车
reduction paste ［印花］稀释浆，冲淡浆
reduction potential 还原电位
reduction production 还原产物
reduction protection agent 还原防止剂《印》
reduction range 缩小倍数
reduction ratio 减速比
Reduction, Recovery, Reuse, Recycle principle〈4R principle〉 4R原则（内部消除、回收、回用、循环，即清洁生产四原则）
reduction resist 还原防染
reduction steamer 还原蒸箱
reduction strip 还原剥色
reduction thickening（＝reduction thickner）冲淡浆
reduction valve 减压阀

reduction wheel 减速轮
reductive discharges 还原拔染
reductive fission 还原开裂
reductive inhibitor 还原抑制剂
red water 铁锈[污染的]水
redwood 红木类树木(用以制染料);红木棕(红棕色)
Redwood second 赖德伍德秒(黏度单位)
Redwood viscosimeter 赖德伍德黏度计
redye 再染色
re-dyeing 复染,重染
reed pulp 芦苇浆
reedy appearance 筘路条痕(织疵)
reedy cloth 筘痕织物,筘痕布
reedy warp 筘痕,筘路(织疵)
reel 纱框;棚架;卷轴;卷绕;退绕
reel dryer 卷筒热风[喷射]烘燥机
reel dyeing 绞盘[绳状]染色
reel dyeing machine 绞盘式绳状染色机
reel dye machine 绞盘[绳状]染色机
reeled off 退绕
reeled silk 绞丝
reeled yarn 绞纱
reel off 退绕
reel stand 纱框架;绷架托脚
reevaporation 再蒸发
re-examination 复验
ref. (=re., reference) 参考,参照;基准;参考资料
reface [花筒]再磨光
refacing [花筒]再磨光
reference ⟨re., ref.⟩ 参考,参照;基准;参考资料
reference beam light 参比测量光束
reference cell 参照电池
reference colour stimuli 参照色刺激

reference cross 基准十字(圆网上的对花标记)
reference diode 参考二极管
reference electrode 参比电极
reference mark 基准标记;参照符号
reference pattern 参考图样
reference pin 基准销(圆网座上的对花标记)
reference plane 基准面
reference point 基准点
reference potential 参考电位
reference sample 参考试样,小样
reference stimulus 原刺激,参照刺激《测色》
reference time 基准时间
reference width 基准门幅,参照门幅
refine 精练,精制,提纯
refinement 精练,精制;净化;改进,改善
refiner 精磨机,精制机,精选机;匀浆机
refining 精练,精制,提纯
refinishing 重新整理,再整理;复修(毛织品在开剪前进行的修补)
reflectance 反射,反射率;反射能力
reflectance curve 反射曲线
reflectance factor 反射因数,反射率因数
reflectance measurement 反射率测定
reflectance photometer 反射式光度计
reflectance spectroscopy 反射光谱法
reflectance-type instrument 反射型测量仪器
reflectance value 反射值
reflected heat 反射热
reflected light 反射光
reflected radiation 反射辐射
reflecting mirror 反射镜
reflection 反射,反映

reflection densitometer 反射光密度计
reflection density 反射密度
reflection electron microscope〈REM〉反射电子显微镜
reflection factor 反射率;反射因数
reflection pigment printing 反光涂料印花
reflection ray 反射[光]线
reflection thickness gauge 反射厚度计
reflective brilliance 反射光辉
reflective index detector 差示折光检测器
reflectivity 反射性;反射比
reflectometer 反光白度计,反射计
reflector 反射器,反射镜
reflect pigment 反光涂料
reflow 逆流,反流
reflux 回流,反流;回流加热
reflux condenser 回流冷凝器
reflux distillation 回流蒸馏
reflux pump 回流泵
refolding 再折叠
reformation 改革,革新,重组
reformee 厚篷帆布
reformer 厚绒布
reforming 重整;改良;再生成
reforming and opening 改革开放
reformulate 再配方,重新配方
reformulation 处方改正
refractance 折射;曲折
refracted light 折射光
refraction 折射;屈折;屈光
refractive exponent 折射率,折射指数
refractive index〈R. I. ,r. i. 〉折射率,折光率,折射指数
refractive index detector〈RI-detector〉差示折光检测器
refractive index increment 折射指数[浓度]增值
refractometer 折射计,折光仪
refractory 耐火;耐火材料
refractory brick 耐火砖
refractory clay 耐火泥
refractory facing 耐火材料表面
refractory fibre 耐火纤维
refractory glass fibre 耐高温玻璃纤维
refractory oxide fibre 耐高温氧化物纤维
refreshing feeling 爽快感
refreshment rate [空气]补充回复率
refrigerant 制冷剂,冷冻剂
refrigerating capacity 冷冻能力,制冷量
refrigerating performance 制冷系数
refrigeration 制冷,冷冻;制冷设备
refrigerator 制冷器,冷冻机;冷藏箱,冰箱
refulling 再缩绒
refuse 回丝,下脚;废料,废物;垃圾
refuse utilization plant 废料利用装置
Reg.(register) 记录表;调风器;寄存器《计算机》;记录器;[通风]节气门;(印花的)对花准确
reg.(regular) 正常;有规则的
Reg.(regulation) 调节,调整;调整率《电》;规则,条例
Reg.(regulator) 调整装置,调节器;调压器;调速器;调节剂
regain 回潮;回潮率;恢复
regain balance 回潮测定秤
regain percentage 回潮率
regain standard 标准回潮率,公定回潮率
regasification 再汽化,再蒸发
regenerable antimicrobial finishing 可再生的抗菌整理
regenerable energy source 可再生能源

regenerated animal fibre 再生动物纤维（指蛋白质纤维等动物纤维）
regenerated cellulose fibre 再生纤维素纤维
regenerated cellulose rayon 再生纤维素人造丝
regenerated fibre 再生纤维
regenerated modified cellulose 再生改性纤维素
regenerated polyester fibre 再生聚酯纤维
regenerated polymer 再生聚合物
regenerated protein fibre 再生蛋白质纤维
regenerated silk yarn 再生绢丝
regenerated wool 再生羊毛
regenerated yarn 再生纤维纱
regenerating column 再生塔，回收塔
regeneration 再生；交流换热[法]
regenerative airheater 回转式空气预热器，再生式空气预热器
regenerative cycle 再生循环
regenerative heat exchanger 再生式热交换器
regenerator 交流换热器
regimental 灰紫蓝色
regimentals 军服
regional pollution 区域性污染
register〈Reg.〉 记录表；调风器；寄存器《计算机》；记录器；[通风]节气门；（印花的）对花准确
register control device [印花]对花控制装置
registered design 注册图案
registered trademark 注册商标（常用®表示，放在商标后）
registering 对花
registering unit [印花]对花装置
register mark [印花]对花记号，十字规矩线
register pin 定位销
register plate 记录板，登记板
register precision 对花精度
register print 双面印花
registration 记录；对花
registration accuracy 对花精度
registration control [滚筒印花]对花装置
Registration, Evaluation, Authorization and Restriction of Chemicals〈REACH〉laws and regulations 化学品注册、评估、授权和限制法规（欧盟法规）
regression analysis 回归分析
regression equation 回归方程
regrinding 重磨，重磨
regular〈reg.〉 正常；有规则的
regular cut 普通式裁剪，常规裁剪
regular dyeing〈RD〉 正常染色
regular interval 常规间隔
regularity 整齐度，均匀度；规整性
regularity and repetition 条理与反复（图案构成的特殊规律）
regularity meter 均匀度试验仪
regular maintenance 日常维修
regular reflection 正反射，镜面反射
regular tenacity〈R. T.，r. t.〉 普通强度
regulated power supply〈RPS〉 稳定电压
regulating 调节
regulating brake 调节闸
regulating constant 调节常数
regulating damper 调节挡板
regulating device 调节装置
regulating fault 调节故障，调节误差
regulating flap 调节挡板
regulating frequency speed governor 变频调

速器(用于印染联合机)
regulating motion 调节运动
regulating shaft 调节轴
regulating spindle 调节轴
regulating technology 调节技术
regulating transformer 调压变压器
regulating valve 调节阀
regulation〈Reg.〉 调节,调整;调整率《电》;规则,条例
regulation nylon 正规锦纶(指锦纶66)
regulation range 调整范围
regulation rod 调节杆
regulator〈Reg.〉 调整装置,调节器;调压器;调速器;调节剂
regulator drive 调速传动
regulators for peroxide bleaching 过氧化氢漂白调节剂,氧漂稳定剂
rehandle 再处理,回修
rehydration 再水化
re-impregnation process 二次浸渍工艺
re-inflate 再充气(小气候试验)
reinforced 加固的,加强的
reinforced composite material 增强复合材料
reinforced concrete〈RC,r.c.〉 钢筋混凝土
reinforced fabric 增强织物
reinforced flexible hose 增强的软管
reinforced frame 加固架
reinforced plastic 增强塑料
reinforced rib [机械部件的]加强筋
reinforced welding 加强焊
reinforcement 加强,增强;加强筋
reinforcement fabrics 增强织物
reinforcement material 增强材料
reinforcer 增强剂

reinforcing agent 增强剂
reinforcing agent to adhesion 黏着强化剂
reinstate 使复原位
rejection 等外品;报废;排斥;阻碍,抑制
rejects 等外品
rejuvenation 恢复过程
related colour 相关色
related shades [深浅不同的]相似色调的色泽
relation 关系,比例关系
relation between the stress amplitude and number of repetition〈S-N〉 应力与反复弯曲次数的关系
relative affinity 相对亲和力
relative colour value 相对色浓度
relative cover 相对覆盖度(指织物密度)
relative density 相对密度
relative diffusion rate 相对扩散速度
relative dyeing rate 相对染色速度
relative equilibrium 相对平衡
relative error 相对误差
relative humidity〈RH,R.H.,rh〉 相对湿度
relative interfacial tension 比表面张力
relative luminous efficiency 比视感度,相对可见度
relative retention 相对保持率
relative spectral power distribution 相对光谱功率分布
relative speed 相对速度
relative tack 相对黏性
relative velocity 相对速度
relative viscosity〈RV〉 相对黏度
relative wall thickness [棉纤维的]相对胞壁厚度
relativity 相关,比较性,相对性

relax 松弛,回缩
relaxation 松弛[作用];弛豫[作用];衰减;回缩
relaxation bath 松弛浴
relaxation compartment 松弛槽
relaxation elasticity 松弛弹性
relaxation machine 松弛精练机(丝绸、化纤织物用)
relaxation mechanism 松弛机理
relaxation of stress 应力弛豫,应力松弛
relaxation performance 松弛性能
relaxation shrinkage〈RS〉 松弛收缩(羊毛);回缩缩率
relaxation shrinkage on wetting 湿后松弛回缩
relaxation spectrum 松弛光谱,缓和光谱
relaxation time 松弛时间
relaxation treatment range 松弛处理机组
relax cylinder dryer 松弛滚筒烘燥机
relaxed fabric 松弛织物
relaxed state 松弛状态
relaxing 松弛作用
relaxing-bulking treatment [膨体纱制品的]膨松处理
relaxing tank 松弛槽
Relax jet dryer 松弛喷射烘燥机(商名)
relax-shrink steaming 松弛及收缩汽蒸
relay 继电器
releasable bond 可松黏合,可除黏合
releasable dog 可松轧头
releasable formaldehyde 释放甲醛
release catch 掣子;卡子;擒纵器
release coated paper 脱膜涂层纸(用于脱膜体系转移印花)
released formaldehyde 释放甲醛
release emulsions 脱膜乳液

release fabric 排液织物(医用)
release film 可释放膜
release input 释放输入(从紧急停车释放出来,恢复启动)
release layer 释放层,脱膜层
release mechanism 松开装置,松压装置
release paper 离型纸,脱膜纸(用于转移涂层)
release switch 释放开关
release time 弛缓时间
release valve 安全阀,泄放阀
releasing 放松;解压,卸压
releasing agent 脱模剂
releasing tension 退解张力
reliability 可靠性,可靠度
reliability test 可靠性试验;耐久性试验
relief 凸纹,浮雕花纹
relief cylinder 凸纹印花滚筒
relief engraving 凸纹雕刻
relief-free 无约束排放,自由排放(水或气)
relief lever 卸压杆,释压杆
relief liquor 放出液
relief offset 凸版胶印,干胶印(印刷)
relief pattern 凸纹花样(阳纹花版)
relief printing 凸纹[滚筒]印花;凸版印刷;发泡印花
relief printing machine 凸纹滚筒印花机
relief spring 保险弹簧,平衡弹簧
relief valve(=relieve valve) 安全阀,保险阀
reload 再装入
relocation 重新定位,重新安置
relustering(=relustring) 再上光整理
REM(reflection electron microscope) 反射电子显微镜

rem.〈remark〉注意

Remacen dyes 雷玛森染料(染聚酯纤维与羊毛混纺织品的染料,商名,德国制)

Remacryl dyes 雷玛克吕染料(阳离子染料,商名,德国制)

Remaflam drying method 雷玛佛莱姆烘燥法

Remaflam dyeing plant 雷玛佛莱姆染色设备

Remaflam process 雷玛佛莱姆工艺(染色和烘燥新工艺,德国制)

remaining soil 残留污垢

Remalan dyes 雷玛兰染料(染羊毛用活性染料,含乙烯砜基团,商名,德国制)

Remalan fast dyes 雷玛兰染料(1∶2金属络合染料,染毛、丝及聚酰胺纤维,商名,德国制)

remanufactured raw materials 再生原料

remark〈rem.〉注意

Remaron dyes 雷玛龙染料(聚酯纤维与纤维素纤维混纺织物印花用的分散性染料和活性染料的混合染料,商名,德国制)

Remazolan dyes 雷玛唑兰染料(染羊毛用活性染料,含乙烯砜基团,商名,德国制)

Remazol dyes 雷玛唑染料(活性染料,内含乙烯砜基团,商名,德国制)

Remazol H dyes 雷玛唑 H 染料(对碱稳定的一种活性染料,商名,德国制)

Remazol Salt FD 雷玛唑缓染盐 FD(在乙烯砜型活性染料印花中作为供碱体,商名,瑞士科莱恩)

remedial measurement 补救措施

remedy 补救,修理;校正

remedy allowance 公差

remnants 零头布,机头布

remobilization 再活化[作用],再运转

remoistening 再润湿

Remol E 染母 E(还原染料用甲法染色时的非离子性匀染剂,商名,德国制)

Remol HT 载体 HT(联苯类化合物,涤纶高温高压染色用,可提高匀染性及移染性,还可作为剥色剂,商名,瑞士科莱恩)

Remol OK 染母 OK(多羟乙基化酯醇,非离子型渗透剂、匀染剂、分散剂、乳化剂,商名,德国制)

remote access 远程访问《计算机》

remote control〈RC, r. c.〉遥控

remote diagnosis 远程诊断《计算机》

remote indicating controller 远程指示控制器

remote indicating instrument 遥示仪表,遥控记录仪表

remote manual control 远程人工遥控

remote reading thermometer 遥测温度计

remote sensing instrument 遥感仪

remote set 遥控设定

remote support contract〈RSC〉远程支持合同(即没有服务工程师的合同,不到现场的服务)

removability 退除性,可去除性

removable disk 可移动磁盘《计算机》

removable insulating plate 可移动的绝热板

removal 除去;移动

removal by filtration 滤除

removal by suction 抽除

removed weft 纬移,浮纬(织疵)

remover 去[涂]膜剂;去污剂

removing dust 除尘
removing stain 去污
rendering hydrophilic 亲水化处理
renewable 再生的
renewal [衣服]翻新；更新
renewal of oil 换油
rennet casein 酶凝酪素；皱胃酪蛋白
renovate 更新，革新
renovation dyeing 翻新染色
reorganization 结构重组，[升温过程中的]结构变化；改组；整顿
reorientation 再取向；重定位
reoxidation 再氧化
reoxygenation 再充氧[作用]
rep.（repair）修理；织补；修补
rep.（repeat）重复；[花纹]循环；单位花样，花回
rep.（=rept.，report）报告
repack 改装，重新包装
repacking 重新包装，改装
repair ⟨rep.⟩ 修理；织补；修补
repair shop 修理工场
repair size 修理尺寸
repeat ⟨rep.⟩ 重复；[花纹]循环；单位花样，花回
repeatability 重演性，再现性，反复性
repeated laundering 重复洗涤
repeated trials 重复试验
repeat length setting system 印花花回长度设定系统
repeat precision 对花精确度
repeat rails in screen table printing 台板印花轨道
repeat setting 花回调整
repeats in the roller circumference 回数（花筒一周占有单元花样的个数）

repeat size 花样循环尺寸
repeat width adjusting 拨针距《雕刻》
repellant 防护剂，驱避剂；相抗的，相斥的
repellants 防水呢，防雨呢（棉经毛纬，美国名称）
repellency 防护性
repellent 防护剂，驱避剂；相抗的，相斥的
repellent treatments 防护处理（织物和服装的防蛀、防水、防霉等防护性整理）
repelling electron group 斥电子基团
Repelotex HE 雷佩洛特克斯 HE（有机硅油乳化液，织物防水剂，商名，美国制）
replacement 取代
replacement parts 备件，备换件
replacement rinse 冷热洗涤
replenisher 补充剂；补充器；调节器
replenishing liquor 补充液
replica 复制品；复型；印模
replica technique 复制技术；复型技术
replication 复制品，拷贝；复制[过程]
report（rep.，rept.）报告
repoussé [印模]敲花
repoussé lace 凸纹花边
representative sampling 代表性取样
repress 抑制，约束（指肥皂中的增强剂抑制肥皂的水解反应）
repressing machine 钢芯排点机
reprocessing product 再加工品，复制品
reproducibility 重现性，再现性
reproducibility of the design already printed 花样复印重现性
reproducible dyeing 能重现的染色
reproducible effect 重现性，重现效果
reproduction 再生产；再生产过程；复制；复制品；翻版

reproduction camera 翻拍机
reprography 复印术
rept.（＝rep.，report）报告
repulsion 推斥，排斥；斥力
repulsion force 反斥力，排斥力
repulsion motor 推斥电动机
repulsion power 排斥力
Reputex 2O 一种抗菌防臭整理剂（商名，英国捷利康）
reqd.（required）要求的，需要的；规定的
requet 漂白亚麻平布（做床单用，法国制）
required〈reqd.〉要求的，需求的；规定的
requisites 必需品
re-raising 再起绒
re-rolling machine 再卷机
rerun 回修；重新运行，再启动；再蒸馏
re-run column 再蒸馏塔
res.（＝R，r.，resis.，resistance）阻力；电阻
research 研究，调查；探测；分析
research and development〈R/D，R&D〉研究与开发，研发
research expenditure 研究费用
research paper 学术论文
reseda［green］木犀草绿（灰光绿色）
resene 氧化树脂
reserve dyeing 防染染色
reserve effect 防染效果，防染效应
Reservehao S 防染剂 S（商名，中国制）
reserve power supply 备用电源
reserve printing 防染印花（指先印花后染色的工艺）
reserve style 防染印花法
reserve tank 储液槽
reserving agent 防染剂

reservoir 储存槽，储存器；蓄水池
reservoir cloth［转移染色］储色布（一种已经轧染分散染料的纯涤纶深色布，应用热轧转移方法，使另一织物中的涤纶得到染色）
re-set 重新调定，重新设置
reset 重新调节，重新安排；复位
reset device 复位装置
reset null 零位再调
reset relay 复位继电器，跳返继电器
re-shaping 再成型（成衣制造）
re-shearing 再剪绒
residence time 停留时间，滞留时间
residential carpet 家用地毯
residential use 居住用途
residual additives 残留添加剂
residual bath 残液；残浴
residual benzene 残留苯
residual chlorine 残留氯
residual crimp 残留皱缩
residual current 残留电流
residual elasticity 剩余弹性，残留弹性
residual elongation 剩余伸长，残留伸长
residual error 残留误差；漏检故障
residual extension 剩余伸长
residual fat 残留脂肪
residual flame 余焰
residual flame time〈RFT〉残余火焰时间，余焰时间
residual formaldehyde 残留甲醛
residual grease content 残留油脂含量
residual liquor minimization 残液最少化
residual moisture 残留水分
residual moisture content 剩余含水量，残留含水量
residual moisture measuring unit 残留含水

测定装置
residual motes 残留棉籽屑（织物前处理后残存物）
residual pectin 残存果胶质（织物前处理后残存物）
residual percentage crimp 残余卷曲率
residual printing pastes 残留印花浆
residual regain 剩余回潮率,残留回潮率
residual regression method 剩余回归法
residual seed coat 残留棉籽壳
residual sericin 残胶量,残胶比
residual shear stress 剩余剪切应力
residual shrinkage 剩余缩[水]率,残留缩[水]率
residual silkgum content 残胶量,丝胶残量
residual size 残浆
residual strain 残留张力
residual stress 残留应力
residual substance 残留物
residual sulfur 残硫量
residue 滤渣,余渣,残渣
residue length 剩余长度；烧剩长度
residue on evapouration 蒸发残渣
residue on ignition 燃烧残渣
resilience 回弹；回弹性能；弹性变形
resilience testing machine [织物]弹性试验机
resiliency 弹性；跳回
resilient elasticity recovery 弹性回复
resilient energy 回弹能
resilient fabric 弹性织物
resilient seal 弹性密封
resiliometer 弹性[测定]计
Resilio roller 雷西利奥辊（德国生产的一种微孔弹性轧辊,由数种吸湿性好的纤维用黏合剂制成薄片,再压成辊筒,

轧水效果极好）
resin 天然树脂；人造树脂
resinate 用树脂浸渍；树脂酸盐
resinated carpet 用树脂处理地毯
resin beat 树脂拍打（一种桃皮绒加工技术）
resin-bonded pigment printing 树脂固着涂料印花
resin composition 树脂组成
resin curing machine 树脂烘焙机
resin dyeing process 染色和整理联合工艺
resin exchanger 离子交换树脂
resin finishing 树脂整理
resin finishing agent 树脂整理剂
resin finishing catalysts 树脂整理催化剂
resin finishing process 树脂整理工艺
resin finishing range 树脂整理机组
resin-fixed pigment printing 树脂固着涂料印花
resin-free finishing 无树脂整理
resinification 树脂化[作用]
resinify 树脂化；树脂处理；浸焦油
resin matrix composites 〈RMC〉树脂基复合材料
resin migration 树脂泳移
resin modified 树脂改性的
resinoid 热固[性]树脂；[已]熟[化]树脂
resinous 有树脂的；含树脂的
resinox 酚—甲醛树脂
resin plate covered rail 衬有树脂润滑板的导轨
resin pocket [层压品夹层内的]淤积树脂
resin precondensate 预缩合树脂
resin resist printing 树脂防染印花法
resin retention 树脂保持性
resin soap 树脂皂,松香皂

resin-soluble dyes 树脂可溶染料
resin stains 树脂污渍
resin streak [层压品表面的]树脂条痕
resin treatment 树脂加工
resin varnishes 树脂漆
resis. (＝R,r.,res.,resistance) 阻力；电阻
resist 防染材料；防腐蚀涂层(用在花版或花筒雕刻制备中)
resist agent 防染剂
resistance ⟨R,r.,res.,resis.⟩ 阻力；电阻
resistance capacitance ⟨RC⟩ [电]阻[电]容
resistance capacitance oscillator 阻容振荡器，RC振荡器
resistance thermometer 电阻温度计
resistance to abrasion 抗磨，耐磨损性
resistance to acid 耐酸性
resistance to ageing 耐老化性
resistance to bending 抗弯曲性，耐弯曲性
resistance to bleeding 抗渗色性，抗渗化性
resistance to cold weather 耐寒性
resistance to creasing 抗皱性，防皱性
resistance to crushing 抗皱性，防皱性
resistance to damage 抗损性(化纤性能)
resistance to fatigue 耐疲劳性
resistance to insects 防蛀性
resistance to mildew 防霉性
resistance to open air 耐外露性，耐暴露性
resistance to plucking [植绒的]抗拔性
resistance to ripping 抗撕裂性
resistance to shrinkage 抗缩性，防缩性
resistance to soiling ⟨RS⟩ 耐污性
resistance to steam heat 耐蒸热性
resistance to tearing 耐撕裂性
resistance to wear 耐穿，耐磨性
resistance to weathering 耐气候牢度
resistance to wrinkling 抗皱性，防皱性
resistance to yellowing 耐泛黄性
resist dyeing 防染染色
Resist H 防染盐H(苯肼磺酸钠，能和纳夫妥AS类化合，阻止显色基的偶合，商名，德国制)
resisting agent 防染剂
resisting agent H 防染剂H
resisting salt S 防染盐S
resistive-capacitive ⟨RC⟩ [电]阻[电]容的
resistivity 抵抗性；电阻系数；电阻率
resistor 电阻器，电阻
resist paste 防染浆
resist printing 防染印花
resists 防染法
resist style 防染印花法
resist with natural resin 天然树脂防染
Resithren dyes 雷齐特伦染料(还原染料与分散的混合染料，商名，德国制)
Resocoton dyes 雷佐科通染料(印聚酯纤维与纤维素纤维混纺织物的分散性及活性混合染料，商名，德国制)
Resofix dyes 雷佐菲克斯染料(铜盐后处理直接染料，商名，瑞士制)
Resolamin dyes 雷索拉明染料(染涤毛混纺纤维的分散染料及酸性染料的混合染料，商名，德国制)
Resolin dyes 雷索林染料(分散染料，商名，德国制)
resolution 分解；分辨力；清晰度；分辨度
resolution power 分辨能力，分辨率
resolver 溶剂，溶媒；分解器
resolving power 溶解能力
resonance 共振[现象]；共鸣
resonance effect 共鸣效果
resonating frequency 共振频率

resorcin 间苯二酚
resorcin latex 间苯二酚胶乳
resorcinol 间苯二酚
resorcinol formaldehyde latex〈RFL〉 间苯二酚甲醛乳胶
resource 资源；设备
respirator 口罩,呼吸器；防毒面具
respiratory protection 呼吸保护器
response 响应；感应；特性曲线；频率特性
response factor 感应因子
Respumit S 雷必美 S(消泡剂,含硅酮,40%活性成分,非离子型,适用于各种染整工艺,商名,德国拜耳)
rest ［扶］架；座；刀架
restart 再起动
resting comfort 静止舒适性
rest position 不工作位置
restrainer 阻制器,定位器
restraining agent 抑染剂,抑制剂(用以降低染料平衡吸尽量)
restraining effect 抑制作用,抑染作用
restraining method 约束［试验］方法(耐燃试验用)
restraint systems 抑制系统
restricted substances list〈RSL〉 限用物质清单
restricted sudsing detergent 低泡洗涤剂
restriction 限制,限定；节气门；扼流圈
restrictor 限制器,节流阀,节气门,定位器
restrike time 再启辉时间
resultant 生成物,反应产物
resultant colour shift 总色相移动《测色》
resultant count 合股支数,总支数
resultant error 综合误差,总误差
resultant yarn number 合股线支数

resulting tension 合张力
resume 摘要,梗概；个人简历
resumption 再取回；再开始,再继续
resuspension 再悬浮［作用］
retail cloth 零售布
retaille 零头碎呢
retail store 零售店
retainer 护圈,挡板,止动装置；定位器
retaining catch 停止挡,闭锁掣子,防退掣子
retaining clip 定位铗
retaining clutch 涨闸式离合器
retaining plate 制动板,挡板
retaining ring 护圈,扣环,挡圈
retaining spring 止动弹簧
retaining washer 弹簧垫圈
retain moisture 滞留湿气
retardancy 阻滞性
retardant 阻滞,延迟；阻化剂,抑制剂；缓染剂
retardation 阻滞［作用］；减速［作用］,减速度
retardation factor value〈Rf value〉 Rf 值,比移值,延迟因素值(纸层析法中用于物质定性分析)
retardation time 阻滞时间
retarded elasticity 延迟弹性
retarded release property 缓释性能
retarded spontaneous recovery 延迟自然恢复
retarder 缓染剂；减速器；阻化剂
retarding agent 阻滞剂,缓染剂,抑制剂
retarding dyeing 缓染
retarding effect 阻滞效应,缓染效应
retarding solvent 缓溶剂
Retardon A 里塔登 A(耐碱及耐硬水的

缓染剂,商名)
retard roll 反转辊
Retargal AN liq. 里塔格尔缓染剂 AN 液(阳离子染料缓染剂,阳离子型,用于腈纶印花,可得光亮白度,商品,瑞士制)
Retargal AW 里塔格尔缓染剂 AW(阳离子型杂环化合物,丙烯腈系纤维及其混纺制品染色匀染剂,阳离子染料印花后的皂洗剂,商名,瑞士制)
Retargal HB liq. 里塔格尔缓染剂 HB 液(阳离子染料缓染剂,用于腈纶印花的清洗,可得洁白的地色,商名,瑞士制)
retentate [超过滤]保留物《环保》
retention 停留,滞留
retentionable screen frame 可保持张力的筛网框
retention period 停留时间
retention power 保留力(气相色谱法术语)
retention time 保留时间(气相色谱法术语)
retention value 保留值(气相色谱法术语)
retention volume 〈RV〉保留体积(气相色谱法术语)
retentivity 保持性
retest 重复测试,复验
retexturing process 再膨化处理
reticulated 网状的;网状织物
reticulate structure 网状结构
retightening 再紧固(如螺丝)
retina 视网膜
retinal illumination 网膜照度《测色》
retouching 手工修理;修版(指印花网版的最后修理)
retouching desk 修描桌《雕刻》
retouching inks 修版色料
retouching stand 修理架(印花圆网或平网手工修网用的支承架子)
retract 收缩,回缩;回程
retractable fibre 收缩性纤维
retracting spring 复位弹簧
retractive force 回缩力,回弹力
retrenchment 紧缩,节省,省略
retrieval 检索
retroaction 逆反应;反力,反馈
retrofitting 式样翻新
retrogradation 退减[作用],倒退,逆行;变稠
retro-reflection 回归反射
retro-reflective coated fabric 回归反射涂层织物
retro-reflective finished textile 回归反射整理纺织品
retro-reflective pigment printing 回归反射涂料印花
return 回collinsbrace 回loops布;纱线染整后占原长度的百分率;返回[操作]
return air 回风
return code 返回代码
return duct 回风管道
returning paste 回收浆料
returning spring 〈RS〉复位弹簧
return paste bucket 回收浆料桶
return paste station 回收浆料(印浆)站
return paste system [印花]回浆系统
return sludge 回流污泥
return valve 回流阀
reusable dyebath media 可回用的染浴介质
reusable routine 可重复使用的程序
reused wool 再生羊毛

reuse of dye bath 染浴再利用
reuse of waste water 废水再利用
rev.(reverse) [织物]背面;反面,反向,回程;可逆的,相反的
Rev.(review) 评论
rev.(revolution) 旋转,转动;转数,周期
Revatol S 利法托尔 S(硝基苯磺酸钠,印花用防染剂,还原拔染中起防染作用和抑制印花浆的还原作用,商名,瑞士制)
reverberation [声音]回响,反响
reversal 反向,倒转
reversal developing(=reversal development) 反转显影
reverse〈rev.〉[织物]背面;反面,反向,回程;可逆的,相反的
reverse colouring 倒色,互换色(花纹颜色与地色相互倒换)
reversed current 逆流
reversed phase 反相
reverse drive 反向传动
reversed sensor polarity 逆向传感器的极化
reverse dyeing method 逆向染色法
reverse effect 反向效果
reverse gear 反向齿轮,倒车挡
reverse knit 针织物反面起花;反面作正面的针织物
reverse micellar system 反胶束体系
reverse osmosis 逆渗透,反向渗透
reverse osmosis process 反渗透法
reverse osmosis treatment 反渗透处理(水处理技术)
reverse printing 反位复印(摄影用)
reverse proteolysis 逆向蛋白水解(通过逆向蛋白水解改变羊绒表面性能)

reverser 复原去渍剂(阻止污渍凝结,并使其溶解);换向开关,换向器
reverse roll coater 逆转辊涂布器
reverse roll coating 逆转辊涂布
reverse side [织物]反面
reverse winder 反向卷绕机
reversibility 可逆性
reversible bonded fabric 双面黏合织物(正反面相同的黏合织物)
reversible cloth 双面织物
reversible colour printing 变色印花
reversible cretonnes 双面印花或提花装饰布(窗帘、家具等用)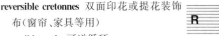
reversible cycle 可逆循环
reversible elongation 可逆伸长
reversible fabric 双面织物
reversible garment 可双面穿织物
reversible homespun 双面粗花呢
reversible mixed-flow pump 可逆式的混流泵
reversible printing 双面印花
reversible reaction 可逆反应
reversibles 双面织物
reversible satin 双面粗棉布(八综缎纹组织)
reversible swelling 可逆溶胀
reversing action 反转
reversing bevel [gear] 换向伞形齿轮
reversing motion 反向运动,换向运动
reversing roller 逆转辊,换向辊
reversing valve 回流阀
review〈Rev.〉评论
revivification of catalyst 催化剂复活
reviving agent 手感改进剂,增效剂
revolution〈rev.〉旋转,转动;转数,周期
revolution counter 转数表,转数计数器

revolutions per minute〈rpm,R. P. M. ,r. p. m.〉 每分钟转数,转/分
revolutions per second〈rps,r. p. s.〉 每秒钟转数,转/秒
revolver （物镜）转换器,旋转器;滚筒
revolver boiling machine 旋转式煮呢机
revolving brush 转刷
revolving expander 旋转式扩幅辊
revolving hank scouring machine 回转式绞纱精练机
revolving input drum 输入转筒
revolving magazine creel 回转复式筒子架
revolving output drum 输出转筒
revolving roller singeing 圆筒烧毛
revolving screen 转动式滤网
revolving singeing machine （＝revolving singer） 圆筒烧毛机
rewetting agent 再润湿[用]剂
rewinding machine 重绕机
reworks 回修品,返工品
RF（＝R. F. ,rf,radio frequency） 射频,无线电频率;高频
RFL（resorcinol formaldehyde latex） 间苯二酚甲醛乳胶
RFT（residual flame time） 残余火焰时间,余焰时间
RFT（right-first-time） [染色]一次准确性
RFT technology 一次准确性技术（新染料的新染色方法）
Rf value（retardation factor value） Rf 值,比移值,延迟因素值(纸层析法中用于物质定性分析)
RF welding 射频焊接
RG（＝rg, RGE,range） 排列;分等;量程,范围,幅度;机组

RGB colour measurement system RGB 测色系统
RGB system（red, green, blue system） 红、绿、蓝三原色
RGE（＝RG, rg. , range） 排列;分等;量程,范围,幅度;机组
RH（＝R. H. ,rh,relative humidity） 相对湿度
RH（＝rh, right hand） 右手;向右的,右旋的
rh（round head） 圆头
rhe 流值(流度单位,黏度单位厘泊的倒数)
rhea 苎麻(印度名称)
rheo.（rheostat） 变阻器,电阻箱
rheodichroism 流变二色性
rheogoniometer 流变测向仪
rheogram（＝rheograph） 流变图
rheologic 流变的
rheological behaviour 流变行为,流变性状,流变性能
rheological curve 流变曲线
rheological equation 结构方程,流变学方程
rheological property 流变性
rheology 流变学
rheology modifier 流变性改进剂《印》
rheometer 流变仪;电流计
rheometric test 流变性试验
rheopectic 震凝的,抗流变的,流凝的
rheopexy 触变性;震凝,摇凝
rheoscan 流变扫描仪
rheostat〈rheo.〉 变阻器,电阻箱
rheostat rocker arm 变阻摇杆臂
rheoviscometer 流变黏度计
rhodamine 若丹明,玫瑰精,碱性桃红

Rhodersil Oil 罗达丝油（柔软剂原料,氨基改性有机硅油,透明黏稠液,用于制作纺织品整理用柔软剂,商名,法国罗地亚）
rhodopsin 视紫红质
rhombic design 菱形图案,菱形花纹
RHR（Rockwell hardness）洛氏硬度
rhs（right hand side）右方,右边
R. H. Twist（right hand twist）反手捻;Z 捻
rhubarb dye 大黄染料（植物染料）
rH-value 氧化还原[电位]值
RH within clothing 衣服内相对湿度
rhythm and metre ［图案设计]节奏与韵律
rhythm crepe 泡泡绉
R. I.（=r. i., refractive index）折射率,折光率,折射指数
ria velvet ［人造丝]利亚绒（中国制）
rib 肋,肋片
riband 饰带
ribbed elastic braid 罗纹松紧带
ribbed fabric 各种经向或纬向棱纹织物的通称
ribbed pipe 肋片管,内螺纹管
ribbed plush 凸条长毛绒;条子长毛绒
ribbed radiator 片式散热器,波形散热器
ribbed tube heating radiator 肋管加热散热器
ribbed velvet 灯芯绒
ribbed velveteen 条子纬绒;灯芯绒
ribbon band 饰带
ribbon crosssection 带状截面,扁形截面
ribbonfil 扁丝
ribbon-like fibre 带状纤维,扁纤维
ribbon type burner 带形火口（烧毛机火口的一种）
ribbon width 长径（纤维和纱线横截面最大直径）
rib knit 罗纹组织,罗纹线圈
ribless corduroy 丰满绒（棉布）,灯芯布
riboflavin 核黄素（维生素 B_2）
ribonucleic acid〈RNA〉核糖核酸（生化）
Ricaguard 脂肪族多元醇类抗紫外整理剂（商名,日本制）
rice flour 米粉
rice starch 米淀粉
rich and colourful 丰富多彩
rich and gaudy 浓艳
Richcel fibre 丽赛纤维
rich colour 丰富的颜色
rich gradation 层次丰富,多层次
ricinoleic acid 蓖麻醇酸
ricinus oil 蓖麻油
rider 游码
rider bar 游码标尺,骑码标尺
RI-detector（refractive index detector）差示折光检测器
ridge design 波纹花纹,凹凸花纹,沟形花纹
ridgy cloth 凹凸不平的布（织疵）
riffle machine 电光机,缎光机
rig 织物对折（布边和布边对合）
rigged cloth 折幅织物,对折织物
rigging and folding machine（=rigging machine）对折机
right elevation 右视图
right-first-time〈RFT〉[染色]一次准确性
right hand〈RH,rh〉右手;向右的,右旋的
righthanded machine 右手机

right hand side〈rhs〉 右方,右边
right hand skew 右手纬斜
righthand thread 右螺纹,顺牙(顺时针方向旋进)
right hand twist〈R. H. twist〉 反手捻;Z捻
right side [织物]正面
right side view 右视图
rigid coupling 刚性联轴节
rigid fabric 硬挺织物
rigidity 刚性;刚度,硬度;稳定性
rigid laminating 刚性层压
rigid pattern 刚性式样
rigid plate handle 刚板状触感,手感硬挺
rigid steel backing 刚性钢板支架
rigid support 刚性支座
rig marks 折痕(缩呢疵)
Rigmel finishing 里梅尔机械防缩整理
Rigmel machine 里梅尔机械防缩整理机
Rigmel shrunk 里梅尔预缩法(缩水率低于1%的机械防缩工艺,也可改善成品的手感和光泽)
rig up 装配
ring 环,圈
ring accumulator 环状储布装置
ring bolt 环头螺栓
ring dyed effect 环染作用
ring dyeing [纤维表面]环染
ring expander 环形扩幅装置
ring forming addition polymerization 成环加成聚合[作用]
ring gear 环形齿轮,齿圈,内啮齿轮
ringing 环移[现象](染料在烘干和染色过程中,向纤维表面泳移所造成的环染现象);[手工去污渍时漫散开的]晕痕

Ringlemann smoke chart 林格尔曼烟雾图表(用于测定烟囱内空气沾染物的色泽、密度及不透明度等)《环保》
ring lubrication 油环润滑
ring [detachment] method 拉环法(测表面张力)
ring motor 环形电动机
ring nozzle 环形喷嘴
ring pipe 环形管
rinsability 可漂洗性,可洗净性
rinse tempering [冷热水]调温冲洗
rinsing 水洗,淋洗,漂洗,漂清
rinsing bath 水洗槽
rinsing booth 冲洗室
rinsing operation 水洗操作
rip 洗涤器;刮板,刮刀;裂幅,撕破
RIP(raster image processor) 光栅图像处理软件(喷墨印花用)
rip cord 触板,挡板(安全装置)
ripcord switch 挡板开关(安全板)
ripe cotton 成熟棉
ripening index 熟成指数《化纤》;[棉纤维的]成熟指数
rip-off selvedge 剥边,撕掉的边
rip-out 拆痕,拆档(织疵)
ripped selvedge [割]破边(疵点)
ripping strength 撕破强力
ripping strength tester [织物]撕破强力试验仪
ripple finish 波纹整理
ripple method 涟波法(测表面张力)
ripple puckers [织物上的]波形小皱纹
ripple washing machine 波动平洗机
rip-stop construction 防裂结构
rip strength 撕破强力
rise in global temperature 地球温度的升

高(温室效应)
riser 上升管;气门;溢水口;浇冒口
rising bubble viscometer 升泡[法]黏度计
rising roll batcher 升降卷布器
rising [tension] roller 松紧辊,张力调节辊,升降辊
risk 危险,冒险
rivel [织物上的]皱纹;褶裥
rivet 铆钉
rivet connection 铆钉连接
riveted flange 铆接法兰,铆合凸缘
riveting 铆接[法]
rivet join 铆接
RMC (resin matrix composites) 树脂基复合材料
RMS (root-mean-square) 均方根
RMV(raising machine version) 起绒机形式
RNA(ribonucleic acid) 核糖核酸(生化)
ro. (recto) 书籍的右页(即单数页)
roans 柔软羊皮
robbing back 起毛背面
robe 长袍;晨衣;浴衣;印花斜纹布
robin's egg blue 知更鸟蛋壳蓝(浅绿光蓝色)
robot 机器人;自动机,遥控机械装置
robotic systems 自控系统
robustness [方法]健全可靠;浓艳性,丰满性
Rochelle 一种轻薄的亚麻织物(作衬衣、床单用)
rochelles shirting 纯亚麻衬衫布(平纹织物)
rocker 摇杆,摇臂
rocker switch 摇杆式开关
rocket 火箭式筒子

rocket packages 火箭式筒装设备(大卷装锥形筒子)
rocking 摇摆轧法
rocking arm 摇臂
rocking bar 摇杆
rocking U-box 摇摆式 U 形堆布箱(溶剂精练设备的部件)
Rockwell hardness〈RHR〉洛氏硬度
Rockwell hardness A-scale〈RA,Ra〉洛氏硬度 A
Rockwell hardness B-scale〈RB,Rb〉洛氏硬度 B
Rockwell hardness C-scale〈RC,Rc〉洛氏硬度 C
rod 棒,杆,棍;连杆,拉杆
rod batcher 辊式卷布机
rodinal 对氨基酚(照相显影剂)
Rodix 罗迪克斯(拔染印花用还原剂,稳定型还原剂混合物,用于真丝、毛织物的酸性染料拔染印花,不会产生白晕渗边现象,颜色明亮,线条清晰,拔染浆稳定性好,商名,意大利米洛)
rod squeegee 刮浆棒
rod teasel machine 拉杆起绒机
roentgen〈R,r.〉伦[琴](辐射剂量单位)
roentgen absorbed dose〈r.,rad〉拉德(吸收辐射剂量单位)
Roica 罗衣卡(氨纶弹力丝,商名,日本旭化成)
Roica BX 罗衣卡 BX(适用于热定形的氨纶)
Roica BZ 罗衣卡 BZ(即氨纶)
Roica HS 罗衣卡 HS(高柔软性氨纶)
Roica SP 罗衣卡 SP(耐氯性氨纶)
roll (用辊)轧,碾;罗拉,辊;辊筒;滚柱;一卷

Roll-a-Belt reaction chamber 辊床式汽蒸箱(法国生产的前处理汽蒸设备)
roll adjusting mechanism 落刀装置
roll bar 飞刀片,打浆辊刀片
roll-batcher [辊式]卷布机
roll bed conveyor 辊式输送带,辊式履带,辊式传送装置
roll bleaching 卷轴漂白
roll blowing 卷轴蒸呢
roll boiling 卷轴煮呢;卷轴煮练
roll calender 滚筒轧光机
roll coating 滚筒涂层
roll core 滚筒芯轴
roll covering machine 滚筒包覆机
roll-decatizing 滚筒蒸呢
rolled end 松紧条痕(织疵)
rolled iron 轧制钢,钢材
rolled section steel 轧制型钢
rolled selvedge 卷边,翻边(织疵)
rolled steel frame 型钢机架
roller 罗拉,辊;滚柱;滚筒
roller ager 导辊式蒸化机
roller bearing (r.b.) 滚柱轴承;滚轴支座;[桁架]伸缩支座
roller bed 辊床(连续漂染加工装置)
roller bed conveyer 辊床式履带,辊式传送装置
roller bed scouring and bleaching range 辊床式练漂机
roller bed steamer 辊床式汽蒸箱
roller-belt accumulator 导辊传送带堆布装置
roller blinds 卷筒式窗帘,卷筒式百叶窗
roller brush 毛刷辊
roller cage [轴承上]滚柱夹圈;滚柱隔离圈;轧辊机座
roller calender 滚筒轧光机
roller card raising machine 钢丝起毛机
roller chain 滚子链
roller chisel 花筒雕刻
roller clutch 自由轮离合器(采用一系列滚子和斜面座配合,只能作单向驱动)
roller conveyor 辊式履带
roller damping 滚筒给湿
roller deformation 轧辊变形
roller developing machine 花筒显影机
roller doctor 花筒印花刮刀
roller dryer 导辊式烘燥机(热风烘燥机,导辊排列有直导辊和横导辊两种)
roller engraving 花筒雕刻
roller engraving machine 花筒雕刻机
roller expander 滚筒式扩幅装置
roller exposing machine 花筒曝光机
roller feed motion 辊式喂给装置
roller finish 滚筒轧纹整理
roller fulling machine 罗拉毡合机;罗拉起毛机
roller heating machine 导辊式焙烘机
roller jet squeezer 滚筒喷射轧液机
roller management 轧辊排列
roller mandrel 滚筒芯轴,花筒轴心
roller melt coating process 热熔辊涂层法
roller milling 滚筒缩绒;花筒刻浅
roller milling machine 滚筒缩呢机;花筒刻线机(滚筒印花雕刻设备)
roller printing 滚筒印花
roller printing machine 滚筒印花机
roller printing pressure adjustment 滚筒印花压力调节
roller recutting 花筒修补
roller scratches 花筒刮伤
roller screen 印花圆网,筛网花筒

roller screen printing machine 圆网印花机
roller seal 滚[筒]封[口]
roller singeing machine 圆筒烧毛机
roller squeegee [网印]刮浆滚筒,磁棒刮刀
roller squeezing apparatus 滚筒轧水器
roller stain 轧辊斑(整理疵)
roller steps 花筒搁座,花筒搁脚
roller storage 花筒储藏
roller strike-off machine 滚筒印花打样机
roller temple 滚筒扩幅
roller thrust bearing 滚柱止推轴承
roller trio 三辊[拉伸机]
roller-type mercerizing machine 滚筒式丝光机
roller vat 卷染机,导辊式染槽
roller washing machine 导辊式水洗机
roller width 滚筒幅宽
rollet 花纹凹槽
roll folding 卷筒(织物装潢方法之一)
roll follower 凸轮转子；成形转子
roll guider 导辊
roll hem 卷边(织疵)
rolling 轧制；辊压
rolling and lapping machine 卷筒卷板两用机(布匹包装)；卷绸机
rolling ball viscometer 落球黏度计
rolling bearing 滚动轴承
rolling-blister finishing process 轧泡整理工艺
rolling friction method 滚动摩擦法
rolling list 荷叶边(织疵)
rolling machine 卷布机
rolling mills 滚筒缩绒机
rolling selvedge 卷边(织疵)
Rollingstatic warp printing machine 罗林斯坦蒂克经纱印花机(利用转移印花技术进行经纱印花,可立即织成印花织物,无需后处理,商名,法国拉美)
rolling-up machine [纱线]卷取机；[布]打卷机
roll mark 辊子印痕
roll rod 圆棒(用于圆网印花或平网印花的磁性刮印机构)
roll setting 落刀
roll squeegee 辊子刮刀
roll-to-roll [织物]卷装进出(指染整加工中进布和出布都是卷装形式)
roll-up mechanism 卷布机构
ROM (read only memory) 只读存储器
Romanthrene dyes 罗马士林染料(还原染料,商名,意大利制)
Rongal A 朗茄尔 A(次硫酸盐的衍生物,高温还原剂,商名,德国制)
Rongalit[e] 雕白粉,吊白块(商名,德国制)
Rongalit[e] C 雕白粉 C,吊白块 C(甲醛化次硫酸钠,直接印花及拔染印花的还原剂,商名,德国制)
Rongalit[e] FD 雕白粉 FD,吊白块 FD(亚磺酸钠盐,直接印花及拔染印花的还原剂,商名,德国制)
Rongalit[e] patash method 雕白粉碳酸钾方法(还原染料印花方法)
rongeant 烂花图案(混纺织物通过化学方法烂成的图案)
röntgen〈R,r.〉 伦[琴](辐射剂量单位)
röntgenogram 伦琴照片；X射线照片
röntgenology 伦琴射线学,X射线学
röntgen ray 伦琴射线,X射线
roof extract ventilation unit 顶部排气通风装置

roof heat recovery system 箱顶热回收系统
roofing washer 瓦形垫圈
roof ventilator 机顶排风风扇
room-size rug 全室地毯
room temperature 车间温度,室温
root-mean-square (RMS) 均方根
root-mean-square deviation 均方根差,标准差
root-mean-square error 均方根误差,标准误差
root-tip difference 毛根—毛尖色差
root-tip-levelling kinetics 毛根—毛尖匀染动力学
ROP (record of production) 生产记录
rope bleaching 绳状漂白
rope bleaching range 绳状漂白联合机(主要由绳状浸轧机、J形汽蒸箱及绳状水洗机组成)
rope chemicking machine 绳状轧漂机
rope continuous scouring and bleaching 绳状连续练漂
rope creases 绳状折痕
rope detwister 绳状退捻装置
rope detwisting and opening machine 绳状退捻开幅机
rope drum 绳轮
rope dyeing 绳状染色
rope dyeing machine 绳状染色机
rope entry 绳状进布
rope expander 绳状开幅机
rope form 绳状
rope form scouring 绳状精练
rope guide wheel 绳状织物导轮
rope [form] hydroextractor 绳状脱水机
rope impregnating machine 绳状浸轧机
rope J-box 绳状J形箱(绳状织物前处理设备)
rope marks 经向长折皱痕;绳状擦伤痕;绳状色条;绳状折皱印,绳状印痕
rope opener 绳状开幅机
rope piler 绳状堆布器
rope plaiter 绳状折布机
rope pulley 绳轮
rope rinsing machine 绳状水洗机,绳洗机
rope saturator 绳状浸轧机;绳状浸渍槽
rope scourer 绳状精练机
rope scouring and bleaching machine 绳状练漂机
rope scouring machine 绳状精练机
rope soaper 绳状洗涤机;绳状皂洗机
rope squeezer (= rope squeezing machine) 绳状轧水机
rope unloader 绳状出布装置(用于绳状染色机等设备)
rope untwister 绳状退捻装置
rope [form] washer 绳状洗涤机
rope washing machine 绳状水洗机,绳洗机
roping 条痕(缩绒疵)
ropy 黏稠的;可拉成丝的;[发酵后]成丝状的
Roracyl dyes 罗拉西尔染料(染丙烯腈系纤维的染料,商名,美国制)
rosamine 红色碱性染料
rosaniline 蔷薇苯胺,品红[碱],玫苯胺
rose 玫瑰色(浅粉红色);喷雾器喷嘴;滤网
Rose-Miles method 罗斯—迈尔斯泡高测定法,罗斯—迈尔斯倾注起泡试验法
rose petal touch 玫瑰花瓣触感(一种新的刷绒技术)
rose pink 玫瑰粉红[色](淡粉红色)

rose red 玫瑰红[色]
rosewood 黄檀棕(浅红棕色)
rosin 松香,松脂
rosin derivatives 松香衍生物
rosin oil 松香油
rosin soap 松香皂
rot. (rotating) 旋转的
rot. (rotation) 旋转,回转,转动;涡流;旋光度
rota- (构词成分)表示"旋转""转动"
rotamer 旋转异构体
rotameter 转子流量计;转子式测速仪;曲线测长计;线曲率测量器
rotary band brushing machine 环带式刷绒机(用于灯芯绒织物加工)
rotary beck dyeing machine 滚动式绳状染色机,转槽式染色机
rotary block printing 旋转式模版印花
rotary blowing machine 回转式蒸呢机
rotary brushing 回转式刷绒
rotary cloth-press 辊式压呢机
rotary compensator 回转式补偿器
rotary counter 旋转式计数器
rotary digester 旋转式蒸煮釜,蒸球
rotary drum drier 转鼓[式]干燥器
rotary drum dyeing machine 转鼓式染色机
rotary drum filter 转鼓[式]过滤器
rotary evapourator 旋转式蒸发器
rotary film printing 滚筒式薄膜印花
rotary fulling mill 滚筒式缩呢机
rotary gig revolving teazel machine 回转式刺果起绒机
rotary kiln 旋转式焚烧炉
rotary knob 旋转钮
rotary lacquered screens 上胶圆网

rotary milling machine 滚筒式缩绒机,滚筒式缩呢机
rotary offset machine 轮转胶印机
rotary precision cutting 旋刀定长切绒(用于植绒)
rotary press 滚筒式压呢机,滚压机
rotary printing machine 滚筒印花机
rotary pump 旋转泵,回转泵
rotary screen coating 圆网涂层
rotary screen dryer 圆网烘燥机
rotary screen exposing machine 圆网曝光机
rotary screen finishing system〈R. S. F.〉圆网整理系统
rotary screen gumming machine 圆网上胶机
rotary screen lacquer coating 圆网涂胶机
rotary screen photoengraving equipment 圆网照相雕刻设备
rotary screen printing〈RSP〉圆网印花
rotary screen printing doctor 圆网印花刮刀
rotary screen printing machine〈RSPM〉圆网印花机
rotary screen printing range 圆网印花联合机(主要由圆网印花机和热风烘燥机组成)
rotary screen roller printing machine 圆网滚筒印花机
rotary screen stencil 圆花网,圆网印花版
rotary screen supporter 圆网托架
rotary screen washing 圆网洗涤
rotary shear 回转式剪毛
rotary sieve 旋转筛,旋转滤网
rotary singeing 圆筒烧毛
rotary singeing machine 圆筒烧毛机

rotary slide valve 回转阀
rotary spherical digester 蒸球
rotary squeegee washer 旋转式加压水洗机
rotary stacker 旋转式叠料器
rotary stretching machine 回转式拉幅机
rotary switch 旋转开关
rotary table 回转工作台
rotary traverse winder 旋转式往复卷绕机
rotary-type heat exchanger 转轮式热交换器
rotary vacuum filtration 旋转式真空过滤
rotary valve 旋转阀
rotary viscosimeter 旋转式黏度计
rotary washer 转鼓式洗涤机
rotatable axle 可旋转的轴
rotatable reception table 回转式承布台
rotate days off 轮休
rotating〈rot.〉旋转的
rotating bar dryer 转棒式烘燥机(绞纱烘燥机的一种,挂棒由输送链传送,且能自转)
rotating bio-disc 生物转盘
rotating biological contactor process 生物转盘法《环保》
rotating carbon slip ring 转动的碳滑环(无油润滑)
rotating dryer 旋转式干燥机
rotating input drum 输入转筒
rotating output drum 输出转筒
rotating rod dryer 回转杆式烘燥机
rotating tensioning device 转动式张力装置
rotation〈rot.〉旋转,回转,转动;涡流;旋光度
rotational velocity 旋转速度
rotational viscometer 旋转式黏度计
rotation speed 回转速度(以 Vr 表示)

rotative screen printing machine 圆网印花机
rotatory dispersion 旋光分散
rotfastness 耐腐牢度,抗腐烂性
rotoconditioner 旋转式给湿机
Rotodye machine 圆筒染色机(用于成衣染色的设备,也可作成衣煮练、漂白用,商名,意大利 MCS 制)
Rotoflush washing machine 旋转冲洗式水洗机(商名,德国制)
rotogravure 照相凹版轮转印刷
rotogravure ink 照相凹版轮转印刷油墨
Rotojet washing machine 旋转喷射式水洗机(商名,德国制)
rotor 旋转器,转子
rotor atomizer 转子喷雾器
rotor damping 转子给湿
rotor-type pump 转子泵(装有内、外两个转子,用作油泵,其工作原理与齿轮泵相似)
rotor washing machine 转筒洗涤机
rotory platform abraser 转台式耐磨仪
Roto-stream dyeing machine 圆筒型染色机(德国生产的一种间歇式喷射染色机)
Rotowa open-width washer 罗脱华平幅洗涤机,转轮式平幅水洗机(商名,德国制)
rot-preventive finish 防腐整理
rotproof finishing 防霉防腐整理
rot proofing 防腐烂,防腐处理
rot proofing agent 防霉剂,防腐剂
rot resistance 抗腐烂性,防腐,耐腐性
rot resistance finish 防腐整理
rot ring pen 绘图钢笔,转轮笔(绘图用)
rot steep [棉布]浸水堆置

rottability of textiles 纺织品腐烂性
rotting 腐败
rotting resistance test 防腐测试
roucou 胭脂树红(浅橙色);橙色植物染料
rouge 红铁粉;胭脂色
rough 布面毛糙(织疵);[花筒表面]局部低陷
rough browns 低级亚麻坯布
rough crepe 粗绉面绸,鸡皮绉
rough cutter 毛刀(割绒疵)
rough cutting 粗切削
rough-dry 不加熨烫的干燥
roughers 呢坯
roughness 粗糙度;毛茸不平
roughness height rating 粗糙度高低值
rough-pile moquette 粗绒头地毯
rough terrain suit 越野工作服
rough test 经验试验
roulette roller 压花纹罗拉;滚花罗拉
round 循环;圆,环;一连串;十足的;环绕;四舍五入成整数
round cross-section 圆形截面
round head ⟨rh⟩ 圆头
round head screw 圆头螺钉
rounding [圆网]复圆;弄圆;倒角;四舍五入成整数
rounding-off method 四舍五入法
round of pattern 花纹循环
round-robin test 循环试验,轮流试验,校对试验
round-trial 轮流试验(校对试验结果)
round wire 起绒杆,圆拉毛杆(织造起毛或起绒织物用)
route ⟨rte.⟩ 线路
routine 程序,例行程序;子程序;例行的

routine analysis 例行分析,日常分析
routine inspection 常规检验,常规检查;定期维修
routine maintenance 例行维修,日常维修
routing 路线,确定的工艺路线
rowdy 条痕,横向或直向条花(织疵)
royal armure 狭幅绉绸
royal blue 品蓝[色];蓝色颜料
royal boucle 高级捻线绒头地毯(绒头线具有永久捻度,耐洗,耐用)
royal cashmere 高级薄呢(英国名称)
royal purple 红紫[色]
royal satin 皇家缎(双股经、单股纬的经面缎)
royal twill 高级斜纹绸(双股丝经,细棉纱纬)
royalty 专利权税,[技术转让]提成费
rpm (= R. P. M., r. p. m., revolutions per minute) 每分钟转数,转/分
r. p. m ratio 转速比
4R principle (Reduction、Recovery、Reuse、Recycle principle) 4R原则(内部消除、回收、回用、循环,即清洁生产四原则)
RPS(regulated power supply) 稳压电压
rps (= r. p. s., revolutions per second) 每秒钟转数,转/秒
RS(relaxation shrinkage) 松弛收缩(羊毛);回缩缩率
RS(resistance to soiling) 耐污性
RS(returning spring) 复位弹簧
RSC(remote support contract) 远程支持合同(即没有服务工程师的合同,不到现场的服务)
R. S. F. (rotary screen finishing system) 圆网整理系统
RS-232 interface RS-232标准串行接口

RSL(restricted substances list) 限用物质清单
RSP(rotary screen printing) 圆网印花
RSPM(rotary screen printing machine) 圆网印花机
RSV (reduced specific viscosity) 比浓黏度
R. T. (=r. t. ,regular tenacity) 普通强度
rte. (route) 线路
RTN process 轧焙烘中性一步固色法(活性染料连续染色方法,英国卜内门开发)
rub [织物]擦伤痕
rubber belt compressive shrinking machine 橡胶毯机械预缩机
rubber belt grinding attachment 修磨橡胶毯装置
rubber blanket [印花机用的]橡皮衬布;橡胶导带
rubber bowl 橡胶滚筒
rubber bowl printing machine 橡胶滚筒印花机
rubber calendar 橡胶毯轧光机,橡胶毯预缩机
rubber cement 橡胶黏着剂
rubber cloth 橡胶布,防水布
rubber coated pressure bowl 橡胶承压辊筒
rubber-coated roller 橡胶辊
rubber-coated textiles 涂橡胶织物
rubber coating 橡胶包覆层
rubber coating waterproofing 涂橡胶防水法
rubber-covered 〈r. c.〉 包橡皮的
rubber-covered pressure bowl 橡胶承压辊
rubber-covered roller 橡胶辊
rubber covering 橡胶包覆层
rubber dough 橡胶黏着剂

rubber elasticity 橡胶弹性
rubber fabric 涂橡胶织物
rubber flinger 橡胶甩水圈
rubber gasket 橡皮衬垫
rubber hardness tester 橡胶硬度测试器
rubberized curved roller 橡胶弯辊
rubberized fabric 涂橡胶织物,橡胶布
rubberizing 上橡胶[工艺],贴胶,涂胶
rubber latex 胶乳,橡胶
rubber proofing [单面]橡胶涂层
rubber roller 橡胶辊
rubber roller covering 橡胶辊包覆层
rubber sheeting 橡皮垫布
rubber shrinking blanket [机械]防缩橡胶毯
rubber silk 涂胶绸
rubber sleeve (预缩机的)橡胶毯
rubber spreading machine 涂胶机
rubber squeegee 橡皮刮刀,橡皮刮[浆]板
rubber substitute 橡胶代用品
rubber varnish 橡胶漆
rubber vevlet 防水棉绒(布上涂以胶液,并植绒,柔软如丝绒)
rubber washer 橡胶垫圈
rubbing 摩擦
rubbing fastness 耐摩擦[色]牢度
rubbing resistance 耐摩擦性
rubbing test 摩擦试验
rubfastness 耐摩擦牢度
rubine 宝石红[色](暗红色)
rub marks 擦痕(洗呢、缩呢疵)
ruby [red] 红宝石;宝石红[色],红玉色
ruby laser 红宝石激光
ruche 褶裥饰边
ruched fabric 褶裥织物
ruching (=ruck) 褶裥饰边;皱,褶;弄皱,

起皱

ruckle 小皱;褶裥

ruffle 褶边;绉纹

ruffle finish [耐久]皱裥整理(使用树脂和机械预缩整理)

rug 小地毯;旅行毯

rug back staining 地毯[对其他材料]返沾色

ruggedness [织物]粗壮性

rugosity 多皱性

rug-over-tufting machine 地毯簇绒机

rule of thumb 凭经验方法,约略的衡量

rules and regulation 规章制度

rules of design 图案法则

rules of operation 操作规程

ruling 摇线

ruling machine [滚模钢芯]摇线机

ruling mill 摇线[滚模]钢芯(花筒雕刻用)

rumchunder 鲁琼德绸(印度各种点子花纹绸的统称)

rumple 皱纹;皱裥;揉皱

run 一个色回的印制量;一个花样的印制总量;一次连续生产总长;[接头]联匹;[卷染]道数;运行,操作;路径,通道

runback 回流管,溢出管

Runic coating 如尼涂层(点涂)

runnability 运行性能

runner 提升辊;导轮;滚子;狭长地毯;走廊地毯;(装饰用)狭长桌布;花边中联线

runner cloth 导布

runner wheel 导轮,辊子提升轮(蒸化机中用)

running 渗化,渗色;运行;行程

running block 动滑车

running characteristic 运转特性,操作特性

running clearance 运行期间的间隙

running-fit 动配合(轴在轴孔内的配合间隙使配合后转动自如,既不松动也不紧阻)

running frequency 转速

running idle 空转;空载运行

running-in 试运转,试车,校车

running-in test 空转试验

running marks 经向长折皱痕;绳状擦伤痕;绳状色条痕

running stability 运行稳定性

running-up 起动

run number 批号,批数

runoff 流出,流量

run-of-the-loom 未经检验的坯布

run-of-the-mill (纺织品的)等外品,次品;未经检验的纱、布

runout 脱开;用完

run proof 防脱散;防抽丝

run resist 防脱散;防抽丝

rupture 破裂,断裂,破坏

rupture strength 抗裂强度,破裂强力,断裂强力

rupture test 破坏试验;断裂试验

rush to repair 抢修

Rusil finish 鲁西[经纹]擦光整理(在钢辊上刻成经向细线纹,进行擦光)

russet brown 黄褐色

rust 铁锈,铁锈色(棕色)

rustic brown 铁锈棕(带红棕色,比铁锈色稍红)

rustless steel 不锈钢

rustling finish 丝鸣整理

rustproof 防锈的
rust resistance 防锈性
rust spot 黄锈渍(织疵)
rust stains 锈斑
rusty scale 锈鳞,锈垢
rutile 金红石(氧化钛矿石名称)
RV(relative viscosity) 相对黏度
RV(retention volume) 保留体积(气相色谱法术语)
Ryudye pigment colours 罗代涂料(印染用涂料,商名,日本制)
Ryudye white binder 罗代罩印白黏合剂系列(适用于深色地印花布的罩印,商名,日本制)

S

s.(secondary) 次级的；第二的；次等的
s.(section) 截面,剖面；区域；部门；部分
s.(silicate) 硅酸盐
s.(spectrometer) 分光计；光谱仪
s.(syn-type) 顺式,顺[基]型
s. a.(sectional area) 横截面积
s. a.(self-acting) 自动的
s. a.(such as) 例如
sabin 塞宾(吸声量单位)
saccharides 糖化物；糖类
saccharification 糖化
saccharimeter 糖检测仪
saccharin 糖精
saccharomycetes 酵母菌
saccharose 蔗糖
sacculus 小囊；球囊
sack degumming 袋练[工艺]
sacrificial 牺牲的
SAD(soil adhesion) 沾污性
sad coloured 深暗颜色的,颜色暗淡的
sadden 变深色
saddening 色泽暗淡处理；后铬媒处理(毛染色用)
saddening agent 消艳剂
saddle 座架；刀架；阀座
Sadtler standard spectra 萨特勒标准[红外]光谱
safari shirt 狩猎衫
safeguard 安全罩,安全装置
safeguards analysis 安全保障分析
safe ironing temperature 安全熨烫温度

safelight [暗室]安全灯
safelocking door 安全锁紧门
safe pressure 安全压力,允许压力
safety 安全,保险；保险装置,安全设备
safety and protective textile 安全防护用纺织品
safety belt 安全带；救生带
safety breaker 安全断流器
safety check list 安全检查表
safety circuitry 安全保护电路
safety clutch 安全离合器
safety coefficient 安全系数
safety colour 安全色
safety concentration 安全浓度
safety controls 保护措施,防护装置
safety coupling 安全联轴器
safety cut-out 安全断流器
safety device 安全装置
safety equipment 安全设施
safety evaluation 安全性评价《环保》
safety factor 安全因数,安全系数
safety fencing 安全围栏
safety flap 安全节气门,安全瓣
safety garment 救生衣
safety glass 安全玻璃
safety guard 安全防护装置,皮带罩壳
safety harness [降落伞]安全背带
Safety, Healthy, Environment consciousness (SHE consciousness) 安全、保健、环境意识
safety helmet 安全帽

safety installation 安全设施
safety instrument 安全仪表
safety lamp 安全灯(俗称行灯)
safety law 安全法
safety lighting 安全照明
safety locking motion 安全保险装置
safety measures 安全措施
safety net 安全网
safety operation 安全操作
safety-oriented 适应安全的
safety pin 安全钉,保险销;安全别针
safety precautions 安全措施,安全预防
safety production 安全生产
safety regulation 安全规则
safety relief valve 安全减压阀
safety roller 安全辊(保护装置)
safety rope 安全绳
safety shield 安全屏障
safety shoes 安全鞋
safety shut-down 紧急停车,安全刹车
safety signs 安全标志
safety sleeve 安全套筒(轧车前安全装置)
safety standard 安全标准
safety stitch 安全线缝
safety stop device 安全停止装置
safety switch 安全开关
safety techniques 安全技术
safety valve 安全阀
safety working pressure 安全操作压力
safflower 红花;红花染料
safflower oil 红花油
saffron 藏红花(植物染料)
saffron yellow 藏红花黄[色](金黄色)
safranine 碱性桃红,[碱性]藏红染料
sag curve 挠度曲线
sage [green] 鼠尾草绿[色](灰绿色)

sagging 下垂,弛垂[现象];倾斜;熔塌;凹陷
sago flour 西谷粉,西米粉
sago starch 西米淀粉
sag resistance 抗下垂性
sakker 绉条纹薄织物,泡泡纱;条格绉绸(可用棉、丝、人造丝等作原料)
sakusan silk 柞蚕丝(日本名称)
SAL (symbolic assembly language) 符号汇编语言
salicylanilide 水杨酰替苯胺(俗名生色精水杨酸,一种优良的防腐剂)
salicylate 水杨酸盐
salicylic acid 水杨酸
saline caustic soda 烧碱食盐液
saline solution 盐溶液
salinometer 盐分计,盐量计,盐液密度计
saliva fastness 耐唾液牢度
sallow 灰黄色
salmon [pink] 鲑肉粉红[色](浅橙色)
sal soda 十水碳酸钠,晶碱
salt accumulation 盐积聚
salt bath 盐浴
salt cake 芒硝
salt content monitor 盐含量监测仪
salt controllability 盐控制性(某些直接染料染色时,用不同的加盐条件控制匀染的性能)
salt-controllable dyes 加盐控制匀染[直接]染料
salt dyes 直接染料
salt effect 盐效应
salt formation 盐层
salt-free dyeing process 无盐染色工艺
salting box 加料槽(绞盘染色机染槽前部的多孔隔板槽)

salting-on 校色(染色至后阶段逐步加小量染料至染液中,以获得标样色泽)
salting-out 盐析,加盐分离
salting-out chromatography 盐析色谱法
salt linkage 成盐结合
salt rejection 盐的排除
salts 盐类
salt sacking 粗糙织物(粗糙似盐袋布)
salt sensitivity 对盐敏感性(指染料染色性能受中性电解质影响的程度,又指染色品受中性电解质的变色反应)
salt shrinking treatment 盐绉缩处理(用于丝织物或丝的交织物)
salt spray test cabinet 盐水喷雾试验箱
salt wash 盐洗(盐析过程)
samardine 法国哔叽
Samaron dyes 舍玛隆染料(染聚酯纤维的分散染料,商名,德国制)
sam-cloth 绣花样本
sample 样品,样本;模型
sample ager 小样蒸化机
sample beam light 样品测量光束
sample blanket 包袱样
sample blender 试样混合器
sample card 样本,样品卡
sample cell 试样池,试样槽
sample cutting device 剪样器,取样器
sampled-data 采样数据
sampled-data control 抽样数据控制
sampled data system 抽样数据系统
sample drying cylinder 小样烘筒
sampled signal 抽样信号
sample dyeing apparatus 小样染色机
sample holder 试样座,试样夹
sample injector 试样注射器
sample loom 织小样机

sample of quality allowance 试样的质量允许
sample picking machine 布样剪齿机
sampler 取样器;取样员;样板;绣花样本
sample room 样品间
sample skein [检验]样丝;样绞
sample skeins test 绞纱试样试验
sampling 制样;抽样[检验];取样
sampling at random 随机取样
sampling container 取样筒
sampling controller 抽样控制器
sampling device 小样罐;取样装置
sampling distribution 样本分布
sampling error 抽样误差
sampling inspection 抽样检验
sampling inspection by attributes 计数抽样检验
sampling inspection by variables 按量抽样检验
sampling machine 制样机,切布样机;(花筒)打样机
sampling plan by variables 计量型抽样方案
sampling point 取样点
sampling rate 抽样速度
sampling schedule 取样程序表
sampling tube 取样管
sampling valve 取样阀
Samuel printing machine 塞氏印花机
SAN(styrene-acrylonitrile copolymer) 苯乙烯—丙烯腈共聚物
Sancowad AN 山科瓦德 AN(阴离子型有机化合物,山科瓦德染法的泡沫发生剂,商名,瑞士制)
Sancowad NI 山科瓦德 NI(非离子型聚乙二醇醚水溶液,山科瓦德染法的泡沫

发生剂,商名,瑞士制)

Sancowad process 山科瓦德节水染色法,山科瓦德微泡沫染色法(瑞士)

sand 沙灰[色](浅黄灰色);沙,砂

Sandacid V liq. 山德酸 V 液(释酸缓冲剂,有机、无机混合物,阴离子型,在热染浴中缓慢释放酸,保持染浴稳定,改善匀染,减少色差,商名,瑞士科莱恩)

sandalwood 檀香木;红木[染料]

sandarac gum 山达脂(香松树胶)

sand bath 砂浴

sand blasting 喷砂[整理](用于牛仔布加工)

Sandene process 山代纤维变性法(商名,瑞士科莱恩开发)

Sandenization 山代纤维变性(纤维素纤维阳离子变性方法,商名,瑞士科莱恩开发)

sanders wood 檀香木;红木[染料]

sand filter 砂滤器

sanding [仿麂皮]起绒工艺;磨毛

sanding machine 磨毛机;砂磨机

sand mill 砂磨

Sandoclean PC liq. 山德净 PC 液(多功能润湿净洗剂,非离子型,具有乳化、分散作用,可生物降解,商名,瑞士科莱恩)

Sandoclean T10 liq. 山德净 T10 环保型前处理助剂(商名,瑞士科莱恩)

Sandocryl B dyes 山德克列尔 B 染料(阳离子染料,商名,瑞士制)

Sandocryl process 山德克列尔法(聚丙烯腈纤维的亚铜离子染色法)

Sandofix 山德士固色系列(固色剂,阳离子型,改善直接染料和活性染料染色和印花的湿牢度,商名,瑞士科莱恩)

Sandofix WE 山德士固色剂 WE(阳离子树脂型含氮缩合物,直接染料固色剂,商名,瑞士科莱恩)

Sandoflex process 山德弗莱克斯冷轧卷丝光法(适用于圆筒针织物,商名,瑞士科莱恩开发)

Sandogen NH 山德近 NH(弱阳离子型含氮缩合物,酸性染料和金属络合染料染聚酰胺纤维匀染剂,商名,瑞士科莱恩)

Sandolan dyes 山德兰染料(适于中性或弱酸染色的酸性染料,商名,瑞士科莱恩)

Sandolan E dyes 山德兰 E 染料(强酸性染料,阴离子型,匀染性和耐日晒色牢度优良,商名,瑞士科莱恩)

Sandolan MF dyes 山德兰 MF 染料(弱酸性染料,阴离子型,良好的移染性,色牢度优良,染色保温时间短,羊毛保护性好,商名,瑞士科莱恩)

Sandolan N dyes (= **Sandolan Milling N dyes**)山德兰 N 染料,山德兰缩绒 N 染料(中性至弱酸性染料,阴离子型,应用于色泽鲜艳的单色和双色拼色,商名,瑞士科莱恩)

Sandolan P dyes 山德兰 P 染料(弱酸性染料,阴离子型,在加入匀染剂和元明粉情况下,相容性和匀染性优良,羊毛成品手感柔软,商名,瑞士科莱恩)

Sandolube NV 山德纳 NV(平滑剂,改良缝纫性,并可作为起绒剂,商名,瑞士科莱恩)

Sandonyl method 山德尼尔法(聚酰胺纤维匀染方法)

Sandopan CBN, BFN 山德潘 CBN, BFN(润湿净洗剂,低泡,适用于纺织品前

Sandopan 2N 山德潘 2N(磺化脂肪酸衍生物,合成洗涤剂,浸湿剂,商名,瑞士制)

Sandoperm 山德平系列(耐久性柔软剂,阳离子型或非离子型,硅弹性体,对各种织物有耐久性柔软效果,商名,瑞士科莱恩)

Sandopur RC liq. 山德飘 RC 液(清洗剂,阴离子型,又是无泡螯合剂、分散剂,保护胶体,可用于染色和印花织物的后清洗,商名,瑞士科莱恩)

Sandospace DPE liq. 山德色 DPE 液(媒染、防染剂,阳离子型,增加阴离子染料对锦纶的亲和力,商名,瑞士科莱恩)

Sandospace MPW 山德色 MPW(地毯置换印花用助剂,商名,瑞士科莱恩)

Sandospace R/S 山德色 R/S(媒染、防染剂,阴离子型,可防止阳离子染料上染锦纶,商名,瑞士科莱恩)

Sandozin NA 山德先 NA(硫酸酯,润湿剂,商名,瑞士科莱恩)

Sandozin NIN 山德先 NIN(烯化氧衍生物,非离子型洗涤剂和润湿剂,商名,瑞士科莱恩)

Sandozin NIT liq. 山德先 NIT 液(润湿剂,非离子型,用于各种染整过程中的润湿及净洗,商名,瑞士科莱恩)

Sandozol dyes 山德素染料(可溶性还原染料,商名,瑞士制)

Sandozol KB 山德素 KB(高度磺化油,渗透剂,具润湿、分散、匀染、柔软作用,商名,瑞士制)

Sandozol NE 山德素 NE(磺化脂肪酸衍生物,渗透剂,商名,瑞士制)

sand paper [金刚]砂纸
sandstone 砂岩棕[色](浅灰棕色)
sandwash finish 砂洗整理
sand washing 砂洗
sandwash silk 砂洗绸
sandwich 叠层制品;多层复合,多层并合
sandwich arrangement 夹层结构
sandwich compound 夹心化合物
sandwich copolymer 嵌段共聚物
sandwich dyeing 夹心染色(合纤染色的一种方法)
sandwiched fabric 叠层织物,夹心织物;衬垫织物(针织);袜子夹底
sandwich effect 夹心效应(毛织物防缩处理中因芯层未处理而产生的不匀现象)
sandwich heating 夹层加热
sandwiching process 夹心[印花]工艺
sandwich laminate 夹心层压
sandwich lamination 夹心黏合,夹层黏合[结构],层压[结构];填充纱黏合
sandwich structure [蜂窝状]夹层结构
sanforize corrugation 预缩皱纹隐痕(织物防缩整理疵)
Sanforized 桑福整理(指织物经机械防缩整理,缩水率在1%以内,商名,美国)
Sanforized Plus 强化桑福整理(指织物经树脂整理后具有较好的防缩防皱等效果,商名,美国)
sanforize pucker 预缩皱痕
sanforizer 预缩整理机
sanforize roughness [过度]预缩布面粗糙(疵点)
sanforizing machine 机械防缩整理机
sanforizing mark 预缩折皱(疵点)
sanforizing oils 预缩用油剂

Sanfor-Knit 桑福尼特整理(针织物防缩整理工艺,该工艺包含机械防缩处理、收缩测试仪器及针织修补技术三个因素,商名,美国桑福)

Sanforlan 桑福伦整理(毛织物的防缩防毡并整理,商名)

Sanforset 桑福瑟特整理(棉或人造丝织物经化学和机械的防缩整理,商名,美国)

sanguine 血红[色]

sani 桑尼(厄瓜多尔及西半球热带地区产的植物染料)

sanitary cotton 卫生棉,脱脂棉

sanitary disposable product 一次性卫生物品

sanitary fibre 抗菌纤维,卫生纤维

sanitary finish 卫生整理(指防汗臭、防腐、杀菌等整理)

sanitary finishing agent 卫生整理剂

sanitary requirement test 卫生性能试验

sanitary wear 卫生服

sanitized 化学防菌整理[的]

sanitized chemical cleaning 化学消毒处理

Sanitized T 山宁泰 T 系列(抗菌整理剂,有机衍生物,阴离子型,有耐久的抗菌性和良好的水洗牢度,商名,瑞士科莱恩)

sanitizer(=**sanitizing agent**) 卫生整理剂,消毒剂

san ramie 野生苎麻(朝鲜商名)

sans envers 双面织物(法国名称)

sansfabrics 无纺布(法国名称)

Sapamine A 色必明 A(N,N-二乙基-N-十八酰基乙二胺乙酸盐,表面活性剂,商名,瑞士制)

Sapamine AK 色必明 AK(阳离子型,脂肪酸酰胺改性的聚合物乳化液,耐洗柔软剂,并有抗静电作用,用于羊毛、聚酰胺纤维及丙烯腈系纤维,商名,瑞士制)

Sapamine AL 色必明 AL(阳离子型,季脂肪醇衍生物,柔软剂,主要用于合成纤维,有抗静电作用,商名,瑞士制)

Sapamine OC 色必明 OC(阳离子型,脂肪酸酰胺衍生物,各种纤维均适用的高效柔软剂,并有抗静电作用,特别适用于砂洗绸及水洗绸,商名,瑞士制)

sapanwood 苏方(一种可提取红色染料的木材)

sap green 树液绿[色](晚绿色);天然绿色染料

sap. No.(**saponification number**) 皂化值

saponide 合成洗涤剂(除肥皂外)

saponifiable matter 可皂化物质

saponifiable wax 可皂化的蜡质

saponification 皂化[作用]

saponification number〈**sap. No. , SN, S. N.**〉皂化值

saponified acetate 皂化醋酯纤维

saponifier 皂化剂

saponin 皂角苷

sapphire [**blue**] 宝石蓝[色]

sapphire 蓝宝石

Sapphire digital printer 蓝宝石数码印花机(商名,斯托克)

Sapriva 莎普里瓦细旦涤纶(适合与羊毛混纺,商名,美国杜邦)

saprogenous system 腐生体系

saprophytes 腐生植物

saprozoites 腐生动物

sarcosine 肌氨酸;N-甲基甘氨酸

saree(=**sari**) 莎丽服;莎丽布

saree printing machine(＝sari printing machine) 莎丽圆筒接版印花机(用两组花网前后拼凑印花,以获得较长的印花图案)

sarong 莎笼,围裙;莎笼布料

sash 饰带

sat.(saturate) 饱和

sat.(saturated) 饱和的

Satanella effect 萨塔尼拉效应(黑地红花或红地黑花的印花)

sateen finish 棉织物充缎子整理

sateen weave [棉织物的]缎纹组织;纬缎组织

satin 缎纹;经缎组织;丝织缎纹织物;缎

satinage 轧光整理,缎光整理

satin alcyonne 表里双色缎

satin back 缎背(指丝织物的背面用缎纹组织);缎背织物

satin berber 缎纹呢(精纺毛织物,经光泽整理)

satin bonjean 精纺缎纹呢

satin drill 贡缎,直贡,斜纹缎,泰西缎

satiner 缎光软整理,软缎整理

satin finish 光泽整理,缎光整理,轧光整理

satin francais 法兰西缎纹呢(经光泽整理,作家具装饰用)

satinize 缎光整理;[棉织物]耐久性光滑整理

satinizing 光泽整理,缎光整理

satinlike(＝satiny) 仿真丝缎纹织物(经轧光上蜡)

satin lissé 印花棉缎;利西棉缎(缎光整理,花纹清楚)

satin lustre 缎纹光泽

satin net 经缎网眼布《针》

satin ribbon 缎带,缎纹丝带

satin weave 缎纹组织

saturant 饱和[浸渍]剂;浸透的;饱和的

saturate〈sat.〉 饱和

saturated〈sat.〉 饱和的

saturated air 饱和空气

saturated aliphatic acid 饱和脂肪酸

saturated aliphatic hydrocarbon 饱和脂肪族烃

saturated calomel electrode〈SCE〉 饱和甘汞电极

saturated colour 饱和色

saturated concentration 饱和浓度

saturated current 饱和电流

saturated ether 饱和醚

saturated hydrocarbon 饱和烃

saturated polyester 饱和聚酯

saturated solution 饱和溶液

saturated state 饱和状态

saturated steam 饱和蒸汽

saturated steam fixation 饱和蒸汽固色

saturated steam setting 饱和蒸汽定形

saturated vapour 饱和蒸汽

saturated water 饱和水

saturated water vapour pressure 饱和水蒸气压力

saturating-bonding 饱和浸渍黏合法

saturating machine 浸湿机

saturating tank 轧槽,浸渍槽

saturation 饱和;饱和度

saturation bath 饱和浴

saturation dye uptake 饱和染着量

saturation factor 饱和因子

saturation point 饱和点

saturation pressure 饱和压力

saturation regain 饱和回潮率(在相对湿

度为 100％、温度为 22℃ 条件下纤维的回潮率)
saturation temperature 饱和温度
saturation value 饱和值
saturator 饱和[浸渍]器；饱和剂
Saueressig roller printing machine 索累西格立式滚筒印花机(商名，德国制)
Savagraph process 萨瓦格拉夫印花法(两相还原染料印花法，商名，美国)
save-all 罩衫；工作服；围裙；围涎
save data 下载数据《计算机》
save recipe 储存处方
Savinese 莎维蛋白酶(用于羊毛制品的后整理，可改善织物的手感和外观，商名，丹麦诺和诺德)
sawtooth effect 锯齿效果《雕刻》
sawtooth roof 锯齿形层顶
SAWTRI (South African Wool Textile Research Institute) 南非毛纺织研究所
saxe blue 灰光浅蓝[色]
Saxon blue 撒克逊蓝(溶于浓硫酸的靛蓝)
Saxony 萨克森地毯(割绒地毯)
Saxony blue 蓝色颜料
Saxony finish 撒克逊整理(粗纺毛织物经缩呢、辊压、拉伸、起绒、剪毛、刷毛与电压后，使呢面有短毛茸的整理)
Saybolt second 赛波特秒[数](用赛波特黏度计测定流体黏度时的表示值)
Saybolt universal viscosimeter 赛波特通用黏度计
Sayl-a-Set 塞阿赛整理(人造丝织物的还原染料染色法，同时起防缩作用，商名)
SBR(styrene butadiene rubber) 丁苯橡胶
SBS (styrene butadiene styrene) copolymer 苯乙烯—丁二烯嵌段共聚物
sc. (scale) 刻度盘；比例尺；鳞片；刻度；秤；锅垢；规模
sc. (＝scil., scilicet) 就是说，即
sc. (＝sci., science) 科学
sc. (＝scr., screw) 螺钉，螺丝
S. C. (short circuit) 短路《电》
sc (silicon controlled) 可控硅
S. C. (silk covered) 丝包的
SCA (secondary cellulose acetate) 二[级]醋酸纤维素(一级醋酸纤维素的熟成物或水解物)
scaffold 脚手架
scaffolding effect 支架效应(例如，涤棉织物在燃烧中棉纤维所起的烛芯支架作用)
scaffolding yarn 伴织纱，支架纱(帮助较弱纱线顺利通过各种工序，织成布匹后可用化学处理将此伴纱除去，也可留在织物上以提高织物的坚牢度)
scald [棉坯布]沸点下煮练
scale〈sc.〉 [羊毛]鳞片；刻度；刻度盘；比例尺；秤；锅垢；规模
scale-down factor 缩小比
scale factor 比例因子，换算系数
scale fibre 有鳞片纤维(黏胶纤维品种之一)
scale frequency [羊毛的]鳞片数
scale hopper 称量加料斗
scale mark 刻度标记
scale masking [羊毛的]鳞片遮蔽
scale model 按比例模型
scale of engraving 雕刻的刻度
scale of hardness 硬度表
scaleover 过刻度
scale pan 天平盘

scale stripping process [羊毛]剥鳞片加工
scale-up 尺度放大,比例放大
scale-up factor 放大比
scaling [图像]缩放
scaling factor 倍率,换算系数
scallop 荷叶边,扇形边,月牙边
scalloped edge 荷叶边
scalloped selvedge 荷叶边,扇形边;脱边（织物拉幅时脱铗疵点）
scaly 鳞片状的,有鳞片的
scan 扫描;搜索
Scandinavian Colour Institute 〈SCI〉 斯堪的纳维亚色彩研究所
scan generator [电子显微镜的]扫描振荡器
scanner 扫描器;扫描程序;析像器
scanner unit 扫描装置
scanning definition 扫描清晰度
scanning disc 扫描盘
scanning electrogen microscope 〈SEM〉 扫描光[照发射]电子显微镜
scanning electron microscope 〈SEM〉 扫描电子显微镜
scanning electron microscopy 扫描电子显微术
scanning element 扫描元件
scanning frequency 扫描频率
scanning head 扫描头
scanning in photo-engraving 照相雕刻扫描
scanning ion microscope 扫描离子显微镜
scanning monitor 扫描监控器
scanning near-field optical microscope 〈SNOM〉 扫描近场光学显微镜
scanning pitch 扫描节距
scanning probe microscope technology 〈SPMT〉 扫描探针显微技术
scanning process 扫描方法
scanning radiometer 〈SR〉 扫描辐射计
scanning rate 扫描速率
scanning roller 扫描辊
scanning speed 扫描速率
scanning station [光电]疵点检验装置
scanning tunneling microscope 〈STM〉 扫描隧道显微镜
scanspot 扫描光点
scarf 围巾;肩上披巾;薄头巾;领巾
scarlet 大红[色],猩红[色]
scarlet corns 虫胭脂,虫红
scatter coating 散布法涂层（用于制领衬布）
scatter coating method 分散涂覆法,散布涂层法（黏合剂施用方法）
scatter dot 散点[喷涂]
scattered flower pattern 散乱花卉图案
scattered light 散射光
scattered motif 散乱图案,散乱花纹
scattering 散射;散布,分散
scattering coefficient 散射系数《测色》
scattering intensity (= scattering strength) 散射强度
scattering theory 散射理论
scatter rug 拼块地毯
scavenge oil 废油
scavenge pipe 吹气清洗管
scavenger 清洁工;清除剂,净化剂;清除器
scavenging agent 清除剂,净化剂
scavenging process 吹扫过程;净化工艺
scavenging time 扫气时间,净化时间
SCBA (self-contained breathing apparatus) 独立(整套)呼吸设施(消防用)

SCE (saturated calomel electrode) 饱和甘汞电极
scenic design 风景图案
scenograph 透视图
scent 香味,气味
SCFD(supercritical carbondioxide fluid dyeing) 超临界二氧化碳流体染色
SCF value (space velocity per cubic foot) SCF值(每立方英尺空间速度值)
schappe 绢丝织物
schapping 发酵脱胶法,腐化脱胶法《丝》
scheduled maintenance 例行维修,预定维修
scheduled overhaul 计划性检修
scheduled stoppage 计划停车率
scheduling 日程安排
schematic cross section 横截面示意图
schematic diagram (= schematic drawing) 原理图;示意图;简图
scheme 方案,规划,计划,设计;草图;平面图;线路图;图表,图解
scherbias drive 串级调速
Schiefer abrasion testing 席费尔磨损试验[法]
Schiffs' reagent 希夫试剂(用于鉴定醛类化学品)
schiller colour 闪色;闪光(矿物的或昆虫的)
Schlafhorst friction coefficient balance 施拉夫霍斯特摩擦系数测定仪
Schollerized 朔勒整理[的](毛织物化学防缩防毡整理,商名)
Schopper's bursting strength tester 肖伯型织物顶破强力机
Schopper's cloth abrasion tester 肖伯型织物耐磨试验仪

schreiner bowl 电光辊
schreiner calender 电光机,缎光机
schreiner finish 电光整理,缎光整理
schreinering 电光工艺,缎光工艺(电光机的钢辊筒刻有极细线纹,棉织物经高压高温轧压后,增强光泽,有如绸缎)
schreinerized 经电光整理,缎光
Schultz number 舒尔兹号数
Schweitzer solution 施韦策尔溶液,施韦策试剂,铜氨溶液
Schweitzer's reagent 施韦策尔试剂(氢氧化铜的氨溶液,分析用,能溶解纤维素纤维)
SCI(Scandinavian Colour Institute) 斯堪的纳维亚色彩研究所
science 〈sc., sci.〉 科学
science fabrics 化学纤维织物
Scientific Design method〈S-D method〉 科学设计法
scientific paper 学术论文
scilicet 〈sc., scil.〉 就是说,即
scintillation method 放火花法,闪光法(激光雕刻术语)
scintillator 闪烁器;闪烁体
scintillometer 闪烁计数器
scission 裂开,裂变,断裂;切开
scission of link 断键
scissors 剪刀
Scitex pattern control system 萨特克斯花样(图案)控制系统(商名,以色列)
sclerometer 硬度计,肖氏硬度计
scleroscope 硬度计
scleroscope hardness 肖氏硬度
SCMC (sodium carboxymethyl cellulose) 羧甲基纤维素钠盐
scollop 荷叶边,扇形边,月牙边

scorching 焦化；[橡胶]过早硫化
scorching mark 烫焦印
scorch mark 黄斑（长丝疵）
scorch resistance 抗烫焦性
scorch test 烫焦试验
scorch tester 烫焦试验器
score 得分记录，比分
Scotch finish 苏格兰整理（指缩呢后紧靠呢面剪毛，仍留有绒毛的整理）
Scotchgard FC 思高洁 FC（拒水拒油整理剂，氟碳树脂，兼有去污效果，商名，美国 3M）
Scotchgard finishing 思高洁整理（织物用含氟化合物的防油拒水整理，商名，美国）
Scotchgard oil repellent finishing 思高洁拒油整理（美国 3M）
Scotchlite 回归反射织物（商名，美国 3M）
Scotch tweed finish 苏格兰粗花呢整理（一种简易的整理方法，包括洗绒、缩绒、洗绒、拉幅、剪毛、热压及蒸呢等工序）
scotopic vision 暗[处]视觉，黄昏视觉
Scott system horizontal cloth strength tester 思高脱卧式织物强力试验仪
Scott system serigraph 思高脱式纱线强力仪
Scott tensile tester 思高脱拉伸强力试验仪（等加负荷，倾斜式）
scourability 洗涤性能
scourable branding fluid 易洗标记液（代替油漆，标记羊身）
scoured silk 熟丝，练熟丝
scoured wool 净洗毛
scouring 煮练，精练；脱胶；洗涤；预浸
scouring agent 精练剂，洗涤剂，净洗剂

scouring and bleaching 练漂工艺
scouring and bleaching machine 练漂设备
scouring and bleaching of cotton 棉练漂
scouring bath 练浴；洗液
scouring bowl 洗毛槽
scouring by ferment 发酵精练
scouring cistern 精练槽
scouring cloth 抹布，拖布
scouring in jigger 卷染机精练
scouring in lap 吊练
scouring in pan 煮布锅煮练
scouring in rope form 绳状煮练
scouring in sack 袋练
scouring loss 煮练损耗
scouring machine 洗涤机，水洗机；煮练机
scouring machine for spun silk 绢丝煮练机
scouring of spun silk 绢丝[纺]煮练
scouring process 洗涤[工艺]，水洗；退浆
scouring shrinkage 精练收缩
scouring spot 练绸印迹
scouring waste 精练废液
scouring yield 精练率
scour-relaxation 精练松弛[工艺]
scourzyme 果胶酶（一种精练酶，丹麦诺维新）
SCP（semi-chemical pulp） 半化学浆
scr.（＝sc. screw） 螺钉，螺丝
SCR（＝S. C. R.，semiconductor-controlled rectifier） 半导体可控整流器
SCR（＝S. C. R.，silicon controlled rectifier） 可控硅整流器
scraf joint 嵌接
scrape coating 刮刀涂层法
scraped mark 擦痕，刮浆印（疵点）
scraper 刮刀，刮板，刮浆刀；刮片（圆网印花用的刮刀两端的塑料刮片）

scraper bar [地毯上浆机的]刮板
scraper ring 刮油环
scraping knife(=scraping squeegee) 刮刀
scratch 抓痕,刮痕,擦痕;划记号
scratch felt 仿驼绒织物
scratchiness 瘙痒感
scratch moiré 刮花波纹整理(多用于人造丝塔夫绸)
scratch proof 耐划的,耐刮的
scratch resistance 耐擦伤性
scratch wash 猫须洗(牛仔裤一种加工方法)
scray 小型 J 形堆布箱
scray box 堆布箱,J 形箱,伞柄箱
SCR drives 可控硅传动
screen 筛网;[网印]网版;荧光屏;幕;光栅;滤光镜;隔离罩
screen analysis 筛网分析
screen angle 网屏角度
screen aperture 筛孔
screen belt 筛网输送带
screen cage 尘笼
screen carrier 圆网托架
screen cassettes 圆网头,圆网盒
screen cleaning device 帘网洗涤装置
screen cleaning tank 帘网清洗槽
screen cloth 筛网布;幕布
screen coating machine 筛网涂胶机
screen coating unit 圆网涂层单元
screen contrast 网屏反差
screen conveyor 网带输送机
screen degreasing 筛网去油脂
screen diagnostics 计算机屏幕诊断
screen disk 网盘;筛盘
screen dot 网点
screen dry 筛式干燥

screened negative 网点负片(经加网版摄制的负片)
screen engraving 筛网雕刻,网版雕刻
screen engraving equipment 筛网雕刻设备
screen exchange 圆网的调换
screen filter 筛滤板
screen frame [筛网]印花网框
screen gauze 筛网(用于平网印花)
screen holder 筛网托座,筛网支架
screening 筛选;遮蔽;筛滤;格网滤渣,粗滤渣
screening agent 掩蔽剂
screening band drier 网带干燥机
screening drum drier 圆网干燥机
screening effect 屏蔽效应
screening fabric 遮蔽布
screening test 屏蔽试验;筛选试验
screening test 筛选试验
screen lacquer 筛网用漆
screen length 圆网长度
screen machine 筛选机
screen making 制版
screen mesh 网目
screen negative 网版负片,加网阴图片
screen perforation 网眼
screen plate 筛选板;过滤板
screen plugging 筛网堵塞
screen positive 网版正片,加网阳图片
screen printing carriage 筛网印花行车
screen printing machine 筛网印花机,网印机,平网机
screen printing with graffiti 粗画风格的网印
screen protective lacquer 筛网保护漆
screen reverse roll coater 圆网逆转辊涂层机

screen ruling 网目线数
screens dimintional stability 筛网尺寸稳定性
screen selecting 筛选
screen shot 截图《计算机》
screen stencil 筛网花纹版,绢网花纹版
screen stretching equipment 绷网设备《雕刻》
screen suction inlet 滤网式吸风口
screen table [印花]筛网台
screen tensioning 圆网拉紧
screen to screen coater 圆网双面涂层机（织物正反面涂层）
screen to screen shading 跳版深浅
screen washer 印花筛网清洗机
screw（sc.，scr.） 螺钉,螺丝
screw axis 螺旋轴
screw bolt 螺栓,螺杆;全螺纹螺栓
screw cap 螺丝灯头;螺钉帽,螺纹帽
screw compressor 螺杆压缩机
screw conveyor 螺旋式输送器
screw die 螺丝绞板
screwed fittings 螺纹连接件
screwed flange 螺纹法兰
screwed rod 螺丝杆
screw expander roll 螺旋扩幅辊
screw feeder 螺杆进料器
screw gauge 螺距规（测量螺丝螺距的规尺,俗称螺丝卡板）
screw lamp socket 螺口灯头
screw pin 螺旋销
screw pump 螺旋泵
screw rod 螺旋杆;丝杆
screw shaped expander 螺旋展幅滚筒,螺旋绷布辊
screw socket bulb 螺口灯泡

screw softening machine 螺旋揉布机
screw spindle 螺旋轴;螺旋柄
screw stirrer 螺旋搅拌器
screw thread 螺纹,螺线
screw valve 螺旋阀
scribbled 混纺纱线（英国名称,用两种以上纤维混纺）
scriber [划线机]划针
scrim back [地毯]组织稀松的底布
scrim fabric 帘子基布,稀松织物
scrimp 折皱,折纹;折皱脱印(印疵)
scrimper bar 扩布板,伸幅板
scrimp rail 伸幅板,[织物的]扩幅[金属]板
scrimps 折皱,折纹(毛织物煮呢、洗呢或缩呢疵);折皱脱印(印疵所致)
scroll 涡管;螺旋线;开幅辊;滚动《计算机》
scroll expander 蜗杆式开幅器,螺旋开布辊
scroll gear 蜗形齿轮
scroll opener 蜗杆式开幅器,螺旋开布辊
scroll pattern 螺旋花纹
scroll pile 卷曲绒毛
scroll roller 螺旋开幅辊
scroop 丝鸣,绢鸣
scroop finish 丝鸣整理,绢鸣整理
scrooping 丝鸣
scrooping agent 丝鸣增效剂
scrooping feel 丝鸣感
scrooping treatment 丝鸣整理
scroopy handle 丝鸣感,绢鸣感
scrub [手工]擦,刷,洗
scrubbed wool 拉绒织物
scrubber 湿式除尘器;洗汽器;刮管器
scrubbers for exhaust air 废气净化器

scrubbing 刷洗
scrub resistance 耐刷洗性
scuff 摩擦痕
scuff-resistance ［植绒织物的］抗擦痕性能
sculptured 浮雕绒头
sculptured design carpet 雕花地毯
sculptured effect 浮雕效应；泡泡纱花纹效应
sculptured pattern 浮雕花纹；凹凸花纹
sculptured pile 雕花绒头
sculptured pile fabric 浮雕绒头织物
sculptured rug 雕花地毯
sculptured wool sweater 浮雕羊毛衫
scum 泡沫；浮色；渣滓
scumbling 薄涂［颜色］
scumming 罩色，留白［处］沾色（机印疵，俗称刮不清）
scum thread 跳纱，浮线（织疵）
scutch 开幅
scutcher 开幅机
scutcher and water mangle 开幅轧水机
scutcher, water mangle and cylinder drying machine 开幅轧水烘燥机
scutching 开［布］幅
s. c. w. (standard copper wire) 标准铜线
S. D. (=s. d., s. dev., standard deviation) 标准差
sd. (seasoned) 风干的，干燥的
S. D. (semi-dull) 半无光
SD (service dress) 制服；军便服
SDC(Society of Dyers and Colourists) 染色工作者学会（英国）
SDC(standard degree of colour fastness) 染色牢度标准等级
SDC (synchron dyeing control) 同步染色控制
SDC system 同步染色控制系统
SDI(sludge density index) 污泥密度指数
SDI(Stork Digital Imaging) 斯托克数码图像制作公司
S-displacement S形弯斜
S-D method (Scientific Design method) 科学设计法
SDO (self directing optimization) 自动对准最佳处理，自动定向最佳处理
s. dev. (=S. D., s. d., standard deviation) 标准差
SDS(service diagnostic system) 服务诊断系统（对设备问题的诊断）
SDS(Synchron Dyeing System) 同步染色系统（商名，德国第斯）
SE(seam efficiency) 缝制效率
S. E. (standard error) 标准误差
sea blue 海蓝色（绿光蓝色）
Seacell 海藻纤维（商名，德国齐默）
sea cell active fibre 海丝活性纤维
sea cell fibre 海丝纤维
seacrest 海浪峰绿［色］（淡绿色）
sea energy 海洋能
seafoam 海沫绿［色］（浅艳绿色）
sea green 海绿色（黄绿色）
sea island cotton 〈SIC〉 海岛棉
seaisland fibre 海岛型纤维
seaisland microfibre 海岛型超细纤维《化纤》
seal 封口；密闭
seal air fan 密封空气风机
sealant 密封剂，密封胶；密封材料；防渗漏剂
sealed ball bearing 封闭式滚珠轴承
sealed bearing 封闭式轴承

sealed edge 封边
sealed-for-life 永久性密封
sealed heat pipe 热导管,热管
sealed jar method 封闭瓶法(织物上游离甲醛测定方法)
sealing bush 密封衬套,密封轴衬
sealing element 密封件
sealing film 密封薄膜
sealing joint 密封接头
sealing regenerative-type air preheater 密封回热式空气预热器
sealing ring 封环(机械部件)
seal packing 密封填料
seal plate 封板
seam 缝,接缝;线缝
seamability 可缝性
seam abrasion machine 线缝磨损机
seam abrasion resistance 缝纫耐磨性
seam automatic device 自动测缝头装置
seam bar marks 缝线条痕(染疵)
seam breakage 缝线断裂
seam cockling 接缝起皱
seam counter timer 缝头计数定时器
seam cracking 缝线开裂
seam detector 线缝探测器
seam efficiency〈SE〉 缝制效率
seamer 缝纫机;缝纫工
seam eye 缝头检出装置
seam finish type 接缝形式
seam flammability 接缝耐燃性
seaming machine 缝纫机
seaming stitch density 缝迹密度
seamless 无缝的;圆形编织
seamless engraved screen 无缝雕刻圆网
seamless engraving 无缝雕刻
seamless pipe 无缝[钢]管

seamless repeat 无缝花型循环
seamless rug 无缝地毯,全幅地毯
seamless tube 无缝管
seamless tubing 无缝管状织物
seamless wear technology 无缝衣着技术
seam mark 缝头色疵;缝头压痕
seam passage [机器上]缝头的通过
seam protection 缝线保护
seam quality 缝制质量
seam security 缝迹牢度
seam sensing system 接缝传感系统
seam signal 缝头信号
seam slippage 脱缝,缝口脱开
seam strength 线缝强力,缝合强力
seam tracing device 缝头探测器
seam welding 线缝焊
seamy side 夹里,里子;反面
search 检索;寻找,检查
searcher 测针,探头
search time 检索时间
sear cloth 蜡布
season colour 应时色,流行色
seasoned〈sd.〉 风干的,干燥的
seasoning 风干;干燥;陈化;时效处理,自然定形
seat 座,座位
seatcover fabrics 椅套织物
sea technology 海洋技术(七大高新技术之一)
seating 椅套起绒织物;基础;支座
seawater desalination 海水淡化
seawater fastness 耐海水牢度
seawater resistance 耐海水性
seaweed 海藻
seaweed green 海藻绿[色](浅灰绿色)
sebacic acid 癸二酸

sec.(second) 秒
sec.(secondary) 次等的;次级的;第二的
SEC(= sec., section) 截面,剖面;区域;部门;部分
seclude 隔绝,隔离,分离
second〈sec.〉 秒
secondary〈s., sec.〉 第二的;次等的;次级的
secondary accelerator 辅助促进剂
secondary acetate 次级醋酯纤维
secondary adsorption 第二类吸附,物理吸附
secondary air 二次风,二次空气
secondary air channel 补风管
secondary amine 仲胺
secondary amine surfactant 仲胺型表面活性剂
secondary antioxidant 辅助抗氧剂
secondary backing [簇绒地毯或非织造地毯的]第二层底布
secondary bio-chemical treatment [印染废水]二级生化处理
secondary bond 次价键
secondary cellulose acetate〈SCA〉 二[级]醋酸纤维素(一级醋酸纤维素的熟成物或水解物)
secondary cell wall 次生胞壁
secondary colours 二次色(由两种原色拼成)
secondary creep 第二潜伸,次级蠕变(纤维受载荷被伸长后,载荷虽经解除,不能恢复至原有的长度)
secondary diamine 仲二胺
secondary dispersion 次级分散;二次色散
secondary drying 二次烘燥
secondary effect 副作用

secondary electron 二次电子;次级电子
secondary energy 二次能源,人工能源
secondary exhaustion 二次上染,二次吸尽
secondary fibre 再生纤维
secondary fibre application 再生纤维应用
secondary fibrillation 二次原纤化
secondary force 二次力
secondary heater 第二加热器;副加热器
secondary hydration 二次水合
secondary ionization 二次电离
secondary layer 次生层
secondary particle 二次粒子
secondary plasticizer 次级增塑剂;助增塑剂
secondary pollution 二次污染
secondary pump 辅助泵
secondary raw material 二次原料;回用原料
secondary reaction 副反应;第二反应
secondary sedimentation basin 二次沉淀池
secondary sewage treatment 二次污水处理;污水二级处理
secondary shades 次色,混合色,调和色(由两种原色拼成,如绿、橘、紫)
secondary steam 二次蒸汽(利用介质加热而产生的蒸汽,或把高压水减压而生成的扩容蒸汽,也称间接蒸汽)
secondary structure 二次结构
secondary tint 柔和色
secondary transition temperature 二次转移温度
secondary treatment of waste water 废水二级处理,废水中级处理
secondary valence 次化合价,副价
secondary wall 次生胞壁
second-generation nylon 第二代锦纶

second-generation synthetic fibre 第二代合成纤维
second grade quality 二级品
second hand wash ［牛仔裤］怀旧洗
second hardening 二级烘焙；二次硬化
second law of thermodynamics 热力学第二定律
second motion shaft 过桥轴
second-order splitting 二级分裂
second-order transition ［第］二级转变
second order transition temperature〈SOTT〉二级转变温度，玻璃化温度
seconds 次品；二等品
section〈s.，SEC，sec.，sect.〉截面，剖面；区域；部门；部分
sectional area〈s.a.〉横截面积
sectional drive 分部传动，分段传动
sectional dryer 分段烘干机（各段温度不同）
sectional elevation 剖视图，断面图
sectional expander 分段扩幅器
sectional interlinings 分段衬
sectional mold 镶合塑模
sectional running 分段运行
sectional side elevation 纵剖图
sectional steel 型钢
sectional type system 拆卸式装置
sectional view 剖视图，断面图
section chief 工段长
sectioning 切片；分段
section iron 型钢
sections〈ss〉截面；区域；部门
section switch box 分配电箱
sector 扇形［面］，扇形体；齿弧；象限；区段，段（计算机磁道的一部分）
sector arm 扇形臂

sector gear 扇形齿轮
security 安全；安全性，可靠性
security control 保护措施，安全技术
Securon 540 赛科朗 540（螯合剂，磷酸组合物，阴离子型，用于漂白和煮练，与有关助剂有协同作用，商名，德国科宁）
sediment 沉淀物，沉积物；渣滓
sedimentation 沉积；淀积；沉降
sedimentation accelarator 沉降促进剂
sedimentation aids 沉淀助剂《环保》
sedimentation basin(＝sedimentation chamber) 沉降槽
sedimentation coefficient 沉降系数
sedimentation constant 沉降常数
sedimentation equilibrium 沉降平衡
sedimentation method 沉淀法
sedimentation pond 沉淀池
sedimentation rate 沉降速率
sedimentation tank 沉淀池（用于生化、物化处理废水）
sedimentation velocity 沉降速度
sediment tank 沉淀池，沉淀槽
seed coat 棉籽壳
seed coat fragments 棉籽壳屑，籽屑
seed coat neps 带籽屑棉结
seeded fabrics 低级棉织物
seed effect 细点子花纹
seed removal 去棉籽壳
seepage 渗出［现象］
seepage wash 渗流水洗
seer 绉条纹薄织物；泡泡纱；条格绉布
seersucker 绉条纹薄织物；泡泡纱（由机织、针织或后整理制成）
seersucker effect 绉条纹效应
seersucker gingham 格子泡泡纱

seersucker printing 泡泡纱印花
see-saw motion 摇摆运动
SEF (self-extinguishing fibre) 自熄纤维
segment〈SEG, seg., segm.〉链段；断片；扇形[体]；(计算机的)程序段，数据段
segmental bonding 局部黏合法
segmental motion 局部运动
segmented copolymer 嵌段共聚物
segment gear 扇形齿轮
segmer 链段
segregated stream 污水，废水
segregated stream treatment 废水处理
segregation 分离；分凝；熔析；偏析
Seikafix binder 塞依卡菲克斯涂料黏合剂(商名，日本大日精化)
Seikafix colour 塞依卡菲克斯涂料色浆(商名，日本大日精化)
sel. (selected) 选择的
Selbana UN 塞芭纳 UN 平滑剂(非离子型，用于各种纺纱过程，具有良好的黏滑平衡特性和抗静电能力，商名，德国科宁)
selected〈sel.〉选择的
selected specular refraction 选择性镜面反射
selection 选择；选址(计算机)；分类
selection criteria 选择标准
selective adsorption 选择性吸收《染》
selective permeable material 选择性透过材料
selective solvent extraction 选择性溶剂抽提[法]，选择性溶剂萃取[法]
selectivity 选择性
selector 选择器；调谐旋钮，波段开关
selector switch 选择开关，选路开关
Selectrol calender 整纬轧光机(用于圆筒针织品，商名，美国制)
seleroprotein 硬蛋白
selerotization 硬化作用
self- (构词成分)表示"自""自动""自身"
self-absorption 自行吸收
self-act 自动
self-acting〈s. a.〉自动的
self-acting control 直接调节，自动控制
self-acting thermostat 自动恒温器，自调恒温器
self-acting valve 自动阀
self-actuated controller 自动控制器
self-adhesion 自行黏着
self-adjusting 自动调节
self-adjusting screen holder 自行调节的筛网支架
self-aligning 自动对准
self-aligning bearing 自定位轴承
self-assembled 自组合
self-association 自缔合[作用]
self-balancing extractor 自动平衡脱水机
self-ballasted mercury lamp 自镇流水银灯
self-bodying 自稠化
self-bonding fibres 自黏合纤维
self-catalysis 自催化[作用]
self-catalyzed reaction 自催化反应
self-centering 自动定中心
self-check 自检，自动检验
self-cleaning 自清洁
self-cleaning filter 自清洁过滤器(自动化水清洁器)
self-colour 本色
self-compensating 自行补偿
self-contained 整套的；内装的
self-contained boiler 整装式锅炉

self-contained breathing apparatus〈SCBA〉独立(整套)呼吸设施(消防用)
self-contained instrument 机内仪表
self-controllable dye 自控性染料
self-crimping fibre 自卷曲纤维
self-crosslinking binder 自交型黏合剂
self-delustering 自身消光
self-diagnosing 自诊断
self-diffusion 自扩散[作用]
self-digest gauze 自消化型纱布
self directing optimization〈SDO〉自动对准最佳处理,自动定向最佳处理
self-edge 布边,纵边
self-electrode 自电极
self-emulsifying 自乳化
self-excitation 自激《电》
self-extinguishing character 自熄性
self-extinguishing fibre〈SEF〉自熄纤维
self-feeding〈s. f.〉自激,自馈;自动给料
self-figure 织花(织成的花纹)
self flame resistance fibre 自阻燃纤维(海藻纤维等)
self-gripping fastener 自闭带扣,尼龙搭扣
self-ignition method 自行点燃法(织物耐燃试验)
self-ignition temperature 自燃温度
self-induction 自感[应]
self-ironing 免烫整理
self-learning 自学的
self-levelling 自动校平,自调平;自匀染
self-levelling dyes 匀染性染料
self-locking nut 自锁螺母
self-lubricating 自动润滑[的]
self-lubricating bearing 自润轴承,含油轴承
self-lubricating slide rails 自润滑导轨

self-luminous colour 自发光色
self-operated regulator 自动调节器
self-oscillating linear induction motor 自振荡直线感应电动机
self-plasticizing action 自增塑作用
self-polymerization 自聚[作用]
self-positioning 自动定位
self-priming pump 自吸泵
self-purification 自净[作用]
self-recording hygrometer 自动记录湿度计
self-recording unit〈S. R. U.〉自动记录器
self-regulation 自动调节,自动调整
self-sealing coupling 自紧联轴节
self-setting bearing 多向调整轴承,自调整轴承
self shade 单色,本色
self-sharpening device 自动磨刀片装置
self-smoothing 免烫
self-smoothing cellulose fabric 免烫纤维素织物
self-smoothing fabrics 免烫织物
self-stripe 织花
self-sustaining reaction 自生[链锁]反应
self-synchronous motor〈self-syn. m.〉自动同步电动机
self-tone 单色调(用深浅同色构成的图案)
self-twist〈ST〉自捻
self-weighting 自重加压
selsyn 自动同步机
selsyn drive 自动同步传动
selsyn generator 自动同步发电机
selvage(=selvedge) 织边,布边;边缘
selvedge booster 布边热增效器(烘干布边专门装置)《针》
selvedge conditioning device 布边滴水给湿

装置
selvedge cutting device 切边装置
selvedge decurler 剥布边器,布边开卷器,防止卷边装置
selvedge defect 边疵
selvedge extracting device 集边装置
selvedge feeler 喂边器
selvedge gassing 烧边器
selvedge gathering device 集边装置
selvedge guider 导布器;吸边器
selvedge gumming device 布边上浆装置
selvedge legend 边印(注明图案、加工厂商或商标名称)
selvedge mark 布边折痕(整理疵);边记,边字
selvedge marking machine 印边机
selvedge monitor 布边监控器
selvedge opener 扩边器
selvedge predryer 布边预烘机
selvedge printing machine 布边印字机,印边机
selvedge-printing press 布边打印器
selvedge-seaming machine 包缝机
selvedge sensor 探边器
selvedge smearing 布边带浆
selvedge spreader 扩边器
selvedge stamping machine 印边机
selvedge tear 破边(织疵)
selvedge trimmer 修边装置;切边器
selvedge trimming 布边修理;切[布]边,修[布]边
selvedge trimming machine 修边机,剪边机
selvedge turndown 翻边(染疵)
selvedge uncurler 防止卷边装置
selvedge unroller 剥边机

selvege trimming device 修剪布边装置
selvege unrolling system 剥边系统
SEM(scanning electrogen microscope) 扫描光[照发射]电子显微镜
SEM (scanning electron microscope) 扫描电子显微镜
seme 满幅点子花纹
semi- (构词成分)表示"半""部分的"""不完全的"以及"[一段时期中]发生两次的"
semi-automatic control 半自动控制
semi-automatic pressing machine 半自动熨烫机
semi-automatic screen printing 半自动筛网印花
semi-automatic screen printing machine 半自动筛网印花机
semi-chemical pulp〈SCP〉半化学浆
semi-chrome leather 混合铬鞣革
semi-circle tester 半圆试验仪(可燃性试验)
semi-circular key 半圆键,月牙键
semi-colloid 半胶质,半胶态
semi-combed 半精梳的
semi-conductor 半导体
semiconductor-controlled rectifier〈SCR,S.C.R.〉半导体可控整流器
semi-conductor integrated circuit 半导体集成电路
semi-conductor laser 半导体激光器
semi-contact dryer 半接触式烘干机
semi-contact heat setting unit 半接触式热定形装置
semi-contact thermosol unit 半接触式热溶装置
semi-continuons bleaching 半连续式漂白

semi-continuons operation 半连续式作业
semi-continuons process 半连续式工艺
semi-continuous activated sludge test 半连续活性污泥试验(测生物降解率)
semi-continuous dyeing range 半连续染色机组
semi-continuous padroll process 半连续轧卷法
semi-cylindrical dye tub 半圆形染槽
semi-decating 半蒸呢[工艺];开式蒸呢
semi-decatizing machine（＝semi-decator）开式蒸呢机
semi-discharge printing 半拔染印花
semi-dress 简式礼服,便礼服
semi-drying oil 半干性油
semi-dull〈S.D.〉半无光
semi-dull silk 半无光丝《化纤》
semi-durable 半耐久性的
semi-durable water repellents 半耐久性防水剂
semi-emulsion 半乳化浆料
semi-emulsion ink 半乳化性油墨(印墨)
semi-enclosed type motor 半封闭式电动机
semi-finish 半光面整理
semi-finished goods 半制品,半成品
semi-finish parts 半成品配件
semi-gloss 半有光;近有光的
semi-manufactured goods 半制品
semi-matt 半无光
semi-mercerizing 半丝光
semi-microanalysis 半微量分析
semi-micro balance 半微量化学平衡
semi-milled finishing 半缩绒整理
semi-permanent decatizing effect 半永久性蒸呢效果
semi-permeable 半渗透性的

semi-permeable membrane 半渗透膜(水处理技术)
semi-pigmentation dyeing 隐色体染色
semi-polar bond 半极性键
semi-portable filter colourimeter 半携带式滤色比色计
semi-quantitative 半定量的;半定量分析
semi-restrained fabric flammability test 半约束式织物阻燃性试验
semi-rotary switch 半旋转式开关
semis 满幅点子花纹
semi-solid fats 半固体脂肪
semi-solvent 半溶剂
semi-synthetic 半合成的
semi-synthetic fibres 半合成纤维
semi-transparency 半透明
semi-tube 半管
semi-wet transfer 半湿法转移[印花]
semi-wet transfer printing process 半湿转移印花法
semi-worsted fabric 半精纺毛织物
send recipe 输送处方
Senegal gum 塞内加尔胶(也称作阿拉伯树胶,可作印花糊料)
Sen-i《纤维》(日本月刊名称)
Sen-i Gakkaishi《纤维学会志》(日本刊物名称)
Sen-i Kako《纤维加工》(日本刊物名称)
Sen-i Kikai Gakkaishi《纤维机械学会志》(日本月刊名称)
senior engineer 高级工程师
Senka-Antifoam 4B 森卡消泡剂4B(硅氧烷混合物,非离子型,印染工序中消泡用,适用温度范围广,商名,日本大祥)
Senka Buffer BA 森卡pH调节剂BA(不含磷的助剂,染整过程中用于pH的

调节,商名,日本大祥)

Senkafix 300 森卡固色剂300(多胺缩合物,阳离子型,用于直接染料或活性染料染色后的固色,不含甲醛,牢度好,商名,日本大祥)

Senkafix 401 森卡固色剂401(季铵盐,不含甲醛,阳离子型,用于活性染料固色,色变少,牢度好,商名,日本大祥)

Senryo to Yokuhin《染料与药品》(日本月刊名称)

sensation unit〈su.〉声感单位(分贝的原名)

sense 读出;检测,断定;感觉

sense organ 感觉器官

Senshoku Kogyo《染色工业》(日本月刊名称)

sensible heat 显热

Sensil 555 瑟锡尔555(抗菌防臭剂,高分子聚合物,阳离子型,应用于各种纤维织物、内衣等,有良好抗菌除臭作用,商名,日本大祥)

sensing 传感;感觉,读出

sensing element(= sensitive element) 传感器,敏感元件

sensitive fabric 敏感织物(如弹力织物等)

sensitive film 感光胶片

sensitive guidance 灵敏的监控[技术]

sensitive guidance of fabric 对织物灵敏的监控

sensitive plate 感光硬片

sensitive to light 对日光敏感的,对日光易褪色的

sensitivity 敏感性,灵敏度;感光性

sensitization 敏化作用,增感作用,激活作用

sensitized 敏化的

sensitized fabric 敏化织物(织物经整理剂浸轧处理并烘燥,但还未焙烘,本身仍具有继续焙固的性能)

sensitized paper 敏化纸;感光纸

sensitized screen 敏化筛网

sensitizer 敏化剂;增感剂;激活剂

sensitizer coating 感光涂胶

sensitizer coating machine 感光涂胶机

sensitizing 敏化[过程];激活[过程]

sensitizing agent 敏化剂;增感剂;激活剂

sensitizing dye 敏化染料

sensitizing effect 敏化效应

sensitometer 感光仪;曝光表

sensitometry 感光度测定;感光学,感光测量学

sensor 传感器;敏感元件

sensor breakage 传感器破损

sensor engineering 传感工程

sensor feeder cable 传感器馈电电缆

sensor function 传感器作用

sensor technology 传感技术

sensory examination 感官检验

sensory perception technology〈SPT〉感官敏感技术

sensory test 感官检验

sentinel 标记

separability 分离性

separable composite fibre 剥离型复合纤维

separable fastening fabric 尼龙搭扣

separated economizer 独立式省煤器

separately excited DC motor 外直流电动机,他激式直流电动机

separating agent 分离剂

separating and joining 分离与连接(成衣制造,包括裁剪)

separating funnel 分液漏斗
separation 分离,离析;分馏;间隔
separation negative 分色底片
separator 分离器;隔板;除尘器
separator roll 分离辊,分离棒
separatory funnel 分液漏斗
sepia 墨色
septa septum 的复数形式
septic plant 污水处理站
septic tank 化粪池
septum 隔板;中隔;隔膜
sequence 序列,数列;程序,次序,顺序
sequence control 程序控制
sequence controlled computer 程序控制计算机
sequence monitor 顺序监测器,顺序控制器
sequence of operation 操作步骤
sequential draw 顺序拉伸,外拉伸
sequester (=sequestering agent) 多价螯合剂
sequestering media 多价螯合介质
sequestrant 多价螯合剂
sequestration 多价螯合作用
SER (=ser., series) 系列
sercious 蚕丝制的;如丝的
serge 哔叽
serged seam 粗缝,疏缝(接缝一般缝两次,用于缝稀松织物)
sergette 轻薄哔叽织物
serge twill 斜纹布
serge weave 哔叽组织
sergine 丝哔叽(日本制)
serging [地毯]包边
serial 连续的,顺序的;串行的,串联的
serial number 顺序号,连续号

serial port 串行接口
serial production 系列生产,成批生产
sericeous 蚕丝制的;如丝的
sericin 丝胶
sericin fixation 丝胶固定
sericin fixing treatment 丝胶[蛋白]固定处理
sericin gumming 丝胶胶着
Sericose 醋酸纤维素(商名)
series ⟨SER, ser.⟩ 系列
series circuit 串联电路
series connection 串联
series DC motor 串激直流电动机
series feedback 串联反馈
series motor 串激电动机
series-parallel connection 串并联,混联《电》
series parallel ⟨S.P., s.p.⟩ 复串,串并联
series winding 串激绕组《电》
serigraphic printing 绢网印刷,筛网印刷
Serilan dyes 塞里兰染料(染聚酯纤维与羊毛混纺品的分散性和酸性混合染料,商名,英国制)
Serilene dyes 塞里林染料(染聚酯纤维的分散染料,商名,英国制)
serine 丝氨酸,3-羟基丙氨酸
serisizing [棉线]上丝液工艺
Serisol dyes 塞里索尔染料(分散染料,商名,英国制)
serpentine coils 蛇形管;螺旋管;盘旋管
serpentine crepe 蛇纹绉(有经向皱纹)
serpentine heater 蛇管加热器,盘管加热器
serrated pin bar 齿形针板
serration 鳞片;细齿
serration scale theory [羊毛毡合]鳞片学

说

serrature 鳞片;细齿
serve component 伺服部件
serve hydraulic 液压伺服装置
service 维修
serviceability 实用性;耐用性
serviceability test 实用性试验,耐用性试验
serviceable life 可使用期,运行寿命
service code 服务代码《计算机》
service condition 工作条件,运行情况
service diagnostic system〈SDS〉 服务诊断系统(对设备问题的诊断)
service dress〈SD〉 制服;军便服
service durability 实用耐久程度
service load 操作负荷,工作负载,实用负载,有效负载
service parts 备用零件
service pump 备用泵,辅助泵
service routine [计算机]服务程序
service soiling method 使用沾污法(沾污测定法之一)
service staff 服务人员;维修人员
service [dye] tank 染料供应槽
service test 穿着试验,试穿,试用
service uniform 军装,制服
service wear index 实用耐磨指数
servicing time 维护检修时间;预检时间
servo 伺服机构,伺服系统
servo-actuator 伺服执行机构
servo-control 伺服控制,继动控制
servo-drive 伺服传动
servo-mechanism 伺服机构,随动机构
servo-motor〈SM, sm, s. m.〉 伺服电动机
servo system 伺服系统,随动系统

servo unit〈su.〉 伺服机构,随动机构
servo valve 伺服阀
Servoxyl VPE 瑟沃克西尔 VPE(聚乙烯乳剂,背面上浆用,商名,荷兰制)
sesquicarbonate of soda 碳酸氢三钠
sessile bubble method 停泡法,固定泡法(测表面张力)
sessile drop method 停滴法,静滴法(测表面张力)
set 定形;调节,调整;安装;[轧辊]加压量;织物经纬密度;批(指原料)
Setamol E 泽塔莫尔 E(高效分散剂,芳香磺酸和羧酸钠混合物,具有优良的生物降解性,商名,德国巴斯夫)
Setamol WS 泽塔莫尔 WS(萘磺酸甲醛缩聚物,染料的胶体保护剂,分散剂,商名,德国制)
setback pin 止退销
set bolt 固定螺栓
set factors 上机参数
set hoop 定位环,紧圈
set length 规定长度
set nut 调节螺母,定位螺母
set off colour 陪衬色
set-point 给定值;凝结点,凝固点
set-point control 设定点控制
set pressure 额定压力,规定压力
set. pt.(setting point) 凝固点;沉淀点;校准点
set roller 定形辊
set screw 固定螺钉,定位螺钉
set screw spanner 止动螺钉扳手
set speed 额定速度,规定速度
set-stretching 拉伸定形
settability 可定形性
setter 给定装置;校验器,调节器;装配工

set time 定形时间;凝固时间,固化时间
setting 定形;安装;调整,调节;隔距
setting agent ［织物］定形剂;硬化剂
setting bath 凝固浴
setting chamber 热定形烘房
setting condition 定形条件
setting effect 定形效应;凝固效应
setting in liquid 液中定形
setting lever 调节杆,操纵杆
setting mechanism 调节机构,校验机构
setting medium 定形介质,凝固介质
setting method 定形方法
setting of gray fabric 坯布定形
setting-on place 开关车横档(织疵)
setting point〈set. pt.〉 校准点;凝固点;沉淀点
setting property of wool 羊毛定形性
setting screw 调节螺钉,定位螺钉
setting time 定形时间
setting treatment 定形处理
setting-up 装配;调整;快速干燥
settle 澄清;沉淀
settlement 固定;沉淀;解决［问题］
settler 沉淀池;澄清器
settling chamber 沉淀装置
settling tank 澄清桶
settling time 沉降时间;稳定时间
sett of fabric 织物经密
set-up 装配,组装;调整;准备;凝固
set-up time 准备时间;安装时间
severe washing test 剧烈洗涤试验,重洗涤试验
Sevron dyes 塞夫隆染料(阳离子染料,商名,美国制)
sewability 可缝纫性
sewability improver(＝sewability improving agent) 缝纫性改进剂
sewability tester 缝纫性试验仪
sewage 污水;下水道
sewage aeration 污水曝气
sewage disposal 废水处理,污水处理
sewage irrigation 污水灌溉
sewage lagoon 污水塘
sewage pump 污水泵
sewage purification 污水净化
sewage sludge 污泥
sewage sludge disposal 污泥处理
sewage treatment 污水处理
sewage treatment works〈STW〉 污水处理厂
sewerage filter 污水过滤器
sewerage filter aid 污水助滤剂
sewer pipe 污水管,沟管
sewing 缝纫;缝头
sewing accessories 成衣附料
sewing cotton 缝纫棉线
sewing damages 缝纫损坏（以针洞计）
sewing finish 提高缝纫性整理
sewing machine 缝纫机
sewing needle 缝纫针
sewing silk 缝纫丝线
sewing table ［缝纫机的］台板
sewing technology 缝纫技术
sewing thread 缝纫线
sewing thread count 缝[纫]线支数
sewing thread lubricants 缝纫线润滑剂
sewing thread on cones 宝塔线
sewing thread on cops 纸芯线,纸纱团
sewing-up 缝头,缝边《针》
sexadentale chelate 六配位体螯合物
sextol 甲基环己醇（羊毛助洗剂）
s. f.(self-feeding) 自激,自馈;自动给料

SF(=S. F., s-f., signal frequency) 信号频率

s. f. (square foot) 平方英尺

SF composite materials (S=spun bonded layer 纺黏布层;F=water proof moisture-permeable film 防水透湿膜)[防水透湿]SF 复合材料

SFD(supercritical fluid dyeing) 超临界流体染色

S-finishing S-整理,表面皂化整理(醋酯纤维织物通过皂化整理,可以改进手感,并减少静电感应)

S folding S 码

sfpm(=s. f. p. m., surface feet per minute) 英尺/分(圆周速度,表面线速度)

SFS(substrate fixation system) 基布固定系统(上胶装置等用)

SFT(supercritical fluid technology) 超临界流体技术

SG(=S. G., s. g., specific gravity) 相对密度

S. H. (Shore hardness) 肖氏硬度,邵氏硬度

s. h. (specific heat) 比热

shabri 白色或银灰色[野山羊]羊绒

shade 色调,色泽,色光;明暗的程度;遮光物

shade bar 色档(织疵)

shade build-up 色泽提升性;染深性

shade card 配色样卡

shade change 色纬档(织疵)

shade cloth 窗帘布

shaded check 渐变色方格花纹

shaded cloth 色泽深浅不匀的织物

shaded designs in Batik 巴蒂克印花中的阴影花纹,爪哇蜡防印花中的阴影花纹

shade-deepening agent [色泽]增深剂

shaded effect 阴影花纹

shade deviation 色泽差异

shaded filling 色纬档,色纬影

shaded satin 阴影缎纹

shaded stripe 阴影条纹

shaded twills 阴影斜纹

shaded yarn 杂色纱线

shadeline 雕刻斜纹线

shade of colour 色光

shade pitching 打色样,拼色

shade reproducibility 色光重现性

shade sorting 色调分类

shade ticket 色泽标签,色泽标样

shade variation 色差

shadiness 影条;晕影(染整疵)

shading 调整色光;[织物]染色差异(指头梢或左右的色差);[绒头织物的]倒顺毛色光改变

shading corrections 修正色光

shading dyes 着色染料(调整色光或增艳用染料)

shading effect 花纹深浅效应

shading-off 色光[调整]过头

shading textiles 遮阳纺织品

shadow barré 影条条痕

shadow cloth 经纱印花纱织物,印经织物,影纹织物

shadow cretonnes 双面影纹花布

shadow dot 影点,阴影部网点

shadow effect 阴阳花纹(隐格或隐条花纹);阴影效应(单色织物中正反捻向纱产生的色光反应)

shadow effect printed towell 隐花印花毛巾

shadow fabric 经纱印花纱织物,印经织

物,影纹织物
shadow green 荫绿[色],黄绿色
shadow mark 印痕
shadow print 虹彩印花,云纹印花,影纹印花(可用花筒雕刻深浅层次、托白浆叠印或经纱印花等方法获得)
shadow series 同色度系列(具有奥斯瓦尔德制相同的色相及纯色与白色含量比例相同的色泽)
shadow silk 闪光绸
shaft 轴
shaft bush 轴衬(机械部件)
shaft coupling 联轴节
shaft gear 主轴齿轮
shafting 轴系
shaft journal 轴颈
shaft seal 轴封
shaft sleeve 轴套(机械部件)
shaggy 起毛的,有毛绒的;粗糙的
shaggy carpet 长毛地毯
shake flask method 振荡瓶法(抗菌试验法)
shake flask test 振荡瓶试验(测生物降解率)
shake willey 振荡式除杂机
shaking apparatus 振荡装置
shameuse 留香绉
shammy 麂皮;麂皮的
shammy fabric 仿麂皮织物
shammy finish [棉针织品]仿麂皮整理
sham plush 假长毛绒
shampooing 皂洗;水洗
shangai 优质柞丝绸
shank 轴;柄;刀柄
Shantung pongee 山东府绸
shaped edge 齿形边,荷叶边(织疵)

shaped steel 型钢
shape memory alloy〈SMA〉形状记忆合金
shape memory composite〈SMC〉形状记忆复合材料
shape memory effect〈SME〉形状记忆效果
shape memory fibre 形状记忆纤维
shape memory functional membrane 形状记忆功能性薄膜
shape memory materials〈SMM〉形状记忆材料
shape memory polymer〈SMP〉形状记忆高分子材料
shape memory polyurethanes〈SMPUs〉形状记忆聚氨酯(用聚氨酯作为形状记忆材料)
shape memory suits 形状记忆服(免烫整理新的名称)
shape of cross section [纤维]横断面形状
shaper 成形装置
shape retention 形状保持性,定形性(指织物制成服装,穿着后仍能保持原形的性质)
shape-retentive finish 保形整理
shape-set 定形
shape-set process 定形方法,定形工艺
shape stability 形状稳定性,形稳性
shape stabilized process〈SSP〉保形工艺(液氨处理与免熨整理相结合的工艺)
shaping 花样;式样
shared electron 共享电子,共价电子
share operating system〈SOS〉共用操作系统
sharp bend 小半径弯管,急弯头
sharp draft 强力通风
sharpening 磨刀(指滚筒印花机刮刀刀片

的磨削);急剧去氧(还原染料染色时,追加少量保险粉使其充分还原)
sharpness 清晰度;精确度
sharpness of creasing 折皱锐度(美国国家干洗研究所术语)
sharpness of doctor 刮刀锐度《印》
sharp odour 剧烈气味
sharp outline 清晰的线条
sharp print 轮廓光洁的印花花纹
sharp soap 含游离碱的肥皂
sharp vat 去氧的还原染液
shattered ends [静电植绒的]受损绒屑
shawl 披巾,披肩
sheaflike structure 束状结构
shear 剪毛机;切变;剪切;安全销
shear deformation 剪切形变
shear-dependence of viscosity 黏度的切变依赖性
shear flow 剪切流动,切变流动
shear gradient 剪切梯度,切变梯度
shearing 刮布;剪毛
shearing bed 剪毛支架
shearing blade 剪毛刀片
shearing brush 剪绒刷
shearing cylinder 剪毛辊
shearing damage 剪毛损伤
shearing defect 剪绒疵;刮布疵
shearing force 剪切力
shearing height adjustment 剪毛高度调节
shearing machine 剪毛机(织物整理用);刮布机
shearing spirals [剪毛机的]螺旋刀片
shearing strain 剪切应变
shearing strength 剪切强力,抗剪强力
shearing stress 切变应力
shearing test 剪切试验,抗剪试验

shear marks 呢绒剪毛疵点;刮痕
shear pin 安全销
shear property 切变性
shear rate 切变速率,剪切速率
shear resistance 抗剪力
shear sensitivity 剪切敏感性
shear thinning characteristics 剪切变稀特性
shear thinning index〈STI〉剪切变稀指数
shear viscosity 切变黏度,剪切黏性
sheath 外皮,外罩;护套屏板;铠装
sheath-core bicomponent fibre 皮芯型双组分纤维
sheath-core fibre 皮芯复合纤维
sheath-core structure 皮芯结构
sheath-core yarn 皮芯[双组分]丝
sheathed heater 夹套加热器;护套加热器
sheathed thermocouple 夹套热电偶
sheath fibre 旋覆纤维(指旋覆在弹性芯丝上形成包芯纱线的纤维)
sheath fibre shrinkage 外皮纤维收缩
sheave 槽轮
SHE consciousness(Safety, Healthy, Environment consciousness) 安全、保健、环境意识
shedding 脱毛
sheen 光彩,光泽;华丽的服装
sheen-type rug 有光地毯(经化学和平整处理,美国制)
sheepskin effect 仿裘皮效应(使织物有卷曲绒头)
sheer 透明薄纱,透明薄织物(统称)
sheer crepe 透明绉,薄绉
sheer fabric 薄纱
sheer gabardine 稀薄轧别丁(一般为丝织),稀薄华达呢
sheerness 透明性;细薄度

Sheerset 希尔瑟整理(棉、黏胶纤维、醋酯纤维薄织物经三聚氰胺甲醛为主的树脂整理,有耐洗、爽挺、弹性、防缩和尺寸稳定等效果,商名)

sheet 薄板,薄片;表,图表;层

sheet dyeing 片经纱浆染法(纱线经轴染色法的一种)

sheet-fed design 单页使用的[转移印花]图纸

sheetings 阔幅平布;被单布

sheeting ticking 阔幅印花被套布

sheet iron〈sh. i., sh. I., shI〉薄铁板,铁皮

sheet lead 铅板,铅皮

sheet metal 金属板;钢皮;薄钢板

sheet molding compound〈SMC〉模压塑料片

sheet polymer 片状聚合物

sheet release 纸片释放

sheet resistance 片电阻

~~sheet-sham~~ 床罩

sheet steel〈Sh. S., shs, sh. st.〉薄钢板

shelf ageing 搁置老化

shelflife 适用期,储存寿命

shell 壳,罩;印花滚筒壳体(不连芯轴)

shellac 虫胶,紫胶

shell boiler 火管锅炉

shell fabric 面料

shell grey 贝壳灰[色](淡黄灰色)

shell roller 印花用铜滚筒(镀铜花筒)

shell sweater 无袖套头女毛衫,贝壳形毛衣

shelter 掩蔽,保护

shepherd check(=shepherd's plaid) 格子花纹织物(一般为黑白格子)

Sherdye pigment colours 秀代涂料(印染用涂料,商名,美国制)

Shetland finish 设得兰整理(使毛织物表面呈毛绒状)

SHF(=S. H. F., shf, s. h. f., superhigh frequency) 超高频

sh. i.(=sh. I., shI, sheet iron) 薄铁板,铁皮

shibori [手工]扎染

shield 防护罩,遮护板;铠装;屏蔽;腋下[吸汗]垫布

shielded cable 屏蔽电缆

shielded wire 屏蔽线

shielding 屏蔽《电》

shielding effect 屏蔽效应

shift 班,轮班;移位,进位;移动;移相;漂移

shift counter 移位计数器

shifter 切换装置,开关

shift factor 移动因子

shifting fork 拨叉

shift operation 交替操作

shift-out 移出

shikonin 紫草色素(植物染料,可染羊毛)

shim 垫片,填隙片;补偿棒,粗调棒

shimmer 微光;闪光

shine 日光;光泽;发油光(指旧织物)

shiners [人造丝织物]极光(疵点)

shipper [皮带]移动装置;[离合器]分离杆

shipping 装货,装运

Shirlastain A 锡拉着色剂 A(用以着染各种纤维,使显出不同色泽以鉴定其类别,商名,英国制)

Shirley cloth 锡莱防水棉布(用长绒丝光棉纱织造,经为合股线,纬为松捻纱,着水后,纤维膨胀,起防水作用)

Shirley flash steamer 锡莱快速蒸化机

Shirley Institute 锡莱研究所（英国的纺织研究所）
Shirley Institute System 锡莱研究所系统
Shirley static eliminator 锡莱静电消除器
Shirm's theory 希尔姆学说（直接染料的直接性与染料分子间的共轭系统长短有关的学说）
shirt ［男式］衬衫，女用［仿男式］衬衫；汗衫
shirt blouse 男式女衬衣
shirting 衬衫料子；本色细平布
shives 碎屑（指天然纤维中包含的植物性杂质）
shock 振动，冲击
shock absorber 减振器；缓冲器
shock ageing 快速蒸化
shock bleaching 快速漂白［工艺］
shock cooling 骤冷
shock curing 快速焙烘
shock drying 快速烘干
shock dyeing 快速染色
shock filter cleaning 过滤器振动清洗
shock freezing 速冻
shock oxidation 快速氧化（还原染料）
shock processing 快速工艺
shock resistance 耐冲击性
shock setting 快速定形
shock soak 快速浸渍
shock test 冲击试验
shock treatment 快速处理
shoe 瓦形物；电热靴；导向板
shoe brake 刹车块，闸块
shoe fold ［织物］两端相向折叠法（按照整匹长度，折成 12～16 层）
shoe interlinings 鞋衬
shoe leather 制鞋皮革

shoe velvet 鞋面丝绒
shop drawing 工作图，生产图，制造图
shop superintendent 车间主任
Shore durometer 肖氏硬度计，邵氏硬度计
Shore hardness〈S. H.〉肖氏硬度，邵氏硬度
Shore sclerscope hardness〈HS，Hs〉肖氏硬度
shorn-pile 剪绒
short 短流的（指厚浆或冻胶的不连续流动性）
shortage 缺货，缺乏，短缺
short ageing 快速蒸化，短时间蒸化
shortage of width 布幅不足
shortbath 小浴比［染液］
short beam measurement 单光束测量
short burnout 短焰燃烧
short circuit〈S. C.〉短路《电》
shortcoming 缺点，不足
short-cure-process 短流程焙烘工艺
shortened process 缩短的工艺流程
shortflow 低切变性（糊料性能）
short grained 细粒的
short liquor 小浴比
short liquor dyeing 小浴比染色
short liquor jets machine 小浴比喷射染色机
short liquor ratio 小浴比
short loop dryer 短环烘燥机
short loop tenter〈S. L. T.〉短环烘燥定形机
shortness 松脆；［金属等的］脆性
short period heat setting 快速热定形
short period steamer 快速汽蒸
short picks 缺纬（织疵）
short piece 短码，缺尺（匹长不足）

shortpiled 起短绒的
short-processing techniques 缩短[的]工艺技术
short range 近程,短程
short-range pH paper 精密 pH 试纸
short rheology 短流变性(印花糊料)
short ripening 快速熟成
short rotor linear induction motor 短转子感应直线电动机
short-run production 小批量生产
short runs 小批量[生产]
shorts [不足规定长度的]短匹
short scan 快扫描
short stator linear induction motor 短定子感应直线电动机
short stick〈S.S.〉短码尺
short-stopper 聚合终止剂
short stopping 急速中止
short-stopping agent 终止剂,速止剂
short-term repeatability 短期重复性
short term test 短期试验
short term use 短期使用
short thickeners 低切变糊料(网印中糊料黏度受刮浆的剪切应力而呈现变化,变化大的称为低切变糊料,变化小的称为高切变糊料)
short-time pad batch process 短时轧堆法
short time test 短期试验,快速试验
short weight 重量不足,低于标准重量
short yards 短码布
shot 注[射]料量
shot blasting 抛丸处理(一种机械表面处理工艺)
shot cloth 闪光绸
shot damage 伤洞(割绒疵)
shot dyeing(=shot-dye) 荧光染色,闪光染色;双色染色
shot effect 闪光效应
shot lisle 闪光莱尔整理产品
shot pattern 闪光图案
shot silk 闪光绸
shot taffeta 闪光塔夫绸
shot weave rayon 人造丝闪光绸
shoulder harness 安全肩带
shovelled 用铲子铲去
shower 喷淋器
shower bleaching 淋漂
shower pipe 淋水管
shower-proof cloth 防雨布
shower-proofing 防雨整理,防淋整理
shower repellency 防雨性能
shower test 防雨试验
shower-washer 淋式水洗机
showery cake bleaching 丝饼淋漂法
show-side [织物的]正面
showy 光亮的,显眼的
shred 零[头]布;碎片,破布;扯碎
shredder 撕碎机,粉碎机,研磨机
shrink 收缩;缩水
shrinkage 缩率;收缩
shrinkage allowance 收缩允许
shrinkage at boiling point 沸点收缩[率]
shrinkage control [织物]缩水率控制
shrinkage control finish 防缩整理
shrinkage controller 防缩剂
shrinkage crack 缩裂
shrinkage crimping 收缩卷曲
shrinkage curve 收缩曲线
shrinkage fit 冷缩配合,红套《机》
shrinkage in boiling water 沸水收缩[率]
shrinkage in length 经向收缩,长度收缩
shrinkage in width 纬向收缩,宽度收缩

shrinkage level 收缩程度
shrinkage machine 预缩机
shrinkage mark 收缩皱纹
shrinkage measurement 缩率测定
shrinkage of cloth 织缩
shrinkage percentage 收缩率，缩［水］率
shrinkage potential 潜在缩率
shrinkage rate 缩［水］率
shrinkage resistance 防缩性
shrinkage shoe 收缩靴
shrinkage test 收缩试验
shrinkage texture 缩绒结构［效应］
shrinkage unit 收缩单元，预缩单元
shrink dryer 收缩烘干机
shrinked weft 缩纬（织疵）
shrink fit 冷缩配合《机》
shrink-fitted shell 热套筒体《机》
shrinking force 收缩力
shrinking meter 收缩比率仪（用于防缩机，指示收缩率）
shrinking treatment 收缩处理
shrink initiation temperature 收缩开始温度
shrink-packaging ［塑料薄膜］热缩包装
shrink percentage 收缩率
shrinkproof 防缩
shrink-proof finish 防缩整理
shrink proofing 防缩整理
shrink proofing effect〈SPE〉防缩效应
shrink-resist agent 防缩剂
shrink resistance 防缩性
shrink resistant finish 防缩整理
shrink tension 收缩张力
shrunk finish 预缩整理
shrunk yards ［布匹］缩后码长
Sh.S.(=shs,sh.st.,sheet steel) 薄钢板
Shu brocade 蜀锦（四川生产的锦类真丝织物）
shuffling roller 推移辊，滑移辊
shunt 分路并联；分流器《电》
shunt circuit 分流电路，并联电路
shunt feedback 并联反馈
shunt motor 并激电动机
shunt voltage 分路电压，分流电压
shut-down 停用，停台
shut-off valve 断流阀，关闭阀
shutter 遮隔板；百叶窗；节气门；断路器《电》
shyer 稀弄（织疵）
SI【法】国际单位制
sialic acid 唾液酸
SIC(sea island cotton) 海岛棉
SIC(=s.i.c., specific inductive capacity)介电常数
siccation 干燥［作用］
siccative 干燥的；催干剂；干料
side-by-side bicomponent fibre 并列型双组分纤维
side-centre-side colour difference 左中右色差
side-centre-side distribution ［加工织物］边中边分布
side chain 侧链，支链
side chain radical 侧链基
side creel 落地纱架
side-effect 副作用
side elevation 侧视图，侧面［图］
side frame ［机器］墙板；边架
side grinding ［针布的］侧磨
side group 侧基；侧基团
side guard 边挡板
side paddle dyeing machine 边桨式染色机
side plate 侧板

side rails 两旁[布铁]导轨
side reaction 支反应;副反应
side seam [地毯]对边拼接;边缝
side shaft 侧轴,边轴
side to centre shade deviation(=side-to-centre shading) [染色]边中色差
side to side shade deviation(=side-to-side shading) [染色]左右色差
side view 侧视[图]
sideways repeat 图案纬向重复
sienna brown 浓黄土棕[色](棕色)
sieve 筛网;筛子;格筛;滤网
sieve analysis 筛析,筛选,筛分
sieve drum drier 筛网圆筒烘干机
sieve method 筛法,过筛淘汰法
sieve nozzle 圆孔式喷嘴
sieve shaker 摇筛机,振动筛分机
sieve texture 筛网组织;筛状结构
Sig(=sig., signal) 信号;信号的
sight glass 观察镜,视镜
sight glass type steam trap 窥视式疏水器
sight hole 检视孔,观察孔
sighting 着色;涂色;指色
sighting agent(=sighting colours) 标志色素,着色剂,指色剂(指易除去的着色物质)
sighting dye 指色染料(暂时着色用)
sight peep 视窗,观察孔
sigma unevenness 总不匀率
sigmoidal curve S形曲线
sign 记号,符号,标记
signal〈Sig, sig.〉信号;信号的
signal bell 信号铃
signal clothing [荧光]信号服装
signal conditioner 信号调节器
signal frequency〈SF, S. F., s-f.〉信号频率

signal generator 信号发生器;测试振荡器
signal lamp 信号灯
signalling alarm 信号报警[装置]
signalling apparatus 信号装置
signalling key 信号键,信号按钮
signal push button 信号按钮
signal response 信号响应(气相色谱仪术语)
signal-to-noise ratio〈SNR, snr.〉信号—噪声功率比
signal wire〈SW〉信号线
significant difference 有意义差别,显著差别
significant digit 有效数[位]
significant figure 有效数[字]
Sigokki machine 雪果开防印印花机(商名,日本制)
silane 硅烷
silane nanosol 含硅纳米溶胶(改善聚酯纤维织物抗静电性能)
silanized support 硅烷化载体
silanol 硅烷醇
silent chain 无声链条
silent gear 无声齿轮
Silesia 西里西亚窗帘布(重浆,轧光整理);西里西亚里子布(经轧光整理,灰色或黑色)
Silhoflec 采用回归反射整理技术的纺织品(商名,荷兰制)
silhouette 线条,轮廓;侧面影像;款式
silica 硅石,二氧化硅
silica-alkoxides 硅醇盐
silica gel 硅胶
silica-gel drier 硅胶干燥剂
silica gel thinlayer chromatography 硅胶薄

层色谱法
silicate〈s.〉 硅酸盐
silicate-free peroxide bleaching 无硅过氧化氢漂白
silicate glass 硅玻璃
silicate ion exchanger 硅酸盐离子交换剂
silicate spots 硅斑,硅垢
siliceous 含硅[的],硅质[的]
Silicia 西里西亚窗帘布(重浆,轧光整理);西里西亚里子布(经轧光整理,灰色或黑色)
silicification 硅化[作用]
silicon carbide fibre 碳化硅纤维
silicon controlled rectifier〈SCR, S. C. R.〉 可控硅整流器
silicon controlled〈sc〉 可控硅
silicon diode 硅二极管
silicone [聚]硅酮;有机硅树脂
silicone antifelting treatment 有机硅防毡缩整理
silicone-based dyes 有机硅染料
silicone-based textile auxiliary 有机硅纺织助剂
silicone catalysts 有机硅催化剂
silicone coating 有机硅涂层
silicone compound 有机硅化合物
silicone cross-linking 有机硅交联
silicone elastomer 有机硅弹性体
silicone finishes 有机硅整理剂
silicone fluid 有机硅流体
silicone fluoride 硅氟化合物
silicone lubricant 有机硅润滑剂
silicone oil 硅酮油,硅油
silicone remover 有机硅去除剂
silicone resin 有机硅树脂,硅氧树脂,硅酮树脂(织物的耐久防雨和柔软整理剂)
silicone rubber roll 硅橡胶辊
silicone softner 有机硅柔软剂
silicone surfactants 有机硅表面活性剂
silicone waterproofing 有机硅防水
silicone water repellent finishes 有机硅拒水整理剂
silicoorganic compound 有机硅化合物
Siligen 适利坚(柔软剂及免烫整理的添加剂,用于纤维素纤维及其混纺织物的柔软整理及免烫整理,给出极佳的柔软、光滑手感,提高织物折皱回复性和弹性,改善缝纫性,商名,德国巴斯夫)
silk 丝,蚕丝;绸缎,丝织品
silk bleaching 丝绸漂白
silk broadcloth 真丝绸
silk-cotton goods 绨;丝棉交织物
silk covered〈S. C.〉 丝包的
silk crepe 真丝绉
silk degumming 生丝精练;丝绸精练
silk dye 丝绸用染料
silkeen 仿绸细棉布
silken 丝的,丝一样的;柔软的;滑润的
silkete 丝光[工艺]
silkete machine 丝光机
silkette 缎纹光亮棉布;丝棉交织里子绸
silk fabric 绸缎,丝织物
silk finish [棉织物]缎光整理,[缝纫线]光滑整理
silk finish on cotton 棉织物的缎光整理
Silkfix 3A 丝绸固色剂 3A(聚胺缩合体为主要成分,阳离子型,用于提高直接染料、酸性染料在丝绸上染色时的湿牢度,商名,日本染化)
silk gauze 绢网

silk ghatpot 真丝绫
silk glue(=silk gum) 丝胶
silk hosiery 真丝袜
silkiness 丝光性,类真丝性
silk inspection and testing 生丝检验
silk jeans 真丝牛仔服
silk knitted article 真丝针织品
silk-like aesthetic 丝样美感
silk-like fabric 仿真丝织物
silk-like fibre 仿真丝纤维
silk-like finish(=silk-like finishing) 仿真丝整理
silk-like handle 丝绸感
silk modifier 丝变性剂,丝改性剂
silk net 丝网,绢网
Silkolene 西科林(仿丝薄棉布,印花轧光)
silk satin 真丝缎
silk scouring 丝绸精练,丝绸脱胶
silk screen 丝绢网(绢网印花用)
silk screen printing 丝网印花
silk scroop 丝鸣,绢鸣
silk scrooping agent 丝鸣增效剂
silk shag 拉绒丝织物
silk union 混合丝
silk velvet 丝绒
silk wadding [蚕茧上的]丝纤维网
silk washing 生丝精练,丝绸精练
silk weighting 丝绸增重[工艺],丝绸增量[工艺]
silkworm protein fibre 蛹蛋白丝
silky 真丝外观
silky touch 真丝感[觉]
sillk boiling off 生丝精练;丝绸精练
silocell 绝热砖
siloxane 硅氧烷(耐久防雨剂和柔软剂的主要成分)
silt 淤泥
Silvatol SO 雪尔污 SO(以非离子型聚乙二醇衍生物为基础,加上阴离子型磺酸盐,可作为织物去油污剂,商名,瑞士制)
silver 银;银色
silver bromide 溴化银
silver coating 涂银,银涂层
silver coating process 涂银工艺
silver compound 银化合物
silver fibre 银纤维
silver grey 银灰色
silver mirror reaction 银镜反应
silver nitrate test 硝酸银试验(用以鉴别蚕丝与羊毛)
silver-plated fibre 镀银纤维
silver powder 银粉
silver powder printing 银粉印花
silver salt 银盐(2-萘酚磺酸钠盐,用于刻色)
silver staining test 银着色试验(测定水解纤维素还原值)
silverstat regulator 接触式调压器
silvertone effect 银饯效应(深色起绒织物,掺用有光白色纤维)
silvertone overcoating 银饯大衣呢
silvery powder 银色粉
similar colour 同类色
similarity 相似性
Simili mercerising 赛丝光法(湿布经高压高温轧光机处理)
simmer ring 轴密封环
simple data-entry unit 简易数据输入装置
simple function 单官能,单功能
simple functional group 单官能团

simple linear polymer 线性聚合物
simple plasticizer 低分子增塑剂
simple polymerisation 纯聚合（共聚合的对称语）
simplex 单程式；单一的，单纯的
simplification 单纯化，简单化
simulate 模拟，仿真
simulated crepe weave 仿绉组织
simulated handknit 仿手工编织
simulating test 模拟试验
simulation 模拟，仿真，模拟物
Simulation of the Consumer's Apparel Purchase 消费者衣物购买[动机]模拟
simulation test 模拟试验
simulator 模拟设备；模拟系统；模拟程序
simultaneous contrast 同时对比
simultaneous dyeing and finishing 同时染色整理
simultaneous dyeing control 同时染色控制
simultaneous finishing processes 同时整理法（多种目的同时整理）
simultaneous motion 同时动作
simultaneous reaction 同步反应
Sinagen dyes 辛纳近染料（不溶性偶氮染料的组分，商名，瑞士制）
sine aqua〈sine aq.〉无水
singe 烧毛，烧绒
singeing 烧毛工艺
singeing and desizing range 烧毛退浆联合机
singeing frame 烧毛机架
singeing machine 烧毛机
singeing plate 烧毛板
single-acting brushing machine 单动式刷呢机，单动式刷布机
single acting raising machine 单动式起绒机
single-bank superheater 单级过热器
single-bath chrome dyeing 一浴法铬媒染色
single bath dyeing 单浴染色
single bath process 一浴法
single beam measurement 单光束测量
single cloth 单层布
single colour effect 单色花纹
single-component coating system 单组分涂层系统
single-contact controller 单触点控制器
single cylinder dryer 单滚筒烘燥机
single cylinder steeping press 单柱式浸渍压榨机
single drum dryer 单滚筒烘燥机
single dye vat 单浴染槽
single-effect evapourator 单效蒸发器
single fancy suiting 单面花呢
single jersey 单面针织物，单面乔赛，汗布
single knit fabric 单面针织物
single-layer stenter 单层拉幅机
single loop compartments 单环式分隔槽（平洗机形式）
single-loop towel 单面毛巾
single nucleotide 单核苷酸
single nucleotide polymorphisms〈SNP〉单核苷酸多态性（检验羊绒、羊毛纤维的技术）
single phase printing process 一相法印花
single-phase〈S.P.〉单相
single-pole〈S.P.〉单板
single sampling 单式抽样
single-screen saturator 单网式饱和浸渍机
single side drying machine 单面烘燥机，单面烘布机

single squeegee 网印单刮刀
single stage pump 单级泵
single-strength brand [染料]普通[力分]牌号
single teasel raising machine 单面刺果起绒机
single width 单幅
single woven 单层织物
single yarn 单纱；单丝
single yarn breaking strength 单纱断裂强力
single yarn tester 单纱强力试验机
sink 刻纹沉陷（雕刻疵）
sinkage 意外损耗
sinter 熔渣
sintered crucible 烧结坩埚
sintered metal 烧结合金
sintered metal bearing 烧结含油轴承，粉末冶金轴承
sintering metal filter 烧结金属过滤器
sintering temperature 烧结温度（用于散布法涂层）
sinter tunnel 烧结隧道（用于散布法涂层）
sinuosity 曲折度，波动性；弯曲
sinuous coil 盘管，蛇形管
Sipex SB 西佩克斯 SB（十二烷基硫酸钠，分散剂，乳化剂，制合成增稠剂时用作稳定剂，商名）
siphon 虹吸；虹吸管
siphonic effect 虹吸效应
siphonic pipe 虹吸管
Sirius dyes 锡利染料（直接耐晒染料，商名，德国制）
Siro FAST（FAST = Fabric Assurance by Simple Testing）赛洛织物客观测试系统（一种织物风格仪，重点用在纯毛、毛混纺产品上，澳大利亚联邦科学与工业研究组织开发）
Sirolan BAP 一种羊毛防缩剂（商名）
Sironized finish 锡罗整理（毛织物的耐久褶裥整理，商名，澳大利亚开发）
Siroset 羊毛织物耐久折裥定形法
Si-Ro-Set process 西罗塞特加工法（毛织物利用还原剂进行耐久褶裥整理）
Siro silky wool 变形截面细羊毛（商名，澳大利亚制）
Siro-spun process 赛络纺纱法（商名）
Siro-spun yarn 赛络纺纱线（商名）
Sirovelle ARW 一种抗静电柔软剂（含氮脂肪酸浓缩液，阳离子型，可提高针织物弹性，赋予合成纤维优良的吸湿抗静电性，可与各种染料同浴使用，也可用于毛织物的防尘吸灰，商名，英国 PPT）
Sirrix AK, AS 舒力斯 AK, AS（全属螯合剂，阴离子型，强碱中稳定，可避免硅及碱土氢氧化物的沉淀，软化水，商名，瑞士科莱恩）
sirve-drum steamer 筛网圆筒汽蒸机
sirve-drum washing machine 筛网圆筒水洗机
SIT（= S. I. T., sit, spontaneous ignition temperature）自燃温度，自燃点
site 现场，工地
site-assembled 工地装配的，散装式的
site erection 现场安装
situation 情况，局面；位置；场所
Sivim rotational abrader 西维姆旋转式织物耐磨仪
size 大小，尺寸；浆，浆液；细度，纤度，条分

size accumulation 浆料积聚（在退浆槽中）
size beck 煮浆桶
size boiling apparatus 煮浆设备
size box 浆槽，浆箱
Size CB, CE 赛爱士 CB, CE（聚丙烯酸酯浆料，阴离子型，适用于各种混纺短纤高支高密经纱的上浆，浆膜弹性好，温湿度范围广，商名，德国巴斯夫）
size cooking pan 煮浆桶
size delivery pump 输浆泵
size deposite 浆垢，落浆
size deviation 尺寸偏差，线密度偏差
size distribution 尺寸大小分布
size film 浆膜
size gum 浆料用胶
size kettle 浆锅
size label 尺码标签
size level control 浆槽液面控制
size loading 上浆率
size mixer 调浆桶；调浆器
size mixing 调浆
size mixing apparatus 调浆设备
size mixture 混合浆料
size of fibre 纤维细度
size of filament 长丝线密度
size pan 煮浆釜
size pick-up 上浆率
size reclaimation device 浆料回收装置
size recovery system 浆料回收系统
size shedding 落浆
size stain 浆斑
size temperature control 浆液温度控制
size trough 浆槽
size variation 支数不匀率
size vat 浆槽，浆箱

sizing 浆纱，上浆
sizing agent 上浆材料，浆料助剂
sizing cloth 浆纱[用]呢；浆辊包布
sizing compatibility 浆料混溶性；浆料互溶性
sizing finish 上浆整理
sizing ingredients 上浆成分
sizing instruction 上浆配方
sizing percentage 上浆率
sizing roller 上浆辊
skein 绞纱；绞丝
skein breaking tenacity 绞纱断裂强度，缕纱强力
skein dyeing 绞纱染色；绞丝染色
skein dyeing machine 绞纱染色机
skeining 成绞
skein printing 绞纱印花
skein reeling machine(＝skein reel) 摇绞纱机；复摇机，扬返机
skein strength 绞纱强力
skein test 绞纱试验
skein washing machine 绞纱水洗机
skeleton 骨架；概略，轮廓
skeleton cylinder [浆纱机]花篮；转笼
skeleton drum 转笼
skeleton roller 主加压辊
sketch 草图；草拟；概略
sketch-making 锌版放样
skew 纬斜(织疵)；歪斜
skew bevel gear 斜伞齿轮
skew-bow fabric straightener [织物]斜纬弓纬整纬器
skew correction rolls 整纬斜辊
skewing (＝skewing of weft) 纬斜
skewness 偏斜度《数》
skewness of fabric 织物的歪斜

skew straightener 整纬器
skew symmetric 对称的纬斜
skew weft 纬斜
ski cloth 滑雪织物
skiing jacket 滑雪上衣
ski knit 滑雪用针织物
skim coating fabric 涂橡胶织物
skimmed milk 脱脂乳(制酪素纤维用)
skimmer [泡沫]分离器;撇乳器,撇渣器
skin 表层
skin and core effect 皮芯效应
skin-care finish 皮肤护理整理
skin coat 皮膜涂层,表面涂层
skin-contact clothing 接触皮肤的衣服(内衣)
skin-core effect 皮芯效应
skin-core structure 皮芯结构
skin cross-linking 皮交联,表面交联
skin effect 皮层效应
skin formation 皮层形成
skin-gas effect 皮层气泡[附着]效应
skin hardness 表面硬度
skin irritant(=skin irritation) 皮肤刺激
skinning 结皮
skin sensitization 皮肤过敏
skin-side comfort 皮肤舒适感
skin temperature 表皮温度
skin tight 贴身
skin tolerance [化学品]对皮肤的刺激程度;皮肤耐受度
skip 跳跃[进位]《自动》;空白指令《计算机》
skips(=skipped thread) 跳花,跳纱(织疵)
skip weld 间断焊缝,跳焊
skirt 侧缘,环,套筒

ski shirt 滑雪衣
ski tricot 滑雪针织物
skitteriness 染色小斑疵点;花筒嵌浆斑
skittery dyeing 斑点染色
skittery of roller printing 滚筒印花疵(斑点)
skiving 切片;削刮
skiwear 滑雪服装
sky blue 天蓝[色]
sky grey 天灰[色](淡灰蓝色)
skying 透风;空气氧化
skying range 透风装置
skying roll 透风滚筒
skylab 天空实验室
SL(=sl, spectrometer lens) 分光计透镜
SL(square law) 平方律
SL(straight line) 直线
slabby 黏稠的;层状的
slack adjuster 空隙调节器,松紧调节器
slack-batch mercerizing 松堆丝光
slack drying 松式烘燥,无张力烘燥
slacken 放松
slack end 松经(织疵);经弓(毛织疵)
slack list 松边(织疵)
slack loop washer 松式环状洗涤机
slack mercerization 松式[碱]丝光
slack mercerizing machine 松式丝光机
slackness [呢绒]不硬挺,无身骨
slack pick 松纬(织疵);纬弓(毛织疵)
slack porosity 膨松性,疏松性;多孔性
slack rope scouring machine 松式绳状煮练机
slack rope washing machine 松式绳状水洗机
slack selvedge 松边
slack suit 一套宽松衣服,便装

slack washing machine 松式绳状水洗机
slack wax 疏松石蜡(经冷却、过滤制得的石蜡)
slag 炉渣,熔渣
slag welding 电渣焊
slag wool [fibre] 渣绒纤维;矿渣棉
slant 倾斜,斜坡
slanted strut 斜撑
slash 横线拉斜;上浆;砑光(布匹最后加工)
slash box 浆槽
slasher dyeing 经轴染色
slasher dyeing process 浆纱机经轴染色法(纱线经轴染色法的一种)
slat-conveyer 平板履带
slate 石板色(紫灰色)
slate blue 石板蓝[色](浅灰蓝色)
slate green 石板绿[色](浅灰绿色)
slate grey 石板灰[色](淡棕灰色)
slazy knitting 布面不清,线圈不匀;稀松针织物
sleaziness 粗纱档(若干根粗节在织物上聚在一起形成的疵点);横向条花(织疵);毛圈的松散性
sleazy 稀松组织
sleazy fabric 稀薄织物;结构不良织物
sleek 轧光斜纹棉布;光滑的
sleeping bags 睡袋
sleeve 套筒,套管,轴套
sleeve bearing 套筒轴承
sleeve coupling 套筒联轴节
sleeve joint 套筒接头
sleeve nut 套筒螺母
sleeving 管状织物
slenderization 细化作用(如羊毛的拉伸细化)

slenderizing wool 细化羊毛
slenderness ratio 细长比,长径比(指长度与直径之比)
slewing angle limitation [链条导轨]回转角的限位
slewing angle of chain rail 链条导轨的回转角
slewing motor 回转式电动机(扫描用)
slicker fabric [上胶]防水织物,雨衣织物
slicking-in roller 轧辊
slidac 滑线电阻调压器
slide 滑板,滑座,滑轨
slide bar 滑杆
slide bearing 滑动轴承
slide bed [剪毛机的]滑动托架,滑动托板
slide block 滑块
slide control 滑动调节
slide damper 闸板
slide gauge 游标卡尺,滑尺
slide glass [显微镜用]载玻片
slide plate 滑板
slider 滑板,滑座,滑轨
slide rule 计算尺
slide switch 滑动开关
slide valve 滑阀,闸阀
slide ways 滑斜面《机》
sliding apron 滑动罩圈
sliding caliper 游标卡尺
sliding clutch 滑动离合器
sliding contact 滑动接触
sliding friction 滑动摩擦
sliding gate pump 叶片泵
sliding poise 游码,滑码
sliding valve 滑动阀
slime spots 浆斑

sliminess 稀黏程度
slinger 抛油环,甩油圈
sling hygrometer 手摇[干湿球]湿度计
sling psychrometer 手摇湿度计
sling tube 悬吊管
slip 滑动;空转;转差率;[衣服的]活络里子;枕套;妇女长衬裙,儿童围裙
slip catch 防止逆转钩,防滑钩,伸缩爪
slip clutch 滑动离合器
slip coupling 滑差联轴节
slip flow 滑流
slippage [皮带]滑溜;经纬滑动;浪纹(染疵、织疵)
slippage effect 滑移效应
slipperiness 稀黏程度
slippery hand 手感光滑
slip-pin 安全钉,保险销;安全别针
slipping clutch 摩擦离合器
slip plane 滑脱面
slip proofing agent 防滑剂
slip resistant 防滑的
slip-retardant finishes 防滑整理
slip ring 滑环
slip-ring motor 滑环式电动机
slip-stick 防滑
slip-stick effect 滑黏效应
slit [狭]缝;裂口;槽
slit burner 狭缝式燃烧器,狭缝式火口
slit cellulose film 纤维素薄膜条
slit film fibre 切膜纤维
slit gas burner 狭缝气体燃烧器
slit heater 槽形加热器,热板(俗称)
slit opener 剖幅机
slit skirt 开衩裙
slitting and opening machine 剖幅开幅机
slitting device 剖幅装置(针织物拉幅机出布区的纵向割布装置)
slitting machine 纵向割布机;剖幅机
slitting resistance 抗撕开性
slitting strength 撕开强力
slit type nozzle 狭缝式喷嘴
sliver printing machine 毛条印花机
SLOE (special list of equipment) 设备清单
slope 斜率,斜度;倾斜
slope roller printing machine 斜式滚筒印花机
slop padding 倾斜浸轧工艺
slops 现成低档衣服;污水
slop starching machine 倾斜浸轧式上浆机
slot 槽,沟,缝,孔隙;存取窗口,插槽《计算机》
slot exhaust unit 狭缝式排风装置
slot jet 扁孔喷嘴
slot liner 槽[绝缘]衬
slot nozzle 狭缝式喷风管
slot scrubber 狭缝式分离器
slot slitting 铣槽
slot squeegee 槽缝式刮浆板
slotted hole 槽孔,长槽孔
slotted seal 长槽形封口
sloughed-off weft 脱纬
slow burning fabric 耐燃织物,慢燃织物
slow cooling 渐冷
slow curing resin 慢焙固树脂
slow-elastic deformation 缓弹性变形
slow elasticity 缓弹性
slowing down cam 减速凸轮
slowing down prop 减速杆,减速撑条
slow motion 减速装置,慢车装置
slow run 低速
slow speed 〈ss〉 低速

slow stock 黏状浆
S. L. T. (short loop tenter) 短环烘燥定形机
slubbing-dyed yarn 粗纱染色纱
slubbing dyeing 粗纱染色《毛》
slubbing printing 毛条印花
slub dyeing 毛条染色
slub effect 竹节效应
sludge 污泥
sludge acclimatization 污泥驯化《环保》
sludge age 泥龄《环保》
sludge bulking 污泥膨胀《环保》
sludge cake 污泥饼
sludge density index〈SDI〉污泥密度指数
sludge dewatering 污泥脱水《环保》
sludge disposal 污泥处置
sludge drying bed 污泥干燥床,干化场
sludge gas 污泥气,沼气
sludge growth index 污泥增殖指数
sludge incinerator 污泥焚烧炉
sludge index 污泥指数《环保》
sludge loading 污泥负荷
sludge seeding 污泥接种
sludge thickening 污泥增稠《环保》
sludge volume index〈SVI〉污泥体积指数
sluice 水门,闸门
sluice line 冲洗管路;排放路线
sluice valve 闸阀
slumber wear 睡衣
slurry 洗毛污水(未脱脂);泥浆,淤浆
slurry filtration 泥浆过滤
slurry pump 淤浆泵
slush 污水,泥浆
slushing 灌泥浆
SM(=sm, s. m. , servo-motor) 伺服电动机

SMA(shape memory alloy) 形状记忆合金
small clip tenter 小布夹拉幅机
small-scale integration〈SSI〉小规模集成电路
small-scale test 小规模试验
small warp ends 织物的经密不足
smalt 大青蓝[色](紫光蓝色);蓝色颜料
smart sound 高音质
smart textile 灵巧纺织品,机敏纺织品;智能性纺织品
SMC(shape memory composite) 形状记忆复合材料
SMC(sheet molding compound) 模压塑料片
SMC(support and maintenance contract) 支持和维护合同(即有服务工程师到现场服务的合同)
SME(shape memory effect) 形状记忆效果
smear 油迹;污斑
smeary 油污的
smelting 熔炼;熔化
SMF(strength management finish) 强力维护整理(使强力不下降的整理)
Smith drum type machine 史密斯滚筒式染色机
Smith's cloth abrasion tester 史密斯织物耐磨试验仪
SMM(shape memory materials) 形状记忆材料
smock 罩衫;工作服
smog 烟雾(用于形容大气中污染的气体)
smog protection 防雾
smoke 烟,烟尘
smoke densitometer 烟雾显像密度计,烟雾浓度计(阻燃试验)

smoke emission 烟雾散发
smoke grey 烟灰色
smoke removal efficiency 烟雾滤除效率
smoke retardants 防烟雾剂,防烟燃剂
smooth agent 平滑剂
smooth backing 光滑背涂
smooth drying 干燥自平,免烫
smooth-drying fabric 免烫织物,干[燥自]平织物
smoothed curve 光滑曲线
smoothing machine 平整机
smooth laminar flow 平滑层流
smoothness 光滑度;滑爽性
smooth run-off 平稳退绕
smooth starting 平稳启动
smooth-touch 手感光滑
smooth type synthetic leather 光滑型合成革
smoulder 发烟燃烧,闷烧;余烬延燃
smouldering 发烟燃烧,闷烧
smoulder proof 助阴燃
smoulder retardants 防阴燃剂
SMP(shape memory polymer) 形状记忆高分子材料
SM process(Stenter Mercerization) 拉幅机丝光工艺(适用于针织品,商名,瑞士科莱恩开发)
SMPUs(shape memory polyurethanes) 形状记忆聚氨酯(用聚氨酯作为形状记忆材料)
smudge 污迹,黑点
smudging 弄脏;涂污
smut 沾脏,沾污;污点,污物
S-N (relation between the stress amplitude and number of repetition) 应力与反复弯曲次数的关系

SN(=S. N., saponification number) 皂化值
snag 擦毛,擦损;钩丝(针织疵)
snag-free finish(=snag-free finishing) (针织品的)防钩丝整理
snagging 钩丝
snag-resistance 防擦毛性,防擦伤性,防钩丝性
snags 钢芯两头的凹凸节齿(雕刻用语,为防止轧制钢芯时左右走动)
snag tester 擦毛试验器,钩丝试验器
snail 涡管;涡轮;涡形板
snapback 迅速回复
snapback fibre 弹性纤维,松紧纤维
snapfastener 按钮,按扣,揿钮,帽钉
snaplock switch 连锁开关,弹簧锁开关
snappers 拖浆,嵌刀(滚筒印花时由于棉屑杂质嵌入刮浆刀口而产生花中间无色,左右有色的条状疵病)
snappy rubber 高弹性橡胶
snap ring 止动环,锁紧环
snaps 拖浆,嵌刀(滚筒印花时由于棉屑杂质嵌入刮浆刀口而产生花中间无色,左右有色的条状疵病)
snarl effect 卷缩效应
SNG (synthetic natural gas) 合成天然气
snip 切口,凹口;切片;切,割
SNOM (scanning near-field optical microscope) 扫描近场光学显微镜
snow ball 起球(织疵)
snowcloth 冬季织物
snowflake 雪花呢
snow shirt(=snow suits) 冬季保暖服装(总称)
snowwhite 雪白
SNP (single nucleotide polymorphisms) 单

核苷酸多态性(检验羊绒、羊毛纤维的技术)

SNP process (sublistic natural printing process) 一种天然纤维转移印花法(德国开发)

SNR(=snr, signal-to-noise ratio) 信号—噪声功率比

snubber 缓冲器;减震器;减声器

snuff[brown] 鼻烟棕[色](黄褐色)

snug fit 服装紧贴合身

soakage 浸湿;吸入量;渗透量

soaker 浸渍剂;浸洗机

soaking 浸渍;泡丝;浸化(指圆网感光后浸在水中,使未感光的胶质软化溶解,以利清洗)

soaking auxiliary 浸吸助剂,浸渍助剂

soaking conveyor belt compartment 浸渍履带式[堆置]箱

soak test 浸泡试验

soap 肥皂

soap and alkali process 皂碱法

soap bath 肥皂液,皂浴

soap boiling 皂煮

soap degumming 皂练《丝》

soaper 皂洗机

soap fastness 皂洗牢度,耐皂性

soaping 皂洗

soaping fastness 皂洗牢度,耐皂性

soaping steamer 皂煮蒸箱

soap lake 肥皂色淀

soap lather 肥皂泡沫

soapless detergent(=soapless soap) 无皂洗涤剂(指合成洗涤剂)

soap lubricant 润滑皂

soap mark 灰伤(真丝绸精练病疵)

soap-mark 皂渍,灰点(绸匹疵)

soap milling 皂洗缩绒

soap-oil emulsion 皂质油乳化液

soap pan 煮皂锅

soap rubbing 皂洗摩擦

soap scum 皂垢

soap shrunk finish [毛织物]皂缩整理

soap softening 肥皂软水法(练丝时用)

soap solubility ratio〈S. R. , S. S. R.〉肥皂溶解度比

soap specks 皂斑

soap spot 皂渍

soap-suds 肥皂沫;起沫皂液

SOAV(solenoid operated air valve) 电磁控制气阀

society〈Soc.〉社会;会,社,协会,团体,公司

Society of Dyers and Colourists〈SDC〉染色工作者学会(英国)

socked stand 电插座板

socket 套节;管套;插座,灯座;管座;接线闸

socket bulb 插口灯泡

socket joint 套筒联接

socket pipe 套接管,承口管

socket pipe for handrail 栏杆套管

socket screw 凹头螺钉

socket spanner(=socket wrench) 套筒扳手

sock fix 短袜蒸汽定形机

sock press 短袜定形机,短袜压烫机

sock treat 短袜处理机

soda ash 纯碱

sodaboil bleaching 碱煮漂白

soda boiling 碱煮

soda cellulose 碱纤维素

soda digestion liquor 碱法蒸煮液

soda lime process 纯碱石灰[软水]法
soda lye 碱液
soda lye causticizing 碱液苛化
sodalye causticizing and precipitating apparatus 淡碱净化设备
sodalye concentrator 淡碱蒸浓设备
soda shrinking 碱缩
soda soap 钠皂,硬皂
soda treatment 碱处理
soda washing 碱洗
sodium acetate 醋酸钠
sodium alginate 海藻酸钠
sodium alkylarylsulfonate 烷基芳基磺酸钠(作渗透剂用)
sodium alkyl naphthalene sulfonate 烷基萘磺酸钠(用作润湿剂和分散剂)
sodium aluminate 铝酸钠(作媒染剂、软水剂、净水剂等用)
sodium amino-triacetate 氨三乙酸钠盐,软水剂 A
sodium azide 叠氮化钠
sodium base grease 钠基润滑脂
sodium bicarbonate 碳酸氢钠,小苏打
sodium bichromate 重铬酸钠,红矾钠
sodium bisulphate 硫酸氢钠
sodium bisulphite 亚硫酸氢钠,重亚硫酸钠
sodium borate 硼酸钠
sodium borohydride 氢硼化钠(还原剂)
sodium bromate 溴酸钠
sodium bromite 亚溴酸钠(氧化退浆剂)
sodium carbonate 碳酸钠,纯碱
sodium carboxymethyl cellulose〈NaCMC,SCMC〉羧甲基纤维素钠盐
sodium cellulose glycolate 羧甲基纤维素钠
sodium chlorate 氯酸钠,氯酸碱

sodium chloride 氯化钠,食盐
sodium chlorite 亚氯酸钠
sodium chlorite bleaching 亚氯酸钠漂白,亚漂
sodium copper chlorophyll 叶绿素铜钠(用于织物染色,起伪装效果)
sodium cyanide 氰化钠
sodium dichloro-isocyanurate 二氯异氰尿酸钠(羊毛防缩整理剂)
sodium dichromate 红矾钠,重铬酸钠
sodium dihydrogen phosphate 磷酸二氢钠
sodium dithionite 保险粉,边二亚硫酸钠
sodium dodecanesulphonate 十二烷基磺酸钠
sodium dodecyl-sulfate 十二烷基硫酸钠(用以漂洗纤维或织物)
sodium dodecyl-sulphosuccinate 十二烷基磺基丁二酸钠(染毛用表面活性剂)
sodium ferrocyanide 黄血盐钠,亚铁氰化钠
sodium fluoride 氟化钠
sodium fluosilicate 氟硅酸钠(防腐剂,杀虫剂)
sodium formate 甲酸钠
sodium hexametaphosphate 六[聚]偏磷酸钠(作软水剂用)
sodium hydrosulphite 连二亚硫酸钠,保险粉
sodium hydroxide 氢氧化钠,烧碱,苛性钠
sodium hypobromite 次溴酸钠(氧化退浆剂)
sodium hypochlorite 次氯酸钠
sodium hypochlorite bleaching 氯漂,次氯酸钠漂白
sodium lamp 钠灯
sodium lauryl sulfate 月桂(十二烷)基硫

酸钠
sodium metanitrobenzene sulfonate 间硝基苯磺酸钠（保护性氧化剂）
sodium metasilicate ［偏］硅酸钠（即水玻璃）
sodium nitrate 硝酸钠
sodium nitrite 亚硝酸钠
sodium oleate 油酸钠
sodium orthophosphate 磷酸二钠，磷酸氢二钠
sodium orthosilicate 原硅酸钠
sodium perborate 过硼酸钠
sodium percarbonate 过碳酸钠
sodium perchlorate 过氯酸钠
sodium peroxide bleaching 过氧化钠漂白工艺
sodium persulfate 过硫酸钠（漂白剂，氧化剂）
sodium phosphate 磷酸钠，磷酸三钠
sodium phosphate glass 磷酸钠玻璃，六偏磷酸钠
sodium polyphosphate 聚磷酸钠
sodium pyrophosphate 焦磷酸钠（软水剂）
sodium pyrosulphite 重亚硫酸钠，亚硫酸氢钠
sodium sesquicarbonate 碳酸氢三钠（$Na_2CO_3 \cdot NaHCO_3 \cdot 2H_2O$）
sodium silicate 水玻璃，硅酸钠
sodium stearate 硬脂酸钠
sodium sulphate 硫酸钠，元明粉，芒硝
sodium sulphide 硫化钠，硫化碱
sodium sulphite 亚硫酸钠
sodium sulphoxylate acetaldehyde 乙醛次硫酸钠（还原剂朗茄尔 A 的化学名称）
sodium sulphoxylate-formaldehyde 甲醛合次硫酸氢钠（雕白粉的化学名称）

sodium thiocyanate 硫氰酸钠
sodium thiosulphate 大苏打，海波，硫代硫酸钠
sodium tripolyphosphate 三聚磷酸钠（软水剂，匀染剂，渗透剂）
sodolin 亚麻、大麻混合制成的织物（意大利制）
Sodyecron dyes 索代克朗染料（分散染料，商名，美国制）
Sodyesul 速得素（预还原液体硫化染料，商名，瑞士科莱恩）
Sofcer relaxing machine 索费瑟松弛机（一种松弛设备，织物夹在上、下两条水平网状导带之间，并有上、下两排喷液管喷液，日本制）
soft acid 软酸（接受电子对能力小、原子体积大、正电荷低或等于零、易极化的酸）
soft agent 软化剂
soft and comfortable clothing 柔软舒适的衣着
soft base 软碱（电负性低、易极化、易失去电子的碱）
soft block 软性嵌段（合成纤维分子结构）
soft bowl 软质轧辊
soft brightening 柔性增艳处理
soft calendering 软轧光
soft cascade 缓和冲流
soft detergent 软性洗涤剂（易生物降解的洗涤剂）
soft dot 软点子，虚点子
softener 柔软剂
softener formulation 柔软剂配方
softening 软化，柔软整理
softening agent 柔软剂
softening agent VS 柔软剂 VS

softening and lubricating agents 柔软平滑剂
softening finishing 柔软整理
softening machine 柔软整理机
softening of hard water by permutit 硬水用泡沸石法软化
softening point (soft. pt.) 软化点
softening point in hot air 干热软化点
softening point in water 水中软化点
softening temperature 软化温度
soft feeling 柔软感；手感柔软
soft fibre 软质韧皮纤维（如亚麻、大麻、黄麻、苎麻等）
soft finish 柔软整理
soft flocks 毛屑，纤维屑
soft flow 缓流（指染液不剧烈的循环流动）
soft flow dyeing machine 缓流染色机（针织物的绳状间歇式染色设备）
soft flow jet dyeing 缓流喷射染色
soft goods 纺织品；匹头
soft handle 柔软手感
soft handle treatment 手感柔软处理
soft jet technique 软喷技术（针织物染色用）
soft laminating 软性层压
softness 柔软度
softness index 柔软度指数
softness tester 柔软度试验机
soft. pt. (softening point) 软化点
soft rubber bowl 软橡胶轧辊
soft segment 软链段（高分子结构）
soft sides 烂纱，断经（织疵）
soft silk 熟丝，脱胶丝
soft soap 软皂，钾皂
soft solid materials 半固体物料，塑性物质

soft stream 缓流
soft stream dyeing system 缓流染色机（一种管道型、软喷式染色设备）
soft stream piece dyeing machine 缓流[绳状]匹染机
soft touch 柔软的触感
software 软件
soft warps 软经纱，轻浆纱（指上浆不足的经纱）
soft water 软水
soft water agent 软水剂
soie batiste 全丝细薄绸（细薄透风）
soil 土壤；污物，污斑；沾污
soilability 易沾污性
soil adhesion〈SAD〉沾污性
soil burial test 土埋试验（纺织品耐腐蚀性试验）
soil conditioner 土壤改良剂
soil covering system 地膜
soil deposition 污垢沉积
soiled ends 污经，油污经纱
soiled swatch 污染的布样，污布样
soiled threads 污丝（生丝疵）
soiled yarn 污纱，污渍纱
soil-hiding 隐渍斑，隐污斑；吸灰，藏灰
soil-hiding quality 藏灰性，吸灰性
soiling 沾污；传色（机印疵，指色浆受另一色浆的传色而变色）
soiling behaviour 易沾污性
soiling medium 污染浴
soiling of the colours [色浆]沾污（滚筒印花的前色浆带入后色浆）
soiling procedure 污染程序，上污程序
soiling test 沾污试验
soil nutrient 土壤营养剂
soil pick-up 吸污，沾污

soil pollution 土壤污染《环保》
soil proof 防污性
soil redeposition 污垢再沉淀,污物再沾污
soil release 去污性
soil release finish〈SR finish〉易去污整理
soil-release testing 去污试验
soil removability 去污能力
soil-removing 去污作用
soil repellency 拒污性
soil repellent 拒污的,防污的;拒污剂,防污剂;防污
soil resistance 抗污性
soil resistance rating 防污等级
soil resistant finish 防污整理
soil-resist fibre 防污纤维
soil retardant 防污整理剂
soil retention〈SR〉污垢保持性;剩余污垢
soil suspending power 污垢悬浮能力
sol 溶胶,液胶
sol.(soluble) 可溶的
sol.(solution) 溶液
Solanthren dyes 索来士林染料(还原染料,商名,德国制)
solar active products 光活化产品,光变色产品
solar active yarn 光变色纱线
Solar Alpha 尼龙碳化锆皮芯纤维(商名)
solar battery 太阳能电池
solar bleaching 日光漂白
solar catalysis 光催化
solar collector 太阳能收集器,太阳能集热器
solar degradation 日光降解
solar drying 日晒干燥
Solar dyes 沙拉染料(直接染料,阴离子型,良好的耐日晒色牢度,后处理用山德士固色剂可改善湿牢度,商名,瑞士科莱恩)
solar energy absorbing fibre 太阳能吸收纤维
solar heat shielding fibre 太阳能屏蔽纤维
solarization 暴晒[作用]
solar optical property 日照性
solar power 太阳能
solar protection 防日晒
solar radiation 太阳能辐射
solar-selective absorbtive fibre 太阳能选择性吸收纤维
solar spectrum 太阳光谱
Solasan fast dyes 沙拉山坚牢染料(染羊毛与黏胶纤维混纺织品的染料,商名,瑞士制)
solder 焊料,焊剂
Soledon dyes 晒丽登染料(可溶性还原染料,商名,英国制)
soleil 棱条缎;斜纹呢;重缎;加光织物(泛称)
soleil velvet 有光平绒
solenoid 电磁线圈,螺线管线圈
solenoid operated air valve〈SOAV〉电磁控制气阀
solenoid relay 电磁继电器
solenoid switch 电磁开关;吸铁开关
solenoid valve 电磁阀,螺线管操纵阀
solenoid valve inkjet printing 电磁阀式喷墨印花
sol-gel 溶胶—凝胶
sol-gel method(=sol-gel process) 溶胶—凝胶法
sol-gel technique 溶胶—凝胶技术
sol-gel transformation 溶胶—凝胶转变

solid 固体;立体;实心;坚固;纯粹,单一;[得色]一致(一般指混纺、交织物中各种纤维都染得同一色泽)

solid acid 固体酸

Solidazol dyes 佐利达措染料(活性染料的一种,含二氯哒嗪酰胺、烷基砜丙酰胺基团,商名,德国制)

Solidazol N dyes 佐利达措 N 染料(活性染料的一种,含乙烯砜基团,商名,德国制)

solid colour 单色,素色

solid coloured cloth 单色织物,素色织物

solid content 固体含量,含固量

solid coupling 刚性联轴节

solid dyeing 单色染色,素色染色

solid effect 色泽一致效果(指混纺、交织物的染色一致)

Solidegal GL 匀染剂 GL(非离子型,用于还原染料、硫化染料染色,还原染色时,可加快染料的还原速度,提高染料溶解度及稳定性,不阻色,不影响色光,商名,瑞士科莱恩制)

solid ground 素地

solidification 凝固[作用],固体化[作用]

solidification heat 固化热

solidification point 凝固点,固化点

solidifying bath 固化浴,凝固浴

solidifying point 〈S.P.〉固化点,凝固点

solidity 织物厚实感

solid layer 坚实层

solid lime 钙皂垢(指钙盐和肥皂生成的悬浮物)

solid line 实线

solid-liquid interface 固体—液体界面

solid negative 单色负片

Solidogen FFL ＋ FGA 索利达近 FFL＋FGA(阳离子季铵化合物,用于直接染料或硫化染料染色的后处理,以增进耐洗牢度,商名,德国制)

Solidogen FR 索利达近 FR(甲醛缩合物,直接染料及硫化染料印染织物的后处理剂,能增进湿牢度,商名,德国制)

Solidokoll K 索利达科尔 K(低黏度增稠剂,用于合纤或混纺织物的轧染浴中,商名,德国制)

solid oxidant 固体氧化剂

solid phase initiation 固相引发[作用]

solid phase polycondensation 固相缩聚[作用]

solid phase polymerization 固相聚合[作用]

solid phase pressure forming 固相压力成型

solid phase reaction 固相反应

solid phase synthesis 固相合成

solid positive 单色正片

solid shade 单色,素色

solid shade dyeing 染成同色(指把含有多种纤维的织物染成一色)

solid solution theory of dyeing 固溶体染色理论

solids-on 加重率

solid sphere 球体

solid-state controller 固态控制器

solid state diffusion 固态扩散

solid state imaging sensor 固体图像传感器《测色》

solid state polycondensation 固态缩聚

solid-state polymerization 固态聚合[作用]

solid-to-solid transition 固—固相转变

solid waste 固体废弃物《环保》

solid woven 紧密的织物;密实编织,多层交织

Solindene dyes 索林登染料(还原染料,商名,意大利制)

Solindolo dyes 索林多洛染料(可溶性还原染料,商名,意大利制)

soliquoid 悬浮[体]

Solochrome dyes 索罗铬媒染料(铬后处理媒染染料,商名,英国制)

solo fibre 单纯纤维,独种纤维

Solo-Matic 索罗麦蒂克(双氧水练漂一步法,商名,美国开发)

Solophenyl dyes 沙拉菲尼尔染料(直接耐晒染料,商名,瑞士制)

solubilisation 增溶[作用];增溶溶解

solubilised sulphur dye 可溶性硫化染料

solubilised vat dye 可溶性还原染料

solubility〈soly〉溶解度;可溶性

solubility parameter 溶解度参数

solubility product 溶度积

solubilizate [被]加溶物

solubilization 可溶化[作用];增溶[作用];增溶溶解

solubilization agent 可溶化剂,助溶剂

solubilizer 增溶剂

solubilizing agent 增溶剂,溶解剂

solubilizing group 助溶基团

solubilizing power 增溶能力

soluble〈sol.〉可溶的

soluble cellulose 可溶性纤维素

soluble conducting polymer 可溶性导电高聚物

soluble dispersion dye 可溶性分散染料

soluble glass 水玻璃

soluble indigo 可溶性靛蓝;靛胭脂,靛红;酸性靛蓝

soluble iron 可溶性铁《环保》

soluble oil 可乳化油,溶性油

soluble starch 可溶性淀粉

soluble starch paste 可溶性淀粉浆

soluble starch thickening 可溶性淀粉糊

soluble vat dyes 可溶性还原染料,溶靛及溶蒽染料

soluble vat esters dye 可溶性酯化还原染料

Solugan 索卢近(变性烷基磺酸盐类,阴离子型洗涤强化剂,具有较好去污作用,一般的干洗溶剂均适用,商名,德国制)

Soluhao B 助溶剂B(商名,中国制)

Solunaptols 索柳纳普妥染料(不溶性偶氮染料偶合组分,商名,英国制)

solute zone 溶质层

solution〈sol.〉溶液

solution adsorption method 溶液吸附法

solution complex 溶液络合物

solution diffusion 溶解扩散

solution dyed 纺前染色的,纺液着色的《化纤》

solution dyed acrylic fibre 纺液着色丙烯腈纤维

solution-dyed yarn 纺前染色纱,纺液着色纱《化纤》

solution dyeing 纺前染色工艺,纺液着色工艺《化纤》

solution pan 液槽

solution polycondensation 溶液缩聚[作用]

solution polymerization 溶液聚合[作用]

solution pressure 溶解压力,溶解压强

solution type binder 溶液型黏合剂

solution type polyacrylic ester copolymer 溶液型聚丙烯酸酯共聚物

solution type PU(PU=polyurethane) 溶液型聚氨酯
solution wash processes 溶液洗涤过程
solutizer 助溶剂；增溶剂
solvation 溶解；溶剂化[作用]
solvatochromism 溶剂化显色[现象]
solvency power 溶解力
solvent adhesion 溶剂黏结
solvent antifelt finishing 溶剂防毡合整理法
solvent-assisted dyeing(=solvent assisted dyeing) 溶剂助染法
solvent-based coating 溶剂型涂层
solvent-based dry coating 溶剂型干法涂层
solvent based PU(PU=polyurethane) 溶剂型聚氨酯
solvent bleaching 溶剂漂白
solvent bonding 溶剂黏合[法]
solvent cleaning machine 溶剂洗涤机
solvent cracking 溶剂裂解
solvent degreasing 溶剂脱脂
solvent distillation 溶剂蒸馏
solvent drum dyeing machine 转筒式溶剂染色机
solvent dyeing 溶剂染色
solvent dyes 溶剂染料(能溶于有机溶剂的染料)
solvent effect 溶剂效应
solvent emulsion 溶剂乳液
solvent extractable matter 溶剂抽出物
solvent extraction 溶剂萃取，溶剂提取
solvent fastness 耐溶剂[色]牢度
solvent finishing 溶剂整理
solvent flammability 溶剂可燃性
solvent-free gel 无溶剂凝胶，干燥凝胶
solvent front 溶剂前沿

solvent hold-up 溶剂滞留
solventless fabric coating 无溶剂织物涂层
solvent milling 溶剂缩绒
solvent-monitoring 溶剂监测
solvent-naphtha 溶剂石脑油
solvent polymerization 溶剂聚合
solvent pumping system 溶剂泵送系统
solvent reclamation 溶剂回收
solvent recovery system 溶剂回收装置
solvent recycling 溶剂循环利用
solvent regenerator 溶剂再生器
solvent reservoir 溶剂储存装置
solvent resistance 耐溶剂性
solvent-resistent fibre 耐溶剂纤维
solvent scouring machine 溶剂精练机
solvent shrink-resist process 溶剂防缩法
solvent-soluble dye 溶剂可溶性染料
solvent tolerance ［溶剂]容许溶解度；[溶剂中的]最大溶解度
solvent treatment 溶剂处理
solvent vapour cure finishing 溶剂气体焙固整理
solvent wide-width-continuous plant 溶剂宽幅连续处理设备
solvent wool scouring 溶剂洗毛
Solvitex BG 索尔维泰克斯 BG(马铃薯淀粉醚，商名，荷兰制)
Solvitose 索尔维托斯系列(能以冷水溶化的淀粉醚，作印花糊料，商名，荷兰艾维贝)
Solvitose C5 索尔维托斯 C5(交联羧甲基马铃薯淀粉印花糊料及上浆剂，商名，荷兰艾维贝)
Solvitose CG 索尔维托斯 CG(交联马铃薯淀粉醚，印花糊料及上浆剂，商名，荷兰艾维贝)

solvolysis 溶剂分解
solvophilic 亲溶剂的
solvophobic 疏溶剂的,憎溶剂的
soly (solubility) 溶解度;可溶性
somber colour 暗淡色,浅黑色
sonde 探测器,探针,探头
sone 宋(响度单位)
Song brocade 宋锦(彩纬显色的锦类真丝织物)
sonic modulus 声波模数
sonic precipitator 声波收集装置
sonic pulse 声脉冲
sonic velocity 声速,音速
sonic velocity method 声速法
sonic washer 声波洗涤机
sonic washing machine 声波洗涤机
soot 烟灰,烟尘,煤烟
soot black 烟炱黑
sooting 烟灰沾污,积灰
sooty 乌黑的,烟炱的,炭黑的
sophisticated design 复杂的花样设计;尖端设计
sophora flower-bud dye 槐米色素(黄色天然染料)
S. O. R. (speed of rotation) 转速
sorbate 吸着物;山梨酸酯
sorbent 吸着剂
sorbitan 山梨糖醇酐,失水山梨糖醇
sorbitan ester surfactants 山梨糖醇酯表面活性剂(乳化剂,纤维用油剂)
sorbitol 山梨糖醇
sorghum red 高粱红(天然植物染料)
Soromine AF 沙罗明 AF(三乙醇胺单硬脂酸甲酸盐,阳离子型表面活性剂,商名,德国制)
Soromine KG 沙罗明 KG(多缩合氧亚烃基的衍生物,非电离性柔软剂,丝鸣剂,商名,德国制)
sorption 吸着[作用]
sorption equilibrium 吸着平衡
sorption hysteresis 吸着滞后现象
sorption isotherm 吸着等温线
sorption of dye 染料吸着
sorption rate 吸着速率
sorption ratio 吸着比
sorptivity 吸着性
sorrel 橘棕色
sort 分类,种类;品级
sorting 拣选,分级,分类
SOS (share operating system) 共用操作系统
SOTT (second order transition temperature) 二级转变温度,玻璃化温度
sound absorbency 吸音性
sound absorption material 吸音材料
sound arresting 防音性
sound fibre 强韧的纤维
sound insulation fabric 隔音布
sound intensity 声音强度
sound isolation 隔声
sound level 噪声水平,噪声程度
sound level gauge 噪声计
sound level meter 声级计《环保》
sound muffler 消声器,减音器
sound pressure 声压
sound pressure level 噪声压力水平,声压水平
sound production 噪声的产生
soundproof 隔音的,防音的
soundproof fabric 隔音布
sound-proofing felts 隔音毡毯
sound sensor 声音传感器

sound wool 拉力强的羊毛
souple silk 半练丝;半练染色丝
soupling 蚕丝半练[工艺]
sour bath 酸浴,酸液
source 源,能源,光源
souring 吃酸,酸化,加酸,酸洗
souring agents 释酸剂
souring cistern 酸洗槽,淋酸槽
South African Wool Textile Research Institute〈SAWTRI〉南非毛纺织研究所
Southern Regional Research Centre〈S. R. R. C.〉南方地区研究中心(美国)
Southern Regional Research Laboratory〈S. R. R. L.〉南方地区研究试验所(美国)
sou'wester 油布长雨衣(尤指在海上风暴时所穿);海员用的防水帽
sow box 浆槽,浆箱
Soxhlet extraction apparatus 索[格利特]氏萃取器
soybean lecithin 大豆卵磷脂(可作乳化剂用)
soybean protein fibre〈SPF〉大豆蛋白质纤维
soybean protein grafting modification 大豆蛋白接枝改性
S. P.(single-phase) 单相
S. P.(single-pole) 单极
S. P.(solidifying point) 固化点,凝固点
sp.(special) 专门的,特殊的
sp.(specific) 专门的,特殊的;比的《物》
SP(spectrophotometry) 分光光度计测定法
S. P.(=s. p., series parallel) 复串,串并联
S. P.(standard pressure) 标准压力

space 空间,太空,宇宙;间隔;空格,存储空间《计算机》
space blanket 太空棉毛毯
space cloth 太空服
space cotton 太空棉
Space Dye 斯佩司染色(将多色花样印花针织物拆开的纱线再行织造,以获得不规则颜色效应,商名,美国制)
space dyed yarn 段染色纱
space dyeing [纱线]段染
space filler 排液器
space flight 航天
space grating 空间光栅
space look 宇航风貌;宇航服款式;太空装
space network polymer 立体网形聚合物
space polymer 立体聚合物
space printer [纱线]间隔印花机
space printing 间隔印花
spacer 间隔物,隔板;垫片,衬垫,管夹;排液器
space requirement 占地面积
spacer-fabric 间隔织物
space rocket 宇宙火箭
spacer tube 间隔管
space shuttle uniforms 航天服
space suits 宇航服
space technology 空间技术(七大高新技术之一)
space vehicle 太空运载工具
space velocity 空间速度
space velocity per cubic foot〈SCF value〉SCF 值(每立方英尺空间速度值)
space wadding 太空棉
spacing collar 间隙调整环
spalling resistance 抗散裂强度,耐散裂性

span 跨距,跨度;变化范围
Span 60 斯盘 60,乳化剂 S-60(失水山梨糖醇硬脂酸酯,商名)
Span 80 斯盘 80,乳化剂 S-80(失水山梨糖醇油酸酯,疏水性表面活性剂,商名)
Spandex 斯潘德克斯(聚氨基甲酸乙酯弹性纤维属名)
Spandex fibre 斯潘德克斯弹性纤维(一种弹性纤维)
Span emulsifiers 斯盘型乳化剂(失水山梨糖醇的高级脂肪醇脂)
spangle [亮晶晶的]金属片
Spanish grain 西班牙皱纹皮革
Spanish green 西班牙绿[色](浅暗绿色)
spanner 扳手,扳头;扳紧器
Spanscour GR 氨纶防黄助剂(抗氧化剂的复配物,非离子型,用于氨纶的精练,防止氧化变色,热定形时也有良好抗黄效果,商名,美国先邦)
span width 净宽
spare [parts] 备[用部]件
spare parts list 备件表
sparge 喷射
spark 火花,电火花
spark ignition 电火花点火
sparking voltage 击穿电压,放电电压
sparkle printing 闪烁印花
sparkle prints 闪烁印花布
sparkling fabric 闪光织物
sparkling nylon 闪光锦纶(多叶形截面长丝)
spark plug 火花塞
spark quencher 火花淬灭剂
sparrow 雀灰[色](浅灰褐色)
spatial 空间的,立体的

spatial arrangement 空间排列,立体排列
spatial structure 立体结构
spatula 刮勺;刮铲;角匙
SPC(statistical process control) 统计过程控制
SPC(stored program control) 存储程序控制《计算机》
SPE(shrink proofing effect) 防缩效应
spec.(specification) 规格,规程;规范,细则,说明书;目录;明细表
spec.(specimen) 样品,试样,样本
spec-finder 光谱检索
special〈sp.〉专门的,特殊的
special binders 特殊[用途]黏合剂
special colour rendering index 特殊显色指数
special finishes 特种整理
special hosiery 特殊功能袜
speciality 特殊产品
special list of equipment〈SLOE〉设备清单
special metameric index 特殊异谱同色指数
special notes 专门用语,术语
special pigment printing 特种涂料印花
special printing 特种印花
special purpose 专用
special refuse 特殊废料
special resin finishing 特种树脂整理
special-shaped 异形的
special steel 特种钢
specialty hair fibres 特种毛发纤维(与羊毛混用,使织物起柔软光泽等特殊效应)
species 种,类;级
specific〈sp.〉专门的,特殊的;比的《物》

specific activity 比活力
specific adhesion 特性黏合；比黏合
specific area 比表面积
specification〈spec.〉 规格，规程，规范，细则，说明书；明细表；目录
specification limit 规格极限
specification of colour difference 色差表示法
specification statement 区分语言（计算机软件）
specific birefringence 双折射率
specific charge 比电荷，荷质比
specific conductance 电导率；电导系数
specific consumption 单位耗量
specific consumption rate 比消耗速率
specific density 相对密度
specific dyeing rate constant 比染色速率常数
specific extinction coefficient 比消光系数
specific gravimeter 比重计
specific gravity bottle 比重瓶
specific gravity〈SG，S.G.，s.g.〉 相对密度
specific heat consumption 单位热耗
specific heat〈s.h.〉 比热
specific humidity 比湿
specific inductive capacity〈SIC，s.i.c.〉 介电常数
specific insulation resistance 比绝缘电阻
specific length 比长度
specific maintenance rate 比维持速率
specific mass 密度
specific rate constant 比速率常数
specific ratio 比率
specific refraction 折射系数，折射度
specific resistance〈SR〉 电阻率，比电阻

specific site 染座
specific speed 比转速
specific strength 比强力，强力系数
specific stress 比应力
specific surface 比表面，表面系数
specific surface area 比表面积（单位体积的表面积）
specific tenacity 比强力
specific thermal conductivity 导热比值
specific viscosity 比黏度，增比黏度
specific volume〈sv〉 比容，体积度（相对密度的倒数）
specific weight〈SW，sw〉 相对密度，比重
specific work of rupture 比断裂功
specified achromatic colour stimulus 白色刺激《测色》
specified achromatic light 标准光《测色》
specified count 规定[织物]密度
specified weight 规定[织物]重量
specimen〈spec.〉 试样，样本，样品
specimen holder 样品座
specimen preparation 试样准备
speck 斑点，污斑；小粒
speck dyeing [毛织物的]斑点染色，斑点涂色（指以直接染料套染毛织物中的植物纤维杂质）
speckiness 染色斑点现象
specking 染斑；修除[毛织物上的]杂质
speckle 染色小斑点，小色渍，色斑，染斑
spectra （spectrum 的复数）谱；光谱；波谱，频谱
Spectra Dyeing 彩虹染色（用喷射染色机染色的新工艺，染色织物呈云斑、彩虹状，德国第斯）
spectral absorption 光谱吸收
spectral analysis 光谱分析[法]

spectral band width 谱带宽
spectral colour 光谱色
spectral concentration 光[谱]密度
spectral curve 光谱曲线
spectral distribution 光谱分布
spectral dye method 光谱染料法(测临界胶束浓度)
spectral luminous efficiency curve 光谱光视效率曲线
spectral match 光谱配色
spectral power distribution curve 光谱能量分布曲线
spectral profile 光谱轮廓图
spectral purity 光谱纯度
spectral radiance factor 光谱辐射亮度因数
spectral radiant power distribution 光谱辐射能量分布
spectral range 波长范围
spectral reflectance 光谱反射值(率),分光反射率,分光反射系数
spectral reflectance curve 光谱反射率曲线
spectral reflectance factor 光谱反射因数
spectral sensitivity 光谱灵敏度
spectral transmittance 光谱透射比,分光透过率,分光透过系数
spectral transmittance curve 光谱透射率曲线
spectral tristimulus values 光谱三刺激值
spectrochemical analysis 光谱化学分析
spectrochemistry 光谱化学
spectrocolourimetry 光谱色度学
spectrofluorimeter 荧光分光光度计
spectro-fluoro-phosphorimeter 荧光磷光分光计
spectrogram 光谱图;频谱图;谱图

spectrograph 摄谱仪;频谱仪
spectrographical identification 光谱鉴定
spectroirradiation 光谱照射,光谱辐射
spectrometer ⟨s⟩ 分光计;光谱仪
spectrometer lens ⟨SL, sl⟩ 分光计透镜
spectrophotometer 分光光度计
spectrophotometric analysis 分光光度分析
spectrophotometric colourimetry 光谱光度测色法
spectrophotometric determination 分光光度定量
spectrophotometric measurement 分光光度测定
spectrophotometric methods 光谱分析法
spectrophotometry ⟨SP⟩ 分光光度计测定法
spectropyrometer 分光高温计
spectroscopic analysis 光谱分析
spectroscopy 光谱学
spectrum 谱;光谱;波谱,频谱
spectrum atlas 光谱图表
spectrum colour 光谱色
spectrum colour stimulus 光谱色刺激《测色》
spectrum finder 光谱检索
spectrum locus 光谱轨迹
specular 反射的,镜子似的
specular exclude 排除镜面反射
specular gloss 镜面光泽
specular include 包括镜面反射
specular reflection 镜面反射
speed 速率,速度
speed change box 变速箱
speed change gear 变速装置;变速齿轮
speed changer 变速器
speed control roller 速度控制辊

speed control switch 速度控制开关
speed control valve 调速阀(具有压力补偿装置的节流阀)
speed counter 转数表
speed discrepancy 速度的差异
speed feedback 速度反馈
speed indicator 示速器,速度计
speed of rotation〈S.O.R〉 转速
speedometer 速度计,测速计;里程计
speed range 速度范围
speed ratio 速比
speed reducer(＝speed reducing unit) 减速器
speed-regulating motor 调速[伺服]电动机
speed steamer [圆筒形针织物的]高速汽蒸机
speed torque 车速转矩,速度转矩
speedtorque characteristics 车速转矩特性曲线
speedup 升速
speed variator 变速器
speed wrench 快速扳手
spent dye liquor 废染液
spent gas 废气
spent lye 废碱液
spent steam 废汽
spermaceti wax 鲸蜡
sperm oil 鲸蜡油
spewing 飞浆,溅浆(印疵)
SPF(soybean protein fibre) 大豆蛋白质纤维
SPF(sun protection factor) 防晒因子(衡量紫外线辐射对人体危害程度的参考指标)
sphere 范围,领域;球状体
spherical bearing 球形支座

spherical boiler(＝spherical digester) 蒸球
spherical roller bearing 球面滚子轴承
spheroidal graphite cast iron 球墨铸铁
spice finish 香味整理
spider 星形轮;带辐条的轮毂;十字叉;多脚架
spider silk 蜘蛛丝
spider weave 蛛网组织,网目组织
spigot 塞子,插头
spigot joint [管端]套筒接合,[管端]轴承接合
spill (＝spillage) 溢出
spillway 溢流道
spinach green 菠菜绿[色](灰光黄绿色)
spindle 轴,芯轴,主轴
spindle batcher 卷轴机
spindle box 主轴箱
spindle carrier 轴支承装置,主轴托架
spindle dyeing machine 锤轴式染色机
spindle oil 锭子油
spindle sleeve 轴套
spindle system dyeing machine [管纱]竖轴式染色机
spin drier(＝spin dryer) 离心式脱水机
spin dyeing 原液着色
spin finishes 纺丝油剂
spinnability 可纺性
spinner 传动给浆辊的齿轮
spinnery 纺纱厂
Spinning 兹平宁(化学纤维原液着色用染料,商名,德国制)
spinning bath additives 纺丝浴添加剂
spinning disk atomiser 旋转式圆盘雾化器
spinning dyes [化纤]原液染色染料
spinning lubricant 纺纱润滑剂(织物煮练时应除去)

spinning oil 纺丝油剂
spin-off 有用的副产品
spins 尖硬纤维
spiral 螺旋,螺旋管;螺线
spiral ager 回绕式[穿布]蒸化机(只在花布背面接触导辊,可免搭浆瑕疵)
spiral blade [剪毛机上的]螺旋刀片
spiral breaker 螺旋辊揉布机
spiral coil 螺旋管,盘香管
spiral condenser 蛇管冷凝器
spiral cutter [剪毛机的]螺旋刀辊
spiral drill 麻花钻
spiral dye beck 绳状染色机,绳状染槽
spiral dyeing 绳状染色
spiral expander 螺旋形扩幅装置
spiral gear 螺旋齿轮,斜齿轮
spiral heat exchanger 螺旋形热交换器
spiral jaw coupling 螺旋形爪式联轴节
spiral knife 螺旋刀
spiralling [狭幅织物的]卷曲
spiral roll breaking machine 螺滚式揉布机
spiral roller 螺旋辊
spiral seaming 螺旋缝合
spiral spring 螺旋弹簧,盘簧
spiral steamer 回绕式[穿布]蒸化机(只在花布背面接触导辊,可免搭浆瑕疵),回转式汽蒸箱
spiral winding 螺旋式卷绕
spiral wound membrane 螺旋状卷绕[过滤]膜(水处理技术)
spiral-wound ultrafiltration membrane 螺旋式超滤膜
spirit level 酒精水准器
spirit soluble dyes 醇溶性染料
spiro-jet 螺旋型喷射染色机

spiropyran 螺吡喃(光致变色化合物)
spissated 凝结了的;浓缩了的
spitalfields 满地小花领带织物
splash board 挡水板
splash guard 防溅护板
splashing 喷溅,飞溅
splash plate 挡水板
splashproof guard 防溅护板
splashproof packing ring 防溅填圈(用以保护轧辊末端的填脂滚珠轴承)
splash-resistant 防溅湿,防淋
splash shield 防溅护板
splicing compound 粘接化合物
spline 花键(轴和孔所开的直牙槽的通称)
split 撕裂,分裂;裂缝;稀弄(织疵)
split-bath degumming 半浴脱胶
split bath dyeing 半浴染色(丝)
split bearing 对开轴承,开缝轴承
split fibre(＝split-film fibre) 裂膜纤维
split jet injector 缝隙式喷嘴
split journal bearing 开缝轴领轴承,对开滑动轴承
split muff coupling 拼合套筒联轴节
split nut 对开螺母,开缝螺母,拼合螺母
split phase motor 分相电动机
split pin 开口销
split rinse 冷热水混合冲洗
split strength test 撕开强力试验(多层胶合织物撕开所需的力)
split suction boxes and weirs 断开式冲吸装置(装于丝光机上)
splittability 易分裂性,裂变性
splitter 分裂器;分解剂;气流分隔片;导流板
splitter plate 导流板,分流板

splitting 分裂;裂解
splitting ability 撕裂性能
splitting of chain 链断裂
splitting of film 裂膜
split up 并幅布分匹
split washer 开口垫圈
splodge(=splotch) 污点,斑点,污迹
splotchy dyeing 染斑
SPMT(scanning probe microscope technology) 扫描探针显微技术
spoke [轮]辐
spoked roller 辐辊
sponge cleaning 海绵清洁法
sponge cloth 揩车布;松软棉布;润湿布
sponge-core package-dyeing 海绵芯筒子染色法(合纤纱线染色时可自由收缩和膨胀,能省去预编和络筒工艺)
sponge effect 点描效果,云纹效果
sponge plastics 多孔[泡沫]塑料
sponge rubber cushion [地毯]橡胶海绵衬垫
sponge silk 海绵绸
sponginess 手感松软
sponging [呢绒]润湿预缩(在裁剪前进行)
sponging machine [毛织物]润湿预缩机
spongy 海绵状的
spongy parenchyma(=spongy tissue) 海绵组织
spontaneous coagulation 自然凝固,自动凝固
spontaneous combustion 自燃
spontaneous crimp 自发卷曲
spontaneous curing 自焙固,自然固化
spontaneous decomposition 自然分解
spontaneous elastic recovery 自发弹性恢复
spontaneous emulsification 自然乳化
spontaneous ignition 自燃
spontaneous ignition temperature (SIT,S. I. T., sit) 自燃温度,自燃点
spontaneous inflammable 自燃的
spontaneous polymerization 自发聚合
spontaneous relaxation 自然松弛
spontaneous selection 自然选择;自发选择
sports and recreation textile 运动及娱乐用纺织品
sports denim 薄软运动服布;褪色劳动布
sport shirt 运动衫
sportswear 运动服装
sport textiles 运动用纺织品
sporty 轻便服装
spot 斑点,斑渍
spot acid wash 斑点酸洗,局部酸洗(用于扎染工艺)
spot analysis 斑点分析,点滴分析
spot and stain removal 去污渍
spot bonding 点状黏合
spot-check 抽查,现场检查
spot cleaner 除污渍喷枪
spot colour 印点色(由点子排列形成的颜色)
spot jamming 选择性干扰,局部干扰
spotless finish 防污渍整理
spot light 聚光灯
spot lighting 局部照明
spot light source 点光源
spot motive 点子花纹
spot of permeation 渗透斑《丝》
spot remover 去斑渍剂
spot resistant 防污渍的
spots 点子花纹棉布(英国名称)
spot test 斑点试验

spotting 去油污渍；蒸呢工艺
spotting agent 去污斑剂（干洗用）
spotting dyeing 斑点染色
spot weld effect 点状黏合效果（羊毛防缩整理用树酯复盖方法）
spot welding 点焊；点状黏合（织物树脂整理时，通过聚合物使纤维间形成点状黏合）
spout 喷口，喷嘴；斜槽；喷射
sprag clutch 制轮离合器
spray 小枝花纹；喷洒，喷雾；喷雾器
spray adhesive laminating machine 喷黏着剂的层压机
spray adhesive system 喷洒黏着剂系统
spray application 喷雾施加法
spray bonding 喷洒黏合
spray booth ［黏合剂的］喷洒室
spray box 喷淋箱
spray coating 喷涂，喷雾涂层，喷洒涂层
spray coating machine ［圆网］喷胶机（用于激光制网）
spray damping machine 喷雾给湿机
spray dish dyeing 转碟喷雾染色
spray drier(＝spray dryer) 喷雾干燥器
spray drying 喷雾干燥
spray dyeing 喷［雾］染［色］
sprayer 喷雾器；喷漆器；喷油器
spray finish 喷洒涂饰剂
spray finishing 喷胶整理，喷浆整理
spray gun 喷枪
spraying apparatus 喷雾器
spraying-bonded wadding 喷浆絮棉
spraying chamber 喷射室
spraying machine 喷雾机
spray jet 喷嘴
spray jet type 喷射式

spray line 喷淋管
spray nozzle 喷孔，喷［雾］嘴
spray painting equipment 喷涂机
spray penetration test 喷淋渗透试验（测织物防水性）
spray pipe 喷淋管
spray printing 喷［雾］印［花］
spray printing machine 喷雾印花机
spray rating tester 喷淋法评级仪
spray resist dyeing 喷雾防染
spray rinsing 喷洗，淋洗
spray sizing 喷雾上浆［工艺］
spray spice finish 喷雾香味整理
spray starching machine 喷雾上浆机
spray test 喷淋试验（测拒水性）
spray test of water repellency 喷淋法防水性试验
spray tube 喷淋管
spray-type cooler 喷淋式冷却器
spray-type hank dyeing machine 喷淋式绞纱染色机
spray/vacuum washing system 喷射/真空水洗机
spray washer 喷洗机
spray water pipe 喷水管
spread 床单；桌布；铺开，伸展；散布
spreadability 覆盖性
spreadable life ［黏合剂］可涂期；［黏合剂］使用期
spread coating 刷涂法；刮涂法
spread defects 散布性疵点，分散性疵点
spreader 喷液器；扩幅器；涂胶机
spreader bar 伸幅杆
spreader knife 上浆刮刀
spreader-steamer ［圆筒形针织物］扩幅汽蒸机

spreading 扩幅,延展;铺[开印]浆
spreading apparatus 扩幅装置,展幅装置
spreading bar 扩幅辊
spreading finish 上浆整理
spreading force 铺展力
spreading knife 刮刀;涂胶刀
spreading machine 扩幅机;刮胶机,涂胶机
spreading mass [铺涂用]浆料;[铺涂用]涂料;[铺涂用]胶料
spreading roll 扩幅辊
spring 弹簧,发条;跳跃
spring back 回弹
spring balance 弹簧秤
spring bed [剪毛机的]弹性托板
spring catch 弹簧挡
spring cloth wind-up motion 弹簧卷布装置
spring coat 弹性涂膜
spring coil 弹簧圈,圈簧;发条盘簧;螺旋弹簧
spring coil fabric 圆簧型[弹性]织物
spring collar 弹簧挡圈
spring green 嫩绿[色]
spring hanger 弹簧吊架
spring head 弹簧头
springiness 弹性
springless tension unit 无弹簧张力装置
spring-loaded tension rail 弹簧支承的张力杆
spring manometer 弹簧式压力表
spring-opposed bellows 弹簧承力波纹管
spring ring 弹簧圈
spring safety valve 弹簧式安全阀
spring strip air preheater 弹簧带式空气预热器

spring support 弹簧支座,弹簧吊架
spring washer 弹簧垫圈(机械部件)
spring-weighted 弹簧加压
springy handle 弹性手感
sprinkled point coating(=sprinkled pastepoint coating) 浆点散粉涂层
sprinkler 淋洒器,洒水器;喷洒头;喷壶
sprinkler system [防火]喷水系统
sprinkling processs 喷洒法
sprinkly dyeing [雪花]留白染色
sprocket [wheel] 链轮
sprocket chain 链轮环链,扣齿链
spruce 云杉;云杉绿[色](浅暗绿色)
SPS(steam press shrinkage) 蒸汽压烫收缩
SP screen 开孔率更大的镍网(适合于涂层加工,商名,荷兰斯托克)
SPT(sensory perception technology) 感官敏感技术
SPT (standard pressure and temperature) 在标准的气压和温度下
spud 煤气喷嘴;销钉;溢水接管
spumescence 泡沫性
spumescent 起泡的
spumous [多]泡沫的
spun bonded backing 纺黏布背衬
spun bonded layer 纺黏布层
spun-bonded non-woven fabric (= spun bonded non-woven film) 纺黏型非织造织物,纺黏法非织造布
spun-coloured(=spun-dyed) 纺前染色的,原液染色的
spun count 纱的实际支数(线密度)
spun dye colour 原液着色剂
spun-dyeing 纺前染色,原液染色
spun fabric 短纤维织物
spun finish 经纹擦光整理

spun glass 玻璃纤维,玻璃丝
spun glass finish ［棉织物］直纹电光整理（轧纹方向与织物经向平行）
spun-laced nonwoven fabric 射流喷网法非织造布
spun-like 仿短纤纱外观;仿短纤纱的
spun-like fabric 仿纱型织物
spun-pigmented 纺前着色的,纺液着色的
Spunray 斯本莱（一种防缩防皱的棉纱和人造丝交织衬衫织物,商名）
spun rayon 人造棉纱,黏胶短纤纱;人造棉织物
spun rayon fabric 人造棉织物,再生短纤维织物
spun rayon velvet 人造棉绒
spun silk 绢丝;绢丝纺［绸］
spun silk fabric 绢丝织物
spun silk mill 绢纺厂
spun silk piece goods 绢丝纺［绸］,绢丝织物
spun silk pongee 绢丝纺［绸］
spun silk yarn 绢丝
spun silk yarn counts 绢丝支数
spun viscose 黏胶短纤纱
spun-with-colour fibre 纺前染色纤维,原液着色纤维
spur gear(＝spur wheel) 正齿轮
spurt pipe 溅散管,多孔喷水管
sputter coating 溅射涂膜
sputter discharge 射流放电（生产等离子体的方法）
sputtering ［金属］喷镀（等离子体处理）
spy ware 窥视器《计算机》
sq.(square) 平方的,正方的,二次幂
SQL(structured query language) 结构化查询语言《计算机》

square〈sq.〉 平方,方形;角尺;平方的,正方的,二次幂
square beater 方形打手
square beater washing machine 方打手绳洗机
square cut ［化纤的］等长切断
square-dot screen 方点网版
square dyeing 方形架染色
square foot〈s. f.〉 平方英尺
square hole nozzle 方孔喷风管
square law〈SL〉 平方律
squareness 垂直度
square nut 方螺母
square screen 平网
square screw 方［牙］螺纹
square tenon joint 方榫接头
square yards equivalent〈SYE〉 平方码当量
squeegee 刮浆器,涂刷器,橡皮滚子,［印花］刮刀
squeegee angle 刮刀角
squeegee angular displacement 刮刀角度的置换
squeegee blade 刮浆刀
squeegee carrier 刮刀架
squeegee coater 刮板式涂布机
squeegee contact angle 刮刀接触角
squeegee grinding machine 磨刮刀机
squeegee holder 刮刀夹
squeegee lip 刀口（指刮刀刀片的未夹入部分）
squeegee passes 刮印次数
squeegee pressure 刮刀压力
squeegee printing machine 刮浆器印花机
squeegee profile 刮刀口剖面
squeegee speed 刮刀速度
squeegee suspension device 刮刀悬置装置

squeegee systems in printing 印花刮浆器体系

squeegee tip 刀口(指刮刀刀片的未夹入部分)

squeegee washer(= squeegee washing machine) 刮刀清洗机

squeeze automatic loading 压浆辊自动调压

Squeeze-jig [松式]轧染卷染二用染色机(可用于轧染、卷染等多功能染整设备,商名,意大利制)

squeeze mangle 轧车

squeeze nip 轧点

squeezer 轧水机

squeeze roller 挤压辊

squeeze suck technique 轧吸技术(一种低给液整理技术,织物经轧点获得普通轧余率后再和相同材料的干织物叠合一起进入第二轧点,干织物的吸液作用使轧液率大大减少)

squeezing 压浆,轧水,压帧

squeezing effect 挤压效果

squeezing hydroextractor 挤压式脱水机

squeezing machine(= squeezing mangle) 轧水机,轧液机

squeezing nip 轧点

squeezing roller 压浆辊;轧水辊;压[液]辊

squeezing/suction technique 挤压/抽吸工艺

SQUID(superconducting quantum interference device) 超导量子干涉仪(测量弱磁场的装置)

SQUID motor 装有超导量子干涉仪的电动机

squirrel-cage motor 鼠笼式电动机

SR(scanning radiometer) 扫描辐射计

S. R. (=S. S. R. ,soap solubility ratio) 肥皂溶解度比

SR(soil retention) 剩余污垢;污垢保持性

SR(specific resistance) 比电阻,电阻率

SRD(standard reference depth) 标准参照浓度

SR finish(soil release finish) 易去污整理

SR finishing(stain resist finishing or soil release finishing) 拒污整理或易去污整理

S-roller 均匀轧车(商名,德国基士特)

S. R. R. C. (Southern Regional Research Centre) 南方地区研究中心(美国)

S. R. R. L. (Southern Regional Research Laboratory) 南方地区研究试验所(美国)

S. R. U. (self-recording unit) 自动记录器

ss(sections) 截面;区域;部门

S. S. (short stick) 短码尺

ss(slow speed) 低速

SS(=ss,switches) 开关;电闸

ss(superheated steam) 过热蒸汽

SS(super soft) 超柔软

SS(suspended solid) 悬浮物质,悬浮固体《环保》

SS-bond(disulphide bond) 二硫[化]键

SSCE(standard saturated sodium chloride calomel electrode) 标准饱和氯化钠甘汞电极

S-S curve(stress-strain curve) 应力—应变曲线

SS durable-press finishing〈SSP〉液氨处理与免烫整理结合的工艺

SSF(super softening finishing) 超柔软整理

SS finishing(super soft finishing) 超柔软整理(日本称液氨整理)

SSI(small-scale integration) 小规模集成电路

SSP(shape stabilized process) 保形工艺(液氨处理与免烫整理相结合的工艺)

SSP(super soft peach) 超柔软桃皮整理

SSP(super soft press, super soft process, SS durable-press finishing) 液氨处理与免烫整理结合的工艺

S.S.R.(=S.R., soap solubility ratio) 肥皂溶解变化

SSR(staining scale rating) 沾色灰卡等级

ST(self-twist) 自捻

ST(standard temperature) 标准温度

St(stoke) 斯[托克斯](运动黏度的物理单位)

st.(stretched) 拉伸的

Stabical A 斯泰比稳定剂 A(双氧水漂白用稳定剂和多价金属螯合剂的混合物,阴离子型,商名,英国制)

Stabical SD new 斯泰比稳定剂 SD 新(多价螯合剂和表面活性剂的混合物,阴离子型,不含硅,用于双氧水漂白,有好的润湿和洗涤作用,商名,英国制)

Stabifix OF 斯泰比固色剂 OF(阳离子型,用于直接染料、活性染料的固色,并提高湿牢度,商名,德国科宁)

Stabifix RF 斯泰比固色剂 RF(不含甲醛的固色剂,改性树脂,用于直接染料、活性染料的固色,并提高湿牢度,商名,德国科宁)

stabilised finish 定形整理,稳定整理(与防缩防皱意义相同)

stabilised power unit 稳定电源设备

stabiliser 稳定剂;稳定器

stability 稳定性,稳定度;复原性,复原力;牢固度,耐久性

stability against laundering 洗烫稳定性

stability constant 稳定常数

stability criterion 稳定[性]判据,稳定[性]评价

stability of breaked bond 断键稳定性

stability of dispersion at high temperature 高温分散稳定性

stability setting 稳定性定形(指干、湿热法)

stability test 稳定性试验

stability to chemicals 药品稳定性

stabilization 稳定[作用]

stabilizator 稳定剂;稳定器

stabilized diazo compound 稳定的重氮化合物

stabilized insoluble azo dyes 稳定不溶性偶氮染料

stabilizer 稳定剂;稳定器

stabilizer against thermal degradation 耐热降解稳定性

Stabilizer CS liq. 稳定剂 CS 液(有机稳定剂,具螯合作用,适用于连续性双氧水漂白,商名,瑞士科莱恩)

Stabilizer SIFA liq. 稳定剂 SIFA 液(有机稳定剂,弱阴离子型,适用于无硅酸盐的双氧水漂白,具抗催化作用,对碱稳定,商名,瑞士科莱恩)

stabilizing oven 定形烘箱(用于纤维热定形或稳定化的烘箱)

stabilizing ring 定位圈,[轴承]止推环

stable 稳定的,坚固的

stable diazonium salt 稳定性重氮盐

stable fabric [尺寸]稳定织物

stable operation 稳定运行

stabler 稳定剂;稳定器
stable state 稳[定状]态
stab-resistant material 防[刀]刺材料,抗[刀]刺材料
stack 堆;堆列;烟囱;通风管;数据表;暂存器《计算机》
stack dryer 堆积式干燥机
stack flue 烟道
stack gas 烟道气
stacking 堆置
stage 工作台;(显微镜的)载物台;级,阶段;程度
stage aeration 阶段曝气
staged filter 分级过滤器
staged reactor 多级反应器
stage evapuration 分段蒸发,分级蒸发
staggered arrangement 交错排列
staggered bank 错列布置管束
staggered gear 交错齿轮
staggered tube 错列管
staging 分段[腐蚀];抑制[腐蚀]
stagnation 停顿,呆滞
stagnation temperature 滞止温度
stain 着色剂,染色剂;污点,斑疵;沾污
stain blocker 污垢阻塞剂;防沾污剂
stained cloth 褪色布;油污布
stained glass 彩色玻璃
stained roving 沾[油]污粗纱(疵点)
stained yarn 油渍纱,油污纱
staining 表面着色;染斑(疵点);沾污
staining method with zinc chloride-iodine 氯化锌和碘溶液的显色法
staining of colour 沾色
staining power 着色力,染色力
staining sample card 沾色样卡
staining scale rating〈SSR〉沾色灰卡等级

staining test 沾污试验;着色试验(鉴别纤维用)
stainless steel 不锈钢
stainless steel belt 不锈钢带
stainless steel bowl 不锈钢[轧]辊(用不锈钢板包覆在铸铁辊体的表面和两侧,经氩弧焊接法焊接而成)
stainless steel calender roll 不锈钢轧辊
stainless steel cistern 不锈钢平洗槽
stainless steel electrode 不锈钢电极
stainless steel guide roller 不锈钢导布辊
stainless steel jacket 不锈钢套
stainless steel J-box 不锈钢 J 形箱
stain-proofing agent 防污剂
stain-release replica 去污评级样卡
stain removal 去污
stain removing agent 除污剂,去污剂
stain removing mark 除污迹痕,去污迹痕
stain repellency 去污性,防污性
stain repellent 去污剂
stain repellent finish 去污整理,防污整理
stain resistance 防沾污性
stain resistant finish 防污渍整理
stain resist finishing or soil release finishing〈SR finishing〉拒污整理或易去污整理
stain warp 渍经(织疵)
stalagmometer [表面张力]滴重计
stale perspiration 陈腐汗液
stall 失速,[因速度不够而]停车
Stalwart-Pickering printing system(=Stalwart printer) 斯塔尔瓦特—皮克林印花机(地毯凸辊印花)
stamping 打印;冲压,锤击;印商标
stamping knife 割刀,冲剪刀
stamping machine 打印机
stamp washer 槌式洗呢机

stanch fibre 止血纤维
stanchion 支柱,撑柱
stand 支柱,支座;墙板
stand-alone 独一无二;独立的
standard〈STD, std.〉标准,基准;规格;[测量]单位
standard absorbed heat 标准吸附热
standard adjacent fabric 标准衬贴织物
standard affinity [染料对纤维的]标准亲和力;标准亲和性
standard atmosphere 标准大气,标准温湿度
standard atmospheric pressure 标准大气压
standard cell 标准电池
standard chemical potential 标准化学电位
standard colour chart 标准色卡
standard colourimetric observer [测色]标准观察者
standard colourimetric reference system CIE CIE[标准]表色制
standard colourimetric system 标准色系统
standard condition 标准状态;标准温湿度条件
standard consistency 标准稠度
standard copper wire〈s. c. w.〉标准铜线
standard degree of colour fastness〈SDC〉染色牢度标准等级
standard depth 标准深度(测试染色牢度等)
standard depth scale [颜色]标准深度分级样卡
standard detergent 标准洗剂
standard deviation〈S. D., s. d., s. dev.〉标准差
standard dyeing affinity 标准染色亲和力
standard dyeing time 标准染色时间

standard dye sample 染料标准品
standard electrode 标准电极
standard entropy of dyeing 标准染色熵
standard error〈S. E.〉标准误差
standard fading booklet 标准褪色卡片
standard fading hours 标准暴晒时间
standard flue gas 标准烟气
standard fuel 标准燃料
standard grade 标准级
standard grey scale 标准灰色分级样卡(染色牢度试验用)
standard heat of combustion 标准燃烧热
standard heat of dyeing 标准染色热
standard humidity 标准湿度
standard hydrogen electrode 标准氢电极
standard illuminant of colourimetry [色度学]标准光源
standard illuminator 标准光源仪
standardisation of measurement 测量标准化
standardization 标准化;校正;标定
standardized curve 标准化曲线,校准曲线
standard length 标准长度
standard light [source] 标准光源
standard method of analysis 标准分析法
standard observer data 标准观测数据(CIE测色系统)
standard of ambient noise 环境噪声标准
standard of environmental sanitation 环境卫生基准《环保》
standard paste 标准[浓]浆(供稀释用)
standard piece glass 织物[标准]分析镜
standard-piece rate 正品率;正布率
standard point 规矩点《雕刻》
standard pressure and temperature〈SPT〉在标准的气压和温度下

standard pressure〈S.P.〉 标准压力
standard reagent 标准试剂
standard reference depth〈SRD〉 标准参照浓度
standard[moisture]regain 标准回潮率(在标准状态下,即相对湿度65%,温度20℃时,织物内自然含水重量与织物的重量之比)
standard sample 标准样
standard saturated sodium chloride calomel electrode〈SSCE〉 标准饱和氯化钠甘汞电极
standard scales 标准样卡
standards of acceptance 验收标准
standard soil 标准污[染]物
standard solution 标准溶液,规定溶液
standard sources 标准光源
standard specification 标准规格
standard strength dyes 标准力分的染料
standard temperature and humidity 标准温湿度
standard temperature and humidity for testing 试验用标准温湿度
standard temperature and pressure〈STP〉 标准温度与压力
standard temperature〈ST〉 标准温度
standard test weight 标准砝码
standard tolerance 标准公差
standard weather meter 标准耐气候色牢度试验仪
standard weave 基本组织,三原组织
standard weight 标准重量(按标准回潮率折算的重量);标准砝码
standard white 标准白度(一般以氧化钡的白度作为标准白度)
standard white plaque 标准白瓷板(测白度用)
standard white tile 标准白板
standard wire 规矩线(点)《雕刻》
stand-by [行动]准备
standby hours 准备小时数,准备时间
Standfast dyeing machine 斯坦法斯特染色机(熔态金属染色机,商名,英国制)
Standfast dyeing process 斯坦法斯特染色法(熔态金属染色法,商名,英国开发)
standing 静置,搁置,储藏
standing bath 续染浴,连缸,套缸
standing dye bath method 染浴续染法
standing time 停台时间;停留时间
stand pipe 竖管
standstill 停止,间歇
stannic chloride [四]氯化锡
stannous chloride 氯化亚锡
Stantex S 6048/7 斯坦特克斯 S 6048/7 纺丝油剂(环保型助剂,用于涤纶和锦纶的高速纺丝,商名,德国科宁)
staple 纤维;纤维长度;切段纤维,[再生]短纤维;U形钉,骑马钉
staple cotton 天然转曲良好的棉纤维;长绒棉
staple cutting 切断成短纤维
staple fibre [再生]短纤维,切段纤维
staple-fibre apparatus 纤维长度试验仪
staple fibre muslin 再生棉细布
staple goods 大路货,主要产品
staple rayon 再生短纤维,黏胶短纤维;人造棉
staples 大路货,主要产品
star ager 星形架蒸化机,星形蒸箱
starch 淀粉;淀粉浆
starch box 浆槽
star checks 星形小格子花纹

starched fabric 上浆织物
starched finish ［淀粉］上浆整理
starched organdy finish 奥甘迪薄浆整理（用淀粉或树脂上浆，使精梳纱细布产生坚挺手感）
starch ester 淀粉酯
starch ether 淀粉醚
starch film 浆膜
starch gum 糊精
starching ［棉布］上浆
starching agent for acetate fibre 醋酯纤维上浆浆料
starching and stentering machine 上浆拉幅机
starching clay 高岭土，白黏土，膨润土
starching finish ［淀粉］上浆整理
starching machine 上浆机
starch iodide paper 淀粉碘化钾试纸
starch lump 浆斑(疵点)
starch mangle 上浆机
starch marks 浆斑(疵点)
starch modification 淀粉改性
starch paste 淀粉浆
starch roller 清水浆滚筒，白浆滚筒
starch-solubilizing enzyme 淀粉溶解酶
starch thickener 淀粉增稠剂
starch thickening 淀粉糊
starch-tragacanth thickening 淀粉龙胶浆
star coupling 万向联轴节，万向接头，星形联接器
star cross section 星形截面
star degumming 星形架精练
star-delta connection 星形—三角形连接法，Y—△连接法《电》
star-delta starter 星形—三角形起动器，Y—△起动器

star-delta switch 星形—三角形转换开关，Y—△转换开关
star dyeing 星形架染色法
star dyeing machine 星形架染色机
starframe 星形架，星形挂框
starframe dyer 星形架染色机
star gear 星形齿轮；［印花机上］传动花筒的大齿轮
starjet 星形喷嘴(烘干机中喷嘴)
star-jet nozzle 星形喷嘴(烘干机所用新型喷嘴)
star pattern 星形花纹
Star print 斯塔式转移印花(一种转移印花法，意大利开发)
star-shaped material carrier 星形架(用于星形架蒸化机)
star-shaped profile 星形截面(指化学纤维)
star steamer 星形架蒸化机，星形蒸箱
start 开动，起动，开始
start button 起动按钮，起步开关
star temple 星形边撑，环形伸幅器
starter 起动器，起动装置
starting bath 头缸，初染浴
starting current 起动电流
starting handle 开关柄，开动手柄
starting lever 起动杆
starting marks 开车痕(织疵，稀弄或密路)
starting ［up］ place 开车痕(织疵，稀弄或密路)
starting ring 起动环，起动圆盘
starting sequence 起动顺序
starting torque 起动转矩
start of text character 正文起始符
start of text〈STX〉正文起始，正文开头

《计算机》
Star Trans scouring & bleaching range 星形练漂机（商名，德国哥拉）
startup 起动，开动；试车，试运转
startup light oil burner 起动用轻油喷燃器
startup procedure 开车工艺规程
startup pump 起动泵
star wheel 星形轮，棘轮
stat. (statistics) 统计学
state 状态，情况
statement 语句；陈述
state of cure 硫化程度；固化程度，焙烘程度
state of disarray 混乱状态，无序状态
state of rawness 粗制状态
state-of-the-art 现代技术；目前工艺水平
state verification 国家鉴定
static absorption test 静态吸收试验（测拒水性）
statical friction coefficient 静摩擦系数
static balance 静平衡
static bioassay test 静态生物测定试验
static characteristics 静态性能
static charge 静电荷
static check 静态检验
static deformation 静［态］形变
static detector head 静态检测头
static dielectric constant 静电介电常数
static dissipation 静电消散
static electricity 静电
static electricity remover 静电消除器
static electrification 带静电［作用］
static elevation 位差
static eliminator 静电消除器
static equilibrium 静平衡

static fatigue 静［态］疲劳
static field 静电场
static flocking 静电植绒
static free 无静电干扰，不受大气干扰
static friction 静电摩擦
static generation 静电产生
static head 静压头
Static Honestometer 奥尼斯特式静电测试仪（商名）
static immersion test 静态浸渍试验（测橡胶防水布的防水性）
static inhibitor 抗静电剂，静电消除剂
staticize 保存（指信息等），静化
static liability 静电倾向
static load 静负载
static mixer 静态混合器
static pressure 静压
static proneness 静电产生倾向
static propensity 静电倾向
static relaxation 静态松弛［作用］
statics 静力学
static shock 静电冲击
static stress 静态应力
static test 静态试验（如抗张试验、弯曲试验、压缩试验等）
static tester 静电测定器
static water absorption 静态吸水
station (stn.) 站，台
stationary 静止的，固定的，稳态的
stationary condition 恒定状态
stationary convex expander roller 固定弯辊
stationary digester 固定式蒸煮釜
stationary fit 静配合；紧配合
stationary liquid phase 液体固定相
stationary phase 固定相
stationary pollution source 固定污染源

stationary state 稳定状态
stationary state diffusion 稳态扩散
statistic 统计的；统计量
statistical analysis 统计分析
statistical bias 统计偏差
statistical method 统计方法
statistical process control〈SPC〉统计过程控制
statistical quality control system 统计的质量控制法
statistics〈stat.〉统计学
statistic-thermodynamic analysis 统计热力学分析
stator 定子《电》
status 状态
Staudinger index 施陶丁格指数（即特性黏度）
stay 撑条；拉条
stay bar 撑杆
stay bolt 撑螺栓，拉撑
stay brace 支撑
stay hook 撑钩
stay pipe 撑管，支持管
stay plate 垫板
st.c.（steel casting）钢铸件
STD（＝std., standard）标准，基准；规格；［测量］单位
steadiness 稳定性，均匀性
steady load 稳定负荷，固定负荷
steady state diffusion 稳态扩散
steady-state flame propagation 火焰稳态蔓延
steady-state viscosity 稳态黏度
steam accumulator 储汽器，汽包
steam admission 进汽
steam admission valve 进汽阀

steam-age 汽蒸
steam ager 蒸化机
steam air ratio 蒸汽空气比
steam atomiser 蒸汽喷雾器
steam baffle 蒸汽挡板
steam black ［苯胺］蒸化黑
steam-blowing 蒸呢
steam blowing machine 蒸呢机
steam boiler 蒸汽锅炉，汽锅
steam brushing machine 汽蒸刷呢机
steam bulking 汽蒸膨化
steam cans 蒸汽圆筒烘燥机
steam chamber 蒸汽室
steam chest 蒸汽夹板；蒸汽柜
steam cleaner 蒸汽清洗机（利用喷射蒸汽来冲洗机件）
steam coil 蛇形蒸汽管，蒸汽盘［香］管
steam column 蒸汽柱
steam condensation 蒸汽凝结
steam condenser 凝汽器
steam consumption 蒸汽耗用量，汽耗
steam content 含汽量
steam content measuring and controlling unit 蒸汽含量测量和控制仪
steam cooking 蒸汽煮练，蒸煮
steam cottage 圆筒蒸箱
steam cracking 蒸汽裂解
steam cure 蒸汽焙烘
steam curing 汽固法
steam cylinder 蒸汽烘筒
steam damping machine 蒸汽给湿机
steam deaerator 热力除氧器
steam demand 蒸汽需要量
steam developing 汽蒸显色
steam-distribution plate 蒸汽分配板，多孔板

steam doctor 蒸汽刮刀,蒸汽刀(强力清洗装置)
steam drain pipe 冷凝水排泄管
steam drier(＝**steam drying machine**) 蒸汽烘燥器;蒸汽烘燥机
steam ejector 蒸汽喷射器
steam envelope 蒸汽夹套
steamer 蒸化机;蒸汽机
steam exhaust 排汽[装置]
steam exhaust port 排汽口
steam filature 缫丝厂,制丝厂
steam fixation 汽蒸固色
steam flow 蒸汽流
steam gage(＝**steam gauge**) 汽压计,压力表
steam generating chamber 蒸汽供应室
steam header 主蒸汽管
steam heat dryer 蒸汽烘燥器;蒸汽烘燥机
steam heated draw roller 蒸汽加热的进布辊
steam heater 蒸汽加热器
steam heating 蒸汽加热,汽暖
steam humidifier 蒸汽给湿器
steam impingement 蒸汽冲击
steaming 汽蒸(指汽蒸定形、汽蒸还原、汽蒸固色等)
steaming calender 汽蒸轧光机(筒状针织物整理设备)
steaming chamber 汽蒸室,汽蒸箱
steaming machine 蒸化机;汽蒸机
steaming of printing 印花蒸化
steaming rate 蒸发率
steaming temperature 汽蒸温度,蒸化温度
steaming tenter 汽蒸拉幅机
steaming unit 汽蒸箱,汽蒸装置

steam inlet pipe 进汽管
steam jacket 汽套,蒸汽夹层
steam jet 汽蒸喷嘴,蒸汽喷射器
steam jet blower 喷汽鼓风机
steam-jet pump 蒸汽喷射泵
steam-jet refrigerating machine 蒸汽喷射致冷机
steam-jet texturing 蒸汽喷射变形[工艺]
steam lance 蒸汽喷枪(热回收装置的清洁器具,能喷射高压蒸汽,冲洗沾污物)
steam lock 汽封
Steam Master 高温高压卧式蒸化机(一种新型卧式蒸化机,主要供化纤织物的高档小批量多品种蒸化用,间歇式生产,较原来的星形架蒸化机有很多优点,商名,日本制)
steam nozzle 蒸汽喷嘴
steam outlet pipe 出气管
steam output 蒸发量
steam perforator 蒸汽喷管
steam permeability 透汽性
steam permeation method 蒸汽渗透法
steam pipe 蒸汽管道,汽管
steam pipeline 蒸汽管路
steam point 沸点
steam preshrinkage finish machine 汽蒸预缩整理机
steam press 汽蒸压烫
steam press shrinkage〈SPS〉 蒸汽压烫收缩
steam pressure regulator 气压调节器
steam puffer 蒸汽喷射烫衣机
steam-purge impregnation system 蒸汽喷射浸渍装置(以蒸汽驱走织物上空气,从而提高吸液率)

Steam purge technique 蒸汽清除[吸液]技术(商名,英国开发)
steam purifier 水汽分离器
steam quantity regulator 汽量调节器(用于各种蒸箱)
steam regulator 汽量调节器(用于各种蒸箱)
steam scrubber 蒸汽清洗装置
steam seal(=steam sealing) 汽封
steam separator 水汽分离器
steam setting 蒸汽定形
steam setting machine 蒸汽定形机
steam-shrunk [呢绒]汽蒸预缩
steam spray 蒸汽喷涂
steam spray pipe 蒸汽喷管
steam stop valve 停汽阀
steam stream 汽流
steam stretching 汽蒸拉伸
steam style 汽蒸显色印花法
steam superheater 蒸汽过热器
steam supply 蒸汽供应[量]
steam supply pipe 供汽管,给汽管
steam throughput 蒸汽流量
steam-to-steam heat exchanger 汽—汽热交换器
steam trap 凝汽阀,阻气排水阀,疏水器
steam tumble 汽蒸圆筒(毛针织物热处理用)
steam tunnel finishing machine 隧道式蒸汽整理机(针织内衣连续自动熨烫设备)
steam turbine 汽轮机
steam type airheater 蒸汽加热空气预热器
steam valve 汽阀,蒸汽阀
steamwater baffle 汽水挡板
steam water ratio 汽水比(织物烘燥过程中汽化织物内一千克水所消耗的蒸汽质量数)
steamwater separation 汽水分离
steapsin 胰脂酶
stearate 硬脂酸盐
stearic acid 硬脂酸,十八[碳][烷]酸
stearyl alcohol 硬脂酰醇,十八酰醇
stearyl amido methyl pyridinium chloride 硬脂酰胺甲基吡啶氯(耐久性防雨剂)
steel blue 钢青色(深蓝色);蓝色颜料
steel bowl 钢滚筒,压光滚筒
steel casting 〈st. c.〉 钢铸件
steel channel 槽钢
steel concrete 钢筋混凝土
steel fibre 钢纤维(用于导电)
steel flat 扁钢
steel framing 钢结构,钢架
steel glass pipe 玻璃钢管
steel grey 钢灰色(紫光灰色);黑色酸性染料
steel reinforced plastics 钢骨塑料
steel sheet 钢板
steelwork 钢结构,钢架
steep bin 浸渍储存箱
steep bleaching process 浸渍漂白法
steeped silk 浸渍丝
steeping 浸渍;浸染;浸泡;前洗[工艺]《丝》
steeping and pressing tank 浸渍压榨机(浴槽式)
steeping in enzyme preparations 酶液浸渍工艺,酵素退浆工艺
steeping liquor 浸渍[碱]液
steeping lye 浸渍碱液
steeping passages 浸渍道数
steeping press 浸渍压榨;浸渍压榨机

Steep Master 斯梯普玛司特液下履带式汽蒸箱(商名,德国制)

Stenacrile dyes 斯泰纳克里尔染料(染丙烯腈系纤维的阳离子染料,商名,意大利制)

stench 恶臭

stencil 刻花模版,刻花纸版

stencil cloth 筛网,筛绢

stencil engraving 型版雕刻

stencil frame [筛网]印花网框

stenciling 刻[花]版;刻版印花

stencil joining 花版接版

stencil lacquer printing 镂[花]版漆印

stencil method [浆印]制版法,刻纸版法

stencil paper 刻花纸版

stencil plate 刻花模版

stencil printing 刻版印花(纸版或锌版印花)

Stenolana dyes 斯泰诺拉纳染料(1∶2金属络合染料,商名,意大利制)

stenter 拉幅机

stenter chain 拉幅链条

stenter chain differential device 拉幅链条差动装置

stenter chain rail 拉幅链条导轨

stenter clip 拉幅布铗

stenter dryer(=stenter drying frame) 拉幅烘燥机

stenter finishing range 拉幅整理联合机

stenter frame 拉幅机

stentering 拉幅[工艺]

stentering allowance 拉幅留量(在花筒雕刻中,为防止花布在拉幅中出现变形而预先留下的余量)

stentering machine 拉幅机

Stenter Mercerization 〈SM process〉拉幅丝光工艺(适用于针织品,商名,瑞士科莱恩开发)

stenter rail 拉幅导轨

step 踏板;位移;间距;节距;托脚杯;级;步骤

step addition polymer 逐步加成聚合物

step aeration process 逐步曝气法,阶段曝气法

step and repeat camera 连拍机

step and repeat machine 连晒机

step bearing 立式止推轴承,竖轴轴承

step bolt 阶梯形螺栓

step brass 轴瓦

step-by-step 步进

step change 有级变速;步进变化,阶跃变化

step-down gearing 减速齿轮

stepless 无级[变速]的

stepless adjustable gear 无级调速装置

stepless change 无级变速;无级变化

stepless control gearing 无级调节传动[装置]

stepless drive 无级变速传动

stepless variable 无级变速

stepmotor 步进电动机

stepped control 有级调节,分级控制

stepping motor 步进电动机

stepping motor drive 步进电机传动

stepping relay 步进继电器

stepping switch 步进开关,阶段开关,分档开关

stepping test 逐步试验,分级试验

step response 阶跃特性,瞬态特性

step switch 步进开关,阶段开关,分档开关

step-up gearing 变速齿轮

step valve 级阀
stepwise cross-linking 分步交联整理
stepwise polymerization 逐步聚合[作用]
stepwise treatment 阶段性处理
ster.(steradian) 球面[角]度
ster.(sterilizer) 消毒器,杀菌器
steradian〈ster.〉 球面[角]度
stereo- （构词成分）表示"立[体]""固[体]"
stereo-block 立构规整嵌段;定向嵌段
stereoblock copolymer 立构规整嵌段共聚物
stereoblock polymer 立[体]构规整嵌段聚合物
stereochemistry 立体化学
stereoisomer 立体异构体
stereometer 立体测量仪,体积计（测纤维比重）
stereomicrograph 立体显微图
stereomicrography 立体显微照相法
stereomicroscopical test 立体显微镜试验
stereoscan electron microscope 立体扫描电子显微镜
stereoscanning 立体扫描
stereoscopic electron micrograph 立体电子显微照相图
stereoscopic microscope 立体显微镜
stereoscopic projector 立体投影仪
steric 空间[排列]的;立体的
steric-crosslinking arrangement 空间交联排列
steric factors 位阻因素
steric hindrance 位阻[现象]
steric isomer 立体异构体
steric retardation 位滞(现象)
steric strain 空间张力

steric structure 空间[排列]结构
sterilant 杀菌剂,消毒剂
sterile 杀菌剂,消毒剂;无[细]菌的
sterile room 灭菌室,无菌室
steriliser 消毒器;灭菌器,杀菌器
sterilization 杀菌[作用],灭菌[作用]
sterilized cotton 消毒棉,药棉
sterilize resistant 耐消毒的
sterilizer〈ster.〉 消毒器,杀菌器
steroid 甾族化合物;类固醇《生化》
sterol 甾醇,固醇
Stevens Water 史蒂文斯水（毛织物防缩整理用剂,商名,美国史蒂文斯）
STI(shear thinning index) 剪切变稀指数
sticker 松紧条痕(织疵)
stickiness 黏着性
sticking 黏搭,胶着
sticking-in 嵌花筒[机印疵],网孔堵塞
sticking places 黏并(浆纱疵)
sticking temperature 黏着温度
stick-lac 树枝虫胶
stick point 软化点《化纤》
stick roller 粘辊
stick-slip 黏滑[现象]（静摩擦力与动摩擦力交替变换时产生）
stick-slip method 黏附滑动法（摩擦系数试验）
sticky 黏性的;黏附感
sticky feeling [穿着]黏附感,不舒服感
sticky point 黏点;液态极限;塑性上限
stiffener 硬挺剂;硬衬;加劲杆;刚性元件
stiffening 硬挺整理
stiffening agent 硬挺整理剂
stiffening cloth 硬衬织物
stiff flow 难流动[性]
stiffness 硬挺性,硬挺度;刚度,刚性;稠

stiffness tester [织物]硬挺度试验仪
stilbene azo dyes 二苯乙烯偶氮染料
stilbene dyes 1,2-二苯乙烯染料
still-air 不流动空气
stimulated elastic restoration 受激弹性回复
stimulation 激励；刺激[作用]
stippled ground 点刻地纹
stipple effect 云纹效应（用细钢钉在印花滚筒上敲出不同深浅的云纹细点,可印出深浅得宜、色彩柔和的云纹）
stipple mill 雪花钢芯
stipple pattern 杂点纹
stipple print 点刻[云纹]印花,细点满地印花
stipple roller 满地细点印花滚筒
stippling 点刻法（印花雕版用）
stippling graver 点纹雕刻刀
stipulation 规定,条款,项目
stirrer 搅拌器
stirring 搅拌,搅动
stitch bonded pile loop fabric 毛圈型缝编织物
stitch bonding 缝编黏合
stitched fabric 缝编织物
stitch lubrication effect 缝纫润滑效果
stitch-out 绗缝
STM(scanning tunneling microscope) 扫描隧道显微镜
stn.(station) 站,台
stochastic 随机的
stochastic sampling 随机抽样
stochastic screening 随机网点[喷墨印花]
stock 原料；存货,备料；台,座
stock dye 原染料,备用染料

stock-dyed fibre 染色散纤维
stock-dyed yarn 散纤维染色纱
stock dyeing 散纤维染色
stock dyeing machine 散纤维染色机
stock fulling mill 捣打式缩绒机
stockinet goods 弹力针织物
stockinette 松紧织物,弹力织物（内衣用）
stocking board 袜子定形板
stock in process 在制品
stock machine 捣打式缩绒机
stock [colour] paste 基本[色]浆,储备[色]浆；原浆
stockpiling [制品或半制品的]积压；堆积
stock reduction paste 备用稀释浆；印花冲淡浆
stock room 原料库；储藏室；材料库,储料间
Stocks' law 斯托克斯定律
stock solution 备用[溶]液,储液
stock thickener(＝stock thickening) 原浆,原糊（作印浆稀释用）
stock vat 还原染料干缸（隐色体浓液）
stock vatting 干缸还原
stoichiometric balance 化学计量平衡
stoichiometry 化学计量法,化学计量学
stoke〈St〉 斯[托克斯]（运动黏度的物理单位）
Stokes' formula 斯托克斯公式
Stoll flex abrasion 斯托尔曲磨（美国测曲磨方法）
stone 英石（重量单位）
stone-finishing real silk 砂洗桑蚕丝绸
stone flax 石棉
stone grey 石头灰[色]（浅橄榄灰色）
stone mangle 石压轧光机,石头轧辊
stone polishing 抛光（滚筒印花）

stoneware 粗陶瓷器
stone wash chemicals 石磨洗化学品
stone washing 石磨洗
stone washing style 石磨风格
stonewash jeans 石磨牛仔裤
Stool-Quartermaster universal weartester 斯托尔—夸特马斯特通用[纱线织物]耐磨测试仪
stop-and-go 时行时止,频繁停止的
stop-and-go device 时行时止装置;停走器
stop-buffers 止动缓冲器
stop button 停车按钮
stop cock 小龙头;管塞
stop collar 紧圈
stop element 停止符号
stop finger 停车指针;停止掣子
stop instruction 停机指令
stop key 止动键
stop lever 制动操作杆,定位杆
stop mark 停车横条(针织疵);停车档;稀档(机织疵)
stop motion 自停装置
stop motion screw 离合螺钉
stop motion wire 停车杆
stop nut 防松螺母,止动螺母
stoppage 停车;停台率
stopper 挡块,挡圈,制动器;阻聚剂
stop pin 止动销,止销,固定销
stopping bolt 止动螺栓
stop plate 制动板
stop ring 止动环
stop screw 止动螺钉
stop valve 截止阀(对液流通道实行启闭控制的阀)
stop watch 停表,秒表
storability 可储藏性

storage 仓库;存储;蓄电;堆布
storage battery 蓄电池
storage battery truck 电瓶车
storage box 积布箱,存布箱
storage condition 储藏条件
storage fastness 耐储藏坚牢度
storage kettle 储浆桶
storage life 适用期;存放期,储存限期
storage period 储藏时间
storage stability 储存稳定性(原液或色浆可保存的时间,印花半制品可存放的时间)
storage tank 储存槽,储存桶
storage vessel 储存容器,储存槽
store 存储;存储器;现成的
stored program control〈SPC〉存储程序控制《计算机》
stored stresses and strains 储存的应力和应变
storehouse 仓库
storeyed drying machine 多层烘布机
Stork Digital Imaging〈SDI〉斯托克数码图像制作公司
storm coat 风雪大衣
Stormer viscometer 斯托默[旋转]黏度计
stout 坚牢的,紧密的,结实的,厚的
stoving 硫熏(丝毛漂白)
stoving fastness 硫熏牢度
STP (standard temperature and pressure) 标准温度与压力
straight chain 直链
straight cutting machine 直条切布机
straightener 正纬器,正纬装置
straightening 矫正,整直
straight line〈SL〉直线
straight-line system 流水作业系统

straightness 伸直度
straight polymer 纯聚合物
straight resin 净树脂
straight run gasoline 直馏汽油
straight run pipe 直段管道
straight shade 单色(用单一染料染成的色泽)
straight soap 纯粹肥皂
straight stay 直撑
straight synthetics 纯合成品;纯合成洗涤剂
straight-through valve 直通阀
straight-walking trial [地毯]直行走步试验
strain 应变;张力;拉紧;浆液绞滤;菌株《环保》
strainer 张紧器,拉紧装置;滤网,粗滤器
straining cloth 滤浆布
straining installation 过滤设备
straining machine 筛[网]滤[浆]机
straining of colour 滤色浆
straining roll 松紧辊,张力辊
strain rebound 应变回弹
strain relaxation 应变松弛
strain roll 张力辊;松紧辊
strap 带,皮带;环,圈;垫片
strap bolt U形钉,蚂蝗钉;带眼螺栓
strap fork 皮带叉,皮带移动器
straps 肩带,背带,打包带
strap uncurler 皮带剥边器(用于针织物热定形机)
stratification 层叠形成,成层[作用]
stratified sampling 分层抽样
stratographic 色层分离的
stratography 色谱法
straw [yellow] 麦秆黄[色](淡黄色)

straw yarn 仿草秆丝[纱];扁平[截面]长丝[纱]
stray heat 散失热
stray light 杂散光,漫射光
streak 刀线,拖刀,拖浆,色条(印疵);条痕;经柳(织疵)
streakiness 条痕,条花[现象](织疵)
Streakmeter 光电式织物条花测定仪(商名)
streak reagent 喷显剂(色层分析用)
streaky 条花,条痕(疵点)
streaky dyeing 色柳,染色条花(染疵)
streaky pattern 横条(染整疵)
stream 气流,液流
stream days 运转日数,生产日数
stream dyeing [循环]液流染色
stream factor 流水作业系数(生产工艺流程中某单一工序在整个生产流程中所占生产时间的比例)
stream feeder 连续给料器
streamline 流水线,流线
streamline flow 层流
stream printing 注液[防染]印花(印好防染浆后以染液喷注染色,用于毛巾、浴衣、国旗等的印花,日本开发)
strength 强力,强度
strength and elongation 强伸度
strength at rupture 断裂强力
strength count product [棉纱]品质指标
strengthening 加固,增强,强化
strengthening agent 增强剂,加固剂
strengthening mechanism 增强机理
strengthing and slenderizing technology 拉伸细化技术《毛》
strength irregularity 强力不匀率
strength loss by light 光照强力损失

strength management finish〈SMF〉 强力维护整理(使强力不下降的整理)
strength measurement and adjustment 力分测定与调节
strength of a dye 染料力分
strength of a dyeing 染浴中的染料量(指染料对被染物重量的百分率)
strength of dyes 染料强度
strength-retention 残留强力
strength tester 强力试验机
strength-to-weight ratio (金属材料的)强度与重量比
stress 应力
stress-concentrator 应力集中点
stress cracking 应力破裂
stress decay 应力衰减
stress distribution 应力分布
stress relaxation 应力松弛,应力弛缓
stress relaxometer 应力松弛仪
stress relief test 应力消除试验(热收缩试验)
stress-strain 应力—应变
stress-strain curve〈S-S curve〉 应力—应变曲线
stress-strain property 应变应变性能,强伸度性能
stress-strain ratio 应力应变比,强力伸长比
stretch [纤维、纱线、织物的]伸张
stretchability 拉伸性
stretch and sag resistance 抗延伸和下垂性
stretch and shrinkage machine 单向伸缩机(印衣制版用)
stretch and shrinkage measuring and control unit 伸缩监控装置(能自动控制拉幅机的超喂量或防缩机的收缩率)

stretch cotton 弹性棉纱
stretch device 拉幅装置
stretched〈st.〉 拉伸的
stretched nylon 弹力锦纶,弹力尼龙
stretched state 紧张状态
stretch elasticity 拉伸弹性
stretcher 伸幅器,张紧器;撑板;绷网机
stretch fabric 弹力织物
stretch growth 第二潜伸,次级蠕变(纤维受载荷被伸长后,载荷虽经解除,不能恢复至原有长度)
stretch hose 弹力袜
stretch [nylon] hosiery 锦纶弹力丝袜
stretch indicator [强力试验机的]伸长指示器
stretching frame 拉幅机
stretching machine 拉幅机;绷网机
stretching roller 扩幅辊
stretching vibration 伸缩振动(红外线)
stretch jersey 弹力平针织物
stretch knit 弹性针织,弹性编织;弹性针织物
stretch nylon 弹力锦纶,弹力尼龙(俗称),锦纶弹力丝
stretch-nylon yarn 弹力锦纶丝
stretch-out view 展开图
stretch pattern 弹性式样
stretch polyester yarn 弹力涤纶丝
stretch potential 潜在伸长
stretch proofing 抗拉伸
stretch tester [针织物]拉伸试验仪
stretch woven fabrics 弹性机织物
stretch yarn 弹力丝[纱]
strié 条痕,条花(织疵,有时指织物花纹)
striated fabric 条纹织物
striation 条痕,条花;沟纹;缝[隙]

strike 瞬染[上色]（指纤维浸入染液初期，染料尚未扩散至其内部时，在纤维表面形成的上色浓度）
strike-back 回胶（织物粘贴时热熔胶回流，造成黏合不牢）
strike levelling test 瞬时匀染试验
strike migration test 瞬时（泳）移性试验，瞬时移染性试验《染》
strike-off （开车时）[印花]打样
strike-off machine [印花]打样机
strike property 瞬染性
strike rate 瞬染[上色]率
strike test 瞬染试验（比较染料上色快慢和匀染性的试验）
strike-through 渗胶（织物粘贴时黏合剂渗入织物）
striking out [绒头织物]起绒整理
string discharge filter 串联过滤器
stringiness 黏稠性
string rug 拼块地毯
strings [布边]色线头（织疵标记）
string washer 松式绳状洗涤机
strip 剥色；抓损疵，擦损疵；条，带
strip chart 带状记录纸
strip cutter 切布样机
strip dyeing 衣片染色（用于运动衫、针织衫等）
stripe 条纹，条子；横档（织疵）
stripe coating cloth 条纹涂布
striped cotton 条子[棉]布
striped crepe 柳条绉
striped fabric 条子织物
stripe finish 条纹呢整理
stripe machine 对条机
stripe matcher 对条机（条纹筒状针织物整理加工的一种附属设备，包括剖幅、条纹理直、再缝合等）
stripe mill 条子钢芯
stripe pattern 条型图案
stripiness 经柳，经向条花（织疵）；[地毯的]条状疵点
striping 经向条子印花
strip method [织物强力试验的]条样法
stripper [圆网]剥胶剂；去色设备；汽提塔
stripping 脱色，褪色，剥色；脱模；剥离
stripping action 剥取作用
stripping agent 反萃（取）剂，褪色剂，剥色剂
stripping bath 剥色浴
stripping device 剥取装置
stripping method 剥色方法，吹脱法
stripping promoter 剥色促进剂
stripping tank 剥胶槽
strip-redye sequence 剥色重染程序
strip test 布条断裂强度测定，织物条样强力试验
Striptex 剥离强度试验仪（商名，瑞士制）
stripy defects 条花疵点
strobophotography 频闪摄影
stroboscope 频闪观测器
stroboscopic lamp 频闪灯
stroke 刮程（手工刮印）；笔划（图形识别用）《自动》；冲程，行程；撞击
stroke length 冲程，刮程（平网印花刮刀的行程）
stroking mechanism 变量机构
strong 浓，深（色彩术语）
strong acid medium dyeing 强酸性染料
stronger [染料]强度（力分）高
strong size 重浆
strong twist yarn 强捻纱

strong viscose rayon 强力黏胶[人造]丝
strontium unit〈su.〉锶单位
structural analysis 结构分析
structural birefrigence 结构双折射，形状双折射
structural colouration 结构生色（物体本身通过对光的散射、干涉、衍射而产生颜色）
structural component 结构元件
structural diagram 结构图
structural fabric 花式织物
structural formula 结构式
structurally coloured fibres 结构生色纤维
structural promoter 结构促进剂，结构助催化剂
structural stability 结构稳定性
structural steel frame 钢结构，钢架
structural support 支承构架
structural viscosity 结构黏度
structure 结构，构造，组织
structure assembling system 组装体系
structured carpet 花式地毯
structured pattern 花式图案，结构花纹
structured query language〈SQL〉结构化查询语言《计算机》
strut 支撑，支柱，支架
strutting 支撑[系统]，加固
strutting piece 支撑件
strychnine 二甲双胍，马钱子碱
STS(substrate transport system) 基布输送系统
stub 轴端；短管，双端螺栓，柱螺栓；短轴；凸头；销子
stubble shearing 剪毛不净
stub tube 短管，管接头
stud bolt 双头螺栓

studded rubber belt 凸头橡胶带
stud-retained coupling 销钉连接器
stud shaft 中间轴；栓轴
stuff colouring 织物上色
stuff goods 毛织品，呢绒
stuffing 填料；填充剂；[织物]填充线
stuffing box 加料槽（绞盘染色机染槽前部的多孔隔板槽）
stuffing box bearing 填料压盖轴承，密封轴承
stuff pump 浆泵
sturdiness 坚实，坚固性
sturdy texture [织物]质地厚实
STW(sewage treatment works) 污水处理厂
STX（start of text）正文起始，正文开头《计算机》
style finish of denim garment 牛仔成衣的风格整理
style of printing 印花类型，印花方式
styling 花样；式样；设计[组织]
stylish 时髦的，时新的，漂亮的
stylus 刻针，钢针；记录针；光笔，指示笔《计算机》
styptic cotton 止血药棉
styptic fibre 止血纤维
styracin 苯丙烯酸苯丙烯酯
styrene 苯乙烯
styrene-acrylonitrile copolymer〈SAN〉苯乙烯—丙烯腈共聚物
styrene-butadiene copolymer 苯乙烯—丁二烯共聚物
styrene butadiene rubber〈SBR〉丁苯橡胶
styrene copolymer 苯乙烯共聚物
styrene-maleic anhydride copolymer 苯乙烯—马来酐共聚物

styrylamine 苯乙烯胺
styryl disperse dye 苯乙烯基型分散染料
su.(sensation unit) 声感单位(分贝的原名)
su.(servo unit) 伺服机构,随动机构
su.(strontium unit) 锶单位
SUB(substitute) 取代,代入;代替物;代替人
SUB(substitute character) 替换字符
subacid 微酸[性]的
subacute dermal toxicity 亚急性皮肤毒性
subacute oral toxicity 亚急性口服毒性
sub-assembly 局部装配;部件,组件;辅助装置
sub cooler 过冷器
subcutaneous water 地下水
sub-ed(sub-editor) 副编辑,助理编辑
subdivision(subd.) 细分类
subdued 柔和的;降低的
subdued lustre 柔和光泽
sub-editor⟨sub-ed⟩ 副编辑,助理编辑
suberin 软木脂,木脂素
subgrade 路基,地基
subgroup movement [分子的]侧链运动
subjective appraisal (= subjective assessment) 主观评定,感官评定
subjective brightness 亮度(英国色度学用语)
subjective characteristic 主观特征,感官鉴定的特征
subjective colour 主观色《测色》
subjective comfort rating 主观舒适度
subjective grading 感官评级
subjective inspection(= subjective test) 主观检验,感官检验
subl.(sublimate) 升华物

subl.(sublime) 升华
Sublaprint 苏布拉普林特转移印花油墨(商名,英国制)
Sublaprint colours 苏布拉普林特染料(转移印花染料,商名,英国制)
sublimability 升华性
sublimable dyes 能升华的染料
sublimate⟨subl.⟩ 升华物
sublimation 升华[作用]
sublimation curve 升华曲线
sublimation diffusion 升华扩散
sublimation energy 升华能
sublimation fastness 升华牢度
sublimation ghosting 升华虚印(转移印花疵)
sublimation [transfer] printing 升华[转移]印花
sublimation transport mechanics 升华机理
sublimative desorption 升华解吸[作用]
sublimatography 升华谱法
sublime⟨subl.⟩ 升华
Sublistatic printing 萨布利斯泰蒂克印花(干热转移印花,商名,法国)
Sublistatic print paper 萨布利斯泰蒂克转移印花纸(商名,法国制)
sublistic natural printing process ⟨ SNP process⟩ 一种天然纤维转移印花法(德国开发)
submerged beam dyeing 液下经轴染色
submerged bleach process 液下浸漂法
submerged centrifuge 浸没式离心机
submerged combustion 水中燃烧法《环保》
submerged fermentation 深层发酵
submerged jet 液下喷射
submerged roll 液下导辊
submerged roll coating 浸没辊涂层法

submerged test 浸渍试验
submerged thermosol process 浸没式热熔加工
submersion 浸渍
subs.(subscription) 预约,订购
subs.(subsidiary) 辅助的,附属的;附属机构
sub-sample 小样,子样,部分样本
subsampling equipment 取小样装置
subscript 下标,注脚,脚号;索引
subscription(subs.) 预约,订购
subsection(subsec.) 细目
subsequent(subseq.) 后来的,其次的
subsidiary〈subs.〉附属机构;附属的,辅助的
subsoil water 地下水
subst.(substantive) 实质的
subst.(substitute) 取代,代入;代替物;代替人
substandard 次级标准,低等级标准;副标准,复制标准;低于标准的,低于定额的
substandard goods 次品,次货
substantive〈subst.〉实质的
substantive colour(=substantive dye) 直接染料
substantive finish 简便有效的整理[技术]
substantive reactant fixable dyes 直接交联固着染料
substantivity 直接[上染]性,亲和性
substation 变电所
substituent 取代基,置换基;替代物
substituent group 取代基团
substituent uniformity 取代基均匀度
substitute〈SUB, subst.〉取代,代入;代替物;代替人

substitute character〈SUB〉替换字符
substituted indigo 靛蓝衍生物
substitute fuel 代用燃料
substituting agent 取代剂
substitution 置换
substitution error 替代误差
substractive colour mixing(=substractive colour mixture) 减法混色,减色法混合
substractive finishing 减法整理(发生质量损失的整理,如精练)
substractive process of antifelting finish 减法防毡缩整理
substrate 被作用物;被染物;基质;底布;地毯底层;衬底
substrate filament 基质长丝(用以生产无机耐高温纤维,采用化学蒸气淀积法,使无机化合物在高温下分解后淀积于基质长丝上)
substrate fixation system〈SFS〉基布固定系统(上胶装置等使用)
substrate heating 基布加热
substrate speed control 基布速度控制
substrate transport system〈STS〉基布输送系统
substratum [促使感光乳剂固着于片基的]底层,胶层
substructure 亚结构
subsurface 表面下的部分;液面下的;地下的
subtractive colour 物体色;减色
subtractive complementary colours 减色法补色
subtractive mixture of colour stimuli 减法混色
subtractive [three] primaries 减法三原色
subtractive process 减色法[混合]

subzero temperature 零下温度
successful acceptance 验收成功
successive contrast 相继对比(凝视单色后出现单色补色的影像)
successive reaction 逐次反应
succinic acid 琥珀酸,丁二酸
succinonitrile 琥珀腈,丁二腈
such as〈s. a.〉例如
sucker 绉条薄织物,泡泡纱;条格绉布;吸管
sucker-weavelike wrinkle 泡泡纱型绉纹
sucrose 蔗糖
sucrose ester surfactant 蔗糖酯表面活性剂
sucrose polyester emulsion 蔗糖聚酯乳液(用于织物柔软整理)
suction 吸;吸力;抽空度
suction box 吸碱装置(用于丝光机);吸尘箱
suction cooling 抽气冷却
suction drum 抽吸圆筒
suction drum backwashing machine 抽吸式转筒复洗机
suction-drum dryer 抽吸式转筒烘燥机
suction-drum hydro-extractor 抽吸式圆网脱水机
suction drum rinsing compartment 吸入式圆网水洗槽
suction drying 真空干燥
suction extraction 抽吸式脱水
suction extractor 真空脱水,抽吸式脱水机
suction fan 抽吸风扇;吸棉风扇
suction filter 吸滤器
suction flask 吸滤瓶
suction gauge 吸力计;真空计

suction hydro-extractor 抽吸式脱水器
suction lift 吸升高度;吸引升力
suction machine 吸尘机,抽吸机
suction main 抽吸总管道
suction nozzle 吸嘴
suction pipe 吸管
suction press 抽吸压烫
suction pump 抽吸泵
suction roll 吸水辊
suction seal 吸力密封
suction slot device 吸气槽装置
suction strainer 吸滤器
suction valve 吸入阀,进气阀
suds 黏稠介质中的空气泡;肥皂水
suds booster 增泡剂
sudsing 起泡
suds return method 洗液回用法
Sudyesul dyes 速得素染料(预先还原的液体硫化染料,应用于浸染和连续染色,重现性好,特别适合于牛仔布染色,商名,瑞士科莱恩)
suede [cloth] 仿麂皮织物
suede effect 绒面效应;仿麂皮效应
suede fabric 仿麂皮织物
suede finish 仿麂皮整理
suede finishing machine 仿麂皮起绒机,磨毛机
suede leather 仿麂皮织物
suede type synthetic leather 仿麂皮合成革
suedine 仿麂皮织物
sueding [织物的]仿麂皮起绒工艺
sueding machine 仿麂皮起绒机
sueding roller 仿麂皮磨毛辊
suit 成套衣服
suitability 适合性,适用性
suiting finish [棉织物]仿亚麻布整理

suitings 衣料
suiting silk 呢(表面少光泽,质地丰厚似羊毛织品的丝织物)
suit pattern 套装图案
sulfamate 氨基磺酸盐,氨基磺酸酯
sulfamic acid 氨基磺酸(酸性染料染深色时的助染剂)
sulfated castor oil 硫酸化蓖麻油,土耳其红油
sulfated tallow 硫酸化牛油脂
sulfate surfactant 硫酸酯型表面活性剂(用于洗毛)
sulfation 硫酸盐化;硫酸酯化
sulfato 硫酸根络
β-sulfatoethylsulfonyl group 乙基砜 β-硫酸酯基团
sulfhydryl [group] 氢硫基,巯基
sulfidation 硫化作用
sulfidation-resistance 抗硫化性
sulfide 硫化物
sulfite 亚硫酸盐
sulfoacylation 磺酰化作用
sulfoalkylation 磺烷化作用
sulfoamidic acid 氨基磺酸
sulfobetaine surfactant 硫[代]内铵盐型表面活性剂
sulfochlorides [烃基]磺酰氯
sulfocyanic ester 硫氰酸酯
sulfoether 硫醚
sulfofatty acid 磺化脂肪酸
sulfoglyceride 磺化甘油酯
sulfonamide dyes 磺酰胺型染料
sulfonatability 磺化性
sulfonate 磺酸盐
sulfonated 磺化的
sulfonated castor oil 磺化蓖麻油

sulfonated dye [含]磺酸基染料
sulfonated [red] oil 磺化红油,土耳其红油
sulfonated soybean oil 磺化豆油(即和毛油)
sulfonate surfactant 磺酸盐表面活性剂
sulfonation 磺化[作用]
sulfone 砜
sulfonic acid group(＝sulfonic group) 磺酸基
sulfonium surfactant 锍型表面活性剂(用作柔软剂)
sulfonyl 磺酰
sulforicinoleic acid 磺化蓖麻油酸
sulfo-sulfonate 磺基磺酸盐(二磺酸盐)
sulfoxide 亚砜
sulfur [yellow] 硫黄黄[色](嫩黄色);硫化黄染料
sulfur compound 硫化合物
sulfur dyeing 硫化染料染色
sulfur dyes 硫化染料
sulfuretted hydrogen 硫化氢
sulfuric acid process [羊毛炭化]硫酸处理(去除羊毛中的草屑等物)
sulfuric ester 硫酸酯
sulfuring 硫熏工艺(丝毛漂白法之一)
sulfurous acid 亚硫酸
sulphide wash 去硫[作用],脱硫[作用]
sulphitolysis 亚硫酸化溶解,亚硫酸盐解(对羊毛的一种前处理方法)
sulphodiethanol 硫代双乙醇,古来辛 A
sulphonated oil 磺化油,土耳其红油
sulphonation 磺化[作用]
sulphone 砜
sulphone chemicals 砜类化学品(与棉纤维起交联作用,用于织物永久性弹性整

理)
sulphonyl dye 磺酰染料(活性染料之一)
sulphosuccinamates 磺基琥珀酰胺酸盐(泡沫,乳化剂)
sulphoxy reducing agent 次硫酸还原剂
sulphur condensed dyes 硫化缩聚染料
sulphur dyes in dispersed form 硫化分散型染料
sulphurization 硫化作用
sulphur liquid dyes 硫化液状染料
sulphur-methyl melamine-formaldehyde resin 磺甲基化三聚氰胺树脂
sulphur removal 去硫,脱硫
sulphur resist [靛蓝]硫黄防染法
sulphurs 硫化染料
sulphur stoving 硫熏[漂白]
sulphur vat dyes 硫化还原染料
sultan 印度绸(印度产,经过柔软、缎纹整理)
sum.(**summary**) 摘要,一览
sumac 苏模鞣料,野葛鞣料,黄栌,盐肤木(制革与染色用)
sumac extract 苏模浸剂,黄栌精,黄栌浸剂
sumaching 单宁处理,上鞣
Sumifix dyes 素米菲克斯染料(活性染料的一种,含乙烯砜基团,商名,日本住友)
Sumifix supra dyes 索米菲克斯超级染料(活性染料,含一氯三嗪和乙烯砜型双活性基团,固色率较高,商名,日本住友)
Sumikaron 素米卡隆(分散染料,商名,日本住友)
Sumikaron dyes 素米卡隆染料(分散染料,商名,日本制)

Sumiplen dyes 素米普伦染料(聚丙烯纤维用染料,商名,日本制)
Sumitex Resin EX-309 素米坦克斯树脂EX-309(低甲醛树脂整理剂,商名,日本住友)
summary〈**sum.**〉 摘要,一览
summit potential 顶点电位
sump 池;潭;槽;污水坑;油盘
sump pump 污水泵;井底水窝水泵
sun and planet gear 行星齿轮
sun bleaching 暴晒漂白
sunburn 日炙棕[色](带红浅棕色)
sunburst pleat 辐射式褶裥
sun cloth 遮阳布
suncrazing 晒裂
sun discolouration 日晒变色
sun drying 暴晒干燥
sun exposure test 日晒试验
sunfast 耐晒的
sun filter 滤阳光窗帘
sunflower[**yellow**] 向日葵黄[色](黄色)
sunflower pectin 向日葵盘果胶(增稠剂)
sunlight degradation 日光降解
sunlight fastness 耐日晒牢度
sunlight resistance 耐日光性
Sunlite 一种中空保暖纤维
sun pleat 扇形褶裥
sun proof 耐晒的
sun protection factor〈**SPF**〉 防晒因子(衡量紫外线辐射对人体危害程度的参考指标)
sun protective finishes 防晒整理
sunscreen index 防晒指数
sun-screening agent 防晒剂
sunshade cloth 阳伞布;窗帘布
sunshine weather-ometer 日晒气候试验仪

sun visors 遮阳板(汽车前玻璃上)
sun wheel 太阳齿轮(周转轮系的中心齿轮)
sup.(superior) 高级的,优良的
sup.(superlative) 最高的,无比的
sup.(supplement) 补充;附加;附录;增刊;补遗;添加物;附加的,补充的,额外的
supatex fabric 非织造织物
superabsorbent 超吸收剂,超吸水剂
super accelerator 超促进剂
super-amphiphilic film 超双亲薄膜(将纳米溶胶涂于玻璃等表面形成亲水、亲油薄膜,起到防雾、清洁等作用)
super-amphiphilic property 超双亲性
superatmospheric pressure 超大气压
Superclear 100-N 索泊克厘尔 100-N(防泳移剂,含有活性羧基和羟基的多糖类大分子化合物,适用于分散、活性、还原等染料连续染色时防止染料泳移,商名,美国制)
super-conductibility 超级传导性能
superconducting quantum interference device 〈SQUID〉 超导量子干涉仪(测量弱磁场的装置)
superconduction 超导《电》
super conduction heat pipe 超导热管
superconductivity 超导性能
super conductor 超导体
supercontraction 超收缩[性]
super contraction of wool 羊毛的超收缩
supercooling 过冷[却]
super crimp 超卷曲
supercritical 超临界的
supercritical carbon dioxide 超临界二氧化碳

supercritical carbon dioxide dyeing 超临界二氧化碳染色
super critical carbondioxide dyeing technology 超临界二氧化碳[介质]染色技术(一种无水染色新技术)
supercritical carbon dioxide fluid dyeing 〈SCFD〉 超临界二氧化碳流体染色
supercritical dyeing 超临界染色
supercritical fluid dyeing 〈SFD〉 超临界流体染色
supercritical fluid extraction 超临界流体萃取
supercritical fluid technology(SFT) 超临界流体技术
superdense 超密的
super drawing 高倍拉伸,超拉伸
super-dry 过干燥
super elasticity 超弹性
superfeed 超喂
superficial expansion 表面膨胀
superficial fault 表面缺陷
superficial structure 表面结构
superfine 超细;特级的
superfine fibre 超细纤维,特细纤维
superfine glass wool 超细玻璃棉
superfine polyester dyeing process 超细涤纶染色方法
super fine wool 超细羊毛
super fluid 超流体
superheated steam drying 过热蒸汽烘燥
superheated steam setting 过热蒸汽定形
superheated steam 〈ss〉 过热蒸汽
superheated vapour 过热蒸汽
superheater 过热器,过热炉
superheater bank 过热器管束
superheater coil 过热器盘管

superheater vent 过热器空气阀
superheating heat 过热热
superheat temperature 过热温度
superhigh frequency〈SHF,S. H. F.,shf,s. h. f.〉超高频
super hydrophobic finishing 超级疏水整理
superintendent 监督员
superior〈sup.〉高级的,优良的
superior grade 超等
superlative〈sup.〉最高的,无比的
supermicro peach fabrics 超细桃皮绒织物
super milling dyes 高耐缩绒染料
supermolecular structure 超分子结构,微胞结构
super molecule 胶束;微胞
supernatant 上清液,上澄液
supernatant liquid 上层澄清液
super natural feeling [新合纤] 超越自然感
super ordinate 最高等级,特级
super phosphate 过磷酸钙;酸性磷酸盐
Super POA 一种增深剂(商名,韩国LG)
superpolymer 高聚物
superpose(=superposition)叠加;复合;使重合
Super-Sat impregnation unit 超饱和浸渍单元(商名,德国)
super-saturated 超饱和,过饱和
supersaturated〈supers.〉过饱和的
supersaturated steam 过饱和蒸汽
supersaturation 过饱和[现象]
super-shrinkage polyester 高收缩聚酯纤维
super soft effect 超柔软效果
super softening finishing〈SSF〉超柔软整理

super soft finishing〈SS finish〉超柔软整理(日本称液氨整理)
super soft peach〈SSP〉超柔软桃皮整理
super soft press〈SSP〉液氨处理与免烫整理结合的工艺
super soft process〈SSP〉液氨处理与免烫整理结合的工艺
super soft〈SS〉超柔软
supersonic cleaner 超声波清洗器
supersonic emulsifier 超声波乳化器
supersonic inspection 超声波探伤法
supersonic noise 超声波噪声
supersonic speed 超声速
supersonic wave 超声波
supersonic wave sewing machine 超声波缝纫机
supersonic wave washing machine 超声波洗涤机
super stiffening finish 特硬手感整理
super stretched wool 超拉伸细化羊毛
superstructure 上部构造
super wash 超级耐洗《毛》
super wash finishing 超级耐洗整理《毛》
super washing 石磨水洗
superwash process [羊毛]超级耐洗法
superwash wool 超级耐洗羊毛
suppl.(supplement)补充,附加;附录;增刊;补遗;添加物;附加的,补充的,额外的
supplement〈sup.,suppl.〉补充,附加;附录;增刊;补遗;添加物;附加的,补充的,额外的
supplier 供应商;厂商
supply 供给,补给;供电,供水;电源
supply chain 供应链
supply duct 送风管道

support（＝support abutment） 支架,支座,支柱,支撑装置
support and maintenance contract〈SMC〉 支持和维护合同（即有服务工程师到现场服务的合同）
support arm 支持臂
support bracket 支架
support coating 支承涂布
supported catalyst 辅助催化剂
supporter 护身,三角带,丁字带；载体
supporting beam 座挡
supporting block 支承垫块
supporting electrode 辅助电极
supporting fabric 辅助织物（里料、衬料等）
supporting frame 支承构架,支架,固定架
supporting plate 托板
supporting roller 承托辊
supporting saddle 支架
support lug 支承撑架
support ring 支承环《试》
support shaft 支架轴
support tube 支承管,悬吊管
suppressant 抑制剂
Supracet dyes 苏普拉塞特染料（分散染料,商名,英国制）
Supraflor 苏普拉弗洛地毯染色机（商名,德国制）
Supralan dyes 苏普拉伦染料（1∶2 金属络合染料,商名,美国制）
Supramin[e] dyes 色派明染料（弱酸性染料,商名,德国制）
supra-molecular structure 超分子结构
surah silk 斜纹绸
surface 面,表面；曲面
surface abrasion 表面磨蚀

surface abrasion resistance 表面耐磨性
surface active agent 表面活性剂
surface active detergent 表面活性洗涤剂
surface active indicator 表面活性指示剂
surface-active ion 表面活性离子
surface activity 表面活[化]度
surface adsorption 表面吸着
surface aeration 表面暴气
surface aeration turbine 表面暴气叶轮《环保》
surface asperity 表面粗糙度
surface attachment 表面依附
surface blush 雾面
surface bonding 表面黏合
surface burning 表面燃烧
surface burning time 表面燃烧时间
surface camber 表面凸度
surface cavity 表面空洞
surface characteristic ［织物］表面特点,［织物］表面风格
surface charge density 表面电荷密度
surface chemistry 界面化学,表面化学
surface cleaning 表面清洁
surface coating 表面涂层
surface coating agent 表面涂层剂
surface colour 表面色
surface concentration 表面浓度
surface conductance 表面传导
surface conductivity 表面导电性
surface contour 表面外形,表面轮廓
surface crack 表面裂纹
surface current 表面电流
surface defect test 表面疵点检验
surface deposition 表面沉淀
surface diffusion 表面扩散
surface-driven batcher（＝surface-driven

winder) 表面传动卷布机
surface dyeing 表面染色(染疵),浮色
surface elastomer [纤维的]表面弹性涂料
surface emerizing machine 表面磨毛机
surface energy 表面能
surface excess 表面过剩
surface expansion 表面膨胀
surface failure 表面破坏
surface feet per minute (sfpm, s. f. p. m.) 英尺/分(圆周速度,表面线速度)
surface fibre 表层纤维,表面纤维
surface finishing 表面处理
surface finishing equipment 表面整理设备
surface-fitting 表面配合,表面平正
surface flash 表面闪燃
surface flaw 表面瑕疵;表面裂隙
surface fluffy 表面毛羽
surface friction 表面摩擦
surface grafting 表面接枝
surface graft modification 表面接枝改性
surface grinding [弹性针布的]平磨;磨面,表面研磨
surface hardness 表面硬度
surface heating 表面加热
surface heat treatment 表面热处理
surface irregularity 表面不规整;表面粗糙
surface layer [表]面层
surface loading 面积负荷
surface membrane 表面薄膜
surface modification 表面变性,表面改性
surface modified fibre 表面改性纤维
surface modified film 表面改性膜
surface moisture content 表面含水量,表面水分
surface morphology [纤维]表面形态

surface nap 表面毛羽
surface negative absorption 表面负吸附
surface pattern effect 表面花式效应
surface phenomenon 界面现象
surface pile density [地毯的]表面绒头密度
surface plasma 表面等离子体
surface poisoning 表面中毒
surface porosity 表面孔隙度
surface positive absorption 表面正吸附
surface pressure 表面压力
surface printing 凸纹花筒印花,阳纹花筒印花
surface printing machine 凸纹滚筒印花机
surface profile 表面轮廓,表面外观
surface property 表面性质
surface reaction control 表面反应控制
surface relief [显微术中的]表面浮雕
surface-repellency test [织物]表面防水试验
surface resin 表面树脂
surface resistance 表面电阻
surface resistivity 表面电阻率
surface roughness 表面粗糙度
surface roughness finish 表面粗糙度整理
surface ruggedness of fabric 织物的粗糙表面
surface scan 表面扫描,表面搜索《计算机》
surface skinning [浆液]表面结皮
surface speed 表面速度,线速度
surface structure 表面结构
surface structure modification 表面结构改性
surface tension 表面张力
surface texture 表面织纹

surface thermometer 表面温度计,贴附温度计
surface treatment 表面处理
surface viscosity 表面黏度
surface water 地表水
surfacing 堆焊;镶面
surfactant 表面活性剂
surfactant analysis 表面活性剂分析
surf green 青竹色
surge tank 调浆槽;调压水槽;中间桶;平衡池
surgical gown 手术衣
surgical operation 外科手术
surgical suture 手术缝合线
surgical textile 外科(手术)用纺织品
surgical wiper 外科(手术)用纱布
surpass 超过,胜过
Surpeba machine 一种纱线连续膨化机(多用于腈纶纱,商名,法国制)
surrounding atmosphere 环境大气
surrounding field 辅助视野《测色》
surrounding of a comparison field 视场背景
surroundings 周围介质;环境
survey 调查;测绘;鉴定;综述,述评
survivability 耐受性;存活率;耐久性
susceptibility 敏感性,感受性,灵敏度;磁化率
susceptible 敏感的,能允许……的
suspended air heater 悬挂式空气加热器
suspended level viscometer 气承液柱黏度计
suspended matter 浮游物,悬浮物,悬浮质
suspended solid〈SS〉悬浮物质,悬浮固体《环保》
suspended weight 吊锤

suspending agent 沉淀防止剂
suspened emulsion 悬浮乳剂
suspension 悬浮体;悬置;吊架,悬架
suspension drying 悬挂烘干
suspension polymerization 悬浮聚合[作用]
suspension system 悬浮法,上浮法《环保》
suspensoid 悬溶胶[体]
sustainability 持续能力
sustainable development 可持续发展
sustained flexing 弯曲疲劳
sustained load 持续负荷
sustained release effect 缓释作用
suture 缝线;缝,接缝;针脚
suture thread 缝合线
sv(specific volume) 比容,体积,度(相对密度的倒数)
SVI(sludge volume index) 污泥体积指数
SVM 杂环芳纶(商名,俄罗斯)
SW(signal wire) 信号线
SW(＝sw,specific weight) 比重,相对密度
SW(＝sw.,switching) 开关;切换;转换
swan-neck bracket 弯脚
swapping [程序的]调动,交换
Sward hardness 斯瓦德硬度
swatch 小块样布;样本;样品
swatch-cutting machine 切布样机
swathe 绷带,包布
swathing cloth 绷带织物;襁褓带子织物
sway bar 平衡杆
swaying cistern 摇框浮洗槽(一种洗涤盛布槽,可以调节作用时间,适用于针织物)
swealing [染料]皱痕泳移(染疵);[手工]洗污渍印

sweat 出汗
sweat absorbent finish 吸汗整理
sweater 运动衫；针织套衫
sweating 渗出；熔化；表面凝水
sweating device 出汗[模拟]装置
sweating manikin 出汗人体模型，出汗假人（用于对服装进行功能性评价的模拟人体，测人体的出汗性能）
sweat retention 汗保持性
sweat secretion 汗分泌量
sweep 刮；扫除；扫描
sweet birch oil 冬青油，水杨酸甲酯
sweet water 淡水，饮料水
swellability 溶胀性
swellant 泡胀剂，溶胀剂
swelling 膨胀，膨化
swelling agent 泡胀剂，溶胀剂
swelling anisotropy 溶胀各向异性
swelling capacity 溶胀量
swelling characteristics 膨化特点
swelling heat 膨胀热，溶胀热
swelling method 膨润法
swelling power 溶胀本领，溶胀能力
swelling pressure 溶胀压
swelling pretreatment 溶胀预处理
swelling property 膨胀性
swelling ratio 溶胀比
swelling-resistant finish 抗膨化整理，抗溶胀整理
swelling testing 膨润试验
swelling value 溶胀值，泡胀值，膨胀值
swift 大滚筒，大锡林；纱框；绡架；丝框
swimming roll[er] padder (= swimming roller mangle) 游动辊轧车
swimming suits and trunks 游泳衣和游泳裤

swimming tub [给浆垫]浮盘（手工模版印花用）
swing 摇摆；振幅
swing arm 摇臂，摇杆，摆动臂
swinging arm compensator 摆臂式松紧调节架
swinging damper 旋转挡板
swinging reach 摆动范围
swing mechanism 摆动机构
swing shaft 摆动轴，摇轴
swirling roller 涡动辊（用于除尘），弯辊
swirl pattern 旋涡型图案
swiss batiste 巴蒂斯特高级薄纱
Swiss crepe organdy 瑞士蝉翼绉（用化学方法整理的薄绉，部分透明，一般为彩色印花）
swissing 滚筒轧光工艺
swissing calender 多辊轧光机（主要用于细薄织物的轧光，轧光时温度较高，压力较大）
swiss muslin 薄细布，薄纱
swiss pongee 充绸丝光棉布（英国名称）
Swiss Textile-Testing Co. Ltd. 〈Testex〉 瑞士纺织品鉴定有限公司
swiss voile 轻量透明薄纱
switch 开关；转接，切换
switch board panel 配电盘，开关配电盘
switch cabinet 开关箱，开关柜
switches 〈SS，ss〉 开关；电闸
switch gear 配电联动器
switch-in(＝switch-on) 开（指电流接通）
switching 〈SW，sw.〉 开关；切换；转换
switching cabinet 开关箱，开关柜
switching instrument 开关仪器
switching mechanism 开关机构
switching time 开关时间

switch lever 开关杆
switch mounting plate 开关座板
switch off 关掉,切断电流,断路
switch-out 关(指电流切断)
switch over 转接,变换(开关变换)
swivel 旋轴;摇臂,摆动杆
swivel arm 旋转臂
swivel bearing 旋转轴承
swivel block 转动滑块
swivel damper 旋转挡板
swivelling device 摆动装置(红外线预烘用)
swivelling movement 旋转运动,摆动
swivelling pressure cloth beam 旋转卷布压辊
swivel pin 转销,回转销子
swivel support 铰接支座
swivel tension device 紧布架
swollen 溶胀的
swollen state 膨润状态
sycamore roll 枫木滚筒
SYE(square yards equivalent) 平方码当量
SYM(=sym.,symmetrical) 对称的
SYMATEX【法】比利时纺织机械制造厂商联合会
symbol 符号,记号,代号
symbolic assembly language〈SAL〉符号汇编语言
symbolic colours 象征色
symbolic notation 符号表示法
symmetrical〈SYM,sym.〉对称的
symmetrical design 对称设计,对称花纹
symmetrical linear structure 对称线型结构
symmetrical pattern 对称花样
symmetry and equilibrium 均齐与平衡(图案构成的基本形式)
symmetry axis 对称轴
Sympatex 新保适(防水防风透气面料,商名,德国制)
syn- (构词成分)表示"共""同"与""同时"
syn(synchronization) 同步;同期
syn(synchronous) 同步的
syn(=sync.,synchronizing) 同步;同步的
syn.(=synth.,synthetic) 合成的
Synacril dyes 西纳克里染料(染聚丙烯腈系纤维用阳离子染料,商名,英国制)
sync.(synchronizing) 同步;同步的
sync.(synchronizing signal) 同步信号
syncancerogenesis 综合致癌作用
synchro 自动同步机,同步器;同步传送;同步的
synchroindented belt 齿形带,同步传动带
synchromesh gear 同步[啮合]齿轮
synchron dyeing control〈SDC〉同步染色控制
Synchron Dyeing System〈SDS〉同步染色系统(商名,德国第斯)
synchronism 同步[性]《电》
synchronization〈syn〉同步;同期
synchronizing〈syn,sync.〉同步;同步的
synchronizing effect 同步染色效应
synchronizing signal〈sync.〉同步信号
synchronous〈syn〉同步的
synchronous-asynchronous motor 同步—异步电动机
synchronous belt 齿形带,同步传动带
synchronous generator 同步发电机
synchronous running 同步运行
synchronous transmission 同步传动

SYNDET (= syndet, syn. det., synthetic detergent) 合成洗涤剂,合成净洗剂

syn-diazosulphonate 顺重氮磺酸盐(活泼的重氮磺酸盐,能与萘酚偶合)

syneresis [胶体]脱水收缩[作用]

synergism 增效作用,协同作用,协合作用

synergist (= synergistic) 增效剂,协合剂,合作剂

synergistic action 增效作用,协合作用

synergistic antioxidant 多效防老化剂

synergistic catalyst 增效催化剂,协合催化剂

synergistic compound 增效复配

synergistic effect 增效效应,协同效应,协合效应

Synferol AH 辛斐罗 AH(渗透剂,匀染剂,耐硬水、酸及弱碱,商名)

synfuel 合成燃料

syngenesis 同生,共生

synopsis(synop.) 对照表,一览表;提要,概要

syntan 合成单宁

Syntapal B 辛塔派尔 B(耐酸、碱及硬水的洗涤剂,商名)

Syntapon CP 辛塔邦 CP(渗透剂,分散机,匀染剂,洗涤剂,商名)

synth. (= syn, synthetic) 合成的

Synthappret A 辛塞普利 A(二氧化硅溶胶,阴离子型,呈弱碱性,适用于各种纤维的防滑剂,商名,德国拜耳)

Synthappret BAP 辛塞普利 BAP(羊毛防毡缩剂,阴离子型,聚异氰酸聚氨基甲酸乙酯的重亚硫酸盐加成物,对羊毛及其混纺织物无需经氯化处理,还有减少起毛起球、抗磨的作用,商名,德国拜耳)

Synthappret LKF 辛塞普利 LKF(含 80％ 聚氨基甲酸乙酯预聚物,在醋酸乙酯溶剂中具有异氰酸酯游离基团,为羊毛防缩防皱剂,并可防起球、防勾丝,商名,德国拜耳)

synthesis 合成[法],综合[法]

synthesis gas 合成气

synthetic 〈syn., synth.〉合成的

synthetic bar 合成固体洗涤剂

synthetic caoutchouc 合成橡胶

synthetic detergent process 合成洗涤剂法

synthetic detergent 〈SYNDET, syndet, syn. det.〉合成洗涤剂,合成净洗剂

synthetic dyes (= synthetic dyestuffs) 合成染料

synthetic elastomer 合成弹性体

synthetic fabrics 合成纤维织物

synthetic fibres 合成纤维

synthetic fuel 合成燃料

synthetic grease 合成润滑脂

synthetic leather 合成革

synthetic leather carrier webs 合成革基布

synthetic natural gas 〈SNG〉合成天然气

synthetic nature dye 合成的天然染料

synthetic pigment 合成颜料

synthetic polymer 合成聚合物

synthetic proteins 合成蛋白质

synthetic pyrethroids 拟除虫菊酯

synthetic reaction 合成反应

synthetic resin 合成树脂

synthetic rubber 合成橡胶

synthetics 化学合成物;合成纤维织物

synthetic segmented copolymer 合成嵌段共聚物

synthetic size 合成浆料

synthetic suede 人造麂皮

synthetic system 综合系统
synthetic textiles 合成纤维纺织品
synthetic thickening 合成增稠剂
synthetic tip 合成尖头（刮刀尖端的材料）
synthetic tragacanth paste 合成龙胶糊
synthetic wax 合成蜡
synthol 合成醇
synthon 合成纤维
Synthrapol RWP 辛塔拉波 RWP（环氧乙烷缩合物，低泡沫、非离子型表面活性剂，羊毛净洗剂，商名，英国制）
syn-type〈s.〉顺式，顺[基]型
syphon 虹吸；虹吸管
syphon pipe type cylinder 虹吸式烘筒（采用虹吸管排除烘筒内的冷凝水）
syringe 注射器

SYS（=sys.，syst.，system）系统；装置制度；方法
systematic analysis 系统分析
systematic error 系统误差
systematic random sample 规律性随机样本
systematic sampling 规律性抽样法
system check 系统检验
system engineering 系统工程
system maker 整机制造厂
system of classification 分类法，分级法
system pressure 系统压力
system（SYS，sys.，syst.）系统；装置；制度；方法
Sytron A 锡特隆 A（金属螯合剂，商名，德国制）

T

t. (temperature) 温度
t. (＝thk., thickness) 厚度；稠度，浓度
t. (torque) 转矩，扭矩；捻矩，扭转
T(twist) 捻度，捻回，加捻；经纱
T_m(mean temperature) 平均温度
T_m(melting temperature) 熔融温度，熔点
TA (time of arrival) 运达时间
tab 标签
tab. (table) 表格
Tabacut II finishing agent 一种去除香烟味的整理剂（商名，日本制）
tabby 平纹绸
tabbying 波纹轧光整理
table〈tab.〉 表格
table adhesives 台板胶
table balance 台秤，托盘天平
table linen 亚麻台布
table printing 台板［绢网］印花
table raising machine 台板式起绒机
tablet 标牌；小块，小片；图形输入板《计算机》
tablet compressing machine 压片机
tablet test ［地毯］小样试验
tablet test at ambient temperature ［地毯］小样燃烧试验
table-vice 台虎钳
table washing device in flat screen printing 平网印花台板洗涤设备
tabs （一米以下的）机头布，短码布，零布
tabular lamp 扁平灯
tabular language 表列语言（计算机软件）

TAC (total air cleaning) 空气总净化
tachodynamo 示速电机
tacho-generator 测速发电机
tachograph 速度记录器，转速表
tachometer 转数计，转速计；流量计
tachometer generator 测速发电机
tachoscope 转速表；手提转速计，手提转数表
tack 图钉；黏着性
tackability 黏着能力；增黏能力
tack-free 不黏手
tackifier 胶黏剂；增黏剂
tackifying ability 黏着能力；增黏能力
tack index 黏性指标
tackiness 胶黏性，黏着性
tackiness agent 黏合剂
tacking ［呢绒整理］缝［边成］筒
tacking adhesion 黏着
tack inhibitor 防黏剂
tackle 滑车，起绒装置
tackoscope 黏度计，测黏仪
tack retention 黏着性保持率；黏着性保持时间
tack-stitching 粗缝，假缝，加固缝
tack strength 胶黏力
tack temperature 发黏温度
tacky 发黏的
tacky fibre 黏结纤维
tactile 触觉的
tactile appraisal 手感评定
tactility 触感

tactometer 触觉测验器
TAED (tetracetylethylene diamine) 四乙酰乙烯二胺（双氧水低温漂白活化剂）
taffeta 塔夫绸
taffeta flannel 轻质条格法兰绒
taffeta fleure 彩色小花卉纹塔夫绸
taffetaline 绢丝塔夫绸
taffetized fabric 仿塔夫绸棉织物
taffety 塔夫绸
tag 标签；[金属的]箍；终端；电缆接头；[计算机]辅助信息
tag board 接线板
Tag closed-cup tester 泰格闭杯试验器
Tag closed tester 泰格密闭闪点试验器
tag cloth 标签布
tagged atom 标记原子，示踪原子
tagged compound 标记化合物
Taglibue closed tester 泰格密闭闪点试验器
Taglibue hydrometer 泰格比重计
Taglibue viscosimeter 泰格黏度计
Tag open-cup tester 泰格开杯试验器
Tag-Robinson colourimeter 泰格—罗宾逊比色计
TAI (Textile Association India) 印度纺织协会
TAI-ash (total alkail-insoluble ash) 非碱溶性灰分总量，非碱溶性物质总量
tail end piece 导布
tailing [染色]头梢色差（染疵）；印花渗色（印疵）
tailing and listing 头梢色差和边深浅
tailing down 花纹拖尾，花纹拖浆（机印疵）
tailing-in 铜刀拖色，小刀铲进（机印疵）
tailings 短码布（美国商业用语）

tailings screen 尾浆筛
tailored fibre 特制纤维（按预定指标设计或制备的化学纤维）
tailoring process 成衣加工
tailoring technology 成衣加工技术
tailor-made fibre 特制纤维（按预定指标设计或制备的化学纤维）
tailor-tacking 粗缝，绗缝
tail pipe [排气]尾管，[泵]吸管
tailstock [机器的]车尾；尾座
takacidin 高杀菌素
TAK dyeing 塔克地毯染色法（将地毯润湿后，利用刮刀将染液引上地毯表面染色）
take-off device 退绕装置；取出机构
take-off roller 送出罗拉，出布辊
take-up 卷取；轧液率，带液率；（缝纫机上提线的）提升装置
take-up after squeezing 轧液率
take-up mechanism 卷取装置
take-up motion 卷取装置；卷取运动
TAK process 多色流染印法（用于地毯染色，商名，德国开发）
talc powder (= talcum powder) 滑石粉
tall oil 妥尔油
tallow [动物]脂，牛油
tallow sulphonates 牛脂磺酸酯
tally 计数；运算；标签
TA-Luft 德国清洁空气条例，德国大气污染防治的技术指导
tamarind gum 罗望子胶，印花糊料
tambour drying machine [夹层]大烘筒烘燥机
tamponing 擦除灰伤（指丝织物用石蜡油等去除擦伤痕）
tamponing machine 擦除灰伤机《丝》

tamtine 绢丝塔夫绸
tan 鞣料;鞣;[鞣革]棕黄色
Tanapal LD-3 泰纳派尔 LD-3(分散染料匀染剂、修色剂,脂肪醇乙氧基化合物及聚丙烯酸酯,非离子型、阴离子型,用于涤纶、锦纶、涤棉织物的匀染,也可用于修色,商名,美国先邦)
Tanaspan ML 泰纳斯潘 ML(匀染剂,阳离子型,在锦纶和氨纶的交织物上能帮助氨纶上色,使织物颜色均匀,色光一致,商名,美国先邦)
Tanasperse MDA 泰纳斯潘斯 MDA(化学退浆助剂,表面活性剂复配物,具有良好缓冲性能的分散剂,用于去除织物上聚酯或丙烯酸衍生物浆料,商名,美国先邦)
Tanaterge REX 泰纳特奇 REX(皂洗剂,矿物质提取物,用于去除水解活性染料,改善色牢度,在硬水和电解质溶液中性能良好,商名,美国先邦)
tandematic decurler 串联式剥边器
tandem fulling machine 双辊缩绒机
tandem jigger 串联卷染机,成对卷染机(俗称上下缸)
tandem position 串联配置,前后排列
tandem rollers 成对轧辊(涂层机)
tandem screen 双片滤网(筛网),串联式滤网(烘房空气通道的过滤网)
tangent 正切;切线《数》
tangential acceleration 切向加速度
tangential displacement 切向位移
tangential firing 切向燃烧
tangential force 切向力
tangential singeing 切向烧毛,弱烧毛
tangential stress 切向应力,切变应力
tangential washing machine 卧式水洗机

tangerine 橘红[色]
tangle 缠结
tangle tentacle 缠结传感器(能探测绳状织物在染色过程中是否打结)
tangling detector 缠车警报装置,缠车检测装置
tangling yarn 交络丝,网络丝
tank 箱,柜;桶;槽;池,罐;库
tankage 储量
tankcar 活动槽车
tank reactor 罐式反应器,釜式反应器
tannase 单宁酸酶,鞣酸酶
Tannex GEO 泰乃克斯 GEO(前处理剂,二性化合物,脂肪醇乙氧基化合物及矿物质,属环保型助剂,低泡双氧水稳定剂,用于棉及混纺织物的间歇漂白,有去除蜡质和油污作用,商名,美国先邦)
Tannex RENA 泰乃克斯 RENA(双氧水稳定剂,无机盐及有机磷化合物,环保型助剂,适用于各种双氧水漂白工艺,商名,美国先邦)
tannic acid 单宁酸,鞣酸
tannin 单宁,二棓酸,单宁酸,鞣酸
tannin boiling 单宁沸煮法(生丝经单宁液沸煮后,可获得较好的光泽和手感)
tannin discharge printing 单宁拔染印花
tannin discharge style 单宁拔染印花法
tannin extract 单宁萃,单宁浸取物
tanning 制革,鞣革
tannin-mordant 单宁媒染剂
tannin-tartar emetic 单宁—吐酒石[媒染剂]
tannin weighting 单宁增重
tap 螺丝攻,丝锥;抽头,分接头《电》
tape 狭幅织物;带;磁带,纸带;带材

tape-controlled electronic programmer 穿孔带控制的电子程序器
tape dyeing and finishing range 带子染整机组
tape-like fibre 带状纤维
tape measure 皮带尺
taperecorder 磁带录音机
tapered selvedge 收缩布边
tapered skirt 开衩裙
tapering 锥度；圆锥形的；圆锥体
taper key 斜键
taper line gratings 锥形线[经纬]密度镜，闪光[经纬]密度仪
taper pin 锥形销[钉]，斜销[钉]
taper reamer 锥形绞刀
taper roll bearing 锥形滚子轴承
tape scouring machine 带式精练机（用于纱线）
tape seam 贴带缝
tape sizing 轴经上浆
tapestry 氆氇（藏族人民手工纺织的毛织物）；花毯；挂毯；提花装饰用毯；像景织物；绒绣
tapestry cloth 挂毯绘画亚麻织物
tapestry painting 挂毯绘画织物
tapestry quilts 织花床单布
tapestry satin 织锦缎
tap funnel 滴液漏斗
tapioca 木薯淀粉
tapis 挂毯，桌毯，地毯
tapissendis 双面印花织物（印度）
tappet 凸盘；挺杆；阀门
tappet clearance 挺杆间隙（阀杆与阀座间的间隙）
tapping 攻螺丝；钻孔；雕刻；雕刻的花纹
tapping and threading 攻绞螺纹

tapping screw 自攻螺钉
tap water 饮用水，自来水
tarare 麻帆布
tare 包装箱；袋皮；皮重
tare weight 袋皮重，皮重
target 目标，指标
target computer 特定程序计算机，结果程序计算机
target language 目标语言《计算机》
target program 目标程序《计算机》
tarnishing 去光泽，[印花]白地沾色（花筒留白部分刮浆不清造成）
tarnishing film 失光泽膜
tarnish inhibitor 晦暗抑制剂
tarnish prevention 防[金属]变色
tarnish resistant flannel 防[银器]变色法兰绒（能吸收硫黄气体，用以包装银器或作银制箱柜的衬里）
tarp(＝tarpaulin) [柏油]帆布，[防水]油布，漆布；[油布]防水衣或防雨帽
tarpaulin coating 防水油布涂层
tar products 焦油产品
tarragon 龙蒿绿[色]（浅暗黄绿色）
tarred canvas 沥青帆布，焦油帆布，防雨帆布
tarring 涂煤焦油；染斑（疵点）
tar spots 柏油污点
tar stain 柏油沾污
tar stains 柏油污渍
tartar 酒石，酒石酸氢钾
tartar emetic 吐酒石，半水合酒石酸氧锑钾
tartaric acid 酒石酸
tartrazine 黄色食用色素
tasarmuga 淡褐色柞蚕丝
task control block 任务控制模块《计算

task dispatcher 任务调度程序《计算机》
task management 任务管理(计算机)
task queue 任务排队(计算机)
tatters 零头布;碎呢
tatty-look 仿旧风格
taupe 豆沙色,灰褐色
taurine 牛磺酸,氨基乙磺酸
taut 拉紧的;整齐的
tautomer(=tautomeride) 互变[异构]体
tautomerism 互变[异构]现象
tautomerization [结构]互变[作用]
tautomerize 互变[异构]
taut pick 紧纬
tawny 黄褐色
taxis 构型规正性;向性,趋性(词尾);排列,次序
TBL (Tebilized) 特比莱防皱整理的
T-blouse 针织圆领衫
TBP(tributyl phosphate) 磷酸三丁酯(消泡剂、抗菌剂)
TBP (true boiling point) 真沸点
TBPA (tetrabromo phthalic anhydride) 四溴邻苯二甲酸酐(阻燃剂)
TBT(tributyltin) 三丁基锡(禁用防腐杀虫剂)
Tc(corrected temperature) 校正温度
Tc(critical temperature) 临界温度
TC(=T. C., tc, t. c., temperature coefficient) 温度系数
TC(=T-controller, temperature controller) 温度控制器
TC(time closing) 延时闭合
TC(total carbon) 总含碳量
TC(transmission control) 传送控制
TCA (Textile Council of Australia) 澳大利亚纺织协会
TCEP [tris (2-carboxyl ethyl) phosphine] 三羧甲基磷酸(羊毛低温染色助剂)
TCM (Traditional Chinese Medicine) 中医学
T-connection T形接线
T-controller (=TC, temperature controller) 温度控制器
TD. (theoretically dry) 绝对干燥
Td. (Titer denier) 旦[尼尔]制纤度
TDA (Textile Distributors Association) 纺织批发商协会(美国)
TDC (Teleprocessing Design Centre) 远程处理设计中心(美国)
TDEA 一种非醇类环氧整理剂(棉、丝绸织物经该类整理剂处理后能提高上色率及干湿弹性)
TDI (toluene diisocyanate) 甲苯二异氰酸酯
TDR(test data report) 试验数据报告
TDR(time delay relay) 延时继电器
TDS (technical data system) 技术数据系统
TDS (total dissolved solids) 总溶解固体[量]
te(totally enclosed) 全封闭的
te(tractive effort) 牵引力
TEA (triethanolamine) 三羟基乙基胺,三乙醇胺(俗名)
tea cloth 茶巾;台布
tea green 茶绿[色](淡暗绿色)
teak [brown] 柚木棕[色](棕色)
teal 凫色(绿光暗蓝色);蟹青[色]
teal blue 凫蓝[色](绿光暗蓝色)
teal duck 水鸭色(绿光暗蓝色)
teams and groups 班组

teams and groups management 班组管理
tear 撕破
teardrop(=teariness) 纬斜,纬纱变位
tearing foil 防爆膜
tearing goods 棉织物与亚麻织物(英国外销到非洲的这类织物的统称)
tearing-out 拔出(指绒头)
tearing strength 撕破强力
tearing test 撕破强力试验
tea rose 茶玫红[色](淡橙红色)
tear resistance 抗撕裂性
tease 起绒,拉绒,起毛
teasel 起绒刺果;刺果起绒机
teasel cloth 起绒织物
teaseling 刺果起绒,刺果起毛;拉绒,起毛
teaseling machine 起绒机
teasel napping machine 刺果起绒机
teasel raising 刺果起绒
teasel raising machine 刺果起绒机
teasing 起绒,拉绒,起毛[工艺]
teazel 起绒刺果;刺果起绒机
teazel finish 起绒整理
teazeling 刺果起绒,刺果起毛;拉绒,起毛
teazle 起绒刺果;刺果起绒机
teazle gig 刺果起毛,刺果起绒
Tebexel process 泰贝凯斯尔树脂整理法(用三硫酸根络乙基锍二钠内盐为交联剂,商名)
Tebilized〈TBL〉 特比莱防皱整理的
Tebolan MDF 塔博伦 MDF(扩散剂,非硅消泡剂,酯类并加乳化剂,商名,德国波美)
Teb-X-Cell finish 泰贝凯锡尔整理(棉织物上多步交联工艺)
Teca 特卡(醋酯短纤维,商名,美国制)
tech.(technical) 技术的

tech.(technics) 技术
techn.(technology) 工艺学;工艺规程;工程技术
technical〈tech.〉 技术的
technical analysis 工业分析
technical atmosphere 工程大气压
technical cloth 工业用布
technical clothing 技术服装
technical data system〈TDS〉 技术数据系统
technical equipment 技术设备
technical expenditure 工程费用;技术费用
technical fabrics 技术纺织品
technical fibre 产业用纤维;工艺[型]纤维
technical file 技术档案
technical gelatine 工业明胶
Technical Information and Documents Unit〈TIDU〉 科技情报文献处(英国)
Technical Information Bureau〈TIB〉 技术情报局
technical innovation 技术革新
technical know-how 专门技术;专有技术
technical licence 技术执照,技术许可证
technical manual〈TM〉 技术指南,技术手册
technical norms 技术标准,技术规范
technical order〈T.O.〉 技术规程,技术命令
technical plastics 工程塑料
technical pure 工业纯《化》
technical regulation〈TR〉 技术规范
technical report〈TR〉 技术报告
technical safety measures 安全技术措施
technical scale 工业规模
technical schedule 工艺规程

technical specification 技术规格;技术说明

technical terminology 技术用语,技术术语

technical textiles 技术纺织品

technician 技术人员;技师

technicolour 彩色印片

technico-scientific periodicals 科技期刊

technics〈tech.〉技术

technique 技术;工程

technique of conveying image 传像技术,导像技术

technological innovation 技术革新

technological process 工艺流程

technological properties 工艺性能

technological transformation 技术改造

technology〈techn.〉工艺学;工艺规程;工程技术

technology transaction 技术转让

Technora 一种多元共聚芳纶(商名,日本帝人)

Tectilon dyes 特克蒂隆染料(染聚酰胺纤维的弱酸性染料,商名,瑞士制)

tecum(= tecuma, tecun) 特肯纤维(棕树叶纤维,制渔网及钓鱼线用,巴西产)

teddy bear cloth 长绒毛织物(羊毛及马海毛制)

Tedecco process 泰代科[液氨丝光]法(商名,挪威)

tee 三通;T形物

tee pipe 三通管,T形管

Teepol 替波尔,页油皂(烯烃混合物,自裂化石蜡经硫酸化而成的合成洗涤剂,商名,日本制)

teeth per inch〈T. P. I.〉每英寸齿数

Teflon 特氟纶(聚四氟乙烯长丝或短纤维,商名,美国制)

Teflon entry and exit seals 聚四氟乙烯进出口封口(用于高温高压的连续染整设备,Teflon 为美国杜邦公司在氟聚合物产品上的商标)

teflonized roller 涂氟滚筒

TEFO【瑞典】纺织研究所

Tegewa desizing scale 退浆效果等级标准(Tegewa 系德国纺织化学品协会名)

Tegewa violet scale 测定退浆率的紫色评级卡

Teijin 帝人(日本商名,亦指该厂所制黏胶长丝)

Teijin-Tetoron [帝人]帝特龙(聚酯纤维,商名,日本制)

tela 蒂拉布(即亚麻布,地中海各国名称)

Telasol 特拉佐(醋酯纤维原液着色用溶剂染料,商名,瑞士制)

tele- (构词成分)表示"远""远距离""电视""电信"

telecommunication 远程通信,电信

telecontrol 遥控,远距离控制

teleindicator 远距离指示器

telemeter〈TLM〉遥测仪,遥测装置

telemetry〈TLM〉遥测技术

telemotor 遥控电动机

telephoto 远距离照相

teleprinter 电传打印机

teleprocessing 远程信息处理

Teleprocessing Design Centre〈TDC〉远程处理设计中心(美国)

teleregulator 远距离调节器

telescoped 伸缩的;[布卷]内层伸出

telescopic baffle 伸缩挡板

telescopic lap splice 套接接头

telescopic tube 伸缩套管

telescoping of processes 工序连续化
teleservice 远程服务
teleservice facility 远程服务设施
telethermometer 遥测温度计
teletron 显像管,电视接收管
teletype(= teletypewriter)〈TTY〉 电传打字[电报]机
television observation system 电视监察系统
television screen 电视荧光屏,电视屏幕
tell-tale 寄存器;计数器;信号装置
telluric treatment 碲处理(用于无污染绿色再生纤维 Tencel 的处理,以获得优良外观和手感)
telogen 调聚剂
telomer 调聚物
telomeric reaction 调[节]聚[合]反应
telomerization 调[节]聚[合]作用(化学高分子制造工艺,用于拒油剂有机氟的制造)
telon 毛麻呢(麻经毛纬粗呢,法国制)
Telon 特纶(聚酯短纤维,商名,美国制)
Telon dyes 天龙染料(聚酰胺纤维及羊毛用的阴离子染料,多数为弱酸性染料,商名,德国德司达)
TEM (transmission electron microscopy) 透射电子显微术
temperature〈TEMP, temp., t.〉温度
θ-temperature θ温度,弗洛里温度
temperature-adaptable fibre and fabrics 调温纤维及纺织品
temperature-adaptable hollow fibres 适温中空纤维
temperature adjustment 温度调节
temperature booster 加热器
temperature buffer 温度缓冲器
temperature coefficient〈TC, T.C., tc, t.c.〉温度系数
temperature control 温度控制;温控器
temperature controllable dye 可控制温度染料
temperature controller〈T-controller, TC〉温度控制器
temperature control valve 温度控制阀
temperature conversion table 温度换算表
temperature correction 温度纠正
temperature dependency 温度依存性
temperature dew-point 露点温度
temperature difference 温差
temperature distribution 温度分布
temperature drop 温度下降,降温
temperature effect 温度效应
temperature feeler 温度探头
temperature fluctuation 温度波动
temperature gradient 温度梯度
temperature-gradient method 分段升温[染色]法
temperature humidity index 温湿度指数
temperature indicating controller〈TIC〉温度指示控制器
temperature indicating paper 测温纸
temperature-indicating pigment 示温颜料
temperature indicating strips 温度指示条(纺织品表面温度测定)
temperature indicator 温度指示器
temperature initial 初洗温度,预浸温度,起始温度
temperature measuring station 温度测位
temperature offset 温度补偿
temperature of initial combustion〈TIC〉开始燃烧问题
temperature of saturation〈tsat〉饱和温度

temperature pick-up 温度传感器
temperature probe 温度传感器,温度探头
temperature profile 温度分布图,温度变化图
temperature programmer 程序升温器
temperature raising 升温
temperature-range properties 温度变化上染性能
temperature recorder〈TR〉 温度记录器
temperature regulated iron 调温熨斗
temperature regulating 调温,温度调节
temperature regulating effect 温度调节效应
temperature regulating fabrics 调温织物
temperature regulating factor〈TRF〉 温度调节因子
temperature regulating function 调温功能
temperature regulating microcapsule 调温微胶囊(用于制造调温纺织品的新技术)
temperature regulator 温度调节器
temperature relay 温度继电器(环境温度变化到预定值时动作的继电器)
temperature scale 温[度]标,温度计刻度
temperature-sensing element 感温元件
temperature sensing probe 温度敏感探针
temperature-sensitive polyurethane〈TS-PU〉 温度感应型聚氨酯
temperature sensitivity 温度敏感性
temperature-time program 温度—时间程序
temperature within clothing 衣服内温度
tempered〈TP,T.P.〉 已回火的;已松开的(指螺钉)
tempering 回火
tempering coil 调温盘管

template 样板,模板
temple 边撑,伸幅器
temporary adhesion 暂时黏着
temporary cross-link 非永久交链
temporary hardness 暂时硬度
temporary hard water 暂时硬水
temporary set [纺织品的]暂时定形;瞬时形变;[高]弹性形变
temporary storage 暂存;暂存存储器
temporary water soluble disperse dyes 水暂溶性分散染料
temporary weaving 假织(供经纱印花)
tenacity 韧度;韧性;强度(单位细度的强力)
Tencel fibre 天丝纤维(无污染绿色再生纤维,商名,英国考陶尔)
Tencel/ramie blended fabrics 天丝苎麻混纺织物
tendel 轧光靛染薄布
tendency 趋势,倾向
tendency to soil 沾污趋向
tender 挡车工,值车工;脆弱;脆化
tender goods 发脆织物
tendering [纤维或织物]脆化,脆损
tendering by carbonization 炭化损伤
tendering of sulfur black 硫化黑[染色织物]的脆化
tendering proof of sulphur colours 硫化染料防脆处理
tendering proof sulphur black 防脆硫化黑
tender lustre 柔和光泽
Ten-o-film 乙氧基化与丙氧基化淀粉(商名)
tenside 表面活性剂
tensile 张力的;抗拉的;可伸展的
tensile brittleness 抗伸

tensile buckling 拉伸变形
tensile elasticity 抗拉伸弹性
tensile elongation 抗拉伸伸长
tensile energy 抗拉能
tensile figure 拉伸指数
tensile force 张力;拉力
tensile hysteresis curve 拉伸滞后曲线
tensile modulus 抗拉模量,抗张模量
tensile product 抗拉积,抗张积
tensile property 拉伸性质
tensile recovery 拉伸回复力
tensile stiffness 拉伸刚性
tensile strain 张力应变
tensile strain recovery 应力恢复
tensile strength 拉伸强力;断裂强力
tensile stress 拉伸应力
tensile stress relaxation 拉伸应力松弛
tensile test 拉伸试验,张力试验
tensile yield point 拉伸屈服点
tensilgraph [纺织材料的]拉伸强力测定器
tensioactive 表面活性的
tensiometer 断裂强度试验机,强力试验机;张力计,张力仪
tensiometric titration 表面张力滴定法
tension 张力,拉力
tension analyzer [动态]张力测定器(用以测定运动中的纱线张力)
tension bar 张力杆;纬挡(织疵)
tension block 张力轮
tension bracket 张力架
tension compensation regulation 张力补偿调节
tension compensator 张力补偿器
tension control 张力控制
tension control dryer 松式烘干机

tension controller 张力控制装置
tension control open-width washer 松式平洗机
tension control rope washer 松式绳洗机
tension cylinder 绷布辊,张力辊
tension device 张力装置
tension disc [摩擦式]张力调节盘;[缝纫机的]夹线板
tensioner 张紧器,紧布器
tensioner brake 张力制动装置
tension feeler 张力探测器
tension-free 无张力的
tension-free state 无张力状态
tension indicator 张力指示仪
tensioning 张紧,绷紧
tensioning block 张力块
tensioning carriage 张力架
tensioning catch 张力扣(拉紧圆网用)
tensioning roller 张力辊,松紧调节辊
tensionless batching machine 无张力卷布机
tensionless calender 无张力轧光机(圆筒针织品用)
tensionless conveyer 无张力输送带,无张力导带
tensionless dryer 松式烘燥机
tensionless dye jigger 无张力卷染机
tensionless framing and steaming machine 无张力拉幅汽蒸机
tensionless handling 无张力操作
tensionless jet drier 无张力热风烘燥机
tensionless open soaper 松式平幅皂洗机
tensionless open-width washing machine 松式平幅水洗机
tensionless setting 无张力定形,松弛定形
tensionless synchronizing drive 松式同步传

动

tensionless washing machine 无张力水洗机

tension measuring device 张力测定仪

tension mercerization 张力丝光

tension meter 张力仪

tension presser 张力压烫机(用以测试织物形态稳定性)

tension pulley 张力盘

tension rail 张力杆

tension regulating roller 张力调节辊

tension regulation 张力控制

tension regulator 张力调整器

tension releasing disc [缝纫机的]松线板

tension roll[er] 张力辊

tension screw seat [缝纫机的]夹线螺钉座

tension screw stud [缝纫机的]夹线螺钉

tension sensitive fabric 张力敏感织物

tension sensitive textiles 张力敏感纺织品

tension set 张力定形;永久变形

tension side [机器]受张力侧

tension slide 张力滑板

tension spring 拉簧;张力弹簧;[缝纫机的]夹线簧

tension structure 紧式结构

tension variator 张力调节器

tension washer 张力垫圈

tension washing machine 紧式水洗机(如普通的直、横导辊式平洗机)

tension weight 张力重锤

tensometer 断裂强度试验机,张力试验机;张力计,张力仪

Tensomodul 动态弹性模数测试仪(商名,瑞士制)

tent 帐幕,帐篷

tentage 帐篷,遮阳篷布;宿营装备

tentative 试验性质的,试用的;试行的,暂行的

tentative experiment 探索性实验

tentative specifications 暂定[技术]规格

tentative standard 暂行标准,试行标准

tentative test 暂行试验法

tent duck 帐篷帆布

tenter [frame] 拉幅机

tenter clip 拉幅布铗

tenter control 拉幅机控制设备

tenter drying 拉幅干燥

tenterette 小型布铗拉幅机

tentering 拉幅

tentering chain 拉幅链条

tentering machine 拉幅机

tentering of weft 纬斜矫正,整纬

tentering roller 扩幅辊

tenter oven 拉幅机烘房

tenter pin 拉幅针板

tenter rail 拉幅导轨

tent fabric 帐篷织物(如帆布等防雨及遮盖用织物)

tenting 帐篷布;帆布

tenuity [空气、流体等的]稀薄度

ter- (构词成分)表示"三""三重""三倍"

Teracoton dyes 特拉科通染料(染涤棉混纺的分散还原混合染料,商名,瑞士制)

Teracron dyes 特拉克隆染料(染涤棉混纺的分散活性混合染料,商名,瑞士制)

Teralan S dyes 特拉兰 S 染料(染涤毛混纺的分散酸性混合染料,商名,瑞士制)

Terasil DI、TI dyes 托拉司 DI、TI 染料(喷墨印花用分散染料印墨,DI 用于直接印花,TI 用于转移印花,商名,瑞士制)

Terasil W dyes 托拉司 W 染料（分散染料，用于涤纶及其混纺织物染色，具有优良的耐洗牢度，商名，瑞士制）

Terasil WW 托拉司 WW 型耐热迁移性分散染料（商名，瑞士制）

teremp 马海长毛绒织物

terephthalic acid bisglycol ester 对苯二甲酸双乙二[醇]酯

terephthalic acid〈TPA〉 对苯二[甲]酸

Tergitol EH 特吉妥尔 EH（乙基己烯磺酸钠，表面活性剂，商名，美国制）

Terindosol dyes 特林多素染料（分散可溶性还原混合染料，商名，瑞士制）

Termanyl 一种高温型淀粉酶（在 85～115℃分解淀粉类浆料，适用于 J 形箱及轧蒸法退浆，商名，丹麦诺和诺德）

terminal 端；接头，接线柱

terminal amino group 末端氨基

terminal block（=**terminal board**）接线盒，分电器接线板

terminal bond 端键

terminal box 接线盒

terminal chain 端链

terminal cover 端盖

terminal diagram 端子图《电》（在生产和维修中用于指导元器件和接插件等的焊接装配）

terminal drying 末期干燥

terminal effect 端基效应

terminal group 端基

terminal hydrophilic group 亲水端基

terminal hydroxy group 羟端基

terminal point 终点，连接点（管道与设备的接点）

terminal potential 终端电位

terminal residue 端余基

terminal temperature 终端温度，最终温度

terminal velocity〈TV〉 终端速度，末速[度]

termination 终止[作用]；终端

termination reaction [链]终止反应

terminator [链增长]终止剂

terminology 术语，专门词汇

Terminox 一种过氧化氢酶（用于氧漂后水洗，可以分解双氧水，简化工序，节约用水，商名，丹麦诺和诺德）

termite-resistant finish 防白蚁整理

terms 术语；条件

ternary 三元；三进制的

ternary acid 三元酸

ternary colour 三元混合色

ternary dye 三元混合染料

ternary mixture 三元混合物，三组分混合物

terpene 萜烯

terpentina 高沸点汽油

terpineol 萜品醇，松油醇

terpolyamides 三元共聚酰胺

terpolymer finish 三元共聚物整理（羊毛织物的定形整理）

terpolymer silicone oil 三元共聚嵌段硅油

terra alba 石膏粉

terra cariosa 钙质细土

terraced 台阶形的

terra cotta 赤土色（赤褐色）

terry 起毛毛圈；毛圈织物；毛巾织物

terry astrakhan 充羔皮[毛圈]织物

terry cloth（= **terry fabric**）毛巾布，毛圈织物

terry-loop goods 毛圈织物

terry pile structure 毛巾组织，毛圈组织；圈毛绒头，未割毛绒头

terry towel 毛巾织物
Tersetile dyes 泰尔塞蒂尔染料(分散染料,商名,意大利制)
tertiary 第三的,第三级的;叔《化》
tertiary amine surfactants 叔胺表面活性剂
tertiary amino alkyl ethers of starch 叔氨基烷基淀粉醚
tertiary amino and sulfonic acid modified starch 叔胺及磺酸变性淀粉
tertiary colours 三拼色,三次色
tertiary treatment 三级处理
tertiary treatment of waster water 废水三级处理
Terylene 一种涤纶(聚对苯二甲酸乙二酯长丝和短纤维,简称聚酯纤维,商名,英国制)
Teryon Super dyes 泰立翁超级染料(分散还原混合染料,商名,日本制)
TES(thermal energy storage) 热能储藏器
tesla 特斯拉(磁通密度单位)
test 试验,检验,测试
testability 可试性
test bench 试验台
test bias 试验偏倚,试验恒定误差
test board 试验盘,试验板,试验台
test bundle 试样
test control 对比用试验标准;[对照]标样
test control specimen 对照标样
test data 检查数据
test data report〈TDR〉试验数据报告
test dyeing 试样染色,染色打样
tester 测试仪,试验器
test error 试验误差
Testex (Swiss Textile-Testing Co. Ltd.) 瑞士纺织品鉴定有限公司
test for crack initiation 龟裂引发试验
testing 试验
testing apparatus 试验装置
testing for stiffness 硬度试验,刚度试验
testing for tension 张力试验
testing items 检验项目
testing method of thickening agent for printing 印花用糊料试验法
testing of colour fastness against washing 染料耐洗牢度试验
test method of dye concentration 染料浓度试验法
test method of dyeing property 染色性能试验法
test method under examination〈TME〉试用的测试方法
test norm 试验标准
test of significance 显著性检验
test OK 正常,无故障
test period 试验时间
test piece 试样;试件;试料
test pressure〈tp〉试验压力
test rig 试验台;试验用具
test run 试车;试验运行;实验程序
test set 试验装置,测试仪器
test specimen 试样
test tube centrifuge 试管离心机
test weaving 试织,小样试织
T. E. S. V.(thermally effective specific volume) 有效比热容
tetra- (构词成分)表示"四"
tetra acetyl ehtylene diamine 四乙烯乙二胺(双氧水漂白活化剂)
tetra base paper 臭氧试纸(在四甲基对苯二胺中浸制)
tetrabromo phthalic anhydride〈TBPA〉四

溴邻苯二甲酸酐（阻燃剂）

Tetracarnit 提特拉卡尼（杂环匀染剂，商名，瑞士制）

tetracetylethylene diamine ⟨**TAED**⟩ 四乙酰乙烯二胺（双氧水低温漂白活化剂）

tetrachloroethylene 四氯乙烯

tetrads 四色相[等间隔]配色

tetra ethyl lead 四乙铅

tetrafluoroethylene ⟨**TFE**⟩ 四氟乙烯

tetrafluoroethylene fibre [聚]四氟乙烯纤维

tetrahedroid(=**tetrahedron**) 四面体

tetrahydrofuran ⟨**THF**⟩ 四氢呋喃（弹力纤维原料）

tetrahydro furfuryl alcohol 四氢糠醇

tetrahydrofurfuryl alcohol ⟨**THFA**⟩ 四氢呋喃甲醇

tetrahydronaphthalene sulfonate 四氢化萘磺酸盐

tetrakisazo dye 四偶氮染料

tetrakis hydroxymethyl phosphonium chloride ⟨**THPC**⟩ 四羟甲基氯化鏻

tetrakis hydroxymethyl phosphonium hydroxide 四羟甲基氢氧化鏻

tetrakis hydroxymethyl phosphonium sulphate ⟨**THPS**⟩ 四羟甲基硫酸鏻

tetralin(=**tetraline**) 四氢化萘（溶剂，除油剂，煮练剂，聚酯纤维的导染剂）

tetramer 四聚[合]物

tetramethylol acetylene diurea 四羟甲基乙炔二脲（耐皂洗和氯漂的棉织物防缩防皱整理剂）

Tetramine dyes 脱曲明染料（染纤维素纤维与羊毛混纺品用的直接染料，商名，瑞士制）

tetrapod test 四撑点式地毯踩踏试验

tetrapod walker 四撑点式地毯踩踏试验仪

tetra sodium pyrophosphate ⟨**TSPP**⟩ 焦磷酸四钠

Tetratex 一种防水透气织物（商名，美国与日本共同开发）

tex 特克斯，特

tex count system 特[克斯]数制

Texhygrometer 相对湿度测定计（商名）

texilscope 混纺[成分]分析仪

Teximat 玻璃纤维毡（绝缘、隔热或过滤用）

texrope belt 三角皮带

text. (**textile**) 纺织的；纺织品；纺织原料

Text. Chem. Col. (**Textile Chemist and Colourist**)《纺织化学家与染色家》（美国月刊）

textile ⟨**text.**⟩ 纺织的；纺织品；纺织原料

textile abrasion tester 织物耐磨试验仪

textile air permeability tester 织物透气性试验机

textile analysing glass 织物密度镜，织物分析镜

Textile Association India ⟨**TAI**⟩ 印度纺织协会

Textileather 皮革[式]织物（商名）

textile auxiliary 纺织助剂

textile bursting strength tester 织物顶破强力试验机

textile chemicals 纺织化学品

Textile Chemist and Colourist ⟨**Text. Chem. Col.**⟩《纺织化学家与染色家》（美国月刊）

textile chemistry 纺织化学

textile commodity 纺织品；纺织原料

textile composite 纺织复合材料

Textile Council of Australia〈TCA〉 澳大利亚纺织协会

Textile Distributors Association〈TDA〉 纺织批发商协会(美国)

textile dressing 织物整理

textile dyeing and finishing industry 织物染整工业

textile dye waste 印染废水

textile engineering 纺织工程

textile fabric 织物(包括交织物、针织物、编结物、非织造织物等)

Textile Fabrics Association〈TFA〉 纺织品组织协会(美国)

textile fibre 纺织纤维

Textile Finishers Association〈TFA〉 纺织品整理家协会(英国)

textile finishing 纺织品整理

textile finishing agent 纺织品整理剂

textile finishing machine 纺织品整理机

textile fire resistance tester 织物阻燃试验机

textile flex-rigidity tester 织物刚度试验机

textile for agriculture application 农业用纺织品

textile for building industry 建筑用纺织品

textile for fishing 渔业用纺织品

textile for health care 保健纺织品

textile for medical use 医疗用纺织品

textile for national defence 国防用纺织品

textile for paper making 造纸用纺织品

textile glass 纺织玻璃纤维

textile hydrostatic pressure tester 织物耐水压试验仪

textile impurities 纺织杂质

Textile Industries〈Text. Industr〉《纺织工业》(美国月刊)

textile industry 纺织工业

textile information retrieval program〈TIRP〉 纺织情报检索程序

Textile Information System〈TIS〉 纺织情况系统(美国麻省理工学院建立的磁带文件系统)

textile insertion [夹胶布的]织物骨架,夹布

Textile Institute and Industry〈Text. Inst. Industr.〉《纺织学会与工业》(英国月刊)

Textile Journal of Australia〈Text. J. Aust.〉《澳大利亚纺织杂志》(月刊)

textile labeling 纺织品标签

textile lubricants 纺织用润滑剂

textile machinery 纺织机械

textile management software〈TMS〉 纺织品管理软件

textile management system〈TMS〉 纺织品管理系统

Textile Manufacturer & Knitting World〈Text. Mfr. Knitt. World〉《纺织制造家和针织世界》(英国月刊)

textile materials 纺织材料;纺织品

textile mill 纺织厂

Textile Month〈Text. Month〉《纺织月刊》(英国)

textile oil 纺织用油(不污染织物,并易洗去)

textile packaging 纺织品包装材料

textile parasite(=textile pest) 纺织品寄生物,纺织品害虫

textile printing 织物印花

textile processing 纺织加工

textile processing fluid〈TPF〉 纺织工艺加工液

textile product for surgery 外科手术用纺织品

Textile Progress〈Text. Prog.〉《纺织进展》(英国季刊)

textile protective agent 纤维保护剂

Textile Research Institute〈TRI〉纺织研究院(美国)

Textile Research Journal〈Text. Res. J., T. R. J.〉《纺织月报》(美国)

textiles for protection from heat and fire 耐高温与防火用纺织品

textiles for sewage filtration 污水过滤用纺织品

textiles in green house 温室用纺织品

textile soap 纺织工业用皂

textile standards 纺织标准

textile structural composite 纺织结构复合材料

Textile Technical Federation of Canada〈TTFC〉加拿大纺织技术协会

textile technician 纺织技术人员,纺织技师

textile technologist 纺织工艺专家,纺织技术专家

textile technology 纺织工艺;纺织科技

Textile Technology Digest〈TTD〉《纺织工艺文摘》(美国月刊)

textile testing 纺织品测试

textile thickness gauge 织物厚度测定计

textile wall papers 纺织品壁纸

textile waste recycling line 废织物回用生产线

textile water repellency tester 织物拒水性试验机

textile water resistance tester 织物耐水试验机

textile weave 织物组织

Textile World〈Text. World〉《纺织世界》(美国月刊)

textilist 纺织专家

Text. Industr.(Textile Industries)《纺织工业》(美国月刊)

Text. Inst. Industr.(Textile Institute and Industry)《纺织学会与工业》(英国月刊)

Text. J. Aust.(Textile Journal of Australia)《澳大利亚纺织杂志》(月刊)

Text. Mfr. Knitt. World (Textile Manufacturer & Knitting World)《纺织制造家和针织世界》(英国月刊)

Text. Month(Textile Month)《纺织月刊》(英国)

textometer 湿度计,测湿计

Text. Prog.(Textile Progress)《纺织进展》(英国季刊)

textralizing 纤维及纱线卷曲工艺

Text. Res. J.(=T. R. J., Textile Research Journal)《纺织学报》(美国月刊)

texture 织物组织;织物密度;织物质地;[材料的]结构;本质,特征

texture back [织物的]反面组织

textured 变形的,使有某种结构的,使有某种特征的

textured carpet 花式地毯

textured [continuous] filament yarn 变形[长]丝

textured yarn fabric 变形丝织物

texture effect [织物的]起绒结构效应(呈粗糙、毛茸或凸纹状)

texture finishing 织物质地整理

texture streaks [地毯]条痕

texture variation 组织变化

texturing 合[成]纤[维]变形工艺
texturing effect 变形效应
texturized polyester 弹力聚酯纤维;卷曲聚酯纤维
texturized yarn 卷曲变形纱(包括天然纤维和化学纤维的卷曲变形纱)
texturizing [纤维或纱线]卷曲工艺,变形工艺
texturizing finishes [织物]变形整理
Texturmat 全自动变形纱卷缩试验仪(商名)
Text. World(Textile World)《纺织世界》(美国月刊)
TF(=t. f., time factor) 时间因数,时间利用系数
TFA(Textile Fabrics Association) 纺织品组织协会(美国)
TFA(Textile Finishers Association) 纺织品整理家协会(英国)
TFE(tetrafluoroethylene) 四氟乙烯
TFI【德】德国地毯研究所
TFT(thin-film transistor) 薄膜晶体管
Tg(=GTT, glass transition temperature) 玻璃化[转变]温度
TGA(thermal gravimetric analysis, thermogravimetric analysis) 热[解]重[量]分析[法]
th.(thermal) 热的,热量的
th.(threshold) 阈,阈值,临界值;界限;[最低]浓度
thackeray washer 止推环
thaw 熔化;融化
Then Airflow dyeing machine 一种气流染色机(商名,德国制)
Then Airflow Synergy dyeing machine 新一代气流染色机(商名,德国制)

theorem 定理
theoretical bleaching 漂白理论
theoretically dry〈TD.〉 绝对干燥
theory 理论,原理
theory of colourimetry 比色的原理
theory of dyeing 染色理论
theory of probability 概率论
theory of washing 净洗理论
therm.(thermometer) 温度计
Thermacol AM 热熔康 AM(高分子聚合物,阴离子型,防泳移剂,用于各种染料的连续轧染工艺,商名)
thermal〈th.〉 热的,热量的
thermal accumulation 热蓄积
thermal activation 热活化[作用]
thermal adaptability 适温性
thermal-adaptable fabrics 适温织物
thermal addition 热力加成作用
thermal after burning 热补燃,热余燃
thermal ageing 加热老成,热力老化
thermal analysis 热分析
thermal analyzer 热分析仪
thermal anemometer 热风速计
thermal balance 热平衡
thermal barrier 绝热层;保温层
thermal bonding 热黏合
thermal capacity 热容量
thermal character [织物]热特征(皮肤接触织物冷暖的感觉)
thermal chrome marking pencil 测热变色铅笔(在织物上涂写记号,温度变化到指示范围外时,记号处即变色或消色),测色常温
thermal cleavage 热裂解,热分裂
thermal cloth 保暖衣
thermal coefficient of expansion 热膨胀系

数

thermal conductivity 导热性；导热系数，导热率

thermal conductivity detector 热导检测器

thermal convection 对流换热

thermal cracking 热裂解

thermal cracking gas chromatography 热裂解气相色谱［法］

thermal-curable system 热固化系统，热固化装置

thermal damage 热损伤

thermal decomposition 热［力分］解，热分解

thermal deformation 热形变

thermal degradation 热降解

thermal dehydration 热脱水

thermal depolymerization 热解聚［作用］

thermal detector 热探测器

thermal diffusion 热扩散

thermal diffusivity 热扩散率

thermal dilation 热膨胀

thermal discolouration 热褪色，热变色

thermal disinfection 热消毒

thermal efficiency 热效率

thermal endurance 耐热性

thermal energy 热能

thermal energy storage〈TES〉热能储藏器

thermal engrgy storage tank 热能储槽

thermal expansivity 热膨胀系数

thermal explosion 热爆炸

thermal feedback 热反馈

thermal fibre 保暖纤维

thermal fixation 热固着（聚酯纤维染色中利用干热把染料分散并固着于纤维内）

thermal fluid heater 热载［流］体加热器

thermal gain 热效益

thermal gravimetric analysis〈TGA〉热［解］重［量］分析法

thermal head 热位差

thermal image 热图像

thermal initiation 热引发［作用］

thermal ink jet〈TIJ〉热喷墨

thermal insulating material 绝热材料

thermal insulation 保温，绝热

thermal insulation test 绝热试验；保暖试验

thermal insulation value〈TIV，T. I. V.〉绝热值；绝热指数；保温率

thermal knit fabric 保暖针织物

thermal loss 热损失

thermally effective specific volume〈T. E. S. V.〉有效比热容

thermally sensitive resistor 热敏电阻

thermally stable polymer 耐热聚合物

thermally treated wool 热处理［过的］羊毛

thermal man 热人仪（耐燃试验用人体模型）

thermal manikin 暖体人体模型，暖体假人（用于服装功能性评价的模拟人体，测量保暖性能）

thermal mechanical analysis 热机械分析

thermal medium oil 热载油

thermal migration 热迁移

thermal oil 热载油

thermal oil installation 热载油装置

thermal oxidative degradation 热氧化降解

thermal performance 热力特性

thermal plastic adhesive 热塑性黏着剂（印花台板用）

thermal plasticity 热塑性

thermal pollution 热污染

thermal post-combustion 二次燃烧
thermal printer 热敏印刷机
thermal property 热学性能;保暖性
thermal protection 防热性,热防护[性]
thermal protective performance ⟨TPP⟩ 保温性能,防热性能
thermal radiation 热辐射
thermal radiation law 热辐射定律
thermal-recovery unit 热回收装置
thermal relaxation 热松弛
thermal relay 热继电器
thermal resistance 耐热性;热阻
thermal retaining property 保温性
thermal retardant material 保温材料
thermal sensitive 热敏的
thermal shock 温度急增,热冲击,热震
thermal shrinkage 热收缩
thermal-shrinkage differential 热收缩差异
thermal spectrum analysis 热光谱分析[法]
thermal spike 热峰
thermal stability 热稳定性,耐热性;耐热度
thermal storage 蓄热
thermal trace 热示踪
thermal transfer dyes 热[升华]转印染料
thermal transfer printing 热转移印花
thermal transition 热转化
thermal transmission 热传递
thermal transmittance test 热传导试验
thermal underwear 保暖内衣
thermal unit ⟨T.U.⟩ 热单位
thermal value 热值
thermal volatilization analysis ⟨TVA⟩ 热挥发分析[法]
thermal wave 热波

thermal woven 保暖织物
thermal yellowing [羊毛]受热泛黄
Thermindex 热指数测试片(商名)
thermistor 热敏电阻
thermite 铝热剂
thermo- (构词成分)表示"热""热电的"
thermoanalysis 热分析
thermobalance 热天平
thermobondability 热黏合性
thermobonded fabric 热黏合非织造布
thermobonding 热黏合
thermobonding calender 热黏合压轧机
thermobonding fibre 热黏合纤维
thermobonding machine 热黏合机
thermobonding powder 热黏合粉
thermobonding process 热黏合法
thermobrush design machine 刷花机
thermo chamber 烘房
thermochemistry 热化学
thermo chromatic printing 热敏变色印花
thermochrome crayon 热变色笔,测色常温笔
thermochrome pencil[s] 热变色笔,测色常温笔
thermochromic 热变色的
Thermochromic colour 热敏变色印花浆(商名,日本松井色素)
thermochromic effect 热变色效应
thermochromism 热变色作用
thermocolour [彩]色温[度]标示;示温涂料
thermocontroller 热敏控制剂
thermocouple 热电偶,温差电偶
thermocouple pyrometer 热电偶高温计
thermocouple thermometer 热电偶温度计
thermocouple wire insulation 热电偶丝的

绝缘套
thermocracking 热裂解
thermodecomposition 热分解
thermodestruction 热破坏[法]
thermodiffusion 热扩散
thermodilatometric analysis 热膨胀分析
thermodurable textile 耐热织物
thermodynamic affinity 热力亲和性
thermodynamical 热力学的
thermodynamical equilibrium 热力平衡
thermodynamic equation 热力学方程
thermodynamics 热力学
thermodynamics of dyeing 染色热力学
thermodynamic steam trap 热动力式疏水器
thermoelastic effect 热弹效应
thermoelasticity 热弹性
thermoelectric couple 热电偶,温差电偶
thermoelectric effect 热电效应,温差电效应
thermoelectric pile 热电偶,温差电偶
thermoelectric sensing element 热电传感元件
thermoelectrometer 热电计
thermo [electric] element 热电偶,温差电偶
thermofixation 热定形[工艺];热固着[工艺]
thermofixing machine 焙烘机(主要用于分散染料在聚酯纤维上的固着)
thermofix process 热固着[工艺]
thermofluid oil boiler 热油锅炉
thermofluid oil circulation heating system 热油循环加热系统
thermoflush system 热冲洗系统(筒子纱染色机的洗涤)

thermofor 蓄热器,流动床
thermoforming 热成形
thermofusible fabric 热熔性织物
thermofusion device 热熔装置
thermogalvanometer 热电电流计,温差电偶电流计
thermogram 温谱图;温度自记曲线;差示热分析图
thermograph 温度记录器,自记式温度计
thermographic process 热成像过程
thermography 热敏成像法
thermogravimetric analysis ⟨TGA⟩ (= thermogravimetry) 热[解]重[量]分析[法]
thermohardening 热定形,热固着;[树脂]热硬化
thermohumidigraphs 温湿度测量器
thermohydrolysis 热水解
thermohydrometer 热比重计
thermohydro setting 湿热定形
thermohydrosetting process 湿热定形工艺
thermohygrograph 温度湿度记录器
thermo insulation value ⟨TIV, T. I. V. ⟩ 绝热值;绝热指数;保温率
thermojunction 热电偶
thermokinetics 热动力学
thermolabile 不耐热的,热不稳定的,受热即分解的
thermolability 不耐热性;热不稳定性
Thermo Light Paper 涂料转印纸(商名,日本松井色素)
thermo light print 热转移耐晒印花
thermolized 表面热处理
thermoluminescence 热致发光;热发光[现象]
thermolysis 热[分]解[作用]
thermoman 热人仪

thermomechanical analysis〈TMA〉 热机械分析[法];温度—形变分析[法]
thermomechanical behavior 热机械性能
thermometer〈therm.〉 温度计
thermometric analysis 热分析
thermometrograph 温度记录器
thermometry 测温法
thermomoter 热电机
thermonegative 吸热的
thermonuclear 热核的
thermonuclear reaction 热核反应
thermo oil 载热油,导热油
thermo oil circulation heating system 载热油循环加热系统
thermo-oxidative degradation 热氧化降解
thermo-oxidative stability 热氧化稳定性
thermo-paint 示温涂料,测温漆
thermopapers 温度试纸,测温纸
thermophilic 耐热性的;适高温的
thermophilic digestion 高温厌氧消化
thermopigment 示温涂料
thermoplast 热塑胶;热塑[性]塑料
thermoplast glue 热熔[台板]胶
thermoplast glueing unit 热塑胶上胶装置
thermoplastic 热塑[性]塑料;热塑[性]的
thermoplastic adhesive 热塑性胶黏剂
thermoplastic composite 热塑复合材料
thermoplastic fibre 热塑性纤维
thermoplasticity 热塑性
thermoplastic plastics 热塑性塑料
thermoplastic polyolefin〈TPO〉 热塑性聚烯烃
thermoplastic polyurethane〈TPU〉 热塑性聚氨酯
thermoplastic resin 热塑性[合成]树脂
thermoplastic rubber〈TPR〉 热塑料橡胶

thermoplast stripper 热熔胶剥离剂
thermopositive 放热的
thermoprinting paper 热转移印花用纸
thermoprinting technology 热转移印花工艺
thermoreaction chamber 汽蒸反应箱(供轧卷练漂汽蒸或轧卷染色汽蒸用)
thermoregulated textile 调温纺织品
thermoregulating colourimeter 有温度调节的比色仪
thermoregulation 温度调节[作用],调温
thermoregulator 温度调节器
thermoretractile 热缩性的
thermoscope 验温器,测温器
thermosensitizing printing 热敏变色印花
thermosensor 热传感器
thermosets 热固性材料
thermosetting 热定形,热固化
thermosetting fibre 热固性纤维
thermosetting machine 热定形机
thermosetting powders 热固性树脂粉末
thermosetting resin 热固[性]树脂
thermoshock process [活性染料]热快速固着法;[合纤]热快速定形法
thermosiphon 热虹吸管
thermosol 热熔胶
thermosol carrier 热熔载体
thermosol dyeing 热熔染色
thermosol dyeing process 热熔染色法
thermosol dyeing range 热熔染色联合机
thermosol fixation process 热熔固着法
thermosoling 热熔胶化
thermosol pad steam process 热熔与浸轧—汽蒸二步法染色工艺
thermosol pad steam range 热熔与浸轧—汽蒸联合机

thermosol process 热熔加工
thermosol unit 热熔设备
thermostability 耐热性,热稳定性
thermostat 恒温器,温度自动调节器
thermostatical control 恒温控制
thermostatic bath 恒温浴
thermostatic bimetal 双金属温度元件
thermostatic control 恒温控制
thermostatic dryer 恒温烘燥箱
thermostatic steam trap 热静力式疏水器
thermotank 恒温箱
thermo tester 耐热色牢度试验仪(测升华牢度)
Thermotex [领衬布]瑟莫脱克斯热收缩测定仪(商名,瑞士制)
thermotherapy 温热疗法
thermotolerant 耐热的;热稳的
thermotropy 热互变
thermowell 热电偶[温度计]套管,温度计套管;温度计插孔
Therolite 中空保暖聚酯纤维(商名,美国杜邦)
theta temperature θ温度,弗路里温度
THF(tetrahydrofuran) 四氢呋喃(弹力纤维原料)
THFA(tetrahydrofurfuryl alcohol) 四氢呋喃甲醇
thi- (构词成分)表示"硫"
thiamine 硫胺素
thiazine 噻嗪
thiazine colouring matters 噻嗪染料
thiazine dyes 噻嗪(结构)染料
thiazole 噻唑,1,3-硫氮杂茂
thiazole dyes 噻唑染料
thick 厚,粗;浓,密,深
thick bar 厚段,密路(织疵)

thickener 印花糊料;增稠剂
thickener gel theory 增稠剂胶凝理论
thickener powder dissolving unit 增稠剂粉末溶解装置
thickener with high dry content 高含固量糊料
thickener with low dry content 低含固量糊料
thickening 变稠;变粗;增厚
thickening agent 印花糊料,稠厚剂
thickening agent for pigment printing 涂料印花增稠剂
thickening efficiency 增稠效率
thickening for reduction 稀释[印花]浆
thickening material 稠厚剂,增稠剂
thickening preparation system〈TPS〉[印花]糊料制备系统
thickness〈t.,thk.〉厚度;稠度,浓度
thickness gauge 厚度测定计;厚薄规,塞尺,卡板
thickness retention ratio 厚度保持率
Thiele silk 锡尔铜氨人造丝
Thies Luft-roto dyeing machine 第斯气流染色机(商名,德国)
thimble 套管
thin 浅色,淡色;稀薄;细
thin-boiling starch 低稠性淀粉
thin-film evaporating unit 薄膜蒸发装置
thin-film psychrometer 薄膜式湿度计
thin-film transistor〈TFT〉薄膜晶体管
thin fluid 稀薄流体
thin-layer chromatography〈TLC〉薄层色谱分析法,薄层色谱分离法
thinner 稀释剂,冲淡剂
thinness [织物的]手感单薄
thinning 稀薄化,稀释化,淡化

thinning agent 稀释剂,冲淡剂
thinning of ozone layer 臭氧层的稀薄(温室效应)
thin paste 薄浆,稀浆
thin shade 浅色调
thio- (构词成分)表示"硫"
thioamide group 硫[代]酰氨基团
thiocarbonic ester 硫代碳酸酯
thiocyanate 硫氰酸盐,硫氰酸酯,硫氰化物
thiodiethylene glycol 硫二甘醇
thioether linkage 硫醚键
thiogenic dyes 硫化染料
thioglycollic acid 巯基乙酸(用以整理毛织物,使产生永褶效应)
thioindigo blue 硫靛蓝
thioindigo dyes 硫靛染料
thioindigo vat dyes 硫靛还原染料
thiol 硫醇,巯基
thiol prepolymer 硫醇预聚物
thiomersal 乙基汞硫代水杨酸钠
thionation 加硫
Thionol dyes 噻翁诺染料(硫化染料,商名,英国制)
Thionol M dyes 噻翁诺 M 型染料(可溶性硫化染料,商名,英国制)
thionone colours 硫酮染料
thiophene 噻吩,硫[杂]茂
thiophene disperse dye 噻吩型[结构]分散染料
thiosemicarbazide 氨基硫脲
thiosulfate 硫代硫酸盐
thiosulfate dyes 硫代硫酸盐染料
Thiotan RS 赛奥坦 RS(苯酚衍生物,防染剂,商名,瑞士制)
Thiotan SRP liq. 赛奥坦 SRP 液(阴离子型,媒染防染剂,提高锦纶羊毛混纺的匀染,商名,瑞士科莱恩)
Thiotan SWN 赛奥坦 SWN(地毯印花用防染剂,商名,瑞士科莱恩)
Thiotan TRN 赛奥坦 TRN(地毯印花用防染剂,商名,瑞士科莱恩)
thiourea 硫脲
thiourea dioxide 二氧化硫脲
thiourea-formaldehyde precondensate 硫脲甲醛预缩物(聚酰胺纤维的阻燃整理剂)
third-generation synthetic fibre 第三代合成纤维
thixotropic effect 触变效应
thixotropy 触变性,摇溶现象
thixotropy modifier 触变性改进剂(印花糊料)
thk. (=t., thickness) 厚度;稠度;浓度
THPC(tetrakis hydroxymethyl phosphonium chloride) 四羟甲基氯化磷
THPS(tetrakis hydroxymethyl phosphonium sulphate) 四羟甲基硫酸鳞
thread 线;丝;螺纹;贯穿
threadbare [织物]露地,露白;陈旧的,破旧的
thread carriers (=thread catchers)导纱器
thread closeness [线的]密实感
thread count 织物经纬密度
threadcount computer [织物]经纬密度计数器
thread counter 织物分析镜,织物密度镜
threadcounting micrometer 织物分析镜,织物密度镜
threaded connection 螺纹连接;螺纹接头
threaded expander roll 螺旋扩幅辊
threaded joint 螺纹接合

threaded rod 螺杆
threaded union 螺纹联结
threader 导布,导布带
threading 穿导布,穿导带
threading roller 导布辊,导辊
thread out 缺经,断经(织疵)
thread pitch 螺距
thread slippage finish 防滑线处理,防滑移处理(防止化学纤维织物中纱线移动变位的整理)
threads per inch〈T. P. I.〉每英寸螺纹数
thread spindle 螺纹轴
threadup 引头;串头子
thready finish 线纹整理
three attributes of colour 色的三属性
three-bath high-temperature dyeing machine 三浴高温染色机
three-bowl friction calender 三辊摩擦轧光机
three-bowl mangle 三辊轧车
three-bowl schreiner calender 三辊电光机
three-bowl water mangle 三辊轧水机
three-colour printing 三[原]色印花
three dimensional 立体的,三维的;三元的,三属性的
three-dimensional body-scanning system 三维人体扫描系统
three-dimensional brightness 三维光泽度;三元光泽度
three-dimensional crosslinked network 三维交联网络,立体交联网络
three-dimensional drape 三维悬垂
three-dimensional dynamic mixer 三维动态混合器
three-dimensional effect 立体效应,三维效应

three-dimensional fabric〈3D fabric〉三维[结构]织物
three-dimensionality 立体性;立体感
three-dimensional net structure 立体型网络结构,三维网络结构
three-dimensional nonwoven 三维非织造
three-dimensional original 立体实样
three-dimensional pleat pressing machine 立体打褶机《针》
three-dimensional polymer 三维聚合物
three-dimensional polymerization 三维聚合;体型聚合
three dimensional print〈3D print〉(＝three dimensional printing〈3D printing〉) 三维立体印刷;三维立体打印;三维立体印花
three-dimensional printing effect 立体印花效应
three-dimensional product 三维产品
three-dimensional space 三维空间
three-dimensional structure 三维[网状]结构;立体结构
three-dimensional textiles 三维纺织品
three E concept（E＝efficiency, economy, ecology）"三 E"理念(效率、经济、生态)
three effects evapouration 三效蒸发
three elemental factors of colours 色彩三要素
three elementary colours 三原色
three-line plug 三线插头
three-line socket 三线插座
three motive design 三原组织(平纹、斜纹、缎纹三个基本组织)
three-necked flask 三口瓶
three origiral colours 三原色

three-phase fourwire system 三相四线制《电》
three-phase induction motor 三相感应电动机
three-phase induction voltage regulator 三相感应调压器
three-phase threewire system 三相三线制《电》
three-phase triangle 三相(水、油、乳液三相)三角
three-phase wound rotor induction motor 绕线转子三相异步电动机
three-piece suit 三件成套服装
three-ply 三股；三层
three primaries (= three primary colours) 三原色
three quarters boiling 七分练
three-roller padding mangle 三辊轧车
three R policy (three R=reduce, reuse, recycle) "三R"政策(还原、再利用、再循环)
three uncurling fingers 三指剥边装置
three-way diverting valve 三通分流阀
three-way valve 三通阀
three-wire system 三线制《电》
threshold〈th.〉阈,阈值,临界值；界限；[最低]限度
threshold concentration 临界浓度,阈浓度(最低限度浓度)
threshold oxidation temperature 极限氧化温度,临界氧化温度
threshold value 阈值,临界值,界限值
thrombin 凝血酶
throstle spun yarn fabric 翼锭纺纱织物；废纺织物
throttle 节流阀,节汽阀；风门调节板

throttle nozzle 节流喷嘴
throttle orifice 节流孔板,节流圈
throttle steam trap 节流孔式疏水器
throttle valve 节流阀；节汽阀；减压阀
throttling cock 节流阀
throttling discharge 节流排出,调节排放
throttling valve 节流阀
through-air dryer fabric 通透式干燥织物
through and through 双面织物(指印花、织花或起绒的双面织物)
through feed 贯穿进给(指物件全面通过)
throughflow 通流,直流
through printing 渗透印花
through-put [原料]通过量；生产量；容许能力；(计算机的)解题能力；吞吐量
through-put capacity 生产能力,流通能力,出力
through-way valve 直通阀
throw-away nappy 一次性尿布
throw off 断开,断路
throwster textured yarn〈TTY〉捻丝厂[生]产[的]变形丝
throwster tints 指示用着色(用于辨认不同批次的纱线、织物等的暂时着色)
thrum 纱头,线头
thrust bar 推杆
thrust bearing 止推轴承
thrust block 止推座
thrust nut 推力螺母
thrust plate 止推板
thrust washer 止推垫圈
thumb nut 蝶形螺母,翼形螺母
thumb pin [压]图钉,按钉
thumb screw 翼形螺钉；指旋螺丝；(缝纫机的)压脚螺钉
thumb wheel switch 手指旋转开关

THV (total hand value) 总手感值

thymol 茴香油,百里[香]酚,麝香草酚,间甲邻丙酚,5-甲基-2-异丙基苯酚

thymol blue 百里酚蓝,麝香草酚蓝,甲异丙酚蓝(酸碱测定用)

thymolphthalein 百里酚酞

thyristor 可控硅;可控硅整流器;半导体开关原件;闸流晶体管

thyristor-controlled motor 可控硅整流器控制电动机

thyristor converter 可控硅变频器;可控硅换流器

thyristor drive system 可控硅传动系统

thyristor installation 可控硅装置

thyristor unit 半导体开关元件;闸流晶体管;可控硅整流器

TI (temperature indicator) 温度指示器

Tiacet dyes 铁爱雪染料(用于锦纶、醋酯纤维的分散染料,并被广泛用于人造花加工,商名,意大利宁柏迪)

Tiacidol dyes 铁雪妥染料(成衣和针织品染色用酸性染料,商名,意大利宁柏迪)

Tiacidol dyes for printing 用于印花的铁雪妥染料(特别挑选的印花用弱酸性染料,具有良好溶解度,适合于丝绸、锦纶、羊毛等印花,包括用于地色拔染与着色拔染,商名,意大利宁柏迪)

Tiacidol P dyes 铁雪妥 P 染料(酸性染料,特别适用于羊毛和毛锦混纺织物的匹染和成衣染色,匀染性好,湿牢度优良,商名,意大利宁柏迪)

Tialanal PF dyes 铁爱兰奴 PF 染料(1∶1 金属络合染料,耐干摩擦牢度好,商名,意大利宁柏迪)

Tialan dyes 铁爱郎染料(混合染料,适用于羊毛、锦纶、氨纶和丝绸染色,色泽鲜艳,牢度优良,匀染性高,商名,意大利柏迪)

tian-xiang damask 天香绢,双纬花绸

Tiasolan L liq. dyes 铁雪郎 L 液体染料(双磺酸型 1∶2 金属络合染料,特别适用于羊毛、锦纶和氨纶的染色,牢度优良,重金属含量低,对环保有利,商名,意大利宁柏迪)

Tiasolan S dyes 铁雪郎 S 染料(单磺酸型 1∶2 金属络合染料,牢度优良,特别适用于毛条及纱线染色,商名,意大利宁柏迪)

TIB (Technical Information Bureau) 技术情报局

TIC (temperature indicating controller) 温度指示控制器

TIC (temperature of initial combustion) 开始燃烧温度

ticket-sewing machine 钉商标机

ticking (=ticks) 枕套及褥罩织物

TIDU (Technical Information and Documents Unit) 科技情报文献处(英国)

tie 布头包纬横线;包扎绳;绑带;领带;打结,捆扎

tie band 扎绞线

tie-coat [合成革]黏接层

tie collar 领圈;围巾

tied carpet 扎结地毯

tie-dyed cotton cloth 扎染棉布

tie dyeing 扎染

tie-dyeing imitation 仿扎染

tie fabric 领带织物

tie point [层状分子结构的]交联点,联结点;结合点

tier drier 多层烘干机

tiering table 印浆盆移动台(手工木版印花用)
tier stenter 多层拉幅机
tie silk 领带绸
tie up thread 扎绞线
tiger cashmere 仿开司米织物(棉经、山羊毛纬交织)
tigering [长毛绒织物]修毛整理
tigering machine 修毛整理机
tigerlilly 卷丹红[色](红橘色)
tiger skin plush 虎皮长毛绒
tight coupling 密封连接；紧接头
tight fit 紧配合
tightness 紧密度，密封度；硬挺度，身骨
tight selvedge 紧边(织疵)
tight-strained fabric guiding 织物张紧导向装置
tight weft 紧纬(织疵)
tigre cashmere 仿羊绒织物(棉经山羊毛纬,英国制)
TIJ(thermal ink jet) 热喷墨
tile 拼块地毯
tile blue 瓦蓝[色](浅灰蓝色)
tile red 瓦红[色](橙色)
tillot 织物包布，包装用织物
tilt 倾斜，斜度
tilting angle 倾斜角
time adjusting device 时限调整装置，定时装置
time and motion study 操作测定，工时研究
time based maintenance 定期维修
time closing〈TC〉 延时闭合
time constant 时间常数
time delay relay〈TDR〉 延时继电器
time factor〈TF, t.f.〉 时间因数，时间利用系数
time interval 时间间隔
timekeeper 定时器；时计
time lag 时间滞后，延时，时滞
time-lag method for diffusion 扩散时滞法，扩散延时法
time-lag relay 延时继电器
time measurement unit〈TMU〉 时间测量单元
time of after flow 阴燃时间
time of arrival〈TA〉 运达时间
time of burning after extinguishing 阴燃时间
time of delivery〈TOD〉 交货时间，送到时间
time of departure〈TOD〉 起程时间，发运时间
time of half dyeing 半染时间(纤维从染液中吸收一半染料量所需的时间)
time of half migration 半泳移时间(布样和其同重量的白布在水内沸煮,至二者染成同样深浅颜色时所需的时间)
time of half reduction 半还原时间
time of half vatting 半还原时间
time-place-occasion〈TPO〉 时间—场所—场合(穿衣三要素)
timeproof 耐久的,耐用的
timer 定时器,记时器；定时延迟继电器
time relay 时间继电器
timesaver system 省时系统
time scale 时标；时间比例；时序表
time-sharing 分时,时间分配
time slicing 时间[分]段
time stopper 定时器
time study 工时测定
time switch 定时开关,控时断路器

time-temperature superposition principle 时[间]—温[度]叠加原理
time to break 断裂时间(强力试验中从开始加重到使试样断裂的时间)
time wage 计时工资
timework 计时工作
timeyield 蠕变
timing 定时,同步
timing belt 同步皮带;牙轮皮带
timing chain 定时控制链条,正时链,定时链
timing corrector 快慢调整器
timing cylinder 绷布辊;透风辊
timing device 计时设备;延时器件;定时器
timing drum 透风滚筒(用以延长处理时间)
timing pulse generator 同步脉冲发生器
timing relay 延时继电器;定时继电器
timing roller 透风辊(主要作用是延长处理时间)
timing valve 延时阀
tin 圆筒;烘筒,烘缸
tin chloride 氯化锡
tin compound 锡化合物
tin crystal 锡晶(氯化亚锡的别名)
tinctorial depth 色浓度
tinctorial power 着色力
tinctorial property 着色性能
tinctorial strength 着色强度
tinctorial substance 着色剂
tinctorial value 色值
tinctorial yield 给色量,得色量,着色量
tincture 色泽,色调;染料,颜料;酊剂;浸剂,浸液;着色
tin cylinder 白铁滚筒

tin discharge 锡拔染法
tin discharge printing 锡拔染印花
tin drum 白铁滚筒
Tinegal MG 坦内格 MG(有机胺衍生物,阳离子染料竣丙烯腈系和聚酯纤维的匀染剂,商名,瑞士制)
Tinegal MR 坦内格 MR[阳离子型缓染剂,特别用作麦西隆(Maxilon)染料配套的缓染剂,有移染性,商名,瑞士制]
Tinegal RWI 坦内格 RWI(部分非离子型脂肪族化合物,染色匀染剂和剥色剂,商名,瑞士制)
Tinegal W 坦内格 W(腈纶染色匀染剂,阳离子型,亦能用于各种纤维的后清洗,改浅色及改善不匀染色,商名,瑞士制)
tineid 蛀虫
tin finish 锡[化合物]增量处理
tinge 淡色调;较淡着色
tingey 廷杰布(优质细密府绸衬衫布,经过仿丝绸整理)
tin mordant 锡[化合物]媒染剂
tinned plate 马口铁,镀锡铁皮
tinning 镀锡;包锡
Tinoclarite CBB 定劳卡维 CBB(膦酸酯衍生物及水溶性高聚物的复配物,阴离子型高效氧漂稳定剂,耐强碱,商名,瑞士制)
Tinoclarite G 定劳卡维 G(阴离子型有机金属络合物,用于双氧水漂白的稳定剂,不含表面活性剂,商名,瑞士制)
Tinoclarite PB 定劳卡维 PB(不含硅的有机复合产品,阴离子型,用于棉及其混纺织物的双氧水冷轧堆漂白,无需汽蒸,白度和毛效均佳,商名,瑞士制)
Tinofast CEL 定劳法 CEL(反应性紫外线

吸收剂,二苯基乙二酰胺,适用于纤维素纤维,商名,瑞士制)

Tinofix LW 添牢弗斯 LW(脂肪族聚季铵盐,直接染料固色剂,商名,瑞士制)

Tinolite pigment colours 天来利特涂料(印花用涂料,商名,瑞士制)

Tinon dyes 天虹染料(靛属还原染料,商名,瑞士制)

Tinopal 天来宝(荧光增白剂的一类,商名,瑞士制)

Tinosan series 定劳生系列(抗菌整理剂,商名,瑞士制)

Tinosol dyes 天虹素染料(可溶性还原染料,商名,瑞士制)

tin phosphate weighting 磷酸锡增重处理
tin pipe 白铁管
tin pulley 白铁滚盘
tin roller 白铁滚筒
tin salt 锡盐(通常指二氯化锡或氯锡酸铵)
tinsel 金属箔;金银线;金银丝交织物
tinsel fabric 金银线织物
tinsel printing 金银丝印花
tint 淡色;极浅着色;易褪着色;明调[含白]
tintage 上色
tinted yarn 着色丝,着色纱
tin tetrachloride 四氯化锡
tinting [花布]浅地染色;着色;标色;浮脏,糊版(印刷用语)
tinting colour 着色[用]染料;指色剂
tinting power(=tinting strength) 着色力
tint mark 色点,色斑
tintometer 色调计,色辉计,比色仪
tin-weighted silk 锡盐增重丝绸
tin weighting 锡增重《丝》

tip 针尖;尖头
tip definition [绒面织物的]表面绒尖整列
tipped [割绒织物]绒尖染色
tipped fabric 绒尖染色绒织物
tippiness 羊毛尖端风化损伤;羊毛尖端较深染色(疵点)
tipping 毛尖染疵
tipping plate 斜板
tip printing 凸纹着色轧花
tippy dyeing 毛尖染色[差异]
tippy wool 尖端受损羊毛
tips 告诫,提示
tip shearing 轻剪绒头(地毯等织物的一种特殊剪毛整理,仅剪去最高一层绒头)
tip switch 触点开关
tire builder fabric 轮胎帘子布
TIRP (textile information retrieval program) 纺织情报检索程序
TIS(Textile Information System) 纺织情报系统(美国麻省理工学院建立的磁带文件系统)
tissu 纺织材料;纺织品(法国名称)
tissue shearing test 织物剪力试验
titanium and fluorine gels 钛氟溶胶(应用于抗紫外拒水整理)
titanium boride fibre 硼化钛纤维
titanium carbide fibre 碳化钛纤维
titanium compound 钛化合物
titanium dioxide 二氧化钛
titanium foil 钛膜
titanium foil as pigment 钛膜涂料
titanium tetrachloride 四氯化钛
titanium white 钛白
titanous sulfate 硫酸钛(拔染剂)

titanox 二氧化钛(消光剂);钛钡白
titer 纤度,线密度;滴定度,滴定率
Titer denier〈Td.〉旦[尼尔]制线密度
titrate 滴定;被滴定液
titration device 滴定装置
titration method 滴定法
titration value 滴定值
titre 线密度;滴定度,滴定率
titrimeter 滴定计
titrimetry 滴定分析[法]
TIV(＝T. I. V., thermal insulation value, thermo insulation value) 绝热值;绝热指数;保温率
tjap printing 木版[金属嵌纹]蜡防印花(东南亚土法)
T-junction box T形套筒,T形接头,三通接头;T形接线匣
TKI 匈牙利纺织研究所
TLC(thin-layer chromatography) 薄层色谱分析法,薄层色谱分离法
TLCA 由多元醇胺和多元羧酸合成的无醛防皱整理剂《丝》
TL84 illuminant TL84 光源
TLm(median tolerance limit) [急毒性]中位容许限度《环保》;半数生存界限浓度
TLM(telemeter) 遥测仪,遥测装置
TLM(telemetry) 遥测技术
TM(technical manual) 技术指南,技术手册
TM(twisting moment) 扭矩,捻矩
TMA(thermomechanical analysis) 热机械分析[法];温度—形变分析[法]
TME(test method under examination) 试用的测试方法
TMS(textile management software) 纺织品管理软件
TMS(textile management system) 纺织品管理系统
TMU(time measurement unit) 时间测量单元
TNO【荷兰】荷兰应用科学研究院
TNT(＝T. N. T., trinitrotoluene) 三硝基甲苯
T. O. (technical order) 技术规程,技术命令
TO(turn over) 周转;翻转;见反面
tobacco [brown] 烟草棕[色]
tobacco cloth 烟草盖布
TOC(total organic carbon) 总有机碳(表示污水中有机物含量的指标)
TOD(time of delivery) 交货时间,送到时间
TOD(time of departure) 起程时间,发运时间
TOD(total oxygen demand) 总需氧量
toddlers 儿童短衫
toggle 触发器;反复电路;手操作开关;肘节;肘节套接
toggle brake 肘节刹车,套环制动器,带伸缩元件的制动器
toggle chuck 肘环套接卡盘
toggle flip-flop 反转触发器
toggle joint 肘形节;弯头结合
toggle lever 肘节杆
toggle plate 摆杆,摆板
toggle switch 拨动式开关,弹簧开关,钮子开关;触发器
togs 特殊服装(如溜冰服装等)
Tog value 托格值(纺织材料和服装隔热性的热阻单位)
toile 薄亚麻织物;帆布;[花边地布上的]

图案

toile a bleuteau 【法】细网眼筛绢（用丝或锦纶制）

toile d'alsace 高级阿尔萨斯薄织物（棉或亚麻女性服装面料,漂白、染色、印花的均有,法国名称）

toile d'argent 银线绸（法国及意大利用银色蚕丝经、银线纬织制）；金银线织物（法国制）

tol. (tolerable) 可允许的

tol. (tolerance) 公差,容限;容许数;耐受度

tolerable〈tol.〉可允许的

tolerable limits 容许范围,容许限度

tolerance〈tol.〉公差,容限;容许数;耐受度

tolerance and fit 公差及配合

tolerance error 公差误差

toluene 甲苯

toluene diisocyanate〈TDI〉甲苯二异氰酸酯

toluene recovery 甲苯的回收

toluol 甲苯

tomato [red] 番茄红[色]（橙红色）

tommy key [印花]对花撬棒

tommy screw 虎钳丝杆,贯头螺丝

tom-tom machine 锤式缩绒机

tom-tom scouring machine 锤式精练机

tom-tom washer 锤式洗涤机

ton 吨;时式,流行

tonality 色调

tone 色调,色光;音

tone colour 色调

tone contrast 色调对比

tone harmony 色调调和

tone in tone 同色调配色

tone in tone dyeing 同色调染色

tone-in-tone effects ［不同类型纤维］同色深浅效应

toneless colour 沉闷色

tone off 色泽渐浅至消失

tone on tone 同系配色（同一色相,不同色调配色）

tone-on-tone dyeing ［不同类型纤维］同色深浅效应染色

tone-on-tone effect 同色深浅效应

toner 调色剂（不含无机粒料的有机染料）;显影剂

toneshade effects 色彩渐变效应

tongs 钳子,夹钳

tongue method 舌形试样法（织物撕破试验）

tongue method of tearing strength 舌形法撕破强力

tongue tear strength 舌形[试样]撕破强力

tongue tear test 舌形试样撕破强力试验

tongue test 舌形法织物撕破试验

toning 调匀颜色

tonnage 吨数,吨位

tontise 彩色毛织物（把毛回丝染成多种颜色纺成毛纱后织成的织物）

tool box 工具箱

toolmaker's microscope 工具显微镜

tool making shop 工具车间

tool steel 工具钢

tooth density 齿密度

tooth depth 齿深

toothed belt 齿形带

toothed conveyor belt 齿形运输带

toothed ring 齿环

toothed rubber belt 齿形橡胶带

tooth pitch 齿距,齿节

top 毛条;化学纤维条;顶端;袜口
top arm weighting 摇臂式加压
topaz 黄玉色(浅棕黄色)
top carbonizing 毛条炭化
topchrome dyeing 铬盐后处理染色,套铬媒染色,后铬媒染色
top coat 上涂;面漆;面层
top colour 罩色
top cover labyrinth 上密封盖
top decatizer 卷状蒸呢机
top drum 高桶(容量高大的桶)
top-dyed 套染的,套色的;毛条染色的
top dyeing 毛条染色
top dyeing machine 毛条染色机
top finish 涂层整理
top grinding 平磨
topical colour 表面着色染料
topical finish 后加工,后整理
topical printing [人物]图形印花
topochemical 局部化学的
topochemical reaction 局部化学反应
topography 形态特征
topometer 外形测试仪
topper [裤子]蒸烫机
topper coat [妇女的]宽松短大衣
topping 套染,套色
topping agent 套色剂,罩色剂
topping denim 牛仔布染色(先染靛蓝后套染硫化染料)
topping finish 套色整理
topping printing 套色印花(先用浅色印花,再用深色套印)
topping-up 套色调节色光
top plate 顶板
top printing machine 毛条印花机
top quality 最高质量,第一流的

top rollers 上排导辊
top scouring 毛条洗涤
top skin finishing 表面整理法
top up 装满
top view 俯视图
top-weight fabric 薄型织物
torn list(＝torn selvedge) 破边(织疵);拉破边(整理疵)
torn-up rag 碎布
toroidal-core transformer 环形核心变压器
torque〈t.〉 转矩,扭矩;捻矩;扭转
torque control 转距控制
torque limitation 转矩极限
torque motor 转矩电动机,力矩电动机
torque wrench 转距扳手
torr 托(低气压压力单位,等于1毫米汞柱)
torsion 转矩;扭力;扭转
torsional deformation 扭转形变
torsional dynamometer 扭力测力计
torsional elasticity 抗扭弹性
torsional fatigue-strength 扭转疲劳强度
torsional force 扭力
torsional rigidity 扭转刚度
torsional strength 抗扭强度
torsional test 扭力测试,扭力试验
torsion balance 扭力天平,扭力秤
torsion denier balance 旦尼尔扭力天平
torsion spring 扭簧
torsion viscometer(＝torsion viscosimeter) 扭力黏度计
tortoise-shell pattern 龟甲花样《印》
tortuosity 曲折;弯扭
tosimeter 微压计
tosylation 甲苯磺酰化
total air cleaning〈TAC〉 空气总净化

total alkali-insoluble ash〈TAI -ash〉 非碱溶性灰分总量,非碱溶性物质总量

total analysis 全分析

total carbon〈TC〉 总含碳量

total correlation coefficient 全相关系数

total deformation 总形变

total dissolved solids〈TDS〉 总溶解固体[量]

total easy care 完全易保养[羊毛]

total emittance 总辐射力

total energy consumption 综合能耗,总能耗

total formaldehyde 总甲醛

total hand value〈THV〉 总手感值

total heat loss 全热损失

total immersion jig 全浸式卷染机

totalizer 加法器,加法计算装置,累加器

total Kjeldahl nitrogen 基耶达定氮总量《环保用语》

totaller 加法器,加法计算装置,累加器

totally enclosed〈te〉 全封闭的

totally enclosed motor 全封闭式电动机

total organic carbon〈TOC〉 总有机碳(表示污水中有机物含量的指标)

total organic halogen〈TOX〉 总的有机卤化物

total oxygen demand〈TOD〉 总需氧量

total print-paste return〈TPR〉 整体印浆回用[系统]

total quality control〈TQC〉 全面质量管理

total reflection 全反射

total resistance to water-vapour transmission 全面抗水蒸汽透过

total solid matters 总固体含量

total solid of waste water 废水总固量

total stitch density 总密度《针》

total suspended solids〈TSS〉 悬浮固体总量(污水处理用语)

total work cost〈t. w. c.〉 总工作费用

touch [纺织原料和产品的]手感

touching up of lines and areas 勾线勾面(电脑分色用语)

touch panel 触摸屏;触摸式控制柜《计算机》

touch roller 接触辊

touch screen 触摸屏《计算机》

touch screen control 触摸屏控制《计算机》

touch-up roller 给液辊(低给液)

toughness 韧性,韧度

toughness index 韧性指数,韧度指数

tour 周转,流通,轮班,巡回

tourmaline 电气石

Tourmaline digital printer 电气石数码印花机(商名,荷兰斯托克)

tow 丝束《化纤》

tow dyeing 丝束[状]染色

tow dyeing machine 毛条轧染机;丝束轧染机

towel 毛巾

towel blanket 毛巾被

towelling 毛巾布,毛巾料

towelling gown 毛巾浴衣;毛巾长晨衣

towelling tea cloth 毛巾茶布

towel suiting 毛巾布料

tower ager 塔式蒸化机(又称快速蒸化机)

tower biological filter 塔式生物滤池《环保》

tower steamer 塔式蒸化机

tower washing machine 塔式水洗机

town gas 城市煤气,家用煤气

TOX(total organic halogen) 总的有机卤化

物

toxic(=toxical) 有毒的;毒剂

toxicity 毒性,毒度

toxicity index 毒性指数

toxicity of pollutants 污染物毒性

toxicological data 毒性数据

toxicology 毒物学

toxic substance control act〈TSCA〉毒物管理条例

toy 蓝黑格子呢;精纺毛织物;精纺毛与丝的交织物

Toyobo 东洋纺(黏胶短纤维,商名,日本制)

Tp(=T. P., tempered) 已回火的;已松开的(指螺钉)

tp(test pressure) 试验压力

TP(=T. P., turning point) 转点,变点

tp(two-pole) 两极的

TPA(terephthalic acid) 对苯二[甲]酸

TPF(textile processing fluid) 纺织工艺加工液

T. P. I.(teeth per inch) 每英寸齿数

T. P. I.(threads per inch) 每英寸螺纹数

TPI(trans-polyisoprene) 反式1,4-聚异戊二烯(属形状记忆聚合物)

T. P. I.(twists per inch) 每英寸捻数

TPO(thermoplastic polyolefin) 热塑性聚烯烃

TPO(time-place-occasion) 时间—场所—场合(穿衣三要素)

TPP(thermal protective performance) 保温性能,防热性能

TPR(thermoplastic rubber) 热塑料橡胶

TPR(total print-paste return) 整体印浆回用[系统]

TPS(thickening preparation system) [印花]糊料制备系统

TPU(thermoplastic polyurethane) 热塑性聚氨酯

TPU film 热塑性聚氨酯薄膜

TQC(total quality control) 全面质量管理

TR(technical regulation) 技术规范

TR(technical report) 技术报告

TR(temperature recorder) 温度记录器

Tr(=tr., Trans, trans, transactions) 学报;会报;论文集

TR(trunk relay) 中继线继电器,中继器

trace 追踪《计算机》;示踪;轨迹;微量

traceability 跟踪能力;示踪能力

trace analysis 痕量分析,微量分析

traced drawing 原图

trace design 描图样

trace element 示踪元素

trace gas technique 微量气体技术(服装小气候试验)

tracer 示踪物;追踪程序;故障检测器《计算机》;描绘器

tracer method 示踪法;触针法

tracing 描图;花纹描绘(将花纹设计转译为数字式,以便喂入计算机)

tracing film 描图薄膜

tracing folio 描样片

tracing paper 描图纸

tracing wheel 描线轮

track 轨道,轨迹

track and jogging suits 运动休闲服

tracking [地毯]条痕

traction roller 牵引辊,拖动辊

traction transmission belt 牵引传送皮带

tractive effort〈te〉牵引力

tractive force 牵[引]力,拉力

trade mark 商标

trade name 商业名称
trade weight 商用砝码；商业重量
Tra. Di. C. (=TRADIC, transistor digital computer) 晶体管数字计算机
Traditional Chinese Medicine〈TCM〉中医学
traditional colour 传统色
traditional colour name 惯用色名，传统色名
tragacanth 龙[须]胶，黄蓍胶
tragacanth gum 龙胶(可作印花糊料)
tragasol gum 角豆树胶(可作印花糊料)
tragon 角豆树胶(高浓粉状)
trailing 拖尾巴
train 机组；轮系；键；序列；顺序
Trans (=trans, Tr., tr., transactions) 学报；会报；论文集
TRANSAC(=Transac, transistor automatic computer) 晶体管自动计算机
transactions〈Tr., tr., Trans, trans〉学报；会报；论文集
trans-addition 反式加成[作用]
transalkylation 烷基转移[作用]
trans-butenediol 反丁烯二醇
trans-cis-isomerism 顺反异构[现象]
Transcolorizer 特拉恩斯克莱转移印花机(连续热转移印花设备，美国制)
transcriber 转录器；再现装置，重复装置
transducer 转换器；传感器；变频器；转录程序
transesterification 酯基转移[作用]，酯交换[作用]
transf. (transformation) 交换，转变，转换，变化
transfer 转移；传送，传输；转换
transferability 可转移性(转移印花的热转移染色性能)
transferability of printing paste 印浆转移性
transfer agent 转移剂(用于链位转移反应)
transfer area 传递面；传热面
transferases 转化酶(用于基团转化反应)
transfer calender 热压转移印花机
transfer catalyst 转移催化剂
transfer coating 转移涂布，转移涂层
transfer coating with release paper 离型纸转移法涂层
transfer coefficient 转移系数
transfer drum 转移滚筒
transfer dyeing 转移染色
transfer efficiency 转移效率
transference number 迁移数
transfer film 贴花膜，转移薄膜
transfer flock printing 转移植绒
transfer function 传递函数，转换函数
transfer gear pump 齿轮输送泵
transfer laminator 转移涂层机；转移胶合机
transfer padding 转移浸轧，转移轧液；转移轧染(指带液布和干布同入轧点而起转移液体作用)
transfer paper storage stability 转移纸储存稳定性
transfer pattern 转印图案
transfer press 转移[印花]压烫机
transfer printing 转移印花
transfer printing machine 转移印花机
transfer printing paper 转移印花纸
transfer printing resist effect 转移印花防印效果
transfer pump 供油泵

transfer ratio 转移系数
transfer reaction 转移反应
transfer reproduction in colour〈TRIC〉颜色转移复制法
transfer roll 移液上浆辊
transfers 转印图案
transfer standard 传递标准
transfer style printing 转移印花
transfer test 移染试验
transfer trolley 运输车
transfer unit [织物] 传送单元
trans-form 反式[异构体]
transformation〈transf.〉交换,转变,转换,变化
transformer 变压器《电》
transformer oil 变压器油
transforming lens 交换透镜
Transforon dyes 转移福隆染料(转移印花用分散染料,商名,瑞士制)
transfusion 倾注;移注;渗流
transgenic gene 转基因
transgenic textile fibres 转基因纺织纤维
transglutaminase 转谷氨酰胺酶(又名 TG 酶)
transient 瞬态,瞬变过程
transient creep 瞬时蠕变
transient state 过渡状态,瞬态
transient temperature 瞬变温度
trans-isomer 反式异构体
trans-isomerism 反式异构[现象]
transistor 晶体管
transistor automatic computer〈TRANSAC, Transac〉晶体管自动计算机
transistor digital computer〈Tra. Di. C., TRADIC〉晶体管数字计算机
transition 转变;过渡;瞬态;转折点
transition curve 过渡曲线
transition element 过渡元素
transition interval [指示剂]变色范围
transition point 转变点,转换点
transition state 过渡[状]态,转变[状]态
transition temperature 转变温度
transit trade 过境贸易
translation 翻译;平移,位移
translator 翻译程序;翻译机,译码器
translite film 透光软片
translucent 半透明的
translucent materials 半透明材料
transmembrane pressure [超过滤]跨膜压《环保》
transmissibility 透过率,透过系数;可传透性
transmission 传动;传动装置;变速器;透射
transmission adapter〈XA〉传输转接器
transmission belt 传动带
transmission densitometer 透光光密度计
transmission electron microscope 透射电子显微镜
transmission electron microscopy〈TEM〉透射电子显微术
transmission factor 透光因数,透射系数
transmission gear 变速齿轮
transmission interface converter〈XIC〉传输接口转换器
transmission line 输电线
transmission measurements 透光度测定
transmission oil 传动润滑油
transmission shaft 传动轴,连接长轴
transmission spectrum 透光光谱
transmission system 传动系统
transmissivity 导水率

transmit〈XMT〉 发送;转换;透射
transmittance(= transmittancy) 透射比,透射率,透光度;透明性;过滤系数
transmitted light 透射光
transmitter 发送器;传递器;传感器
transmitting ratio 传动比
transmutation 蜕变
transparence(= transparency) 透明性;透光度
transparency to electronic detection 隐形防电子侦察
transparent binder 透明黏合剂
transparent body 透明体
transparent covering 透明罩膜
transparent film 透明薄膜,玻璃纸;透明胶片
transparent finishing 透明整理
transparent printing 透明印花
transparent quartz tube 透明石英管
transparent thread 透明线
transplantation 移植
trans-polyisoprene〈TPI〉 反式1,4-聚异戊二烯(属形状记忆聚合物)
transport 移动;运输
transportation fabric 车辆坐垫织物
transportation textile 交通运输用纺织品
transport equipment 传输设备
transport eyes 运输用的索眼
transport systems 传输系统
transport textiles 交通工具用纺织品
transposed pattern [互]换位[置]图案
transposition [互]换位[置];移位
transprints paper 转移印花用纸
trans-trans isomer 反—反式[同分]异构体
transudate 渗出液

transverse 横[向]的;横梁
transverse bearing 径向轴承
transverse brushing machine 横向刷绒机(灯芯绒织物加工用)
transverse contraction 横向收缩
transverse cross-linking 横向交联
transverse relaxation 横向松弛
transverse section 横截面
transverse strength 横向[挠曲]强度
transverse tension 横向张力
transversing roller 横向移动辊
transverter 交换器;变频器;换能器
trap 凝汽筒;护罩,吸尘罩;收集;陷波电路;疏水器
trapezoidal thread 梯形螺纹
trapezoid method 梯形试样法(织物撕破强力试验)
trapezoid method of tearing strengh 梯形法撕破强力
trapezoid tear strength [织物]梯形试样撕破强力
trapezoid test 织物梯形撕破试验
trapping 收集;截留,捕捉
trash 杂质
trash chamber 除杂室,除尘室
trash content 含杂率
trash extraction 除杂
trash meter 含杂[光电]测定仪
travel 动程;运动;输送
travelling counting glass 游动式织物密度镜
travelling crane 行车,移动式起重机
travelling wince piler 往复堆布器
travel plaiter 移动式落布架
travers 横向条纹;横条花
traverse arm [连接印花机刮浆刀的]横向

往复连杆
traverse bar 横动杆
traverse cam 横动[成形]凸轮
traverse guide pulley 往复导轮
traverse length 往复长度
traverse motion 横向往复运动
traverse net 六角网眼
traversing lift 往复升降
traversing roller 横向移动辊
tray dryer 盘式烘燥机
treadle 踏板
tread test 轮胎里程试验；踩踏试验（地毯耐用性试验）
treated water 净化水，处理过的水
treatment condition 处理条件
treatment effect 处理效果
treatment vessel 处理箱，反应箱
treble cloths 三层织物
treble column cylinder drying machine 三柱烘筒烘燥机
tree bark 树皮绉
treebark stripe 树皮条纹（花型名称）
trefoil bur 三叶形草刺
trellised pleat pressing machine 格花打褶机《针》
trellis work type design 格花刺绣花样
trestle 支架
trevet(=trevette) [丝绒]割绒刀
Trevira 特雷维拉（聚对苯二甲酸乙二酯纤维，商名，德国制）
Trevira Finesse 特雷维拉聚酯微细旦纤维（商名，德国制）
Trevira Micronesse 特雷维拉聚酯超细旦纤维（商名，德国制）
TRF(temperature regulating factor) 温度调节因子

tri- （构词成分）表示"三""三次""三倍"
TRI (Textile Research Institute) 纺织研究院（美国）
triacetate 三醋酸酯；三醋酯纤维
triacetate rayon 三醋酯人造丝
triacetate staple fibre 三醋酯短纤维
triacetyl cellulose 三乙酰基纤维素
triad 三[单]元组；三价基；三价元素；三合一的
triads 三色相[等间隔]配色
trial and error [method] 反复试验[法]，逐次逼近[法]
trial-manufacture(=trial-production) 试制
trial running 试运转，试车
trial sale 试销
trial site [地毯]试用地点
trial speed 试验速度
triangle knife 三角刮刀
triangle profile(=triangular profile) 三角异形[纤维]；三角形截面
Trianized 锦纶针织物防皱整理（商名，美国）
trianizing 锦纶针织物热定形
triarylmethane dyes 三芳甲烷类染料
Triatex minimum application 〈Triatex-MA〉 一种低给液工艺（商名，瑞士）
triaxial 三维的，三轴向的，三向的
triaxial fabrics 三轴向[平面]织物
triazine 三嗪，三氮杂苯
triazine dyes 三嗪染料（活性染料的一种）
triazinone(=triazone) 三嗪酮（棉织物的耐氯、防缩、防皱树脂整理剂）
tribasic 三元的，三碱的
tribasic acid 三元酸
tribasic alcohol 三元醇
tri-blend fabric 三合一织物

tribo-charging 摩擦充电
triboelectric 摩擦带电的
triboelectricity 摩擦电
triboelectrification 摩擦起电
tribometer 摩擦计
tributyl phosphate〈TBP〉磷酸三丁酯(消泡剂,抗菌剂)
tributyltin〈TBT〉三丁基锡(禁用防腐杀虫剂)
TRIC (transfer reproduction in colour) 颜色转移复制法
tricel 三醋酯纤维
trichloroacetic acid 三氯醋酸,三氯乙酸
2,4,6-trichloroanisole 2,4,6-三氯苯甲醚(聚酯纤维的染色载体)
trichlorobenzene 三氯苯
trichlorocarbanilide 三氯均二苯脲(杀菌剂,用于织物消毒整理)
trichlorocyanuric acid 三氯氰酸
trichloroethylene 三氯乙烯(去除织物油脂蜡质的不燃性溶剂)
trichlorofluoromethane 一氟三氯甲烷(干洗剂)
trichloropyrimidine dyes 三氯嘧啶[型]染料(活性染料的一种)
trichoderma 绿色木霉
trichospoum penicillatum 青霉变种霉
trichroism 三原色[现象]
trichromatic 三原色的;三色版的
trichromatic coefficient 三原色系数
trichromatic colourimeter 三原色比色计
trichromatic colour separating 三原色分色
trichromatic coordinates 三原色[色度]坐标
trichromatic matching 三原色比色
trichromatic printing 三原色印花(用红黄蓝三种原色套印而成的多套色效应)
trichromatic system 三原色表色系统
trickling 滴流,细流
trickling bed 生物滤池
trickling filter 生物滤池;滴滤池
Trico Dim Test Instrument 织物三维变形测试仪(商名,匈牙利制)
tricot 经编织物
tricot cut 一卷(或一匹)经编织物
tricot de laine 羊毛针织物(做海军服用)
tricot guider 圆盘吸边器(主要用于经编针织物)
tricot heat setting machine 经编织物热定形机
tricot knit corduroy 经编灯芯绒
tricot knitting 经编
tricresyl phosphate 磷酸三甲苯酯(增塑剂)
tri-dimensional 三维的,三度的
triethanolamine〈TEA〉三羟基乙基胺,三乙醇胺(俗名)
trifoil crosssection 三叶形截面
trifunctional reactive dyes 三官能团活性染料
trigger 闸柄;扳机;起动线路;触发线路;触发器
triglyceride 三甘油酯
trihydric acid 三价酸
trilateral cross section 三角形截面,三边形截面
trilobal 三叶形[的]
Trilon B,TB 曲力龙 B,TB(螯合剂,乙二胺四醋酸化合物,商名,德国巴斯夫)
trim 装饰,整修,调整;切毛边
trimer 三聚[合]物
N,N',N'-trimethylmelamine N,N',N'-三

甲基三聚氰胺(耐皂洗耐氯的树脂整理剂)
trimethyl methylol melamine 三甲基[三]羟甲基三聚氰胺(防缩防皱树脂整理剂)
trimethylol melamine 三羟甲基三聚氰胺(防缩防皱树脂整理剂)
trimmed selvage motion 修边装置
trimming 装饰
trimming machine 布边修剪机
trimming stand 修剪架(圆网附属设备,将圆网套在架上,修剪其两端至规定长度)
trinitrobenzene 三硝基苯
trinitrocellulose 三硝基纤维素(俗称火药棉)
trintrotoluene〈TNT, T. N. T.〉 三硝基甲苯
trio 三辊[拉伸机]
triode 三极管
trioxan[e](=trioxymethylene) 三噁烷,三聚甲醛,聚甲醛
trip 自停机构;跳闸装置;分离机构
tri-pad [针织物]松式开幅脱水处理机
triphasic 三相的
triphenylmethane dye 三苯甲烷染料
triple bond 三键
triple cloths 三层织物
triple effect evapourator 三效蒸发器
triple point 三相[交]点(固、液、气三相)
triple pressure zone calander 三辊轧光机
triple sheer crepe 半透明薄绉
triple shift 三班制
tripped 跳闸《电》
tripping mechanism 脱扣机构,断路机构
tripping point 跳闸点《电》

trip-set 行程调定
trip thread 引燃线
trip wheel 棘轮
tris[N-methylol-2-carbamoylethyl] amine 三[N-羟甲基-2-氨基甲酰乙]胺(棉织物的耐酸洗、碱洗和氯损的防缩防皱交联剂)
tris-aziridinyl phosphine oxide〈APO〉 三乙烯亚氨基膦化氧(棉织物阻燃整理用)
trisazo compound 三偶氮化合物
trisazo dye 三偶氮染料
triskelion cross section 三弯叶形截面,三弯脚形截面
trisodium phosphate 磷酸三钠
tris[polyhaloaliphatic] phosphate 三[多卤化脂肪族]磷酸酯(纤维用阻燃剂)
tris(2-carboxyl ethyl)phosphine〈TCEP〉三羧甲基磷酸(羊毛低温染色助剂)
tristimulus colourimeter 三原色比色计
tristimulus coordinates 三原色[色度]坐标
tristimulus values(X, Y, Z) 三刺激值(X、Y、Z)
tritanopia 丙型色盲,蓝黄色盲
tritanopic chromaticity confusion 丙型色盲错乱(蓝、黄色盲)
Triton 曲力通系列(精练剂,润湿剂,净洗剂,商名,美国制)
Triton X 100 曲力通 X 100(烷基芳基聚醚醇,用作净洗剂、精练剂、乳化剂、润湿剂,商名,美国制)
triturate 磨碎;磨碎物
tritylation 三苯甲基化
trivalent chromium 三价铬
trivat(=trivet, trivette) [丝绒]割绒刀
trivial name 俗名

T. R. J. (= Text. Res. J., Textile Research Journal) 《纺织学报》(美国月刊)
trochoid 摆动
trolley 集电器;滚轮,触轮《电》;手推车,小车
trolley head 触轮,集电头《电》
trolley system 馈电装置
trolley wheel 滚轮,触轮
trombone cooler 长度可调式冷却器
trommel 滚筒筛
trona 天然碱
tropal 衬里[用]织物,里子布
tropical 薄型外衣织物,夏令织物;热带的
tropical cotton 薄型棉布(热带服装用)
tropical design 热带风情图案
tropical fancy suiting 薄花呢
tropical suiting 夏令织物;薄型精纺呢
tropical wear 轻便服装;热带服装
trouble 故障,事故
trouble-free 无故障
trouble-free functioning 无故障功能
trouble-free operation 正常操作
trouble-free running 正常运转
trouble light 事故信号灯
trouble shooting 查找故障;故障检修
trouble shooting guide 故障排查指南,故障检修指南
trough 槽,盆,缸,池;托盘;储槽
trouserings 裤料
trousers 裤子
Trouton viscosity 特鲁通黏度
Trubenizing 托律本硬挺整理[工艺](将含醋酯纤维的棉衬布夹入两层棉布间,一同浸入丙酮液内,使醋酯纤维溶化,然后轧压,三层布黏合为一,用以制衬衫领、袖口等,商名)
truck 布车,布箱,手推车;卡车;运货车
truck dryer 推车式烘燥机
Tru-colour jet printer 彩色喷墨印花机(用于打样布,商名,荷兰斯托克)
true boiling point〈TBP〉真沸点
true colour 真色度(去除悬浮固体后的溶液色度)
true colour unit 真色度单位(1.0毫克/升氯铂酸盐离子的色度)
true complement 补码《计算机》
true form 原码;原码形式《计算机》
true gauge length [强力试验机上铗子钳口间的]正确隔距长度
true selvedge 光边
true silk 真丝,天然丝,蚕丝
true solution 真溶液
true tensile strength 最大拉伸强力
true-up 校准
trumpet-shaped duct 喇叭口管道(绳状染色机)
truncate 截尾;截顶;截断
trunk 箱;干路;干管;中继线
trunke 运动短裤;平脚短泳裤
trunk relay〈TR〉中继线继电器,中继器
trypsin 胰蛋白酶
tsat(temperature of saturation) 饱和温度
TSCA(toxic substance control act) 毒物管理条例
T-screen T形圆网(用以印制转印纸的特制圆网)
T-shirt printer T恤衫印花机
TSPP(tetra sodium pyrophosphate) 焦磷酸四钠
TS-PU(temperature-sensitive polyurethane) 温度感应型聚氨酯

T-square 丁字尺
TSS (total suspended solids) 悬浮固体总量(污水处理用语)
tsumugi 绵绸(日本名称)
TTD (Textile Technology Digest)《纺织工艺文摘》(美国月刊)
TTFC (Textile Technical Federation of Canada) 加拿大纺织技术协会
T-topper 圆领衫
TTY (teletype, teletypewriter) 电传打字[电报]机
TTY (throwster textured yarn) 捻丝厂[生]产[的]变形丝
T. U. (thermal unit) 热单位
tub 盆,桶,槽
tubbable silk 耐洗丝绸
tub-dip 小量染纱;小量浆纱
tube 管;纱管;电子管;(簇绒地毯机的)植绒管
tube clip 管卡
tube conveyor 空[筒]管输送带
tube dyeing machine 管式染纱机
tube fittings 管子零件,管配件
tube fuse 管状保险丝
tube opening machine 圆筒形针织物开幅机
tube spanner 管子扳手,管子钳
tube-type heat exchanger 列管式热交换器
tube-type jet dyeing machine 管式喷射染色机
tube wrench 管子扳手,管子钳
tub fabric 耐洗织物
tub-fast 耐洗耐烫的
tubing 管状织物
Tubivis ECO 400 一种环保型增稠剂(商名,瑞士 CHT)

tub-liquoring 小量染纱;小量浆纱
tub silk 耐洗丝绸
tubular 圆筒形的,管状的
tubular continuous digester 管式连续蒸煮器
tubular dryer 管式干燥器
tubular fabric 圆筒织物
tubular fabric electronic slitter 圆筒织物电子开幅机
tubular fabric singeing machine 圆筒织物烧毛机
tubular fabric slitting machine 圆筒织物剖幅机
tubular fabric turning machine 圆筒织物翻布机
tubular fabric vertical drying machine 圆筒织物立式烘干机
tubular form 圆筒状
tubular hosiery 圆筒形针织物
tubular ionizator 管状电离器(用于静电消除)
tubular knit mercerizing 筒状针织布丝光
tubular knit mercerizing machine 筒状针织物丝光机
tubular knitted fabric 圆形针织物
tubular membrane ultrafiltration 管状膜超过滤(水处理技术)
tubular ozonizer 管状臭氧发生器
tubular radiator 管式散热器(散热器的芯子是圆管、方管或椭圆管)
tub washer 盆式洗毛机
tub washing machine 洗桶机,桶式洗衣机
tucks 褶皱织物,褶裥织物
tucks stain 折皱污迹(织疵)
tuck weave 起皱组织,褶裥组织
tucum (=tucun) 特肯纤维(棕树叶纤维,

制渔网及钓鱼线用,巴西产)
tuft 毛圈;簇绒;毛茸;绒头
tuft definition 簇绒清晰度
tuft-detuft process 簇绒法间隔染色
tufted blanket 簇绒毯
tufted broadloom carpet 簇绒宽幅机织地毯
tufted carpet 簇绒地毯
tufted fabric 簇绒织物,栽绒织物
tufted pile 簇绒
tufted rug 簇绒地毯,绒头地毯
tufteds 簇绒织物,栽绒织物
tufted yarn 簇绒纱
tufting 簇绒;簇绒法
tufting machine 簇绒机,栽绒机
tufting-out 绒头脱落,绒头磨蚀
tufting substrate 簇绒基底
tuft mockado 充丝绒(全毛或丝毛交织的拉绒织物)
tuft pull strength 绒头抗拔强力
tufts 簇绒产品(指织物)
tuft withdrawal force [地毯的]绒头拔出力
tuft withdrawal tensometer [地毯]绒头拔出力试验仪
tulle baran 印花绒布(英国制)
tumble 转筒缩绒;翻滚,滚动
tumble abrasion 翻转式耐磨[试验]
tumble dry 转笼干燥
tumble dryer 转笼烘燥机(主要用于服装干燥)
tumble dyeing machine 转笼式染色机
tumble jar dynamic absorption test 转动瓶动态吸收试验(测抗水性)
tumbler 倒扳开关;转臂
tumbler-effect 转笼式效果

tumbler finishes 转笼烘干整理剂
tumbler shrinkage 转笼烘干收缩
tumbler washer 转笼式洗衣机
tumbling barrel process 转涂法;转筒翻滚法(将纺织品在筒内互相摩擦产生光洁表面)
tumbling machine 揉搓滚光机(合成革加工)
tumbling mixer 转鼓式混合器
tumbling pilling tester 翻滚式起球仪
tungsten-bromine lamp 溴钨灯
tungsten compound 钨化合物
tungsten-halogen lamp 卤钨灯
tungsten-iodine lamp 碘钨灯
tuning 调整,校准;调谐
tunnel drier for bobbins 隧道式[筒管丝]干燥机
tunnel drying machine 隧道式烘燥机
tunnel shrinking machine 隧道式收缩机
tunnel steamer 隧道式汽蒸机
tunnel stoving machine 隧道式[羊毛]熏白机
tunnel test 隧道试验(测试地毯阻燃性)
turbidimeter 浊度计
turbidimetric analysis 比浊分析
turbidimetric method 浊度测定法;比浊法
turbidimetric titration 浊度滴定[法]
turbidimetry 浊度测定法;比浊法
turbidity 浊度;混浊[性]
turbidity point 浊度点
turbidometer 浊度计
turbine 涡轮机,叶[汽或水]轮机
turbine pump 涡轮泵,叶轮泵
turbine-type impeller agitator 涡轮型叶轮搅拌器
turboblower 涡轮式鼓风机

turbo electrofinisher 涡轮式电加热压烫机
turbofan 涡轮通风机
turbogenerator 汽轮发电机
turbo humidifier 涡轮式给湿机
turbomixer 叶轮式混合器
turbonator 振荡器
Turbo setter 吐波式定形机（直接成条的短纤维，经减压蒸汽热处理，使纤维成为无残余收缩的装置）
turbo spray washer 旋转式喷淋洗涤机
turbulator 扰流器
turbulence 湍流，紊流
turbulent diffusion 湍流扩散
turbulent dyeing process 湍流染色法
turbulent flow 紊流，湍流
Turkey blue 土耳其蓝［色］（蓝色）
Turkey red 土耳其红［色］（红色）；红色颜料
Turkey red oil 土耳其红油，磺化蓖麻油，太古油
turmeric 郁金姜黄［色］（黄色）；黄色植物染料（俗称姜黄）
turmeric paper 姜黄［试］纸
turn 转；圈；翻倒；转变；捻回
turn around time 生产周期
turnbuckle ［松紧］螺丝扣；紧线器《机》
Turnbull's blue 滕［部尔］氏蓝［色］（深蓝色）
turned down (= turned off) ［花筒］车削（车削除旧花纹）
turned over edge 卷边，翻边（织疵）
turning device 转动装置
turning lathe 车花筒机《印》
turning machine 车削花筒机；（圆筒形针织物的）翻布机
turning moment 转动力矩

turning point〈TP，T. P.〉转点，变点
turning site 转弯部位（地毯试验）
turning tool 车刀；切削刀
turnmeter 转速计；回转速度指示器
turn over〈TO〉周转；翻转；见反面
turn-over figure 两面相同的图案
turnover point ［曲线］转折点，倒转点
turns ［纱线］捻数，捻度；转数
turntable 转盘，转台
turpentine 松节油（印花助剂）
Turpex KM 吐尔佩克斯 KM（阴离子型活性聚合物，能增强纤维素纤维针织品的强力与耐磨性能，对合纤有防抽丝效果，商名，瑞士制）
Turpex NP 吐尔佩克斯 NP（非离子型的油改性三聚氰胺甲醛缩合物，耐水洗的合纤及其混纺织物柔软剂，商名，瑞士制）
turquoise blue 绿松石蓝［色］（绿光艳蓝），湖蓝［色］，翠蓝［色］
turquoise green 绿松石绿［色］（绿光艳蓝，比绿松石蓝稍绿）
turret ring gear 转塔环形齿轮
tussah 柞蚕丝；柞蚕丝织物
tussah cloth 柞丝绸
tussah fabric 柞丝织物
tussah silk 柞蚕丝；柞蚕丝织物
tussah velvet 柞蚕丝绒（以柞蚕丝作绒头）
tussah wool 仿柞丝呢
tussores 罗缎（丝光细经粗纬棱条布）；柞蚕丝，野蚕丝
tussur silk 柞蚕丝
Tuterette process 一种棉/涤织物机械收缩工艺
TV(terminal velocity) 终端速度，末速

[度]

TVA（thermal volatilization analysis）热挥发分析[法]

TW（Textile World）《纺织世界》（美国月刊）

Twaddell（＝Twaddle）特沃德尔（比重计）

Twaddell degree〈Tw〉特沃德尔度（液体比重度数）

Twaddle hydrometer 特沃德尔比重计

twalle 人[造]丝素绸（统称）

t. w. c.（total work cost）总工作费用

tweed 粗呢,粗花呢

tweed jersey 仿毛平纹针织物；粗呢运动衫

tweel 斜纹

Tween 吐温（失水山梨糖醇脂肪酸酯与环氧乙烷的加成物,属非离子型乳化剂、溶解剂等,商名,英国制）

Tween 80 吐温 80,乳化剂 T-80（失水山梨糖醇油酸酯聚氧乙烯醚,亲水性表面活性剂,商名）

Tween emulsifiers 吐温型乳化剂

tweezers 捏钳,镊子

twill 斜纹

twill crepe 斜纹绉

twilled satin 直贡呢；棉直贡呢

twilled swansdown 凸纹布,绒布

twill habutai 斜纹绸

twills 斜纹织物,哔叽

twill weave 斜纹组织

Twin air 中空保暖聚酯纤维（商名,日本旭化成）

twin air cylinders 双气缸

twin fabric 组合织物

twin helical gear 人字齿轮

twin jig 成对卷染机（上下缸染色用）

twinpad dye tank 双浸渍染槽

twin-print 双面印花

twin roller stretching machine 双辊拉伸机

twinset 两件式服装

twin star dyeing machine 双星形架染色机

twist〈T〉捻度,捻回,加捻；经纱

twist-and-contraction meter 捻度和捻缩试验仪

twist break method 解捻法

twist carpet 加捻纱地毯

twist coefficient 捻度系数

twist counter 捻度试验仪

twist crimped yarn 加捻卷曲纱[线],热定形卷曲纱[线]

twist curled yarn [假捻]卷曲变形纱

twisting 加捻

twisting force 扭力

twisting moment〈TM〉扭矩,捻矩

twist per metre 每米捻数（以 T_m 表示）

twist regularity meter 光电式[纱线]捻度均匀度测定仪

twist run back 退捻,回捻

twist-set-untwist method 加捻—热定形—解捻法（即假捻法,变形纱制造法之一）

twists per inch〈T. P. I.〉每英寸捻数

twist stress relaxation 扭曲应力松弛,扭力松弛

two bath dyeing 两浴染色

two bath process 二浴法

two-bath union dyeing 交织物二浴染色法

two-bowl embossing calender 两辊拷花轧压机

two-bowl squeezer 两辊轧车

two colour dyeing 二色染色（织物上两种纤维同浴染色产生两种颜色）

two-colour effect 双色效应,两色效应
two-component mixing jets 双组分混合喷嘴
two dips and two nips 二浸二轧
two-disc tensioner 双[摩擦]盘张力器
two-drum boiler 双汽包锅炉
two-faced plush cloth 双面绒布
two-faced terry cloth fabric 双面毛圈织物
two feed 双喂[系统]
two-fold yarns 双股线
two-layer fabric 双层织物
two-layer laminates 双层层压制品
two-loop 双环,双层,双幅
two-mode method 二阶段荧光测色方法
two-monochrometer method 二分光器荧光测色法
two-phase cleaning 两相净洗(即乳液净洗)
two phase flash-ageing method 两相快速蒸化法
two phase laminating process 二相法层压工艺
two-phase mercerizing process 二相法丝光工艺
two-phase polycondensation 二相缩聚
two-phase printing process of vat colours 还原染料两相印花法
two-phase process 两相法
two-phase structure 二相结构
two-phase vat discharge prints 二相法还原染料拔染印花
two-phase wet fixation process 二相法湿固着工艺
two-piece bearing 双开轴承;拼合轴承
two-piece dress 两件套服装
two-ply 双股;双层;叠织[起毛]织物

two-pole〈tp〉 两极的
two-roll padder 两辊轧车
two-sided effect 正反面效应,阴阳面效应
two-sided knit goods 双面针织物
two side in one 双面合一
two sideness [织物]两面色差,正反面差异性
two-speed drive 双速转动
two-stage printing(＝two-stage printing process) 两步法印花,两相法印花(指先印染料和糊料,后用化学品处理得色)
two-stage reserve printing 两步法防染印花
two-stage rope scouring and bleaching plant 双程绳状练漂机
two-step system 两步法(如对混纺织物进行两次染色)
two-tone 双色调(浓淡或明暗色调)
two-tone colour 同色不同色调;二并色
two-tone dyeing 双色调染色
two-tone effects 同色深浅效应;[织物]正反面同色深浅效应;闪色效应
two-tone stripe 双色条纹(两种不同颜色交替相间的条纹)
two water action [洗衣机的]双向水流洗涤方式
two-way coat 两用外套
two-way design 双面花样
two-way elastic fabrics 双向弹力织物
two-way print 双面印花
two-way stretch woven fabrics 双向(经纬)拉伸机织物
tying 接结;捆扎
Tylose 泰罗斯(纤维素醚,商名,德国制)
Tyndall effect 丁达尔效应(光线的反射、

散射理论）
Tyndall phenomenon 丁达尔现象
typed cloths 标准家用布（如餐巾、茶巾、盘垫、杯垫布等，布上常织有用途名称）
type sample 标[准]样[品]
type test 典型试验
typewriter cloth（＝**typewriter ribbon**） 打字[机]带
typha fibres 香蒲纤维（一种韧皮纤维）

typical formula 代表处方，典型处方
typical value 代表值《数》
typographic block [活版]印[刷]版
typography 活版印刷术，凸版印刷术
tyre cord fabric 轮胎帘子布
tyre fabric 轮胎用织物
Tyrian purple 泰尔红紫（从海螺中获得的紫色染料）
tyrosine 酪氨酸

U

UA（ultra-audible） 超音速的
U-bar 槽钢
Ubbelodhe viscosimeter 乌伯娄德黏度计
U-bend U形弯头
U-bolt U形螺栓,马蹄螺栓
U-box U形槽（染整连续加工设备）
UBS（urea-bisulphite solubility） 尿素亚硫酸氢盐溶解度（用于测定羊毛的碱损）
Ucarsil TE-24 欧卡西尔 TE-24（硅酮柔软剂,可增强树脂整理织物的耐久熨烫性,商名,美国制）
UCC（Union Carbide Corporation） 联合碳化物公司（美国）
UCL（=U.C.L., upper control limit） ［质量］控制上限

U-clamp U形夹
UCMTF【法】法国纺织机械制造商协会
UCR（under colour removal） 地色去除
UCS（uniform colour scale） 均距色标度
UCS diagram（uniform chromaticity scale diagram） 均距色度图
UD（ultradeep dyeing） 超深染色性
udell 接受［冷凝水汽］器
UF（=U.F., U/F, u.f., urea formaldehyde） 脲醛［树脂］
UF（=u.f., ultra-fine） 超细度;超细的,特细的
UF（ultrafiltration） 超滤［作用］
UF（utilization factor） 利用系数
U/F ratio（urea-to-formaldehyde ratio） 脲与甲醛［克分子］比

U-gauge U形压力表
UHF（=uhf, ultra high frequency） 超高频
UHMWPE fibre（ultra-high molecular weight polyester fibre） 超高分子量聚乙烯纤维（优良抗腐蚀,高吸热性）
u.i.（ut infra） 如下所述,如下所示
UL（Universal League） 世界联盟
ulbricht sphere 积分球《测色》
ULPA filter（ultra low penetration air filter） 超级低穿透率空气过滤器
ult.（ultimate） 极限的
ult.（ultimo） 上月的
ultiliy uniform 公共事业人员制服
ultimate〈ult.〉 极限的
ultimate bending strength 极限弯曲强力
ultimate compression strength 极限耐压强力
ultimate consumer 最终消费者
ultimate elongation 极限伸长,断裂伸长
ultimate load 极限载荷,最大负荷
ultimate newtonical viscosity 极限牛顿黏度
ultimate oxygen demand〈UOD〉 ［污水］理论需氧量;极限需氧量
ultimate strength 极限强度,极限强力
ultimate stress 极限应力
ultimate tensile strength〈UTS, U.T.S.〉 极限抗拉强力
ultimate yield 最终得率
ultimo〈ult.〉 上月的
ultra- （构词成分）表示"极端""超""过"
ultra-audible〈UA〉 超音速的

ultracentrifugal analysis 超离心分析
ultracentrifuge 超离心机
ultra-deep-dyeable fibre 超深染性纤维
ultradeep dyeable yarn 超深染色纱
ultradeep dyeing〈UD〉超深染色性
ultradeeping 超深染色
ultra-disperse dyes 超细分散染料
ultrafibrillen 超[细]原纤
ultrafilter 超滤器
ultrafiltration〈UF〉超滤[作用]
ultrafiltration membranes 超滤膜（水处理技术）
ultrafiltration process 超滤法（水处理技术）
ultrafiltration technology 超滤技术（水处理技术）
ultrafiltration treatment 超滤处理
ultra-fine〈UF, u. f.〉超细度；超细的，特细的
ultrafine denier fibre(=ultrafine fibre) 超细纤维《化纤》
ultra high frequency〈UHF, uhf〉超高频
ultra-high molecular weight polyester fibre〈UHMWPE fibre〉超高分子量聚乙烯纤维（优良抗腐蚀，高吸热性）
ultra low penetration air filter〈ULPA filter〉超级低穿透率空气过滤器
ultramarine 群青（如人造群青、天然群青）
ultramicro 超微量，超微
ultramicroanalysis 超微量分析
ultramicro balance 超微量天平
ultramicro coating 超微粒子涂层
ultramicro determination 超微量测定
ultramicro emulsion 超微乳液
ultramicro fibre 超细纤维

ultramicro film 超微粒子膜
ultramicron 超微细粒
ultramicro sample 超微试样
ultramicroscope 超[倍]显微镜
ultramicrostructure 超微结构
ultramicrotome 超薄切片机
ultra-oscilloscope 超短波示波器
Ultraphil 一种透湿散热功能性助剂（商名，瑞士制）
Ultraphil HCT 一种吸湿排汗剂（商名，瑞士制）
Ultraphor 乌特拉福（荧光增白剂，商名，德国巴斯夫）
ultraporous structure 超微孔结构
ultrarapid accelerator 超速加速器
ultra-rapid ager 超速蒸化机
ultra-rapid developer 超速显色剂
ultrarapid frequency 超射频率
ultra-γ ray 超γ射线，宇宙线
ultra-red 红外的
ultra-red ray 红外线
Ultra Release Teflon 多功能型易去污特氟龙（商名，美国杜邦）
ultrashort wave〈USW, usw〉超短波
ultrasoft 超柔软
Ultrasoft CW 超柔软整理剂CW（有机硅类羊毛防毡缩剂，兼有柔软作用，可省却氯化处理工序，商名，英国PPT）
ultrasonic air-floating method 超声气浮法《环保》
ultrasonic apparel machinery 超声波成衣机械
ultrasonic bath 超声波浴
ultrasonic bleaching 超声波漂白
ultrasonic bonding 超声波黏合
ultrasonic bonding process 超声波黏合工

ultrasonic calender 超声波压轧机(用于层压多层纺织品或其他材料)
ultrasonic cleaner 超声波除垢器;超声波清洗装置
ultrasonic control 超声控制
ultrasonic cutting 超声波切割
ultrasonic degradation 超声波降解[作用]
ultrasonic [flaw] detector 超声波探伤仪
ultrasonic dispersion 超声波分散
ultrasonic dyeing 超声波染色
ultrasonic eco-pretreatment 超声波生态前处理(酶处理)
ultrasonic finishing 超声波整理
ultrasonic generator 超声波发生器
ultrasonic humidifying system 超声给湿系统
ultrasonic inspection 超声波检验
ultrasonic irradiation 超声波辐射
ultrasonic laminating unit 超声波层压机
ultrasonic melt-joining 超声波熔接
ultrasonic pulse technique 超声波脉冲技术
ultrasonic purification 超声波净化《环保》
ultrasonic quilting machine 超声绗缝机
ultrasonic radiation 超声[波]辐射[作用]
ultrasonic sensor 超声波传感器
ultrasonic shearing 超声波剪切
ultrasonic thermo-bonding 超声波加热黏合
ultrasonic vibration dyeing 超声波振荡染色
ultrasonic washing 超声波洗涤
ultrasonic wave 超声波
ultrasonic welding 超声波焊接
ultrasonic wool scouring 超声波洗毛

ultra speed 超速
Ultratex ESU, EMJ 欧特斯 ESU、EMJ(多功能自交联耐久性柔软剂,硅酮弹性体,适用于浸渍加工,商名,瑞士制)
Ultratex FSA, FSL, FMK 欧特斯 FSA、FSL、FMK(多功能耐久性柔软剂,硅酮弹性体,适用于浸轧加工,能改善织物回弹性抗皱性,商名,瑞士制)
ultra-thin film 超薄薄膜
ultrathin section (电子显微术的)超薄切片
ultraviolet〈UV, u/v, uv, u. v.〉 紫外线;紫外[线]的;光谱紫外区
ultraviolet absorber〈UV absorber〉 紫外线吸收剂;紫外线吸收器
ultraviolet absorption spectrum 紫外线吸收光谱
ultraviolet activation 紫外线活化[作用]
ultraviolet and visible spectrum analysis 紫外线和可见光谱分析
ultraviolet curing of polymers 聚合物紫外固化
ultraviolet cut-off filter 紫外线滤光器
ultraviolet degradation 紫外线降解
ultraviolet detector〈UVD〉 紫外线探测器
ultraviolet electronic energy spectrum 紫外光电子能谱
ultraviolet filter 紫外线滤色镜
ultraviolet fluorescence meter 紫外线荧光仪
ultraviolet fluorescence of textiles 纺织品紫外线荧光性
ultraviolet indicator 紫外线指示器
ultraviolet inhibitor 紫外线抑制剂
ultraviolet irradiation〈UVI〉 紫外线辐射
ultraviolet lamp 紫外光灯

ultraviolet light 紫外光[线]
ultraviolet light absorber 紫外线吸收剂
ultraviolet luminescence 用紫外线照射发光
ultraviolet microscope 紫外光显微镜
ultraviolet protection agent 紫外线防护剂
ultraviolet [ray] protection factor(UPF) 紫外线防护因子,紫外线辐射防护系数
ultraviolet protection finish 防紫外线整理
ultraviolet radiation 紫外线辐射[作用]
ultraviolet radiation absorber 紫外线辐射吸收器
ultraviolet ray 紫外线
ultraviolet ray curing coating 紫外线焙固涂层
ultraviolet ray curing ink 紫外线焙烘印墨
ultraviolet ray proof(= ultraviolet ray protection) 防紫外线
ultraviolet ray source 紫外光源
ultraviolet ray stabilizer 紫外线稳定剂
ultraviolet-reflectance agent 紫外线反射剂
ultraviolet screener 紫外线屏蔽剂
ultraviolet sensitizer 紫外敏化剂
ultraviolet shield agent 紫外线屏蔽剂
ultraviolet shielding ratio 紫外线屏蔽率
ultraviolet shield technology 紫外线屏蔽技术
ultraviolet spectrophotometer 紫外线分光光度计
ultraviolet spectrophotometry 紫外光谱测定法
ultraviolet spectrum 紫外光谱
Ultravon LX 欧特拉冯 LX(阴离子型净洗分散剂,用作煮练助剂,洗涤功能佳,分散作用大,商名,瑞士制)
umber 赭色,浓[红]茶色;棕土(一种天然颜料)
umbrella cloth 伞布
umbrella hank stand 伞形绞纱架
umbrella silk 伞绸
UMIST (University of Manchester Institute of Science and Technology) 曼彻斯特大学科学技术学院(英国)
Umstätter viscometer 乌[姆施泰特尔]氏黏度计
unavailable energy 无用能
unbacked carpet 未衬底地毯
unbalance 不平衡,失衡
unbiased variance 无偏方差,均方差
unbleached 未漂的
unbleached linen 未漂亚麻布
unbleached material 坯布
unboiled silk 未脱胶丝,未煮练丝
unbranched chain 直链,不分支的链
uncertainty principle 不确定性原理
unclipped carpet 不剪绒地毯
uncoiling 解卷曲;伸直
unconventional energy sources 新能源
uncreasable 不皱的
uncrushable carpet 耐压地毯(紧捻卷曲绒头)
uncurling action 去卷[曲]作用;剥边[作用]
uncurling apparatus 剥边装置
uncurling finger 剥边螺旋辊
uncut pile 未剪绒毛
uncut pile carpet 未剪绒毛地毯
uncut worsted 毛哔叽(美国名称);精纺毛绒斜纹织物;轻缩绒精纺毛织物
undecomposed〈undec.〉 未分解的
undecylenic acid 十一碳烯酸
underbaked 焙烘不足

under base 打底
under base colour 地色
under basing 打底处理
underbed 底架,底座,托板
under blade [剪毛机上的]底刀,平刀
underbracing 下支撑
undercapacity 容量不足;功率不足
undercloth 印花[机]衬布,垫布
underclothes 内衣,衬衣
undercoat 下涂层
under coated fabric 内涂层织物
under colour addition 地色增益(印花分色用术语)
undercoloured 染色不足
under colour removal 〈UCR〉 地色去除(指彩色图像中减少三原色光量,印花分色用术语)
undercooling 过冷,冷却过度
undercure 焙烘不足,固化不足;硫化不足
under cut 浮雕《雕刻》
underexposure 曝光不足
underfeed 反超喂[装置];喂不足
underjigger 液下卷染机
under lap 延展线
underlay [地毯]衬垫
under line 分线《雕刻》
under liquor dwell unit 液下堆放单元
under liquor scouring and bleaching 液下连续练漂
under liquor scouring and bleaching range 液下(履带式)练漂机
underliquor squeezing device 液下轧压装置
underplate 底板,底座,垫板
underpressure 压力不足;负压
underpressure distillation 减压蒸馏

underrefining 精练不足,欠精练
under refining yarn 生纱
under rest 汗背心
under screen 漏底
undershrinking 收缩不足
undersize 缩小尺寸
underskirt 衬裙;汗衫
under structure 里层结构
undertone 地彩;淡色,浅色
undervoltage relay 〈uvr〉 低压继电器
underwater degumming machine 液下精练机
under water jig(＝under water jigger) 液下卷染机
underwear 内衣
under weight 重量不足
under width 幅宽不足
undischargeable 不可拔性
undistorted 不失真
undress 便装,便服
undressed 未加工的,未经处理的
undressed finish 轻缩绒整理(微露地纹)
undressed worsted 缩绒精纺毛织物
undue heat 过度热量;过分热量
undulate 使成波纹
undyed 未染色的
undyed cloth for staining 沾色白布(作沾色试验用)
uneasily biodegradable auxiliary 不易生物降解的助剂
Unel 尤奈尔(有锦纶6、锦纶66和聚氨基甲酸酯纤维等品种,商名,加拿大制)
unequal-arm scale 不等臂秤
UNESCO (United Nations Educational, Scientific and Cultural Organization) 联合

国教育、科学及文化组织,联合国教科文组织
unetched fibre 未侵蚀纤维
uneven 不匀(如印染色泽)
uneven ageing 老化不匀
uneven bleaching 漂白不匀
uneven blotch 满地不匀
uneven bottoming 打底不匀,底染不匀;预媒染不匀
uneven calendering 轧光不匀
uneven cloth 布面不匀[的]织物
uneven compression 挤压不匀(预缩整理疵)
uneven cooking 斑煮,煮练不匀
uneven cover 布面不匀(织疵)
uneven covering [地色]刮浆不匀
uneven craping 皱斑;皱疵;起绉不匀
uneven crimping 皱缩不匀,卷曲不匀
uneven developing 显色不匀
uneven discharge 拔染不匀
uneven drying 烘燥不匀
uneven dyeing 染色不匀
uneven dyeing for fashion effects 花式染色(不均匀的时尚染色效果)
uneven filling 横档,纬档
uneven finishing 云斑(丝绸精练漂白疵);整理不平整
uneven fixing 固着不匀
uneven ground 地色深浅不匀
uneven lustre 光泽不匀
uneven mark 印花不匀(印疵,指同位花纹色泽有差异)
uneven mercerization 丝光不匀
uneven milling 缩呢不匀,缩绒不匀
uneven napping [布面]起绒不匀
unevenness 不匀率,不匀度;不匀性

unevenness of bleaching 漂白不匀
uneven printing 印花不匀,印花斑(疵点)
uneven printing of repeats 接版深浅(手工印花疵)
uneven raising 起绒不匀,拉绒不匀,刮绒不匀
uneven resin finishing 树脂整理不匀
uneven rolling 轧痕,轧印
uneven scouring 练斑(绸匹各部分精练程度不一)
uneven selvege 不匀布边
uneven setting 定形不匀
uneven singeing 烧毛不匀
uneven stitch length 缝线长度不匀
uneven structure 不均匀组织
uneven water proofing 防水不匀
uneven whiteness 白度不匀
unexposed film 未曝光的胶片
unfast dyeing 不坚牢染色
unfilled 未填充的;未增重的
unfinished 未整理的
unfinished worsted 轻微起绒整理的精纺毛织物
unfixed dye 未固着染料
unflammability 不燃性
unfolding 打开,摊开
ungreenable black 不泛绿黑元
ungummed silk 熟丝,精练蚕丝;精练丝织物
unhomogeneity 不均一性
uni- (构词成分)表示"单""一"
unicoloured 单色的;染单一色的
unicolour paper 单色[转印]纸
unidimensional 线性的,一维的,一元的
unidirectional composite 单向[增强]复合材料

unidirectional flow 单向流动
unidirection diffusion 单向扩散
UNIDO (United Nations Industrial Development Organization) 联合国工业发展组织
Unidyne TG，410H，411 尤尼得 TG，410H，411(防油防污剂,乳液,商名,日本大金)
uniflow 单流,直流
uniflow principle in drying 单流烘干原理
uniform 均匀的;制服,军服
uniform acceleration 匀加速度
uniform chromaticity scale diagram 〈UCS diagram〉均距色度图
uniform cloth 制服呢
uniform colour scale (= uniform colour space)〈UCS〉UCS 表色系统,均匀色空间;均距色标度
uniform diffusion 均匀扩散
uniform dispersion 均匀分散体
uniform dyeing 匀染,均匀染色
uniformity 均匀度,整齐度
uniform load distribution [轧辊]均压效果
uniform squeezer 均匀轧车
Uni-hotflue dryer 尤尼热风烘燥机(一种横导辊式热风烘燥设备,日本制)
unilateral 单方向的,一面的
uninflammable 不易燃的
uninterruptible power system〈ups〉不间断供电设备(停电时也可保证供电的设备)
UNIO (United Nations Information Organization) 联合国情报组织
union 混纺织物;交织织物
union broadcloth 防雨厚毛毯;军用防雨披风

Union Carbide Corporation〈UCC〉联合碳化物公司(美国)
union colour 一致色泽,混染同色(指把含多种纤维的织物染成一色)
union coupling 联管节,联轴节
union damask 交织装饰布
union dyed fabric 染成同色的交织物
union dyeing 混纺交织物染色
union dyes 混染染料(能把混合纤维制品染成同色)
union dyes for wool nylon blends 毛锦混纺用中性染料(1∶2 金属络合染料)
union elbow 弯头[接头]
union joint 管接头,联管节
union nut 联管螺母,管接螺母
union towel 交织毛巾
union yarns 混纺纱
Unipad padder 尤尼派德均匀轧车(一种新型轧车,荷兰制)
uni-paper 单色[转移]纸
Uniperol AC 乌尼佩罗 AC(氧乙基化合物,羊毛、锦纶及其混纺织物的匀染剂,弱阳离子型,商名,德国巴斯夫)
Uniperol KA 乌尼佩罗 KA(腈毛混纺织物的匀染剂,非离子型,商名,德国巴斯夫)
Uniperol SE 乌尼佩罗 SE(氧乙基化合物,专用作聚酰胺纤维和羊毛的匀染剂,商名,德国巴斯夫)
Uniperol W 乌尼佩罗 W(聚乙二醇醚磺化产物,弱阴离子型,羊毛和合纤的染色助剂,具有分散、匀染和保护胶体的作用,商名,德国巴斯夫)
uniplaner wave [毛纤维的]平面波形卷曲
uniqueness 唯一性
unique style 独特风格

UNIS（United Nations Information Service）联合国情报服务处
unispace dyeing 等间隔段染（纱线染色）
Unisperse colours 尤尼斯派司新型水性涂料色浆（含低黏度涂料分散体，商名，瑞士制）
unit 单元；单位；零件，部件；装置，机构；机组，组合
unit area 单位面积
unitary backing ［地毯］单一涂胶
unitary hue 单一色相（指色立体中红、黄、绿、蓝四个单一色）
unit drive 单独传动
United Nations Educational, Scientific and Cultural Organization〈UNESCO〉联合国教育、科学及文化组织，联合国教科文组织
United Nations Industrial Development Oranization〈UNIDO〉联合国工业发展组织
United Nations Information Organization〈UNIO〉联合国情报组织
United Nations Information Service〈UNIS〉联合国情报服务处
united pieces 联匹
United States of America Standards Institute〈USASI〉美国标准协会
United States Patent Office〈USPO〉美国专利局
United States Patent〈USP, U.S. Pat.〉美国专利
United States Standard〈USS, U.S. St.〉美国标准
uniterm ［专利］单元名词
Unithron dyes 尤尼思朗染料（分散和还原混合染料，商名，日本制）

Unithron K dyes 尤尼思朗 K 染料（分散还原复合染料，涤棉混纺织物染色用，商名，日本住友）
unit operation 单元操作
unit package 最小单位包装
unit process 单元作业，单元过程
unit-repeating design 单一花纹循环
unit wiring diagram 单元布线图（用于指导整机的装配，亦称线路装配图）
Univ.（university）大学
Univadine 3FLEX 尤尼维典 3FLEX（匀染剂，非离子型，适用于涤纶染色，有匀染、分散和移染作用，商名，瑞士制）
Univadine PA 尤尼维典 PA（匀染剂，阴离子型，适用于锦纶染色，有特殊覆盖条花的作用，商名，瑞士制）
univalence（=univalency）一价，单价
Universa Colour System 尤尼弗萨多用途加工系统（一种用于小批量生产的轧蒸设备，主要进行连续染色，德国制）
universal calender 通用轧光机
universal chuck 万能卡盘，自动定心卡盘
universal coupling 万向联轴节，万向连接器，万向接头
universal cropping machine 通用剪毛机
universal dividing head 万能分度头
universal grinder 往复磨辊
universal indicator 通用指示剂，万能指示剂
universal joint 万向接头，万向节
Universal League〈UL〉世界联盟
universal loop ager 通用悬挂式蒸化机
universal motor 交直流两用电动机，通用电动机
universal permanent decatizing machine "Decoclav" 德科克拉夫通用型耐久罐

蒸机(商名,德国制)
universal pH indicator pH 通用试纸
universal singeing machine 通用烧毛机
universal stains 通用着色剂(鉴别纤维用)
universal standard 通用标准
universal strength tester 通用强力试验机
universal tester 万能试验机,通用试验仪
universal vertical roller printing machine 通用立式滚筒印花机
universal wear tester 通用式耐磨试验仪
university〈Univ.〉大学
University of Manchester Institute of Science and Technology〈UMIST〉曼彻斯特大学科学技术学院(英国)
Uniwash 高效平幅水洗机(商名,意大利制)
unknown loss 未知损失
unknown sample 未知样品
unlevel dyeing 染色不匀
unlevelling dyes 不匀性染料
unlevel shade 不匀色泽
unlimited pattern area 无限制花型范围
unloader 减压器
unloading port 出布口
unlocking 解开;开锁
unmodified resin 原树脂
unnapped 未起绒的
unoriented fibre 未取向纤维
unoriented rigion 非取向区,未取向区域
unoxidizable 不可氧化的
unpacking 开箱
unpaired electron 不成对电子
unperturbed 无干扰
unpinning device 脱针装置(针板拉幅机出布处的附属装置)

unpinning monitor 脱针监控器
unpleasant odor 令人不愉快的气味
unprereduced printing 非预还原印花(印花浆内不含烧碱、保险粉,只含染料、吊白粉、碳酸钾)
unprocessed fabric 未加工布
unravelling 解开,拆散;阐明,解决
unreacted dye 未反应的染料
unregularity 不规则性,无规律性
unrelated colour 非相关色
unrolling 退卷
unrolling surface 退卷表面
uns.(＝unsym., unsymmetric[al]) 不对称的
unsaponifiable 不能皂化的
unsaturated aliphatic compound 不饱和脂肪族
unsaturated bond 不饱和键
unsaturated polyester 不饱和聚酯
unsaturated state 不饱和状态
unsaturated steam 不饱和蒸汽
unsaturation 不饱和[现象]
unscoured silk 未脱胶丝,未煮练丝
unsharp 轮廓不明显的,不清晰的,模糊的
unshrinkable finish 防缩整理
unshrinkable wool 防缩羊毛
unshrinking 防缩整理
unspool 退绕
unsquare pattern 斜花
unstability 不稳定性
unstable 不稳定的;不坚牢的,褪色的(指染料)
unstable equilibrium 不稳定平衡
unstable radical 不稳游离基(或自由基)
unstationary state(＝unsteady state) 非稳

态

unsupported film 无支承薄膜

unsymmetric[al]〈uns.，unsym.〉不对称的,非对称的

unsymmetry 不对称,非对称

untarnishable thread 不变暗的金属线

untwisted 解捻的,退捻的；无捻的

untwisted silk 未捻丝

untwisted yarn 无捻丝,无捻纱

untwist-retwist method 退捻加捻法（捻度试验）

ununiformity 不均一性

ununiform weft 纬密不匀（织疵）

ununiform weft bar 粗细纬档（织疵）

U-nut U 形螺母

unvatted 未还原的

unwashable fabrics 不宜水洗[的]织物（只适于干洗）

unwashable suits 不宜水洗[的]服装

unweaving mark 拆布痕迹,拆痕

unwinding 退卷,退绕

unwinding process 退卷过程,退绕过程

unwoven fabric 非织造织物,无纺织物

UOD（ultimate oxygen demand）[污水]理论需氧量；极限需氧量

up and down motion（＝up and down movement）上下运动

up curve 上升曲线

update 现代化；更新

updated equipment 新式设备

UPF（ultraviolet [ray] protection factor）紫外线防护因子,紫外线辐射防护系数

upfield 高磁场

upflow 向上流动；上流

upgrade production 升级换代产品

up-grading 提高等级,提高质量

upholstery 室内装潢,室内装饰；室内装潢业

upholstery fabric 家具装饰织物,家具布

upholstery fabric finishing 家用纺织装饰物整理

upholstery twill silk for painting 裱画绫（绫类丝织物用作装裱）

U-pipe U 形管

upkeep 维修,保全保养

UP padder UP 轧车（一种均匀轧车,其轧辊有一个油压腔,可以补偿辊筒的挠度,日本制）

upper binder [多层织物的]上层接结

upper bushing 上轴瓦

upper control limit〈UCL，U.C.L.〉[质量]控制上限

upper plastic limit 黏结点；塑性上限

upright boiling kier 立式煮布锅

upright pile 立绒

upright pile finish 立绒整理

upright shaft 立轴

UPS（uninterruptible power system）不间断供电设备（停电时也可保证供电的设备）

upset 翻转,倒转

upstream 上游；溯流的,逆流的

uptake 吸收,吸取量

up-to-date technique 新技术

upward view 仰视图

uranium hexafluoride 六氟化铀

urea 脲,尿素,碳酰二胺

urea-bisulphite solubility〈UBS〉尿素亚硫酸氢盐溶解度（用于测定羊毛的碱损）

urea formaldehyde〈UF，U.F.，U/F，u.f.〉脲醛[树脂]

urea formaldehyde condensate 脲醛树脂缩

合体
urea formaldehyde precondensate 脲醛树脂初缩体
urea formaldehyde resin 脲醛树脂,尿素甲醛树脂(织物防缩防皱整理剂)
urea-free printing 无尿素印花
urea glutarate 戊二酸脲(爱尔新类染料的助剂)
urea method of printing 尿素印花法
urea peroxide 过氧化尿素(新型漂白剂)
urea resin 尿素树脂
urease method 脲酶法(尿素树脂测定方法)
urea-to-formaldehyde ratio〈U/F ratio〉脲与甲醛[克分子]比
Ureol P 衣里罗 P(水溶性脲醛树脂初缩体,织物的防缩防皱整理剂,商名,瑞士制)
urethane 氨基甲酸酯
urethane coated fabric 氨基甲酸酯涂层织物
urethane elastic fibre 氨基甲酸酯弹性纤维(即氨纶)
urethane elastomer 聚氨酯弹性体
urethane foam 聚氨酯泡沫
urethane foam cushion 聚氨酯[地毯]泡沫衬垫
urethane laminating machine 氨基甲酸酯层压机
urethane leather 聚氨酯人造革
urethane rubber 聚氨酯橡胶
urgent shutdown 紧急停车
uric acid 尿酸
uron 乌龙(与甲醛缩合可作棉织物的防缩防皱交链剂)
Ursol dyes 乌搔染料(染毛皮用氧化色基,商名,德国制)
Uruguay wool 乌拉圭羊毛(以交配种细毛为主)
usable life 适用期,可用寿命
usage 使用,用法
USASI〈United States of America Standards Institute〉美国标准协会
USASCII〈USA Standard Code for Information Intercharge〉美国信息交换标准代码
USA Standard Code for Information Intercharge〈USASCII〉美国信息交换标准代码
USB〈universal serial bus〉通用串行总线《计算机》
USBS(＝U.S.B.S.,United States Bureau of Standards)美国标准局
USDA〈United States Department of Agriculture〉美国农业部
used goods 旧货,旧纺织品
useful efficiency 有用效率
usefulness index 实用性指数
user 用户,客户
user ecology 用户生态学,消费者生态学
user interface 用户接口《计算机》
use value 使用价值
use-yellowing 使用泛黄[现象]
U-shaped U-形的
U-shaped manometer U形压力表
U-shaped trough U形槽
U siphon U形虹吸管
U-Sofcer U形索弗瑟松弛槽(日本生产的一种新型松弛设备,由原来的水平式改进成立式U形槽)
Usometer 织物耐磨牢度测定器(商名)
USP(＝U.S.Pat.,United States Patent)

美国专利
USPO（United States Patent Office） 美国专利局
USS（＝U. S. St., United States Standard） 美国标准
U-steel 槽钢
usual maintenance 小修,日常维修《机》
USW（＝usw,ultrashort wave） 超短波
utensil 器皿,用具
Utiliscope 工业电视装置
utilities 公用设施;实用性
utilizational coefficient 利用系数
utilization factor〈UF〉 利用系数
utilization of capacity 开工率
ut infra〈u. i.〉 如下所述,如下所示
UTS（＝U. T. S., ultimate tensile strength） 极限抗拉强力
U-tube gauge U形管压力计
UV（＝U/V, uv, u. v., ultraviolet） 紫外线;紫外[线]的;光谱紫外区
UV A 长波紫外线
UV absorber 紫外线吸收剂;紫外线吸收器
UV B 中波紫外线
UV C 短波紫外线
UV cure coating agent 紫外线焙烘涂层剂
UV cured coating 紫外线焙烘涂层
UV-cut proceeded fabrics 防紫外线加工织物
UVD（ultraviolet detector） 紫外线探测器
UV excimer radiator 紫外激发辐射
UV exclude 排除紫外线成分
UV filter 紫外线滤光器

UV fluorescent dyes 紫外荧光染料
UVI（ultraviolet irradiation） 紫外线辐射
Uvibond inks 尤维邦德油墨(紫外线焙固油墨,商名,英国制)
UV include 包括紫外线成分
Uvinul D-49 尤维纳 D-49(二苯甲酮衍生物,紫外线吸收剂,能增进合成纤维的染色耐日晒牢度,商名,美国制)
Uvinul N-35 尤维纳 N-35(取代的丙烯腈化合物,非极性聚合物的紫外线吸收剂,能增进合成纤维的染色耐日晒牢度,商名,美国制)
Uvitex BAC 尤辉得 BAC(荧光增白剂,阳离子型,适用于腈纶,耐亚氯酸钠漂白,白度鲜艳,提升率高,牢度高,商名,瑞士制)
Uvitex EBF 尤辉得 EBF(荧光增白剂,非离子型,用于涤纶,低能量型,商名,瑞士制)
Uvitex WG 尤辉得 WG(荧光增白剂,阴离子型,适用于羊毛、丝绸,商名,瑞士制)
UV lamp 紫外灯
UV-Laser 紫外激光(准分子激光)
UV-Laser on textile surface modification 紫外激光在纺织品上的表面改性处理
UV marking pen 紫外记号笔
UV-no textile 防紫外线系列纺织品
uvr（undervoltage relay） 低压继电器
UV radiation 紫外辐射
UV-resistance 防紫外辐射
UV-resistant 抗紫外的,不受紫外线作用的

V

V(vacuum tube) 真空管；电子管
v(valve) 阀,活门；电子管
v.(via) 经过
v.(vide) 参见
v.(viscose rayon) 黏胶[人造]丝
v(viscosity) 黏度；黏[滞]性
V(voltage) 电压
V(voltmeter) 伏特表,电压表
VA(＝V.A.,value analysis) 价值分析
VA(＝va,volt-ampere) 伏安
Vac.(＝vac.,vacuum) 真空
VAc(vinyl acetate) 醋酸乙烯酯
Vacumat 瓦丘梅真空式转移印花机（德国制）
vacuole 液泡；空[液]泡；空穴；析稀胶粒
vacuole formation 形成空泡,形成空腔,空穴成形
vacuometer 真空计,低压计
vacu-pad 真空[抽吸]轧染,真空[抽吸]浸轧
vacuum〈Vac.,vac.〉 真空
vacuum bonding 真空黏合[工艺]
vacuum breaker valve 真空中断阀
vacuum cleaner 真空吸尘机
vacuum cleaning 真空吸尘,真空除尘
vacuum coat 真空涂覆；真空涂层
vacuum defoamation(＝vacuum degassing) 真空脱泡
vacuum dehydrating and overextending system 真空脱水和过伸长系统
vacuum deposition 真空沉积

vacuum drum dryer 真空转鼓烘燥机
vacuum drum washing range 真空转鼓水洗机
vacuum dryer 真空烘干机
vacuum drying chamber 真空干燥室
vacuum drying oven 真空干燥[烘]箱
vacuum dyeing 真空染色,减压染色
vacuum evapouration 真空蒸发
vacuum extractor 真空萃取器；真空脱水机
vacuum film coating 真空薄膜涂层
vacuum filter 真空滤器,[空]吸滤器
vacuum finishing 真空处理
vacuum gauge 真空计
vacuum hood 真空罩
vacuum hydroextracting 真空脱水
vacuum hydroextractor 真空脱水机
vacuum impregnating core mercerizing 真空浸碱透芯丝光
vacuum impregnating machine 真空浸渍机
vacuum impregnation 真空抽吸浸渍
vacuum manometer 真空压力计
vacuum mercerization 真空丝光过程
vacuum metalizing 真空金属化处理
vacuum plasma 真空等离子体
vacuum plating velum 真空镀膜
vacuum pump 真空泵
vacuum relief valve 真空安全阀
vacuum rotary dryer 旋转式真空烘燥机
vacuum seal 真空封口
vacuum sieving machine 真空筛分机

vacuum slot extractor 狭缝式真空脱水装置
vacuum steam autosetter 真空及蒸汽自动定形机
vacuum steamer 真空汽蒸箱
vacuum strainers 真空过滤器
vacuum suction 真空吸水
vacuum suction roll 真空吸[水]辊
vacuum take-off 真空抽吸
vacuum transfer printing 真空转移印花
vacuum transfer printing machine 真空转移印花机
vacuum tube〈V〉真空管；电子管
vacuum tube rectifier 真空管整流器
vacuum tumbler dryer 真空转筒式烘燥机
vacuum valve 真空阀
vagabond effect 浪子效果（牛仔服一种处理方法的效果）
vagabond look 浪子风尚（衣着时尚）
val.（value）值，数值
valence 化合价，原子价
valence angle 价角
valence bond theory 价键理论
valence shell 价电子层
valency 化合价，原子价
valency isomerism [化合]价异构[现象]
valentia 棉毛丝交织格子织物
valent weight 当量
valeric acid 戊酸
valetine 凡立丁
validity 有效性；确实性
validity check 有效性检查
valine 缬氨酸；α-氨基异戊酸
valley 波谷；最低点
value〈val.〉值；大小；价值；能力；浓淡色度；亮度，明度

γ value γ值；酯化度
value analysis〈VA，V.A.〉价值分析
value of vapour diffusion resistance 抗蒸气扩散值
valve〈v〉阀，活门；电子管
valve base 阀座
valve body 阀体
valve bonnet 阀帽
valve disc 阀盘，阀座
valve holder 阀座；灯座，管座
valve lever 阀门操纵杆
valve operating mechanism 阀门操作机构
valve seat 阀座
valve spindle 阀轴，阀杆
valve stem 阀杆
vanadium 钒
vanadium black 钒盐精元，钒盐氧化元
vanadium dichloride 二氯化钒
vanadium pentoxide 五氧化二钒
Van der Waals bonding 范德瓦耳斯键合
Van der Waals force 范德瓦耳斯力
vanilla 香草黄[色]（淡杏黄色）
vanillin 香草醛
vanishing 涂清漆
Van Wees wet-brushing machine 范威湿刷起绒机
Van Wyk system 范维克系统
Van Wyk theory 范维克定理（反映纤维压缩性能）
Vapocol method 瓦波科尔染色法（合纤经浸轧染液后烘干，再用三氯乙烯汽蒸，使合纤膨化，商名，美国）
Vapojet dryer 瓦波过热蒸汽烘燥机（商名）
vapor【美】= vapour
Vaporloc machine 瓦波洛克平幅高温高

Vaporloc 压连续汽蒸机(商名,英国制)
Vaporloc range 瓦波洛克高温高压平幅连续练漂联合机(分成煮练和氧漂两部分,特点是使用高温高压反应罐,英国制)
vapour 蒸汽;汽化物
vapour barrier 隔蒸汽层
vapourization 蒸发,汽化
vapourization-condensation sequence 蒸发—凝结的反复转化
vapourization efficiency 蒸发效率,汽化效率
vapourization heat 蒸发[潜]热
vapourizer 汽化器,蒸发器
vapourizing unit 汽化装置
vapour-jet dryer 蒸汽喷射干燥机
vapour lift pump 蒸汽升液泵
vapour lock 汽封
vapour permeability 蒸汽透过性
vapour phase〈VP〉气相
vapour phase chemical finishing 气相化学整理
vapour phase chromatography〈VPC, V.P.C.〉气相色谱法
vapour phase condensation 气相缩合
vapour phase crosslinking 气相交联
vapour phase durable press finishing 气相压烫整理
vapour phase dyeing 气相染色
vapour phase esterification 气相酯化
vapour phase finishing 气相整理
vapour phase heating system 气相加热系统
vapour phase inhibitor〈VPI, V.P.I〉气相[氧化]抑制剂
vapour phase preparation 气相制备法

vapour phase 3 process〈VP3 process〉气相交联法
vapour phase radiation grafting 气相辐射接枝
vapour phase treatment 气相处理
vapour phase xanthation 气相黄化
vapour pressure〈V.P., vp, v.p.〉蒸汽压力
vapour pressure curve 蒸汽压力曲线
vapour [phase] transfer printing 气相转移印花
vapour transmission 汽化传递
var 乏(电抗功率单位)
var.(variable) 变数,变量;可变的,变化的
var.(variance) 方差,偏差;差异;变化;分散;色散
var.(variant) 变形,变体,变种;不同的,变异的
variability 变异率,变率;变异性,可变性,易变性
variable〈var.〉变量,变数;可变的,变化的
variable capacitor〈VC〉可变电容
variable connector 可变连接点;多路开关
variable drop size〈VDC〉可变的液滴尺寸(喷墨印花术语)
variable frequency drive〈VFD〉变频传动《电》
variable head viscosimeter 可变压差黏度计
variable reluctance〈VR〉可变磁阻
variable resistance 可变电阻
variable resistor〈VR〉可变电阻器
variable speed clutch 变速离合器
variable speed control unit 变速控制装置

variable speed drive 变速驱动装置
variable-speed drive in synchronism 同步调速传动
variable-speed drive of induction motor by adjusting stator voltage 异步电动机调压调速
variable-speed motor〈VS motor〉调速电动机,变速电动机
variable speed reel 变速卷轴
variable-speed scanning 变速扫描
variable temperature dyeing processes 可变温度染色工艺
variable timesetting relay 可调时间继电器
variable transformer 可调变压器
variable traverse motion 变幅往复运动
variac 自耦变压器
Variacrol 凡丽阿克罗儿光敏变色印花浆(商名,意大利制)
variance〈var.〉方差,偏差;差异;变化;分散;色散
variance analysis 方差分析
variant〈var.〉不同的,变异的;变体,变种,变形
variate 变值
variation 偏差;变异;变分
variation and regularity 变化与统一
variation coefficient 变异系数
variator 变速器;聚束栅;伸缩[接]缝
vari-bow expander 可调式扩幅弯辊
vari-coloured 多色的
varied cut presser 可变花压板
variegated 杂色的;多样化的(指纺织品的色泽)
variegated yarn 多色纱(通常用印花法制造)
variegation 扎染

variety 品种;变种
variety of colours and designs 花色品种
Variflex dye pad 凡丽弗莱克斯挠度可调染色轧车(一种均匀轧车,轧点压力左、中、右可调,德国制)
Variogen 凡丽华近(能与金属络合的色基,商品)
Variolux colour matching chamber 梵里勒克斯比色(测色)箱(标准光源箱,商名,丹麦制)
variometer 变压器;变压表
vario nozzle 多级可调喷嘴(喷射染色用喷嘴,可随加工织物种类调节其直径)
Vario press application system 一种泡沫施加装置(商名)
varistor 变阻器;可变电阻;非线性电阻
varnish 清漆,凡立水
varnished cloth 漆布
varnishing 涂蜡,上蜡
varnishing machine 上蜡机(花筒雕刻设备,供花筒上固体或液体蜡);上清漆机(平网印花机的附属设备)
varnish making 制蜡
vat 瓮,槽,池,缸,桶;还原染缸;[染料]还原体
vat acid [还原染料]隐色酸
vat ageing 还原蒸化
vat ager 还原蒸箱
Vat-Craft process 瓦脱—克拉夫特[还原染料]染色法(用同位素催化还原,再用紫外线照射显色)
vat discharge paste 还原染料拔染浆
vat-dyed 还原染料染色的
vat dye fine dispersions 还原染料超细分散体
vat [colour]**dyeing** 还原染料颜色

vat dyeing faults 还原染料染疵
vat dye instability 还原染料不稳定性
vat dye printing 还原染料印花
vat dyes 还原染料
vat ester dyes 酯化还原染料(可溶性还原染料)
vat fermentation [靛青]发酵还原
vat padding 还原染料轧染
vat pigment continuous dyeing range 还原染料悬浮体连续染色机组
vat-pigment printing process 还原染料两相法印花(还原染料印花后,轧碱性还原液,再进行快速蒸化)
vat powder 保险粉
Vatrolite 伐妥来特(保险粉,商名,美国制)
vattability 可还原性
vatting 还原;干缸,养缸(还原染料染色前的预还原操作)
vatting condition 还原条件
vatting in a long liquor 全浴还原
vat yellow paper 还原黄试纸
V-belt V形皮带,三角皮带
V-belt pulley 三角皮带轮,三角皮带盘
V-block 元宝铁,V形铁
VC(variable capacitor) 可变电容
VC(vinyl chloride) 氯乙烯
VC(voltcoulomb) 伏特库仑
VCD(video compact disk) 视频光盘
VCM(vinyl chloride monomer) 氯乙烯单体
VDC(variable drop size) 可变的液滴尺寸(喷墨印花术语)
VdCh 【德】德国化学工作者协会
VDI 【德】德国工程师论文
VDT(visual display terminal) 图像显示终端《计算机》
VDU(volumetric dispensing unit) [印浆]容量式调配单元
vector(vec.) 向量,矢量
vector control of induction motor 异步[感应]电机矢量控制
vector diagram 矢量图,向量图
Vectran 一种共聚类纤维(高强高模纤维,低温时耐磨,尺寸稳定性好)
veer 调向,旋转,转向
VEF(viscoelastic fibre) 黏弹性纤维
vegetable cork 软木衬料(隔热用)
vegetable dyes 植物染料
vegetable fibre 植物纤维
vegetable globulin 植物球蛋白
vegetable gum 植物胶
vegetable hair 植物纤维(包括棉、木棉、松叶纤维、马利筋属植物纤维等,做填充料用)
vegetable matter [羊毛中]植物性杂质(草籽、干草等)
vegetable oil 植物油
vegetable protein 植物蛋白质
vegetable protein fibre 植物性蛋白质纤维
vegetable silk 黏胶纤维
vegetable tanning 植物鞣法;植物鞣革
vegetable waxes 植物蜡
vegetable weighting 植物性增重《丝》
vegetable wool 黏胶短纤维充毛棉布
vehicle 载体,载色剂;媒介物;运载工具;车辆
vein 破洞,裂缝(织疵)
vel.(velocity) 速度
velam fibre 维兰纤维
Velan finish 维兰防水整理(耐久性防水整理,商名)

Velan PF 维兰 PF(硬脂酰胺甲吡啶氯化物,阳离子防水整理剂、柔软剂,商名,英国制)

Velesta pigment colours 韦莱斯塔涂料(印花及轧染用涂料,商名,意大利制)

velludo 经立绒织物,经起绒织物

velocimeter 速度计

velocity〈vel.〉速度

velocity coefficient 速度系数

velocity gradient 速度梯度

velocity ratio 速比

velograph 速度计

velour 丝绒,天鹅绒;棉绒;拉绒织物;维罗呢

velour finish [粗呢]起绒整理

velour overcoating 起绒大衣呢

velours 丝绒,天鹅绒;棉绒;拉绒织物;维罗呢

velours de France 法国丝绒(彩印拷花)

velours ecrase 异向平绒(整理时,将绒头朝不同方向压倒)

velours embosse 叠花丝绒,拷花丝绒

velours frappe 拷花丝绒(用热辊压成)

velours paon 平绒

velours printing 绒面印花

velours venetian 威尼斯丝绒(大花花纹,印花或拷花)

velure 天鹅绒;仿天鹅绒织物

Velustrol 维露斯系列(平滑剂,聚乙烯乳液,阴离子型、非离子型,适用于浸轧、浸渍工艺,能改善纤维缝纫性,改善树脂整理中纤维强力损伤,商名,瑞士科莱恩)

velutum 丝绒,天鹅绒

velvet 丝绒,经绒,立绒,天鹅绒

velvet article 绒头织物

velvet bath towel 绒面浴巾

velvet carpet 割绒地毯

velvet/corduroy cutting machine 天鹅绒/灯芯绒割绒机

velvet dyeing 丝绒染色

velveteen 纬绒,棉绒,平绒

velveteen finish 平绒整理

velveteen rug 平绒地毯

velvet finish 仿麂皮整理;起绒整理《毛》

velvet flat embossing calender 丝绒拷花机

velveting 丝绒织物,精细绒头织物

velvet knife 割绒刀

velvet-on-velvet 叠花丝绒

velvet pile 割绒

velvet-plain 平绒

velvet raising 丝绒起绒

velvet shearing machine 丝绒剪毛机

velvet velludo 经立绒织物,经起绒织物

velvety stuff 天鹅绒织物

veneer board 胶合板

Venetian 直贡呢;棉直贡[呢];威尼斯缩绒呢;威尼斯精纺细呢

Venetian finish 威尼斯[轧纹]整理(低档织物经轧纹后可产生高度光泽)

Venetian pink 威尼斯淡粉红[色]

Venetian red 威尼斯红[色](红光棕色)

vent 通风孔;出口

vent.(**ventilating**) 通风的

vent.(**ventilation**) 通风,换气

vent.(**ventilator**) 通风器,通风机;通风管;电扇

vent hood 通气罩

ventilated drying oven 通风烘箱

ventilating〈vent.〉通风的

ventilating fan 排气风扇

ventilation〈vent.〉通风,换气

ventilation cover 排风罩
ventilation duct 通风管道
ventilation grille 通风格栅
ventilation system 通风系统
ventilator〈vent.〉通风器,通风机;通风管;电扇
Ventile 文泰尔防雨布(商名)
venting 通风;排放
venting passage 通风通道
vent port 排气孔
Venturi [tube] 文丘里流量管(统称文氏管)
Venturi absorber 文丘里吸收器
Venturi feed tube 文丘里喂入管
Venturi flowmeter 文丘里流量计
Venturi meter 文丘里测量管
vent valve 放气阀,空气门
ver.〈versus〉……对……,与……相比
verdigris [green] 铜绿[色](浅暗绿色);绿色颜料
verge 边,边缘;圈,环;接近;毗连
verification 校准,校验,鉴定
Vermatex process 费麦特克斯工艺(一种液氨处理工艺)
vermiculite hydrosol 蛭石水溶液(用于阻燃隔热整理)
vermillion 朱红;银硃,朱硃;红色颜料(指硫化汞)
vermin-proof 防虫的
vermin proofing 防虫加工,防虫整理
vernier 游标,游标尺
vernier calipers 游标卡尺
vernier condenser 微调电容器
vernier motor 微调电动机
vernier rheostat 微调电阻器
Verofix dyes 费罗菲克斯染料(染羊毛的活性染料,含一氯二氟嘧啶基团,商名,德国制)
versatility [用途]多面性,多方面适合性
versicolour 闪色;虹彩
version 型式,版本,译文
versus〈ver.〉……对……,与……相比
vertex 顶,顶点
vertical blind 立式遮阳帘
vertical board 立式配电盘
vertical burning test 垂直燃烧测试
vertical centrifugal hydroextractor 立式离心脱水机
vertical centrifugal machine 立式离心机
vertical cloth run 直导辊穿布[方式]
vertical compensator 立式松紧架
vertical cylinder dryer 立式滚筒干燥机
vertical duplex rotory screen printing machine 立式双面圆网印花机
vertical duplex screen printing machine 立式双面筛网印花机
vertical dyers 联合染整厂
vertical end-ring glueing device 立式[圆网]闷头胶黏设备(圆网印花机的附属设备)
vertical filter 立式滤池
vertical flammability test 垂直法可燃性试验
vertical flash ager 立式快速蒸化机,立式闪蒸机(用于二相法印花)
vertical flocculator 立式絮凝池
vertical gallery camera 立式制版照相机
vertical goods guidance 织物垂直运行
vertical heat setting machine 立式热定形机
vertical injector 立式喷液器
vertical interval〈VI〉垂直间距;等距离;

等高线间距
vertical kier 立式蒸布锅,立式煮锅
vertical loop steamer 立式成环蒸化机
vertical migration 垂直迁移法,垂直泳移
vertical open-width washing machine 立式平幅水洗机
vertical organization 全能[生产]机构,全能工厂(指纺织印染整理联合工厂)
vertical padder 立式轧车
vertical paper press 立式纸板压机
vertical pattern 立式图案
vertical perforated drum dryer 立式圆网烘燥机
vertical plant 联合工厂(如纺织印染联合厂)
vertical press 立式压呢机
vertical raising machine 立式起绒机
vertical return chain 立式往返链(拉幅机)
vertical return pin chain 立式往返针链(拉幅机针板链条的一种新的往返形式)
vertical roller printing machine 立式滚筒印花机
vertical rotary screen printing machine 立式圆网印花机(还可用于双面印花)
vertical screw 立式螺杆
vertical shaft 立轴
vertical shearing machine 立式剪毛机
vertical space dyeing printing unit 立式间隔染色印花机
vertical spindle dyeing machine 立式锭轴染色机(用于染纱线)
vertical squeezer 立式轧车
vertical star dyeing machine 立式星形架染色机
vertical streak 纵向条痕(织疵)

vertical stripes 直条纹
vertical sueding machine 立式磨毛机;立式仿鹿皮起绒机
vertical test method 垂直试验法(醇焰法)
vertical type printing machine 立式印花机
vertical view 俯视图
very soluble 〈V.S., v.s.〉易溶[解]的
vesicant 糜烂剂;糜烂[性]的
vessel 容器;器皿
vest 背心;汗衫;内衣,衬衣
VFD(variable frequency drive) 变频传动《电》
VGA(video graphics adapter) 视频图形显示卡《计算机》
VH(Vickers hardness) 维氏硬度
VHN(Vickers hardness number) 维氏硬度值
VI(vertical interval) 垂直间距;等距离;等高线间距
VI(viscosity indicator) 黏度指示器
VI(=V.I., viscosity index) 黏度指数
via 〈v〉经过
viable 可行的;能生存的;能活的
Vialon Fast dyes 菲阿隆染料(染聚酰胺用1:2金属络合染料,商名,德国制)
Vibatex AN 维勃坦克斯 AN(聚乙烯树脂的乳化液,耐洗的防滑剂,防钩丝剂,商名,瑞士制)
Vibatex HKN 维勃坦克斯 HKN(硬挺整理剂,非离子型,适用于领衬布,好的耐洗性,商名,瑞士制)
Vibatex SL-MC 维勃坦克斯 SL-MC(乙烯衍生物的共聚物,手感改善剂,具防披[裂]性能,商名,瑞士制)
vibrating jet method 振荡射流法(测动表面张力)

vibrating roller 摆动辊,振动辊
vibration 振动
vibration absorbent 减振
vibration-damper 减振器
vibration-free 防振;无振动
vibration-free running 无振运转
vibration isolation 隔振,防振
vibration isolator 防振装置
vibrations per minute〈VPM, vpm〉每分钟振动数
vibration washer 振荡式水洗机
vibrator 振动器
vibratory felting machine 振动制毡机
vibrograph 振动记录器,示振器
Vibro-jigger 振荡卷染机(染色设备,内有振荡装置,可以提高染色效果,日本制)
vibrometer 振动计,振荡计
Vibrotex machine 一种振荡洗涤机(商名,德国制)
vibro washer 振荡洗涤机
vice 台钳,[老]虎钳
vice versa〈VV, vv, v. v.〉反过来也一样,反之亦然

vicinal 邻位的,连[位]的
Vickers hardness〈VH〉维氏硬度
Vickers hardness number〈VHN〉维氏硬度值
Vicol 维科尔(丙烯酸类浆料,有 A、N_{40} 等产品编号,商名,英国制)
Vicontin I liq. conc. 维固定 I 浓液(促染剂,非离子型,可改善染料溶解度,促进坚牢素 SF 染料在轧蒸法中的固色,商名,瑞士科莱恩)
victoria 维多利亚绸;维多利亚布;维多利亚双重斜纹呢

Victoria Blue 维多利亚蓝[色](蓝色);蓝色碱性染料
Victoria Violet 维多利亚紫[色]
vicuna finish [毛织物]仿骆马绒整理
vicuna wool 骆马毛织品
vide〈v.〉参见
videlicet〈viz.〉即,就是
video 电视;视频;影像的
video camera 摄像机
video compact disk〈VCD〉视频光盘
video display 视频显示(可用于连拍机、电子分色仪等)
video graphics adapter〈VGA〉视频图形显示卡《计算机》
video screen 影像荧光屏;电视荧光屏
video tape 电视录像带,录像磁带
video telephone 可视电话
vide supra〈v. s.〉参见上文
vieley cloth 色纱绉布(英国制)
view 视图;形式,式样
viewing angle 探测角
viewing aperture 测量孔
vignette 渐晕[图案];晕映
vignette print 渐晕印花[花纹]
vignetting 渐晕;光阑阻
vignetting effect 渐晕效应
vignetting stop 格晕光阑(摄影用语)
vigor finishing 健康整理,活力整理
vigoureux-dyed top 印花毛条
vigoureux printing 毛条印花
vigoureux printing machine 毛条印花机
vigoureux yarn 毛条印花毛纱
Vinal 维纳尔(聚乙烯醇系纤维的属名,美国)
vinometer 酒精比重计
vinyl 乙烯基

vinyl acetate〈VAc〉 醋酸乙烯酯
vinyl acetate-acrylic ester copolymer 醋酸乙烯酯丙烯酸酯共聚物
vinyl acetate-crotonic acid copolymer 醋酸乙烯酯丁烯酸共聚物
vinyl alcohol 乙烯醇
vinylal fibre 乙烯醇系纤维
vinylation 乙烯[基]化[作用]
vinyl benzene 苯乙烯
vinyl chloride〈VC〉 氯乙烯
vinyl chloride monomer〈VCM〉 氯乙烯单体
vinyl coating 乙烯基涂料;乙烯基涂层
vinyl compound 乙烯化合物
vinyl cyanide 乙烯基氰,丙烯腈
vinyl ether 乙烯醚
vinyl fibre 乙烯基系纤维
vinyl film 乙烯薄膜
vinyl foam coated fabrics [聚]氯乙烯涂料海绵布,泡沫聚氯乙烯涂层织物
vinyl group 乙烯基
vinylidene 聚偏[二]氯乙烯纤维;亚乙烯基
vinylidene chloride 偏二氯乙烯
vinyl monomer 乙烯单体(用于丝的接技聚合)
vinyl morpholine 乙烯吗啉
Vinylon 维尼纶(聚乙烯醇缩醛纤维的统称,商名,日本制)
vinyl polymer 乙烯基聚合物
vinyl pyridine 乙烯基吡啶
vinyl pyrrolidone 乙烯吡咯烷酮
vinyl sponge leather [聚]氯乙烯涂料海绵布,聚氯乙烯海绵人造革
vinyl stearyl ether 十八烷基乙烯基醚(防水整理剂)

vinyl sulfone〈VS〉 乙烯砜(活性染料的活性基团)
vinylsulfone dyes 乙烯砜染料
vinylsulfone reactive dyes 乙烯砜型活性染料
vinyl tile 乙烯地砖
violaceous 紫罗兰色
violanthrone 蒽酮紫(紫色还原染料)
violet 紫色,紫罗兰色
violet scale 紫色评级卡(碘溶液测织物上退浆率)
virgin curve 起始曲线,初始曲线,新曲线
virgin silk 坯(蚕)丝,原丝
viridine green 艳黄绿色;绿色颜料
viridine yellow 浓黄绿色
virtual impedance 有效阻抗
virtual machine environment〈VME〉 虚拟计算环境
virtual value 有效值
virus 病毒
vis.(viscosity) 黏度;黏[滞]性
Visa 维萨涤棉布(具有易去油污及耐久压烫性能,商名,美国制)
visc.(viscosity) 黏度;黏[滞]性
viscid 黏滞的,黏的,半流体的
viscidity 黏性
Viscobalance 用于不透明液体的落球式黏度计(商名,德国)
Visco dyes 菲斯科染料(直接染料,商名,瑞士制)
visco-elastic 黏[滞]弹性的
viscoelastic behaviour 黏弹性状
visco-elastic body 黏弹体
viscoelastic cross effect 黏弹性交叉效应,韦森堡效应
viscoelastic deformation 黏弹性变形

viscoelastic fibre〈VEF〉 黏弹性纤维
viscoelastic fluid 黏弹性流体,马克斯韦尔流体
visco-elasticity 黏弹性
viscolloid 黏性胶体
viscometer 黏度[测定]计
viscometry 黏度测定法
Viscontrol VA [连续测定用的]VA 旋转式黏度计(商名,德国制)
viscose 黏胶液;黏胶纤维
viscose cellulose 黏胶纤维素
viscose cord fabric 黏胶帘[子]布
viscose fibre 黏胶纤维
viscose filament yarn 黏胶长丝
viscose film 黏胶薄膜;玻璃纸,透明纸
viscose process 黏胶法
viscose rayon〈v.〉 黏胶[人造]丝
viscose rayon cake dyeing 黏胶丝饼染色
viscose rayon fibre 黏胶人造纤维
viscose staple fibre 黏胶短纤维
viscosimeter 黏度[测定]计
viscosimetric analysis(=viscosimetry) 黏度测定[法]
viscosity〈v, vis., visc.〉 黏[滞]性;黏度
viscosity abnormality 黏度反常性
viscosity anomaly 黏[滞]性反常,黏性异常
viscosity coefficient 黏度系数
viscosity constant 黏度常数
viscosity control 黏度控制
viscosity conversion 黏度换算
viscosity exponent 黏度指数
viscosity gradient 黏度梯度
viscosity index〈VI, V.I.〉 黏度指数
viscosity indicator〈VI〉 黏度指示器
viscosity modifier 黏度改进剂(印花糊料)

viscosity number 黏数;比浓黏度(增比黏度与溶液浓度之比)
viscosity ratio 相对黏度,黏度比
viscosity recorder〈VR〉 黏度记录器
viscosity stabilizer 黏度稳定剂
viscosity-temperature coefficient〈VTC〉 黏度—温度系数
Viscotes(Visual Communication Technology System) 维司科特数字[码]印花系统(商名,日本开发)
viscous 黏[滞]的
viscous drag 黏性阻力
viscous flow 黏[性]流(纤维大分子)
viscous fluid 黏[性]流体
viscous force 黏[滞]力
vise 台钳,[老]虎钳
visibility 能见度;视程;视野
visibility function 视见函数
visibility scale 可见性标度
visible absorption spectrum 可见吸收光谱
visible defect 外观疵点
visible outline 外形线
visible portion of the electromagnetic spectrum 电磁光谱的可见部分
visible ray 可见光
visible spectrum 可见光谱
visible yarn (地毯的)绒头部分
Visil fibre 一种耐高温阻燃黏胶纤维(商名,芬兰制)
vision 视觉
visual acuity 视觉敏锐度
visual angle 视角
visual appearance 外观质量
visual assessment 目测,目光鉴定
visual colourimeter 目测比色计
visual colourimetry 目视比色法

Visual Communication Technology System〈Viscotes〉 维司科特数字[码]印花系统(商名,日本开发)
visual control instrument 目视控制仪,目视控制装置
visual depth 目测[色泽]深度
visual difference 视差
visual discrimination 目测鉴别
visual display 可见显示
visual display system 直观显示系统,目测显示系统
visual display terminal〈VDT〉 图像显示终端《计算机》
visual display unit 显示屏幕
visual effect 视觉效应
visual evaluation 目光鉴定,目光评价
visual examination 目测,表观检验
visual field 视野
visual grading 目光判定
visual impact 视觉效应;视觉刺激
visual inspection 目测,肉眼检查
visualization 显像;目视观察
visualizing masks for colour monitors 彩色监控设备
visualizing reagent 显示试剂
visual light scattering photometer 目视式散射光度计
visual measurement 目测
visual observation 目光观察
visual perception 视觉
visual persistence 视觉暂留
visual photometry 视觉光度学
visual presentation 视觉展示
visual purple 视紫质
visual rating method 肉眼评级法
visual ratings 目测等级

visual readout 可见读出,目测读出
visual sensation 视感觉
visual signal 视频信号
visual standard 目测标样
visual stimulus 视刺激
visual test 目光检验,肉眼检验
vitality 活力;持久性
vitaminised cloth 维生素布,维他命布(布经维生素P处理后,制成贴身衫裤,有助血管收缩)
vitreous fibre 玻璃质纤维,透明纤维
vitreous luster 玻璃光泽
vitreous polymer 玻璃状聚合物
vitreous state 玻璃状态,透明状态
vitrification 玻璃化[作用]
vitrification point 玻璃化温度
vitrify 玻璃化
vitriol 硫酸盐,矾;硫酸
vitriol oil [浓]硫酸
Viveral 胃微儿(胰酶退浆剂,商名,德国制)
vivid [色、光]强烈的,鲜明的,鲜艳的
vivid colour 鲜艳明亮色泽
vividness 鲜艳度,鲜明度
viz.(videlicet) 即,就是
VM(=vm,v.m., volatile matters) 挥发性物质
VM(voltmeter) 伏特计,电压表
VME(virtual machine environment) 虚拟计算环境
V.O.C.(volatile organic chemicals) 挥发性有机化合物
vogue 时尚,流行,时髦
void 孔隙,空间
void content 孔隙含量;孔隙率
voided-structure fibre 空隙结构纤维

void ratio 孔隙比
voile 巴里纱(一种轻量透明薄纱,用特别紧捻纱织造,并经烧毛,又称华而纱)
vol.(volume) 容积,体积,容量;声音,音量;卷,册
volatile 挥发[性]的
volatile component 挥发成分
volatile matters〈VM,vm,v.m.〉挥发物质
volatile organic chemicals〈V.O.C.〉挥发性有机化合物
volatile residue 挥发残存物
volatile solvent 挥发性溶剂
volatility 挥发性;挥发度
volatilization 挥发[作用]
volt 伏[特](电压单位)
voltage〈V〉电压
voltage drop 电压降
voltage feedback 电压反馈
voltage generator 电压发生器
voltage loss 电压损失
voltage range 电压范围
voltage regulated synchronizing system 调压同步系统
voltage regulator 稳压器,调压器
voltage stabilizer 稳压器
voltage stabilizing diode 稳压二极管
voltaic battery 原电池,伏打电池
volt-ampere〈VA,va〉伏安
voltcoulomb〈VC〉伏特库仑
voltmeter〈V,VM〉伏特表,电压表
voltohmmeter〈VOM,V.O.M.〉伏欧表,电压—电阻表
volume〈vol.〉体积,容积;容量;声量,音量;卷,册
volume conductivity 单位体积导电率

volume density 体积密度;表观密度
volume expansion 体积膨胀
volume fashion 大众化流行款式
volume percentage 体积百分比
volume production 大批量生产
volume resistivity 体积电阻率
volume specific heat 定容比热
volume swelling 体积溶胀
volumeter 体积计
volumetric analysis 容量分析
volumetric cylinder 量筒
volumetric dispensing unit〈VDU〉[印浆]容量式调配单元
volumetric flask 容量瓶,量筒
volumetric loading 水力负荷
volumetric solution〈VS〉滴定标准液
volumetric specific heat 容积比热
voluminous crimping 膨松卷曲
voluminous fleecy handle 膨松起绒手感
voluminous yarn 膨体纱
volute spring 锥形簧,涡旋弹簧
VOM(=V.O.M.,voltohmmeter) 伏欧表,电压—电阻表
Von der Wehl steamer 冯德韦尔蒸箱(用于还原染料印花的卷轴式蒸箱)
vortex roller 涡流式滚筒
vortex scrobber 涡流式洗涤器(用于清洁烘干机中废气)
vortex scrubber 涡流式洗涤器
vortex-type seal 涡旋式密封垫
VP(vapour phase) 气相
V.P.(=vp,v.p.,vapour pressure) 蒸汽压力
VPC(=V.P.C.,vapour phase chromatography) 气相色谱法
VPI(=V.P.I.,vapour phase inhibitor)

气相[氧化]抑制剂
V piece V形块,三角板
VPM(=vpm, vibrations per minute) 每分钟振动数
VP3 process(vapour phase 3 process) 气相交联法
VR(variable reluctance) 可变磁阻
VR(variable resistor) 可变电阻器
VR(viscosity recorder) 黏度记录器
v. s. (vide supra) 参见上文
VS(vinyl sulfone) 乙烯砜(活性染料的活性基团)
VS(volumetric solution) 滴定标准液
V. S.(=v. s., very soluble) 易溶[解]的
V-screen V形圆网
V-shaped trough V形槽
VSM 【德】瑞士机械工业协会

VS motor (variable-speed motor) 调速电动机,变速电动机
V-squeegee V形刮刀
VTC(viscosity-temperature coefficient) 黏度—温度系数
Vulcafor process 威尔凯福化学防缩整理法(人造棉防缩整理剂,主要使用聚异氰酸盐,商名,英国)
vulcanite roller 硬橡胶辊
vulcanization 硫化[作用];硬化[作用]
vulcanized oil 硫化油
vulcanized rubber 硫化橡胶
vulcanizing agent 加硫剂
vultex 硫化橡浆,硫化胶乳
VV(=vv, v. v., vice versa) 反过来也一样,反之亦然

W

w. (weft) 纬,纬纱
W. A. (width average) 宽度平均值
wadded cloth 衬垫经(或纬)织物
wadded double cloth 衬垫双层织物
wadding 填絮,填料;衬垫
waffle backing 凹凸面涂底,蜂窝状背涂(地毯用)
waffle surface 凹凸表面;蜂窝形表面
wale streak 经柳(织疵);直条花(针织疵点)
walkometer 缩绒检测器
walk test 踩踏试验,行走试验(地毯)
walkway 过道,通道
wall carpet 壁毯
wall cloth 糊壁花布
wall cooling 回壁冷却
wall covering 壁纸
wall hangings 壁毯;壁布
wall paper 壁纸
wall surface heat losses 壁面热损失
wall to wall [地毯]满铺
wall-to-wall carpet 全室地毯
walnut [brown] 胡桃壳棕[色](浅棕色)
wandering 漂移,游离
W and F (warp and filling) 经向加纬向
WAN-dried fibre 溶剂[交换]烘干纤维(W指水,A指醇类,N指苯)
W and W (=W & W, wash and wear) 洗可穿,免烫
Ward-Leonard drive 渥特—勒奥那尔特式传动

wardrobe 衣服(总称)
ware 制品,同类物品
warehouse 仓库
warm colour 暖色(指红、黄、橙等)
warm cupboard 保温箱
warm hue 暖色相
warmness 温暖(常指手感)
warm-retention property 保暖性
warmth retaining tester 保暖试验仪
warmth retention property 保暖性能
warmth tester [织物]保暖[性]测试仪
warm tone 暖色调
warm up [仪器的]预热
warm white fluorescent lamp 暖白色荧光灯《测色》
warner 警报器
warning colour 警戒色
warning colouration 涂(染)有警戒色
warning devices 报警装置
warning lamp 报警灯,指示灯
warning shield 警告护牌
warning signal (=warning symbols) 报警信号;警告符号
warp 经,经纱
warpage 翘曲,弄歪
warp and filling 〈W and F〉 经向加纬向(如经向和纬向物理性能相加的数据)
warp beam 经轴
warp beam dyeing machine 经轴染色机
warp beam hydroextractor 经轴脱水机
warp dyeing 经纱染色

warp effect 经面花纹
warp falling 缺经,断经,头路(织疵)
warp fault 错经;经纱疵
warp floats 经向跳花
warp holding place 松紧条痕,经吊痕
warp indigo dyeing machine 经纱靛蓝染色机
warping 整经
warp knitted cauterising fabric 经编烂花织物
warp-knitted fabric 经编针织物
warp knitted net 网孔经编织物
warp knitted tablecloth 经编台布
warp knitting 经编
warp pattern 经向图案
warp printed cretonne 印经装饰布
warp printing 经纱印花
warp printing machine 经纱印花机
warp printing technology 经纱印花技术
warp-rebeaming machine 倒轴机,并轴机
warp seconds 经纱织疵
warp sett ［织物］经密
warp shrinkage 经缩
warp sizing and dyeing 经纱上浆与染色
warp stain 渍经
warp streak 经向条花,分经路,经柳
warp stretch woven fabrics 经向拉伸织物
warp stripe 经向条花,纵向条斑,经挡
warp tension 经向张力
warp thread 经纱
warp-wise 经向
warpwise defects 经向疵点
warpwise stretch 经向伸张
warp yarn 经纱,经线
warranty test 保证试验,认可试验,验收试验

WAS(wash-active substance) 洗涤剂,洗涤活性物质
Wascator 羊毛防缩试验仪(国际羊毛局)
washability 耐洗性
washable 可洗的,耐洗的
washable blanket 可洗的橡皮衬布(指印花机上随印随洗烘)
washable colour 耐洗的坚牢染料
washable foil 可洗涤的箔片（印花用）
wash-active substance〈WAS〉洗涤剂,洗涤活性物质
wash and starch 浆洗
wash and use 免烫,洗可穿
wash and wear〈W and W,W & W〉免烫,洗可穿
wash and wear cycle 洗可穿周期
wash and wear finish 免烫整理,洗可穿［防缩防皱］整理
wash and wear process 洗可穿整理法,免烫整理法
wash and wear rating 洗可穿评级
wash-bath 洗槽
washboards 洗后扭斜(针织疵)
wash bowl(＝wash cistern) 水洗槽
wash-crease resistant 防洗皱的
wash drum 洗鼓
washed carpet(＝washed rug) 有光地毯
washer 垫圈;洗净器;洗涤机,洗衣机
washer breaks 洗涤折痕(整理疵)
washer cloth 水洗绉
washer mark 洗呢斑
washer wrinkle fabric 水洗布(有绉纹)
washer wrinkles 洗呢经向折痕;洗涤折痕（染整疵）
washes 洗涤废水;洗涤液
wash fast 耐洗的

wash fast finish 耐久整理
wash fastness 耐洗牢度
wash-finishing fabric 水洗织物
wash goods 耐洗的衣服或织物
washing 洗净
washing activity 洗涤性能
washing agent additives 洗涤添加剂
washing agent dosing 洗涤剂用量
washing agent enzymes 洗涤添加酶
washing assistant 洗涤助剂
washing bottle 洗涤瓶
washing centrifuging machine 水洗离心机
washing cycle 洗涤周期（全工序洗涤一次的周期）
washing detergent 洗涤剂
washing detergent analysis 洗涤剂分析
washing lye 洗用碱水
washing machine for wollen yarn 绒线洗涤机
washing mechanics 洗涤机理
washing off properties 洗净性能
washing powder 洗衣粉，粉状洗涤剂
washing process 洗涤工艺
washing resistance 耐洗性
washing shrinkage ［织物的］缩水率
washing soda 洗涤碱；十水［合］碳酸钠
washing test 耐洗［牢度］试验
wash leather 揩拭用［鹿］皮
wash performance 水洗性能
wash permanency 耐洗性
wash-proof finish 耐洗整理
wash-resistant antistatic agent 耐洗抗静电剂
wash run （一道或一次）洗涤过程
wash-wear 免烫，洗可穿
wash-wear rating 洗可穿等级

wash wheel 转轮洗涤装置
wash wrinkle 洗涤皱
wash-wrinkle resistant 防洗皱的
waste 回丝，废纱头；废料，污物
waste acid 废酸
waste alkali 废碱
waste and hazardous parts 废料和危险材料
waste boiler 废热锅炉
waste cotton 废棉
waste disposal method 废料处理法
waste fibre 下脚纤维（总称）
waste gas 废气
waste gas emission standard 废气排放标准
waste gas pollution control 废气污染控制
waste heating 废气加热
waste heat recovery unit 废热回收装置
waste liquor 废液
waste lye 废碱液
waste management 废物处置
waste oil 废油《环保》
waste plains 低级平布
waste recovering 废料回收
waste residue 废渣
waste sorter 废料分类工；废料分类机
waste stabilization pond 污水稳定塘
waste steam 废汽
waste treatment 废物处理《环保》
waste water and sludge treatment 废水和废渣处理
waste water degasification 废水除气
waste water dye removal 废水脱色
waste water evaluation 废水鉴定；废水评估
waste water heat exchanger 污水热交换器
waste water lagoon 废水氧化塘

Waste Water Levy Act 废水处理法案(德国)
waste water load 废水负荷
waste water pit 废水池
waste water pollution 废水污染
waste-water purification 污水净化,废水净化
waste water reclamation 废水回收;废水再生
waste water regulation 废水排放的相关法规
waste water renovation 废水净化回收;废水重复使用
waste water stabilization pond 污水稳定塘
waste waterstream 污水流
waste-water treatment 污水处理
waste water treatment unit 污水处理单元
waste water ultrafiltration treatment 废水的超滤处理(水处理技术)
waste water wool scouring 废水洗毛
watch-dog 监控设备,监视器
watch-dog timer 监视计时器
water absorbability 吸水性,吸湿性
water-absorbent finish 吸水整理
water-absorbing fibres 吸水性纤维
water absorption 吸水率
water absorptive softening agent 吸水性柔软剂
water and wind proof property 防水、防风性能
water atomizer 水喷雾器
water balance 水平衡
water-based dry coating 水基干法涂层
water-based ink 水溶性油墨
water-based wet coagulation coating 水基湿法凝固涂层

water bath 水浴
water bloom 水华(水中浮游生物猛增时发生亮光)
water blue 水蓝[色](浅暗绿蓝色)
water-borne bacteria 水生细菌
water borne coating 水性涂层
water borne soil 水中带来的污垢
water-borne stains 水媒[介]污渍
water-bottom [蒸箱]底部水层
water calender 轧水机
water calorimeter 水量热器
water capacity [含]水容量,持水量
water cartridge filter 芯式滤水器
water circulating pump 水环泵(液环泵的一种,其能量转换的介质是水,常用作真空泵)
water cistern 水洗槽
water clarification 水的净化
watercolour 水彩颜料;水彩画
water column 水柱
water conditioner 软水剂,水质改善剂
water consumption 水耗量
water content 含水量,含水率
watercooled 水冷的
water-cooled footstep 水冷却轴承,水冷却托杯
water cooled roll 冷水辊,冷却辊
water-cooled suit 水冷服装
water-cooling 水冷却
water correction plant 水质改善设备,水净化设备
water course 水道,水路;水流
water damage 水渍
water damping 水给湿
water decarbonization 脱去水中碳酸盐工艺(软水处理)

water degasification 水的除气工艺
water demanganizing 水的去锰工艺
water deoxidation 水的脱氧工艺
water desalination 水[的]脱盐[作用]
water discharge legislation 水的排放条例
water dispersion 水分散液
water dispersion type PU 水分散型聚氨酯
water distilling apparatus 蒸馏水装置
water drop penetrating test 水滴渗透性试验
water economizer 省水器
water economy 节水
watered 波纹的,云纹的
watered gauze 香云纱
water emulsion 水乳胶;水[外相]乳状液
water-endangering substances 水污染物
water evapouration 蒸发量
water extraction machinery 脱水机械
waterfast 耐水的;不透水的
water fastness 耐水牢度
water feed valve 给水阀
water filter 滤水器
water flow 水流
water for living 生活用水
water-free breaking elongation 干态断裂伸长
water gas 水煤气
water gauge〈W.G.〉水[位]标[尺],水位指示器;水表
water glass 水玻璃,硅酸钠
water green 水绿[色](浅暗黄绿色)
water hardening salts 导致水硬度提高的盐
water hardness 水硬度
water head 水位差,水头
water-heat set 热水定形

water holding capacity 最大保水容量
water holding property 保水性
water hose 水龙带
water imbibition 吸水性
water impermeability 不透水性
water-impermeable finishes 不透水整理,拒水整理
water incrustation 水垢
watering 波纹,云纹;松板印,云斑
watering calender 波纹轧光机
watering cloth 波纹布,云纹布
watering machine 波纹轧光机
water in oil 水油相,油包水
water-in-oil emulsion 水油相乳化液,油包水乳化液(水为内相,油为连续外相)
water-in-oil system 油包水系统
water insoluble dye 水不溶性染料
water jacket 水套
water-jacketed condenser 水套冷凝器
water-jet air pump 喷水空气泵
water-jet scrubber 喷水洗涤器
water jet vacuum pump 水射流真空泵
water leakiness 透水度;漏水性
waterless 无水的,干的
waterless dyeing 无水染色(指溶剂染色)
waterless dyeing process 无水染色工艺
water-level 水平面;水准器
water level regulator 水平面调节器,水准调节器
water lock 水封口,水封闭
waterlogged 浸透水的
water-logged soil 浸透水的污垢
water loving 亲水的
water lute 水封,液封
water mangle 轧水机
water mangling and cylinder drying machine

轧水烘燥机
watermark 水痕,水渍,水印(疵点);水位标
water mark calender 波纹轧光机
water marked finish 波纹轧光整理
water matrix 水基质
water meter 水表
water milling 水缩绒;水磨
water miscibility 水混溶性
water mould 水霉菌
water of condensation 缩合水;冷凝水
water of constitution 化合水,结构水
water of crystallization 结晶水
water of hydration 结合水,结晶水
water-oil repellent finish 防水、防油整理
water, oil, stain proofing and soil release finish 防水、防油、防污和易去污整理
water penetration test 水浸透试验
water percolation test 水渗滤试验(测织物防水性)
water permeability 渗透性,渗水性
water permeability test 渗水性试验(测织物防水性)
water pick up 吸液率
water pistols 水枪
water pollution 污水
Water Pollution Control Law 污水管理条例《环保》,污水管理法
water power load 水力负荷
water pressure resistance 耐水压性
water-pressure vapour 水蒸气压力
waterprint design 水纹图案
waterproof 防水的,不透水的
water proof and breathable fabric〈WBF〉防水透气织物
water proof and moisture permeable fabrics 防水透湿织物
water proof and moisture permeating finish 防水透湿整理
water proof and moisture permeating finishing agent 防水透湿整理剂
water proof and permeability 防水透湿性
water proof breathable finishing 防水透气整理
water proof breathable laminate 防水透气层压
water proof breathable microporous membrane 防水透气微孔膜
waterproof cloth 防水布
water proof duck 防水帆布
water proof finish 防水整理
water proofing 防水工艺;防水的,不透水的
waterproofing〈WP, wp〉防水的,不透水的
waterproofing agent 防水剂
water proof moisture permeable film 防水透湿膜
waterproof sealing 防水密封
water proof, wind proof and breathable fabrics〈WWB〉防水、防风、透气织物
water protection 水质保护
water pump 水泵
water purification 水的净化,净水[法]
water quality classification 水质分类
water quality monitoring 水质监测
water ratio 浴比
water recirculation pump 水循环泵
water recirculator 循环水冷却器
water recovery apparatus 水回收设备
water recycling [已处理的]水的循环使用
water regain 水分率,回潮率

water removal 脱水
water repellency〈WR〉拒水性
water repellency test 拒水性试验
water repellency treatment 拒水性处理
water repellent 拒水的；拒水剂
water repellent agent 拒水整理剂
water repellent finish 拒水整理
water resistance 抗水性
water resistance adhesive 抗水性黏着剂
water resistance tester 抗水性试验仪
water resistant 抗水的
water-resistent grease 防水油脂
Water Resources Law 水资源法
water retention 保水率，轧液率（用来衡量轧液效果，轧液率越小，表示轧液效率越高）
water retention on oven-dry weight 干基轧液率（湿织物重减去绝干织物重与绝干织物之比）
water-retention property 保水性
water-retention value〈WRV〉水分保持值，含湿量
water roller 清水轧辊
water sampler 取水样器
water sampling device 取水样器
water-saving pump flushing mechanism 节水泵的冲洗机理
water seal 水封口，水封闭
water seeping 渗水性
water sensitive 水敏感性的
water separator 汽水分离器；水分离器
water shedding 脱水
water-shedding properties 排水性能
water shrinkage ［织物］缩水率
water softener 软水剂；软水器
water softening 水的软化［作用］

water softening agent 软水剂
water solubility 水溶性
water soluble dye 水溶性染料
water soluble fibre 水溶性纤维
water soluble gum 水溶性胶
water solution 水溶液
water spot 水滴，水渍
water spot resistant 防水渍的
water spotting 水渍［变色］试验
water spotting fastness 水渍牢度《丝》
water spray 喷水，洒水；注水
water standard 标准水质《环保》，水质标准
water table 地下水面
water tank 水箱，水槽，水桶
water test 水压试验
water thickness swell ［毡呢的］水湿增厚
water-tight 不透水的，不漏水的
water-tight clothing 防水衣
water-to-air heat exchanger 水—气热交换器
water-to-goods ratio 水—织物重量比（即浴比）
water tolerance 耐水度
water transport 水分转移
water trap 脱水器；聚水器
water treatment 水处理
water-tube boiler 水管式锅炉
water-turbine 水轮机
water vapour diffusion resistance 抗水蒸气扩散
water vapour permeability〈WVP〉水蒸气透过性，透湿性能
water vapour permeability value〈WVP value〉透湿性能值
water vapour pressure〈WVP〉水蒸气压

力,汽压

water vapour transmission〈WVT〉 水汽传递,水蒸气透过性能

water vapour transmission test 水汽传递试验,水蒸气透过[性]试验

water-wash cloth 水洗布

water-water heat exchanger 水—水热交换器

water wicking 导水;导水物(如利用织物导汗)

watery fusion 结晶熔化

watt〈wt,wt.〉 瓦[特](功率单位)

wattage 瓦[特]数

watt-hour〈WH,whr〉 瓦特小时

watt-hour meter 瓦时计,电度计

wattmeter〈WM,wm〉 瓦特计,电力表

watt[s] per candle〈w/c,wpc,w.p.c.〉 瓦[特]/烛光

wave 波,波形,云纹;仿羔皮粗纺呢

wave angle 波程角

waveband 波段,频带

wave channel 波道

wave crest 波峰

waved pique 波纹凹凸组织

waveform 波形

wave front 波前;波峰;波阵面

waveguide 波导,波导管

waveguide bend 弯波导

wave-length 波长

wave-length band pass 波长带通

wave-length calibration 波长校正

wave-length interval 波长间隔

wave-length resolution 波长分辨率

wave-length spectrometer 波长分光计,波长分光仪

wave-length spectrum 波[长]谱

wave mechanics 波动力学

wave number 波数

wave pattern 波浪形图案

wave roller 波形辊,起波辊(辊筒表面成棘爪形,在运转中可使布起波动作用)

wave-shade defects 松板印(绸缎练染疵)

wave stitch 波形缝

waviness 波纹

waviness phenomena 波纹现象

wavy-line screen 波纹网版

wavy selvedge 松边,木耳边,起伏不平的布边(织疵)

wavy staple 卷曲纤维

wax 蜡

wax cloth 蜡布;油布(英国名称)

wax content 棉蜡含量

wax emulsion 蜡乳液

wax fill in 嵌蜡《雕刻》

wax finish 上蜡整理

waxing 上蜡

waxing and lustering machine 上蜡上光机

waxing machine 上蜡机

wax jet machine 喷蜡机(圆网印花制网用)

wax-like handle 蜡状手感

wax mark 蜡斑

wax printing 蜡防印花

wax resist printed fabric 蜡防印花布

wax resist printing 蜡防印花

wax-soluble dyes 蜡溶性染料

wax yellow 蜜蜡黄[色](暗黄色)

waxy finish 蜡状整理

waxy substance 蜡质物

waxy touch 蜡状手感

ways 色位(印花织物同一花样的几种不同配色)

Wb（weber） 韦伯（磁通量单位）
w. b.（wet bulb thermometer） 湿球温度计
WB（white balance） 白平衡（光学术语）
WBF（water proof and breathable fabric） 防水透气织物
WBS（waterproof breathable sorptive）material 防水透气吸附材料
w/c（watt per candle） 瓦［特］/烛光
WC（=W/C,wc,w. c.,with care） 小心
WC（wet combustion） 湿燃［法］
WCA（wet crease recovery angle） 湿折皱回复角
wcy（wincey） 上等绒布，棉毛绒布
WD（wiring diagram） 接线图；线路图，布线图
wd（wood） 木材，木质
weak acid dyes 弱酸性染料
weak acid medium dyeing 弱酸性染料
weak colour 浅色，弱色（染料用量很小时表现的相对色）
weaker ［染料］强度低，染料力分低
weak spot 脆弱点（织疵）；浅色斑
wear 穿着；戴；磨损
wearability 服用性能，穿着性能；耐磨损性
wearable computing 可穿着的计算机技术（计算机技术与服装的结合）
wearable intelligence 可穿着的智能产品（计算机技术与服装的结合）
wear behaviour 服用性能
wear-comfort 穿着舒适感
wear fastness properties 服用色牢度性能
wearing apparel 服装，衣服
wearing comfort 穿着舒适性
wearing of cloth 布的耐磨性；织物服用性
wearing performance 穿着性能

wearing value 耐穿性
wear-life ［织物的］穿着寿命，可服用期
wear-off 脱落
wear out 磨损，穿破；用旧
wear-prone parts 易损零件
wear-resistance 耐磨性，耐服用性
wear resistant compound material 耐磨复合材料
wear test 穿着试验
wear testing 耐磨试验；穿着试验
wear to backing ［地毯］磨损到背衬
wear trial 穿着试验
weatherability 耐气候性，耐风吹雨打性
weather ageing 大气老化作用，气候风化作用
weather-all coat（=weather coat） 晴雨大衣
weather cloth 防雨帆布，篷帆布
weathered piece 风渍布匹
weathered wool 风化损伤羊毛
weather fastness 耐气候性，耐气候坚牢度
weathering 风化，风蚀，老化
weathering ageing 气候老化
weathering damage 风化损伤
weathering quality 耐气候性，耐风蚀性
weathering test 耐气候牢度试验，风蚀试验
weather mark 风印，风渍（印染疵）
weatherometer 耐气候牢度试验仪
weatherproof〈WP〉 防风雨的；不受气候影响的
weather proof and winter clothing 防风雨和防寒服装
weather-proof finish 耐气候整理，抗风蚀整理
weather protection 防风雨，抗风化

weather resistance 耐气候性
weather-stained 受气候变色的,风渍[变色]的
weather strip [门、窗的]挡风雨嵌条
weavability 可织性
weave 织造,组织,织纹
weave analysis 织物组织分析,织纹分析
weave contraction 织造收缩
weave design 织纹设计
weave structure 织物结构
weaving 织,机织,织造;机织工艺,织造工艺
weaving bar 横挡,纬挡(织疵)
weaving mill 织造厂
weaving shed 织造工场,织造车间
weaving slack 经纱张力不匀(织疵)
web 纤维网,网状物;织物;金属薄片
web bonding effect 胶黏作用
web centre guiding 布匹居中导辊
web control equipment 织物控制器(如剥边、整纬、开幅、松紧装置等)
weber〈Wb〉韦伯(磁通量单位)
web guiding device 导布装置
web scouring method [针织物]网压平幅精练法
website 网站
Webstar 威勃司达自动导布器(自动跟踪导布装置,日本制)
wedding gown 新娘服
wedge 楔子;楔形物
wedge angle [刮刀]楔角
wedge joint 楔接合
wedge key 楔形键
wedge ruler(=wedge scale) 楔尺[卡](测试纤维直径)
wedgeshaped 楔形的

wedge shaped belt V形带,三角带
wedge slot 楔形槽
wedge spring 楔弹簧
wedge-type seal 楔形密封
Wedgewood(=Wedgwood) 威基伍花布(白色浮雕和彩纹的印花织物)
weed 杂草
weed killer 除草剂,除莠剂
weepage 渗漏,滴出
weeping [印花]残浆;渗出,滴漏现象
weft〈W.〉纬;纬纱
weft bar 纬挡,通纬挡(织疵)
weft bow 弓纬,纬弧
weft density control 纬密控制
weft distortion 纬斜
weft filament fabric 纬长丝织物
weft-knitted fabric 纬编针织物
weft printing 纬纱印花
Weftrol 威夫特罗尔整纬器(自动整纬装置,英国制)
weft shift 纬移
weft shrinkage 纬向收缩
weft straightener 正纬器;[织物]扩幅装置
weft strainer 纬斜防止装置
weft stretch woven fabrics 纬向拉伸织物
weft tight 紧纬(织疵)
weftwise defects 纬向疵点
weft yarn 纬纱
weigh average molecular weight 重均分子量
weighbridge 台秤,计量台
weighed inside oven [烘]箱内称重
weighed outside oven [烘]箱外称重
weighing 称重,加权
weighing bottle 称量瓶

weighing capacity 称重容量
weighing container [烘箱]烘篮,烘盘
weighing machine 称量机,磅秤
weighing scale 称量天平
weighing stations [染料]称重工作站
weigh pan 称量盘,天平盘,称量装置
weight〈wt〉权;重量;锤;砝码
weight-average degree of polymerization〈Xw.〉重量平均聚合度
weight commercial 商业重量
weight deduction [碱]减量处理
weight disc(=weight disk) 重锤圆盘,加重圆盘
weighted black 增重黑《丝》(用单宁硫酸亚铁增重处理的黑色)
weighted cloth(=weighted fabric) 加重织物
weighted lever 加压杠杆
weighted ordinate method 等间隔波长法,加权坐标法
weighted silk 重磅真丝,增重丝绸(用金属盐增加重量的丝绸)
weighter 增重剂,增量剂
weighting [织物]增重;称重;加权
weighting agent 加重剂,增重剂
weighting cotton fabric 重浆棉布
weighting dyeing 增重染色
weighting finishing [织物]增重整理《丝》
weighting percentage 加重率,增重率
weighting substances 加重物质,加重剂
weighting to par 脱胶丝增重至原来重量《丝》
weight in wet base 含水重量,湿重
weight lever 重锤杆,加压杆,均重杆
weight linear meter 每米长重量
weight loaded compensator 重锤加压[摆式]松紧架
weight loss 失重
weight loss in scouring and bleaching 练漂损耗
weight method 称重法
weight of boil-off gum content [精练]脱胶量
weightograph 自记式称重仪
weightometer 重量计;自动[称重]秤
weight percent〈w/o〉重量百分比
weight per foot〈wt per ft, wt per ft.〉每英尺重量
weight per unit area 单位面积重量
weight reduction finishing 减量整理
weight spring 重锤弹簧
weirs 冲碱装置(用于丝光机)
weld 焊接;焊缝
welded-on 堆焊
welded pipe 焊接管,有缝管
welded seam 焊缝
welding 焊接[法],焊接工艺
welding electrode 焊条
welding goggles 护目镜(焊接劳保用品)
welding jig 焊接夹具
welding machine 焊机
welding procedure 焊接工艺
welding spatters 焊渣,焊溅物
welding torch 焊枪,焊炬
welington 惠灵顿防雨呢(英国制)
well covered cloth 布面丰满[的]织物
well fit 合身(服装)
wellness fabrics 健康织物
well suited 成套
wenshang grosgrain 文尚葛(葛类丝织物)
WET(whole effluent toxicity) 整体污水毒性

WETA(water, energy, time, auxiliary) factors 在染整中的四要素(水、能源、时间、助剂)
wet adhesive bonding 湿黏合
wet adhesive process 湿黏工艺(层压)
wet after cleaning treatment [干洗后]湿清洁处理
wet and dry bulb hygrometer 干湿球湿度计
wet and dry bulb thermometer 干湿球温度计
wet applicator 喷雾器,雾化器
wet blending 湿掺和
wet blowing 湿蒸呢
wet blowing machine 湿蒸呢机
wet bonding 湿法黏合
wet brushing 湿刷绒
wet brushing machine 湿刷绒机(灯芯绒整理设备)
wet bulb depression 湿落度数(即干湿球温差)
wet bulb hygrometry method 湿球测湿法
wet bulb temperature 湿球温度
wet bulb thermometer 湿球温度计
wet carbonization 湿[式]炭化
wet-chlorination process 湿氯化防缩工艺(羊毛整理)
wet cleaning 湿洗
wet coagulating coating 湿法凝固涂层
wet coagulation 湿凝固法
wet coating 湿法涂层
wet colour 湿润色
wet combustion〈WC〉湿燃[法]
wet condition 湿润条件
wet crease recovery 湿皱回复性
wet crease recovery angle〈WCA〉湿皱回复角
wet crease resistance 抗湿皱性
wet crocking fastness 湿摩擦牢度
wet crosslinking 湿交联
wet-crosslinking process 湿交联法
wet curing 湿烘焙
wet decating(=wet decatizing) 湿蒸呢
wet decatizing machine 湿蒸呢机
wet developing(=wet development) 湿显色
wet dust collector 湿式除尘器
wet dwell processe 湿堆置工艺
wet fading 湿褪色
wet fastness 湿牢度
wet-fastness properties 湿牢度性能
wet finishing 湿整理
wet fire extinguisher 水灭火器
wet fixation process 湿固着法
wet fixation transfer printing 湿固着转移印花
wet heat setting 湿热定形
wet impregnation 浸渍,浸轧
wet iron bleeding 湿熨褪色
wet laid 湿法成网
wet laminate process 湿层压法
wet light fastness 湿日光牢度
wet-memory effect 湿定形效应
wet mercerization 湿丝光
wet migration process 湿泳移法
wet modulus 湿润弹性率
wetness 潮湿,湿度
wet on dry [process] 湿罩干[印花]法
wet-on-dry finishing 干布湿整理
wet on wet [process] 湿罩湿[印花]法;湿布轧液加工法
wet on wet resist printing 湿防印花;潮布

轧液加工
wet-oxidation 湿式氧化《环保》
wet paraffining 石蜡润滑
wet pickup〈WPU〉 纤维吸液率;湿浸轧率
wet press 湿轧辊;湿压机
wet printing system 湿印花系统
wet processing 湿加工整理,湿处理
wet raising 湿起毛,湿起绒
wet recovery 湿抗皱性,湿折皱恢复
wet relaxation 湿松弛复原
wet rubbing fastness 湿摩擦牢度
wet scrubber 湿式擦洗器,湿法除尘器
wet setting 湿[热]定形
wet shrinkage finishing 湿态收缩整理
wet-softening agent 浴中柔软剂
wet-soiling 湿污染(污垢在洗液中从一织物转移到另一织物上)
wet spreader and extractor 湿布扩幅脱湿机
wet steam 湿蒸汽
wet steaming 湿汽蒸
wet stoving 湿硫熏漂白(用二氧化硫溶液进行丝、毛织物的漂白)
wet strength 湿强力
wet stretch process 湿拉伸法
wet sueding machine 湿磨毛机
wettability 可湿性;润湿度,吸湿度
wettability power 吸湿能力,润湿率
wettability test 吸湿剂处理效果试验,吸湿性试验
wetted surface〈W.S.,ws〉 潮湿表面,湿面
wet tenacity 湿强度
wet tenacity and elongation 湿强度及断裂伸长率

wet tensile strength 湿强力,湿拉力;湿抗拉强度
wetter 润湿剂
Wetter NLF 惠脱 NLF(渗透剂,环氧乙烷缩聚物的混合物,非离子型,高效、低泡、耐氯,特别适合于毛条连续防缩克络因处理工艺,也可用于散毛、纱线的防毡缩工艺,商名,英国 PPT)
wet test 润湿试验
Wettihao BN 渗透剂 BN(商名)
wetting 润湿
wetting agent 润湿剂,浸湿剂
wetting agent 润湿剂
Wetting Agent FL 63 润湿剂 FL 63(染色用快速渗透剂,用于还原、硫化染料的染色,商名,瑞士科莱恩)
wetting angle 润湿角,[纤维的]湿吸附角
wetting heat 润湿热
wetting-out 浸湿,湿透(指轻度碱煮,使织物易于吸湿)
wetting-out agent 润湿剂,浸湿剂
wetting power 润湿力,浸湿力
wetting process 润湿过程
wetting promoters 润湿促进剂
wetting property 润湿性
wetting roller 浸渍辊
wetting time test 润湿时间试验(测毛织物防水性)
wet to dry strength ratio 干湿强度比
wet transfer printing 湿[态]转移印花法
wet-transfer printing machine 湿[态]转移印花机
wet transmission 湿传递
wet treatment 湿处理,湿加工
wet water 润湿水(含有强润湿剂的水)
wet-without-dry recovery 湿[无干]折皱恢

复,湿[无干]弹性恢复(指织物经某种树脂整理后,有湿折皱恢复性能,而没有或只是稍有干折皱恢复性能)

wet wrapper damping machine 卷布给湿机(预缩整理用)

wet wrinkle fastness 湿折皱牢度

wet wrinkle recovery 湿皱回复性

W. G. (water gauge) 水表;水[位]标[尺],水位指示器

WH(=whr, watt-hour) 瓦特小时

wheat flour 小麦粉

wheat starch 小麦淀粉

Wheatstone bridge 惠斯登电桥

wheel 轮;齿轮

wheel barrow 手推车

wheel box 齿轮箱

wheel chuck 阶梯弹簧夹头

wheel guard 齿轮防护罩

whet 磨;磨锐(指尖端)

WHG【德】德国水体保护法

whip-cords(=whipcord) 马裤呢

whipping 锁边;搅打;打成泡沫

whirler 旋转器;离心式滤气器;离心浇板机(锌板上浇感光胶用)

whirling machine [起绒织物]卷绒机

whiskering 毛绒(疵点)

whisper blade 均匀刮刀(涂层机用)

white-back duck 劳动布

white balance〈WB〉白平衡(光照术语)

white binder 白涂料黏合剂

white cast iron 白口铸铁

white content 白色量《测色》

white copperas 皓矾(即硫酸锌)

white core 白芯(靛蓝染色中纱线表面染上色泽而内部仍呈现白色的现象)

white dextrin 白糊精

white dischargeability 拔白性

white discharge printing 拔白印花,雕白印花

white discharge style 拔白印花,雕白印花

white-dotted cotton 白点印花[棉]布

white dyeing 增白,加白;上蓝

white dyes 荧光增白剂

white finish 增白处理

white finishing room 漂白整理间

white fluorescent lamp 白荧光灯

white goods 漂白织物

white lead 铅白,碱式碳酸铅

white lime 熟石灰,白涂料,白刷料

white liquor 烧碱液

whiteness 白度

whiteness Berger 拜尔格白度

whiteness CIE 国际照明协会标准白度

whiteness formula 白度公式

whiteness Harrison 哈里森白度

whiteness Hunter 亨特白度

whiteness index 白度指数

whiteness meter 白度计

whiteness or yellowness index of wool 羊毛白度或泛黄指数

whiteness reduction 白度降低

whiteness reflectency 白度(光度计测出的织物白度值)

whiteness retention 白度保持性

whiteness Stensby 斯坦斯比白度

whitening 加白,增白;漂白

whitening agent 增白剂

white oil 白油,轻油

white-on-white 白织提花织物;白布上印无光白

white paste 白浆

white pigments 白色涂料

white pile 白毛(绸缎印疵)
white point 白色点,非彩色点,中性点,中和点(色度学用语)
white resist paste 防白浆
white resist printing(=white resist style) 白色防染,防白印花
white scale quality label 白度样卡质量标签
white selvedge 白边
white sheeting 漂白布
white-soil effect 白污效应(施用无色粒状阻污剂于毛毯,产生白污效应,可降低有色尘污物的黏附)
white souring 漂后酸洗[工艺]
white specks 白斑丝(生丝疵)
white spirit 石油溶剂,漆用汽油
white standard 白度标准,白色标准
white vitriol 皓矾,七水[合]硫酸锌
white wash 水洗(指漂白酸洗后的第二次水洗)
white water〈w.w.〉白水
Whitex 惠泰克斯(荧光增白剂,商名,日本制)
whiting 漂白;白粉
whitney finish 惠特尼大衣呢起绒波纹整理[法]
whizzer 脱水机;离心机
WHO(World Health Organization) 世界卫生组织
whole effluent toxicity〈WET〉 整体污水毒性
whole metabolism 全代谢《环保》
wholesale 批发
whole sale price 批发价
whole saler 批发商
whole-water-based synthetic thickner 全水型合成增稠剂
wick 芯,灯心;油绳
wickability 导液性,油绳吸液性;毛细管性,芯吸性
wickable hydrophilic finishing 芯吸亲水整理
wicking 渗化《印》;芯吸
wicking belt of fibre 纤维毛细吸湿区域
wicking effect 芯吸效应
wicking height 毛细升高值;吸水上升高度
wicking property 芯吸性能
wicking rate 芯吸速率,毛细渗湿率
wicking tendency 渗化倾向《印》
wick lubricator 油绳润滑杯
wick oiler 油绳润滑杯
wick yarn 灯芯纱
wide board filter 宽频带过滤
widening 扩幅;扩布
wide range control 宽范围调节
wide range speed-adjustable DC motor 广调速直流电动机
width 宽度;[织物]幅宽
width altering device 调幅装置
width average〈W.A.〉宽度平均值
width barrier 横档疵点
width control 幅宽控制;调幅装置(烧毛机火口的火焰幅度或拉幅机轨道幅度的调节装置)
width difference 幅宽偏差
width holder 撑布架,幅撑
widthing 定幅
width milling 纬向缩绒
width of bottom 底布幅宽
width of cloth in loom 在机[织物]幅宽
width of dyeing and finishing machine 染整

机器宽度
width of gray cloth 坯布幅宽
width stretching 扩幅
width tolerance [织物]幅宽公差
widthways shade variations 纬向色差（染疵）
widthwise 纬向，横向
wild flax 野麻（罗布麻）
wildness 表面起毛（纱条由于干燥或静电而造成的表面起毛）
wild silk 野蚕丝
wild yak 野牦牛
Willesden process 一种防水方法（用于厚重织物）
Williams Unit 威廉氏槽（蒸汽夹层加热的多格插块狭槽，用于连续平幅加工，特点是用液量少，对比轧染作用时间长）
Wilton carpet 威尔顿地毯
wince crease 绳状折皱（疵点）
wincey〈wcy〉 上等绒布，棉毛绒布
winch 绞盘，绞车；六角盘
winch and cylinder drying machine 带透风滚筒的圆筒单面烘燥机
winch beck 绞盘[绳状]染槽
winch dyeing 绞盘绳状染色
winch dyeing machine 绞盘绳状染色机
winch mark 绞盘[染色机]擦伤痕（染疵）
wincing [绒头织物]热水浸泡除散毛整理法
wind 卷绕；缠绕；弯曲
Windbar 风雨布（商名，美国）
windblown system 鼓风植绒法
windbreaker 风雪衣
windbreaker cloth 防风布；防风厚呢
wind break screening 挡风屏

windcheater 防风衣
wind chill 风冷
winder roller 卷布辊
wind gage 风速计
winding arbour 卷布轴
winding click motion 卷绕离合器；扇形离合器
winding-off 退绕，退卷
winding oil 卷绕油剂
winding regulator 卷绕调节装置
winding roller 卷取辊，卷取罗拉，卷绕辊
winding tension 卷绕张力
winding unit 卷绕装置，卷布装置
window 窗口，视窗《计算机》
windowblind holland（= window holland） 窗帘布
window shade cloth 遮光窗帘布
wind proof agent 防风剂
wind protection 防风
wind protect net 防风网
wind screen（= windshield） 挡风板，风屏，风挡，挡风玻璃
wind shield cover 挡风罩盖
Windsor brilliant 温莎耐洗光亮棉衣料（英国制）
Windsor duck 温莎印花帆布（做夏天衣服用，英国制）
Windsor louisine 温莎优质耐洗印花棉衣料（英国制）
Windsor tie 温莎式阔领带（打成松散蝴蝶结式）
wind speed〈WS〉 风速
windtunnel balance 风洞天平
wind velocity〈w/v〉 风速
wine red 酒红[色]（深红色）
wine yellow 酒黄[色]（浅灰黄色）

wing nut 蝶形螺母

wingrip tear test 织物翼形［试样］撕破强力试验

winter damage ［棉织物洗后］冬季在室外晾干的损伤

wiper 刮器,刮油器；电刷,弧刷《电》

wiper bar(=wiper blade) 刮板,刮片,刮液刀

wipe[r] roll 剥离辊

wiping 刮浆

wiping and cleaning of machine 揩车

wiping cloth 揩布

WIRA(=W. I. R. A., Wira, Wool Industries Research Association) 羊毛工业研究协会(英国)

wire brush 钢丝刷

wire brushing roller 钢丝刷辊

wire card 钢丝针布,钢丝梳理

wire card raising machine 钢丝梳理起绒机

wire frame ［筒子染色用］弹簧筒管

wire gauge 线规

wire gauze 线网；铁丝网；滤网

wire raising machine 钢丝起绒机

wire velvet 天鹅绒(纬向织入竹丝或金属丝,使绒经成为绒头的丝绒织物)

wiriness 手感粗硬

wiring diagram〈WD〉接线图；布线图,线路图

Wirk. Stric. Tech. 【德】《针织技术》(德国刊物)

wiry end 紧捻,过捻(织疵)

wiry wool 粗硬毛

wiry yarn 硬纱,坚韧纱

wistaria(=wisteria) 紫藤色(浅紫色)

with care〈WC, W/C, wc, w. c.〉小心

withdraw ［纱线］退绕

withdrawal roller 退绕辊

within-laboratory 试验室内的

within-laboratory component 试验室内测试的数据或组分

with-scale 顺鳞片《毛》

Witney finish 威特尼［毛毯］整理

WM(=wm, wattmeter) 瓦特计,电力表

WMC (Wool Manufacturer's Council) 毛纺织品制造商协会

woad 菘蓝(植物靛青染料)

wobbling 摇晃；不稳定运转

W/O(water in oil) emulsion 水油相乳化液,油包水乳化液

Wofafix S special 沃法菲克斯S特号(直接染料及硫化染料染色的后处理剂,能增进耐洗、耐晒牢度,商名,德国制)

Wofapon AH 沃法邦AH(烷基磺酸钠,合成洗涤剂,商名,德国制)

wollen paper bowl 羊毛纸辊

Wonder Print 奇妙印花品(喷墨印花产品,商名,日本制)

wood〈wd〉木材,木质

wood alcohol 甲醇

wood bowl 木质轧辊

wood cellulose 木纤维素

wooden guide rail 导布木条

wooden pallet 堆布木板

wooden pin 导布栓

wooden plug 木塞

wooden roller ［卷布］木辊

wood-filled bowl 层压木纤维轧辊

wood washer 垫木

woof 纬,纬纱

wool 羊毛

wool acidifying 羊毛浸酸［处理］

wool blended fabric 混纺毛织物
wool brightening agents 羊毛增艳剂
wool bunting 旗纱(稀松平纹毛织物)
wool chemistry 羊毛化学
wool chlorinating machine 毛条氯化机
wool cortex 羊毛角质层
wool count 毛线支数
wool cream 和毛油
wool crepe 绉纹呢
wool denim 毛牛仔布
wool dye 羊毛用染料
wool dyeing 羊毛染色
wool-dyeing viscose staple fibre 能用上染羊毛的染料染色的黏胶短纤维
woolen【美】= woollen
wool fabric 毛织物,呢绒
Wool Fastea 105 羊毛固色剂 105(阴离子型,毛织物酸性染料染色时能增进色牢度,商名)
wool fat 羊毛脂
wool fat extraction 羊毛脂萃取
wool felt 毛毡
wool grease 羊毛脂
wool grease recovery 羊毛脂回收
wool hosiery 羊毛袜
Wool Industries Research Association〈WI-RA, W. I. R. A.,Wira〉羊毛工业研究协会(英国)
wool keratin 羊毛角蛋白
wool labeling 毛织物质量标签(美国规定的质量表示法,其中包括织物中新羊毛和回用毛的含量以及其他纤维的含量等)
woolled skin 仿绵羊皮
woolen 羊毛的;粗纺的,粗梳的《毛》;粗纺毛织物,粗梳毛织物

woollen blanket 毛毯
woollen cloth(＝woollen fabric) 粗纺毛织物
woollen finish for cottons 棉织物仿法兰绒整理,棉织物拉绒整理
woollen finishing 粗纺毛织物整理
woollen fleece 长毛大衣呢
woollen goods 呢绒,粗纺毛织物
woollenization 羊毛化
woollen lady's cloth 女大衣呢
woollen overcoating 大衣呢
woollen suiting 粗花呢
woollen sweater 绒线衫;羊毛衫
woollen sweater embossed pattern finishing 羊毛衫浮纹整理
woollen sweater embossing finish 羊毛衫浮雕整理
woollen sweater printing 羊毛衫印花
woollen thread 粗纺毛线
woollen type〈W type〉毛型(指化纤)
woollen yarn 粗纺毛线
woollie finish 仿毛整理
woollie nylon 弹力锦纶,仿毛尼龙(变形纱线)
wool-like aesthetic 羊毛样美感
wool-like fabric 仿毛织物
wool-like fibre 仿毛型纤维
wool-like finishing 仿毛整理
wool-like handle 毛型手感
wool-like polyester filament 仿毛涤纶长丝
woolliness[似羊毛]柔软性;毛线状;绒毛性
wool lubricant 和毛油
woolly 羊毛状的
woolly finish 毛形整理
Wool Manufacturer's Council〈WMC〉毛

Woolmark 纺织品制造商协会
Woolmark 羊毛制品防缩合格商品标记
wool modification 羊毛改性作用
wool modifying 羊毛改性
wool oil 和毛油
wool pest 羊毛蛀虫
wool polyester fabric 毛涤薄花呢
wool printing 羊毛印花
wool protection against insect attack 羊毛防蛀
wool protection agent 羊毛保护剂
wool quality 羊毛品质
wool reactive dyes 羊毛用活性染料
Wool Record and Textile World〈Wool Rec. Text. World〉《羊毛记录和纺织世界》(英国月刊)
Wool Research Organization of New Zealand〈WRONZ〉新西兰羊毛研究组织
wool-rich blends(＝wool-rich materials) 羊毛为主的混纺织物
wool scale 羊毛鳞片
Wool Science Review〈Wool Sci. Rev.〉《羊毛科学评论》(英国刊物)
wool scouring 洗毛
wool scouring machine 洗毛机
wool scouring range 洗毛机组
Wool Seal [国际羊毛秘书处的]防缩处理标记
wool serge 全毛哔叽
wool setting 羊毛定形
wool sheer 轻薄毛织物
wool shrinkage control 羊毛缩水率控制
wool slenderization 羊毛细化
wool slenderization technology 羊毛细化技术
wool/spandex woven fabrics 毛/氨纶机织物
wool structure 羊毛结构
wool touch 毛型手感
wool type〈W type〉毛型(指化纤)
wool velvet 海虎绒
wool washing 洗毛
wool yellowing 羊毛泛黄
words per minute〈WPM, wpm, w.p.m.〉每分钟字数,字/分
workability 可加工性,加工性能
workability test 加工性试验
work beam 卷取辊;卷布辊
work capacity 生产能力,工作量
work elastic recovery 弹性回复功能
work factor 做功系数
workhardening 加工硬化
workholding device 工件夹持装置
work in 插进,引进;织入
working 加工;运转;工作;液体内操作(特别指纱线在液体内进行染整加工)
working area 暂[时]存[储]区,中间结果存储区《计算机》
working-cloth width 织物阔幅
working drawing 工作图
working environment 工作环境
working life 适用期,使用寿命
working loss [纺织品]加工损耗(重量减轻或缩水)
working order 作业命令
working plan 工作计划;工作程序图
working position 工作位置
working substance 工质(工作质的简称),工作介质
working width 工作幅阔,工作宽度
work in three 三班制
work of adhesion 黏着功能

work of rupture 断裂功
work overtime 加班
work place illumination 工作区域照明
work recovery 回复功能
workshop 车间,工场,工厂;工艺;专题研究组
work-to-break 断裂功
World Health Organization〈WHO〉 世界卫生组织
World Textiles Agreement〈WTA〉 世界纺织品协会
World Trade Organization〈WTO〉 世界贸易组织
worm 蜗杆;蛇形管,盘香管
worm gear 蜗轮
worm gearing 蜗轮传动装置
worm gear reducer 蜗轮减速器
worm mark 虫渍;污渍
worm pump 蜗轮泵
worm reducer 蜗杆减速器
worm transmission 蜗轮传动
worm wheel 蜗轮
worn denim look 牛仔布仿旧风格
worn look 用旧了的外观;穿旧了的外观
worn part 磨损的零件
worsted〈wstd.〉 精纺毛织物,精梳毛织物;精梳毛纺的,精纺的
worsted cloth 精纺毛织物,精梳毛织物
worsted counts 精纺毛纱支数,精梳毛纱支数
worsted fabric 精纺毛织物
worsted finishing 精纺毛织物整理
worsted flannel 精纺法兰绒
worsted gabardine 精纺轧别丁,精纺华达呢
worsted goods 精纺毛织品

worsted melton 精纺麦尔登呢
worsted serge 精纺毛哔叽
worsted-type spun rayon cloth 毛型黏胶纤维织物
worsted yarn count system 精纺毛纱支数制,精梳毛纱支数制
wound coverage 伤口贴,伤口覆盖物
wound dressing material 创伤被覆材料(医用材料)
wound insulation 缠绕绝热[层]
wound rotor induction motor 绕线转子交流电动机
woven backing [地毯]底布
woven designs 织造花样,织物花纹
woven fabric 机织[织]物(与针织、编结、黏合等类织物相区别)
woven hose 管袋类梭织物;水龙带
woven in waste 杂物织入
woven jacketed hose 水龙带
woven like 仿机织物外观
woven looks 类机织物外观(指针织物)
woven narrow fabric 机织带子;机织狭幅织物
woven stretch fabrics 弹力织物(一般用弹力纱线织制)
WP(weatherproof) 防风雨;不受气候影响的
WP(=wp, waterproofing) 防水的,不透水的
wpc(= w. p. c., watts per candle) 瓦[特]/烛光
WPM(= wpm, w. p. m., words per minute) 每分钟字数,字/分
WPPL(water-proof paper-lined) bag 防水纸衬里[丙纶]袋
WPU(wet pickup) 湿浸轧率;纤维吸液率

WR(water repellency) 拒水性
WR(wrinkle resistant) 防皱
WRA(wrinkle recovery angle) 折皱回复角
wrangler shade 威格色,中蓝色牛仔裤
wrap 包,缠;围巾;外套;包身布;[包裹物的]一层,[环绕的]一圈
wrappage 毯子;围巾;头巾;宽松女服,女子便服;包装材料
wrapper 包布(蒸呢、煮呢用);睡衣;化妆衣;童装长斗篷
wrapper fibre [气流纺的]包缠纤维
wrapping 包装
wrapping angle 抱合角,包围角
wrapping paper 包装纸
wrapping top 绕辊[的]粗毛条;轧水辊包覆毛条
wraps 外衣;围巾,头巾;披肩;毯子;手帕
wrench 扳头,扳手;对花撬棒《印》
WRG(wrinkle resist garment) 防皱服
wringable laundry 可轧水洗涤方式
wringer 压榨机,绞干机,轧水机
wringing 轧水,压榨,绞干
wringing fit 轻迫配合,轻打配合
wringing machine 压榨机,绞干机,绳状轧水机
wring machine 绞干机,绳状轧水机
wring pump 叶轮泵
wrinkle 皱;皱纹;折皱;折叠
wrinkle fabric 折皱布
wrinkle-free finishing 免烫整理;不皱整理
wrinkle intensity 单位面积折皱数(美国国家干洗研究所术语)
wrinkle mark 皱痕;接缝痕(整理疵)
wrinkleometer 测皱仪

wrinkle on calendering 轧光折皱
wrinkle on process 加工折皱
wrinkle pattern 折皱标样(美国国家干洗研究所规定的折皱图样)
wrinkle-proof 抗皱褶
wrinkle recovery 折皱回复度,回能性
wrinkle recovery angle〈WRA〉 折皱回复角
wrinkle recovery degree 折皱回复度
wrinkle recovery ratio 折皱回复率
wrinkle recovery tester 折皱回复试验仪
wrinkle-recovery tester 折皱回复度试验仪,弹性回复度试验仪
wrinkle resistance 防皱性,耐折皱性
wrinkle resistant(WR) 防皱
wrinkle resistant finish 防皱整理
wrinkle resist garment(WRG) 防皱服
wrinkling 起皱
wrinkly cloth 皱面织物
wrist pin 十字头销,活塞销,肘节销
write-off [因严重损失而]报废,完全无用;销账
wrong colour 错色
wrong colour of weft 纬向错色
wrong craping 皱疵,皱斑(织疵)
wrong end 错经(织疵)
wrong filling(=wrong pick)错纬(织疵)
wrong side [织物]反面,背面
wrong timing 错误计时,错误调速
wrong wefting 错纬(织疵)
wrong width 幅宽不符
WRONZ(Wool Research Organization of New Zealand) 新西兰羊毛研究组织
Wronz scouring 朗兹洗毛(新西兰羊毛研究所开发的洗毛方法)
wrought iron 熟铁,锻铁

WRV (water-retention value) 水分保持率,含湿量

WS (wind speed) 风速

W. S. (=ws, wetted surface) 潮湿表面,湿面

wstd. (worsted) 精纺毛织物,精梳毛织物;精梳毛纺的,精纺的

wt (weight) 权;重量;锤;砝码

wt (=wt., watt) 瓦[特]

WTA (World Textiles Agreement) 世界纺织品协定

WTO (World Trade Organization) 世界贸易组织

wt per ft (=wt per ft., weight per foot) 每英尺重量

W type (woollen type, wool type) 毛型(指化纤)

Wurknit process 沃克尼特毛织品防缩整理法(界面聚合相间交链法)

Wurlan 沃兰(羊毛防缩剂,商名,美国制)

Wurlan process 沃兰毛织品永久性压烫整理法(界面聚合法)

w/v (percent weight in volume) 单位容积中重量百分比

w/v (wind velocity) 风速

WVP (water vapour permeability) 水蒸气透过性,透湿性能

WVP (water vapour pressure) 水蒸气压力,汽压

WVP value (water vapour permeability value) 透湿性能值

WVT (water vapour transmission) 水汽传递,水蒸气透过性能

w. w. (white water) 白水

WWB (water proof, wind proof and breathable fabrics) 可呼吸织物;防水、防风、透气织物

WW (wash and wear) finish 洗可穿整理,免烫整理

WW (wash and wear) process 洗可穿整理法,免烫整理法

XA (transmission adapter) 传输转接器
X-Age 一种电磁波屏蔽服的牌号(商名,日本钟渊纺)
xanthane 苍耳烷
xanthan gum 合成生物聚合胶,黄原胶(能耐热、耐强酸和强碱的合成胶质)
xanthate 黄原酸盐
xanthating (=xanthation) 黄[原酸]化[作用]
xanthene dyes 呫吨染料
xanthin 黄色天然色素(来自黄花)
xanthine oxidase 黄质氧化酶;醛酶
xanthophyll 叶黄素
xantho-protein reaction 黄蛋白反应
x-axis x轴[线],Ox轴;横轴,横坐标轴
x-axle x轴,Ox轴
X-brace 交叉支撑
xenon-arc lamp 氙弧灯
xenon electric-flash lamp 氙气电子闪光灯
xenon fadeometer 氙光耐晒[色]牢度试验仪
xenon lamp 氙灯
Xenotest 氙灯式耐晒[色]牢度试验
Xenotest light fastness tester 氙光耐晒[色]牢度试验器
xerogel 干燥凝胶
xerograph 静电复印术,干印术
xerographic printing 静电印刷;静电复印
xerography 静电复印术,干印术
xeror 静电复印机;静电复印件
xeroxing machine 静电复印机

XIC (transmission interface converter) 传输接口转换器
XIO (execute input/output) 执行输入/输出
XL (extra large) 特大号
X-linkable 可交联[的]
XMT (transmit) 发送;转换;透射
xpln. (explanation) 解释,说明
XPS (X-ray photoelectron spectra) X射线光电子光谱
XPS (X-ray photoelectron spectroscopy) X射线光电子光谱学
XRA (X-ray absorption) X射线吸收
X-ray X射线,X光
X-ray absorption ⟨XRA⟩ X射线吸收
X-ray analysis X射线分析
X-ray detection X射线检查
X-ray diagram X射线图
X-ray diffraction X射线衍射
X-ray diffractometry X射线衍射法
X-ray electronic energy spectrum X光电子能谱
X-ray fluorescence ⟨XRF⟩ X射线荧光
X-ray fluorescent analysis X射线荧光分析
X-ray fluorometry X射线荧光分析法
X-ray fluoroscopy ⟨XRF⟩ X射线荧光分析[法]
X-ray pattern X射线图
X-ray photoelectron spectra ⟨XPS⟩ X射线光电子光谱

X-ray photoelectron spectroscopy〈XPS〉 X射线光电子光谱学
X-ray proofing agent 防X射线剂
X-ray proofing fabrics 防X射线织物
X-ray proofing function 防X射线功能
X-ray proofing laminated fabrics 防X射线的层压织物
X-ray spectrum analysis X射线光谱分析
X-ray tester X射线试验器
XRF(X-ray fluorescence) X射线荧光
XRF(X-ray fluoroscopy) X射线荧光分析[法]
X-shaped filament X形[截面]长丝
XTAL(＝X-tal, crystal) 晶体
X-unit X单位(波长单位)
xyl.(xylene) 二甲苯
xylanase 木聚糖酶(去除半纤维素, 用于纸浆漂白等)
xylene〈xyl.〉 二甲苯
Xylene dyes 柴林染料(酸性染料, 商名, 瑞士制)
Xylene recovery 二甲苯的回收
xylene sulfonate 二甲苯磺酸盐
xylenol 二甲苯酚
xylography 木刻术; 木版[印刷]术; 木版画印画法
xylol 混合二甲苯(各种二甲苯的混合物, 俗称工业二甲苯)
xylon 木质, 木纤维
xylylene chloride 苯二甲基氯
xylylene diamine 苯二甲胺
xylylene glycol 苯二甲醇
XYZ colourimetric system XYZ表色系

Y

Ya-Jiang tussah pongee 鸭江绸(传统的柞蚕丝织物)
yamamai silk 山蚕丝,天蚕丝
yard〈yd,yd.〉 码(长度单位,1码等于91.44厘米);场
yardage 匹头,布匹
yardage clock(=yardage counter) 计码表,码分表
yardage goods 匹头;按码出售[的]织物
yardage production 匹头布生产
yardage roll 布卷,成卷的布;卷布辊,布轴
yardage yield 以码为单位的产量
yarder 码布工
yard goods 匹头;按码出售[的]织物
yard goods knitting 匹头针织物;按码计算的针织物
yards〈yds,yds.〉 码数
yardstick 码尺
yarn 纱,纱线
yarn abrader 纱线耐磨试验仪
yarn appearance 纱线外观
yarn ballistic test 纱线冲击强力试验
yarn bleaching machine 纱线漂白机
yarn break 断纱,断线(针织疵)
yarn bulk and steaming unit 纱线膨化汽蒸装置
yarn bundle cohesion 纱束抱合性,丝束抱合性
yarn conditioning machine 纱线给湿机
yarn count 纱线支数(线密度)

yarn defect 纱疵
yarn drier 纱线烘干机
yarndye(Y.D.) 纱线染色
yarn dyed 色纱线
yarn-dyed fabric(=yarn dyed woven fabric) 色织布,色织物
yarn dyehouse 染纱厂,染纱间
yarn dyeing 纱线染色
yarn dyeing automation 纱线染色自动化
yarn dyeing fabric 色织布
yarn dyeing in package 筒子纱染色
yarn dyeing in skein 绞纱染色
yarn finishing 纱线整理
yarn finishing applicator 纱线整理仪器
yarn gassing machine 纱线烧毛机
yarn humidifying agent 纱线增湿剂
yarn knitted into tubular form 筒状针织纱
yarn liquoring 小量染纱;小量浆纱
yarn lubricant 纱线润滑剂
yarn lustering machine 纱线上光机
yarn mercerization 纱线丝光
yarn mercerizing machine 纱线丝光机
yarn number 纱线支数(线密度)
yarn package 纱线卷装
yarn package dyeing machine 筒子纱染色机
yarn package preparation 筒装纱准备
yarn package wring machine 络筒机
yarn polishing machine 纱线上光机
yarn printing 纱线印花
yarn printing machine 纱线印花机

yarn raising 纱线起毛
yarn singeing machine 纱线烧毛机
yarn size 纱线支数
yarn sizing 纱线上浆
yarn spray-dyeing machine 纱线喷射染色机
yarn stacking installation 堆纱装置
yarn stand 筒子架;筒管支持器
yarn steaming 蒸纱
yarn steaming apparatus 蒸纱箱
yarn steaming chamber 蒸纱箱;蒸纱室
yarn strength tester 纱线强力试验仪
yarn strength utilization [织物中]纱线强力利用系数
yarn tensile strength tester 纱线拉伸强力试验仪
yarn texturing machine 化纤丝卷曲变形处理机
yarn twist 纱线捻度
yarn tying 纱线接结
yarn unevenness 纱线不匀
yarn variation 横路(针织疵)
yarn waste 废纱线
Y-axis Y 轴线,纵坐标轴《数》
Y. D. (yarndye) 纱线染色
yd(＝yd., yard) 码(长度单位,1 码等于 91.44 厘米);场
yds(＝yds., yards) 码数
year book〈Yr. B., Yr. bk〉年鉴,年报
yeast 酵母
yellow (yel.) 黄色,黄色的;变黄
yellow dextrin 黄糊精
yellower [色光]偏黄
yellowing 泛黄,发黄
yellowing and browning 泛黄变褐(指丝)
yellowing during wear 穿着变黄

yellowing of wool 羊毛泛黄
yellowing on storage 储藏泛黄
yellowing percentage 泛黄率
yellowing resistant finish 防泛黄整理
yellowish 带黄色的
yellowish ironing 烫焦(疵点)
yellowish pink 肉色
yellowness 黄度
yellowness factor 泛黄系数,泛黄度
yellowness index〈YI〉泛黄指数
yellow ochre 赭石黄[色](浅暗橘黄色);黄色颜料
yellow prussiate of potash 黄血盐钾,亚铁氰化钾
yellow prussiate of soda 黄血盐钠,亚铁氰化钠
yellow red 黄红[色],带黄光的红色
YI (yellowness index) 泛黄指数
yield 得量;收获率;一磅染整品的码数
yield behaviour 屈服性能
yield control 产量控制
yielding strain 永久应变;塑性应变
yield limit 屈服点,屈服极限
yield per pass 每次收率,循环一次的收率
yield point〈YP, Y. P.〉屈服点;流动点;击穿点
yield point value 塑变值;屈服[点]值
yield scales 得率标度
yield strain 屈服应变
yield strength 屈服强度
yield stress 屈服应力
yield temperature 屈服温度;流动温度
yield value 屈服值;塑变值
yokel colour 乡土色彩
yolk 羊毛汗脂
yolk yellow 蛋黄色

Yoracryl dyes 约阿克里染料(阳离子染料,商名,英国制)
Youhao acid chrome dyes 友好酸性媒介染料(商名)
Youhaocol dyes 友好涂料色浆(印染用涂料,商名)
Youhao direct and direct fast dyes 友好直接及直接耐晒染料(商名)
Youhaologen dyes 友好酞菁素染料(商名)
Youhaoneuter dyes 友好中性染料(1∶2金属络合染料,商名)
Youhao reactive dyes 友好活性染料(商名)
Youhao solvent dyes 友好溶剂染料(商名)
Youhao sulfur and Youhaodron dyes 友好硫化及友好昌染料(硫化及硫化还原染料,商名)
Youhaothol dyes 友好妥染料(不溶性偶氮染料,商名)
Youhaothrene dyes 友好士林染料(蒽醌还原染料,商名)
young fashion 青年款式
young fustic 染料木(黄色植物染料)
Young's modulus 杨氏模量
YP(＝Y. P., yield point) 屈服点;流动点;击穿点
Y-pipe 叉形管
Yr. B.(＝Yr.-bk, year book) 年鉴,年报
Y-shaped filament Y形[截面]长丝
yun brocade 云锦
Yuzen printing 友禅印花(早期手工印花法,日本)

Z

Zahn cup 赞恩杯(杯底有孔,用来测定杯内液体流尽所需时间,以确定黏度)

Zahn viscosimeter 锥盘黏度计,锥杯黏度计

Zarts solution Zarts 溶液(区分铜氨纤维与黏胶纤维)

Z-axis Z 轴《数》

Zaza-printing method Zaza 纱线印花法

ZD(zero defect) 无缺陷

zebra stripes 斑马条纹

zein 玉米蛋白

zein fibre 玉米蛋白纤维

Zeiss Elrepho photometer 蔡司埃里福测光计(比色用)

Zeiss Spectrophotometer for colour matching 蔡司配色用分光光度计

Zeiss universal microscope 蔡司通用显微镜

Zelan finish 泽伦整理(棉、麻、人造丝、丝织物的防水防污整理,商名)

Zelan S 泽伦 S(聚合蜡分散液,用作防水剂,商名,美国制)

Zelan water repellent 泽伦防水剂(商名,美国制)

Zener diode 齐纳二极管

zenith [blue] 天顶蓝[色](淡紫光蓝色)

zeolite 泡沸石(软水剂)

zeolite process 泡沸石软水法

Zepel 泽泼尔(含氟化合物,织物防污防水剂,商名,美国制)

zephyr 细薄织物(各种轻薄、柔软织物的统称);轻薄色织直条平布(棉为主);轻薄色织席纹呢(以毛为主);高级细软毛线(刺绣、钩针织物、针织品等用)

zephyr construction 轻薄织物结构

zephyr shirting 薄纱衬衫料

zerk 加油嘴

zero 零;零点;零度

zero adjuster 零位调节器

zero adjustment 归零;调零;零位调整[装置]

zero axis 零位轴线

zero burette 自满滴定管

zero check [仪器的]调零,零点调整

zero condition 调零条件

zero defect〈ZD〉 无缺陷

zero discharge 零排放;无出料;空转

zero drift 零点漂移

zero error 零位误差

zero fibre 零纤维(低聚合度,含有杂质的纤维素短纤维)

zero hardness [water] 软水

zeroing 调整零点

zerol gear 弧齿伞齿轮

zero line〈ZL〉 基线差,零线

zero mark 零度,零点标志

zero point 零点

zero point energy 零点能

zero point of raising 起毛零点

zero point raising 零点起绒

zero point shifting 零点漂移

zero pollution 零污染

zero position 零位
zero potential 零电位
zero pulse 零脉冲
zero reading 零点读数
zero set control 分段控制
zero setting 调到零点
zero shear viscosity 零剪切黏度
zero shift 零点漂移
zero state 零[状]态
zero-twist 无捻的
zero valance 零价
zero water 蒸馏水
Zeset TP 泽赛特 TP(羊毛防缩剂,商名,美国杜邦)
zeta 希腊字母 ζ
zeta potential ζ-电势,ζ-电位;动电势,动电位
zigzag 曲折的,之字形的,人字形的,锯齿形的
zigzag arrangement 交错排列
zigzag chain 曲折链
zigzag pattern 曲折图案
zigzag seam(=zigzag stitch) 曲折缝
Zimmer flatbed printing machine 齐默[地毯]平板网印机(奥地利制)
zinc acetate 醋酸锌
zincate caustic solution 锌酸盐碱性溶液
zinc bloom 锌华,氧化锌
zinc borate 锌硼酸盐
zinc borofluoride 氟硼酸锌(环氧树脂、APO 树脂等整理用的强酸性催化剂)
zinc chloride 氯化锌,锌氯粉
zinc chloride resist dyeing 氯化锌防染
zinc chrome 锌黄
zinc compound 锌化物
zinc corbonate 碳酸锌

zinc cutting 刻锌板(雕刻用语)
zinc discharge printing 锌盐拔染印花
zinc dust 锌粉
zinc dust resist paste 锌粉防染浆
zinc finish 锌辊轧光整理,锌辊轧花整理
zinc fluoroborate 氟硼酸锌
zinc green 锌绿[色]
zinc hydroxide 氢氧化锌
zincing 镀锌
zinc-lime-vat 锌—石灰—还原染浴
zinc naphthenate 环烷酸锌(织物的杀菌剂)
zinc nitrate 硝酸锌
zinc oxide powder 锌氧粉,锌白粉,氧化锌
zinc plate 锌板,锌版
zinc plate cutting machine 锌板雕刻机
zinc plate cutting table 锌板雕刻台(滚筒印花的一种雕刻设备,供锌板雕刻及锌板上色用)
zincplating 镀锌
zinc powder 锌粉
zinc sheet 锌板,锌片
zinc soap 锌皂
zinc stannate 锌锡酸盐
zinc stearate 硬脂酸锌
zinc sulfoxylate formaldehyde 甲醛合次硫酸锌(强还原剂,剥色拔染用)
zinc sulphate 硫酸锌,皓矾,锌矾
zinc[lime] vat 锌瓮;锌石灰还原染液
zinc white 锌白
zinc yellow 锌黄
zine sulfide 硫化锌
zipper 拉链
Zirchrome process 锆铬染色法(一浴矿物染色法,染后织物具有杀菌效果)

zirconia element 氧化锆元件（传感器）
zirconium compound 锆化合物
zirconium silicate fibre 锆石纤维，硅酸锆纤维
zirconium soaps 锆皂
zirconyl acetate 醋酸氧锆（防霉防腐剂）
zirconyl ammonium carbonate 碳酸氧锆铵（灰绿色泽矿物染料）
Zirpro antiflaming agent 齐普罗[羊毛]阻燃剂
Zirpro procedure 齐普罗[羊毛阻燃]整理法
Zirpro process 齐普罗整理（羊毛阻燃整理）
ZL（zero line） 零线，基准线
zonal 带状的
zonal sampling 区域抽样
zone 带；区；层；晶带；色区；[计算机]存储区；三行区（穿孔卡顶部的三行）
zone map 区域图案，部分图案（电子雕刻用语）
zone melting 区域熔化
zone of inhibition 抑菌区
zoning 分区取样（从各区内取小样，拼成有代表性的样品）；分区，分区制
zoobenthos 底生殖物《环保》
zoom 图像放大，缩小；变焦距
zoom 变焦[距]
zooming 图像缩小放大；变焦[距]
zoom microscope 变焦[距]显微镜
zoomorphic design 兽形花纹，动物花纹
Zwick tensile tester 兹维克强力试验仪
zwitter ion 两性电解质离子，两性离子
zymase 酿酶，酒化酶
zyme 酶
zyme desizing 酶退浆
zymochemistry 酶化学
zymohydrolysis 酶解[作用]
zymolysis 发酵；酶解作用
zymosis 发酵[作用]

主要参考书目

[1] Hans-Karl Rouette. Springer 纺织百科全书(注译本)[M]. 北京:中国纺织出版社,2008.

[2] 黄故. 新英汉纺织词汇[M]. 北京:中国纺织出版社,2007.

[3] 托托拉,默克尔. 仙童英汉双解纺织词典(第七版)[M]. 黄故等,译. 北京:中国纺织出版社,2004.

[4] 梅自强. 纺织辞典[M]. 北京:中国纺织出版社,2007.

[5] 姚虎卿. 化工辞典[M]. 5版. 北京:化学工业出版社,2014.

[6] 纺织行业生产力促进中心. 第五届功能性纺织品及纳米技术研讨会论文集[C]. 北京:2005.

[7] 纺织行业生产力促进中心. 第二届功能性纺织品及纳米技术研讨会论文集[C]. 北京:2002.

[8] 中国纺织工程学会染整专业委员会. 国际涂料应用和特种印花学术交流会论文集[C]. 上海:2004.

[9] 中国纺织科技信息研究所. 纺织导报[J]. 北京:2002－2014(部分).

[10] 中国纺织信息中心. 毛纺科技[J]. 北京:2002－2014(部分).

[11] 全国印染科技信息中心. 印染[J]. 上海:2002－2014(部分).

附录 国家与地区名单
（包括参加奥运会及邮票发行）

英文名称	中文名称	货币	文字
Afghanistan	阿富汗	Afghani	PE
Actutaki	艾图塔基（大洋洲中一小岛）	New Zealand $	E
Atland	阿兰群岛（又称奥兰群岛，芬兰属）	Markka	Fin
Alaska	阿拉斯加（美国）	US $	E
Alderney	奥尔德尼岛（英吉利海峡，近法国）	EP	E
Algeria	阿尔及利亚	Dinar	A,F
Albania	阿尔巴尼亚	Rupee	E
American	美属萨摩亚（南太平洋）	US $	E
Andorra(＝Andorre)	安道尔（欧洲）	Peseta	F,S
Angola	安哥拉（非洲）	Kwanza	Pu
Antigua & Barbura	安提瓜和巴布达（中美，西印度群岛）	dollar	E
Anguilla	安圭拉岛（英，中美）	dollar	E
Antilles	荷属安的列斯群岛（中美，西印度群岛）	Guilder	H
Argentina	阿根廷（南美）	Austral	S
Armenia	亚美尼亚（独，亚洲）		
Aruba	阿鲁巴岛（安的列斯群岛，荷属）	Guilder	H
Ascension. Is	阿森松群岛（英，大西洋，非洲西）	EP	E
Australia	澳大利亚	dollar	E
Australian Antarctic Territory	澳属南极领地	dollar	E
Austria(＝Usterreich)	奥地利		
Azania	阿扎尼亚（南非）		
Azerbaijan(＝Azarbaycan)	阿塞拜疆		
Azores(＝Acores)	亚速尔群岛	Escudo	Pu
Bahamas Is	巴哈马群岛（中美，加勒比海岛国）	dollar	E
Bahrain (＝Ajman, Capital Manama)	巴林（中亚）	Dinar	A,E
Balearic Is	巴利亚利群岛（西班牙东部群岛，属西班牙）		
Bangladesh	孟加拉	Taka	M,E

续表

英文名称	中文名称	货币	文字
Barbados	巴巴多斯(拉美,近加勒比海)	dollar	E
Belarus	白俄罗斯		
Belgium(=Belgique)	比利时	Franc	F
Belize	伯利兹(拉美,原隶属洪都拉斯)	dollar	E
Benin	贝宁(非洲,原称达荷美)	Franc	F
Bequia	贝基亚岛(加勒比海)	EC dollar	E
Bermuda	百慕大群岛(英属,大西洋)	dollar	E
Bhutan	不丹(亚洲)	Ngultram	E
Bolivia	玻利维亚(南美)		S
Bosnia & Herzegovina	波斯尼亚和黑塞哥维那(原南斯拉夫)		
Bophuthatswana	博普塔茨瓦纳(南非)	SA Rand	E
Bootswana	博茨瓦纳(非洲)	Pula	E
Brazil	巴西(南美)	Cruzeiro	Pu
British Virgin Is	英属维尔京群岛(拉美,加勒比海诸岛)		E
Brunei(=Darussalam)	文莱(太平洋岛国,马来西亚附近)	dollar	E
Bulgaria	保加利亚	Leba	Bu
Burma(=Myanmar)	缅甸	Kyat	E
Burkina Faso	布基纳法索(非洲)	Franc	F
Burundi	布隆迪(非洲)	Franc	F
British Antarctic territory	英属南极领地	EP	E
Cambodia(=Kampuchea)	柬埔寨	Riel	F
Cameron	喀麦隆(非洲)	Franc	F
Canada	加拿大	dollar	E,F
Canary Is	加那利群岛(西班牙,大西洋东北部)		
Cape verde	佛得角(非洲)独立国	Escudo	Pu
Cayman Is	开曼群岛,鳄鱼岛(加勒比海,近古巴)	dollar	E,F
Central African Republic	中非共和国	Franc	F
Ceylon(=Srilanka)	锡兰(即斯里兰卡)		
Chad(=Tchad)	乍得(非洲)		
Chile	智利	Peso	S
Christmas Is	圣诞岛(南太平洋,属基里巴斯)	dollar	E
People Republic of China	中华人民共和国	CNY	C
Ciskei	西斯凯(南非)	SA Rand	S

续表

英文名称	中文名称	货币	文字
Colombia	哥伦比亚	Peso	S
Comoros	科摩罗群岛(东非海岸)	Franc	F
Congo(＝People Republic of Congo)	刚果人民共和国(刚果布)	Franc	F
Congo	刚果民主共和国(刚果金)	Franc	E
Cook Is	库克或科克岛(英大洋洲,属新西兰)	New Zealand $	E
Cocos[Keeling]Is	科科斯群岛(沃州地区)	Australia $	E
Costa Rica	哥斯达黎加	colon	S
Cote Devoire	科特迪瓦(西非)	Franc	F
Croatia(＝Hrvatska)	克罗地亚(原南斯拉夫地区)		
Cuba	古巴	Peso	S
Curacao	库拉索岛(荷属,加勒比海岛国)		
Cyprus(＝Kibris)	塞浦路斯(西亚)	Pound	E
Czech Republic(＝Ceska)	捷克共和国	Kwanza	Cz
Denmark	丹麦	Krone	Den
Djibouti	吉布提(西非,法属索马里)	Franc	F
Dominica republic	多米尼加共和国(中美)	dollar	E
Dominicana	多米尼加共和国	Peso	S
Ecundor(＝Galanagos Is)	厄瓜多尔(南美)	Secre	S
Egypt	埃及	Pound	A,F,
El Salvador	萨尔瓦多(中美)	Coler	S
Espana	西班牙	Peseta	S
Eire	爱尔兰	Pound	E
Equatorial Guinea	赤道几内亚(非洲)	Peseta	S
Estonia	爱沙尼亚		
East Timor	东帝汶(原印度尼西亚属地)		
Ethiopia	埃塞俄比亚(阿比西尼亚)	Birr	E
Eritrea	厄立特里亚(东非)		
Fiji	斐济(南太平洋)	dollar	E
Finland(＝Suomi,Lappland)	芬兰	marka	Fin
France	法国	Franc	F
Faroe Is	法鲁群岛(丹属,英国北)	Krone	Den
Folkland Is	福克兰群岛(英,南美,即玛尔维纳斯群岛)	Pound	E
Gabon(＝Gabonaise)	加蓬(非洲)	Franc	F
Gambia	冈比亚(非洲)	palasi	E

续表

英文名称	中文名称	货币	文字
Georgia	格鲁吉亚(独,近土耳其,亚洲)		
Germany(＝Deutschland)	德国		
Ghana	加纳(非洲)	Cedi	E
Gibraltar	直布罗陀	Pound	E
Great Britain	英国	EP	E
Greece(＝Hellas)	希腊		
Greenland	格棱兰(丹麦)	Krone	Den
Grenada	格林纳达(中美)	EC dollar	E
Guam	关岛(美)		
Guatemala	危地马拉(中美)	Quatzal	S
Guinea	几内亚(非洲)	Franc	F
Guinea Bissau	几内亚比绍(西非)	Peso	Pu
Guyana	圭亚那(南美)	dollar	E
Guyana	圭亚那(法属,南美)		
Haiti	海地(中美)	gourde	F
Hawaii	夏威夷(美)		
Honduras	洪都拉斯(拉丁美洲,中美)	lempira	S
Hongkong	香港(中国)		
Hungary(＝Magyar)	匈牙利	Forint	Hu
Icelang(＝Island)	冰岛	ISK	
India	印度	Rupee	E
Indonisia	印度尼西亚	Rupiah	Indo
Iran	伊朗	Rial	E,F
Iraq	伊拉克	Dinar	E
Ireland(＝Eire)	爱尔兰(见前页)		
Israel	以色列	Shekei	E
Italy	意大利	Lira	I
Ivory Coast	象牙海岸(非洲,非国家)		
Isla of Man	马恩岛(英国,加勒比海)	EP	E
Jamaca	牙买加(拉丁美洲,中美)	dollar	E
Japan(＝Nippon)	日本	JPY	
Jersey	泽西岛(英国,近法国海岸)	EP	E
Jordan	约旦	Dinar	E
Jugoslavija	南斯拉夫	dinaer	
Kazakstan	哈萨克斯坦(独)		
Kenya	肯尼亚(非洲)	Shilling	E

续表

英文名称	中文名称	货币	文字
Korea DPR	朝鲜	KPW	
Korea Rep	韩国	KRW	
Kuwait	科威特	Dinar	E
Kirghizastan(=Kyrgyzstan)	吉尔吉斯坦(独,亚洲)		
Kiribati	基利巴斯(南太平洋)	Australia $	E
Kuzey Kibris(=Turk Cumhuriyeti)	北塞浦路斯土耳其共和国	Lira	Turk
Laos	老挝	Kip	F
Latvia	拉脱维亚		
Lebanon	黎巴嫩	Pound	F
Lesotho	莱索托(南非)	Maluti	E
Libya	利比亚	Dinar	E
Leichtenstein	列支敦士登	Swiss Franc	G
Lithunia(=Lietuva)	立陶宛		
Liberia	利比里亚(非洲)	dollar	E
Luxembourg	卢森堡	Franc	F
Madagascar(=Malagasy)	马尔加什共和国(东非岛国)	Franc	F
Madeira	马德拉群岛(非洲,东大西洋)	Escuda	Pu
Malawi	马拉维(东南非)	Kwanza	E
Malaysia	马来西亚	Ringgio	Ma
Macau	澳门(中国)		
Maldivi Is	马尔代夫群岛(印度洋岛国)	Rufiyea	E
Mali	马里(非洲)	Franc	F
Malta	马耳他(地中海岛国,欧洲)	Pound	E
Mauritius	毛里求斯(东非小岛)	Rupee	E
Mauritania	毛里塔尼亚(非洲)	Quguiya	F
Macedonia(=Makedonija)	马其顿(原南斯拉夫地区)		
Mariana Is.	马里亚纳群岛(西太平洋)		
Marshall Is.	马绍尔群岛(太平洋,大洋洲)		
Micornesia	密克罗尼西亚(南太平洋,大洋洲)	US $	E
Mexico	墨西哥	Peso	S
Monaco	摩纳哥(德法之间小国)	Franc	F
Mogolia	蒙古		
Morocco	摩洛哥	Dirham	F
Montserrat	蒙特塞拉特(加勒比海)	EC dollar	E
Mozambique(=Mocambique)	莫桑比克(非洲)	Metical	Pu

续表

英文名称	中文名称	货币	文字
Myanmar(=Burma)	缅甸	Kyat	E
Moldova	摩尔多瓦(乌克兰,罗马尼亚间小国,有中国村)		
Namibia	纳米比亚(西南非)	SA Rand	E
Nauru	瑙鲁(西太平洋,大洋洲)	Australia $	E
Nepal	尼泊尔	Ropee	E
Netherland	荷兰	Guilder	H
New Caledonia(=Nouvelle)	新喀里多尼亚(法属,南太平洋)	Franc	F
New Zealand	新西兰	dollar	E
Nevis(=St. Christopher)	尼维斯(东加勒比海)	EC dollar	E
Nicaragua	尼加拉瓜(拉美)	CorDOBA	S
Niger	尼日尔(非洲)	Franc	F
Nigeria	尼日利亚(非洲)	Naira	E
Norfolk Is	诺福克岛(澳属,南太平洋)	Aus $	E
Norway(=Norge)	挪威	Krone	N
Niuafo'ou	纽阿福欧(汤加属)	Pa,anga	E
Niue	纽埃(新西兰属)	New Zealand $	E
Omen	阿曼(亚洲)	Rial	E
Pakistan	巴基斯坦	Rupee	E
Palastan	巴勒斯坦(亚洲)		
Palau	帕劳(贝劳,南太平洋小岛,大洋洲)	US $	E
Panama	巴拿马(中美)	Ballboa	E
Papua New Guinea	巴布亚新几内亚(南太平洋,大洋洲)	Kina	E
Paraguay	巴拉圭(南美)	Guarani	S
Peru	秘鲁(南美)	Inti	S
Philippines	菲律宾	Peso	E
Poland(=Polska)	波兰	Zioty	Po
Polynesia	玻利尼西亚(中太平洋)	Franc	F
Polytugal	葡萄牙	Escudo	Pu
Puerto Rico	波多黎各岛(加勒比海诸岛)		
Pitcairn Is	皮特克恩群岛(英属,南太平洋)	New Zealand $	E
Penthyn Is	彭林岛(新西兰属)	New Zealand $	E
Qatar	卡塔尔(亚洲,沙特旁小国)	Riyal	E
Rhodesia	罗得西亚(即津巴布韦,非洲)		
Ras-Al-Khajmah	基里文角(阿拉伯联合酋长国)		

续表

英文名称	中文名称	货币	文字
Reunion	留尼汪岛(法属,东非,在马尔加什旁小岛)		
Romania	罗马尼亚	Leu	Ro
Russian Fedration	俄联邦	R	Rus
Rwanda(＝Rwandaise)	卢旺达(非洲)	Franc	F
St. Christopher &. Nevis	圣基茨和尼维斯联邦(东加勒比海)	EC dollar	E
St. Heleno	圣赫勒拿(英属,大西洋近非洲)	EP	E
Sahara o. c. c	非洲撒哈拉(见 West Sahara)		
St. Kitts	圣基茨(即尼维斯,在加勒比海)	EC dollar	E
St. Lucia	圣卢西亚	EC dollar	E
St. Pierre &. Miqueion	圣皮埃尔和密克群岛(法属)	Franc	F
St. Vincent &. Grena-dines	圣文森特群岛与格林纳达群岛(在加勒比海)	EC dollar	E
Sam Marina	圣马力诺(欧)	Lira	I
Samoa	萨摩亚群岛(南太平洋,大洋洲)	Tala	E
Sao Tome &. Principe	圣多美和普林西比(非中西部岛国)	Dobra	P
Saudi Arabia	沙特阿拉伯	Riyal	E,F
Senegal	塞内加尔(西非)	Franc	F
Seychelles(＝Sesel)	塞舌尔群岛(非洲,西印度洋小岛)	Rupee	E
Sierra Loone	塞拉利昂(非洲)	Leone	E
Sikkim	锡金(尼泊尔旁小国)		
Singapore	新加坡	dollar	E,C
Slovakia(＝Slovensko) re-pubilc	斯洛伐克共和国		
Slicenia(＝Slovenija)	斯洛文尼亚(原南斯拉夫地区)		
Solomon Is	所罗门群岛(大洋洲)	dollar	E
Somalia	索马里(非洲)	Shilling	S,O
South Africa	南非	Rand	E
South Georgia(＝and South Sandwich Is)	南乔治亚岛(英属,南非)	Pound	E
Spain(＝see Espana)	西班牙		
Srilanka	斯里兰卡	Rupee	E
Sudan	苏丹(非洲)	Pound	E
Suriname	苏里南(南美)	Guilder	H
Swaziland	斯威士兰(南非)	Lilangeni	E
Sweden(＝Sverige)	瑞典	Krone	Sw

续表

英文名称	中文名称	货币	文字
Switzerland(＝Helvetia)	瑞士	Franc	F,G,E
Syrian Arab(＝Syria)	叙利亚	Pound	E,F
Serbia Montonegro	塞尔维亚(原南斯拉夫)和黑山共和国		
Tapi	台北(中国)		
Tahiti	大溪地(又称塔西提岛,南太平洋)		
Tajikistan	塔吉克斯坦(独)		
Tanzania	坦桑尼亚联合共和国(非洲)	Shilling	E
Tchad(＝Chad)	乍得(非洲)	Franc	F
Thiland	泰国	Baht	E,Thi
Timor	东帝汶(印度尼西亚群岛中)	New Zealand $	E
Togo	多哥(非洲)	Franc	F
Tokelau	托克劳群岛(新西兰)		
Tonga	汤加(太平洋,大洋洲)		
Transdnestra	外特尼尔斯坦共和国(欧洲北,近乌克兰)		
Transkei	特水斯凯(南非)	SA Rand	E
Tristan da cunha	特里斯坦-达库尼亚(非洲南端,大西洋)	EP	E
Trinided & Tobago	特立尼特和多巴哥(加勒比海)	dollar	E
Tunisia	突尼斯(非洲)	Dinar	F
Turkey	土耳其	Lira	Turk
Turkmenistan	土库曼斯坦(独)		
Terres Australes	法属南方及南方领地	Franc	E
Turks & Caicos Is.	特克斯和凯科斯群岛	US $	E
Tuvalu	图瓦卢群岛(南太平洋)	Sust. $	E
United Kingdom	英国	EP	E
Uganda	乌干达(非洲)	Shilling	E
Ukraine	乌克兰		
United Arab Emirates	阿联酋(亚洲)	Dirham	E
United Nations	联合国	US $	E
Union Is	尤宁群岛(加勒比海)	EC	E
Uruguay	乌拉圭(南美)	Peso	S
USA	美国	US $	E
Upper Volta	上奥尔特(非洲,即布基纳法索)		
Uzbekistan	乌兹别克(独)		
Vanantu	瓦努阿图(又称温纳图,南太平洋)	Vatu	E,F

续表

英文名称	中文名称	货币	文字
Vatican City State	梵蒂冈	Lira	Latin,I
Venda	文达(南非)	SA Rand	E
Venezuela	委内瑞拉(南美)	Balivar	S
Vietman	越南	Gilder	Vi
Virgin Is(独立)	维尔京群岛(加勒比海)		
Weatern Somoa	西萨摩亚(太平洋,大洋洲)		
West Sahara(=Sahara OCC)	西撒哈拉(非洲)		
Yemen A. R. (=Arab Republic of Yemen)	阿拉伯也门共和国	Riyal	E
Yugoslavia	南斯拉夫		
Yemen Peoples Democratic Republic	也门人民民主共和国		
Zarie	扎伊尔(即刚果金)		
Zambia	赞比亚(非洲)	Kwanza	E
Zimbabwe	津巴布韦(非洲)	dollar	E
Zil Elwagne(=Seset)	塞舌尔外岛(群岛,非洲东端,近印度洋)	dollar	E

注：
A=阿拉伯文　　　　　Bu=保加利亚文　　　C=中文　　　　　　Cz=捷克文
Den=丹麦文　　　　　E=英文　　　　　　　F=法文　　　　　　Fin=芬兰文
G=德文　　　　　　　H=荷兰文　　　　　　Hu=匈牙利文　　　　I=意大利文
Latin=拉丁文　　　　 M=孟加拉文　　　　　Mo=马来西亚文　　　N=挪威文
Po=波兰文　　　　　 Pu=葡萄牙文　　　　　Ro=罗马尼亚文　　　Rus=俄文
S=西班牙文　　　　　So=索马里文　　　　　Sw=瑞典文　　　　　Thi=泰文
Turk=土耳其文　　　 Vi=越南文　　　　　　P=普什图文　　　　　EP=英镑
EC dollar=东加勒比元　　　　　　　　　　　SA Rand=南非兰特

简介 / Archroma. 传承值得信赖历史的崭新品牌

创始于Clariant纺织、纸张和乳液三大业务单元，Archroma为更好地满足本土市场客户的需求，为您度身定制优异性能的产品和色彩解决专案。

基于数十年从事色彩和化学制品的改革创新、世界级的质量标准、高水平服务及提高成本收益和可持续性发展的承诺，Archroma不断发展新科技，提升您日常生活用品的美感和性能。

同一个世界，同一个愿望
更美好的生活
Archroma将不遗余力打造
能满足您"情""感"需求的产品。

Archroma. Life enhanced.

www.archroma.com

青岛瑞恩博化工有限公司（Qingdao Rainbow Chemical Co ., Ltd）作为全球知名品牌昂高化工（中国）有限公司〔原科莱恩化工（中国）有限公司〕纺织部北方区总代理，为纺织品企业提供"优质、高效、环保"的染料和助剂。公司以"科研、创新、满足客户需求"为发展目标，始终坚持"客户至上"的服务宗旨和"质量、诚信、与客户共赢"的经营理念，成为众多上市公司和大、中型企业的长年签约供应商。相信我们会成为您最有价值的合作伙伴。

源于**瑞**士，感**恩**生活，**博**采众长

地址：青岛市城阳区宝陆莱路
电话：0532-85920099
传真：0532-85927272
邮箱：admin@qdrainbow.com.cn